Applications of Encapsulation and Controlled Release

Encapsulation and Controlled Release Series

Series Editor

Dr. Munmaya K. Mishra

The field of encapsulation, especially microencapsulation, is a rapidly growing area of research and product development. The ***Encapsulation and Controlled Release*** Series covers the entire field, with its volumes presenting the fundamental processes involved and exploring how to use those processes for different applications in industry. Titles are written at a level comprehensible to non-experts, each is a rich source of technical information, and many address current practices in research and industry.

Handbook of Encapsulation and Controlled Release
Edited by Munmaya Mishra

Applications in Encapsulation and Controlled Release
Edited by Munmaya Mishra

For more information about this series, please visit: https://www.crcpress.com/Encapsulation-and-Controlled-Release/book-series/CRCENCCONREL

Applications of Encapsulation and Controlled Release

Edited by
Munmaya K. Mishra

CRC Press is an imprint of the
Taylor & Francis Group, an **informa** business

CRC Press
Taylor & Francis Group
6000 Broken Sound Parkway NW, Suite 300
Boca Raton, FL 33487-2742

First issued in paperback 2022

ISBN: 978-1-03-240117-1 (pbk)
ISBN: 978-1-138-11878-2 (hbk)
ISBN: 978-0-429-29952-0 (ebk)

DOI: 10.1201/9780429299520

Publisher's Note
The publisher has gone to great lengths to ensure the quality of this reprint but points out that some imperfections in the original copies may be apparent.

Library of Congress Cataloging-in-Publication Data

Names: Mishra, Munmaya K., editor.
Title: Applications of encapsulation and controlled release / [edited by] Munmaya K. Mishra.
Description: Boca Raton : CRC Press, Taylor & Francis Group, 2019. | Includes bibliographical references.
Identifiers: LCCN 2019021452 | ISBN 9781138118782 (hardback : alk. paper)
Subjects: LCSH: Controlled release technology. | Drug delivery systems.
Classification: LCC TP156.C64 A67 2019 | DDC 660/.28--dc23
LC record available at HYPERLINK "https://protect-us.mimecast.com/s/k2A4CjRvnlfnopEylfRjza6?domain=lccn.loc.gov" https://lccn.loc.gov/2019021452

Visit the Taylor & Francis Web site at
http://www.taylorandfrancis.com

and the CRC Press Web site at
http://www.crcpress.com

To my family.

Also, to those who made and will make a difference through polymer research for improving the quality of life!

Contents

Preface

The field of encapsulation, especially microencapsulation, is a rapidly growing area of research and product development. The first edition of the book, *Handbook of Encapsulation and Controlled Release*, covered the entire field, presenting the fundamental processes involved and exploring how to use those processes for different applications in industry. The book also laid the groundwork for further advancements in encapsulation technologies and controlled release applications. As the field is evolving, especially the area of controlled release applications, there is a clear need for a source of technical information that has sufficient broad coverage and is written at a level comprehensible to non-experts. With these considerations in mind, the decision to publish the second edition of the book, entitled *Applications of Encapsulation and Controlled Release*, was apparent.

This book is designed for scientists and engineers working in various industries, including food, consumer products, pharmaceuticals, medicine, agriculture, nutraceuticals, dietary supplements, cosmetics, flavors, and fragrances. It offers a broad perspective on a variety of applications and processes, providing research information, figures, tables, illustrations, and references. Catering to professionals, researchers, students, and general readers in academia, industry, and research institutions, *Applications of Encapsulation and Controlled Release* is a much-needed reference for continued research and development in controlled release applications.

I feel honored to undertake the important and challenging endeavor of developing the second edition of *Applications of Encapsulation and Controlled Release*, which will cater to the needs of many who are working in the field. I would like to express my sincere gratitude and appreciation to the authors for their excellent professionalism and dedicated work. I would like to thank the entire management of the book program of the Taylor & Francis Group (CRC Press), including Ms. Barbara Knott and Ms. Danielle Zarfati, who made this possible.

Munmaya K. Mishra
Editor

Editor

Munmaya Mishra, PhD, is a polymer scientist who has worked in the industry for more than 30 years. He has been engaged in research, management, technology innovations, and product development. Currently, he is the principal at Encapcr, LLC and serves as the chief consultant in the field of encapsulation/controlled release and polymer applications. He held research and technical management responsibility at Altria Research Center. He serves as the editor-in-chief of three renowned Taylor & Francis polymer journals and has contributed immensely to multiple aspects of polymer applications, including encapsulation and controlled release technologies. He has authored and coauthored hundreds of scientific articles, is the author or editor of eight books, and holds over 75 U.S. patents, over 50 U.S. patent-pending applications, and hundreds of world patents. He has received numerous recognitions and awards. He is the editor-in-chief of the recently published 11-volume *Encyclopedia of Biomedical Polymers and Polymeric Biomaterials* and the *Encyclopedia of Polymer Applications.* He founded the International Society of Biomedical Polymers and Polymeric Biomaterials as well as a scientific meeting titled *Advanced Polymers via Macromolecular Engineering*, which has gained international recognition and is under the sponsorship of the International Union of Pure and Applied Chemistry.

Contributors

B. A. Aderibigbe
Department of Chemistry
University of Fort Hare, Alice Campus
Eastern Cape, South Arica

Muhammad Sajid Hamid Akash
Department of Pharmaceutical Chemistry
Government College University Faisalabad
Faisalabad, Pakistan

Farooq Azam
Incharge Pharmacy Department
University of Agriculture Faisalabad
Faisalabad, Pakistan

Saleheen Bano
Department of Polymer and Process Engineering
Indian Institute of Technology Roorkee
Roorkee, India

Argyro Bekatorou
Department of Chemistry
University of Patras
Patras, Greece

Anuradha Biswal
Department of Chemistry
Veer Surendra Sai University of Technology
Burla, India

Luisa F. Cabeza
GREiA Research Group
INSPIRES Research Centre
University of Lleida
Lleida, Spain

Yahya E. Choonara
Wits Advanced Drug Delivery Platform Research Unit
Department of Pharmacy and Pharmacology
School of Therapeutic Sciences
Faculty of Health Sciences
University of the Witwatersrand
Johannesburg, South Africa

Marcos Akira d'Ávila
Department of Chemical Processes
School of Chemical Engineering
University of Campinas—UNICAMP
Campinas, Brazil

Frédéric Debeaufort
Food and Wine Physical Chemistry Lab
Université de Bourgogne Franche Comte
Dijon, France

Farha Deeba
Department of Polymer and Process Engineering
Indian Institute of Technology Roorkee
Roorkee, India

Stéphane Desobry
Université de Lorraine
ENSAIA, LIBio
Vandœuvre Cedex, France

Francesco Donsì
Department of Industrial Engineering
University of Salerno
Fisciano, Italy

Lisa C. du Toit
Wits Advanced Drug Delivery Platform Research Unit
Department of Pharmacy and Pharmacology
School of Therapeutic Sciences
Faculty of Health Sciences
University of the Witwatersrand
Johannesburg, South Africa

Elisabeth Dumoulin
UMR Ingénierie Procédés Aliments AgroParisTech, INRA
Université Paris-Saclay
Massy, France

Tahir Farooq
Department of Applied Chemistry
Government College University Faisalabad
Faisalabad, Pakistan

Giovanna Ferrari
Department of Industrial Engineering
and
ProdAl scarl
University of Salerno
Fisciano, Italy

Ana Figueiras
Department of Pharmaceutical Technology
and
REQUIMTE/LAQV, Group of Pharmaceutical Technology
Faculty of Pharmacy
University of Coimbra
Coimbra, Portugal

Arijit Gandhi
Department of Pharmaceutics, Gupta College of Technological Sciences
Ashram More
Asansol, India

Jessica Giro-Paloma
Department of Materials Science and Physical Chemistry
Universitat de Barcelona
Barcelona, Spain

Vikas V. Gite
Department of Polymer Chemistry
School of Chemical Sciences
Kavayitri Bahinabai Chaudhari, North Maharashtra University
Jalgaon, India

Arruje Hameed
Department of Biochemistry
Government College University Faisalabad
Pakistan

Zikhona Hayiyana
Wits Advanced Drug Delivery Platform Research Unit
Department of Pharmacy and Pharmacology
School of Therapeutic Sciences
Faculty of Health Sciences
University of the Witwatersrand
Johannesburg, South Africa

Hira Ijaz
Faculty of Pharmacy
University of Sargodha
Sargodha, Pakistan

A. Inés Fernández
Department of Materials Science and Physical Chemistry
Universitat de Barcelona
Barcelona, Spain

Sougata Jana
Department of Pharmaceutics, Gupta College of Technological Sciences
Ashram More
Asansol, India
and
Department of Health and Family Welfare
Directorate of Health Services
Kolkata, India

P. Kaushik
ICAR-Indian Agricultural Research Institute
New Delhi, India

Pariksha J. Kondiah
Wits Advanced Drug Delivery Platform Research Unit
Department of Pharmacy and Pharmacology
School of Therapeutic Sciences
Faculty of Health Sciences
University of the Witwatersrand
Johannesburg, South Africa

Pierre P. D. Kondiah
Wits Advanced Drug Delivery Platform Research Unit
Department of Pharmacy and Pharmacology
School of Therapeutic Sciences
Faculty of Health Sciences
University of the Witwatersrand
Johannesburg, South Africa

Anurag Kulshreshtha
Department of Polymer and Process Engineering
Indian Institute of Technology Roorkee
Roorkee, India

Bijender Kumar
Department of Polymer and Process Engineering
Indian Institute of Technology Roorkee
Roorkee, India

Jitendra Kumar
ICAR—Indian Agricultural Research Institute
New Delhi, India
and
ICAR—Institute of Pesticide Formulation Technology
Gurugram, India

Prabhat Kumar
Advanced Instrumentation Research Facility
Jawaharlal Nehru University
New Delhi, India

Pradeep Kumar
Wits Advanced Drug Delivery Platform Research Unit
Department of Pharmacy and Pharmacology
School of Therapeutic Sciences
Faculty of Health Sciences
University of the Witwatersrand
Johannesburg, South Africa

Mariana Magalhães
Department of Pharmaceutical Technology
and
REQUIMTE/LAQV, Group of Pharmaceutical Technology
Faculty of Pharmacy
University of Coimbra
Coimbra, Portugal

Mahendra S. Mahajan
Department of Polymer Chemistry, School of Chemical Sciences
Kavayitri Bahinabai Chaudhari, North Maharashtra University
Jalgaon, India

S. Majumder
ICAR-Indian Agricultural Research Institute
New Delhi, India
and
ICAR-Indian Institute of Vegetable Research
Varanasi, India

Athanasios Mallouchos
Food Science and Human Nutrition Department
Agricultural University of Athens
Athens, Greece

Thashree Marimuthu
Wits Advanced Drug Delivery Platform Research Unit
Department of Pharmacy and Pharmacology
School of Therapeutic Sciences
Faculty of Health Sciences
University of the Witwatersrand
Johannesburg, South Africa

Mònica Martínez
Department of Materials Science and Physical Chemistry
Universitat de Barcelona
Barcelona, Spain

Yuvraj Singh Negi
Department of Polymer and Process Engineering
Indian Institute of Technology Roorkee
Roorkee, India

Viness Pillay
Wits Advanced Drug Delivery Platform Research Unit
Department of Pharmacy and Pharmacology
School of Therapeutic Sciences
Faculty of Health Sciences
University of the Witwatersrand
Johannesburg, South Africa

Stavros Plessas
Department of Agricultural Development
Democritus University of Thrace
Orestiada, Greece

Ruchir Priyadarshi
Department of Polymer and Process Engineering
Indian Institute of Technology Roorkee
Roorkee, India

Kalyani Prusty
Department of Chemistry
Veer Surendra Sai University of Technology
Burla, India

Junaid Qureshi
Department of Pharmacy
Bahauddin Zakariya University
Multan, Pakistan

K.V. Radha
Department of Chemical Engineering, A.C.Tech. Campus
Anna University
Chennai, India

Maria Inês Ré
Université de Toulouse
IMT Mines d'Albi
Albi, France

Kanwal Rehman
Institute of Pharmacy, Physiology and Pharmacology
University of Agriculture Faisalabad
Faisalabad, Pakistan

Ufana Riaz
Materials Research Laboratory, Department of Chemistry
Jamia Millia Islamia
New Delhi, India

Maria Helena Andrade Santana
Department of Biotechnological Processes
School of Chemical Engineering University of Campinas—UNICAMP
Campinas, Brazil

Ana Cláudia Santos
Department of Pharmaceutical Technology
and
REQUIMTE/LAQV, Group of Pharmaceutical Technology
Faculty of Pharmacy
University of Coimbra
Coimbra, Portugal

S. Saranya
Department of Chemical Engineering, A.C.Tech. Campus
Anna University
Chennai, India

Dhruba Jyoti Sarkar
ICAR—Central Inland Fisheries Research Institute
Barrackpore, India
and
ICAR—Indian Agricultural Research Institute
New Delhi, India

Niladri Sarkar
Department of Chemistry
Veer Surendra Sai University of Technology
Burla, India

Kalyan Kumar Sen
Department of Pharmaceutics, Gupta College of Technological Sciences
Ashram More
Asansol, India

Mariarenata Sessa
Department of Industrial Engineering
University of Salerno
Fisciano, Italy

Nilesh Shah
The Dow Chemical Company
Collegeville, Pennsylvania

Najam Akhtar Shakil
ICAR-Indian Agricultural Research Institute
New Delhi, India

Sauraj Singh
Department of Polymer and Process Engineering
Indian Institute of Technology Roorkee
Roorkee, India

Sarat K. Swain
Department of Chemistry
Veer Surendra Sai University of Technology
Burla, India

Ume Ruqia Tul-Ain
Faculty of Pharmacy
University Of Sargodha
Sargodha, Pakistan

Christelle Turchiuli
UMR Ingénierie Procédés Aliments AgroParisTech, INRA
Université Paris-Saclay
Massy, France
and
Université Paris Sud
IUT d'Orsay
Université Paris-Saclay
Orsay, France

Francisco Veiga
Department of Pharmaceutical Technology
and
REQUIMTE/LAQV, Group of Pharmaceutical Technology
Faculty of Pharmacy
University of Coimbra
Coimbra, Portugal

Anurakshee Verma
Materials Research Laboratory, Department of Chemistry
Jamia Millia Islamia
New Delhi, India

Fanwen Zeng
The Dow Chemical Company
Collegeville, Pennsylvania

1

Multi-Cyclodextrin Supramolecular Encapsulation Entities for Multifaceted Topical Drug Delivery Applications

P. D. Kondiah, Yahya E. Choonara, Zikhona Hayiyana, Pariksha J. Kondiah, Thashree Marimuthu, Lisa C. du Toit, Pradeep Kumar, and Viness Pillay

CONTENTS

1.1 Introduction

Topical delivery systems are one of the most popular applications of controlled, encapsulated drug delivery platforms. This is due to the ease of administration and the direct application, which translates to site specificity, enhanced therapeutic effectiveness, and elimination of systemic side effects, such as hepatotoxicity, through bypassing the hepatic metabolic processes (Park et al. 2017). Topical formulations have to be compatible with the body surface at the intended area of administration, since interactions of the formulation with the surface will affect the pharmacokinetic properties of bioactives. This includes permeation across the biological membranes, drug release behavior, solubility, absorption, and the overall pharmacological activity (Choudhury et al. 2017). Various drug delivery systems are constantly being developed to improve the pharmacokinetics of topically delivered drug molecules (Choudhury et al 2017). Recent researches have revealed the capability of cyclodextrins (CDs) in enhancing drug permeability, solubility, and bioavailability without severely altering the biological barrier as well as the drug functionality (Xu et al. 2018). CDs form one of the most versatile technologies that have the capability to encapsulate a wide range of drug molecules for a variety of applications (Sherje et al. 2017). This chapter aims to highlight the application of CDs in topical drug delivery and how their benefits can be modified through the formulation of multi-cyclodextrin entities. The topical administration routes discussed in this chapter are divided into three categories: surface of tissues (ocular, oral cavity, and tooth surface routes), skin (dermal route), and mucous membranes (buccal, sublingual, and nasal). The current applications of CDs in these routes are elaborated and the substantial benefit of multi-cyclodextrin entities proposed.

1.2 Cyclodextrin Properties That Confer Suitability in Drug Delivery Applications

CDs are cyclic oligosaccharides composed of glucopyranose rings co-joined by the α-1, 4 glycosidic bonds (Xiao and Grinstaff 2017). There are three common types used in drug delivery: α- (six rings), β- (seven rings), and γ-cyclodextrins (eight rings) (Gautam et al. 2017). Cyclodextrin forms are differentiated by size and diameter of the central cavity, solubility, molecular weight, and toxicity profiles. All the configurations of cyclodextrins are water soluble with some limitations (Zhu, Yi and Chen 2016). Important features of CDs that make them ideal for drug delivery are their amphipathic nature, with an inner hydrophobic central cavity and a hydrophilic outer surface, and the ability to form inclusion and non-inclusion complexes with other compounds. The lipophilicity of the central cavity is imposed by the presence of the glycosidic oxygen bridges and hydrogen atoms, while the hydrophilicity of the outer surface is due to the presence of numerous primary and secondary hydroxyl groups (Sherje et al. 2017). When CDs incorporate lipophilic, low–molecular weight, and accurately sized compounds within their central cavity through non-covalent interactions, an improvement in permeability and aqueous solubility, stability (chemical, hydrolytic, thermal, etc.), dissolution rate, site specificity, and bioavailability is obtained (da Silva Mourão et al. 2016). Cyclodextrins are generally non-toxic; however, different administration routes could impose certain adverse effects: for example, the parenteral route of administration causes nephrotoxicity due to renal clearance (Azzi et al. 2018). This nephrotoxicity is most prominent on parenteral administration of α- and β-cyclodextrins (Azzi et al. 2018) and oral administration of some β-cyclodextrin derivatives (Lucio et al. 2018). Cyclodextrins have been reported to exert hemolytic activity at high concentrations (Yang et al. 2018). The toxicity profile of cyclodextrins is summarized in Table 1.1.

Host–guest chemistry (inclusion complex formation) is the process of incorporating a chemical agent within another; in this case, the cyclodextrin molecule is the host, and the lipophilic compound, usually a bioactive agent, functions as the guest (Barman and Roy 2018a). The product formed is known as an

TABLE 1.1

Characteristic Features of α, β, and γ-Cyclodextrins

Cyclodextrin Molecule	Central Cavity Diameter (Å)	Molecular Weight (g/mol)	Solubility in Water (mg/mL)	Contraindicated Routes of Administration	Specific Toxicity
α-cyclodextrin	4.5	972	145	Intravenous	Hemolysis and nephrotoxicity Li et al. (2004), Lockwood, O'Malley and Mosher (2003)
β-cyclodextrin	7.0	1135	18.5	Intravenous	Hemolysis and nephrotoxicity Li et al. (2004), Lockwood, O'Malley and Mosher (2003)
γ-cyclodextrin	8.5	1297	232	N/A	Hemolysis Li et al. (2004)

inclusion complex. Inclusion complexes are supramolecular cages composed of active molecules incorporated within the cyclodextrin core in a 1:1 interaction ratio (Barman and Roy 2018b), as displayed in Figure 1.1 (Kayaci and Uyar 2011; Kida et al. 2003), with adequate water solubility and complex reversibility (Cai, Huang, and Li 2017). Inclusion complex formation is best carried out in aqueous solutions, where enthalpy-rich water molecules are released from the cyclodextrin cavity, allowing inclusion or non-inclusion interactions with the guest compound (Tambe et al. 2018). The inclusion or non-inclusion interaction is largely influenced by the size, molecular weight, polarity, ionic charge, solubility, and stability of the guest molecule in aqueous solutions. These properties, therefore, determine the selectivity of cyclodextrins to include certain compounds over others (Suvarna, Gujar, and Murahari 2017). The formation of inclusion complexes may significantly alter the biological and physicochemical parameters of the incorporated compound (Pápay et al. 2016).

The binding of CD molecules to cholesterol and other phospholipid components of the biological membrane after topical administration improves membrane permeability (Duchêne and Bochot 2016). The extraction of these biological components leads to the opening up of pores, which allows greater permeation of the formulation than would be expected in the absence of CDs (Ryzhakov et al. 2016). The aqueous solubility of lipophilic compounds is improved due to the bonding between the hydroxyl groups of CDs and the substituent molecules (Mayank and Rajashree 2018) as well as the presence of external surface hydroxyl groups. Since CDs form stable complexes in certain solutions, this enables their use as solubilizing encapsulating agents (Ol'khovich et al. 2017). In addition, the formation of an inclusion complex improves the stability of the included compound by protecting it against chemical degradation, hydrolysis, and oxidation due to the resultant minimal interaction between the compound and the external environment (Ho et al. 2017). The cumulative result of all these processes leads to enhanced dissolution rate and bioavailability of the guest compound at the targeted site, followed by controlled release (Zhang et al. 2017). These functions have been experimentally investigated within drug delivery parameters with good outcomes. CDs have become a subject of interest in gene delivery. One of the early used polymers in gene delivery is polyethylenimine. The limitations associated with using this compound, such as high toxicity, have led to the investigation of other compounds that could serve the same functions in more efficient ways. CDs have been tested for localized gene delivery, stabilization of labile proteins, DNA, and RNA, and increasing the transfection efficiency of nucleic acids. CDs do not elicit immune responses *in vivo* and thus have low toxicity (Biscotti et al. 2018). Bellocq et al. (2004) reported a synthetic biocompatible cyclodextrin-based delivery system with localized and sustained delivery of growth factors for wound healing therapy. A β-cyclodextrin–poly(ethylenimine) complex in nanoparticulate form was used as the controlled delivery vehicle for nucleic acids. The overall improvement achieved by this cyclodextrin-based carrier was high affinity of binding to the target site, as confirmed in a study conducted by Park and co-workers (Park et al. 2006).

1.3 Types of Multi-Cyclodextrin Entities: Properties and Applications

Multi-cyclodextrin entities are amphiphilic cyclodextrin derivatives that are formed through the assembly of cyclodextrin molecules. They possess unique mechanical and stimuli-responsive properties and significant drug encapsulation and controlled release features. The multi-cyclodextrin moieties of interest include cyclodextrin-based polyrotaxanes, cyclodextrin-based nanosponges, and cyclodextrin polymers (Liu et al. 2018).

1.3.1 Cyclodextrin-Based Polyrotaxanes

Polyrotaxanes (PRs) are formed when the bucket-shaped cyclodextrin molecule is threaded onto a compound with a linear geometry, such as polyethylene glycol (PEG) (Bai et al. 2018). The occurrence of threading is encouraged by bond formation, which takes place between the oligosaccharide and the linear polymeric molecules as a result of non-covalent interactions. PRs can be used as host–guest-based controlled delivery systems for drugs, peptides, proteins, and enzymes because of their ability to form inclusion and non-inclusion complexes. They undergo spontaneous de-complexation in

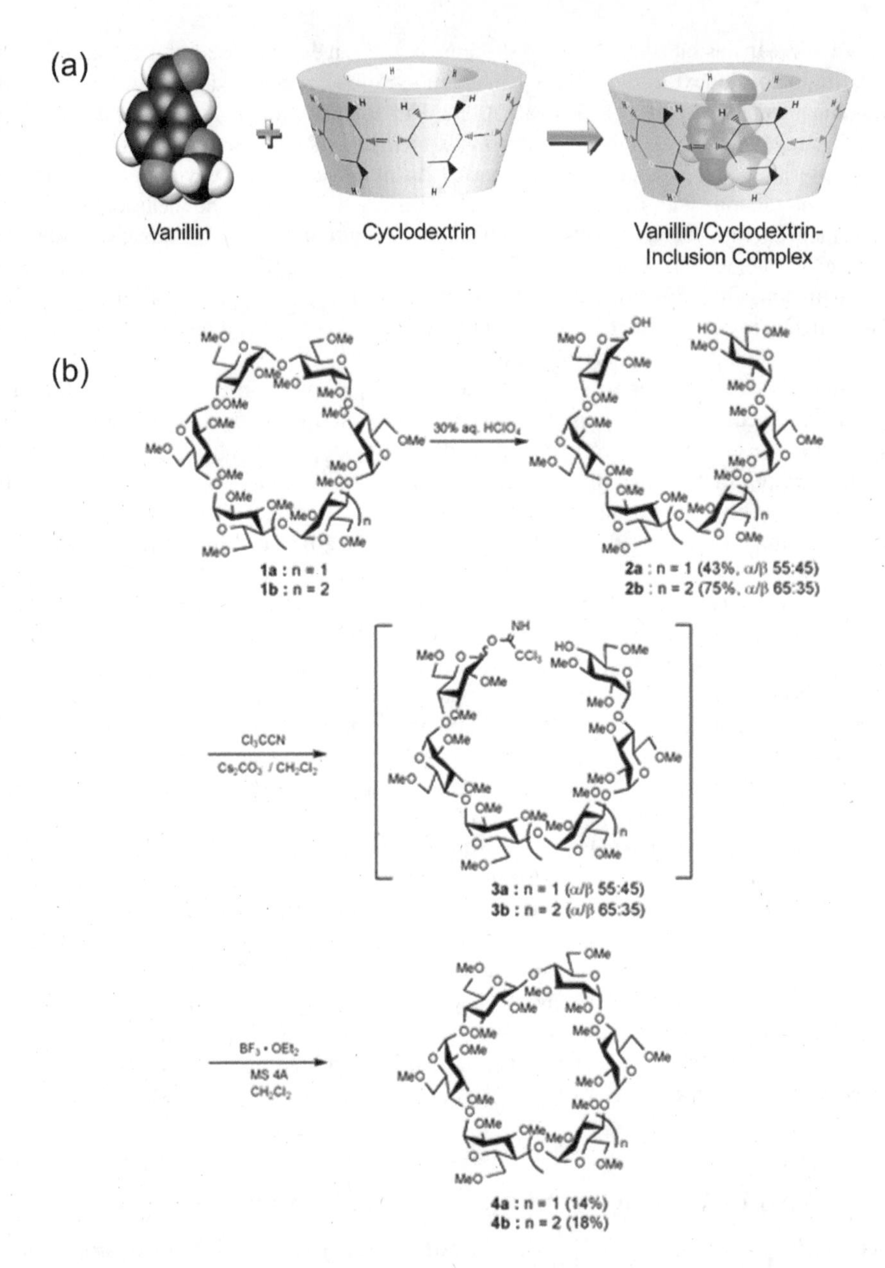

FIGURE 1.1 Schematic representation of (a) inclusion complex; (b) novel cyclodextrin derivatives incorporating one β-(1,4)-glucosidic bond, with the skeleton-modified cyclodextrin fitting the size and shape of the guest molecule. ([a] Reprinted with permission from Kayaci, F. and T. Uyar. 2011. Solid inclusion complexes of vanillin with cyclodextrins: their formation, characterization, and high-temperature stability. *Journal of Agricultural and Food Chemistry* 59, no. 21: 11772–11778. Copyright 2011 American Chemical Society. (b) From Kida, T., A. Kikuzawa, Y. Nakatsuji, and M. Akashi. 2003. A facile synthesis of novel cyclodextrin derivatives incorporating one β-(1,4)-glucosidic bond and their unique inclusion ability. *Chemical Communications* no. 24: 3020–3021. Reproduced by permission of The Royal Society of Chemistry.)

aqueous solutions and physiological media under specified temperature conditions. PRs possess a unique architecture, which can be ideal for use in encapsulated and targeted drug delivery (Resmerita et al. 2017). They function as stimuli-responsive supramolecular architectures by attaching to compounds with target-associated activity, such as peptides, proteins, etc., which respond to the stimulus at specific receptor sites. Different factors can be manipulated to ensure targeted controlled drug delivery, such as pH, temperature, and the attachment of antibodies to the polyrotaxane structure (Gómez et al. 2018). A prednisolone–α-cyclodextrin-star poly (ethylene glycol) poly-pseudopolyrotaxane with delayed pH-sensitive behavior was synthesized for targeted drug delivery. Prednisolone is an anti-inflammatory gluco-corticosteroid used for various autoimmune diseases. This class of drugs has very serious side effects, and targeted delivery could potentially decrease the dose needed to produce the pharmacological activity. Encapsulation of prednisolone within the pseudopolyrotaxane provides structural protection against acid hydrolysis and could potentially reduce the dose and frequency of administration (Resmerita et al. 2017).

Molecular recognition and self-assembly of supramolecular architectures are achieved through binding or encapsulation (Qin et al. 2017). The molecular recognition effect of PRs can be enhanced by conjugating other functional agent(s) (biological/chemical compounds) to the polyrotaxane structure (Adeoye and Cabral-Marques 2017), and this also functions to impart improved water solubility. For example, Choi and co-workers revealed that the addition of saccharide groups to cyclodextrin molecules threaded in a polyrotaxane increased the biomolecular recognition of lectin due to the mobility of the CDs affecting the intermolecular interactions (Choi et al. 2006), while in another study, a bis-(β-cyclodextrin) was synthesized and possessed multiple binding sites and recognition abilities, whereby a hydrophobic agent was incorporated, and the other site was determined to be suitable for metal binding (Cheng et al. 2003).

1.3.2 Cyclodextrin-Based Nanosponges

Cyclodextrin-based nanosponges (CDNS) are cyclodextrin cross-linked supramolecular architectures. The cross-linking is stabilized by non-covalent interactions such as hydrogen bonding, van der Waal's forces, ionic interactions, etc. They are dichotonic, hyper-cross-linked, porous, colloidal nanosystems (Caldera et al. 2017). They function as nanocarriers for a range of hydrophilic and lipophilic compounds with marked encapsulation efficiency (Morin-Crini et al. 2018). The encapsulation efficiency is markedly affected by the nature of the cross-linking agent; therefore, the fine tuning of the cyclodextrin to cross-linker ratio can improve drug loading, solubility, and controlled release kinetics. Nanosponges have a higher number of cyclodextrin molecules interconnected as compared with pure cyclodextrins; this is the reason for their significant swelling in aqueous solutions and a common mode of controlled drug release platforms (Gabr, Mortada, and Sallam 2018). They can be transported through aqueous media and improve the aqueous solubility of compounds with high lipophilicity (Caldera et al. 2017).

CDNS have found wide applications as novel excipients in drug delivery (Pushpalatha, Selvamuthukumar, and Kilimozhi 2018). This is due to the availability of various forms of nanosponges due to the cross-linking agent used, which is responsible for imparting new functional groups that provide novel structure–activity relations. Cross-linking agents such as diphenylcarbonate, pyromellitic dianhydride, epichlorohydrin, etc. have been extensively researched in the development of innovative nanosponge systems. These nanosponge forms are useful in numerous drug administration routes, as explained in the following paragraphs.

Carboxylated CDNS have been synthesized and used in the oral delivery of acyclovir, an antiviral drug used to treat herpes simplex virus (Lembo et al. 2013). The challenges that surround the delivery of acyclovir lie in the fact that it possesses a low oral bioavailability of about 10–30% because of slow and incomplete absorption in the gastrointestinal tract. The encapsulation of acyclovir in carboxylated CDNS resulted in prolonged release properties of the acyclovir drug and efficient enhancement of its antiviral efficacy *in vitro* on an HSV-1-based cell line. CDNS were also synthesized using diphenylcarbonate as a cross-linker (Minelli et al. 2012). These nanosponges were further used to incorporate camptothecin, an efficacious antitumor agent for hematological and solid tumors. This agent is highly labile in physiological environments due to the presence of the lactone ring in each chemical structure, which makes it highly susceptible to hydrolysis. Therefore, it is prone to chemical degradation and has very low aqueous solubility. The encapsulation resulted in prolonged release of camptothecin with an increase in concentration over time. The *in vitro* efficacy of camptothecin-loaded diphenylcarbonate CDNS was improved, as cytotoxicity was achieved at lower concentrations compared with commercial camptothecin formulations. The anti-proliferative effect and DNA damage were more pronounced, and proper stabilization of the drug was achieved. A ternary mixture of β-cyclodextrin nanosponges, itraconazole, and copolyvidonum was formulated by Swaminathan and co-workers (Swaminathan et al. 2007). Itraconazole is a biopharmaceutical classification system class II drug with dissolution rate–limited bioavailability. Copolyvidonum is a water-soluble polymer used in this mixture to enhance the encapsulation efficiency of the nanosponges (Swaminathan et al. 2007). The ternary mixture provided great improvement in the solubility of itraconazole, evidenced by the phase solubility studies, and therefore overcame the challenge of low bioavailability, as expected.

1.3.3 Cyclodextrin Polymers

Cyclodextrin-based polymers were the first of the supramolecular architectures to be fabricated. They are chemically linked cyclodextrin molecules produced using cross-linking agents such as epichlorohydrin or modification with other polymers (Chang et al. 2018). They can be either soluble or insoluble in water, depending on the chemicals used for their formation, and this directly affects their applications (Li et al. 2018). Cyclodextrin polymers have found a wide variety of applications in controlled drug delivery due to their low toxicity profiles by numerous administration routes. Cyclodextrin polymers are usually modified with other polymeric agents such as polyethylenimine, chitosan, hyaluronic acid, etc. Cyclodextrin-based polymers have been used extensively in gene and drug delivery applications. Zawko and co-workers prepared hyaluronic acid–based hydrogels functionalized with the methacryloyl derivative of β-cyclodextrin for the delivery of hydrocortisone (Zawko, Truong, and Schmidt 2008). Functionalization with a cyclodextrin contributes inclusion complex–forming capability to the hyaluronic acid hydrogel, and the system can be used in various drug delivery routes and dosage forms. A similar study was undertaken by Das and Subuddhi, in which preformed cyclodextrin and naproxen complexes were incorporated in a pH-responsive hydrogel made from chitosan and polyvinyl alcohol, cross-linked with glutaraldehyde as a colon-targeted delivery system, which was successful in delivering drug only in the intestinal fluid (Das and Subuddhi 2013). Thus, this delivery system could function well as a controlled drug-encapsulated platform intended to act specifically in the large intestine and for localized disease treatment.

1.4 Approaches to the Production of Cyclodextrins: Synthesis and Derivatives

Generally, cyclodextrins are produced by the enzymatic degradation of starch, specifically the amylase component. This reaction is catalyzed by the glucosyltransferase enzyme (Zhang et al. 2017). The production of α- and β-cyclodextrins occurs naturally in the presence of the bacterium *Bacillus macerans* (Pal, Gaba, and Soni 2018).

1.4.1 Cyclodextrin-Based Polyrotaxanes

Polyrotaxanes are synthesized through the threading of cyclodextrin molecules on linear inclusion compounds and/or the rotaxanation of pseudopolyrotaxanes (Li et al. 2016). The polyrotaxane yield is dependent on temperature control, reaction time, type of inclusion compound, molecular weight, and solvent type.

1.4.1.1 Threading Process

Cyclodextrin molecules are threaded onto a linear polymeric inclusion compound such as polydimethylsiloxane (PDMS) in the presence of a polar aprotic solvent (dimethyl formamide [DMF], dimethyl sulfoxide [DMSO], or acetonitrile), and the pseudopolyrotaxane (Figure 1.2a and b) precipitates are collected by centrifugation, filtration, or lyophilization (Iguchi et al. 2013). The threading process is driven by thermodynamic variations. The first reported synthesis involved the use of cyclodextrins threaded on PEG (Poudel et al. 2018). The product obtained can be either single or multiple cyclodextrin molecules

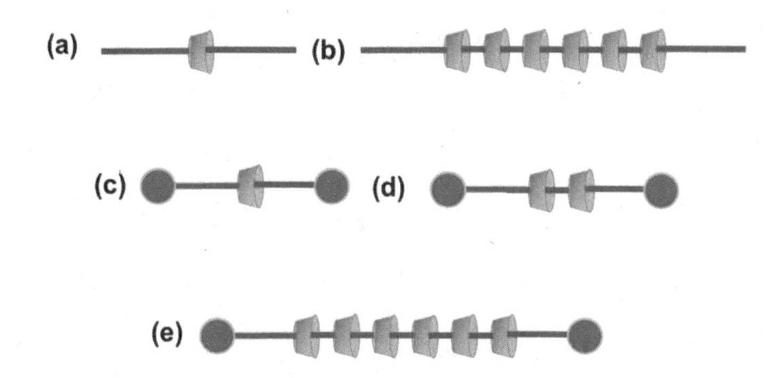

FIGURE 1.2 Diagrammatic representation of the nomenclature of cyclodextrin-based pseudopolyrotaxanes (a) single ring-like cyclodextrin molecule threaded on a linear compound, (b) multiple ring-like cyclodextrin molecules threaded on a linear compound and polyrotaxanes, and (c, d, and e) one, two, and multiple ring-like cyclodextrin molecules threaded on a linear compound capped with a stopper molecule to prevent de-threading.

threaded on the linear compound, as seen in Figure 1.2a and b. All this is a result of attenuation in the reaction parameters.

1.4.1.2 Rotaxanation of Pseudopolyrotaxanes

To prevent the occurrence of de-threading of the cyclodextrins, an end-capping molecule is added at both ends of the linear compound. This process is known as *rotaxanation*. The end-capping molecules should have very concrete specificity for the linear molecule to prevent reaction with the cyclodextrin molecules (Alvarez-Lorenzo, García-González, and Concheiro 2017). The rotaxanation of pseudopolyrotaxanes produces polyrotaxanes (Figure 1.2c–e). The difference between pseudopolyrotaxanes and polyrotaxanes is that pseudopolyrotaxanes easily de-thread in aqueous solutions due to the absence of end-capping molecules and easily separate in aqueous solutions, while polyrotaxanes remain insoluble (Higashi et al. 2017).

1.4.2 Cyclodextrin-Based Nanosponges

Nanosponges are synthesized through the cross-linking of cyclodextrins and further reduction to a nanorange size (Singh et al. 2018). Their shapes may vary, ranging from spherical to rod-like structures. The techniques employed in synthesis include solvent techniques and ultrasound synthesis. Solvent techniques, cross-linking cyclodextrins in the presence of a polar aprotic solvent such as DMF or DMSO, and the ultrasound-assisted method ensure the proper cross-linking of cyclodextrins (Leudjo Taka, Pillay, and Yangkou Mbianda 2017). The products are collected and washed with water to remove the non-reacted residues and dried in an oven at 60 °C, which is followed by a purification process (soxhlet extraction) using ethanol. The chemical nature of the cross-linker offers novel and peculiar properties to the CDNS. Moreover, the controlled drug release behavior from the complex varies according to the concentration of the cross-linkers employed during the synthesis. Similar reports were discussed by Caldera and co-workers (Caldera et al. 2017), evaluating the degree of cross-linking with CD nanosponges employing the popular cross-linker pyromellitic dianhydride. Spectroscopic evaluation proved that the highest degree of cross-linking was obtained using a molar ratio of 1:6 (CD: pyromellitic dianhydride). Higher molar ratios showed a decrease in the degree of cross-linking, possibly due to steric hindrance from the cross-linker linked to the CD units. Figure 1.3 thus explains the synthesis reaction, employing CDs as controlled release nanosponges, with significant influence of the cross-linker in relation to the optimum physico-chemical properties of the system (Caldera et al. 2017).

Cyclodextrin polymers are synthesized via various methods: cross-linking with other reagents, polymerization of acrylic monomers with cyclodextrins, and linking to a polymeric backbone through either covalent bonding or physical encapsulation in a polymeric chain (Thomsen et al. 2017). The most common of these methods is cross-linking bifunctional reagents such as di-epoxides.

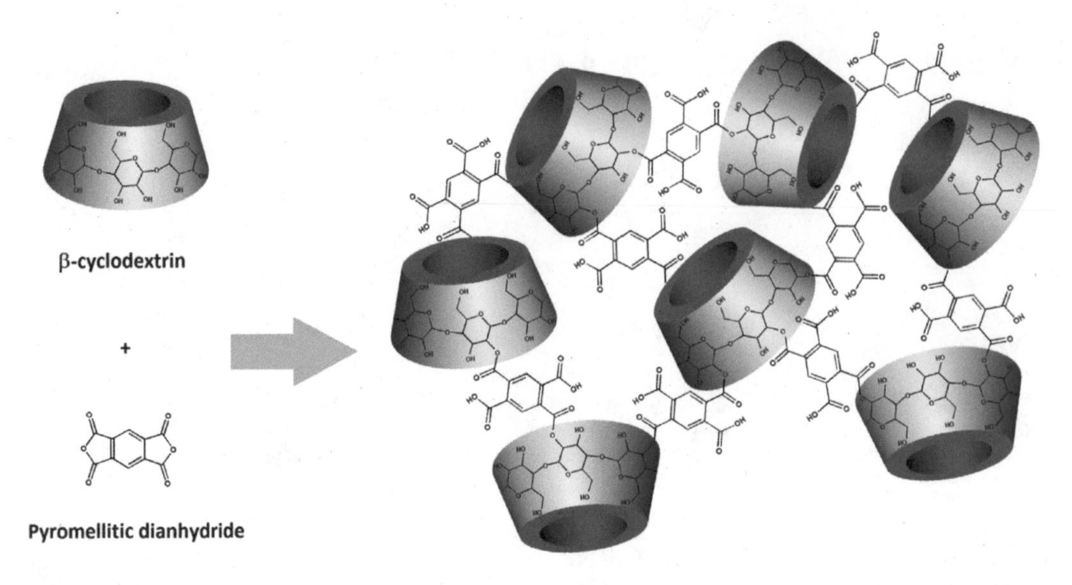

FIGURE 1.3 Schematic representation of cyclodextrin-based nanosponges employing a pyromellitic cross-linker for controlled release kinetics. (From *International Journal of Pharmaceutics*, 531, Caldera, F., M. Tannous, R. Cavalli, M. Zanetti, and F. Trotta, Evolution of cyclodextrin nanosponges, 470–479, Copyright (2017), with permission from Elsevier.)

1.5 Cyclodextrins and Multi-Cyclodextrins Applied for Topical and Targeted Drug Delivery

Topical and localized drug delivery is the direct administration of pharmaceuticals for local effects. This mode of drug delivery is advantageous due to site targeting, elimination of systemic effects, and improved drug encapsulation efficiency. This route of drug delivery presents inherent benefits. These include bypassing hepatic clearance, which improves drug bioavailability, limited systemic adverse events due to toxic drug blood levels, and prevention of unwanted effects in other body organs. Cyclodextrins, after topical administration, increase the amount of drug available at the surface of biological membranes such as the surface of tissues, skin, and mucosal membranes. This improves the drug's bioavailability by ensuring that the drug is available for absorption. It has been suggested in previous work that cyclodextrins themselves do not penetrate biological membranes (Shelley and Babu 2018). After delivering the drug to the membrane surface, cyclodextrins remain extracellular, as seen in Figure 1.4. This also means that possible intracellular damage is avoided. By attaching to the membrane surface, cyclodextrins may also improve the membrane permeability by binding with membrane components. Cyclodextrins are non-irritant to the surfaces of tissue, skin, and mucosa.

The applications of cyclodextrins for topical drug delivery are described primarily under three classifications: the surface of tissues and the dermal and mucosal routes. According to this classification, the tissue surface route of administration includes the eye, oral cavity, and tooth surfaces. The mucosal route includes buccal, sublingual, and nasal. Cyclodextrin-based formulations used for application in these routes are discussed in the following subsections.

1.5.1 Tissue Surface Applications

1.5.1.1 Ocular Surface Applications

Many ocular diseases are treated by surgery as well as administration of therapeutics by means of topical routes, intravitreal injection, or implantation. However, these administration routes are generally accompanied by complications that could have potential detrimental effects on the eye, which could lead

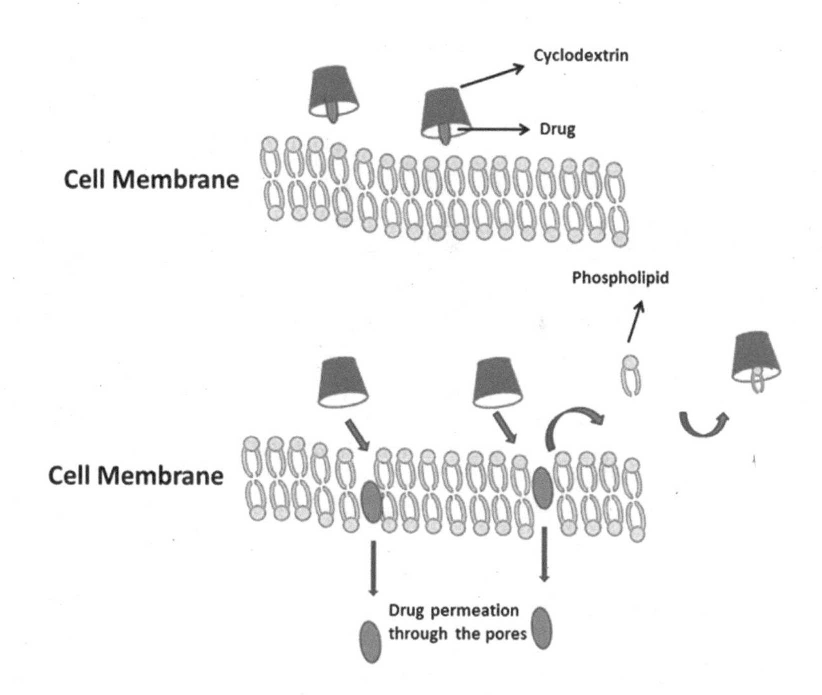

FIGURE 1.4 Diagrammatic representation of the pathway taken by cyclodextrin complexes after topical administration to biological membranes. Biological membranes have components such as phospholipids and cholesterol; their locations are the primary sites of entry of the drug. The cyclodextrin extracts these components to form inclusion complexes, which remain in the extracellular space until excretion takes place. The drug diffuses in through these new temporary pores. Drug diffusion also occurs through membrane transport pathways.

to severe vision impairment or blindness (Loftsson and Stefánsson 2017). The challenges associated with topical drug delivery systems for ocular diseases mainly revolve around penetration, absorption, solubilization, and bioavailability limitations. These are exerted by the natural makeup of the eye as well as the disease state. The ocular biofactors that significantly affect therapeutic effectiveness are the biological membranes, such as the cornea, sclera, and blood–retinal barrier, by notably affecting drug permeation and fluid drainage mediated through lacrimation, which may lead to premature drug elimination. The chemical degradation of drugs by ocular enzymes alters their absorption, therapeutic activity, and bioavailability (Jóhannsdóttir et al. 2017). The ideal properties of compounds intended for topical administration to the eyes are that they should be lipophilic enough to cross the ocular membranes and hydrophilic enough for stability and solubility in the aqueous humor; ultimately, they should be amphiphilic in nature. The use of cyclodextrins and multi-cyclodextrins achieves this, as they are amphiphilic compounds with a hydrophobic central cavity and an outer hydrophilic surface. A number of lipophilic drugs exist with applications in ocular disease treatment that have poor aqueous solubility and low bioavailability; cyclodextrins combat this through their outer hydrophilic surface, which provides stability in aqueous media and therefore, enhanced aqueous solubility (Jansook, Ogawa, and Loftsson 2018).

Jansook and co-workers reported that the use of γ-cyclodextrin modified with hydroxypropyl methylcellulose (HPMC) for the delivery of dorzolamide produced controlled drug release and improved the permeation and bioavailability of the drug to the posterior segment of the eye (Jansook et al. 2010). The major drawback of this group of drugs is its side effects. The γ-cyclodextrin/HPMC system could offer the possibility of reduced dosage with appropriate therapeutic success and delivery for drugs that can cause extreme adverse events. Another study aimed at improving the delivery of carbonic anhydrase inhibitors using a hydrophilic acrylic hydrogel with a built-in pendant cyclodextrin. The application of this delivery system achieved adequate penetration and bioavailability for acetazolamide and ethoxzolamide. Ribeiro and co-workers reported that the delivery system could be used for other antiglaucoma drugs (Ribeiro et al. 2012). According to Tanito and co-workers, cyclodextrins were well tolerated when applied topically to the eye. The study involved the use of cyclodextrins in a dexamethasone eye drop formulation. Clinical

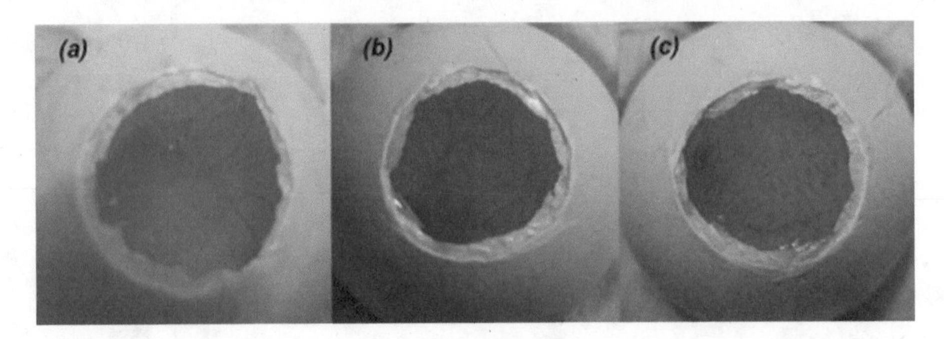

FIGURE 1.5 HET-CAM test result of (a) γ-cyclodextrin: HPMC nanogel, (b) negative control (saline), and (c) positive control (0.1 N NaOH). The test gave an irritation score close to zero for the nanogel and negative control, confirming that the nanogel is non-irritant. (From *International Journal of Pharmaceutics*, 441, Moya-Ortega, M., T. Alves, C. Alvarez-Lorenzo, A. Concheiro, E. Stefánsson, M. Thorsteinsdóttir, and T. Loftsson, Dexamethasone eye drops containing γ-cyclodextrin-based nanogels, 507–515, Copyright (2013) with permission from Elsevier.)

trials demonstrated that the active component, dexamethasone, maintained its efficacy against diabetic edema, decreased central macular thickness, and greatly enhanced visual acuity (Tanito et al. 2011). Alvarez-Lorenzo and co-workers reported the development of γ-cyclodextrin-based drug-loaded nanogels by cross-linking γ-cyclodextrin using an emulsification/solvent evaporation process. These aqueous controlled release eye drops were specifically studied for *in vivo* topical administration to rabbits. Moreover, the study revealed that the formulation showed reduced appearance and severity of side effects such as irritation, redness, and localized lethal effects (Figure 1.5) (Moya-Ortega et al. 2013). These nanogels can be applied to treat diseases that require an extended duration of treatment.

The use of cyclodextrin-based polyrotaxanes and nanosponges in future can offer the delivery of both hydrophilic and lipophilic agents and macromolecules and achieve targeted and controlled drug delivery (Sherje et al. 2017). Cyclodextrin-based polyrotaxanes can be used as transport-targeted delivery systems of prodrugs to overcome chemical degradation and can be linked to the prodrug activating agents to ensure complete therapeutic activity at the site of action. The targeting of ion channels/transporters is also possible. These cyclodextrin-based supramolecular architectures are amphiphilic in nature, which is ideal for permeation of the extracellular matrix and the inner endothelium layer (Li et al. 2016). Their physical nature bypasses melanin binding in the uvea; therefore, there is no disruption in drug pharmacological activity, as melanin is most likely to bind to lipophilic agents, and these supramolecular architectures have an outer hydrophilic layer. Since cyclodextrin-based supramolecular architectures can entrap both hydrophilic and hydrophobic drugs, they can improve the delivery of these agents to both the anterior and the posterior segments of the eye (Figure 1.6).

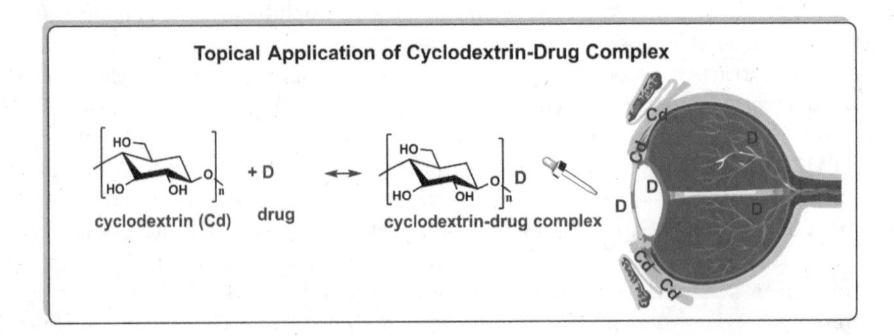

FIGURE 1.6 Diagrammatic representation of the mechanism of permeability enhancement exerted by cyclodextrin-based drug delivery systems. A reversible inclusion complex of a cyclodextrin and drug molecule forms on topical administration to the eye. The cyclodextrin extracts phospholipids and cholesterol from the biological membrane, which opens up temporary pores that function as sites of entry for the drug molecules, while the cyclodextrins themselves are excreted through the uveoscleral pathway.

1.5.1.2 Oral Cavity and Tooth Surface Applications

Loftsson and co-workers investigated the effects of β-cyclodextrins and their hydrophilic derivatives on silica-based triclosan toothpaste (Loftsson et al. 1999). The study found that the addition of β-cyclodextrins enhanced triclosan solubilization. The combination of triclosan, β-cyclodextrin, and carboxymethylcellulose increased the bioavailability, duration of action, and substantivity of triclosan. The tooth surface is covered by tooth enamel, whose main component is hydroxyapatite (HA). Lee and co-workers developed a biomineral-binding conjugate of alendronate–β-cyclodextrin (Liu et al. 2007). The conjugate showed a very strong binding affinity for HA. This conjugate holds great potential as a novel delivery system for oral cavity–localized diseases such as tooth decay, periodontitis, etc.

1.5.1.3 Dermal Applications

Skin has a prominent barrier function, and hence, the need to develop an optimized drug delivery formulation that overcomes these barriers is indispensable; otherwise, the bioactives may achieve only limited bioavailability (Bianchi et al. 2018). The properties required for an appropriate skin drug delivery chemical enhancer are: (i) it must be pharmaceutically inert, (ii) non-toxic and immediate and reversible in action, (iii) chemically and physically compatible, and (iv) cosmetically acceptable. Excipients such as cyclodextrins, when incorporated in pharmaceutical formulations, can extract dermal lipids, opening pores, which improves bioactive permeation and bioavailability for better therapeutic efficiency and efficacy (Sakulwech et al. 2018).

An attempt to overcome the challenges of dermal drug delivery using cyclodextrins has been made with success in the past. These include an increase in absorption, dissolution and solubility, skin permeability, and controlled release. Randomly methylated β-cyclodextrin (RAMEB) and 2-hydroxypropyl β-cyclodextrin (HP-β-CD) were reported earlier to incorporate a central dopaminergic agonist, piribedil, and were found to extract all the major classes of lipids, thus being able to open up more pores for penetration. Furthermore, they enhanced the percutaneous absorption of S-9977 hydrochloride (a cognition-enhancing drug) (Legendre et al. 1995). The combination of RAMEB with oleic acid caused the flux of S-9977 to increase about 30-fold. This delivery system could benefit other drugs with limited penetration. In another study, β-cyclodextrin and HP-β-CD were complexed with ketoprofen, a non-steroidal anti-inflammatory and analgesic agent, which is insoluble in aqueous media. The complexes were incorporated in multilamellar vesicle liposomes and are expected to positively influence drug penetration enhancement and improve drug efficacy (Maestrelli et al. 2005). In a similar study performed by Maestrelli and co-workers, the cyclodextrin complexation improved drug solubilization and entrapment efficiency in the aqueous phase of the liposome structure with extended release kinetics (Maestrelli et al. 2006). The effect of pH on the solubility of the drug–HP-β-CD complex was determined by using different concentrations of the complex at varying pH values. The ionized drug complex had a 2.5-fold increase in solubility as compared with the unionized complex and a much better intrinsic permeability (Sridevi and Diwan 2002). Sigurðoardottir reported on the effect of polyvinylpyrrolidone on cyclodextrin complexation of hydrocortisone and its diffusion through hairless mouse skin (Sigurðoardóttir 1995). The study revealed that with a gradual increase in the cyclodextrin concentration and a constant hydrocortisone concentration, a flux of hydrocortisone was observed until all its molecules were in solution following a flux decrease, resulting in increased cyclodextrin concentration. The addition of oleic acid to the system further enhanced the permeation capacity.

The application of multi-cyclodextrins in topical skin drug delivery would provide a more pronounced benefit as compared with unmodified cyclodextrins. Based on the applications of multi-cyclodextrin entities for other drug administration routes, by use of appropriate concentrations and further modification, can produce the desired mode of drug release. Multi-cyclodextrin entity based delivery systems can be manipulated to form different dosage forms including patches which achieve controlled and prolonged drug release. One of the challenges of dermal drug delivery is that most drugs, especially hydrophilic drugs, do not possess a suitable partition coefficient to interact with lipids (Sakulwech et al. 2018). Hence, the use of multi-cyclodextrin entities such as nanosponges would enable the encapsulation of these agents and achieve enhanced and targeted drug delivery when using polyrotaxanes.

1.5.2 Mucosal Membrane Applications

The mucous membranes cover internal organs (buccal, nasal, and sublingual mucosae) and form linings in cavities exposed to external organs. The nasal mucosa–targeting drugs are commonly formulated as nasal sprays, allowing maximum absorption for immediate therapeutic benefits. Thus, these routes of administration are essential for bypassing hepatic metabolism, allowing significant concentrations of encapsulated drug to be absorbed due to the profusely rich blood supply.

1.5.2.1 Buccal Mucosa

Buccal drug delivery is advantageous due to its high patient acceptance as compared with other non-oral drug administration routes (Mura et al. 2016). Therapeutic benefits are produced through the following mechanisms: bypassing liver clearance, direct access to systemic circulation, and therefore, enhanced bioavailability. The disadvantages are that the buccal membranes possess low penetrability and surface area. Cyclodextrins have been proved to provide various benefits in buccal drug delivery, such as improvement in permeation, stability, solubility, absorption, muco-adhesiveness, and the rate of dissolution. Figueiras and co-workers used methylated cyclodextrins as penetration enhancers for omeprazole, a hydrophobic agent that is highly unstable in neutral conditions (Figueiras et al. 2009). When the drug was complexed with pure β-cyclodextrins and methyl-β-cyclodextrins separately, both its stability and its penetration capacity were enhanced. A successful delivery system for carvedilol was reported by using methyl-β-cyclodextrins. The challenges in the use of carvedilol are its high metabolic clearance and poor solubility. Methyl cyclodextrin–based buccal tablets enhanced its muco-adhesive strength, permeation, and bioavailability (Mura et al. 2016). Another cyclodextrin derivative, HP-β-CD, has been reported for the effective delivery of darifenacin through the buccal route. Darifenacin is a urinary antispasmodic with significantly poor solubility and bioavailability of about 15–19% (Jagdale et al. 2013). Complexation with cyclodextrins produced an enhanced dissolution rate as compared with the free drug powder. This was administered using a patch dosage form, bypassing hepatic metabolism, with HP-β-CD significantly enhancing buccal absorption.

1.5.2.2 Sublingual Mucosa

Mannila and co-workers used α-cyclodextrins and chitosan derivatives to deliver low-solubility peptides intra-orally (sublingually) using cyclosporine, a hydrophobic agent, as the model peptide. This study confirmed enhanced solubility of the α-cyclodextrin–cyclosporine complexes as compared with the pure peptide agent (Mannila et al. 2009). The complexes had improved sublingual bioavailability. The addition of chitosan derivatives did not significantly affect the pharmacokinetic properties, substantiating α-cyclodextrin as a plausible drug delivery system alone. When a formulation is applied sublingually, it undergoes absorption due to profuse blood capillaries in the connective tissue under the epithelium. This route of drug delivery has received great attention as a solution in the delivery of labile peptides and proteins. As indicated, the major challenge in the delivery of hydrophobic drugs is their erratic and insufficient oral bioavailability. Researchers have reported the use of pure and modified cyclodextrins as mucosal membrane–localized delivery systems with improved solubility, penetration, and absorption. Hydrophobic drugs and labile peptides and proteins also displayed enhanced bioavailability using cyclodextrins (Jagdale et al. 2013).

1.5.2.3 Nasal Mucosa

The nasal route of drug delivery has been used for localized interests as well as the delivery of drugs to other parts of the body (e.g., the circulatory system) (Yalcin et al. 2016). Methyl-β-cyclodextrin, a hydrophilic cyclodextrin derivative, was found to act as a potent nasal absorption enhancer for peptides. This research can be dated a while back to studies undertaken by Romeijn and co-workers (Romeijn et al. 1990). The methylated β-cyclodextrin derivative is also known to function as a paracellular permeation enhancer, opening up the tight junctions in the nasal epithelium. Hydrophilic derivatives of cyclodextrins such as HP-β-CD and RM-β-CD were assessed for nasal toxicity and concentrations as high as 20% w/v, displaying no signs of toxic activity (Yalcin et al. 2016). Zhang and co-workers investigated the nasal delivery of insulin using a hyper-branched polyglycerol-functionalized β-cyclodextrin. The results proved that the insulin and the β-cyclodextrin derivative readily passed through the nasal mucosa with significant absorption (Zhang et al. 2011). Thus, CD derivatives and multi-cyclodextrin entities could be a new area of research in the future via the nasal drug delivery route.

1.6 Conclusion

Cyclodextrins have shown great benefits in topical drug delivery systems. This encourages their application for the efficient absorption and encapsulation of drugs for the treatment of locally manifested diseases. After delivering the drug to the membrane surface, cyclodextrins remain extracellular. This also means that possible intracellular damage is avoided. By attaching to the membrane surface, cyclodextrins may also improve the membrane permeability by binding with membrane components. Cyclodextrins are non-irritant to the surfaces of tissue, skin, and mucosa. They have also been proved to provide various benefits in tissue and mucosal membrane absorption, improving permeation, stability, and solubility. Hydrophobic drugs and labile peptides and proteins also displayed enhanced bioavailability using cyclodextrins. Since cyclodextrin-based supramolecular architectures can entrap both hydrophilic and hydrophobic drugs, they have significant potential for tissue surface absorption and membrane-targeted controlled drug delivery.

Conflict of Interest

The authors declare no conflict of interest.

Acknowledgments

This study has been funded by the National Research Foundation (NRF) of South Africa.

REFERENCES

Adeoye, O., and H. Cabral-Marques. 2017. Cyclodextrin nanosystems in oral drug delivery: a mini review. *International Journal of Pharmaceutics* 531, no. 2: 521–531.

Alvarez-Lorenzo, C., C. García-González, and A. Concheiro. 2017. Cyclodextrins as versatile building blocks for regenerative medicine. *Journal of Controlled Release* 268: 269–281.

Azzi, J., A. Jraij, L. Auezova, S. Fourmentin, and H. Greige-Gerges. 2018. Novel findings for quercetin encapsulation and preservation with cyclodextrins, liposomes, and drug-in-cyclodextrin-in-liposomes. *Food Hydrocolloids* 81: 328–340.

Bai, S., M. Hou, X. Shi, J. Chen, X. Ma, Y. Gao, Y. Wang, P. Xue, Y. Kang, and Z. Xu. 2018. Reduction-active polymeric prodrug micelles based on α-cyclodextrin polyrotaxanes for triggered drug release and enhanced cancer therapy. *Carbohydrate Polymers* 193: 153–162.

Barman, B., S. Barman, and M. Roy. 2018a. Inclusion complexation between tetrabutylphosphonium methanesulfonate as guest and α- and β-cyclodextrin as hosts investigated by physicochemical methodology. *Journal of Molecular Liquids* 264: 80–87.

Barman, S., B. Barman, and M. Roy. 2018b. Preparation, characterization and binding behaviors of host-guest inclusion complexes of metoclopramide hydrochloride with α- and β-cyclodextrin molecules. *Journal of Molecular Structure* 1155: 503–512.

Bellocq, N., D. Kang, X. Wang, G. Jensen, S. Pun, T. Schluep, M. Zepeda, and M. Davis. 2004. Synthetic biocompatible cyclodextrin-based constructs for local gene delivery to improve cutaneous wound healing. *Bioconjugate Chemistry* 15, no. 6: 1201–1211.

Bianchi, S., B. Machado, M. da Silva, M. da Silva, L. Bosco, M. Marques, A. Horn et al. 2018. Coumestrol/hydroxypropyl-β-cyclodextrin association incorporated in hydroxypropyl methylcellulose hydrogel exhibits wound healing effect: in vitro and in vivo study. *European Journal of Pharmaceutical Sciences* 119: 179–188.

Biscotti, A., C. Barbot, L. Nicol, P. Mulder, C. Sappei, M. Roux, I. Déchamps-Olivier, F. Estour, and G. Gouhier. 2018. MRI probes based on C6-peracetate β-cyclodextrins: synthesis, gadolinium complexation and in vivo relaxivity studies. *Polyhedron* 148: 32–43.

Cai, H., Y. Huang, and D. Li. 2017. Biological metal–organic frameworks: structures, host–guest chemistry and bio-applications. *Coordination Chemistry Reviews*.

Caldera, F., M. Tannous, R. Cavalli, M. Zanetti, and F. Trotta. 2017. Evolution of cyclodextrin nanosponges. *International Journal of Pharmaceutics* 531, no. 2: 470–479.

Chang, D., W. Yan, D. Han, Q. Wang, and L. Zou. 2018. A photo-switchable dual-modality linear supramolecular polymer based on host–guest interaction of cyclodextrin and pseudorotaxane. *Dyes and Pigments* 149: 188–192.

Cheng, J., K. Khin, G. Jensen, A. Liu, and M. Davis. 2003. Synthesis of linear, β-cyclodextrin-based polymers and their camptothecin conjugates. *Bioconjugate Chemistry* 14, no. 5: 1007–1017.

Choi, H., A. Takahashi, T. Ooya, and N. Yui. 2006. Molecular-recognition and binding properties of cyclodextrin-conjugated polyrotaxanes. *Chemphyschem* 7, no. 8: 1668–1670.

Choudhury, H., B. Gorain, M. Pandey, L. Chatterjee, P. Sengupta, A. Das, N. Molugulu, and P. Kesharwani. 2017. Recent update on nanoemulgel as topical drug delivery system. *Journal of Pharmaceutical Sciences* 106, no. 7: 1736–1751.

Das, S., and U. Subuddhi. 2013. Cyclodextrin mediated controlled release of naproxen from pH-sensitive chitosan/poly(vinyl alcohol) hydrogels for colon targeted delivery. *Industrial & Engineering Chemistry Research* 52, no. 39: 14192–14200.

da Silva Mourão, L., D. Ribeiro Batista, S. Honorato, A. Ayala, W. de Alencar Morais, E. Barbosa, F. Raffin, and T. de Lima e Moura. 2016. Effect of hydroxypropyl methylcellulose on beta cyclodextrin complexation of praziquantel in solution and in solid state. *Journal of Inclusion Phenomena and Macrocyclic Chemistry* 85, no. 1–2: 151–160.

Duchêne, D., and A. Bochot. 2016. Thirty years with cyclodextrins. *International Journal of Pharmaceutics* 514, no. 1: 58–72.

Figueiras, A., J. Hombach, F. Veiga, and A. Bernkop-Schnürch. 2009. In vitro evaluation of natural and methylated cyclodextrins as buccal permeation enhancing system for omeprazole delivery. *European Journal of Pharmaceutics and Biopharmaceutics* 71, no. 2: 339–345.

Gabr, M., S. Mortada, and M. Sallam. 2018. Carboxylate cross-linked cyclodextrin: a nanoporous scaffold for enhancement of rosuvastatin oral bioavailability. *European Journal of Pharmaceutical Sciences* 111: 1–12.

Gautam, S., S. Karmakar, R. Batra, P. Sharma, P. Pradhan, J. Singh, B. Kundu, and P. Chowdhury. 2017. Polyphenols in combination with β-cyclodextrin can inhibit and disaggregate α-synuclein amyloids under cell mimicking conditions: a promising therapeutic alternative. *Biochimica et Biophysica Acta (BBA) – Proteins and Proteomics* 1865, no. 5: 589–603.

Gómez, E., S. Anguiano Igea, J. Gómez Amoza, and F. Otero Espinar. 2018. Evaluation of the promoting effect of soluble cyclodextrins in drug nail penetration. *European Journal of Pharmaceutical Sciences* 117: 270–278.

Higashi, T., N. Ohshita, T. Hirotsu, Y. Yamashita, K. Motoyama, S. Koyama, R. Iibuchi et al. 2017. Stabilizing effects for antibody formulations and safety profiles of cyclodextrin polypseudorotaxane hydrogels. *Journal of Pharmaceutical Sciences* 106, no. 5: 1266–1274.

Ho, S., Y. Thoo, D. Young, and L. Siow. 2017. Inclusion complexation of catechin by β-cyclodextrins: characterization and storage stability. *LWT* 86: 555–565.

Iguchi, H., S. Uchida, Y. Koyama, and T. Takata. 2013. Polyester-containing α-cyclodextrin-based polyrotaxane: synthesis by living ring-opening polymerization, polypseudorotaxanation, and end capping using nitrile N-oxide. *ACS Macro Letters* 2, no. 6: 527–530.

Jagdale, S., P. Mohanty, A. Chabukswar, and B. Kuchekar. 2013. Dissolution rate enhancement, design and development of buccal drug delivery of darifenacin hydroxypropyl β-cyclodextrin inclusion complexes. *Journal of Pharmaceutics* 2013: 1–11.

Jansook, P., N. Ogawa, and T. Loftsson. 2018. Cyclodextrins: structure, physicochemical properties and pharmaceutical applications. *International Journal of Pharmaceutics* 535, no. 1–2: 272–284.

Jansook, P., E. Stefánsson, M. Thorsteinsdóttir, B. Sigurdsson, S. Kristjánsdóttir, J. Bas, H. Sigurdsson, and T. Loftsson. 2010. Cyclodextrin solubilization of carbonic anhydrase inhibitor drugs: formulation of dorzolamide eye drop microparticle suspension. *European Journal of Pharmaceutics and Biopharmaceutics* 76, no. 2: 208–214.

Jóhannsdóttir, S., J. Kristinsson, Z. Fülöp, G. Ásgrímsdóttir, E. Stefánsson, and T. Loftsson. 2017. Formulations and toxicologic in vivo studies of aqueous cyclosporin A eye drops with cyclodextrin nanoparticles. *International Journal of Pharmaceutics* 529, no. 1–2: 486–490.

Kayaci, F., and T. Uyar. 2011. Solid inclusion complexes of vanillin with cyclodextrins: their formation, characterization, and high-temperature stability. *Journal of Agricultural and Food Chemistry* 59, no. 21: 11772–11778.

Kida, T., A. Kikuzawa, Y. Nakatsuji, and M. Akashi. 2003. A facile synthesis of novel cyclodextrin derivatives incorporating one β-(1,4)-glucosidic bond and their unique inclusion ability. *Chemical Communications* no. 24: 3020–3021.

Legendre, J., I. Rault, A. Petit, W. Luijten, I. Demuynck, S. Horvath, Y. Ginot, and A. Cuine. 1995. Effects of β-cyclodextrins on skin: implications for the transdermal delivery of piribedil and a novel cognition enhancing-drug, S-9977. *European Journal of Pharmaceutical Sciences* 3, no. 6: 311–322.

Lembo, D., S. Swaminathan, M. Donalisio, A. Civra, L. Pastero, D. Aquilano, P. Vavia, F. Trotta, and R. Cavalli. 2013. Encapsulation of acyclovir in new carboxylated cyclodextrin-based nanosponges improves the agent's antiviral efficacy. *International Journal of Pharmaceutics* 443, no. 1–2: 262–272.

Leudjo Taka, A., K. Pillay, and X. Yangkou Mbianda. 2017. Nanosponge cyclodextrin polyurethanes and their modification with nanomaterials for the removal of pollutants from waste water: a review. *Carbohydrate Polymers* 159: 94–107.

Li, J., H. Xiao, J. Li, and Y. Zhong. 2004. Drug carrier systems based on water-soluble cationic β-cyclodextrin polymers. *International Journal of Pharmaceutics* 278, no. 2: 329–342.

Li, L., X. Chen, Q. Xia, X. Wei, J. Liu, Z. Fan, and M. Guo. 2016. Formation and characterization of pseudo-polyrotaxanes based on poly(p-dioxanone) and cyclodextrins. *Carbohydrate Polymers* 142: 82–90.

Li, X., M. Zhou, J. Jia, and Q. Jia. 2018. A water-insoluble viologen-based β-cyclodextrin polymer for selective adsorption toward anionic dyes. *Reactive and Functional Polymers* 126: 20–26.

Liu, T., X. Wan, Z. Luo, C. Liu, P. Quan, D. Cun, and L. Fang. 2018. A donepezil/cyclodextrin complexation orodispersible film: effect of cyclodextrin on taste-masking based on dynamic process and in vivo drug absorption. *Asian Journal of Pharmaceutical Sciences.*

Liu, X., H. Lee, R. Reinhardt, L. Marky, and D. Wang. 2007. Novel biomineral-binding cyclodextrins for controlled drug delivery in the oral cavity. *Journal of Controlled Release* 122, no. 1: 54–62.

Lockwood, S., S. O'Malley, and G. Mosher. 2003. Improved aqueous solubility of crystalline astaxanthin (3,3′-dihydroxy-β, β-carotene-4,4′-dione) by Captisol® (sulfobutyl ether β-cyclodextrin). *Journal of Pharmaceutical Sciences* 92, no. 4: 922–926.

Loftsson, T., N. Leeves, B. Bjornsdottir, L. Duffy, and M. Masson. 1999. Effect of cyclodextrins and polymers on triclosan availability and substantivity in toothpastes in vivo. *Journal of Pharmaceutical Sciences* 88, no. 12: 1254–1258.

Loftsson, T., and E. Stefánsson. 2017. Cyclodextrins and topical drug delivery to the anterior and posterior segments of the eye. *International Journal of Pharmaceutics* 531, no. 2: 413–423.

Lucio, D., M. Martínez-Ohárriz, Z. Gu, Y. He, P. Aranaz, J. Vizmanos, and J. Irache. 2018. Cyclodextrin-grafted poly(anhydride) nanoparticles for oral glibenclamide administration. In vivo evaluation using *C. elegans. International Journal of Pharmaceutics* 547, no. 1–2: 97–105.

Maestrelli, F., M. González-Rodríguez, A. Rabasco, and P. Mura. 2005. Preparation and characterisation of liposomes encapsulating ketoprofen–cyclodextrin complexes for transdermal drug delivery. *International Journal of Pharmaceutics* 298, no. 1: 55–67.

Maestrelli, F., M. González-Rodríguez, A. Rabasco, and P. Mura. 2006. Effect of preparation technique on the properties of liposomes encapsulating ketoprofen–cyclodextrin complexes aimed for transdermal delivery. *International Journal of Pharmaceutics* 312, no. 1–2: 53–60.

Mannila, J., K. Järvinen, J. Holappa, L. Matilainen, S. Auriola, and P. Jarho. 2009. Cyclodextrins and chitosan derivatives in sublingual delivery of low solubility peptides: a study using cyclosporin A, α-cyclodextrin and quaternary chitosan N-betainate. *International Journal of Pharmaceutics* 381, no. 1: 19–24.

Mayank, P., and H. Rajashree. 2018. Multicomponent cyclodextrin system for improvement of solubility and dissolution rate of poorly water soluble drug. *Asian Journal of Pharmaceutical Sciences*.

Minelli, R., R. Cavalli, L. Ellis, P. Pettazzoni, F. Trotta, E. Ciamporcero, G. Barrera, R. Fantozzi, C. Dianzani, and R. Pili. 2012. Nanosponge-encapsulated camptothecin exerts anti-tumor activity in human prostate cancer cells. *European Journal of Pharmaceutical Sciences* 47, no. 4: 686–694.

Morin-Crini, N., P. Winterton, S. Fourmentin, L. Wilson, É. Fenyvesi, and G. Crini. 2018. Water-insoluble β-cyclodextrin–epichlorohydrin polymers for removal of pollutants from aqueous solutions by sorption processes using batch studies: a review of inclusion mechanisms. *Progress in Polymer Science* 78: 1–23.

Moya-Ortega, M., T. Alves, C. Alvarez-Lorenzo, A. Concheiro, E. Stefánsson, M. Thorsteinsdóttir, and T. Loftsson. 2013. Dexamethasone eye drops containing γ-cyclodextrin-based nanogels. *International Journal of Pharmaceutics* 441, no. 1–2: 507–515.

Mura, P., M. Cirri, N. Mennini, G. Casella, and F. Maestrelli. 2016. Polymeric mucoadhesive tablets for topical or systemic buccal delivery of clonazepam: effect of cyclodextrin complexation. *Carbohydrate Polymers* 152: 755–763.

Ol'khovich, M., A. Sharapova, G. Perlovich, S. Skachilova, and N. Zheltukhin. 2017. Inclusion complex of antiasthmatic compound with 2-hydroxypropyl-β-cyclodextrin: preparation and physicochemical properties. *Journal of Molecular Liquids* 237: 185–192.

Pal, A., R. Gaba, and S. Soni. 2018. Effect of presence of α-cyclodextrin and β-cyclodextrin on solution behavior of sulfathiazole at different temperatures: thermodynamic and spectroscopic studies. *The Journal of Chemical Thermodynamics* 119: 102–113.

Pápay, Z., Z. Sebestyén, K. Ludányi, N. Kállai, E. Balogh, A. Kósa, S. Somavarapu, B. Böddi, and I. Antal. 2016. Comparative evaluation of the effect of cyclodextrins and pH on aqueous solubility of apigenin. *Journal of Pharmaceutical and Biomedical Analysis* 117: 210–216.

Park, I., H. von Recum, S. Jiang, and S. Pun. 2006. Supramolecular assembly of cyclodextrin-based nanoparticles on solid surfaces for gene delivery. *Langmuir* 22, no. 20: 8478–8484.

Park, S., E. Lih, K. Park, Y. Joung, and D. Han. 2017. Biopolymer-based functional composites for medical applications. *Progress in Polymer Science* 68: 77–105.

Poudel, A., F. He, L. Huang, L. Xiao, and G. Yang. 2018. Supramolecular hydrogels based on poly (ethylene glycol)-poly (lactic acid) block copolymer micelles and α-cyclodextrin for potential injectable drug delivery system. *Carbohydrate Polymers* 194: 69–79.

Pushpalatha, R., S. Selvamuthukumar, and D. Kilimozhi. 2018. Cross-linked, cyclodextrin-based nanosponges for curcumin delivery—physicochemical characterization, drug release, stability and cytotoxicity. *Journal of Drug Delivery Science and Technology* 45: 45–53.

Qin, Q., X. Ma, X. Liao, and B. Yang. 2017. Scutellarin-graft cationic β-cyclodextrin-polyrotaxane: synthesis, characterization and DNA condensation. *Materials Science and Engineering: C* 71: 1028–1036.

Resmerita, A., K. Assaf, A. Lazar, W. Nau, and A. Farcas. 2017. Polyrotaxanes based on PEG-amine with cucurbit[7]uril, α-cyclodextrin and its tris- O -methylated derivative. *European Polymer Journal* 93: 323–333.

Ribeiro, A., F. Veiga, D. Santos, J. Torres-Labandeira, A. Concheiro, and C. Alvarez-Lorenzo. 2012. Hydrophilic acrylic hydrogels with built-in or pendant cyclodextrins for delivery of anti-glaucoma drugs. *Carbohydrate Polymers* 88, no. 3: 977–985.

Romeijn, S., M. Deurloo, W. Hermans, N. Schipper, J. Verhoef, and F. Merkus. 1990. Absorption enhancement of intranasally administered insulin by sodium tauro-24,25-dihydrofusidate in rabbits and rats. *Journal of Controlled Release* 13, no. 2–3: 320–321.

Ryzhakov, A., T. Do Thi, J. Stappaerts, L. Bertoletti, K. Kimpe, A. Sá Couto, P. Saokham et al. 2016. Self-assembly of cyclodextrins and their complexes in aqueous solutions. *Journal of Pharmaceutical Sciences* 105, no. 9: 2556–2569.

Sakulwech, S., N. Lourith, U. Ruktanonchai, and M. Kanlayavattanakul. 2018. Preparation and characterization of nanoparticles from quaternized cyclodextrin-grafted chitosan associated with hyaluronic acid for cosmetics. *Asian Journal of Pharmaceutical Sciences* 13, no. 5: 498–504.

Shelley, H., and R. Babu. 2018. Role of cyclodextrins in nanoparticle-based drug delivery systems. *Journal of Pharmaceutical Sciences* 107, no. 7: 1741–1753.

Sherje, A., B. Dravyakar, D. Kadam, and M. Jadhav. 2017. Cyclodextrin-based nanosponges: a critical review. *Carbohydrate Polymers* 173: 37–49.

Sigurðoardóttir, A. 1995. The effect of polyvinylpyrrolidone on cyclodextrin complexation of hydrocortisone and its diffusion through hairless mouse skin. *International Journal of Pharmaceutics* 126, no. 1–2: 73–78.

Singh, P., X. Ren, T. Guo, L. Wu, S. Shakya, Y. He, C. Wang, A. Maharjan, V. Singh, and J. Zhang. 2018. Biofunctionalization of β-cyclodextrin nanosponges using cholesterol. *Carbohydrate Polymers* 190: 23–30.

Sridevi, S., and P. Diwan. 2002. Optimized transdermal delivery of ketoprofen using pH and hydroxypropyl-β-cyclodextrin as co-enhancers. *European Journal of Pharmaceutics and Biopharmaceutics* 54, no. 2: 151–154.

Suvarna, V., P. Gujar, and M. Murahari. 2017. Complexation of phytochemicals with cyclodextrin derivatives—an insight. *Biomedicine & Pharmacotherapy* 88: 1122–1144.

Swaminathan, S., P. Vavia, F. Trotta, and S. Torne. 2007. Formulation of betacyclodextrin based nanosponges of itraconazole. *Journal of Inclusion Phenomena and Macrocyclic Chemistry* 57, no. 1–4: 89–94.

Tambe, A., N. Pandita, P. Kharkar, and N. Sahu. 2018. Encapsulation of boswellic acid with β- and hydroxypropyl-β-cyclodextrin: synthesis, characterization, in vitro drug release and molecular modelling studies. *Journal of Molecular Structure* 1154: 504–510.

Tanito, M., K. Hara, Y. Takai, Y. Matsuoka, N. Nishimura, P. Jansook, T. Loftsson, E. Stefánsson, and A. Ohira. 2011. Topical dexamethasone-cyclodextrin microparticle eye drops for diabetic macular edema. *Investigative Ophthalmology & Visual Science* 52, no. 11: 7944.

Thomsen, H., G. Benkovics, É. Fenyvesi, A. Farewell, M. Malanga, and M. Ericson. 2017. Delivery of cyclodextrin polymers to bacterial biofilms—an exploratory study using rhodamine labelled cyclodextrins and multiphoton microscopy. *International Journal of Pharmaceutics* 531, no. 2: 650–657.

Ungaro, F., R. d'Emmanuele di Villa Bianca, C. Giovino, A. Miro, R. Sorrentino, F. Quaglia, and M. La Rotonda. 2009. Insulin-loaded PLGA/cyclodextrin large porous particles with improved aerosolization properties: in vivo deposition and hypoglycaemic activity after delivery to rat lungs. *Journal of Controlled Release* 135, no. 1: 25–34.

Xiao, R., and M. Grinstaff. 2017. Chemical synthesis of polysaccharides and polysaccharide mimetics. *Progress in Polymer Science* 74: 78–116.

Xu, J., Y. Xue, G. Hu, T. Lin, J. Gou, T. Yin, H. He, Y. Zhang, and X. Tang. 2018. A comprehensive review on contact lens for ophthalmic drug delivery. *Journal of Controlled Release* 281: 97–118.

Yalcin, A., E. Soddu, E. Turunc Bayrakdar, Y. Uyanikgil, L. Kanit, G. Armagan, G. Rassu, E. Gavini, and P. Giunchedi. 2016. Neuroprotective effects of engineered polymeric nasal microspheres containing hydroxypropyl-β-cyclodextrin on β-amyloid (1–42)–induced toxicity. *Journal of Pharmaceutical Sciences* 105, no. 8: 2372–2380.

Yang, L., M. Li, Y. Sun, and L. Zhang. 2018. A cell-penetrating peptide conjugated carboxymethyl-β-cyclodextrin to improve intestinal absorption of insulin. *International Journal of Biological Macromolecules* 111: 685–695.

Zhang, J., D. Liu, Y. Shi, C. Sun, M. Niu, R. Wang, F. Hu, D. Xiao, and H. He. 2017. Determination of quinolones in wastewater by porous β-cyclodextrin polymer based solid-phase extraction coupled with HPLC. *Journal of Chromatography B* 1068–1069: 24–32.

Zhang, S., H. Zhang, Z. Xu, M. Wu, W. Xia, and W. Zhang. 2017. *Chimonanthus praecox* extract/cyclodextrin inclusion complexes: selective inclusion, enhancement of antioxidant activity and thermal stability. *Industrial Crops and Products* 95: 60–65.

Zhang, W., H. Zhou, X. Chen, S. Tang, and J. Zhang. 2009. Biocompatibility study of theophylline/chitosan/β-cyclodextrin microspheres as pulmonary delivery carriers. *Journal of Materials Science: Materials in Medicine* 20, no. 6: 1321–1330.

Zhang, X., X. Zhang, Z. Wu, X. Gao, C. Cheng, Z. Wang, and C. Li. 2011. A hydrotropic β-cyclodextrin grafted hyperbranched polyglycerol co-polymer for hydrophobic drug delivery. *Acta Biomaterialia* 7, no. 2: 585–592.

Zhu, G., Y. Yi, and J. Chen. 2016. Recent advances for cyclodextrin-based materials in electrochemical sensing. *TrAC Trends in Analytical Chemistry* 80: 232–241.

2

Nanocomplexes for Topical Drug Delivery

Sougata Jana, Arijit Gandhi, and Kalyan Kumar Sen

CONTENTS

2.1 Introduction

Topical drug delivery has been shown to be a most attractive alternative to the conventional drug delivery methods of oral/injection administration. As well as non-invasiveness and convenience, the skin also acts as a "reservoir" that sustains delivery over a period of days. It offers multiple sites to avoid toxicity and local irritation, yet it can also concentrate drugs at local areas to avoid undesirable systemic effects (Hadgraft and Lane 2005; Díaz-Torres 2010). However, today, the clinical use of drug delivery through the skin is limited by the fact that very few drugs can be delivered topically at a viable rate. This problem is due to the stratum corneum of skin acting as an efficient barrier that limits the penetration of drugs through the skin, and some non-invasive methods are known to significantly increase the penetration of this barrier. To increase the availability of drugs for topical delivery, the use of nanocomplexes is interesting and has emerged as a valuable alternative for delivering hydrophilic and lipophilic drugs through the stratum corneum for the treatment of many diseases with the possibility of having a systemic or local effect. These nanocomplexes (such as nanoparticles, nanostructured lipid carriers, nanosuspensions, solid lipid nanoparticles [SLN], etc.) can be formulated using materials of many different types, and they are very different in their structure and chemical nature. They are too small to be detected by the immune system, and they can deliver the drug to the target sites using lower drug doses to reduce side effects. Fabricated nanocomplexes designed for drugs and other materials on the nanoscale (1–500 nm) can modify the basic bioactivity and properties of materials. Improvements in surface area, solubility, controlled release, and site-specific targeted delivery are some important characteristics that nanotechnology can design into drug delivery systems. Nanocarriers are attractive alternatives for systemic drug delivery via topical application (Rizwan et al. 2009; Escobar-Chávez et al. 2010; Papakostas et al. 2011). The technology provides high concentrations of drug to permeate through the skin and functionally create a drug depot in the epidermis and stratum corneum. This route of drug delivery provides similar advantages to transdermal patch technology in avoiding direct contact with the gastrointestinal tract (GIT) and hepatic first-pass effects, and it is also cosmetically more acceptable to many patients. The enhancement of physical permeation has led to more interest in topical drug delivery via the skin. Some of these novel advanced technologies for the permeation enhancement include iontophoresis, ultrasound, electroporation, microneedles to open up the skin, and more currently, the use of topical nanocarriers (Desai et al. 2010; Schroeter et al. 2010; Neubert 2011). Various research and reviews have been published containing detailed discussions on various aspects of transdermal nanocarriers. The present chapter focuses on an updated overview of the use of nanocomplexes in the pharmaceutical field, specifically in the area of topical and transdermal drugs. The design of nanocomplexes for application through the skin in the cosmetic and pharmaceutical fields has been emerging and interesting in the last few decades.

2.2 Skin Morphology and Barrier Functions

The skin serves as an excellent biological barrier in the human body. The skin composes 4% of body weight and is less permeable (102–104 times) than a blood capillary wall. The outer part of the skin (epidermis) is generally 0.02–0.2 mm thick in humans. The stratum corneum comprises the outermost 5–20 μm (Wertz and Downing 1989). The region below the epidermis is called the dermis. It is composed of the outer papillary and the inner reticular dermis, which taken together, are thicker (usually 5–20 times) than the epidermis and thus, normally measure up to 2 mm in thickness. These two inner skin parts contain only a few cells, predominantly fibroblasts and adipocytes, but a lot of collagen (~70%). The latter forms fibrous bundles, many micrometers long, to shelter as well as support glands, some immunologically active cells, nerves, and dermal lymph and blood capillaries. The segmental area of cutaneous blood vessels peaks ~100–150 μm below the skin surface (i.e., within the papillary dermis embracing the upper cutaneous blood plexus) (Braverman 1997). The total area of blood vessels then decreases quasi-exponentially to a depth of ~0.75 mm with a site- and skin condition–dependent decay length of ~0.1 mm; further permeability sink decay is about six times less steep (Cevc and Vierl 2007). The cutaneous micro-vasculature represents the secondary skin barrier by acting as a sink for the molecules that had diffused across the primary skin barrier. The papillary dermis is composed of relatively thin elastic elaunin fibers, which are broadly perpendicular to the skin surface. These fibers merge with the microfibrillar cascade (oxytalan fibers) and then intercalate into the dermal–epidermal junction. The collagen fibers in the reticular dermis are much thicker and largely parallel to the skin surface. Dermal collagens form a continuous, elastic network and provide mechanical elasticity to the skin from the reticular and papillary dermis and up to the epidermis (Kielty et al. 2002). The dermis transitions seamlessly into the hypodermis, which together with the dermal fatty deposit, helps to absorb mechanical shocks that might otherwise endanger the skin vasculature and nerves.

Although hair follicles, reaching from the outermost parts of the skin to the blood vessels, generally allow direct access of drug/bioactive compounds, such agents predominantly penetrate and permeate the skin by passing the horny layer, specifically using the tortuous intercellular pathway between the corneocytes (Barry 2002). Therefore, skin absorption varies with the physicochemical characteristics of the substances and the anatomical region of the application area. Generally, compounds having a molecular mass ˂0.6 kDa, adequate solubility in oil and water, and a high partition coefficient are well absorbed (Moseret et al. 2001). A partially damaged barrier, as observed in many skin diseases (e.g., psoriasis and atopic eczema), can favor drug permeation through the horny layer (Anigbogu et al. 1996), So, adequate skin penetration is still an important challenge in the development of topical dermatologicals. Conventional drug carrier systems, such as ointments and creams, result in drug uptake of only a few percent, which moreover, as a rule, is linked to a rather high inter-individual variation of uptake rates. Subsequently, drug concentration levels within the diseased skin may be sub-therapeutic in some patients while inducing unwanted systemic or local side effects in others. However, highly lipophilic compounds can penetrate via the hair follicle (Münster et al. 2005). Follicular uptake may occur in the case of rigid crystalline particles and particulate carriers with a size range of 3–10 μm. Apart from the horny layer, micrometer size does not exclude penetration into follicular orifices (Otberg et al. 2004). Besides acting as a permeation barrier, the horny layer is also a reservoir for topically applied compounds (Teichmann et al. 2005).

2.3 Purpose of Topical Preparation

To formulate an effective and efficient topical preparation, consideration must be given to the intended purpose. This is directly concerned with the site of action and the desired effect of the preparation. Topical preparations may be used for

- *Surface effects:* cosmetic (enhancement of appearance), protective (prevention of moisture loss, sunscreen), antimicrobial (reduction of infection).
- *Stratum corneum effects:* protective (e.g., sunscreens that penetrate this layer), keratolytic (a sloughing of the skin, useful in the treatment of psoriasis), protective (moisturizing).
- *Viable epidermal and dermal effects:* several classes of drugs may penetrate to these layers (anti-inflammatory, anesthetic, antipruritic, antihistamine). Although it is difficult for drugs to penetrate the stratum corneum, once they are in the dermis, they can diffuse into the general circulation. It is difficult to formulate a drug with only a local effect without subsequent uptake by the blood.
- *Systemic effects:* a few drugs, such as scopolamine, nitroglycerin, clonidine, and estradiol, have been formulated in a manner to achieve systemic effects.
- *Appendage effects:* some classes of drugs are intended to exert their action in these portions of the skin (antimicrobial and antiperspirant). Infection remains a major cause of morbidity and mortality following the shock phase in the burn patient. Measures to reduce the risk of wound infection and subsequent sepsis include early excision where possible and the use of topical antimicrobial creams such as silver sulfadiazine. The patient suffering major burns is at risk from both cutaneous and systemic infection (Hajare et al. 2014).

2.4 Nanotechnology for the Skin

Since the 1970s, skin has been used for the topical delivery of drugs or bioactive compounds. Nanoparticulate delivery systems to the skin are being increasingly used to develop systemic and local therapies. Generally, nanoparticle-mediated drug delivery into the epidermis and dermis without barrier modification has met with little success (Schneider et al. 2009). Where the barrier is compromised, however, such as in diseased or aged skin, there may be potential for increased particle penetration. Squamous cell (ulcerated) carcinoma is one example. The opportunities of and obstacles to nanoparticle drug delivery are only just beginning to be explored in ongoing clinical trials. For example, nanoparticles containing capsaicin are being used in the treatment of pain associated with diabetic neuropathy. Advances in particulate engineering and formulation science and an improved understanding of nanoparticles and skin interactions will undoubtedly lead to important, clinically relevant improvements in transdermal drug delivery. From the safety point of view, it must be recognized that non-biodegradable nanoparticles could be taken up and retained by the reticuloendothelial system (Erdogan 2009). Furthermore, there is the potential for local toxicity: recent research showed that some nanoparticles induce keratinocyte apoptosis, whereby the subtle relationship between shorter and longer phosphatidylcholine chain lengths makes the difference between life and death for keratinocytes (Liang and Chou 2009). Particles may interact with skin at a cellular level as adjuvants. These nanoparticle–skin interaction approaches can be used to enhance immune reactivity for the application of topical vaccines (Prow et al. 2010). Another nanoparticle–skin interaction approach is the topical use of silver nanoparticles as over-the-counter antimicrobial agents (Tian et al. 2007), where the nanoparticles showed slow release of silver ions, which have antimicrobial and wound healing properties. Silver ions from the nanoparticles can not only inhibit microbial proliferation but also accelerate wound healing. This formulation results in controlled release of silver ions while the nanoparticles remain on the skin surface.

2.5 Different Nanocarriers for Topical Delivery

The different advantages and applications of nanocarriers in topical drug delivery are given in Table 2.1.

2.5.1 Nanoparticles

During the last few years, there has been much more research interest in the field of nanoparticulate-based drug delivery systems as vehicles for various large and small molecules (Pal et al. 2007). Nanoparticles are colloidal/solid particles, where particle size ranges from 1 to 1000 nm. The major aims in designing nanoparticles as delivery systems are to control surface properties, particle size, and release of pharmacologically active compounds to achieve site-specific drug action at the therapeutically optimal rate and dose regimen. Among them, polymeric nanoparticles have attracted more attention due to their ease of surface modification, specificity, and stability (Liu et al. 2008). These colloidal vehicles are expected to develop adhesive interactions within the mucosa and remain in the GIT while preventing the entrapped

TABLE 2.1

Summary of Advantages and Applications of Nanocarriers in Topical Drug Delivery

Nanocarrier	Advantages	Topically delivered drugs	References
Nanoparticles	They are able to avoid the immune system due to their size. They can include antibodies in their surface to reach target organs. They achieve site-specific action.	Minoxidil, Triptolide, Dexamethasone phosphate, Testosterone	Mei et al. (2003)
Nanoemulsions	Easily applied to skin and mucous membranes. They are non-toxic and non-irritant.	Ketoprofen, Celecoxib	Kim et al. (2008); Baboota et al. (2007)
Liposomes	Control release based on natural lipids. High biocompatibility. Simple manufacture.	Amphotericin B, Estradiol, Methotrexate	Manosroi et al. (2004)
Niosomes	Biodegradable and low toxicity. Easy to prepare. Can encapsulate both hydrophilic and lipophilic moieties. Able to target organs for drug delivery.	Minoxidil, Hydroxyzine	Junyaprasert et al. (2012)
Dendrimers	They are easily prepared and functionalized. They increase the bioavailability of drugs. They covalently associate drugs. They also act like solubility enhancers, increasing the permeation of lipophilic drugs.	5-Fluorouracil, Diflunisal, Indomethacin	Yiyun et al. (2007); Venuganti and Perumal (2008)
SLN	Excellent biocompatibility. Improve stability of pharmaceuticals. High and enhanced drug content. Control and/or target drug release. High long-term stability.	Ketoconazole, Isotretinoin, Doxorubicin	Liu et al. (2007)

drug from enzymatic degradation until the release of the entrapped drug is either released or absorbed in an intact particulate form. Some reports suggest that the favored site for uptake is the lympho-epithelial M cells in the Peyer's patches (PP). It has been stated that microparticles remain in the PP, whereas nanoparticles are disseminated systemically (Hans and Lowman 2002). The different methods for preparation of nanoparticles include coacervation/precipitation, emulsion cross-linking, the emulsion coalescence method, spray drying, ionic gelation, and the reverse micelle method.

The advantages of using nanoparticles as a drug delivery system include that the surface characteristics and particle size of nanoparticles can be easily fabricated to achieve both active and passive drug targeting after administration. They sustain and control the release of the drug during transportation and at the site of localization, altering organ distribution of the drug and subsequent clearance so as to achieve an increase in the drug's therapeutic efficacy and reduced side effects. Controlled release and particle degradation properties can be readily modulated by the choice of matrix components. Drug loading is relatively high, and drugs can be incorporated into the systems without any chemical modification; this is an important factor for preserving the drug's activity. Site-specific drug targeting can be obtained by attaching targeting molecules to the surface of particles or the use of magnetic guidance (Kommareddy et al. 2005; Lee and Kim 2005).

Epicutaneous pharmacological applications are all based on organic particles. They sometimes focus on decreasing the depth of cutaneous penetration as well, but otherwise, seek to enhance skin permeation (Pardeike et al. 2009), mainly via a (semi)occlusive superficial film formation (Uner 2005).

2.5.2 Nanoemulsions

Nanoemulsions generally contain 20–500 nm large droplets, which are stabilized by surfactants. Nanoemulsions are isotropic dispersed systems of two non-miscible liquids, normally consisting of an oily system dispersed in an aqueous medium or an aqueous medium dispersed in an oily medium but forming droplet or other oily phases of nanometer size. They are thermodynamically non-stable systems, in contrast to microemulsions, because nanoemulsions need high energy to produce them. They are susceptible to Oswald ripening and as a consequence, susceptible to flocculation, creaming, and other physical instability problems related to emulsions. Despite this, they can be stable (metastable) for long periods due to their extremely small size and the use of sufficient surfactants. Hydrophilic and hydrophobic drugs can be fabricated in nanoemulsions, because it is possible to make oil/water or water/oil nanoemulsions. Shearing, especially in the high concentration range, accelerates the physical deterioration of any nanoemulsion. This may be problematic if a concentrated nanoemulsion is squeezed through a nanoporous membrane, such as skin. First introduced for dermal use in the early 1990s, nanoemulsions have since been explored extensively for this purpose (Wu 2001; Khandavilli and Panchagnula 2007; Shakeel 2008). Nanoemulsions can be formulated by three methods: microfluidization, high-pressure homogenization, and phase inversion temperature.

Khurana et al. developed a nanoemulsion (NE) gel formulation containing meloxicam (MLX) for transdermal drug delivery and showed maximum sustained and controlled drug release capacity. An *in vitro* drug release study showed that ~92% of MLX was released within 7 h from MLX solution, whereas ~95% was released within 24 h from MLX–NE gel, as depicted in Figure 2.1.

Percutaneous absorption studies showed higher permeation of MLX from NE gel than from drug solution. Differential scanning calorimetry (DSC) and Fourier transform infrared (FTIR) studies revealed stratum corneum lipid extraction as a possible penetration enhancer mechanism for MLX–NE gel. Confocal laser scanning microscopy (CLSM) studies showed permeation of the NE gel to the deeper layers of the skin (down to 130 μm). The MLX–NE gel turned out to be biocompatible and non-irritant and showed maximum inhibition of paw edema in rats over 24 h compared with MLX solution (Figure 2.2) (Khurana et al. 2013).

Ahmed et al. formulated alginate core-coated chitosan nanoparticles (NP) containing rabeprazole (RP) by a water-in-oil (W/O) nanoemulsion technique. The formulation of transdermal patches containing RP–NP avoids the first-pass effect and drug peroral acid sensitivity. The formulated patches showed minimal patch-to-patch variation. Patches loaded with RP–NP exhibited substantial skin permeability and controlled drug release and showed Fickian diffusion (Ahmed and El-Say 2014). The application

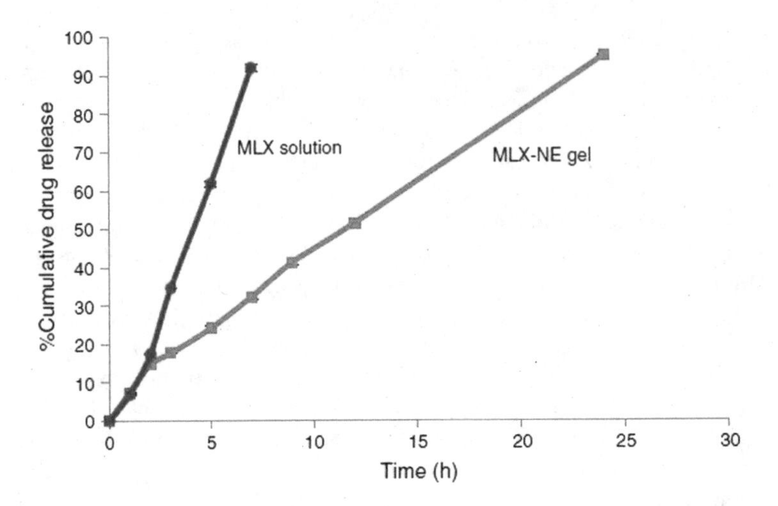

FIGURE 2.1 In vitro release profile of meloxicam from meloxicam-containing nanoemulsion gel (MLX–NE gel) and meloxicam (MLX) solution. (Reprinted from *Life Sciences*, 92, Khurana, S., N.K. Jain, P.M.S. Bedi, Nanoemulsion based gel for transdermal delivery of meloxicam: physico-chemical, mechanistic investigation, 383–392, Copyright (2013), with permission from Elsevier.)

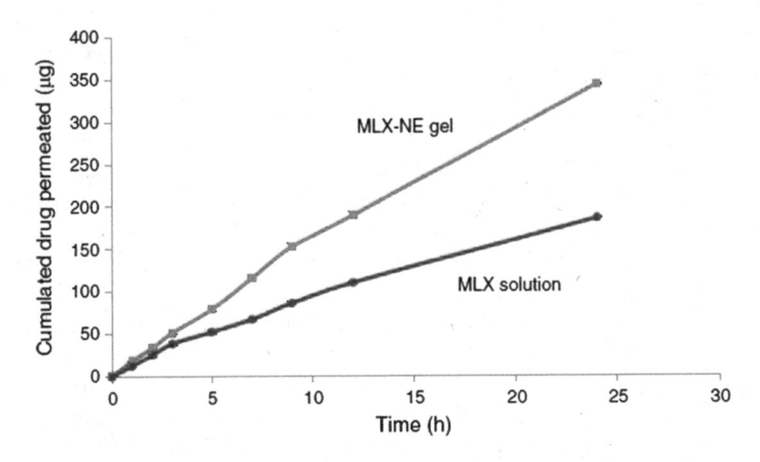

FIGURE 2.2 *Ex vivo* skin permeation profile of meloxicam from meloxicam-loaded nanoemulsion gel (MLX–NE gel) and meloxicam (MLX) solution across excised rat skin. (Reprinted from *Life Sciences*, 92, Khurana, S., N.K. Jain, P.M.S. Bedi, Nanoemulsion based gel for transdermal delivery of meloxicam: physico-chemical, mechanistic investigation, 383–392, Copyright (2013), with permission from Elsevier.)

of nanoemulsions for topical administration of ketoprofen was reported in some research publications. Sakeena et al. prepared nanoemulsions of palm oil esters by an emulsification method for the delivery of ketoprofen in the carrageenan-induced rat hind paw edema model. The nanoemulsions showed good drug release through methyl acetate membrane *in vitro* and comparable efficacy as compared with Fastun® gel *in vivo* (Sakeena et al. 2010). In another experiment, Kim et al. formulated nanoemulsions of ketoprofen with favorable stability and a high skin permeation rate (Kim et al. 2008). Baboota et al. investigated the potential of nanoemulsions for transdermal delivery of celecoxib. It was shown that the steady-state flux and permeability coefficient increased significantly compared with the gel formulations. In addition, the anti-inflammatory effects on carrageenan-induced paw edema in rats were higher (Baboota et al. 2007). Wu et al. explained the preparation of water-in-oil nanoemulsions containing expression plasmid DNA that appear to facilitate follicular transfection following topical application *in vivo*. The nanoemulsions without DNA had mean particle sizes of 14.6–42.3 nm, while the nanoemulsions with DNA showed sizes of 20.2–32.1 nm. Expression plasmids encoding chloramphenicol acetyltransferase (CAT) or human interferon-2 cDNA were formulated in water-in-oil nanoemulsions and applied to murine skin. It was shown that the deposition site of plasmid DNA was primarily in follicular keratinocytes. The transgene expression was optimal at 24 h following the topical application of a single dose of water-in-oil nanoemulsion containing plasmid DNA as quantified by reverse transcription polymerase chain reaction (RT-PCR) and enzyme-linked immunosorbent assay (ELISA). The efficiency of nanoemulsion-mediated transfection was most effective in the context of normal versus atrophic hair follicles. The results of this study suggested that the efficiency of transfection and the dynamics of transgene expression appear to be augmented by the use of a nanoemulsion vehicle. This may be the result of simple physical protection of plasmid DNA from endogenous deoxyribonucleases present in the skin. It is also possible that this effect is a consequence of alterations in cell membrane fluidity or membrane integrity that result from the presence of non-ionic detergents in the oil phase of the nanoemulsion. Other possibilities include undefined components of the organic plant oil that may facilitate transfection (Wu et al. 2001). Several mechanisms have been suggested for the enhancement of skin permeation from nanoemulsion formulations following topical administration. Surface charge–modified nanoemulsion droplets could have a significant influence on the binding affinity of droplets to the skin; nanoemulsions could act as drug reservoirs; the high concentration of drugs in nanoemulsions could result in a high concentration gradient; or due to the small droplet size of nanoemulsions, the oily droplets might become embedded into the stratum corneum, and the drug molecules could be directly delivered from the oily droplets into the stratum corneum.

2.5.3 Solid Lipid Nanoparticles (SLNs)

SLNs are submicron solid/colloidal carriers, particle size range 50–1000 nm, which are composed of physiological lipid and dispersed in water or in aqueous surfactant solution. SLNs show various properties such as small size, high drug loading, large surface area, and the interaction between phases at the interface and are attractive for their potential to increase the performance efficacy of pharmaceutical formulations (Gandhi et al. 2012). These lipid nanoparticles modify the release, body distribution, and kinetics of associated drugs. SLNs show great promise for controlled and site-specific drug targeting and hence, have attracted wide attention from researchers. The major advantages for topical products are the protective properties of SLNs for chemically labile drugs against degradation and the occlusion effect due to film formation on the skin. A major step forward was the development of ISLNs ("intelligent" SLNs), which can be used not only for topical but also for all other routes of delivery. In ISLNs, the SLN releases the incorporated bioactive compound/drug in a controlled manner after it receives a stimulating impulse. Such stimulating impulses are an increase in temperature or loss of water from an SLN-containing cream or SLN dispersion (Mei et al. 2003). SLNs have been used for topical application for various drugs such as DNA, tropolide, imidazole, ketoconazole, anticancers, antifungals, vitamin A, isotretinoin, flurbiprofen, and glucocorticoids. The penetration of podophyllotoxin–SLN into the stratum corneum along with the skin surface showed epidermal targeting. By using glyceryl behenate, vitamin A–loaded nanoparticles can be formulated. Various processes are useful for the improvement of penetration with sustained release. Isotretinoin-loaded lipid nanoparticles were formulated for topical drug delivery. Tween 80 and soyabean lecithin are used for the hot homogenization method for this purpose. The methodology is useful because of the increase of accumulative uptake of isotretinoin in skin. Flurbiprofen-loaded SLN gel for topical application has the advantage of delivering the drug directly to the targeted sites, which will produce higher tissue concentrations (Mei et al. 2005; Liu et al. 2007).

Anti-androgens may be applicable topically for acne using lipid nanoparticles. Mainly, loading of a lipophilic ester prodrug of the nonsteroidal anti-androgen RU 58841 has been reported. The active androgen is rapidly released by keratinocytes, dermal papilla cells, fibroblasts, and sebocytes (SZ95 cells). Interestingly, no drug has been detected in the acceptor medium following RU58841 myristate application to pig skin and even more importantly, reconstructed human epidermis (Münster et al. 2005). This should mean a clear reduction in systemic side effects, since the skin model has proved to be over-predictive as compared with human skin. Loading dye (Nile red) into SLNs allows the agent to be detected in high concentrations within the hair follicle infundibulum to a depth of about 650 μm.

Therefore, SLNs also have the potential to deliver not only anti-acne drugs but also agents for male baldness to the site of disease, which has also been claimed for liposome-encapsulated RU 58841. Unloaded and loaded SLNs have already been investigated for possible use in the application of ultraviolet (UV) absorbers. Cetyl palmitate nanodispersions act both as particulate UV blockers themselves and as carriers for UV-absorbing agents (2-hydroxy-4-methoxy benzophenone; Eusolex® 4360). These results showed a threefold increase in UV protection, which allows a reduction of the concentration of the UV absorber (Müller et al. 2002). Currently, similar effects have been explained when testing 3,4,5-trimethoxybenzoylchitin-loaded SLN. An increased effect is seen if tocopherol is added (Song et al. 2005). Moreover, microparticular preparations forming a UV-protecting film covering the skin appear interesting with respect to sunscreens (Lademann et al. 2004).

Raza et al. (2013) prepared isotretinoin-loaded SLNs for the topical treatment of acne. The drug entrapment efficiency (DEE) of the optimized formulation was ~89% and the size ~75 nm.

2.5.4 Nanostructured Lipid Carriers (NLCs)

NLCs are formulated by mixing liquid lipids (oils) and solid lipids. To obtain a uniform mixture for the particle matrix, solid lipids are mixed with liquid lipids, preferably in a ratio from 70:30 to 99.9:0.1. NLCs have been produced as novel systems consisting of physiological lipid materials suitable for dermal, topical, and transdermal applications. Many methods are used for the fabrication of lipid nanoparticles (NLCs). These methods are the microemulsion technique, high-pressure homogenization, emulsification–solvent evaporation, emulsification–solvent diffusion, solvent injection (or solvent displacement), phase inversion, the multiple emulsion technique, ultrasonication, and the membrane contractor technique. Among them, high-pressure homogenization is the most favorable method due to the many advantages it has compared with the other methods: e.g., the possibility of production on a large scale, the avoidance of organic solvents, and the short production time (Müller et al. 2002; Müller et al. 2007). The release of drug from lipid particles is observed to occur by lipid particle degradation and diffusion in the body. Ideally the drug release should be stimulated by an impulse when the particles are administered. NLCs accommodate drugs due to their highly unordered lipid structures. By applying a stimulation impulse to the matrix to convert it into a more ordered structure, a desired burst of drug release can be started. NLCs of certain structures can be stimulated in this way, for example, when applying the particles to the skin incorporated in a cream. The increase in temperature and water evaporation lead to an increase in the release rate of the drug (Schäfer-Korting et al. 2007).

Pinto et al. developed and assessed the potential of NLCs loaded with methotrexate as a new approach for topical therapy of psoriasis. Methotrexate-loaded NLCs were prepared via a modified hot homogenization combined with ultrasonication techniques using either polysorbate 60 (P60) or 80 (P80) as surfactant. The produced NLCs were within the nanosize range (274–298 nm) with a relatively low polydispersity index (<0.25). The entrapment efficiency of methotrexate in NLC–P60 and –P80 was approximately 65%. Cryo-scanning electron microscopy (SEM) images showed the spherical shape of the empty and methotrexate-loaded NLCs. The *in vitro* release of methotrexate from the NLCs followed a fast release pattern, reaching approximately 70% in 2 h. An *in vitro* skin penetration study demonstrated that methotrexate-loaded NLC–P60 had higher skin penetration when compared with free methotrexate, suggesting a significant role of drug–nanocarriers in topical administration. The results indicate the potential of NLCs for the delivery of methotrexate as topical therapy for psoriasis (Pinto et al. 2014).

Gainza et al. demonstrated the *in vivo* effectiveness of the topical administration of epidermal growth factor (rhEGF)–loaded lipid nanoparticles in healing-impaired mice. They reported the effectiveness of rhEGF–NLCs in a more relevant preclinical model of wound healing, the porcine full-thickness excisional wound model. The rhEGF–NLCs showed a particle size of around 335 nm and a high encapsulation efficiency of 94%. *In vivo* healing experiments carried out in large white pigs demonstrated that 20 μg of rhEGF–NLCs topically administered twice a week increased the wound closure and percentage of healed wounds by day compared with the same number of intralesional administrations of 75 μg free rhEGF and empty NLCs. Overall, these findings demonstrated that topically administered rhEGF–NLCs may generate *de novo* intact skin after full-thickness injury in a porcine model, thereby confirming their potential clinical application for the treatment of chronic wounds (Gainza et al. 2014). An NLC topical formulation was also designed as an alternative to oral and parenteral (IM) delivery of artemether (ART), a poorly water-soluble drug. A study was conducted by Nnamani et al. in which a Phospholipon 85G–modified Gelucire 43/01–based NLC formulation containing 75% Transcutol was loaded with ART by gradient concentration. ART-loaded NLCs were stable for up to 1 month of storage and polydispersed with a size range of 247–530 nm. *Ex vivo* evaluation showed detectable ART amounts after 20 h, which gradually increased over 48 h, achieving approximately 26% cumulative amount permeated irrespective of the applied dose. This proves that ART permeates excised human epidermis, where the current formulation acts as a reservoir to gradually control drug release over a prolonged period of time (Nnamani et al. 2014). Chen et al. investigated the potential of RPV–NLCs (ropivacaine-loaded NLCs) as a transdermal delivery system. RPV–NLCs were prepared by the method of emulsion evaporation–solidification at low temperature. The DEE of the optimized RPV-NLC formulation was 81.45 ± 2.16%. The average particle size was 203.5 ± 1.2 nm. The *in vitro* permeation study stated that RPV–NLCs could promote the permeation of RPV through skin to obtain transdermal delivery. In the mice writhing test, RPV–NLCs provided an analgesic effect by reducing the writhing response with an inhibition rate of 89.1% compared with the control group. In addition, the mechanism of permeation enhancement for NLCs investigated by histopathological study and DSC analysis revealed that NLCs could interact with stratum corneum, weaken the barrier function, and facilitate drug permeation (Chen et al. 2015). Joshi and Patravale formulated an NLC-based topical gel of celecoxib for the treatment of inflammation and allied conditions. NLCs prepared by the microemulsion template technique were characterized by photon correlation spectroscopy (PCS) for size and by SEM studies. The nanoparticulate dispersion was suitably gelled and assessed for *in vitro* release and *in vitro* skin permeation using rat skin. The efficacy of the NLC gel was established using a pharmacodynamic study in the Aerosil-induced rat paw edema model. The skin permeation and rat paw edema pharmacodynamic studies were carried out in comparison with a micellar gel that had the same composition as the NLC gel except for the solid lipid and oil. The NLC-based gel described in this study showed faster onset and elicited prolonged activity until 24 h (Joshi and Patravale 2008).

2.5.5 Liposomes

Liposomes are microscopic vesicles composed of one or more lipid bilayers arranged in a concentric fashion enclosing an equal number of aqueous compartments. Various amphipathic molecules have been used to form the liposomes, and the method of preparation can be tailored to control their size and morphology. Drug molecules can either be encapsulated in the aqueous space or intercalated into the lipid bilayer; the exact location of a drug in the liposome will depend on its physicochemical characteristics and the composition of the lipids (Gregoriadis, 1976). Liposomes, i.e., phospholipid vesicles, are widely applied for the topical treatment of diseases in dermatology. Many drugs encapsulated into liposomes show enhanced skin penetration. Because of their ability to provide sustained and controlled release of the loaded material, liposomes also have a potential for being applied vaginally. The major disadvantage of using liposomes topically and vaginally lies in the liquid nature of the preparation. To achieve the viscosity desirable for application, liposomes should be incorporated into a suitable vehicle. It has been well established that liposomes are fairly compatible with viscosity-increasing agents (methylcellulose and polyacrylic acid [Carbopol]) (Vyas and Khar 2001).

Liposomes are similar to the epidermis with respect to their lipid composition, which enables them to permeate the epidermal barrier to a large degree compared with other dosage forms. Where liposomes differ is that they are thought to act as "drug localizers," not only as "drug transporters," i.e., to enhance the accumulation of drug at the target site of administration. By virtue of the penetration of individual phospholipid molecules or non-ionic ether surfactants into the lipid layers of the stratum corneum and epidermis, they may serve as penetration enhancers and facilitate dermal delivery, leading to higher localized drug concentrations (Bangham et al. 1995).

Nishihata et al. reported that the *in vitro* penetration of diclofenac, formulated as a gel, through rat dorsal skin was poor. This lack of transport was due to the poor permeability of the drug through the stratum corneum. However, the co-administration of hydrogenated soya phospholipid liposomes increased the concentration of drug in the subcutaneous tissue as well as its permeation through the skin (Nishihata et al. 1987). Jacobes et al. tested the effect of pre-treating skin with phosphatidylcholine (PC), applied as a liposomal suspension, on the bioavailability of four formulations of corticosteroids with different inherent potencies. At the end of the second week of corticosteroid application, the blanching response to all four formulations on the PC–liposome-treated arms was significantly higher than on the control-treated arms. They suggested that the applied phospholipids either supplement the lipid content of the skin or provide a thin film in intimate epidermal contact. It was argued that such a film may promote hydration of the stratum corneum and also provide an environment into which corticosteroids initially partition before a subsequent, more controlled release to the underlying tissue (Jacobes et al. 1988). Rodney et al. fabricated the therapeutic and immunologic effects of the topical application of a liposome preparation containing both a macrophage activator, muramyltripeptidylethanolamine, and a recombinant antigen, glycoprotein D of HSV-1. This formulation was tested *in vitro* for its ability to stimulate peripheral blood lymphocytes and *in vivo* for the control of recurrent herpes genitalis in guinea pigs. They found that the liposomal antigen adjuvant preparation was capable of enhancing antigen-specific lymphocyte stimulation, which may be related to the observed 75% suppression of the frequency and severity of reactivation of recurrent HSV-2 genitalis compared with those of placebo controls (Rodney et al. 1989).

Various anti-inflammatory compounds have been useful topically as anti-aging remedies. Manconia et al. formulated C-phycocyanin liposomes and proved that the protein was mainly localized in the stratum corneum, while there was no permeation through the whole skin; 2% C-phycocyanin liposomes showed drug accumulation higher than that of the corresponding free 2% C-phycocyanin gel. Liposomal encapsulation also increased the anti-inflammatory activity of C-phycocyanin (Manconia et al. 2009). A study conducted by Yarosh et al. demonstrated that ursolic acid incorporated into liposomes increases both the ceramide content of cultured normal human epidermal keratinocytes and the collagen content of cultured normal human dermal fibroblasts. In clinical tests, the ceramide content of human skin was increased over an 11 day period. Liposomes had an effect on keratinocyte differentiation and dermal fibroblast collagen synthesis similar to that of retinoids (Yarosh 2000). In another study, Wen et al. formulated arbutin liposomes for enhancing skin-whitening activity. It was reported that although the permeation rate of arbutin in the liposome formulations decreased compared with arbutin solution, the deposition amount of arbutin in the epidermis/dermis layers increased in the liposomal formulation. These results revealed that a liposomal formulation could enhance the skin deposition of hydrophilic skin-whitening agents, thereby enhancing their activity (Wen et al. 2009). Fang et al. evaluated the effect of liposomal composition on the efficiency of transdermal catechin delivery. Catechins, which possess significant antioxidant activity, were encapsulated in liposomes using ethanol and anionic surfactants. The authors suggested that incorporation of anionic surfactants such as deoxycholic acid and dicetyl phosphate in the liposomes in the presence of 15% ethanol increased the catechin permeation by five- to sevenfold as compared with the control. Skin permeation studies were performed on mouse skin. The flexibility of bilayers was stated as an important factor governing the enhancing effect of liposomes (Fang et al. 2006).

2.5.6 Niosomes

Liposomes are simple microscopic lipid vesicles ranging from 20 nm to several micrometers in size. They are composed of a lipid bilayer, which presents with an aqueous volume entirely enclosed by a membrane composed of lipid molecules in such a way that both lipophilic and hydrophilic drugs can be successfully entrapped (Biju et al. 2006). Lipophilic drugs are entrapped within the bilayer membrane, whereas hydrophilic drugs are entrapped in the central aqueous core of the vesicles (Anwekar et al. 2011). Liposomes can be used for both topical and oral drug delivery. They act by attaching to cellular membranes and appear to fuse with them, releasing their contents into the cell. Sometimes, they are taken up by the cell, and their phospholipids are incorporated into the cell membrane, by which the drug trapped inside is released. In the case of phagocytic cells, liposomes are taken up, the phospholipid walls are acted on by organelles called lysosomes, and the entrapped drug is released. A number of components are present in liposomes, with phospholipids and cholesterol being the main ingredients. The most commonly used natural phospholipid is phosphatidylcholine (PC). Some other phospholipids, such as phosphatidyl serine (PS), phosphatidyl inositol (PI), phosphatidyl ethanolamine (PE), and phosphatidyl glycerol (PG), can also be used (Gandhi et al. 2012).

An increase in the penetration rate has been observed due to the transdermal delivery of drugs incorporated in niosomes. Jayraman et al. examined the topical delivery of erythromycin from various formulations, including niosomes, to the hairless mouse. From the studies and confocal microscopy, it was seen that non-ionic vesicles could be formulated to target pilosebaceous glands (Jayaraman et al. 1996). The conventional oral administration of hydroxyzine hydrochloride (an antihistamine) causes central nervous system (CNS) sedation, dry mouth, and tachycardia, whereas topical use in the form of semisolid dosage forms would lead to systemic side effects (Elzainy et al., 2003). A modified proniosomal gel of hydroxyzine hydrochloride was prepared by a coacervation–phase separation technique with different combinations of phospholipids and non-ionic surfactants (Spans and Tweens). Statistical experimental design was applied to optimize the various formulation variables. The optimized formulations were evaluated by *in vitro* and *ex vivo* permeation, skin irritation, skin deposition, and stability studies. Stability for 3 months at refrigeration temperature and quite high encapsulation efficiency (95%) and drug deposition in the stratum corneum in 24 h (90%) were found for the optimized formulation. The results showed that modified proniosomal formulations of hydroxyzine hydrochloride were appropriate for a topical drug delivery system for the treatment of localized urticaria (Rita and Lakshmi, 2012).

2.5.7 Dendrimers

In 1985, Donald A. Tomalia synthesized the first family of dendrimers. The word "dendrimer" originated from two words: the Greek word *dendron*, meaning tree, and *meros*, meaning part (Jana et al. 2012). They are synthetic nanomaterials that are approximately 2–10 nm in diameter. They are hyperbranched and monodisperse three-dimensional (3-D) molecules with defined molecular weights, large numbers of functional groups on the surface, and well-established host–guest entrapment properties. They are made up of layers of polymer surrounding a central core. The dendrimer surface contains many different sites to which drugs may be attached and also attachment sites for materials such as polyethylene glycol (PEG), which can be used to modify the way the dendrimer interacts with the body (Babu et al. 2010; Padilla et al. 2002).

Since 1979, dendrimers have generally been prepared using two major strategies. The first was introduced by Tomalia, called the *divergent method*, in which the growth of dendrimers originates from a core site. The second method, pioneered by Hawker and Fréchet, follows a *convergent growth process*, in which several dendrons are reacted with a multifunctional core to obtain a product. Dendrimers have narrow polydispersity; the nanometer size range of dendrimers can allow easier passage across biological barriers. All these properties make dendrimers suitable as hosts, binding guest molecules either in the interior or on the periphery of the dendrimers. Dendrimers have found applications in transdermal drug delivery systems. Generally, bioactive drugs have hydrophobic moieties in their structure, resulting in low water solubility, and as a result, efficient delivery into cells is inhibited. Dendrimers, designed to be highly water soluble and biocompatible, have been shown to be able to improve drug properties such as solubility and plasma circulation time via transdermal formulations and to deliver drugs efficiently (Jana et al. 2012). Chauhan et al. studied the transdermal ability of polyamidoamine (PAMAM) dendrimers using indomethacin as the model drug. *In vitro* permeation studies showed an increase in the steady-state flux with increase in the concentration of all three types: G4-NH_2, G4-OH, and G-4.5 PAMAM dendrimers. For the *in vivo* pharmacokinetic and pharmacodynamic studies, indomethacin and dendrimer formulations were applied to the abdominal skin of Wistar rats, and blood was collected from the tail vein at predetermined time intervals. The indomethacin concentration was significantly higher with PAMAM dendrimers when compared with the pure drug suspension. The results showed that an effective concentration could be maintained for 24 h in the blood with the G4 dendrimer–indomethacin formulation. Therefore, the data suggested that the dendrimer–indomethacin-based transdermal delivery system was effective and might be a safe and efficacious method for treating various diseases (Chauhan et al. 2003).

2.6 Delivery of Drugs through Nanocarriers

2.6.1 Anti-Inflammatory Drugs

There are various research results on nanoparticle-delivered drugs with anti-inflammatory properties for topical use, including aceclofenac, flufenamic acid (FFA), betamethasone-17-valerate, celecoxib, clobetasol propionate, flurbiprofen, corticosterone, ketoprofen, glycyrrhetic acid, naproxen, prednicarbate, nimesulide, and triptolide. Kuntsche et al. (2008) investigated the potential for SLN (tripalmitate), smectic nanoparticles (cholesteryl myristate and cholesteryl nonanoate), and cubic nanoparticles (glycerol monooleate) to deliver corticosterone to the skin (Kuntsche et al. 2008). Nanoparticle–skin interactions were evaluated with fluorescence microscopy. Storage of these nanoparticles for >15 months did not dramatically change the diameter or polydispersity of any of the formulations. Permeation studies were performed in excised human skin and a rat epidermal keratinocyte organotypic culture model mounted in Franz-type diffusion cells. Cumulative corticosterone in the receptor was measured over 48 h. These results showed that the phosphate buffered saline (PBS) control had the highest cumulative permeation at 1.5%, cubic nanoparticles at 1.0%, smectic nanoparticles at 0.4%, and SLNs at 0.1% in human epidermis. Luengo et al. described topical FFA delivery with poly-lactic-co-glycolic acid particles loaded with FFA (Luengo et al. 2006). The particles were characterized with atomic force microscopy. Cumulative penetration (area under the penetration curve) was determined for free and gel particle formulations. After 24 h, the particulate group showed significantly more FFA (6.7 μg/cm^2) in the deep skin layers than the non-particulate control (3.8 μg/cm^2)—an 1.8-fold increase.

Jana et al. 2014 investigated chitosan-based nanoparticles in Carbopol gel containing aceclofenac. The formulated drug-loaded nanoparticles were analyzed by the FTIR, DSC, field emission SEM (FE-SEM), and powder X-ray diffraction (P-XRD) methods. The *in vitro* drug release from nanoparticles revealed sustained aceclofenac release over 8 h. The formulated nanoparticles were composed of chitosan (200 mg), egg albumin (500 mg, and sodium tripolyphosphate (2% w/v). Drug entrapment was 96.32%, average particle size was 352.90 nm, and zeta potential was −22.10 mV. The nanoparticle-loaded Carbopol 940 gels were applied to rat skin and compared with the commercial formulation (Figure 2.3).

2.6.2 Anti-Photoaging Drugs and Antioxidants

Retinoids are generally used for anti-aging and topical acne purposes in topical nanoparticle delivery (Eskandar et al. 2009). Controlled release and improvement of stability for the formulated nanoparticles are the primary focal points. Castro et al. formulated SLNs containing tretinoin with an efficiency of 94% by using a stearylamine (a lipophilic amine). The ion pairing interaction occurs between the lipophilic amine and tretinoin to increase the lipophilicity of the tretinoin, which causes increased drug encapsulation efficiency (Castro et al. 2009). Tretinoin derivatives, vitamin A, isotretinoin, and retinol have also been delivered to skin using SLNs. The SLNs containing isotretinoin described by Liu et al. were in the particle size range of 30–50 nm, and the pharmacokinetic parameters were evaluated in excised abdominal rat skin. The DEE of formulated nanoparticles was 99% depending on the concentration of surfactant (4.5% Tween 80 and 8% soybean lecithin). Stability was evaluated up to 3 months at 2–8 °C with no changes in isotretinoin levels noted. The steady-state flux for the non-SLN control (0.06% isotretinoin) was 0.76 ± 0.3 μg/cm^2 h, but there was no isotretinoin found in the receptor chambers after 8 h. However, the skin retention of the control (2.81 μg) and the maximum SLN (3.65 μg) formulation was comparable. This was an increase of 30% from the non-SLN control (Liu et al. 2007). Vitamin A palmitate is converted into retinol in the skin and finally, to tretinoin. Vitamin A palmitate has been fabricated into SLNs ~350 nm in diameter by Pople and Singh. The pharmacokinetics of this formulation was examined using human cadaver skin with Keshary Chien cells. After 24 h, the release of drug was 67.5% from the SLNs and 54.4% from the gel control, implying that the flux was unusually higher with the SLN formulation. SLN formulations usually result in lower flux with longer-term release (Pople and Singh 2006). Zhao et al. stated the benefits of a hydrofluoroalkane foam to aid tocopheryl acetate release from lipid nanoparticles on contact with the skin. The nanoparticles were prepared with an oil-in-water

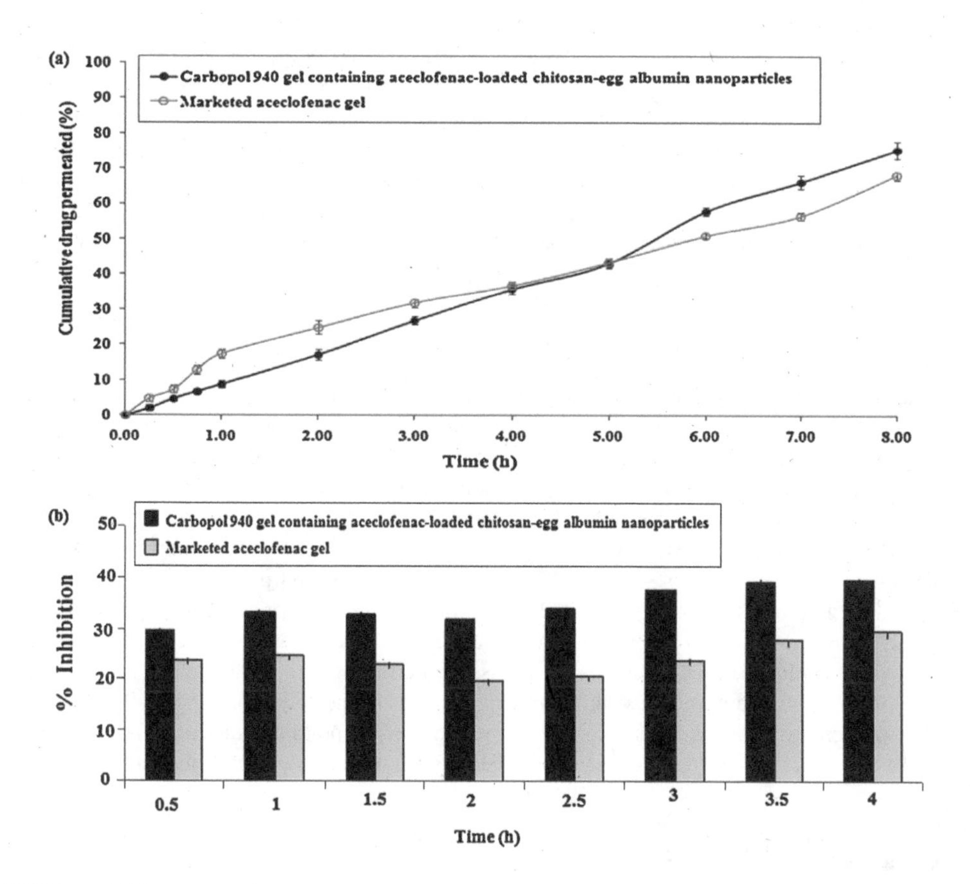

FIGURE 2.3 (a) The comparative *ex vivo* drug permeation from Carbopol 940 gel containing aceclofenac-loaded chitosan–egg albumin nanoparticles and a marketed aceclofenac gel through excised mouse skin (mean ± S.D.; n = 3); (b) comparative percentage inhibition profile of paw edema for Carbopol 940 gel containing aceclofenac-loaded nanoparticles and marketed aceclofenac gel at various time intervals in carrageenan-induced rat model for anti-inflammatory activity evaluation. (Reprinted from *Colloids and Surfaces B: Biointerfaces*, 114, Jana, S., S. Manna, A.K. Nayak, K.K. Sen, S.K. Basu, Carbopol gel containing chitosan-egg albumin nanoparticles for transdermal aceclofenac delivery, 36–44, Copyright (2014), with permission from Elsevier.)

emulsion, and the particle size range of the loaded nanoparticles was 53–57 nm. The drug encapsulation efficiency was 90%. Permeation into human skin was performed by treating skin with 400 μg/cm^2 tocopheryl acetate for 24 h (Zhao et al. 2009).

2.6.3 Antimicrobial Agents

Several experiments have been conducted to prepare topical nanoparticle formulations for antimicrobial delivery (Jain et al. 2009). These experiments involved two types of nanoparticles: SLN for the delivery of antifungal drugs and silver nanoparticles (Sanna et al. 2007; Sanna et al. 2009).

SLN particles containing econazole nitrate were characterized, and permeation studies were done to validate the system (Sanna et al. 2007). The particle size was 140–154 nm, and the DEE was 97–102%. The cumulative percentage econazole release from non-particle-containing gel into porcine skin was 124.72 μg/cm^2 after 24 h, and from the formulated gel (particle containing), it was 48.46 μg/cm^2. This is explained by the drug leaving the outer particle surface and leaving a high concentration of drug in the core. Tian et al. 2007 described silver nanoparticle topical delivery and wound healing evaluation in mouse models. The particle size of the silver nanoparticles was 14 nm. Silver nanoparticle–impregnated dressings were changed daily, making treatment continuous for the course of the experiments. Thermal injury took 35.4 days to heal without intervention and took nine fewer days (26.5 days) with silver nanoparticle treatment. Injury evaluated with the same concentration of silver, but without nanoparticles (silver sulfadiazine), took 37.4 days to heal. Samberg et al. evaluated an *in vivo* experiment in porcine skin (Samberg et al. 2010), in which the skin was treated topically with silver nanoparticles (particle size range 20–80 nm) for 14 days. The authors evaluated the effects of washing the nanoparticles and carbon coating the nanoparticles before application. Their results revealed that the sites of application showed focal inflammation, and none of the groups penetrated the stratum corneum. These *in vitro* and *in vivo* data revealed that carbon-coated silver nanoparticles were less toxic in nature.

2.6.4 Anti-Proliferative Agents

Several antiproliferation and anticancer drugs, such as 5-fluorouracil (5FU), 5-aminolevulinic acid (ALA), podophyllotoxin, and paclitaxel, were used in the form of drug-loaded nanoparticles (Venuganti and Perumal 2008; 2009). Podophyllotoxin is a topical antiproliferative drug used in first-line treatment for genital warts, but there is systemic absorption, and side effects can be dangerous. To overcome this type of problem, Chen et al. formulated a potential topical SLN–podophyllotoxin delivery to the skin surface (Chen et al. 2006). SLNs were prepared from soybean lecithin, tripalmitin, polysorbate 80, and poloxamer 188 in an aqueous solution. The prepared SLNs showed particle sizes between 44 and 194 nm, and all had negative zeta potential. Podophyllotoxin has a fluorescence property and can be excited at 290 nm with emission at 633 nm, so the endogenous fluorescence property was used to visualize podophyllotoxin localization after treating excised porcine ear skin. Podophyllotoxin fluorescence accumulation can be found in furrows and hair follicles. The encapsulation of podophyllotoxin in SLNs resulted in an increase in the cumulative podophyllotoxin. But, no podophyllotoxin could be detected in the receptor chamber of SLN-treated skin, showing that most of the drug was contained in the SLNs on the outer layers of the skin.

2.7 Conclusion

Topical drug delivery systems (TDDS) are self-administered and non-invasive. Due to their decreasing of side effects, they have received increased attention during the last few years. The important challenge to topical drug delivery is the barrier nature of skin, which protects against the entry of most active compounds. Today, nanocomplexes have been tried and tested to overcome the stratum corneum barrier to obtain higher transdermal permeation, and they have been fabricated to avoid immune system rejection and to reach specific target sites. However, the routes these nanocomplexes follow are very different. The main advantages of using nanocomplexes arise from their characteristic parameters, such as their high surface energy, particle size, architecture, and composition and attached molecules. The recent advances with regard to materials, techniques, and fabrication methods facilitate the development of new and better nanocarriers. This chapter is focused on several research reports, and their outcomes and results have been cited here in a concise manner. We hope this chapter will be helpful to new researchers for further investigations.

REFERENCES

Ahmed, T.A., K.M. El-Say. 2014. Development of alginate-reinforced chitosan nanoparticles utilizing W/O nanoemulsification/internal crosslinking technique for transdermal delivery of rabeprazole. *Life Sciences* 110: 35–43.

Anigbogu, A.N., A.C. Williams, B.W. Barry. 1996. Permeation characteristics of 8-methoxypsoralen through human skin; relevance to clinical treatment. *Journal of Pharmacy and Pharmacology* 48: 357–366.

Anwekar, H., S. Patel, A.K. Singhai. 2011. Liposome-as drug carrier. *International Journal of Pharmacy and Life Sciences* 2: 945–951.

Baboota, S., F. Shakeel, A. Ahuja, J. Ali, S. Shafiq. 2007. Design, development and evaluation of novel nanoemulsion formulations for transdermal potential of celecoxib. *Acta Pharmaceutica* 57: 315–320.

Babu, V.R., V. Mallikarjun, S. Nikhat, G. Srikanth. 2010. Dendrimers: a new carrier system for drug delivery. *International Journal of Pharmaceutical and Applied Sciences* 1: 1–10.

Bangham, A.D., M.M. Standish, J.C. Watkins. 1995. Diffusion of univalent ions across the lamellae of swollen phospholipids. *Journal of Molecular Biology* 13: 238–252.

Barry, B.W. 2002. Drug delivery routes in skin: a novel approach. *Advanced Drug Delivery Reviews* 54 (Suppl. 1): S31–S40.

Biju, S.S., S. Talegaonkar, P.R. Mishra, R.K. Khar. 2006. Vesicular systems: an overview. *Indian Journal of Pharmaceutical Sciences* 68: 141–153.

Braverman, I.M. The cutaneous microcirculation: ultrastructure and microanatomical organization. *Microcirculation* 4: 329–340.

Castro, G.A., A.L. Coelho, C.A. Oliveira et al. 2009. Formation of ion pairing as an alternative to improve encapsulation and stability and to reduce skin irritation of retinoic acid loaded in solid lipid nanoparticles. *International Journal of Pharmaceutics* 381: 77–83.

Cevc, G., U. Vierl. 2007. Spatial distribution of cutaneous microvasculature and local drug clearance after drug application on the skin. *Journal of Controlled Release* 118: 18–26.

Chauhan, A.S., S Sridevi, K.B. Chalasani et al. 2003. Dendrimer-mediated transdermal delivery: enhanced bioavailability of indomethacin. *Journal of Controlled Release* 90: 335–343.

Chen, H., X. Chang, D. Du et al. 2006. Podophyllotoxin-loaded solid lipid nanoparticles for epidermal targeting. *Journal of Controlled Release* 110: 296–306.

Chen, H., Y. Wang, Y., Zhai, G. Zhai, Z. Wang, J. Liud. 2015. Development of a ropivacaine-loaded nanostructured lipid carrier formulation for transdermal delivery. *Colloids and Surfaces A: Physicochemical and Engineering Aspects* 465: 130–136.

Desai, P., R.R. Patlolla, M. Singh. 2010. Interaction of nanoparticles and cell-penetrating peptides with skin for transdermal drug delivery. *Molecular Membrane Biology* 27: 247–259.

Díaz-Torres, R., 2010. Transdermal nanocarriers. In Escobar-Chávez, J.J., Merino, V., editors. *Current Technologies to Increase the Transdermal Delivery of Drugs*. Bussum, the Netherlands, Bentham Science, pp. 120–140.

Elzainy, A.A.W., X. Gu, F.E.R. Simons, K.J. Simons. 2003. Hydroxyzine from topical phospholipid liposomal formulations: evaluation of peripheral antihistaminic activity and systemic absorption in a rabbit model. *The AAPS Journal* 5: 41–48.

Erdogan, S. 2009. Liposomal nanocarriers for tumor imaging. *Journal of Biomedical Nanotechnology* 5: 141–150.

Eskandar, N.G., S. Simovic, C.A. Prestidge. 2009. Nanoparticle coated emulsions as novel dermal delivery vehicles. *Current Drug Delivery* 6: 367–373.

Fang, J.Y., T.L. Hwang, Y.L. Huang, C.L. Fang. 2006. Enhancement of the transdermal delivery of catechins by liposomes incorporating anionic surfactants and ethanol. *International Journal of Pharmaceutics* 310: 131–138.

Gainza, G., D.C. Bonafonte, B. Moreno et al. 2015. The topical administration of rhEGF-loaded nanostructured lipid carriers (rhEGF-NLC) improves healing in a porcine full-thickness excisional wound model. *Journal of Controlled Release* 197: 41–47.

Gandhi, A., A. Paul, S. Sheet, A. Banerjee. 2012. Solid lipid nanoparticles: a promising approach in drug delivery system. *American Journal of PharmTech Research* 2: 168–189.

Gandhi, A., S.O. Sen, A. Paul. 2012. Current trends in niosome as vesicular drug delivery system. *Asian Journal of Pharmacy and Life Science* 2: 339–353.

Gregoriadis, G. 1976. The carrier potential of liposomes in biology and medicine. *New England Journal of Medicine* 295: 704–710.

Hadgraft, J., M.E. Lane. 2005. Skin permeation: the years of enlightenment. *International Journal of Pharmaceutics* 305: 2–12.

Hajare, A.A., S.S. Mali, A.A. Ahir et al. 2014. Lipid nanoparticles: a modern formulation approach in topical drug delivery systems. *Journal of Advanced Drug Delivery* 1: 30–37.

Hans, M.L., A.M. Lowman. 2002. Biodegradable nanoparticles for drug delivery and targeting. *Current Opinion in Solid State & Materials Science* 6: 319–327.

Jacobes, M., G.P. Martin, C. Marriott. 1988. Effects of phosphatidylcholine on the topical bioavailability of corticosteroids assessed by human skin blanching assay. *Journal of Pharmacy and Pharmacology* 40: 829–833.

Jain, J., S. Arora, J.M. Rajwade, P. Omray, S. Khandelwal, K.M. Paknikar. 2009. Silver nanoparticles in therapeutics: development of an antimicrobial gel formulation for topical use. *Molecular Pharmaceutics* 6: 1388–1401.

Jana, S., A. Gandhi, K.K. Sen, S.K. Basu. 2012. Dendrimers: synthesis, properties, biomedical and drug delivery applications. *American Journal of PharmTech Research* 2: 32–55.

Jana, S., S. Manna, A.K. Nayak, K.K. Sen, S.K. Basu. 2014. Carbopol gel containing chitosan-egg albumin nanoparticles for transdermal aceclofenac delivery. *Colloids and Surfaces B: Biointerfaces* 114: 36–44.

Jayaraman, C.S., C. Ramachandran, N. Weiner. 1996. Topical delivery of erythromycin from various formulations: an in vivo hairless mouse study. *Journal of Pharmaceutical Sciences* 85: 1082–1084.

Joshi, M., V. Patravale. 2008. Nanostructured lipid carrier (NLC) based gel of celecoxib. *International Journal of Pharmaceutics* 346: 124–132.

Junyaprasert, V.B., P. Singhsa, J. Suksiriworapong, D. Chantasart. 2012. Physicochemical properties and skin permeation of Span 60/Tween 60 niosomes of ellagic acid. *International Journal of Pharmaceutics* 2012: 303–311.

Khandavilli, S., R. Panchagnula. 2007. Nanoemulsions as versatile formulations for paclitaxel delivery: peroral and dermal delivery studies in rats. *Journal of Investigative Dermatology* 127: 154–162.

Khurana, S., N.K. Jain, P.M.S. Bedi. 2013. Nanoemulsion based gel for transdermal delivery of meloxicam: physico-chemical, mechanistic investigation. *Life Sciences* 92: 383–392.

Kielty, C.M., M.J. Sherratt, C.A. Shuttleworth. 2002. Elastic fibres. *Journal of Cell Science* 115: 2817–2828.

Kim, B.S., M. Won, K.M. Lee, C.S. Kim. 2008. In vitro permeation studies of nanoemulsions containing ketoprofen as a model drug. *Drug Delivery* 15: 465–469.

Kommareddy, S., S.B. Tiwari, M.M. Amiji. 2005. Long-circulating polymeric nanovectors for tumor-selective gene delivery. *Technology in Cancer Research & Treatment* 4: 615–625.

Kuntsche, J., H. Bunjes, A. Fahr et al. 2008. Interaction of lipid nanoparticles with human epidermis and an organotypic cell culture model. *International Journal of Pharmaceutics* 354: 180–195.

Lademann, J., A. Rudolph, U. Jacobi et al. 2004. Influence of nonhomogeneous distribution of topically applied UV filters on sun protection factors. *Journal of Biomedical Optics* 9: 1358–1362.

Lee, M., S.W. Kim. 2005. Polyethylene glycol-conjugated copolymers for plasmid DNA delivery. *Pharmaceutical Research* 22: 1–10.

Liang, C.H., T.H. Chou. 2009. Effect of chain length on physicochemical properties and cytotoxicity of cationic vesicles composed of phosphatidylcholines and dialkyldimethylammonium bromides. *Chemistry and Physics of Lipids* 158: 81–90.

Liu, J., W. Hu, H. Chen, Q. Ni, H. Xu, X. Yang. 2007. Isotretinoin-loaded solid lipid nanoparticles with skin targeting for topical delivery. *International Journal of Pharmaceutics* 328: 191–195.

Liu, Z., Y. Jiao, Y. Wang, C. Zhou, Z. Zhang. 2008. Polysaccharide-based nanoparticles as drug delivery systems. *Advanced Drug Delivery Reviews* 60: 1650–1662.

Luengo, J., B. Weiss, M. Schneider et al. 2006. Influence of nanoencapsulation on human skin transport of flufenamic acid. *Skin Pharmacology and Physiology* 19: 190–197.

Manconia, M., J. Pendás, N. Ledón, T. Moreira, C. Sinico, L. Saso. 2009. Phycocyanin liposomes for topical anti-inflammatory activity: in-vitro in-vivo studies. *The Journal of Pharmacy and Pharmacology* 61: 423–430.

Manosroi, A., L. Kongkaneramit, J. Manosroi. 2004. Stability and transdermal absorption of topical amphotericin B liposome formulations. *International Journal of Pharmaceutics* 270: 279–286.

Mei, Z., H. Chen, T. Weng, Y. Yang, X. Yang. 2003. Solid lipid nanoparticle and microemulsion for topical delivery of triptolide. *European Journal of Pharmaceutics and Biopharmaceutics* 56: 189–196.

Mei, Z., Q. Wu. 2005. Triptolide loaded solid lipid nanoparticle hydrogel for topical application. *Drug Development and Industrial Pharmacy* 31: 161–168.

Moser, K., K. Kriwet, C. Froehlich, Y.N. Kalia, R.H. Guy. 2001. Supersaturation: enhancement of skin penetration and permeation of a lipophilic drug. *Pharmaceutical Research* 18: 1006–1011.

Müller, R.H., A. Hommoss, J. Pardeike, C. Schmidt. 2007. Lipid nanoparticles (NLC) as novel carrier for cosmetics—special features & state of commercialisation. *SÖFW* 9: 40–46.

Müller, R.H., M. Radtke, S.A. Wissing. 2002. Solid lipid nanoparticles (SLN) and nanostructured lipid carriers (NLC) in cosmetic and dermatological preparations. *Advanced Drug Delivery Reviews* 54 (Suppl. 1): S131–S155.

Münster, U., C. Nakamura, A. Haberland et al. 2005. RU 58841-myristate-prodrug development for topical treatment of acne and androgenetic alopecia. *Pharmazie* 60: 8–12.

Neubert, R.H. Potentials of new nanocarriers for dermal and transdermal drug delivery. *European Journal of Pharmaceutics and Biopharmaceutics* 77: 1–2.

Nishihata, T., K. Kotera, Y. Nakano, M. Yamazaki. 1987. Rat percutaneous transport of diclofenac and influence of hydrogenated soya phospholipids. *Chemical and Pharmaceutical Bulletin* 35: 3807–3812.

Nnamani, P.O., S. Hansen, M. Windbergs, C. Lehr. 2014. Development of artemether-loaded nanostructured lipid carrier (NLC) formulation for topical application. *International Journal of Pharmaceutics* 477: 208–217.

Otberg, N., H. Richter, H. Schaefer, U. Blume-Peytavi, W. Sterry, J. Lademann. 2004. Variations of hair follicle size and distribution in different body sites. *Journal of Investigative Dermatology* 122: 14–19.

Padilla, O., H. Ihre, L. Gagne, J. Fréchet, F. Szoka. 2002. Polyester dendritic systems for drug delivery applications: in vitro and in vivo evaluation. *Bioconjugate Chemistry* 13: 453–461.

Pal, D., A.K. Nayak. 2010. Nanotechnology for targeted delivery in cancer therapeutics. *International Journal of Pharmaceutical Sciences Review and Research* 1: 1–7.

Papakostas, D., F. Rancan, W. Sterry, U. Blume-Peytavi, A. Vogt. 2011. Nanoparticles in dermatology. *Archives of Dermatological Research* 303: 533–550.

Pardeike, J., A. Hommoss, R.H. Muller. 2009. Lipid nanoparticles (SLN, NLC) in cosmetic and pharmaceutical dermal products. *International Journal of Pharmaceutics* 366: 170–184.

Pinto, M.F., C.C. Moura, C. Nunes, M.A. Segundo, S.A. Costa Lima, S. Reis. 2014. A new topical formulation for psoriasis: development of methotrexate-loaded nanostructured lipid carriers. *International Journal of Pharmaceutics* 477: 519–526.

Pople, P.V., K.K. Singh. 2006. Development and evaluation of topical formulation containing solid lipid nanoparticles of vitamin A. *AAPS PharmSciTech* 7: 91–96.

Prow, T.W., X. Chen, N.A. Prow et al. 2010. Nanopatch-targeted skin vaccination against West Nile virus and chikungunya virus in mice. *Small* 6: 1776–1784.

Raza, K., B. Singh, P. Singal, S. Wadhwa, O.P. Katare. 2013. Systematically optimized biocompatible isotretinoin-loaded solid lipid nanoparticles (SLNs) for topical treatment of acne. *Colloids and Surfaces B: Biointerfaces* 105: 67–74.

Rita, B., P.K. Lakshmi. 2012. Preparation and evaluation of modified proniosomal gel for localised urticaria and optimisation by statistical method. *Journal of Applied Pharmaceutical Science* 2: 85–91.

Rizwan, M., M. Aqil, S. Talegaonkar, A. Azeem, Y. Sultana, A. Ali. 2009. Enhanced transdermal drug delivery techniques: an extensive review of patents. *Recent Patents on Drug Delivery & Formulation* 3: 105–124.

Rodney, J.Y., L.B. Rae, C.M. Thomas. 1989. Antigen-presenting liposomes are effective in treatment of recurrent herpes simplex virus genitalis in guinea pigs. *Journal of Virology* 63: 2951–2958.

Sakeena, M.H., S.M. Elrashid, F.A. Muthanna et al. 2010. Effect of limonene on permeation enhancement of ketoprofen in palm oil esters nanoemulsion. *Journal of Oleo Science* 59: 395–400.

Samberg, M.E., S.J. Oldenburg, N.A. Monteiro-Riviere. 2010. Evaluation of silver nanoparticle toxicity in skin in vivo and keratinocytes in vitro. *Environmental Health Perspectives* 118: 407–413.

Sanna, V., E. Gavini, M. Cossu, G. Rassu, P. Giunchedi. 2007. Solid lipid nanoparticles (SLN) as carriers for the topical delivery of econazole nitrate: in-vitro characterization, ex-vivo and in-vivo studies. *Journal of Pharmacy and Pharmacology* 59: 1057–1064.

Sanna, V., A. Mariani, G. Caria, M. Sechi. 2009. Synthesis and evaluation of different fatty acid esters formulated into Precirol ATO-based lipid nanoparticles as vehicles for topical delivery. *Chemical and Pharmaceutical Bulletin* 57: 680–684.

Schäfer-Korting, M., W. Mehnert, H.C. Korting. 2007. Lipid nanoparticles for improved topical application of drugs for skin diseases. *Advanced Drug Delivery Reviews* 59: 427–443.

Schneider, M., F. Stracke, S. Hansen, U.F. Schaefer. 2009. Nanoparticles and their interactions with the dermal barrier. *Dermato-Endocrinology* 1: 197–206.

Schroeter, A., T. Engelbrecht, R.H. Neubert, A.S. Goebel. 2010. New nanosized technologies for dermal and transdermal drug delivery: a review. *Journal of Biomedical Nanotechnology* 6: 511–528.

Shakeel, F. 2008. Celecoxib nanoemulsion: skin permeation mechanism and bioavailability assessment. *Journal of Drug Targeting* 16: 733–740.

Teichmann, A., U. Jacobi, H.J. Weigmann, W. Sterry, J. Lademann. 2005. Reservoir function of the stratum corneum: development of an in vivo method to quantitatively determine the stratum corneum reservoir for topically applied substances. *Skin Pharmacology and Physiology* 18: 75–80.

Tian, J., K.K. Wong, C.M. Ho et al. 2007. Topical delivery of silver nanoparticles promotes wound healing. *ChemMedChem* 2: 129–136.

Uner, M. 2005. Skin moisturizing effect and skin penetration of ascorbyl palmitate entrapped in solid lipid nanoparticles (SLN) and nanostructured lipid carriers (NLC) incorporated into hydrogel. *Pharmazie* 60: 51–755.

Venuganti, V.V., O.P. Perumal. 2008. Effect of poly(amidoamine) (PAMAM) dendrimer on skin permeation of 5-fluorouracil. *International Journal of Pharmaceutics* 361: 230–238.

Venuganti, V.V., O.P. Perumal. 2009. Poly(amidoamine) dendrimers as skin penetration enhancers: influence of charge, generation, and concentration. *Journal of Pharmaceutical Sciences* 98: 2345–2356.

Vyas, S.P., R.K. Khar. 2001. Formulation aspects of liposomes. *Advances in Liposomal Therapeutics* 1: 128–129.

Wen, A.H., M.K. Choi, D.D. Kim. 2006. Formulation of liposome for topical delivery of arbutin. *Archives of Pharmacal Research* 29: 1187–1192.

Wertz, P.W., D.T. Downing. 1989. Stratum corneum: biological and biochemical considerations. *Transdermal* 35: 1–22.

Wu, H. 2001. Topical transport of hydrophilic compounds using water-in-oil nanoemulsions. *International Journal of Pharmaceutics* 220: 63–75.

Wu, H., C. Ramachandran, A.U. Bielinska et al. 2001. Topical transfection using plasmid DNA in a water-in-oil nanoemulsion. *International Journal of Pharmaceutics* 221: 23–34.

Yarosh, D.B. 2000. Liposomal ursolic acid (merotaine) increases ceramides and collagen in human skin. *Hormone Research* 54: 318–321.

Yiyun, C., M. Na, X. Tongwen. 2007. Transdermal delivery of nonsteroidal anti-inflammatory drugs mediated by polyamidoamine (PAMAM) dendrimer. *Journal of Pharmaceutical Sciences* 96: 595–602.

Zhao, Y., M. Moddaresi, S.A. Jones, M.B. Brown. 2009. A dynamic topical hydrofluoroalkane foam to induce nanoparticle modification and drug release in situ. *European Journal of Pharmaceutics and Biopharmaceutics* 72: 521–528.

3

Polymeric Hydrogels for Controlled Drug Delivery

Hira Ijaz, Farooq Azam, Ume Ruqia Tul-Ain, and Junaid Qureshi

CONTENTS

3.1 Introduction

Hydrogels are an upcoming class of hydrophilic, polymeric three-dimensional networks with significant absorbing capacity for water and biological fluids [1, 2]. Their stimulation of living tissues is attributed to soft consistency, water content, and porosity [2, 3]. Hydrogels are chemically stable and biodegradable; they substantially disintegrate as well as dissolve [1]. Biomedical applications of hydrogels have been evident since 1960, when it was reported in the literature that Wichterle and Lim cross-linked poly (hydroxyethyl methacrylate) (PHMA) [1]. Initially, a hydroxyethyl methacrylate (HEMA)–based hydrogel was formulated using ethylene glycol dimethacrylate (EGDMA) as a cross-linker. However, in 1970, many reports made claims about these novel physical and chemical assemblies. In 1980, modifications of hydrogels were carried out, and calcium alginate–based microcapsules were formulated by Lim and Sun for tissue engineering [3].

Hydrogels are of two types. (1) Physical gels are also known as "reversible gels," as they have molecular complexation [4, 5] and chemical interactions such as ionic bonding, hydrogen bonding, and hydrophobic interactions [3, 6] Reversible hydrogels have the property of swelling and dissolving with slight modifications in pH, temperature, and ionic strength. (2) Chemical gels are also known as "permanent gels," as they employ covalent bonding for macromolecule condensation with polymeric cross-linking in solution as well as in the dry state [8]. However, functionalized groups attributed to charged and uncharged ions. pH-dependent swelling is attributed to charged hydrogels that undergo shape changes with respect to an electric field [9–11].

Two methods are employed for the preparation of chemical hydrogels. (1) Three-dimensional polymerization employs hydrophilic monomers that polymerize in the presence of a polyfunctional cross-linking agent or by direct cross-linking of hydrophilic polymers. Free radicals are generated via an initiator such as potassium persulfate (KPS) [12, 13], ammonium persulfate (APS) [14], benzoyl peroxide, 2-2,azo-bisisobutyronitrile (AIBN), and ammonium peroxodisulfate [14] or by employing electron beam, gamma, or ultraviolet (UV) radiation [14, 15]. Furthermore, the three-dimensional polymeric network results in residual monomer, which is toxic and subsequently leached out. Purification is carried out with the aid of water via extraction to eliminate unreacted monomer for up to several weeks [14].

Different approaches are employed to eliminate and avoid unreacted monomer. One approach employs monomer conversion by condensation into a three-dimensional polymeric network followed by post-polymerization (thermal and radiation treatment of monomers) [14]. Another approach to overcome toxicity is to use non-toxic monomers such as macromolecules (polyethylene glycol [PEG] dimethylacrylate) [14].

Water-soluble hydrophilic polymers include poly(acrylic acid) [16], polyvinyl alcohol (PVA) [17], polyvinyl [18] pyrrolidone, PEG [19], polyacrylamide, and polysaccharides [20]. They are water soluble, non-toxic polymers for biological and pharmaceutical applications. The simultaneous formulation and sterilization of the hydrogels are carried out by radiation-induced cross-linking using gamma rays [10, 16].

Microwave radiation has been newly introduced for hydrogel formation [21]. In this method, a hydrophilic polymer such as PVA and poly(methyl ether-alt-maleic anhydride) are mixed at room temperature, and cross-linking is carried out using temperature treatment under high pressure by employing a microwave/autoclave, which is safe and economical and does not require purification [21]. Several research papers, academic reviews, reports, monographs, and databases focus on the synthesis, preparation, and application of hydrogels [21] (Figure 3.1).

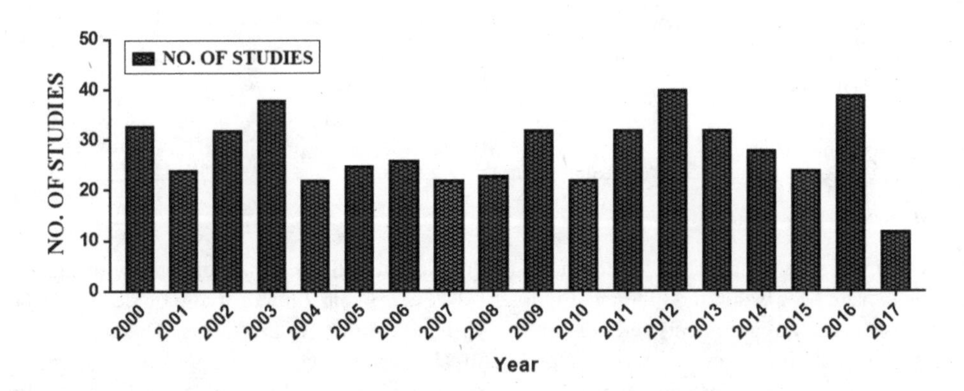

FIGURE 3.1 Number of studies conducted on hydrogels.

3.2 Method of Synthesis

Both chemical and physical methods employ cross-linking methods for the formation of hydrogels [14]. Covalent bonds between polymer chains are accountable for chemical cross-linking, whereas physical methods employ physical interactions to avert drug release [14].

3.2.1 Physical Methods

Physical methods are more acceptable, as they do not require a cross-linker.

(1) *Hydrogen bonding:* Poly methacrylic acid (PMAA) and polyacrylic acid (PAA) form a complex with PEG by hydrogen bonding between the carboxylic group of PAA and PMAA and the oxygen of PEG [14, 20]. Hydrogen bonding is present between PMAA and PEG as well as between polymethacrylic acid grafted ethyl cellulose (PMAA-g-EG). However, carboxylic acid protonation is responsible for pH-dependent swelling [14, 22]. (2) *Amphiphilic graft and block polymerization* undergoes self-building in aqueous media to formulate polymeric micelles as well as hydrogels [23]. *Diblock hydrophilic polymers* furnish a lamellar phase. Multiblock hydrophobic networks have hydrophilic grafting and a hydrophilic backbone with attached hydrophobic segments [24]. (3) In *crystallization*-based cross-linking, PVA forms a gel at room temperature after freeze-thawing [25]. (4) In *ionic interaction*, cross-linking employs a calcium salt at physiological pH at room temperature. The polymer acts as a carrier for protein delivery and cell encapsulation [26, 27].(5) In *protein interaction*–based cross-linking, an additional cross-linking agent is employed to graft rabbit IgG to cross-linked polyacrylamide. The hydrogel swells in the presence of antigen, and further antibody release occurs due to decreased cross-linked density [28] (Figure 3.2).

3.2.2 Chemical Cross-linking

Chemical cross-linking has the added advantage of excellent mechanical strength. (1) A complementary *chemical reaction* employs hydrophilic groups in hydrophilic polymers: amine carboxylic acid, isocyanate-OH/NH_2, and Schiff base to recognize covalent bonds. [28]. (2) *High-energy radiation* such as gamma rays, electron beams, etc. is employed to polymerize unsaturated substances [29]. (3) *Free radical polymerization* is employed to produce cross-linked hydrogels. The polymerizable groups of hydrophilic polymers are cross-linked with monomers [30]. Natural, synthetic, and semi-synthetic

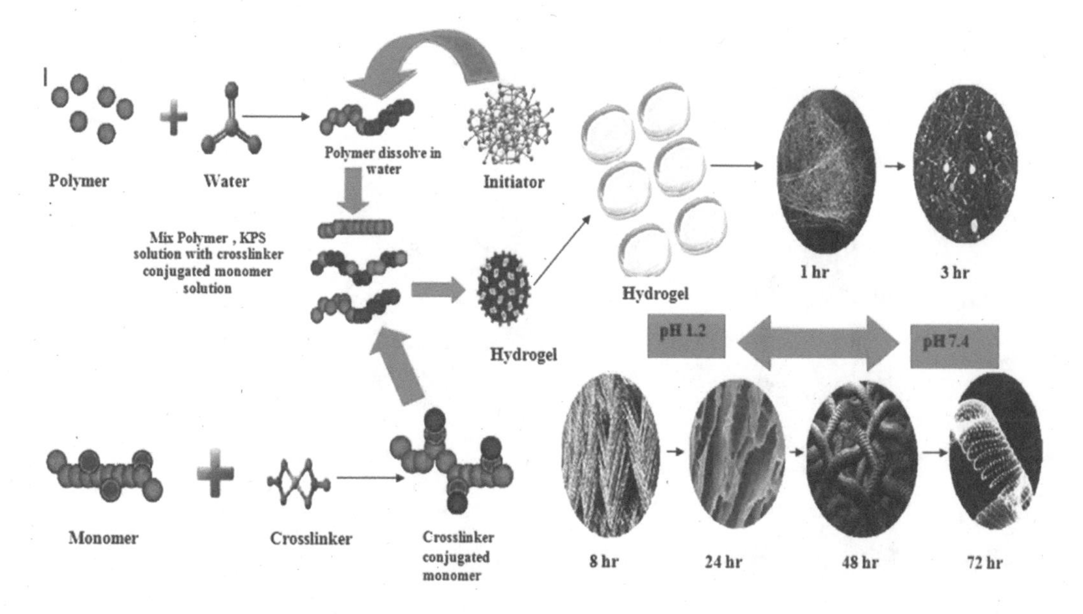

FIGURE 3.2 Formulation of hydrogel via polymerization.

polymers are employed. Methacrylic acid groups are coupled with monosaccharides and disaccharides to formulate hydrogels. Furthermore, UV-polymerization is employed to produce a planned structure. A photo-responsive system is planned to release drug in response to UV [30]. (4) A PEG-based hydrogel has been formulated by using an *enzyme glutaminyl* group functionalized with tetrahydroxy polyethylene glycol [31].

3.3 Hydrogels for Oral Drug Delivery

Due to their vital physical and chemical characteristics, hydrogels aid in drug delivery [32]. Porosity can be controlled by cross-link density in matrix and water affinity, which in turn, aid in drug entrapment and release [33]. Their sustained release properties enable controlled drug release, which results in therapeutic drug concentration at the site of action for a considerable time period [34]. Drug loading and release follow several mechanisms, such as chemically controlled, swelling controlled, environmentally controlled, and diffusion controlled [32, 33].

Diffusion control can be illustrated by both reservoir and matrix systems [35–37]. Diffusion through hydrogel depends on the mesh and the hydrogel pores. The reservoir contains core drug, coated with hydrogel membrane (capsule, cylinder, sphere, and slab). A higher central drug concentration leads to constant drug release, whereas in the matrix, the drug is homogeneously disseminated and dissolved within the three-dimensional polymeric network [36]. Porosity and macromolecular mesh size are related to drug release. The initial drug release is directly related to the square root of time, as compared to the reservoir, in which drug release is independent of time and concentration [35, 27, 37–39] (Figure 3.3).

In swelling control, a glassy polymer contributes to matrix formation, and swelling occurs as it comes into contact with physiological fluids, such as blood, urea, urine, etc. [40]. The matrix expands apart from boundary, and polymer chain expansion occurs, which leads to drug release. This is known as *Case III transport* and relates to time-dependent and constant kinetics. In *anomalous transport*, swelling control is coupled with diffusion [41] (Figure 3.4).

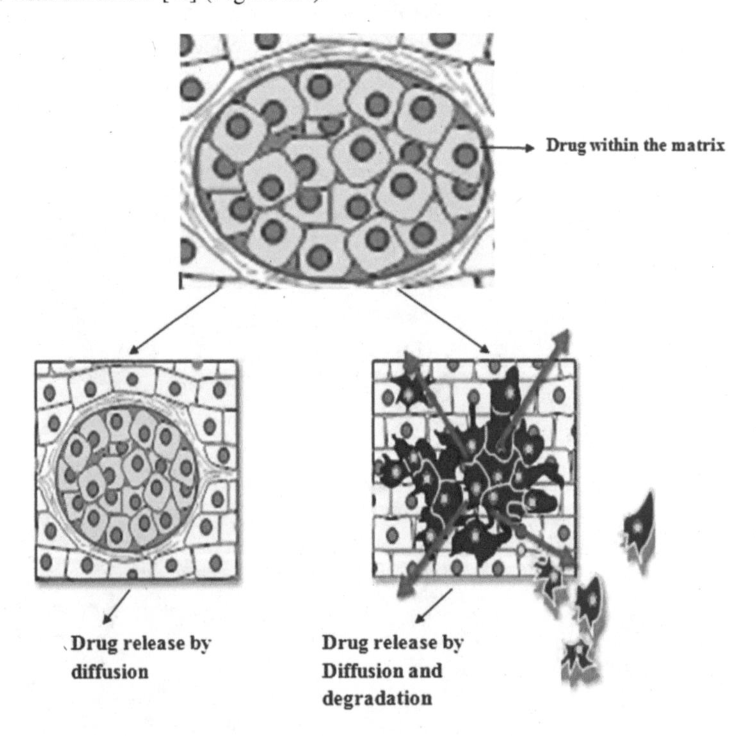

FIGURE 3.3 Mechanism of drug release from hydrogel.

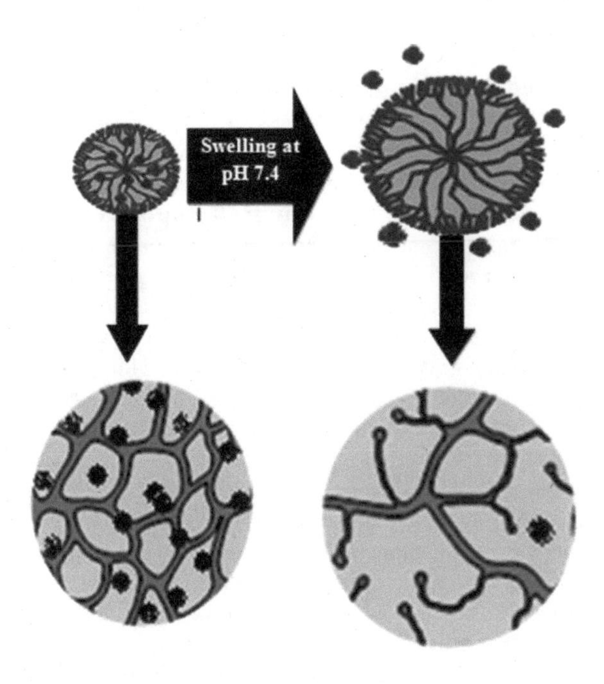

FIGURE 3.4 Drug within hydrogel matrix in swelling and drug release.

A diffusion gradient prevails between drug in the hydrogel matrix and in the surrounding environment, and drug release occurs from the region of higher concentration to lower concentration [1].

$$J = -Dda / Dx$$

- J is the "diffusion flux," of which the dimension is amount of substance per unit area per unit time, so it is expressed in units such as mol m^{-2} s^{-1}. J measures the amount of substance that will flow through a unit area during a unit time interval.
- D is the diffusion coefficient or diffusivity. Its dimension is area per unit time, so typical units for expressing it would be m^2/s [25].
- a (for ideal mixtures) is the concentration, of which the dimension is amount of substance per unit volume. It might be expressed in units of mol/m^3.
- x is position, the dimension of which is length. It might thus be expressed in the unit m.
- D is proportional to the squared velocity of the diffusing particles, w.

Hydrogels are employed for the controlled drug delivery of antihypertensives such as metoprolol, lisinopril, and non-steroidal anti-inflammatory drugs (NSAIDS). The proteolytic degradation of active moieties, proteins, peptides, vitamins, and minerals via proteolytic enzymes can be overcome by hydrogels. Insulin-loaded hydrogels have been developed by cross-linking poly (methacrylic acid-g-ethylene glycol) using tetraethylene glycol dimethacrylate as a cross-linker [39]. Hydrogen bonding between ether and carboxyl groups on the grafted series stabilizes the matrix at acidic pH. pH-responsive swelling occurs at high pH (in the upper portion of the small intestine), and the polymeric complex/nexus dissociates, thereby increasing pore size, which in turn, leads to insulin release. However, strong adhesion of the hydrogel to the intestinal mucosa improves the release and absorption of drugs. [20].

In the recent era, hydrogel-based three-dimensional polymeric nexuses improve controlled drug delivery at predetermined rates. Significant dissolution and release characteristics can be achieved by employing different hydrophilic and hydrophobic polymers. Drug release occurs in response to different physiological conditions such as cancer and diabetes mellitus [23].

3.4 Polymers Employed for Preparation of Hydrogels

3.4.1 Polyhydroxyethyl Methacrylate

Polyhydroxyethyl methacrylate is of great pharmaceutical importance. It was modified naturally as well as synthetically in 1955, thereby improving drug delivery. The method of preparation, the polymer concentration, the nature of the polymer, the nature and concentration of the cross-linking agent, and the temperature also affect the characteristics of the hydrogel [23]. Rao et al. (2013) formulated a polyhydroxyethyl methacrylate (PHEMA) and acrylamidoglycolic acid–based hydrogel for the controlled drug delivery of 5-fluorouracil (FU) using N,N′-methylene bis acrylamide (MBA) as a cross-linking agent. KPS was employed as the initiator. *In vitro* release studies indicated the release of FU for more than 12 h [42]. Furthermore, Suhag et al. (2008) synthesized polymeric hydrogels containing salbutamol sulfate for transdermal drug delivery. HEMA was cross-linked with methacrylic acid (MAA) and N-[3 (dimethylamino)propyl]methacrylamide (DMAPMA) using EGDMA. N,N',N'-tetramethylethylenediamine (TEMED) and APS were used as the initiator [43]. Mahkam and Allahverdipoor (2004) formulated PHEMA and MAA–based hydrogels for the sustained drug delivery of glibenclamide (Figure 3.5). Covalent cross-linking between terephthalic acid and HEMA was carried out using citric acid as the cross-linker. Drug release depends on the degree of swelling and cross-linking [44]. However, subsequently, an amount of water is also employed for formulation of the hydrogel. The resulting product/hydrogel matrix is washed with an ethanol/water mixture to remove unreacted, toxic monomer, as it has a negative impact on living cells. PHEMA is also employed for artificial skin synthesis and burn dressings, as it has excellent healing properties. 1,1,1-trimethylol propane dimethylacrylate has been employed as a cross-linker for the synthesis of PHEMA hydrogels, which are soft, containing 30–40% water [44].

3.4.2 Polyethylene Glycol (PEG) and Its Derivatives

PEG, also known as poly(oxyethylene)/poly(ethylene oxide), is employed for medical and biomedical applications. Polyethylene glycol dimethacrylate (PEGDMA) and polyethylene glycol methacrylate (PEGMA) are also employed for the preparation of polymeric hybrids (Figure 3.6). PEG-based hydrogels are non-toxic, biocompatible, biodegradable, and hydrophilic, which makes them excellent for drug delivery [22]. PEG-based hydrogels are physically, chemically, and biologically stable and stimuli responsive [34]. Due to these novel properties, they are called *smart/intelligent gels.* Physical stimuli include temperature, pressure, light, solvent, radiation, and electric and magnetic fields. Chemical stimuli include pH, ions, and molecule recognition [32].

Chemically cross-linked PEG hydrogels are synthetic and biologically active; they act as a scaffold for protein/recombinant and tissue production. Furthermore, PEG-based hydrogels are also employed as drug carriers and efficient drug delivery systems [45]. Nikouei et al. (2016) formulated temperature/pH-sensitive hydrogels containing PEG and functionalized poly (ε-caprolactone) for the controlled drug

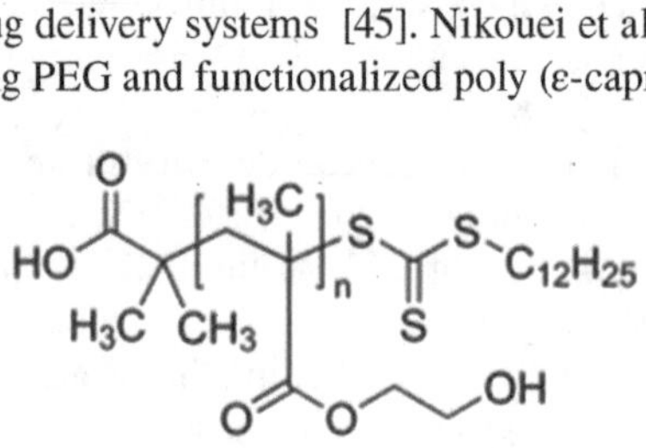

FIGURE 3.5 Structure of polyhydroxyethyl methacrylate.

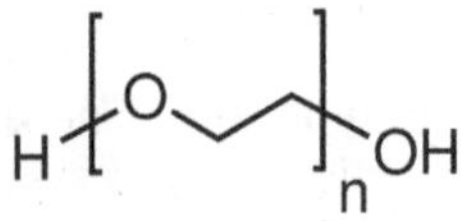

FIGURE 3.6 Structure of polyethylene glycol (PEG).

delivery of macromolecules. The ring opening polymerization technique was employed for the preparation of hydrogels [46]. In addition, Nehls et al. (2016) employed the host/guest relationship between an azobenzene (guest) and a β-cyclodextrin (host) for the preparation of a PEG-based hydrogel. Azobenzene isomerization occurs in the presence of UV light irradiation, thereby decreasing complex formation and accelerating drug release [47].

3.4.3 Polyvinyl alcohol (PVA)

PVA-based hydrogels have good water absorption and gas permeation and are flexible as well as biocompatible. Excellent water absorption contributes to their good mechanical properties. PVA-based hydrogels are used for the preparation of contact lenses, cartilage reconstitution, artificial organ regeneration, and drug delivery [24, 48]. A cycle of alternate freezing and thawing results in a hydrogel with efficient mechanical strength. Crispim et al. (2012) formulated a PVA–GMA-based hydrogel. PVA–GMA is PVA functionalized with vinyl groups (glycidyl methacrylate [GMA]); it is employed as a biomaterial and a drug delivery carrier [49] (Figure 3.7).

3.4.4 Polyvinyl Pyrrolidone

These hydrogels have broad biomedical applications and are employed for wound dressing, as they have single-step formulation and sterilization, thereby decreasing cost and improving softness with the ability to store a large quantity of water [19, 33] (Figure 3.8).

3.4.5 Polyacrylates

Polyacrylates include PAA, employed in agriculture and biomedical science. Elliott et al. (2004) employed a free radical polymerization technique to formulate a polyacrylic acid–based hydrogel for pH-dependent drug delivery [50] (Figure 3.9).

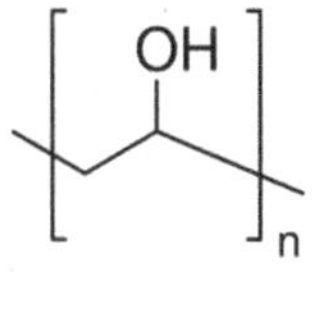

FIGURE 3.7 Structure of polyvinyl alcohol.

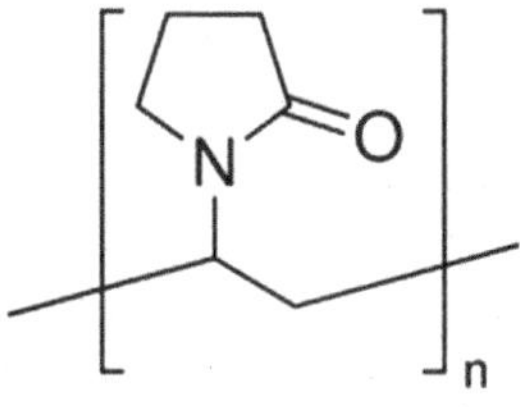

FIGURE 3.8 Structure of polyvinyl pyrrolidone.

$$[-CH_2-CH-]_n$$
$$|$$
$$COO-R$$

FIGURE 3.9 Structure of polyacrylates.

3.4.6 Xanthan Gum

Xanthan gum is a microbe-based, branched-chain, high–molecular weight extracellular heteropolysaccharide obtained industrially under aerobic conditions from *Xanthomonas campestris* employed as thickener in cosmetics and in food [51, 52, 53]. It has the novel feature of formulating a "synergistic hydrogel" with gluco- or galactomannans (interaction between two polysaccharides) [51, 52]. D-glucuronyl, D-glucosyl, and D-mannosyl acid residues are present in a molar ratio of 2:2:1 with a variable ratio of O-acetyl and pyruvyl residues [51, 52]. Mannose (-1,4) glucuronic acid (-1,2) attached alternative to glucose residues by -1,3 linkages in the backbone which constitute trisaccharide side chain [51]. A pyruvic acid moiety is joined by a ketal linkage on half the terminal mannose residues. 6-O-substituted acetyl groups are present on mannose residues (Figure 3.10). High temperature or low ionic strength results in a disordered and flexible network. However, high ionic strength or low temperature results in an ordered network (single or double helix configuration). Deprotonation of O-acetyl and pyruvyl residues occurs at pH >~4.5, thereby amplifying charge density across the xanthan chain, which in turn, aids in Ca^{2+} ion–mediated physical cross-linking [51]. The high acceptability of xanthan gum is attributed to its stability, biocompatibility, and safety. It has been used in combination with other polymers, such as chitosan for xylan immobilization and alginate to encapsulate urease [51, 52].

3.4.7 Hydroxyethyl Cellulose (HEC)

Hydroxyethyl cellulose (HEC) is a crucial spinoff cellulose polysaccharide, which has wide application in drug delivery. It is highly water soluble and non-ionic. It is employed as a thickening agent in emulsions and suspension and for film formation and water retention, and is therefore used for formulating hydrogels for controlled drug delivery. The water retention capacity of HEC is twice that of alkyl polysaccharides, thereby leading to a stronger colloid [54] (Figure 3.11). HEC is biocompatible and biodegradable with film-forming ability. It is employed as a thickening and viscosity-increasing agent (viscosifier) in the food, cosmetic, biophysical, and biotechnological industries [55].

3.4.8 Pectin

Pectin is a complex natural biopolymer that has found wide application in the biotechnology and pharmaceutical industry [56]. Heteropolysaccharides are found in large amounts in the primary cell walls

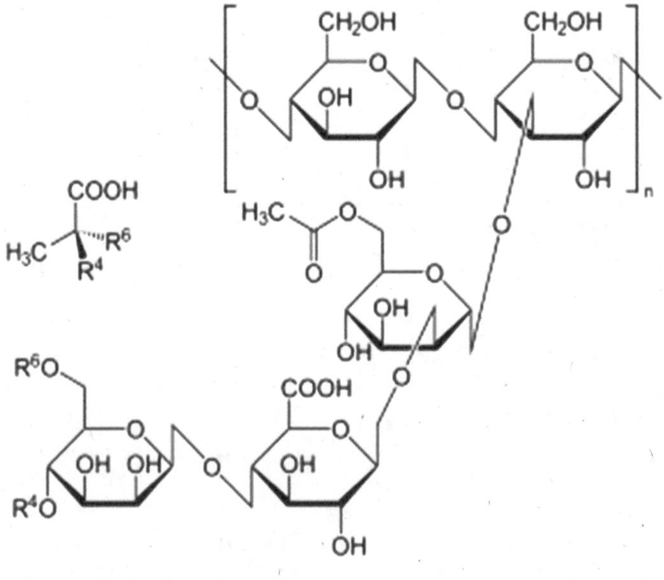

FIGURE 3.10 Structure of xanthan gum.

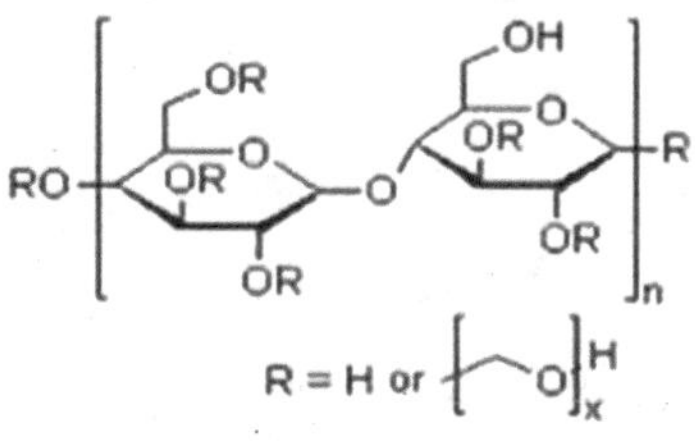

FIGURE 3.11 Structure of hydroxyethyl cellulose.

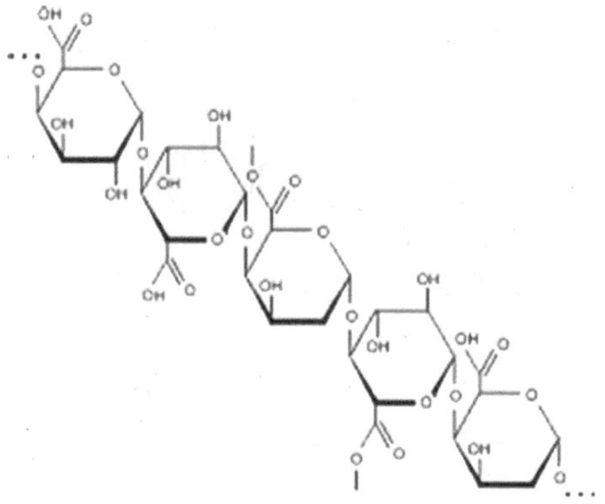

FIGURE 3.12 Structure of pectin.

of dicotyledons and play vital roles in growth and development [56]. Important structural elements are the homogalacturonan (HG) and type I rhamnogalacturonan (RG-I) regions, which are "smooth" and "hairy" regions, respectively [56]. HG regions have (1→4) linked α-D-GalpA residues, which are partially methylated at C-6 and partially acetyl-esterified at O-2 and/or O-3 [56] (Figure 3.12). Pectin is used as a thickening agent, a gelling agent, and a colloidal stabilizer. It is employed as a matrix for drug encapsulation and the delivery of protein, peptides, cells, and drugs [57]. Cross-linked pectin (hydrogel) absorbs and retains water (hundreds of times its own weight) and is therefore also called "superabsorbent" [56, 57]. This remarkable feature is attractive to technologists and researchers in various fields, such as drug delivery systems (DDS) and agriculture. A blend of biodegradable pectin and hydrophilic acrylic polymer exhibits vital features depending on its composition, the interaction between polymers, and hydrogen bonding [56, 57] (Table 3.1).

3.5 Monomers Employed for Formation of Hydrogels

3.5.1 Acrylic Acid (AA)

AA is an electric- and pH-sensitive polymer with the innate ability to formulate complexes with polybases [65]. It is widely employed for targeted drug delivery in the gastrointestinal tract (GIT) [66]. This pH-sensitive electrolyte has significant superabsorbent properties. Its pH-dependent features can be altered by altering the polymer. pH-responsive interpenetrating networks (IPN) have gained wide acceptance due to their reported pH-sensitive, thermosensitive and electrosensitive properties [66]. Furthermore, it has been reported in the literature that PVP-co-AA has been employed for pH-dependent drug delivery, ocular drug delivery, and the formulation of mucoadhesive microspheres and polymer–ceramic composites. PVP/AA hydrogels have good gel strength and TiO_2 immobilization and are excellent chemical sensors. A PVA/AA-based hydrogel has been formulated for the controlled drug delivery of vitamin B_{12} [65]. The pKa value of AA ranges from 4.5 to 5.0, and swelling at physiological pH (7.4) is attributed to carboxylic acid group ionization [66].

TABLE 3.1

Various Polymers Used in Formulation of Hydrogels

Drug	Polymer	Chemical Structure	Reference
Ketoprofen	Tamarind seed polysaccharide	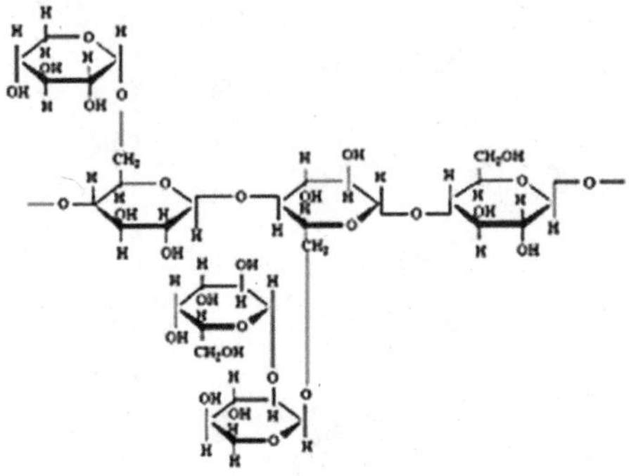	[58]
Dexamethasone	Polyethylene glycol	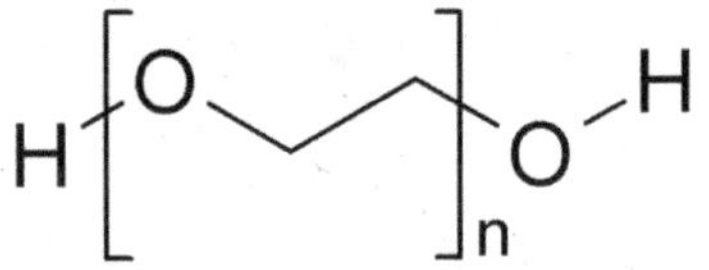	[59]
Indomethacin (IND)	Sodium alginate	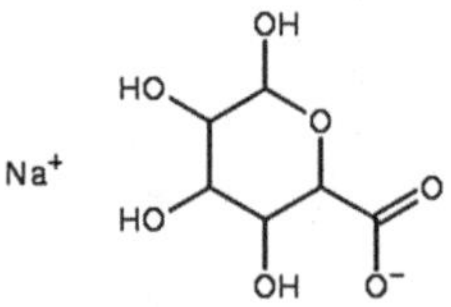	[60]
Protein delivery	Chitosan (CS)	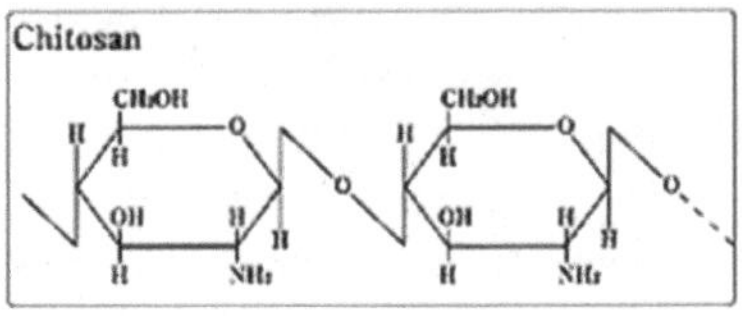	[61]
Protein	Alginate	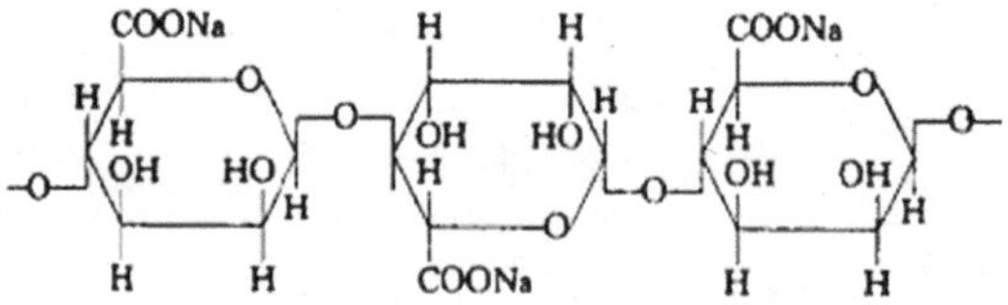	[62]
Theophylline and 5-fluorouracil (5-FU)	*N*-vinyl pyrrolidone (NVP), polyethylene glycol diacrylate (PAC) and chitosan	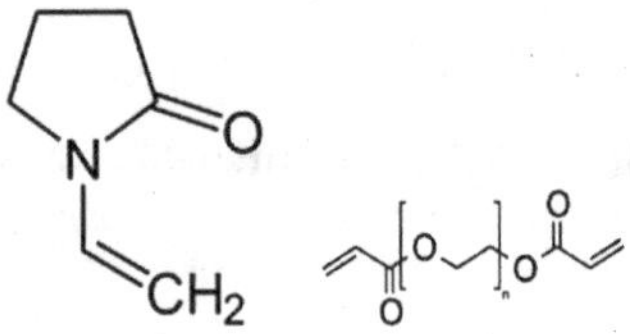 ***N*-vinyl pyrrolidone (NVP)**, polyethylene glycol diacrylate (PAC)	[63]
Ketoprofen	Xanthan gum and sodium carboxymethyl cellulose (NaCMC)	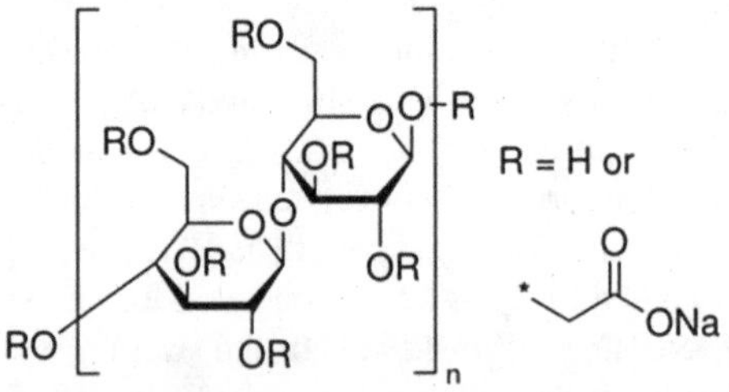 **sodium carboxymethyl cellulose**	[64]

3.5.2 Methacrylic Acid (MAA)

MAA is employed for enteric drug delivery. It is also employed for controlled drug delivery. As the concentration of MAA increases, the swelling of the hydrogel increases after a certain period of time in the intestine/alkaline media. Singh et al. (2008) formulated MAA-based hydrogel for the controlled delivery of fungicide (Triram) [67]. In addition, Sajeesh et al. (2006) formulated a MAA-based cyclodextrin/insulin complex for the oral controlled delivery of insulin [68] (Table 3.2).

3.5.3 Initiator Employed for Hydrogels

An initiator is a chemical species with the innate ability to react with monomer, thereby forming an intermediate product, which reacts with other monomers within polymeric blends. KPS (also called potassium peroxodisulfate), APS, and sodium persulfate are commonly employed initiators in polymerization reactions. Initiators generate free radicals and accelerate the radical reaction. They mostly contain weak bonds with a small energy of dissociation. Wang and Wang (2010) formulated a pH-responsive sodium alginate/poly(sodium acrylate) and PVP–based hydrogel for controlled drug delivery by employing APS as initiator [10, 11] However, Ramanan and his coworkers formulated a temperature-sensitive hydrogel for controlled delivery of BSA (bovine serum albumin) by a free radical polymerization technique employing KPS as initiator [77] (Table 3.3).

3.5.4 Cross-linkers Employed for Preparation of Hydrogels

The two most commonly employed polyfunctional cross-linkers are N,N′-MBA and EGDMA. They employ ionic and covalent bonds between polymeric chains. Pourjavaidi et al. (2004) polymerized AA and kappa-carrageenan to formulate a superabsorbent composite by employing MBA as cross-linker [81] However, Lei et al. (2013) formulated a PEG dimethacrylate–based hydrogel by using EGDMA as cross-linker [82] (Tables 3.4 through 3.6).

3.5.5 Applications

Hydrogels have become very popular due to their unique properties, such as high water content, softness, flexibility, and biocompatibility. Natural and synthetic hydrophilic polymers can be physically or chemically cross-linked to produce hydrogels. Their resemblance to living tissue opens up many opportunities for applications in biomedical areas. Currently, hydrogels are used for manufacturing contact lenses, hygiene products, tissue engineering scaffolds, DDS, and wound dressings.

3.5.6 Subcutaneous Drug Targeting

Polymeric nexuses are used in implants and subcutaneous drug targeting. Implants should be biodegradable, compatible, and not susceptible to inflammation so that they do not require follow-up post-operative intervention [115]. These use physical and chemical interactions for loading therapeutic drugs [116].

3.5.7 Transdermal Delivery

Low–molecular weight drugs can be delivered by a transdermal drug delivery system (TDDS); hydrogels have an increased water content, which gives a comforting feeling to the patient's skin, leading to better patient compliance [117]. Glimepiride (antidiabetic) and sulfonylurea, with solubility issues, are potential candidates for delivery as hydrogels [117].

3.5.8 Delivery of Growth Factors

Hydrogel implants aid in the delivery of glycosaminoglycan (GAG) (a growth factor) in developing bones, joints, cartilage, and nerve tissues. GAG also aids in lesion healing. Both GAG and chondroitin

TABLE 3.2

Various Monomers Used in Formulation of Hydrogels

Drug	Monomer	Chemical Structure	Reference
Ofloxacin (OFL)	Acrylic acid (PAC)	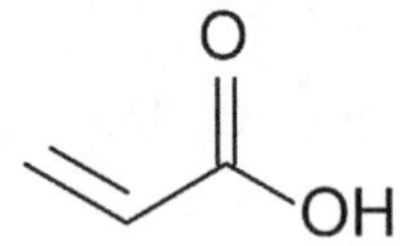 O, OH	[69]
Riboflavin (RF)	Polyglycidyl methacrylate	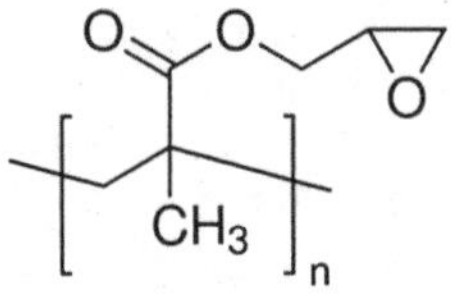 O, O, O, CH_3, n	[70]
Losartan potassium	Poly(acrylic acid) (PAA) and poly(acrylamide) (PAAM)	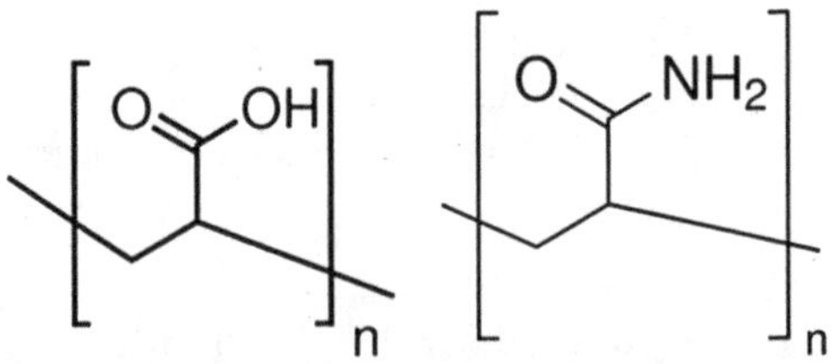 O, OH, n; O, NH_2, n **Poly(acrylic acid) (PAA), poly(acrylamide) (PAAM)**	[71]
Doxorubicin	Acrylic acid (AA)	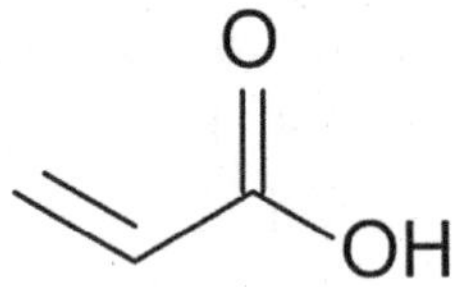 O, OH	[72]
Gliclazide	Poly (acrylic acid)	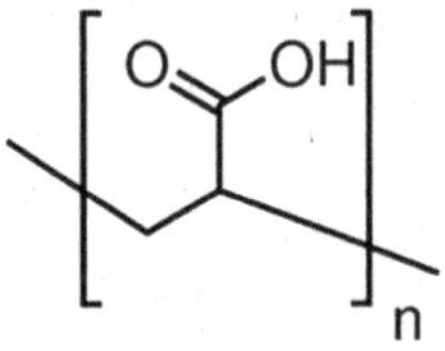 O, OH, n	[73]
Diethylcarbamazine citrate	Hydroxyethyl methacrylate	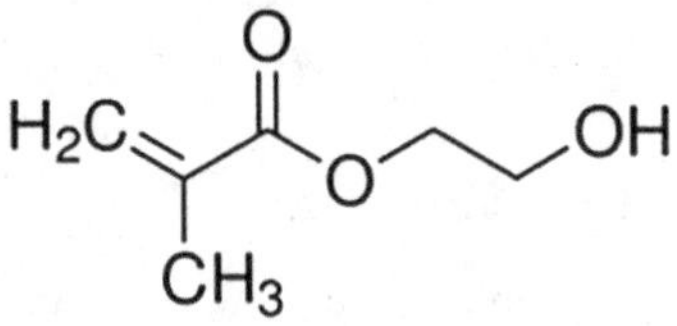 O, H_2C, O, OH, CH_3	[74]
Oseltamivir phosphate	Hydroxyethyl methacrylate	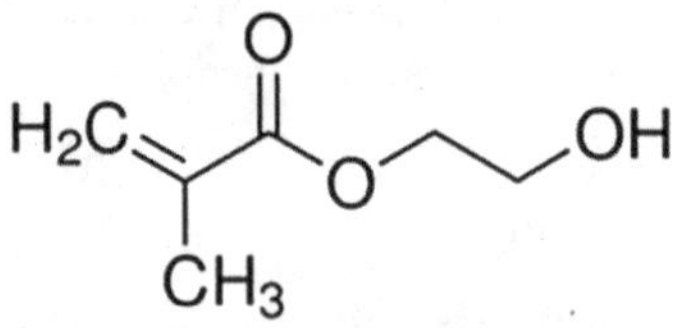 O, H_2C, O, OH, CH_3	[75]
Monoclonal antibodies (MAbs)	Methacrylic acid	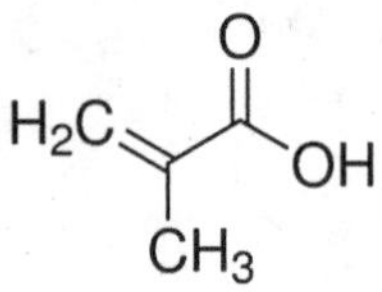 O, H_2C, OH, CH_3	[76]

TABLE 3.3
Various Initiators Used in Formulation of Hydrogels

Drug	Initiators	Chemical Structure	Reference
Nifedipine	Benzoyl peroxide	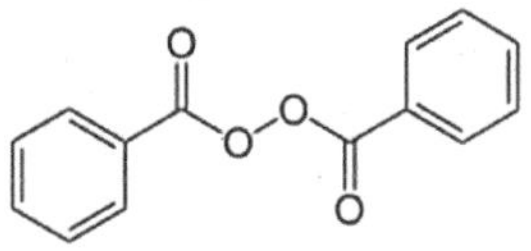	[78]
Ciprofloxacin and ornidazole	Potassium persulfate (KPS)	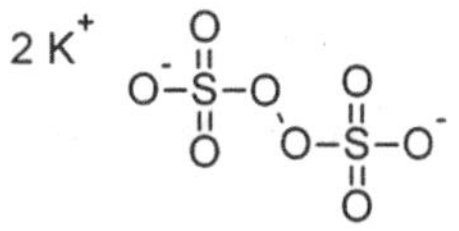	[79]
–	Ammonium persulfate	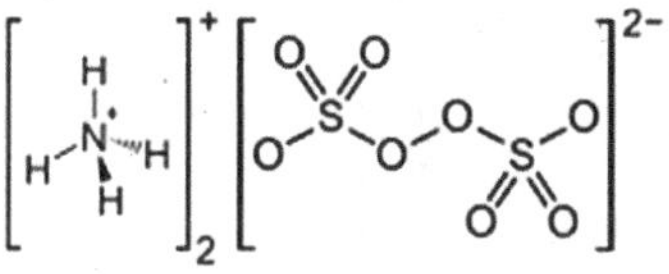	[80]

TABLE 3.4
Polyfunctional Cross-linkers

Drug	Cross-linker	Chemical Structure	Reference
Pheniramine maleate	EGDMA	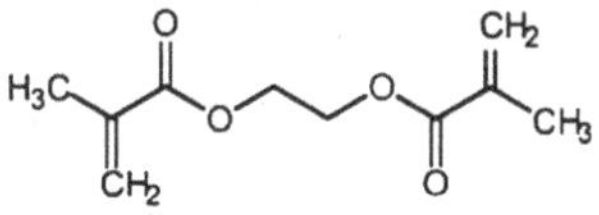	[83]
Tramadol hydrochloride	EGDMA	–	[83]
Tramadol HCl	MBA	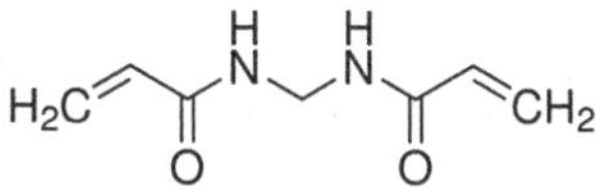	[65]

sulfate are used in polymeric nexuses for controlled drug delivery (Lee et al., 2011). Pegylated growth factor–loaded nexuses induce osteoinduction. It has also been illustrated in the literature that mesenchymal stem cells and BMP-2 cells aid in bone formation [118].

3.5.9 Cancer

Today, injectable doxorubicin-based hydrogels are available for the treatment of cancer. They remarkably inhibit osteolysis, lung cancer, and osteosarcoma (primary and secondary). A thermosensitive and photosensitive nexus was formulated via UV irradiation for paclitaxel drug delivery for the treatment of angiogenesis, Lewis lung cancer (3LL) cells, and solid tumors. An intratumorally injected (EMT-6 tumors) polymeric nexus had comparable effectiveness to that of Taxol injection [119].

Fundamental approaches for the treatment of cancer are surgery, chemotherapy, and radiotherapy. Hydrogels are used for both chemotherapy and radiotherapy approaches. Brachytherapy is radio-therapeutic implantation in adjacent target tissue, which aids in the delivery of a higher dose of drug but involves invasive procedures [120]. Furthermore, hydrogel radiolabeling aids in controlled exposure. Moreover, a 131I norcholesterol (131I-NC)–loaded hydrogel aids in the treatment of breast cancer. Traditionally available microcapsules, gels, and microspheres that undergo passive drug loading have limited drug loading. Sometimes, they involve invasive procedures [121].

TABLE 3.5
Patents

Patent	Reference
Compliant yet tough hydrogel systems as ultrasound transmission agents	[84]
Hydrogel compositions	[85]
Hydrogel encapsulated cell patterning and transferring method and cell-based biosensor using the same	[86]
Hydrogel encapsulated cells and anti-inflammatory drugs	[87]
Antimicrobial hydrogel formulation	[88]
Compositions and methods relating to an occlusive polymer hydrogel	[89]
Colored hydrogel contact lenses with lubricious coating thereon	[90]
Aminoglycoside hydrogel microbeads and macroporous gels with chemical crosslink, method of preparation and use thereof	[91]
Bioadhesive and injectable hydrogel	[92]
Hydrogel contact lenses with lubricious coating thereon	[93]
Method for using hydrogel sheet for treating wound	[94]
Hydrogel Beads With Self-Regulating Microclimate pH Properties	[95]
Formulation for silicone hydrogel, silicone hydrogel and method for forming silicone hydrogel	[96]
Hydrogel capsules and process for preparing the same	[97]
Composition comprising a hydrogel and pesticides	[98]
Hydrogel-forming material, premix, and hydrogel formation method	[99]
Hydrogel composition	[100]
Process for the preparation of nanoparticles of noble metals in hydrogel and nanoparticles thus obtained	[101]
Shaped article made of porous hydrogel, manufacturing process therefor and use thereof	[102]

3.5.10 Wound Dressing

Ideally, wound dressing should be biocompatible and aid in protection from bacterial infection. A hydrogel nexus provides quick wound healing, smooth scarring, and improved vascularization by supplying chitooligomers at the lesion site [122]. The polymeric nexus has a reparative nature and aids in providing a therapeutic amount of drug at the wound site [123].

3.5.11 Oral Treatment

Various diseases of the oral cavity include stomatitis, periodontal diseases, fungal and viral diseases, and oral cancers. They have the advantages of reduced first-pass hepatic metabolism and reduced susceptibility to acid and proteolytic degradation [124]. Their mucoadhesive properties help to improve penetration, thereby aiding in improving the therapeutic effect by maintaining a therapeutic level of the antimicrobial agent. Paracellular permeability via the mucosal epithelia aids in the remarkable transfer of transforming growth factor-β (TGF-β) across the porcine mucosa [125]. Hydrogels have been formulated for local drug delivery in oral mucosa. They prohibit the adhesion of *Candida albicans* to buccal cells. A chlorhexidine gluconate–loaded nexus aids in sustained drug release. However, an ipriflavone–loaded nexus helps increase bone density in periodontal pockets [124].

3.5.12 Ophthalmic Delivery

Conventional drug delivery leads to simultaneous elimination from the ocular cavity. Poor absorption from the eye leads to poor bioavailability. These novel systems aid in prolonging drug retention and penetration via employing penetration enhancers and bioadhesive polymers [125]. These systems are promising in terms of enhanced corneal resident time. *In situ* gels are also promising, as they have a prolonged retention time. A Timolol-based hydrogel has a greater ability to reduce intraocular pressure as compared with a conventional dosage form [126].

TABLE 3.6
Different Drugs Loaded in Hydrogels

Drugs	Route of administration	Type	Method of Preparation	Polymer	Monomer	Cross-linker	Reference
Perindopril Erbumine	Oral	pH responsive	Free radical polymerization	Xanthan gum	Acrylic acid	Methylene bisacrylamide	[74]
Bovine serum albumin (BSA)	Oral	Thermo/pH double sensitive	Thermosensitivity	Chitosan	–	–	[103]
Cilnidipine	Oral	pH responsive	Physical cross-linking	Chitosan–alginate	–	–	[104]
–	Injectable	Thermo sensitive	De-cross-linkable	Dextrin, starch	Acrylamide	–	[105]
Polymyxin B	Oral	–	Physical cross-linking	Polycarbonates	–	–	[106]
Albumin	Oral	Thermosensitive	Free radical polymerization	N-isopropylacrylamide		Tetramethylethylenediamine (TEMED), methylene bisacrylamide (MBA)	[107]
Acid orange 8 (AO8) and bovine serum albumin (BSA)	Oral	pH responsive	Free radical polymerization	Hydroxyethyl methacrylate	Methyacrylic acid	Tri(ethylene glycol) dimethacrylate (TEGDMA)	[108]
Ibuprofen	Oral	pH responsive	Free radical polymerization	Guar gum	Methacrylic acid	Ethylene glycol dimethacrylate (EGDMA)	[109]
Amoxicillin and meloxicam	Oral	pH responsive	Free radical polymerization	Chitosan	Acrylic acid	Maleic anhydride	[110]
Ofloxacin	Oral	Light and temperature responsive	Free radical polymerization	Hydrogen tetrachloroaureate	Acrylic acid	N,N-methylenebisacrylamide	[111]
Bovine testicular hyaluronidase, indomethacin, pilocarpine, diclofenac sodium, hydrocortisone, 6a-methyl-prednisolone, cortisone, corticosterone, dexamethasone	Oral	Temperature sensitive	Free radical polymerization	Hyaluronic acid	Poly(ethylene glycol)	1-Ethyl-3-[3-(dimethylami no)-propyl]carbodiimide (EDCI)	[112]

(Continued)

TABLE 3.6 (CONTINUED)
Different Drugs Loaded in Hydrogels

Drugs	Route of administration	Type	Method of Preparation	Polymer	Monomer	Cross-linker	Reference
Bovine serum albumin (BSA)	Oral	–	–	N,O-carboxymethylchitosan and alginate	Genipin	–	[61]
Diclofenac sodium and procaine HCl	Oral	Temperature responsive	Free radical polymerization	Poly(N-isopropylacrylamide)–poly(vinylpyrrolidinone)	–	Poly(ethylene glycol) dimethacrylate	[113]
Acetylsalicylic acid and Theophylline	Oral	pH sensitive	Free radical polymerization	Hemicellulose	Acrylic acid	N,N-methylenebisacrylamide	[114]

3.5.13 GIT Targeting

Drug delivery in the GIT involves certain formidable barriers. Certain drugs degrade in response to the acidic environment of the GIT, and limited residence time limits their therapeutic effectiveness. Colon targeting is also useful for the treatment of various diseases such as carcinoma, infection, Crohn's disease, and ulcerative colitis. Non-specific targeting requires an increased dose of drug and leads to side effects [127]. pH-sensitive drug delivery allows targeting to selected areas of GIT, which overcomes the issue of non-targeted drug delivery. It has been reported in the literature that the polymeric nexus swells in response to an acidic environment, and drug release occurs in response to pH [128, 129]. Amoxicillin and metronidazole have been delivered via this type of nexus. Similarly, 5-fluorouracil and insulin are selectively released in the intestine. Nitrofurantoin (NF)–based microcapsule depicted pH-dependent swelling. Polypeptides have been protected from intestinal degradation via enzyme inhibition. Serine proteases and metallopeptidases have been inhibited by the covalent attachment of competitive inhibitors, such as the Bowman–Birk inhibitor, and chelating moieties, such as EDTA [130].

3.6 Conclusion

Polymeric nexuses have advantages over other polymer-based DDS in achieving the desired targeted controlled action of drugs. The formulation of this kind of system requires a good understanding of the physical and chemical nature of the substances. They are good delivery systems for drugs administered orally with low molecular weight, proteins, and peptides. This delivery system has a greater carrier capacity for hydrophilic and hydrophobic drugs, increased stability, and greater possibility of different routes of administration. The future of polymeric DDS products is promising due to the keen interest of formulation scientists.

REFERENCES

1. Kabiri, K., Omidian, H., Zohuriaan-Mehr, M. J., & Doroudiani, S. (2011). Superabsorbent hydrogel composites and nanocomposites: a review. *Polymer Composites*, *32*(2), 277–289.
2. Richter, A., Paschew, G., Klatt, S., Lienig, J., Arndt, K. F., & Adler, H. J. P. (2008). Review on hydrogel-based pH sensors and microsensors. *Sensors*, *8*(1), 561–581.
3. Ganji, F., Vasheghani-Farahani, S., & Vasheghani-Farahani, E. (2010). Theoretical description of hydrogel swelling: a review. *Iranian Polymer Journal*, *19*(5), 375–398.
4. Boucard, N., Viton, C., Agay, D., Mari, E., Roger, T., Chancerelle, Y., & Domard, A. (2007). The use of physical hydrogels of chitosan for skin regeneration following third-degree burns. *Biomaterials*, *28*(24), 3478–3488.
5. Loh, X. J., Peh, P., Liao, S., Sng, C., & Li, J. (2010). Controlled drug release from biodegradable thermo-responsive physical hydrogel nanofibers. *Journal of Controlled Release*, *143*(2), 175–182.
6. An, L., Zhao, T. S., & Zeng, L. (2013). Agar chemical hydrogel electrode binder for fuel-electrolyte-fed fuel cells. *Applied Energy*, *109*, 67–71.
7. Rosales, A. M., & Anseth, K. S. (2016). The design of reversible hydrogels to capture extracellular matrix dynamics. *Nature Reviews Materials*, *1*, 15012.
8. Arifuzzaman, M., Behrend, C., DesJardins, J., & Anker, J. N. (2016). A smart polymer hydrogel as a chemical sensor on biomedical implant. *Chemistry Annual Research Symposium*, 11. https://tigerprints.clemson.edu/cars/11.
9. Wang, J., & Li, X. (2015). Synthesis of polyacrylamide/modified silica composite hydrogels for synergistic complexation of heavy metal ions. *Desalination and Water Treatment*, *53*(1), 230–237.
10. Wang, M., Zhang, X., Li, L., Wang, J., Wang, J., Ma, J., Yuan, Z., Lincoln, S. F., & Guo, X. (2016). Photo-reversible supramolecular hydrogels assembled by α-cyclodextrin and azobenzene substituted poly (acrylic acid)s: effect of substitution degree, concentration, and tethered chain length. *Macromolecular Materials and Engineering*, *301*(2), 191–198.
11. Wang, W., & Wang, A. (2010). Synthesis and swelling properties of pH-sensitive semi-IPN superabsorbent hydrogels based on sodium alginate-g-poly (sodium acrylate) and polyvinylpyrrolidone. *Carbohydrate Polymers*, *80*(4), 1028–1036.

12. Ren, X. Y., Yu, Z., Liu, B., Liu, X. J., Wang, Y. J., Su, Q., & Gao, G. H. (2016). Highly tough and puncture resistant hydrogels driven by macromolecular microspheres. *RSC Advances*, *6*(11), 8956–8963.
13. Strachota, B., Hodan, J., & Matějka, L. (2016). Poly (N-isopropylacrylamide)–clay hydrogels: control of mechanical properties and structure by the initiating conditions of polymerization. *European Polymer Journal*, *77*, 1–15.
14. Ahmed, E. M. (2015). Hydrogel: preparation, characterization, and applications: a review. *Journal of Advanced Research*, *6*(2), 105–121.
15. Zhang, S., Shi, Z., Xu, H., Ma, X., Yin, J., & Tian, M. (2016). Revisiting the mechanism of redox-polymerization to build the hydrogel with excellent properties using a novel initiator. *Soft Matter*, *12*(9), 2575–2582.
16. Calixto, G., Yoshii, A. C., Rocha e Silva, H., Stringhetti Ferreira Cury, B., & Chorilli, M. (2015). Polyacrylic acid polymers hydrogels intended to topical drug delivery: preparation and characterization. *Pharmaceutical Development and Technology*, *20*(4), 490–496.
17. Yang, Y., Chen, J., Yang, L., Chen, B., Sheng, Z., Luo, W., Sui, G., Lu, X., & Chen, J. (2016). Effect of D-(+)-glucose on the stability of polyvinyl alcohol Fricke hydrogel three-dimensional dosimeter for radiotherapy. *Nuclear Engineering and Technology*, *48*(3), 608–612.
18. Sheikholeslami, P., Muirhead, B., Baek, D. S. H., Wang, H., Zhao, X., Sivakumaran, D., & Hoare, T. (2015). Hydrophobically-modified poly (vinyl pyrrolidone) as a physically-associative, shear-responsive ophthalmic hydrogel. *Experimental Eye Research*, *137*, 18–31.
19. Yu, D., Sun, C., Zheng, Z., Wang, X., Chen, D., Wu, H., Wang, X., & Shi, F. (2016). Inner ear delivery of dexamethasone using injectable silk-polyethylene glycol (PEG) hydrogel. *International Journal of Pharmaceutics*, *503*(1), 229–237.
20. Fitzgerald, M. M., Bootsma, K., Berberich, J. A., & Sparks, J. L. (2015). Tunable stress relaxation behavior of an alginate-polyacrylamide hydrogel: comparison with muscle tissue. *Biomacromolecules*, *16*(5), 1497–1505.
21. Caló, E., & Khutoryanskiy, V. V. (2015). Biomedical applications of hydrogels: a review of patents and commercial products. *European Polymer Journal*, *65*, 252–267.
22. Gulrez, S. K., Phillips, G. O., & Al-Assaf, S. (2011). *Hydrogels: Methods of Preparation, Characterisation and Applications*. INTECH Open Access Publisher.
23. Qi, X., Hu, X., Wei, W., Yu, H., Li, J., Zhang, J., & Dong, W. (2015). Investigation of Salecan/poly (vinyl alcohol) hydrogels prepared by freeze/thaw method. *Carbohydrate Polymers*, *118*, 60–69.
24. Voorhaar, L., & Hoogenboom, R. (2016). Supramolecular polymer networks: hydrogels and bulk materials. *Chemical Society Reviews*, 45(14), 4013–4031.
25. Ma, X., Guo, L., Ji, Q., Tan, Y., Xing, Y., & Xia, Y. (2016). Physical hydrogels constructed on a macro-cross-linking cationic polysaccharide with tunable, excellent mechanical performance. *Polymer Chemistry*, *7*(1), 26–30.
26. Stroganov, V., Al-Hussein, M., Sommer, J. U., Janke, A., Zakharchenko, S., & Ionov, L. (2015). Reversible thermosensitive biodegradable polymeric actuators based on confined crystallization. *Nano Letters*, *15*(3), 1786–1790.
27. Dehghan-Niri, M., Tavakol, M., Vasheghani-Farahani, E., & Ganji, F. (2015). Drug release from enzyme-mediated in situ-forming hydrogel based on gum tragacanth–tyramine conjugate. *Journal of Biomaterials Applications*, *29*(10), 1343–1350.
28. Ranjha, N. M., Mudassir, J., & Akhtar, N. (2008). Methyl methacrylate-co-itaconic acid (MMA-co-IA) hydrogels for controlled drug delivery. *Journal of Sol-Gel Science and Technology*, *47*(1), 23–30.
29. Fekete, T., Borsa, J., Takács, E., & Wojnárovits, L. (2016). Synthesis of cellulose-based superabsorbent hydrogels by high-energy irradiation in the presence of crosslinking agent. *Radiation Physics and Chemistry*, *118*, 114–119.
30. Kim, B., Hong, D., & Chang, W. V. (2015). Kinetics and crystallization in pH-sensitive free-radical crosslinking polymerization of acrylic acid. *Journal of Applied Polymer Science*, *132*(27).
31. Anjum, F., Lienemann, P. S., Metzger, S., Biernaskie, J., Kallos, M. S., & Ehrbar, M. (2016). Enzyme responsive GAG-based natural-synthetic hybrid hydrogel for tunable growth factor delivery and stem cell differentiation. *Biomaterials*, *87*, 104–117.
32. Gupta, P., Vermani, K., & Garg, S. (2002). Hydrogels: from controlled release to pH-responsive drug delivery. *Drug Discovery Today*, *7*(10), 569–579.

33. Peppas, N. A., Wood, K. M., & Blanchette, J. O. (2004). Hydrogels for oral delivery of therapeutic proteins. *Expert Opinion on Biological Therapy*, *4*(6), 881–887.
34. Hamidi, M., Azadi, A., & Rafiei, P. (2008). Hydrogel nanoparticles in drug delivery. *Advanced Drug Delivery Reviews*, *60*(15), 1638–1649.
35. Ijaz, H., Qureshi, J., Danish, Z., Zaman, M., Abdel-Daim, M., Hanif, M., Waheed, I., & Mohammad, I. S. (2015). Formulation and in-vitro evaluation of floating bilayer tablet of lisinopril maleate and metoprolol tartrate. *Pakistan Journal of Pharmaceutical Sciences*, *28*(6), 2019–2025.
36. Qureshi, J., Ijaz, H., Sethi, A., Zaman, M., Bashir, I., Hanif, M., & Azis, M. (2014). Formulation and in vitro characterization of sustained release matrix tablets of metoprolol tartrate using synthetic and natural polymers. *Latin American Journal of Pharmacy*, *33*(9), 1533–1539.
37. Zaman, M., Qureshi, J., Ejaz, H., Sarfraz, R. M., ullah Khan, H., Sajid, F. R., & ur Rehman, M. S. (2016). Oral controlled release drug delivery system and characterization of oral tablets; a review. *Pakistan Journal of Pharmaceutical Research*, *2*(1), 67–76.
38. Wang, Y., Yang, X., Qiu, L., & Li, D. (2013). Revisiting the capacitance of polyaniline by using graphene hydrogel films as a substrate: the importance of nano-architecturing. *Energy & Environmental Science*, *6*(2), 477–481.
39. Wang, Q., Yang, Z., Ma, M., Chang, C. K., & Xu, B. (2008). High catalytic activities of artificial peroxidases based on supramolecular hydrogels that contain heme models. *Chemistry–A European Journal*, *14*(16), 5073–5078.
40. Yew, Y. K., Ng, T. Y., Li, H., & Lam, K. Y. (2007). Analysis of pH and electrically controlled swelling of hydrogel-based micro-sensors/actuators. *Biomedical Microdevices*, *9*(4), 487–499.
41. Huang, X., & Brazel, C. S. (2001). On the importance and mechanisms of burst release in matrix-controlled drug delivery systems. *Journal of Controlled Release*, *73*(2), 121–136.
42. Rao, K. M., Mallikarjuna, B., Krishna Rao, K. S. V., Sudhakar, K., Rao, K. C., & Subha, M. C. S. (2013). Synthesis and characterization of pH sensitive poly (hydroxy ethyl methacrylate-co-acrylamidoglycolic acid) based hydrogels for controlled release studies of 5-fluorouracil. *International Journal of Polymeric Materials and Polymeric Biomaterials*, *62*(11), 565–571.
43. Suhag, G. S., Bhatnagar, A., & Singh, H. (2008). Poly (hydroxyethyl methacrylate)-based co-polymeric hydrogels for transdermal delivery of salbutamol sulphate. *Journal of Biomaterials Science, Polymer Edition*, *19*(9), 1189–1200.
44. Mahkam, M., & Allahverdipoor, M. (2004). Controlled release of biomolecules from pH-sensitive network polymers prepared by radiation polymerization. *Journal of Drug Targeting*, *12*(3), 151–156.
45. Hennink, W. E., & Van Nostrum, C. (2012). Novel crosslinking methods to design hydrogels. *Advanced Drug Delivery Reviews*, *64*, 223–236.
46. Nikouei, N. S., Ghasemi, N., & Lavasanifar, A. (2016). Temperature/pH responsive hydrogels based on poly (ethylene glycol) and functionalized poly (e-caprolactone) block copolymers for controlled delivery of macromolecules. *Pharmaceutical Research*, *33*(2), 358–366.
47. Nehls, E. M., Rosales, A. M., & Anseth, K. S. (2016). Enhanced user-control of small molecule drug release from a poly (ethylene glycol) hydrogel via azobenzene/cyclodextrin complex tethers. *Journal of Materials Chemistry B*, *4*(6), 1035–1039.
48. Ijaz, H., Qureshi, J., Danish, Z., Zaman, M., Abdel-Daim, M. O., & Bashir, I. (2015). Design and evaluation of bilayer matrix tablet of metoprolol tartrate and lisinopril maleate. *Advances in Polymer Technology*, *36*(2), 152–159.
49. Crispim, E. G., Piai, J. F., Fajardo, A. R., Ramos, E. R. F., Nakamura, T. U., Nakamura, C. V., Rubira, A. F., & Muniz, E. C. (2012). Hydrogels based on chemically modified poly (vinyl alcohol)(PVA-GMA) and PVA-GMA/chondroitin sulfate: preparation and characterization. *Express Polymer Letters*, *6*(5), 383–395.
50. Elliott, J. E., Macdonald, M., Nie, J., & Bowman, C. N. (2004). Structure and swelling of poly (acrylic acid) hydrogels: effect of pH, ionic strength, and dilution on the crosslinked polymer structure. *Polymer*, *45*(5), 1503–1510.
51. Bueno, V. B., Bentini, R., Catalani, L. H., & Petri, D. F. S. (2013). Synthesis and swelling behavior of xanthan-based hydrogels. *Carbohydrate Polymers*, *92*(2), 1091–1099.
52. Ciolacu, D., & Cazacu, M. (2011). Synthesis of new hydrogels based on xanthan and cellulose allomorphs. *Cellulose Chemistry and Technology*, *45*(3), 163.

53. Paradossi, G., Finelli, I., Cerroni, B., & Chiessi, E. (2009). Adding chemical cross-links to a physical hydrogel. *Molecules, 14*(9), 3662–3675.
54. Rhim, J. W., Hong, S. I., Park, H. M., & Ng, P. K. (2006). Preparation and characterization of chitosan-based nanocomposite films with antimicrobial activity. *Journal of Agricultural and Food Chemistry, 54*(16), 5814–5822.
55. Peppas, N. A., Bures, P., Leobandung, W., & Ichikawa, H. (2000). Hydrogels in pharmaceutical formulations. *European Journal of Pharmaceutics and Biopharmaceutics, 50*(1), 27–46.
56. Morris, G. A., Kök, S. M., Harding, S. E., & Adams, G. G. (2010). Polysaccharide drug delivery systems based on pectin and chitosan. *Biotechnology and Genetic Engineering Reviews, 27*(1), 257–284.
57. Sadeghi, M. (2011). Pectin-based biodegradable hydrogels with potential biomedical applications as drug delivery systems. *Journal of Biomaterials and Nanobiotechnology, 2*(01), 36.
58. Boppana, R., Kulkarni, R. V., Mohan, G. K., Mutalik, S., & Aminabhavi, T. M. (2016). In vitro and in vivo assessment of novel pH-sensitive interpenetrating polymer networks of a graft copolymer for gastro-protective delivery of ketoprofen. *RSC Advances, 6*(69), 64344–64356.
59. Dong, K., Dong, Y., You, C., Xu, W., Huang, X., Yan, Y., Zhang, L., Wang, K., & Xing, J. (2016). Assessment of the drug loading, in vitro and in vivo release behavior of novel pH-sensitive hydrogel. *Drug Delivery, 23*(1), 174–184.
60. Işıklan, N., & Küçükbalcı, G. (2016). Synthesis and characterization of pH-and temperature-sensitive materials based on alginate and poly (N-isopropylacrylamide/acrylic acid) for drug delivery. *Polymer Bulletin, 73*(5), 1321–1342.
61. Chen, S. C., Wu, Y. C., Mi, F. L., Lin, Y. H., Yu, L. C., & Sung, H. W. (2004). A novel pH-sensitive hydrogel composed of N, O-carboxymethyl chitosan and alginate cross-linked by genipin for protein drug delivery. *Journal of Controlled Release, 96*(2), 285–300.
62. Eldin, M. M., Omer, A., Wassel, M., Tamer, T., Abd-Elmonem, M., & Ibrahim, S. (2015). Novel smart pH sensitive chitosan grafted alginate hydrogel microcapsules for oral protein delivery: II. Evaluation of the swelling behavior. *International Journal of Pharmacy and Pharmaceutical Sciences, 7*(10), 331–337.
63. Shantha, K. L., & Harding, D. R. K. (2000). Preparation and in-vitro evaluation of poly [N-vinyl-2-pyrrolidone-polyethylene glycol diacrylate]-chitosan interpolymeric pH-responsive hydrogels for oral drug delivery. *International Journal of Pharmaceutics, 207*(1), 65–70.
64. Kulkarni, R. V., & Sa, B. (2008). Evaluation of pH-sensitivity and drug release characteristics of (poly acrylamide-grafted-xanthan)–carboxymethyl cellulose-based pH-sensitive interpenetrating network hydrogel beads. *Drug Development and Industrial Pharmacy, 34*(12), 1406–1414.
65. Sohail, K., Khan, I. U., Shahzad, Y., Hussain, T., & Ranjha, N. M. (2014). pH-sensitive polyvinylpyrrolidone-acrylic acid hydrogels: impact of material parameters on swelling and drug release. *Brazilian Journal of Pharmaceutical Sciences, 50*(1), 173–184.
66. Bukhari, S. M. H., Khan, S., Rehanullah, M., & Ranjha, N. M. (2015). Synthesis and characterization of chemically cross-linked acrylic acid/gelatin hydrogels: effect of pH and composition on swelling and drug release. *International Journal of Polymer Science, 15*(1), 1–15.
67. Singh, B., Sharma, D. K., & Gupta, A. (2008). In vitro release dynamics of thiram fungicide from starch and poly (methacrylic acid)-based hydrogels. *Journal of Hazardous Materials, 154*(1), 278–286.
68. Sajeesh, S., & Sharma, C. P. (2006). Cyclodextrin–insulin complex encapsulated polymethacrylic acid based nanoparticles for oral insulin delivery. *International Journal of Pharmaceutics, 325*(1), 147–154.
69. Cuggino, J. C., Contreras, C. B., Jimenez-Kairuz, A., Maletto, B. A., & Alvarez Igarzabal, C. I. (2014). Novel poly (NIPA-co-AAc) functional hydrogels with potential application in drug controlled release. *Molecular Pharmaceutics, 11*(7), 2239–2249.
70. El-Ghaffar, M. A., Hashem, M. S., El-Awady, M. K., & Rabie, A. M. (2012). pH-sensitive sodium alginate hydrogels for riboflavin controlled release. *Carbohydrate Polymers, 89*(2), 667–675.
71. Kaith, B. S., Jindal, R., Kumar, V., & Bhatti, M. S. (2014). Optimal response surface design of Gum tragacanth-based poly [(acrylic acid)-co-acrylamide] IPN hydrogel for the controlled release of the antihypertensive drug losartan potassium. *RSC Advances, 4*(75), 39822–39829.
72. Hu, X., Wei, W., Qi, X., Yu, H., Feng, L., Li, J., Wang, S., Zhang, J., & Dong, W. (2015). Preparation and characterization of a novel pH-sensitive Salecan-g-poly (acrylic acid) hydrogel for controlled release of doxorubicin. *Journal of Materials Chemistry B, 3*(13), 2685–2697.

73. Bajpai, S. K., Chand, N., & Soni, S. (2015). Controlled release of anti-diabetic drug Gliclazide from poly (caprolactone)/poly (acrylic acid) hydrogels. *Journal of Biomaterials Science, Polymer Edition*, *26*(14), 947–962.
74. Kamisetti, R. R., Vilasitha, W., Muvvala, S., Alluri, R., & Bhavani, D. (2015). Formulation and in vitro evaluation of modified release oral hydrogel driven drug delivery systems of diethylcarbamazine citrate. *RGUHS Journal of Pharmaceutical Sciences*, *5*(2), 10–19.
75. Rajeswari, K. R., Reddy, R. R., Himabindu, A., Sudhakar, M., & Ramesh, A. (2015). Article Details Design, development and in vitro evaluation of modified release hydrogel based drug delivery systems of oseltamivir phosphate. *Indian Drug*, *52*(10), 34–39.
76. Carrillo-Conde, B. R., Brewer, E., Lowman, A., & Peppas, N. A. (2015). Complexation hydrogels as oral delivery vehicles of therapeutic antibodies: an in vitro and ex vivo evaluation of antibody stability and bioactivity. *Industrial & Engineering Chemistry Research*, *54*(42), 10197–10205.
77. Ramanan, R. M. K., Chellamuthu, P., Tang, L., & Nguyen, K. T. (2006). Development of a temperature-sensitive composite hydrogel for drug delivery applications. *Biotechnology Progress*, *22*(1), 118–125.
78. Razzaq, R., Ranjha, N. M., & Rashid, Z. (2018). Preparation and evaluation of novel pH-sensitive poly (butyl acrylate-co-itaconic acid) hydrogel microspheres for controlled drug delivery. *Advances in Polymer Technology*, *37*(1), 247–255..
79. Das, D., Das, R., Mandal, J., Ghosh, A., & Pal, S. (2014). Dextrin crosslinked with poly (lactic acid): a novel hydrogel for controlled drug release application. *Journal of Applied Polymer Science*, *131*(7), 217–230.
80. Ranjha, N. M., Hanif, M., Naz, A., Shah, M. S., Abbas, G., & Afzal, Z. (2016). Synthesis and characterization of cetirizine-containing, pH-sensitive acrylic acid/poly (vinyl alcohol) hydrogels. *Journal of Applied Polymer Science*, *133*(19).
81. Pourjavadi, A., Harzandi, A. M., & Hosseinzadeh, H. (2004). Modified carrageenan 3. Synthesis of a novel polysaccharide-based superabsorbent hydrogel via graft copolymerization of acrylic acid onto kappa-carrageenan in air. *European Polymer Journal*, *40*(7), 1363–1370.
82. Lei, J., Mayer, C., Freger, V., & Ulbricht, M. (2013). Synthesis and characterization of poly (ethylene glycol) methacrylate based hydrogel networks for anti-biofouling applications. *Macromolecular Materials and Engineering*, *298*(9), 967–980.
83. Bukhari, S. M. H., Khan, S., Rehanullah, M., & Ranjha, N. M. (2015). Synthesis and characterization of chemically cross-linked acrylic acid/gelatin hydrogels: effect of pH and composition on swelling and drug release. *International Journal of Polymer Science*, *15*(3), 1–15.
84. Zhao, X., Lin, S., & Yuk, H. (2018). U.S. Patent No. 9,878,506. Washington, DC: U.S. Patent and Trademark Office.
85. Sershen, S. R., Pai, S. S., & Gong, G. (2018). U.S. Patent No. 9,861,701. Washington, DC: U.S. Patent and Trademark Office.
86. Kim, T., & Choi, W. S. (2018). U.S. Patent No. 9,869,616. Washington, DC: U.S. Patent and Trademark Office.
87. Anderson, D. G., Langer, R. S., & Dang, T. T. (2018). U.S. Patent No. 9,867,781. Washington, DC: U.S. Patent and Trademark Office.
88. Eddy, P. E. (2018). U.S. Patent Application No. 15/819,381.
89. DePinto, J. T., Templer, D. A., Nikitenko, A. A., Gamerman, G., Waller, D. P., Bolick, D., & Lissner, E. (2018). U.S. Patent Application No. 15/833,959.
90. Tucker, R. C., Pruitt, J. D., & Qiu, Y. (2018). U.S. Patent Application No. 15/704,188.
91. Rege, K., Grandhi, T. S. P., & Potta, T. (2018). U.S. Patent Application No. 15/547,173.
92. Bian, L., Feng, Q., Kongchang, W. E. I., Li, G., & Sien, L. I. N. (2018). U.S. Patent No. 9,889,086. Washington, DC: U.S. Patent and Trademark Office.
93. Qiu, Y., Pruitt, J. D., & Nelson, J. (2018). U.S. Patent Application No. 15/704,205.
94. Inamoto, Y., Kamakura, T., Takahashi, M., & Ohura, T. (2018). U.S. Patent No. 9,877,871. Washington, DC: U.S. Patent and Trademark Office.
95. McClements, D. J., Zhang, Z., & Zhang, R. (2018). U.S. Patent Application No. 15/690,748.
96. Chien, H. W. (2018). U.S. Patent No. 9,896,531. Washington, DC: U.S. Patent and Trademark Office.
97. Lei, Y., & Xu, L. (2018). U.S. Patent Application No. 15/722,022.
98. Ludan, Z., Hanssen, J. H. L., & Tweehuysen, R. (2018). U.S. Patent Application No. 15/510,729.

99. Matsumoto, K., & Kashino, T. (2018). U.S. Patent No. 9,872,907. Washington, DC: U.S. Patent and Trademark Office.
100. Kakinuki, K. (2018). U.S. Patent Application No. 15/539,728.
101. Costa, A. L., & Blosi, M. (2018). U.S. Patent Application No. 15/547,662.
102. Oka, T., Ishiodori, A., & Yoshihara, M. (2018). U.S. Patent No. 9,868,840. Washington, DC: U.S. Patent and Trademark Office.
103. Bai, X., Bao, Z., Bi, S., Li, Y., Yu, X., Hu, S., Tian, M., Zhang, X., Cheng, X., & Chen, X. (2018). Chitosan-based thermo/pH double sensitive hydrogel for controlled drug delivery. *Macromolecular Bioscience*, *18*(3), 1700305.
104. Umaredkar, A. A., Dangre, P. V., Mahapatra, D. K., & Dhabarde, D. M. (2018). Fabrication of chitosan-alginate polyelectrolyte complexed hydrogel for controlled release of cilnidipine: a statistical design approach. *Materials Technology*, *8*(3), 1–11.
105. Li, S., Xia, Y., Qiu, Y., Chen, X., & Shi, S. (2018). Preparation and property of starch nanoparticles reinforced aldehyde–hydrazide covalently crosslinked PNIPAM hydrogels. *Journal of Applied Polymer Science*, *135*(5), 10–18.
106. Obuobi, S., Voo, Z. X., Low, M. W., Czarny, B., Selvarajan, V., Ibrahim, N. L., Yang, Y. Y., & Ee, P. L. R. (2018). Phenylboronic acid functionalized polycarbonate hydrogels for controlled release of polymyxin B in *Pseudomonas aeruginosa* infected burn wounds. *Advanced Healthcare Materials*, *7*(13), 1701388.
107. Zhang, X. Z., Wu, D. Q., & Chu, C. C. (2004). Synthesis, characterization and controlled drug release of thermosensitive IPN–PNIPAAm hydrogels. *Biomaterials*, *25*(17), 3793–3805.
108. He, H., Cao, X., & Lee, L. J. (2004). Design of a novel hydrogel-based intelligent system for controlled drug release. *Journal of Controlled Release*, *95*(3), 391–402.
109. Seeli, D. S., & Prabaharan, M. (2017). Guar gum oleate-graft-poly (methacrylic acid) hydrogel as a colon-specific controlled drug delivery carrier. *Carbohydrate Polymers*, *158*, 51–57.
110. Wang, Y., Wang, J., Yuan, Z., Han, H., Li, T., Li, L., & Guo, X. (2017). Chitosan cross-linked poly (acrylic acid) hydrogels: drug release control and mechanism. *Colloids and Surfaces B: Biointerfaces*, *152*, 252–259.
111. Amoli-Diva, M., Sadighi-Bonabi, R., & Pourghazi, K. (2017). Switchable on/off drug release from gold nanoparticles-grafted dual light-and temperature-responsive hydrogel for controlled drug delivery. *Materials Science and Engineering: C*, *76*, 242–248.
112. Luo, Y., Kirker, K. R., & Prestwich, G. D. (2000). Cross-linked hyaluronic acid hydrogel films: new biomaterials for drug delivery. *Journal of Controlled Release*, *69*(1), 169–184.
113. Geever, L. M., Cooney, C. C., Lyons, J. G., Kennedy, J. E., Nugent, M. J., Devery, S., & Higginbotham, C. L. (2008). Characterisation and controlled drug release from novel drug-loaded hydrogels. *European Journal of Pharmaceutics and Biopharmaceutics*, *69*(3), 1147–1159.
114. Sun, X. F., Wang, H. H., Jing, Z. X., & Mohanathas, R. (2013). Hemicellulose-based pH-sensitive and biodegradable hydrogel for controlled drug delivery. *Carbohydrate Polymers*, *92*(2), 1357–1366.
115. Kurisawa, M., Chung, J. E., Yang, Y. Y., Gao, S. J., & Uyama, H. (2005). Injectable biodegradable hydrogels composed of hyaluronic acid–tyramine conjugates for drug delivery and tissue engineering. *Chemical Communications*, (34), 4312–4314.
116. Hoare, T. R., & Kohane, D. S. (2008). Hydrogels in drug delivery: progress and challenges. *Polymer*, *49*(8), 1993–2007.
117. Hurkmans, J. F. G. M., Bodde, H. E., Driel, L. V., Doorne, H. V., & Junginger, H. E. (1985). Skin irritation caused by transdermal drug delivery systems during long-term (5 days) application. *British Journal of Dermatology*, *112*(4), 461–467.
118. Lee, K., Silva, E. A., & Mooney, D. J. (2011). Growth factor delivery-based tissue engineering: general approaches and a review of recent developments. *Journal of the Royal Society Interface*, *8*(55), 153–170.
119. Seo, S. H., Han, H. D., Noh, K. H., Kim, T. W., & Son, S. W. (2009). Chitosan hydrogel containing GMCSF and a cancer drug exerts synergistic anti-tumor effects via the induction of CD8+ T cell-mediated anti-tumor immunity. *Clinical & Experimental Metastasis*, *26*(3), 179–187.
120. Emoto, S., Yamaguchi, H., Kamei, T., Ishigami, H., Suhara, T., Suzuki, Y., Ito, T., Kitayama, J., & Watanabe, T. (2014). Intraperitoneal administration of cisplatin via an in situ cross-linkable hyaluronic acid-based hydrogel for peritoneal dissemination of gastric cancer. *Surgery Today*, *44*(5), 919–926.

121. Wolinsky, J. B., Colson, Y. L., & Grinstaff, M. W. (2012). Local drug delivery strategies for cancer treatment: gels, nanoparticles, polymeric films, rods, and wafers. *Journal of Controlled Release*, *159*(1), 14–26.
122. El Fawal, G. F., Abu-Serie, M. M., Hassan, M. A., & Elnouby, M. S. (2018). Hydroxyethyl cellulose hydrogel for wound dressing: fabrication, characterization and in vitro evaluation. *International Journal of Biological Macromolecules*, *111*, 649–659.
123. Wu, D. Q., Zhu, J., Han, H., Zhang, J. Z., Wu, F. F., Qin, X. H., & Yu, J. Y. (2018). Synthesis and characterization of arginine-NIPAAM hybrid hydrogel as wound dressing: in vitro and in vivo study. *Acta Biomaterialia*, *65*, 305–316.
124. Chen, C. C., Baikoghli, M. A., & Cheng, R. H. (2018). Tissue targeted nanocapsids for oral insulin delivery via drink. *Pharmaceutical Patent Analyst*, *7*(3), 1–8.
125. Mohamed, H. A., Radwan, R. R., Raafat, A. I., & Ali, A. E. H. (2018). Antifungal activity of oral (Tragacanth/acrylic acid) Amphotericin B carrier for systemic candidiasis: in vitro and in vivo study. *Drug Delivery and Translational Research*, *8*(1), 191–203.
126. Ruel-Gariepy, E., & Leroux, J. C. (2004). In situ-forming hydrogels—review of temperature-sensitive systems. *European Journal of Pharmaceutics and Biopharmaceutics*, *58*(2), 409–426.
127. Nanjawade, B. K., Manvi, F. V., & Manjappa, A. S. (2007). RETRACTED: in situ-forming hydrogels for sustained ophthalmic drug delivery. *Journal of Controlled Release*, *122*(2), 119–134.
128. Ma, L., Liu, M., Liu, H., Chen, J., Gao, C., & Cui, D. (2010). Dual crosslinked pH- and temperature-sensitive hydrogel beads for intestine-targeted controlled release. *Polymers for Advanced Technologies*, *21*(5), 348–355.
129. Jain, S. K., Jain, A., Gupta, Y., & Ahirwar, M. (2007). Design and development of hydrogel beads for targeted drug delivery to the colon. *AAPS PharmSciTech*, *8*(3), E34–E41.
130. Lee, K. Y., Peters, M. C., Anderson, K. W., & Mooney, D. J. (2000). Controlled growth factor release from synthetic extracellular matrices. *Nature*, *408*(6815), 998.

4

Polymer–Drug Conjugates as Drug Delivery Systems

Sauraj Singh, Ruchir Priyadarshi, Bijender Kumar, Saleheen Bano, Farha Deeba, Anurag Kulshreshtha, and Yuvraj Singh Negi

CONTENTS

4.1 Introduction

The successful delivery of a drug to its site(s) of action is a challenging prospect in light of the many physicochemical, biopharmaceutical, and pharmacokinetic barriers it may face. The polymer–drug conjugate offers several significant advantages over the parent drug molecule, including improved aqueous solubility of hydrophobic drugs, enhanced stability of drugs, improved bioavailability, prolonged plasma life, etc. In 1975, Ringsdorf proposed an ideal model for a polymer–drug conjugate, shown in (Figure 4.1), in which bioactive agents were attached to a hydrophilic polymer directly or through biologically responsive linkers.

The covalent attachment of a drug to a hydrophilic polymer provided a means to overcome the physicochemical and pharmacokinetic challenges associated with small molecule therapeutics. Through this conjugation, the drug's pharmacokinetic profile becomes that of the polymer, thereby improving the solubility, preventing rapid renal clearance, and altering the tissue distribution characteristics.

In recent years, nanotechnology has opened new perspectives for biological and biomedical applications to improve the selective delivery of anticancer compounds to their site of action. Nanoparticles (NPs) can be preferentially delivered to the tumor site because of the enhanced permeation and retention (EPR) effect. In addition to that, nanotechnologies can improve drug properties in several ways: by controlling release and distribution, by enhancing drug absorption, and by protecting the drug from degradation. Many nanodrug delivery systems, such as micelles, liposomes, polymersomes, and polymeric NPs, have been used for the delivery of chemotherapeutic agents in cancer therapy.

Amphiphilic polymer–drug conjugates can self-assemble in water to form NPs that offer several distinct advantages in the field of anticancer drug delivery. In this chapter, we discuss nanosystems based on

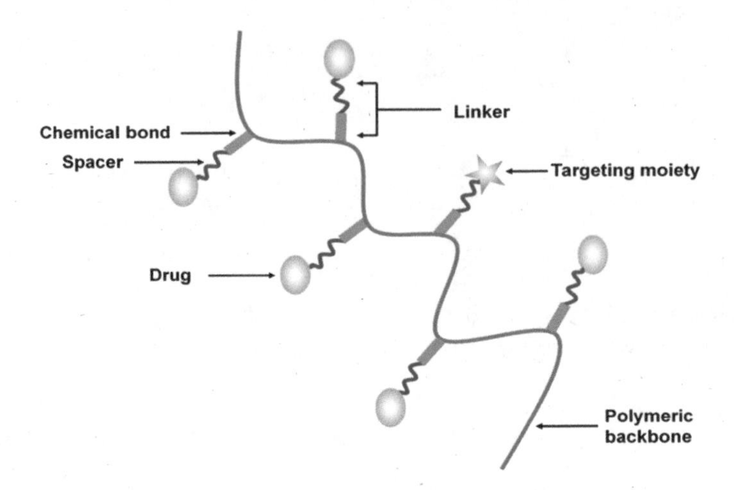

FIGURE 4.1 Schematic presentation of polymer–drug conjugates (prodrugs). (Reprinted from *Journal of Controlled Release*, 222, Pang X, Jiang Y, Xiao Q, Leung AW, Hua H, Xu C., pH-responsive polymer-drug conjugates: Design and progress, 116–129, Copyright (2016), with permission from Elsevier.)

polymer–drug conjugates for cancer therapy. In addition, we discuss new trends in the field of prodrug-based nanoassemblies that enhance the delivery efficiency of anticancer drugs, with special emphasis on smart stimuli-triggered drug release.

4.2 Self-Assembly Behavior of Amphiphilic Polymer–Drug Conjugates

Self-assembly is the spontaneous disorder-to-order transition of molecules driven by noncovalent interactions such as the Van der Waals and hydrogen bonding forces, and the hydrophobic effect (2). Amphiphilic block copolymers can spontaneously self-assemble into core/shell nanostructured micelles in aqueous solution at concentrations above the critical micelle concentration (CMC). Hydrophobic interactions are the most extensively studied noncovalent interactions and the driving force for the spontaneous self-assembly of amphiphilic diblock copolymers into micelles in water (3).

Over the past few decades, the concept of self-assembly has been successfully used in the design of drug delivery systems (DDSs) that are capable of releasing bioactive compounds at therapeutically effective rates and concentrations (4).

4.3 Self-Assembled Polymer–Drug Conjugates

Various types of biocompatible polymers, including synthetic and block copolymers, dendritic polymers, and natural polysaccharides, have been used in the synthesis of polymer–drug conjugates for cancer therapy. The advantage of these polymers is that they have a large number of active functional groups on the polymer skeleton, and anticancer drug moieties are usually coupled onto polymer chains as side groups for high drug-loading efficiency (Table 4.1.)

4.4 Stimuli-Responsive Self-Assembled Polymer–Drug Conjugates

Tumor tissue exhibits different intracellular microenvironments, such as acidic pH (5, 6), a high concentration of glutathione (GSH) enzyme (2–10 mM), and more reductive environments as compared with normal tissue (35–42). These biological features were employed to design stimuli-responsive polymer–drug conjugates or other nanotherapeutics for intracellular drug delivery. The pH-responsive

TABLE 4.1
List of Self-Assembled Polymer–Drug Conjugates

Categories	Polymers	Drug coupled	Refs
Polyethylene glycol (PEG)	PEG-40k	Camptothecin	(5)
	PEG-2k	Paclitaxel	(6)
	Multi-hydroxyl polyethylene Glycol derivative	5-Fluorouracil	(7)
PEG block copolymers	PEG-polymerized block of camptothecin	Camptothecin	(8)
	PEG–block-dendritic polylysine	Camptothecin	(9)
	mPEG–b-P(LA-co-MCC-OH)	Cisplatin	(10)
	mPEG–b-PAMAM	Doxorubicin	(11)
	PEG–b-P(HEMA-co-EGMA)	Doxorubicin	(12)
	PEG–poly(aspartic acid) copolymer	Oxaliplatin	(13)
	Poly(ethylene oxide)-block-polyphosphoester	Paclitaxel	(14)
	PEG–b-poly(acrylic acid) block copolymers	Paclitaxel	(15)
	mPEG–b-P(LA-co-MCC)	Platinum	(16)
Polysaccharides	Pullulan	Doxorubicin	(17)
	Hyaluronic acid (HA)	Doxorubicin	(18)
	Chitosan	Doxorubicin	(19)
	Heparin	Paclitaxel	(20)
	Xylan	Curcumin	(21)
	Cholesterol-modified hyaluronic acid (HA)	Curcumin	(22)
	PEGylated carboxymethyl cellulose	Docetaxel	(23)
	Xylan	5-Fluorouracil	(24)
	Pectin	5-Fluorouracil	(25)
	Dextran	Curcumin	(26)
	Dextran	Budesonide	(27)
	beta-Cyclodextrin	5-Fluorouracil	(28)
	Carboxymethyl cellulose	Tetrahydrocurcumin	(29)
	Xylan	Ibuprofen	(30)
Others	HPMA copolymer	Doxorubicin	(31)
	Poly(L-g-glutamylglutamine) (PGG)	Paclitaxel	(32)
	Polyisoprene (PI)	Gemcitabine	(33)
	Poly(methyl methacrylate) (PMMA)	Gemcitabine	(34)

polymer–drug conjugates were developed by incorporating pH-sensitive bonds between the polymer and the drug. Over the last few years, numerous pH-sensitive polymer–drug conjugates have been synthesized using various pH-sensitive bonds (43–46). The most commonly used pH-sensitive bonds and their corresponding degradation products are represented in Table 4.2.

In addition to the pH stimuli, redox response is another promising stimulus for intracellular drug delivery. GSH is a primary antioxidant and a thiol-containing tripeptide, which is found in higher concentration in tumor cells (2–10 mM) than in normal cells (2–10 μM) (47–51). These sharp differences in GSH levels between tumor and normal cells develop the possibility of designing redox-sensitive polymer–drug conjugates. Due to the poor oral bioavailability of polymer–drug conjugates, they are generally administrated by the intravenous route. After intravenous administration, the polymer-drug conjugate enter the blood circulation, and the stimuli-linkage between the polymer and the drug in the conjugate remains stable, which ensures the stability of the polymer–drug conjugate during circulation. On arrival within the tumor site, the polymer–drug conjugate will enter into the cell through the endocytosis process, where the pH-sensitive bond will be destroyed in the acidic intracellular endosomal compartments (pH 5–6), and drug release takes place inside the tumor cells. Polymer–drug conjugates with

TABLE 4.2

Examples of pH- and Redox-sensitive Chemical Bonds and Their Degradation Products

Name	Chemical structure		Degradation products
Hydrazone		H^+ →	+
Schiff base		H^+ →	+ $H_2N—R$
Acetal		H^+ →	+ R—OH
Cis- Aconityl		H^+ →	+ $H_2N—R$
β- Thiopropionate		H^+ →	+ R—OH
Disulfide bond		DTT →	+

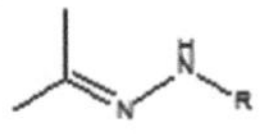

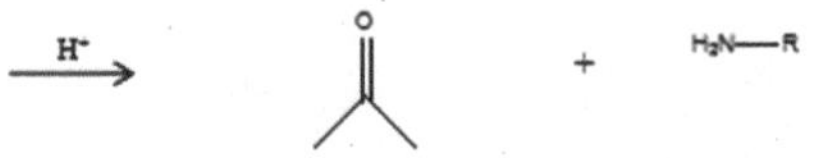

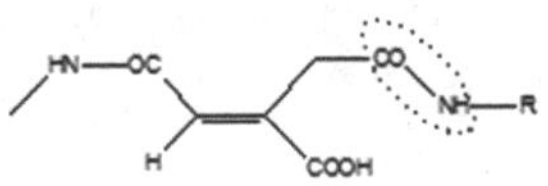

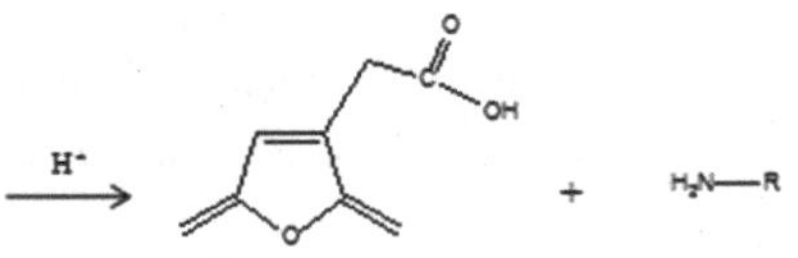

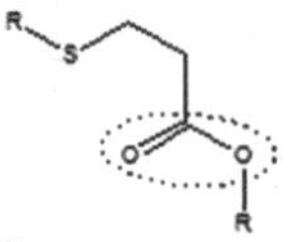

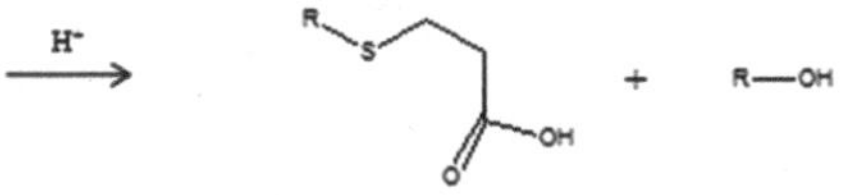

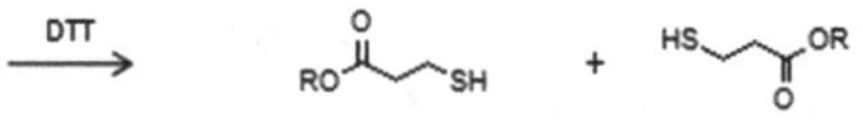

redox-sensitive bonds, i.e., a disulfide linkage, release drug intracellularly in the presence of glutathione (GSH) enzyme, as shown in (Figure 4.2) (52–61).

4.5 pH-Responsive Self-Assembled Polymer-Drug Conjugates

In recent years, various stimuli-responsive polymer–drug conjugates have been developed using a variety of pH/redox-responsive chemical bonds or linkers. Of these, pH-sensitive (hydrazone, Schiff base, acetal, cis-acrotinyl, and β-thiopropionate bonds) and redox-sensitive (disulfide) bonds have been frequently used in the preparation of stimuli-responsive polymer–drug conjugates.

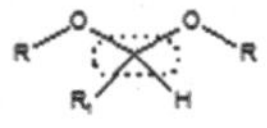

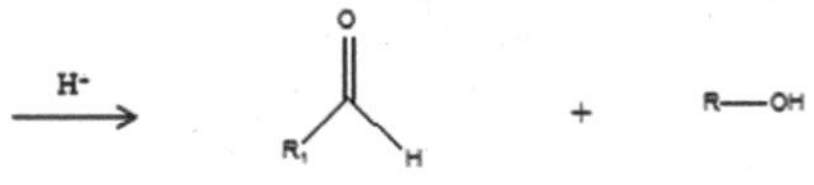

To date, numerous polymer–drug conjugates have been developed using the various pH-labile chemical bonds between polymer and drug.

4.5.1 Polymer-Drug Conjugates Based on Hydrazone Bond Cleavage

Doxorubicin (DOX)-conjugated poly (L-lactic acid)–methoxy poly (ethylene glycol) (PLLA–mPEG–DOX) copolymers have been reported by Park et al. (62), where DOX was conjugated with a pH-sensitive hydrazone bond (Figure 4.3a). The amphiphilic polymer–drug conjugate formed self-assembled nanoparticles in aqueous solution with the average diameter 89.1 nm. Most drug was released in acidic conditions (pH 5.0) from the (PLLA–mPEG–DOX) conjugate, which indicated that the bond between polymer and drug in the conjugate is highly pH sensitive in nature. The DOX–PLLA–mPEG micelles exhibited better cytotoxic activity than free DOX.

Similarly, adriamycin (ADR) was linked to block copolymer of poly (ethylene glycol)–poly-(aspartate hydrazone adriamycin) (PEG–PASP–ADR) through an acid-sensitive hydrazone linker (Figure 4.3b). The amphiphilic (PEG–PASP–ADR) block copolymers formed micelles in aqueous medium, which

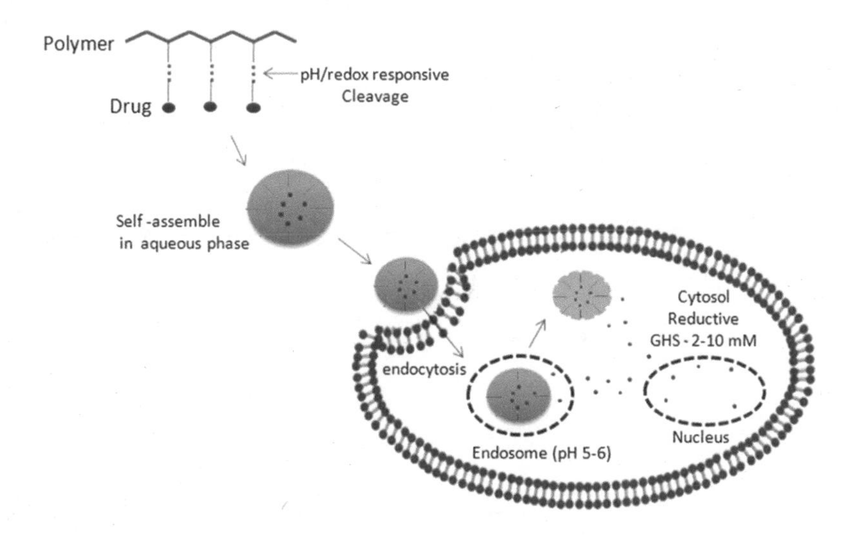

FIGURE 4.2 Schematic illustration of intracellular environments and drug release mechanisms.

shows stability in physiological conditions (pH 7.4) but cleavage in acidic conditions (pH 5–6). The (PEG–PASP–ADR) conjugate micelles improve the therapeutic efficacy of ADR due to selective intracellular pH-triggered drug release (63).

A multifunctional folate-decorated polymeric micelle based on [FA–PEG–P(Asp-Hyd-ADR)] was developed by Kataoka et al (64) for enhancing the intracellular delivery of the anticancer drug ADR. The (ADR) was conjugated with folate containing [FA–PEG–P(Asp-Hyd-ADR) copolymer with a pH-sensitive hydrazone bond (Figure 4.3c). The [FA–PEG–P(Asp-Hyd-ADR)] showed a pH-sensitive drug release profile. A surface plasmon resonance (SPR) study demonstrated that [FA–PEG–P(Asp-Hyd-ADR)] has a strong folate-binding proteins (FBP) tendency. Furthermore, the cell growth inhibitory activity of the [FA–PEG–P(Asp-Hyd-ADR)] was enhanced significantly due to enhanced cellular uptake.

Jing et al. (65) developed a novel mixed micelles DDS based on (mPEG-b-P(LA-co-DHP)) block copolymer, in which DOX and folic acid were attached to copolymers via a pH-sensitive hydrazone bond and a carbamate linkage, respectively (Figure 4.3d). The mixed micelles had a well-defined core-shell structure with the size ranging from 70 to 100 nm. The micelles showed pH-dependent release behavior, and a greater amount of the drug was released in acidic medium relative to physiological pH (7.4).

A pH-sensitive polymeric micellar system based on (PEO-b-PAGE) was prepared for the delivery of DOX (66). The drug was linked to copolymers with a highly pH-sensitive hydrazone bond (Figure 4.3e). The polymeric micellar system showed pH-dependent drug release behavior; i.e., 43% DOX was released at acidic pH (5.0), whereas only 16% DOX was released at physiological pH (7.4).

4.5.2 Polymer–Drug Conjugates Based on Schiff Base Bond Cleavage

An acid-labile Schiff base bond containing polymer–drug conjugate based on polylactide-graft-doxorubicin (PLA-g-DOX) was synthesized by Cheng et al. The DOX was attached to PLA through an acid-sensitive Schiff base linkage (Figure 4.4a). The PLA-g-DOX formed self-assembled nanoparticles in phosphate-buffered saline (PBS) (7.4) with average diameter 100 nm. The PLA-g-DOX NPs showed acid-sensitive drug release behavior due to the presence of Schiff base conjugation linkage. Cellular uptake studies against MCF breast cancer cells showed higher antitumor activity of PLA-g-DOX NPs than free drug (67).

Recently, Schiff base bond linkage pH-sensitive prodrug NPs based on an oxidized sodium alginate–DOX conjugate were developed by Wang et al. for the simultaneous delivery of DOX and curcumin (Figure 4.4b). The oxidized sodium alginate–DOX conjugate forms NPs in aqueous solution, and curcumin is encapsulated into the core of the prodrug NPs through hydrophobic interaction. Cur-DOX NPs are highly pH sensitive in nature and efficiently released the drugs. The Cur-DOX NPs showed potent antitumor activity compared with free drug (68).

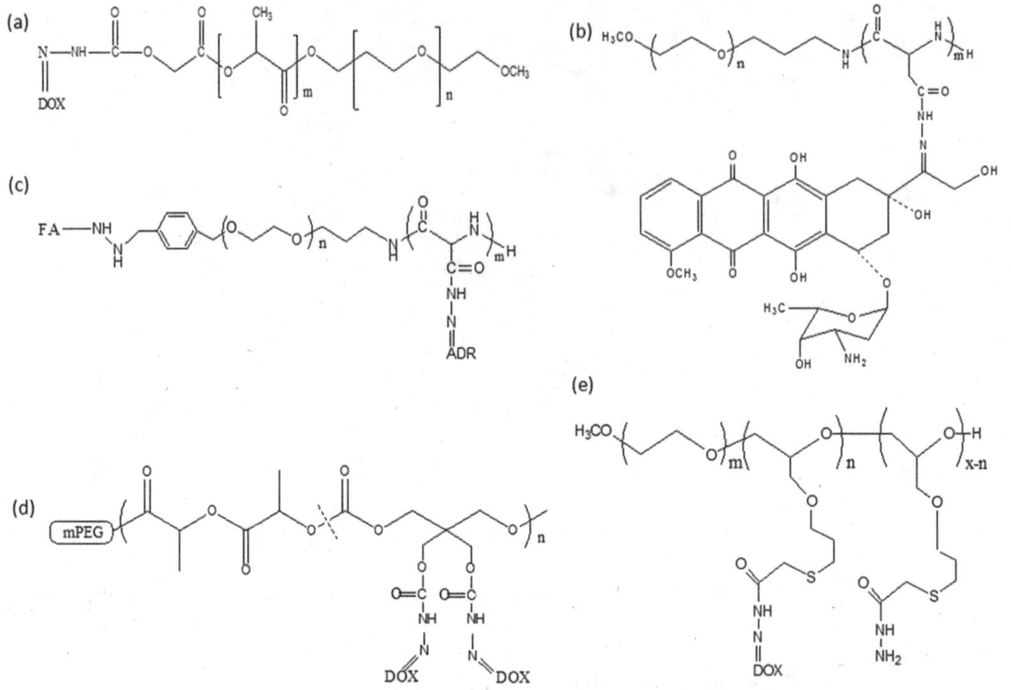

FIGURE 4.3 (a) Hydrazone-linked (PLLA–mPEG–DOX) conjugate. (b) Hydrazone-linked (PEG–PASP–ADR) conjugate. (c) Hydrazone-linked [Fol–PEG–P(Asp-Hyd-ADR)] conjugate. (d) DOX-conjugated (mPEG-b-P(LA-co-DHP)) copolymers via hydrazone bond. (e) Hydrazone-linked (PEO-b-PAGE–DOX) conjugate.

Similarly, Zhang Y. et al. developed poly (ethylene glycol)–DOX conjugated nanoparticles (PEG-DOX NPs) for the co-delivery of DOX and curcumin, where DOX was connected with PEG through a Schiff base reaction (Figure 4.4c). Subsequently, curcumin was loaded into the PEG-DOX NPs. The majority of the drug was released at acidic pH (5.0), relative to pH 7.4, due to the presence of the acid pH-sensitive Schiff base bond. The PEG-DOX-Cur NPs released most of the drug in acidic conditions (pH 5.0), whereas a negligible amount of the drug was released in physiological conditions (pH 7.4). An *in vivo* study demonstrated that the PEG-DOX-Cur NPs showed better antitumor activity compared with free drugs (69).

A diblock copolymer of poly (ethylene glycol) (PEG) and polycarbonate (PC) was synthesized by Yang et al. (70) for the intracellular delivery of DOX. The DOX was covalently attached to aldehyde groups of polycarbonate through the acid-labile Schiff base linkage (Figure 4.4d). The DOX-containing PEG-PC-DOX block copolymers formed self-assembled NPs in PBS (pH 7.4) with a particle size of 100 nm. The results obtained from an *in vitro* drug release study demonstrated the pH-dependent release behavior of DOX from NPs. More of the drug was released at pH 5.0 than pH 7.4. A cytotoxic and cellular uptake study of the DOX-containing NPs against human breast cancer MCF-7cells showed that the DOX-containing NPs killed the cells more efficiently than free drug.

4.5.3 Polymer–Drug Conjugates Based on Acetal Bond Cleavage

Curcumin-conjugated methoxy poly (ethylene glycol)–poly (lactic acid) (mPEG–PLA) copolymer-based micelles were developed to enhance the therapeutic efficacy of curcumin in cancer therapy. The curcumin was conjugated with copolymer via a pH-sensitive acetal bond as well as by an ester bond (control) (Figure 4.5a). The curcumin-conjugated amphiphilic mPEG–PLA copolymers formed self-assembled micelles with the hydrodynamic size 91.1 ± 2.9 nm. The micelles displayed pH-dependent drug release behavior, and most of the drug was released in acidic pH (5.0) relative to physiological pH (7.4). The micelles exhibited better cytotoxic activity than free drug (71).

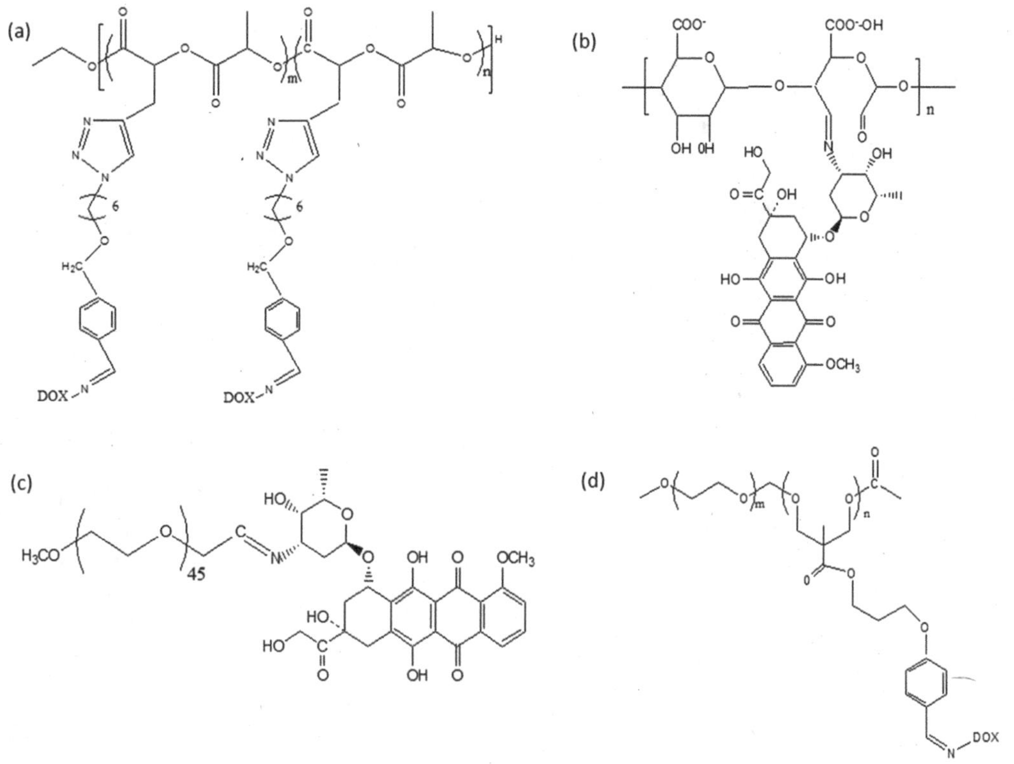

FIGURE 4.4 (a) Schiff base–linked (PLA-g–DOX) conjugate. (b) Schiff base–linked sodium alginate–DOX conjugate (c) Schiff base–linked (PEG–DOX) conjugate. (d) Schiff base–linked (PEG–PC–DOX) conjugate.

An acid-labile polymer–drug conjugate based on DOX-*acetal*-PEG-*acetal*-DOX conjugate was reported by Wang et al., in which DOX was attached to PEG through acetal and carbamate linkages (Figure 4.5b). The dissociation of prodrug micelles in acidic medium was observed by monitoring the change in size through dynamic light scattering (DLS) analysis. The prodrug exhibited pH-triggered DOX release behavior at acidic pH (5.0) rather than at physiological pH (7.4). A cell cytotoxic and intracellular drug release study indicated that the prodrug exhibited excellent antitumor activity compared with free drug (72).

4.5.4 Polymer–Drug Conjugates Based on cis-Acrotinyl Bond Cleavage

Kakinoki et al (73) synthesized poly (vinyl alcohol)–doxorubicin (PVA–DOX) conjugates, in which the drug (DOX) was conjugated with PVA through an acid-labile acrotinyl bond in cis and trans configurations (Figure 4.5c). Both PVA-cis-acrotinyl-DOX and PVA-trans acrotinyl-DOX isomers showed stability at physiological pH (7.4) but released the drug (DOX) at acidic pH (5.0). The drug release profile showed that the cis isomer released the drug more rapidly than the trans isomer. The cytotoxicity of both isomers against J774.1 cells indicated a significant difference in antitumor activity; PVA-cis-acrotinyl-DOX exhibited higher antitumor activity than PVA-cis-acrotinyl-DOX.

An acid-sensitive polymer–drug conjugate based on glycol-chitosan–DOX (GC–DOX) was synthesized by Jeong et al. The DOX was conjugated with glycol-chitosan (GC) via a pH-sensitive N-cis-acrotinyl bond (Figure 4.5d). The DOX-conjugated GC forms nanoaggregates with uniform size ranging from 250 to 300 nm. The DOX was further entrapped into the hydrophobic core of glycol-chitosan (GC DOX) nanoaggregates to enhance the drug content. The GC–DOX nanoaggregates showed pH-dependent release behavior at acidic pH (5.0), whereas at neutral pH (7.0), the drug release was almost negligible. The *in vivo* antitumor activity showed that the GC–DOX nanoaggregates showed antitumor effect similar to free drug (DOX) with a longer circulation time and fewer side effects (74).

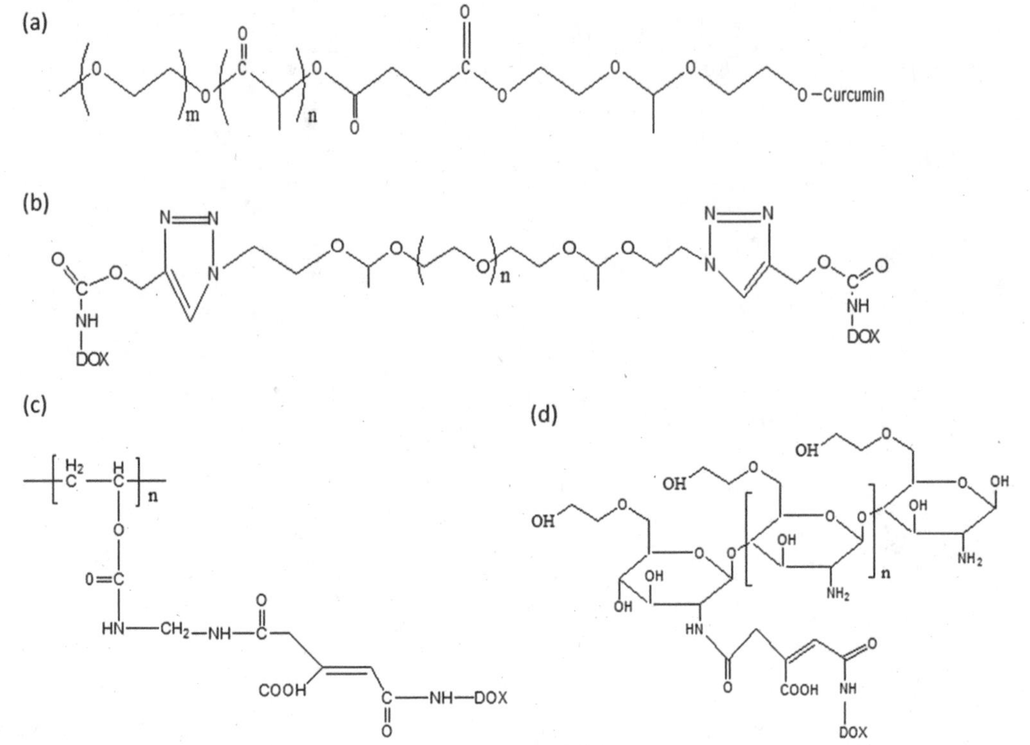

FIGURE 4.5 (a) Acetal-linked (mPEG–PLA–CUR) conjugate. (b) Acetal-linked (DOX-*acetal*-PEG-*acetal*-DOX) conjugate. (c) Cis-acrotinyl-linked (PVA–DOX) conjugate. (d) Cis-acrotinyl-linked (GC–DOX) conjugate.

4.5.5 Polymer–Drug Conjugates Based on β-Thiopropionate Bond Cleavage

A folate-decorated dual (pH and redox)-responsive prodrug micellar drug delivery system based on (FA-PEG-b-P ((PTX-SS-CL)-*co*-CL)) conjugates was developed by Rezaei et al. (75), in which paclitaxel (PTX) was linked through acid- and redox-cleavable linkages (Figure 4.6a). The amphiphilic conjugates formed nanosized (95 mm) micelles in PBS buffer solution. The prodrug micelles were more stable at physiological pH (7.4) than acidic pH (5.0). The prodrug micelles released most of the drug at acidic pH (5.0). The folate-decorated conjugate prodrug micelles displayed enhanced therapeutic efficacy compared with free drug with fewer side effects (75).

An acid-labile polymer–drug conjugate based on poly (ethylene oxide)-*b*-polyphosphoester-*graft*-PTX drug conjugate (PEO-*b*-PPE-*g*–PTX) was developed by Wooley et al (76), in which PTX was connected with the polymer backbone through an acid-labile β-thiopropionate linkage (Figure 4.6b). The PTX-conjugated copolymer forms NPs in aqueous medium with an average diameter of 114 nm. The NPs show pH-triggered drug release in acidic conditions (pH 5.0) compared with neutral physiological conditions (pH 7.4). The cytotoxic study demonstrated that NPs show five to eight times higher antitumor activity than free drug (PTX).

4.6 Polymer–Drug Conjugates Based on Disulfide Bond Cleavage

Cao Y. et al. (3) synthesized a redox-sensitive polymer–drug conjugate micelle system based on methoxy poly (ethylene glycol)–poly (lactic acid) copolymer (mPEG–PLA), in which curcumin was linked through a disulfide bond and an ester bond (control), respectively (Figure 4.7a). The self-assembled micelles showed a hydrodynamic diameter of 115.6 ± 5.9 nm. A drug release study was performed in the presence of 10 mM GSH enzyme, which showed that the rate of drug release from disulfide-linked micelles is three times higher than control.

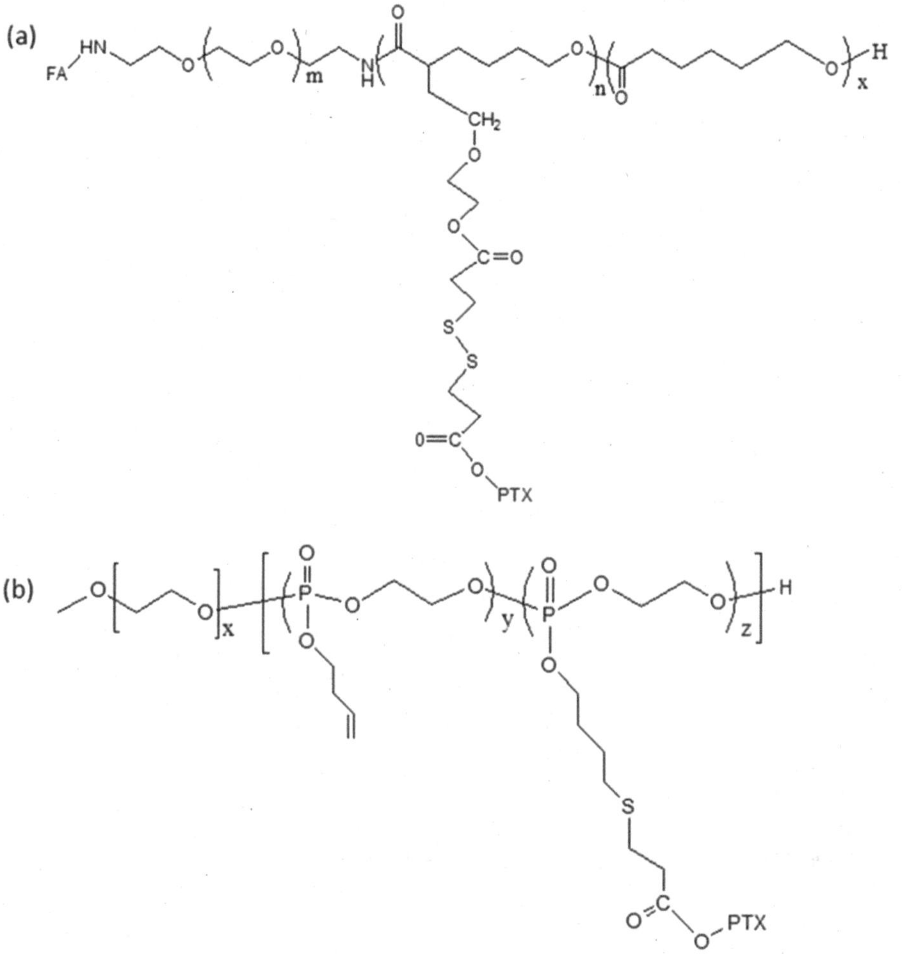

(a)

(b)

FIGURE 4.6 (a) PTX-linked (FA-PEG-b-P ((PTX-SS-CL)-*co*-CL)) conjugate via β-thiopropionate linkage. (b) β-thiopropionate-linked (PEO-*b*-PPE-*g*–PTX) conjugates.

Bai et al. (77) developed a dual-responsive DDS for the safe and efficient delivery of the anticancer drug DOX. A redox-sensitive prodrug based on poly (ethylene glycol)–DOX conjugate (PEG-SS-DOX) was developed, which was subsequently cross linked with Cu^{2+} to enhance the stability of the prodrug to the external environment. The release study demonstrated that prodrug NPs were quite stable at physiological pH (7.4), and most of the drug was released under simulated intracellular conditions.

Mixed micelles based on mPEG-SS-PTX and mPEG-SS-DOX were developed by Luan al. (78), in which both PTX and DOX were connected with methoxy poly (ethylene glycol) through a redox-sensitive disulfide linkage (Figure 4.7b and c). Both (mPEG-SS-PTX) and (mPEG-SS-DOX) form mixed micelles in aqueous medium with an average diameter of 93 nm. The mixed micelles show pH-triggered drug release in acidic conditions (pH 5.0) compared with neutral physiological conditions (pH 7.4). The cytotoxic study demonstrated that the mixed micelles exhibit better antitumor activity than the free drugs paclitaxel (PTX) and doxorubicin (DOX).

Hu et al. (79) synthesized a redox-responsive hydroxyethyl starch-doxorubicin (HES-SS-DOX) conjugate in which drug was attached through the redox-responsive disulfide bond (Figure 4.7d). The stability studies indicated that the HES-SS-DOX conjugate was quite stable in extracellular conditions, whereas it showed a sharp redox response in stimulated reducing conditions. The *in vivo* antitumor activity study

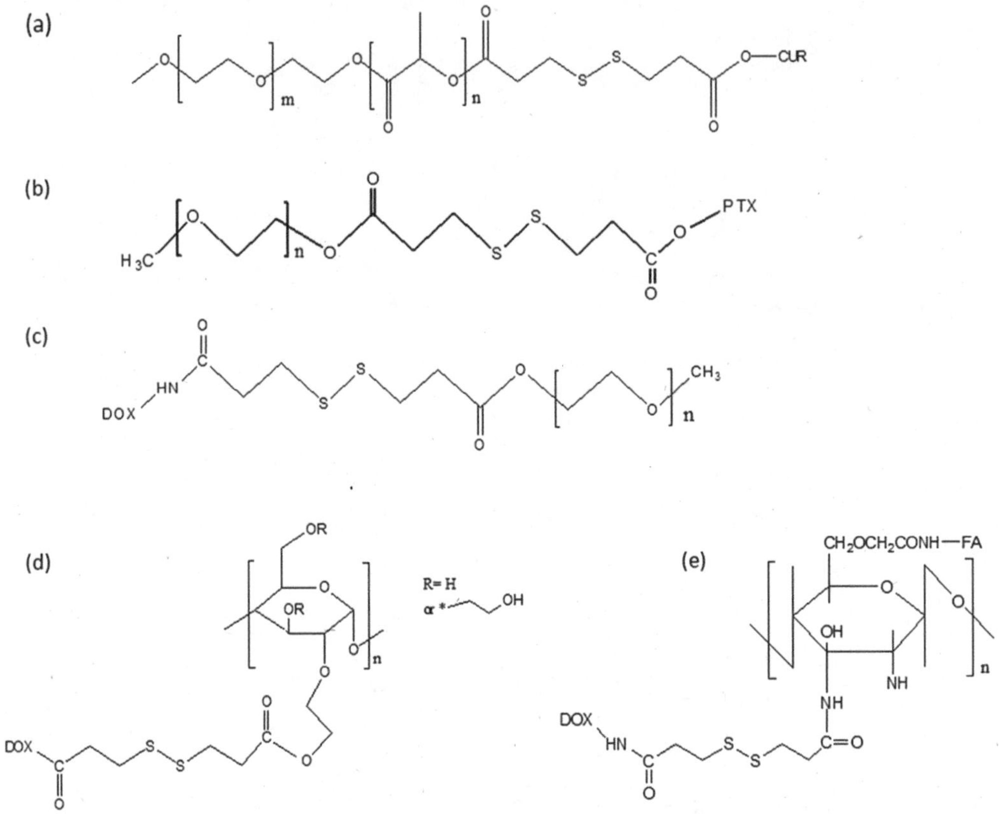

FIGURE 4.7 (a) Disulfide-linked (mPEG-PLA-CUR) conjugate. (b) Disulfide-linked (mPEG-SS-PTX). (c) Disulfide-linked (mPEG-SS-DOX) conjugate. (d) Disulfide–linked (HES-SS-DOX). (e) Disulfide-linked carboxymethyl chitosan–DOX conjugate.

demonstrated that HES-SS-DOX showed better antitumor efficacy and reduced toxicity as compared with free DOX.

Zhang X et al. designed cisplatin (CDDP)-cross-linked redox-sensitive micelles based on carboxymethyl chitosan for the combination delivery of cisplatin and DOX, in which DOX was conjugated to carboxymethyl chitosan via a disulfide bond (Figure 4.7e). The micelles were stable at physiological pH (7.4), whereas the *in vitro* release study demonstrated that that the micelles were highly glutathione sensitive, and the drug release depended on the concentration of GSH. The cytotoxicity study revealed that the cisplatin-cross-linked micelles loaded with DOX exhibited enhanced therapeutic efficacy compared with free drugs (DOX, CDDP). Cellular uptake and intracellular release revealed that the cisplatin-cross-linked micelles loaded with DOX could efficiently deliver and release DOX into the cancer cells (80).

A redox-responsive hyaluronic acid–paclitaxel (HA-ss-PTX) conjugate was developed by Yin et al. for dual tumor targeting as well as selective intracellular drug delivery, in which PTX was linked through a highly reductive disulfide bond (Figure 4.8a). The disulfide linkage between polymer and drug prevents premature drug release during blood circulation and allows drug release in the reductive environment of the cell. The cellular uptake of the HA-ss-PTX toward MCF-7 cells indicated that HA-ss-PTX exhibited rapid intercellular drug release (81).

Cheewatanakornkool K. et al. (82) synthesized thiolated pectin-DOX conjugates, in which DOX was attached with a reductive disulfide bond (Figure 4.8b). The drug release study confirmed the cleavage of disulfide bonds in a reductive environment. The thiolated pectin–DOX conjugates showed significantly higher cytotoxicity activity than free drug DOX against CT26 cell lines (82).

Yigang Su et al. developed redox-responsive polymer–drug conjugates, in which DOX was connected with stearic acid–grafted chitosan oligosaccharide through a disulfide bond (Figure 4.8c).

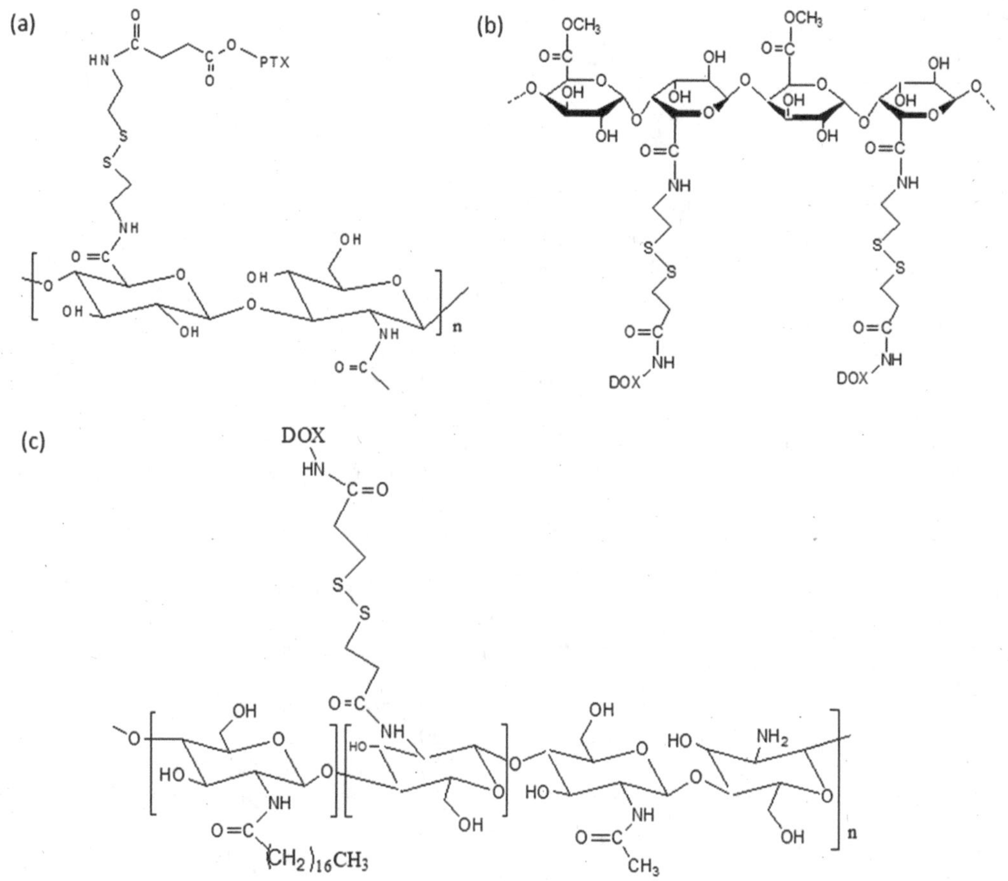

FIGURE 4.8 (a) Disulfide-linked (HA-ss-PTX) conjugate. (b) Disulfide-linked thiolated pectin–DOX conjugate. (c) Disulfide-linked stearic acid–grafted chitosan oligosaccharide (CSO-SA-DOX) conjugate.

The amphiphilic polymer–drug conjugates formed nanosized (62.8 nm) micelles in an aqueous system. The drug release study showed that DOX-SS-CSO-SA micelles were highly glutathione sensitive (83).

4.7 Conclusion and Perspectives

As discussed in this chapter, stimuli-responsive polymer–drug conjugates have been widely investigated. These studies demonstrated that stimuli-responsive polymer–drug conjugates have great potential to release the bioactive agent at the tumor site with reduced systemic side effects. Although much progress has been achieved over the past decades, a large number of stimuli-responsive polymer–drug conjugates still face severe challenges due to the slight difference in pH values among the various endosomes of diverse tumors. Therefore, to overcome the complexity of intracellular drug delivery and further enhance the therapeutic efficacy of anticancer drugs, multiresponsive polymer–drug conjugates with various biological and physiological features will have an effective impact on cancer therapy in the near future.

Acknowledgments

The authors gratefully acknowledge the Ministry of Human Resource Development (MHRD), New Delhi for financial support to conduct the study.

REFERENCES

1. Pang X, Jiang Y, Xiao Q, Leung AW, Hua H, Xu C. pH-responsive polymer-drug conjugates: Design and progress. *J Control Release.* 2016;222:116–29.
2. Cheetham AG, Chakroun RW, Ma W, Cui H. Self-assembling prodrugs. *Chem Soc Rev.* 2017;46(21):6638–63.
3. Cao Y, Gao M, Chen C, Fan A, Zhang J, Kong D, et al. Triggered-release polymeric conjugate micelles for on-demand intracellular drug delivery. *Nanotechnology.* 2015;26(11):115101.
4. Larson N, Ghandehari H. Polymeric conjugates for drug delivery. *Chem Mater.* 2012;24(5):840–53.
5. Li XQ, Wen HY, Dong HQ, Xue WM, Pauletti GM, Cai XJ, et al. Self-assembling nanomicelles of a novel camptothecin prodrug engineered with a redox-responsive release mechanism. *Chem Commun.* 2011;47(30):8647–9.
6. Zhu L, Wang T, Perche F, Taigind A, Torchilin VP. Enhanced anticancer activity of nanopreparation containing an MMP2-sensitive PEG-drug conjugate and cell-penetrating moiety. *Proc Natl Acad Sci USA.* 2013;110(42):17047–52.
7. Man L, Zhen L, Xun S, Tao G, Zhirong Z. A polymeric prodrug of 5-fluorouracil-1-acetic acid using a multi-hydroxyl polyethylene glycol derivative as the drug carrier. *PLoS One.* 2014;9(11):e112888.
8. Hu X, Hu J, Tian J, Ge Z, Zhang G, Luo K, et al. Polyprodrug amphiphiles: Hierarchical assemblies for shape-regulated cellular internalization, trafficking, and drug delivery. *J Am Chem Soc.* 2013;135(46):17617–29.
9. Zhou Z, Ma X, Jin E, Tang J, Sui M, Shen Y, et al. Linear-dendritic drug conjugates forming long-circulating nanorods for cancer-drug delivery. *Biomaterials.* 2013;34(22):5722–35.
10. Aryal S, Hu CJ, Zhang L. Polymer-cisplatin conjugate nanoparticles for acid-responsive drug delivery. *ACS Nano.* 2010;4(1):251–8.
11. Zhan F, Chen W, Wang Z, Lu W, Cheng R, Deng C, et al. Acid-activatable prodrug nanogels for efficient intracellular doxorubicin release. *Biomacromolecules.* 2011;12(10):3612–20.
12. Talelli M, Iman M, Varkouhi AK, Rijcken CJF, Schiffelers RM, Etrych T, et al. Core-crosslinked polymeric micelles with controlled release of covalently entrapped doxorubicin. *Biomaterials.* 2010;31(30):7797–804.
13. Xiao H, Li W, Qi R, Yan L, Wang R, Liu S, et al. Co-delivery of daunomycin and oxaliplatin by biodegradable polymers for safer and more efficacious combination therapy. *J Control Release.* 2012;163(3):304–14.
14. Gu Y, Zhong Y, Meng F, Cheng R, Deng C, Zhong Z. Acetal-linked paclitaxel prodrug micellar nanoparticles as a versatile and potent platform for cancer therapy. *Biomacromolecules.* 2013;14(8):2772–80.
15. Xiao H, Song H, Yang Q, Cai H, Qi R, Yan L, et al. A prodrug strategy to deliver cisplatin(IV) and paclitaxel in nanomicelles to improve efficacy and tolerance. *Biomaterials.* 2012;33(27):6507–19.
16. Xiao H, Noble GT, Stefanick JF, Qi R, Kiziltepe T, Jing X, et al. Photosensitive Pt(IV)-azide prodrug-loaded nanoparticles exhibit controlled drug release and enhanced efficacy in vivo. *J Control Release.* 2014;173(1):11–17.
17. Li H, Bian S, Huang Y, Liang J, Fan Y, Zhang X. High drug loading pH-sensitive pullulan-DOX conjugate nanoparticles for hepatic targeting. *J Biomed Mater Res A.* 2014;102(1):150–9.
18. Oommen OP, Garousi J, Sloff M, Varghese OP. Tailored doxorubicin-hyaluronan conjugate as a potent anticancer glyco-drug: An alternative to prodrug approach. *Macromol Biosci.* 2014;14(3):327–33.
19. Chen C, Zhou JL, Han X, Song F, Wang XL, Wang YZ. A prodrug strategy based on chitosan for efficient intracellular anticancer drug delivery. *Nanotechnology.* 2014;25(25):255101.
20. Wang Y, Xin D, Liu K, Zhu M, Xian J. Heparin-paclitaxel conjugates as drug delivery system: Synthesis, self-assembly property, drug release, and antitumor activity. *Bioconjugate Chem.* 2009;20:2214–21.
21. Sauraj, Kumar SU, Kumar V, Priyadarshi R, Gopinath P, Negi YS. pH responsive prodrug nanoparticles based on xylan-curcumin conjugate for the efficient delivery of curcumin in cancer therapy. *Carbohydr Polym.* 2018;188:252–9.
22. Wei X, Senanayake TH, Warren G, Vinogradov SV. Hyaluronic acid-based nanogel-drug conjugates with enhanced anticancer activity designed for the targeting of CD44-positive and drug-resistant tumors. *Bioconjug Chem.* 2013;24(4):658–68.
23. Ernsting MJ, Murakami M, Undzys E, Aman A, Press B, Li SD. A docetaxel-carboxymethylcellulose nanoparticle outperforms the approved taxane nanoformulation, Abraxane, in mouse tumor models with significant control of metastases. *J Control Release.* 2012;162(3):575–81.

24. Sauraj, Kumar SU, Gopinath P, Negi YS. Synthesis and bio-evaluation of xylan-5-fluorouracil-1-acetic acid conjugates as prodrugs for colon cancer treatment. *Carbohydr Polym.* 2017;157:1442–50.
25. Wang QW, Liu XY, Liu L, Feng J, Li YH, Guo ZJ, et al. Synthesis and evaluation of the 5-fluorouracil-pectin conjugate targeted at the colon. *Med Chem Res.* 2007;16(7–9):370–9.
26. Raveendran R, Bhuvaneshwar GS, Sharma CP. Hemocompatible curcumin-dextran micelles as pH sensitive pro-drugs for enhanced therapeutic efficacy in cancer cells. *Carbohydr Polym.* 2016;137:497–507.
27. Varshosaz J, Emami J, Tavakoli N, Fassihi A, Minaiyan M, Ahmadi F, et al. Synthesis and evaluation of dextran-budesonide conjugates as colon specific prodrugs for treatment of ulcerative colitis. *Int J Pharm.* 2009;365(1–2):69–76.
28. Udo K, Hokonohara K, Motoyama K, Arima H, Hirayama F, Uekama K. 5-Fluorouracil acetic acid/β-cyclodextrin conjugates: Drug release behavior in enzymatic and rat cecal media. *Int J Pharm.* 2010;388(1–2):95–100.
29. Plyduang T, Lomlim L, Yuenyongsawad S, Wiwattanapatapee R. Carboxymethylcellulose-tetrahydrocu rcumin conjugates for colon-specific delivery of a novel anti-cancer agent, 4-amino tetrahydrocurcumin. *Eur J Pharm Biopharm.* 2014;88(2):351–60.
30. Daus S, Heinze T. Xylan-based nanoparticles: Prodrugs for ibuprofen release. *Macromol Biosci.* 2010;10(2):211–20.
31. Yang Y, Pan D, Luo K, Li L, Gu Z. Biodegradable and amphiphilic block copolymer-doxorubicin conjugate as polymeric nanoscale drug delivery vehicle for breast cancer therapy. *Biomaterials.* 2013;34(33):8430–43.
32. Yang D, Liu X, Jiang X, Liu Y, Ying W, Wang H, et al. Effect of molecular weight of PGG-paclitaxel conjugates on in vitro and in vivo efficacy. *J Control Release.* 2012;161(1):124–31.
33. Harrisson S, Nicolas J, Maksimenko A, Bui DT, Mougin J, Couvreur P. Nanoparticles with in vivo anticancer activity from polymer prodrug amphiphiles prepared by living radical polymerization. *Angew Chemie Int Ed Engl.* 2013;52(6):1678–82.
34. Wang W, Li C, Zhang J, Dong A, Kong D. Tailor-made gemcitabine prodrug nanoparticles from well-defined drug-polymer amphiphiles prepared by controlled living radical polymerization for cancer chemotherapy. *J Mater Chem B.* 2014;2(13):1891–901.
35. Huo M, Yuan J, Tao L, Wei Y. Redox-responsive polymers for drug delivery: From molecular design to applications. *Polym Chem.* 2014;5(5):1519–28.
36. Wen H. Redox sensitive nanoparticles with disulfide bond linked sheddable shell for intracellular drug delivery. *Med Chem.* 2014;4(11):748–55.
37. Zhang X, Han L, Liu M, Wang K, Tao L, Wan Q, et al. Recent progress and advances in redox-responsive polymers as controlled delivery nanoplatforms. *Mater Chem Front.* 2017;1(5):807–22.
38. Chen W, Zhong P, Meng F, Cheng R, Deng C, Feijen J, et al. Redox and pH-responsive degradable micelles for dually activated intracellular anticancer drug release. *J Control Release.* 2013;169(3):171–9.
39. Cheng R, Meng F, Deng C, Klok HA, Zhong Z. Dual and multi-stimuli responsive polymeric nanoparticles for programmed site-specific drug delivery. *Biomaterials.* 2013;34(14):3647–57.
40. Khorsand B, Lapointe G, Brett C, Oh JK. Intracellular drug delivery nanocarriers of glutathione-responsive degradable block copolymers having pendant disulfide linkages. *Biomacromolecules.* 2013;14(6):2103–11.
41. Feng Q, Tong R. Anticancer nanoparticulate polymer-drug conjugate. *Bioeng Transl Med.* 2016;1(3):277–96.
42. Wang Y, Luo Q, Zhu W, Li X, Shen Z. Reduction/pH dual-responsive nano-prodrug micelles for controlled drug delivery. *Polym Chem.* 2016;7(15):2665–73.
43. Kanamala M, Wilson WR, Yang M, Palmer BD, Wu Z. Mechanisms and biomaterials in pH-responsive tumour targeted drug delivery: A review. *Biomaterials.* 2016;85:152–67.
44. Gao W, Chan J, Farokhzad OC. pH-responsive nanoparticles for drug delivery. *Mol Pharm.* 2010;7(6):1913–20.
45. Wei H, Zhuo RX, Zhang XZ. Design and development of polymeric micelles with cleavable links for intracellular drug delivery. *Prog Polym Sci.* 2013;38(3–4):503–35.
46. Chang M, Zhang F, Wei T, Zuo T, Guan Y, Lin G, et al. Smart linkers in polymer–drug conjugates for tumor-targeted delivery. *J Drug Target.* 2016;24(6):475–91.
47. Guo S, Lv L, Shen Y, Hu Z, He Q, Chen X. A nanoparticulate pre-chemosensitizer for efficacious chemotherapy of multidrug resistant breast cancer. *Sci Rep.* 2016;6(1):21459.

48. Meng F, Cheng R, Deng C, Zhong Z. Intracellular drug release nanosystems. *Mater Today*. 2012;15(10):436–42.
49. Kim HC, Kim E, Ha TL, Jeong SW, Lee SG, Lee SJ, et al. Thiol-responsive gemini poly(ethylene glycol)-poly(lactide) with a cystine disulfide spacer as an intracellular drug delivery nanocarrier. *Colloids Surfaces B Biointerfaces*. 2015;127:206–12.
50. Zhang Q, Re Ko N, Kwon Oh J. Recent advances in stimuli-responsive degradable block copolymer micelles: Synthesis and controlled drug delivery applications. *Chem Commun*. 2012;48(61):7542.
51. Jiang M, Zhang R, Wang Y, Jing W, Liu Y, Ma Y, et al. Reduction-sensitive paclitaxel prodrug self-assembled nanoparticles with tetrandrine effectively promote synergistic therapy against drug-sensitive and multidrug-resistant breast cancer. *Mol Pharm*. 2017;14(11):3628–35.
52. Lee MH, Sessler JL, Kim JS. Disulfide-based multifunctional conjugates for targeted theranostic drug delivery. *Acc Chem Res*. 2015;48(11):2935–46.
53. Wen H, Dong H, Liu J, Shen A, Li Y, Shi D. Redox-mediated dissociation of PEG–polypeptide-based micelles for on-demand release of anticancer drugs. *J Mater Chem B*. 2016;4(48):7859–69.
54. Kumar A, Lale SV., Mahajan S, Choudhary V, Koul V. ROP and ATRP fabricated dual targeted redox sensitive polymersomes based on pPEGMA-PCL-ss-PCL-pPEGMA triblock copolymers for breast cancer therapeutics. *ACS Appl Mater Interfaces*. 2015;7(17):9211–27.
55. Jia L, Li Z, Zhang D, Zhang Q, Shen J, Guo H, et al. Redox-responsive catiomer based on PEG-ss-chitosan oligosaccharide-ss-polyethylenimine copolymer for effective gene delivery. *Polym Chem*. 2013;4(1):156–65.
56. Sun H, Guo B, Li X, Cheng R, Meng F, Liu H, et al. Shell-sheddable micelles based on dextran-SS-poly(ε-caprolactone) diblock copolymer for efficient intracellular release of doxorubicin. *Biomacromolecules*. 2010;11(4):848–54.
57. Li F, Chen WL, You BG, Liu Y, Yang SD, Yuan ZQ, et al. Enhanced cellular internalization and on-demand intracellular release of doxorubicin by stepwise pH-/reduction-responsive nanoparticles. *ACS Appl Mater Interfaces*. 2016;8(47):32146–58.
58. Deng C, Jiang Y, Cheng R, Meng F, Zhong Z. Biodegradable polymeric micelles for targeted and controlled anticancer drug delivery: Promises, progress and prospects. *Nano Today*. 2012;7(5):467–80.
59. Tong R, Lu X, Xia H. A facile mechanophore functionalization of an amphiphilic block copolymer towards remote ultrasound and redox dual stimulus responsiveness. *Chem Commun*. 2014;50(27):3575–8.
60. Li X-Q, Wen H-Y, Dong H-Q, Xue W-M, Pauletti GM, Cai X-J, et al. Self-assembling nanomicelles of a novel camptothecin prodrug engineered with a redox-responsive release mechanism. *Chem Commun*. 2011;47(30):8647–9.
61. Luo C, Sun J, Liu D, Sun B, Miao L, Musetti S, et al. Self-assembled redox dual-responsive prodrug-nanosystem formed by single thioether-bridged paclitaxel-fatty acid conjugate for cancer chemotherapy. *Nano Lett*. 2016;16(9):5401–8.
62. Yoo HS, Lee EA, Park TG. Doxorubicin-conjugated biodegradable polymeric micelles having acid-cleavable linkages. *J Control Release*. 2002;82(1):17–27.
63. Bae Y, Nishiyama N, Fukushima S, Koyama H, Yasuhiro M, Kataoka K. Preparation and biological characterization of polymeric micelle drug carriers with intracellular pH-triggered drug release property: Tumor permeability, controlled subcellular drug distribution, and enhanced in vivo antitumor efficacy. *Bioconjug Chem*. 2005;16(1):122–30.
64. Bae Y, Jang W-D, Nishiyama N, Fukushima S, Kataoka K. Multifunctional polymeric micelles with folate-mediated cancer cell targeting and pH-triggered drug releasing properties for active intracellular drug delivery. *Mol Biosyst*. 2005;1(3):242–50.
65. Hu X, Liu S, Huang Y, Chen X, Jing X. Biodegradable block copolymer-doxorubicin conjugates via different linkages: Preparation, characterization, and in vitro evaluation. *Biomacromolecules*. 2010;11(8):2094–102.
66. Hrubý M, Koňák Č, Ulbrich K. Polymeric micellar pH-sensitive drug delivery system for doxorubicin. *J Control Release*. 2005;103(1):137–48.
67. Yu Y, Chen CK, Law WC, Weinheimer E, Sengupta S, Prasad PN, et al. Polylactide-graft-doxorubicin nanoparticles with precisely controlled drug loading for pH-triggered drug delivery. *Biomacromolecules*. 2014;15(2):524–32.

68. Gao C, Tang F, Gong G, Zhang J, Hoi MPM, Lee SM-Y, et al. pH-responsive prodrug nanoparticles based on a sodium alginate derivative for selective co-release of doxorubicin and curcumin in tumor cells. *Nanoscale.* 2017;12533–42.
69. Zhang Y, Yang C, Wang W, Liu J, Liu Q, Huang F, et al. Co-delivery of doxorubicin and curcumin by pH-sensitive prodrug nanoparticle for combination therapy of cancer. *Sci Rep.* 2016;6(1):21225.
70. Ke XC, Coady DJ, Yang C, Engler AC, Hedrick JL, Yang YY. pH-sensitive polycarbonate micelles for enhanced intracellular release of anticancer drugs: A strategy to circumvent multidrug resistance. *Polym Chem.* 2014;5(7):2621–8.
71. Li M, Gao M, Fu Y, Chen C, Meng X, Fan A, et al. Acetal-linked polymeric prodrug micelles for enhanced curcumin delivery. *Colloids Surfaces B Biointerfaces.* 2016;140:11–18.
72. Wang H, He J, Cao D, Zhang M, Li F, Tam KC, et al. Synthesis of an acid-labile polymeric prodrug DOX-acetal-PEG-acetal-DOX with high drug loading content for pH-triggered intracellular drug release. *Polym Chem.* 2015;6(26):4809–18.
73. Kakinoki A, Kaneo Y, Ikeda Y, Tanaka T, Fujita K. Synthesis of poly(vinyl alcohol)-doxorubicin conjugates containing cis-aconityl acid-cleavable bond and its isomer dependent doxorubicin release. *Biol Pharm Bull.* 2008;31(1):103–10.
74. Son YJ, Jang JS, Cho YW, Chung H, Park RW, Kwon IC, et al. Biodistribution and anti-tumor efficacy of doxorubicin loaded glycol-chitosan nanoaggregates by EPR effect. *J Control Release.* 2003;91(1–2):135–45.
75. Tabatabaei Rezaei SJ, Sarbaz L, Niknejad H. Folate-decorated redox/pH dual-responsive degradable prodrug micelles for tumor triggered targeted drug delivery. *RSC Adv.* 2016;6(67):62630–9.
76. Zou J, Zhang F, Zhang S, Pollack SF, Elsabahy M, Fan J, et al. Poly(ethylene oxide)-block-polyphosphoe ster-graft-paclitaxel conjugates with acid-labile linkages as a pH-sensitive and functional nanoscopic platform for paclitaxel delivery. *Adv Healthc Mater.* 2014;3(3):441–8.
77. Bai L, Wang X-H, Song F, Wang X-L, Wang Y-Z. "AND" logic gate regulated pH and reduction dual-responsive prodrug nanoparticles for efficient intracellular anticancer drug delivery. *Chem Commun.* 2014;51(3):3–6.
78. Zhao D, Wu J, Li C, Zhang H, Li Z, Luan Y. Precise ratiometric loading of PTX and DOX based on redox-sensitive mixed micelles for cancer therapy. *Colloids Surfaces B Biointerfaces.* 2017;155:51–60.
79. Hu H, Li Y, Zhou Q, Ao Y, Yu C, Wan Y, et al. Redox-sensitive hydroxyethyl starch-doxorubicin conjugate for tumor targeted drug delivery. *ACS Appl Mater Interfaces.* 2016;8(45):30833–44.
80. Zhang X, Li L, Li C, Zheng H, Song H, Xiong F, et al. Cisplatin-crosslinked glutathione-sensitive micelles loaded with doxorubicin for combination and targeted therapy of tumors. *Carbohydr Polym.* 2017;155:407–15.
81. Yin T, Wang J, Yin L, Shen L, Zhou J, Huo M. Redox-sensitive hyaluronic acid–paclitaxel conjugate micelles with high physical drug loading for efficient tumor therapy. *Polym Chem.* 2015;6(46):8047–59.
82. Cheewatanakornkool K, Niratisai S, Manchun S, Dass CR, Sriamornsak P. Thiolated pectin–doxorubicin conjugates: Synthesis, characterization and anticancer activity studies. *Carbohydr Polym.* 2017;174:493–506.
83. Su Y, Hu Y, Du Y, Huang X, He J, You J, et al. Redox-responsive polymer-drug conjugates based on doxorubicin and chitosan oligosaccharide- g -stearic acid for cancer therapy. *Mol Pharm.* 2015;12(4):1193–202.

5

Cross-Linked Polymers for Drug Delivery Systems

B. A. Aderibigbe

CONTENTS

5.1 Introduction

Cross-linked polymers exhibit three-dimensional networks. They are prepared from synthetic polymers, carbon-based materials such as fullerenes, graphene oxide, carbon nanotubes, and natural polymers. An *in situ* gel is a good example of a cross-linked polymer. They are easy to prepare, easy to administer, and undergo a sol-gel transition when administered. *In situ* gels offer several advantages that make them useful as drug delivery systems, such as reduced side effects, improved patient compliance, extended residence time, enhanced bioavailability, etc. [1]. They are administered via the nasal, ocular, oral, vaginal, and intraperitoneal routes. Their unique properties make them potential systems for drug delivery. This chapter will be focused on the design and the therapeutic efficacies of the recently reported *in situ* gels.

5.2 *In Situ* Gels in Drug Delivery

In situ gels loaded with bioactive agents are administered via the nasal, ocular, oral, vaginal, and intraperitoneal routes. They are used to deliver therapeutics for the treatment of infectious and non-infectious diseases.

5.2.1 *In Situ* Gels for Ocular Administration

The delivery of therapeutics to the eye with topical eye drops is characterized by low bioavailability. Several constraints, which could be classified as anatomical and physiological, such as tear turnover, nasolacrimal drainage, reflex blinking, and ocular static and dynamic barriers hinder deeper ocular drug permeation [2, 3]. Less than 5% of the topically applied dose reaches the deep ocular tissues (Figure 5.1). Therapeutics are delivered to the posterior segment ocular tissues by intravitreal injections, periocular injections, and systemic administration. However, intravitreal injection is the most common route of drug administration to treat posterior ocular diseases. This involves repeated eye puncture, resulting in

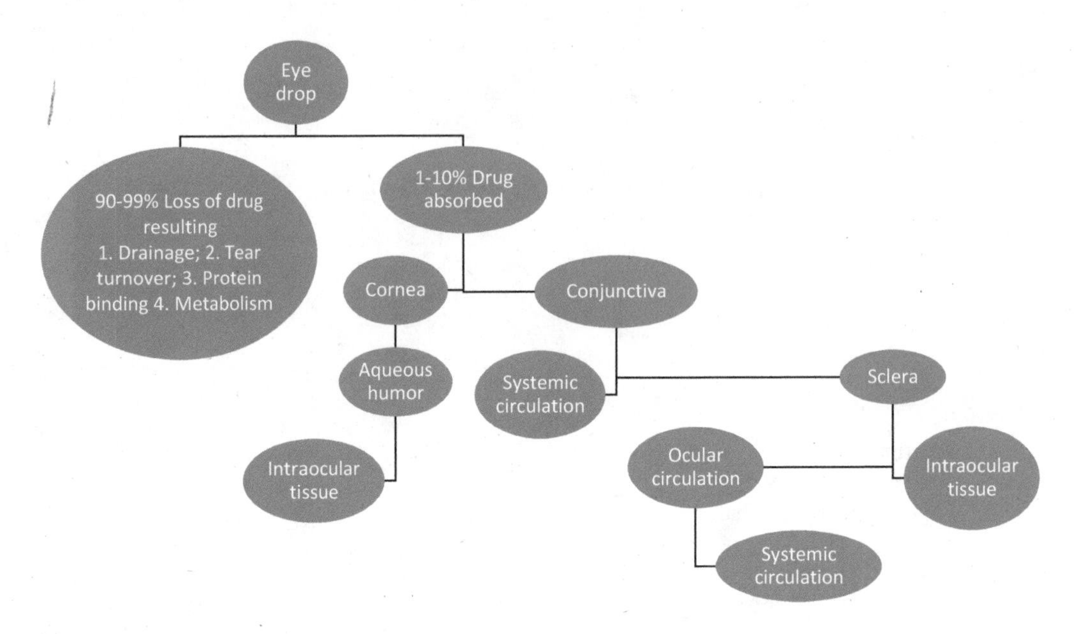

FIGURE 5.1 A schematic representation of ocular drug uptake.

severe side effects such as endophthalmitis, hemorrhage, retinal detachment, and poor patient tolerance [2, 3]. To overcome the barriers to ocular drug delivery and enhance ocular drug bioavailability, several systems have been designed, including *in situ* gels. Several bioactive agents have been loaded onto *in situ* gels for ocular administration (Figure 5.2).

Ahmed et al. developed ketoconazole poly(lactide-co-glycolide) (PLGA) nanoparticles (NPs) by a nanoprecipitation technique, followed by loading onto *in situ* gel formulations for ocular drug delivery [4]. Ketoconazole is used for the treatment of fungal eye infections. However, it is limited by short duration of action due to its elimination half-life of 19 and 43 min in the aqueous humor and cornea, respectively [4, 5]. Its high molecular weight also hinders its transport. The drug entrapment efficiency and particle size of the formulation and the NPs, respectively, increased with an increase in the concentration of PLGA. The particle size and the zeta potential value were 331 ± 24 nm and 4.11 ± 0.34 mV, respectively. The drug permeation in human epithelial cells was high and sustained from the formulation, resulting from the small particle size and the drug entrapment inside the polymeric matrix. The *in situ* gel formulations also exhibited enhanced antifungal activity when compared with the free drug. An alginate-chitosan *in situ* gel formulation loaded with optimized ketoconazole NPs exhibited higher drug permeation through epithelial cell lines when compared with other formulations. The *in situ* gel formulation was reported to be a potential ophthalmic formulation for the treatment of fungal eye infections [4].

Makwana et al. developed an *in situ* gel eye drop formulation from sodium alginate and hydroxypropylmethyl cellulose (HPMC) as a viscosity enhancer for sustained drug delivery and loaded it with ciprofloxacin hydrochloride [6]. The pH of the *in situ* gel solution was in the range of 6.49–6.58, and the gelling capacity of the gels was immediate over an extended period. The optimum viscosity of the gels resulted in easy instillation into the eye, in which a rapid sol to gel transition occurred. The gels revealed their potential to enhance ocular drug bioavailability and are patient compliant [6]. Bhowmik et al. developed *in situ* gels for the delivery of ketorolac tromethamine. The gel was prepared from methylcellulose and HPMC. HPMC increased the viscosity of the gels and controlled the drug release over a period of 4 h. The hydration rate of the *in situ* gels was lower when compared with the normal tear flow rate, suggesting that the gels are compatible with the eye and do not have the potential to cause eye irritation induced by excessive dehydration of the aqueous ocular cavity. The *in situ* gels were stable over a period of 90 days and retained clarity, with no significant changes observed in gelation temperature, viscosity, *in vitro* drug release profile, and salt leaching properties [7]. Puranik et al. prepared a voriconazole ophthalmic *in situ* gel for the ophthalmic delivery of an antifungal agent. The gel was prepared from sodium alginate

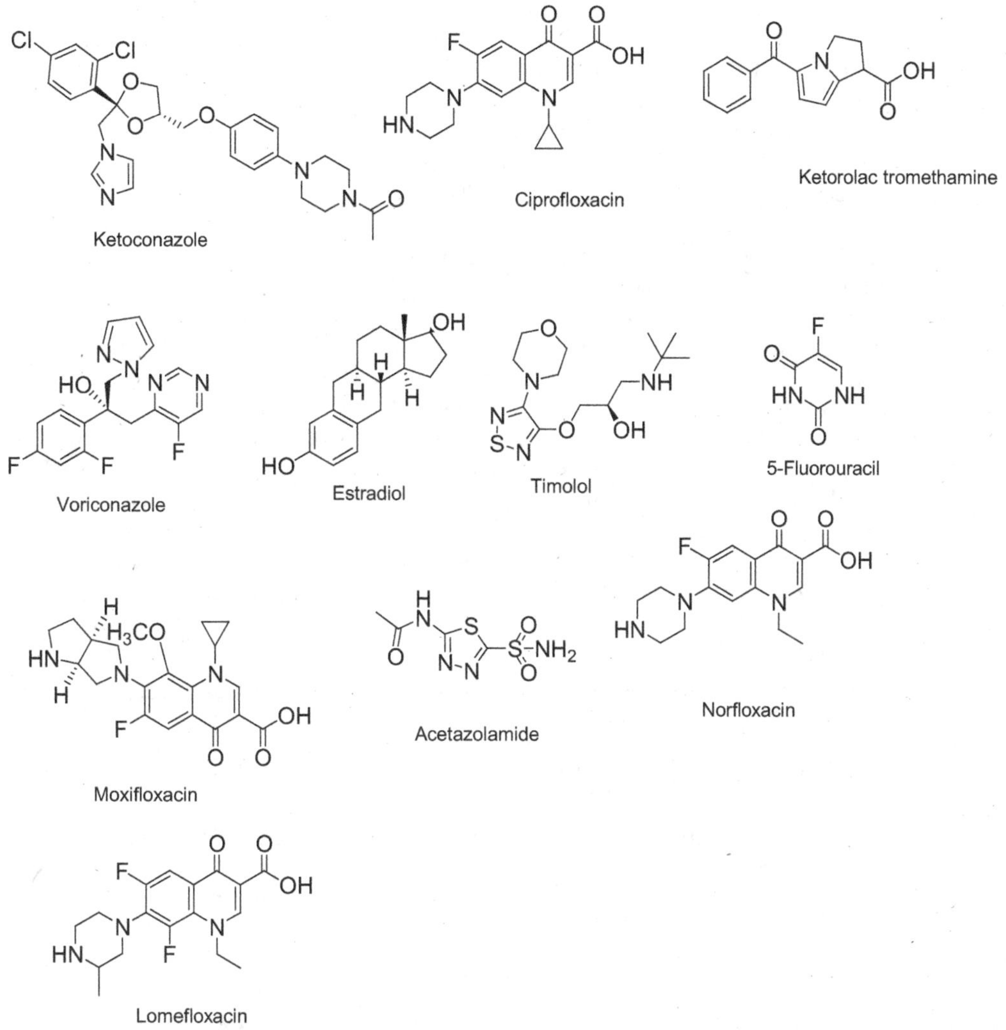

FIGURE 5.2 Examples of selected drugs loaded into *in situ* gels for ocular administration.

and HPMC. The developed formulations exhibited sustained drug release over a period of 8 h. The pH of the formulations was in the range of 6.7 to 7.2 with a good gelling capability that is sustained over an extended period of time. The *ex vivo* permeation effect of the gel using goat cornea revealed a 52% drug release in the goat cornea in 5 h [8].

Kotreka et al. prepared an ion-activated *in situ* gel loaded with estradiol eye drops for the prevention of age-related cataracts. The *in situ* gels were prepared from gellan gum, polysorbate-80, mannitol, potassium sorbate, and edetate disodium dihydrate (EDTA). Drug release from the gel followed a non-Fickian mechanism over a period of 8 h with 80% drug release. The drug release was by diffusion due to polymer erosion. The formulation was stable over a long period of 6 months. The apparent viscosity of the formulation was in the range of 12.5–23.2 cps. The formulations exhibited phase transition when in contact with the simulated tear fluid cations. The absorption of the drug into the eyes in the aqueous humor was 250-fold higher than in the systemic circulation, resulting from the viscoelastic nature of the gel, which promoted extended precorneal drug residence time and consequential reduction in drug drainage through the nasolacrimal duct. There was an absence of significant irritation or toxicity in the ocular tissues *in vivo* [9]. Mohan et al. prepared *in situ* gels for the delivery of ciprofloxacin from poly acrylic acid (Carbopol 940), HPMC, Pluronic F-127, and gellan gum. The formulations were liquid at room temperature and underwent rapid gelation at a pH of 7.4 and at a temperature of 37°C. The antimicrobial activity of the loaded ciprofloxacin in the formulations prepared was retained, and the formulations were stable over an extended period of 7 weeks [10].

Patil et al. formulated norfloxacin *in situ* gel for the treatment of conjunctivitis from Carbopol-940 and HPMC. The drug encapsulation in the formulations was found to be in the range of 98.30–99.97%. The pH was in the range of 5.4–7.2 with a drug content of 98.0–99.7%. The nature of the gel formed was dependent on the concentration of the polymers used. The *in vitro* release studies in simulated tear fluid showed prolonged drug release, suggesting slow diffusion of the drug via the polymer matrix [11]. Cao et al. reported poly(*N*-isopropylacrylamide)-chitosan thermosensitive *in situ* gel loaded with timolol maleate for ocular drug delivery. *In vivo* studies on rabbits showed that the C_{max} of timolol maleate in aqueous fluid for the formulation was two-fold higher than for conventional eye drops. The formulation also had the capacity to reduce the intraocular pressure when compared with the conventional eye drops over a period of 12 h. The MTT assay showed an insignificant cytotoxic effect of the formulation at a concentration range of 0.5–400 μg/ml, revealing its biocompatibility [12]. Gupta et al. reported a chitosan and gellan gum–based novel *in situ* gel system loaded with timolol maleate, a drug used for the treatment of glaucoma. Ocular retention of the formulation was enhanced with no signs of irritation *in vivo* [13].

Gratieri et al. reported an *in situ* gel composed of poly (ethylene oxide)–poly (propylene oxide)–poly (ethylene oxide) with Poloxamer and chitosan. The mechanical strength and texture properties of the formulations were enhanced by the addition of chitosan. A 10 min instillation of the formulation with a composition of Poloxamer/chitosan 16:1 in human eyes resulted in 50–60% of the gel remaining in contact with the corneal surface, which was a fourfold increase in retention when compared with the conventional solution [14]. Nagarwal et al. prepared *in situ* gel loaded with 5-fluorouracil for the treatment of conjunctival/corneal squamous cell carcinoma. Drug-loaded polylactic acid (PLA) nanoparticles were dispersed in sodium alginate solution to form *in situ* systems. The nanoparticles were spherically shaped with a size range of 128–194 nm. The drug release was diffusion controlled with a high burst effect. *In vivo* studies showed a high drug level in the aqueous humor resulting from the extended retention time of the formulation [15].

Gao et al. investigated a poly-(DL-lactic acid-co-glycolic acid)-polyethylene glycol formulation for the ocular delivery of dexamethasone acetate. *In vivo* studies showed that the C_{max} of the drug in the anterior chamber for the PLGA–PEG–PLGA solution was sevenfold higher when compared with eye drops. The results revealed that the *in situ* gel improved the bioavailability of the eye drug [16]. Liu et al. reported an *in situ* gel prepared from a combination of Gelrite and alginate. The concentrations of Gelrite and alginate solutions suitable for the ocular drug delivery system were 0.3% and 1.4% (w/w), respectively. The concentration of matrine in the precorneal area was high when compared with matrine-containing simulated tear fluid. The Gelrite/alginate formulation retained the loaded drug when compared with either Gelrite or alginate alone. The formulation was non-irritant [17]. Mandal et al. prepared *in situ* gels from alginate and HPMC for the delivery of moxifloxacin hydrochloride. The antibacterial studies revealed inhibition values against *S. aureus* (28.66 mm) and *Escherichia coli* (30.99 mm). *In vivo* studies in rabbits indicated the absence of eye redness, swelling, and watering of eyes. No ocular damage or abnormality in the cornea, iris, or conjunctiva was observed. The formulations were stable over a period of 1 month with no change in the appearance of the gels. The drug release was in the range of 78–94% over a period of 10 h [18].

Yu et al. developed *in situ* polyethylene glycol (PEG) hydrogels for sustained release of avastin for the treatment of corneal neovascularization. *In vitro* cytotoxicity studies on L-929 cells over 7 days of incubation showed the absence of cytotoxic effect. The drug release was within a period of 14 days [19]. Morsi et al. loaded acetazolamide, an antiglaucoma drug, onto an ion-induced nanoemulsion-based *in situ* gel for ocular delivery. The nanoemulsion formulations were prepared from peanut oil, Tween, cremophor EL, gellan gum, xanthan gum, Carbopol, HPMC, and propylene glycol. The drug release profile from the gel was sustained, and the gel was stable at all studied temperatures. However, the formulation of gellan/Carbopol exhibited partial drug precipitation during storage. Formulations of gellan gum/xanthan gum and gellan gum/HPMC exhibited high therapeutic efficacy with extended intraocular pressure–lowering effect when compared with commercial eye drops and oral tablets. The formulation is a promising therapeutic for the treatment of glaucoma with reduced systemic side effects of acetazolamide due to the local inhibition of carbonic anhydrase. The formulation also exerted an extended therapeutic effect, which can improve patient compliance resulting from reduced frequency of administration [20].

Upadhayay et al. extended the corneal contact time of norfloxacin in the treatment of extra-ocular diseases by designing a pH-triggered *in situ* gel system. The system was designed by an ionotropic gelation method using chitosan cross-linked by sodium tripolyphosphate. The pH of the *in situ* gels was in the range of 5.84–6.25, revealing that it can be tolerated by the human eye. Increasing the concentration of Carbopol increased the gelling capability of the gels, resulting from increased ionization of the functional groups present in Carbopol 934P, leading to increased electrostatic repulsion between the adjacent –COOH groups and the subsequent expansion of the polymer network. The formulation also displayed prolonged antimicrobial efficacy when compared with the marketed formulation. Ocular irritation was not observed *in vitro* [21]. Gadad et al. developed temperature-sensitive *in situ* gel formulations, which could undergo a transition from liquid to semisolid gel when exposed to the physiological temperature of the eye. Lomefloxacin hydrochloride, a fluoroquinolone antibiotic used to treat various conjunctival infections, was loaded onto the gel prepared from Pluronic F127, Pluronic F68, and sodium alginate. The pH was in the range of 7–7.5, and the viscosity was 1590–3370 cps at a physiological temperature of 37°C. The stability of the formulation was short-term, and 4°C was the suitable storage temperature for the formulations. The permeation of the formulation through goat cornea was 79%. The antibacterial activity of the loaded drug was retained. No sign of damage to the cornea was observed *in vivo*. The formulation reduced the infection of bacterial conjunctivitis within 5 days [22].

5.2.2 *In Situ* Gels for Vaginal Administration

Delivery of therapeutics to the vaginal area is challenging due to vast physiological variations, such as differences in pH, microflora, changes during menstruation, and cervical mucus [23, 24]. The vaginal epithelium has a high activity of enzymes, which affect the stability of intravaginal delivery systems and devices [24]. Absorption of drugs from vaginal delivery systems occurs by drug dissolution in the vaginal lumen followed by membrane penetration (Figure 5.3) [24]. The human vagina is an S-shaped fibromuscular collapsible tube that extends from the cervix of the uterus [24]. The wall of the vagina is composed of three layers: the epithelial layer, the muscular coat, and the tunica adventitia [24]. The thickness of the vaginal epithelial cell layer provides support and also increases the surface area of the vaginal wall. The loose connective tissue of the tunica adventitia further increases the elasticity of the vagina. There are abundant arteries, blood vessels, and lymphatic vessels in the walls of the vagina. Drugs absorbed from the vagina do not undergo first-pass metabolism, because blood leaving the vagina enters the peripheral circulation, which empties into the internal iliac veins. The vagina secretes a large amount of fluid, which includes cervical secretion, transudation from the blood vessels, and secretions from the endometrium and the fallopian tubes. The amount of the vaginal fluid also changes throughout the menstrual cycle [24]. However, the delivery of drugs via the vagina offers several advantages, such as eliminating side effects of nausea and vomiting, preventing high-clearance hepatic first-pass elimination, and avoiding contact with the digestive fluid, thereby preventing the enzymatic degradation of some drugs [24]. Different bioactive agents are incorporated into *in situ* gels for vaginal administration (Figure 5.4).

Rençber et al. prepared a mucoadhesive *in situ* gel formulation loaded with clotrimazole for the treatment of vaginal candidiasis [25]. It was prepared from a mixture of Poloxamer 407 and 188 and HPMC. A formulation containing 20% PLX 407, 10% PLX 188, and 0.5% HPMC was found to be suitable for vaginal administration. HPMC is a mucoadhesive polymer, which was added to enhance the

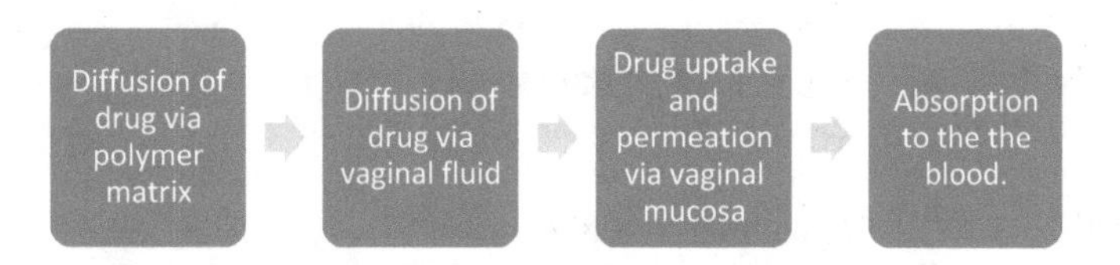

FIGURE 5.3 A schematic representation of drug uptake via vaginal administration.

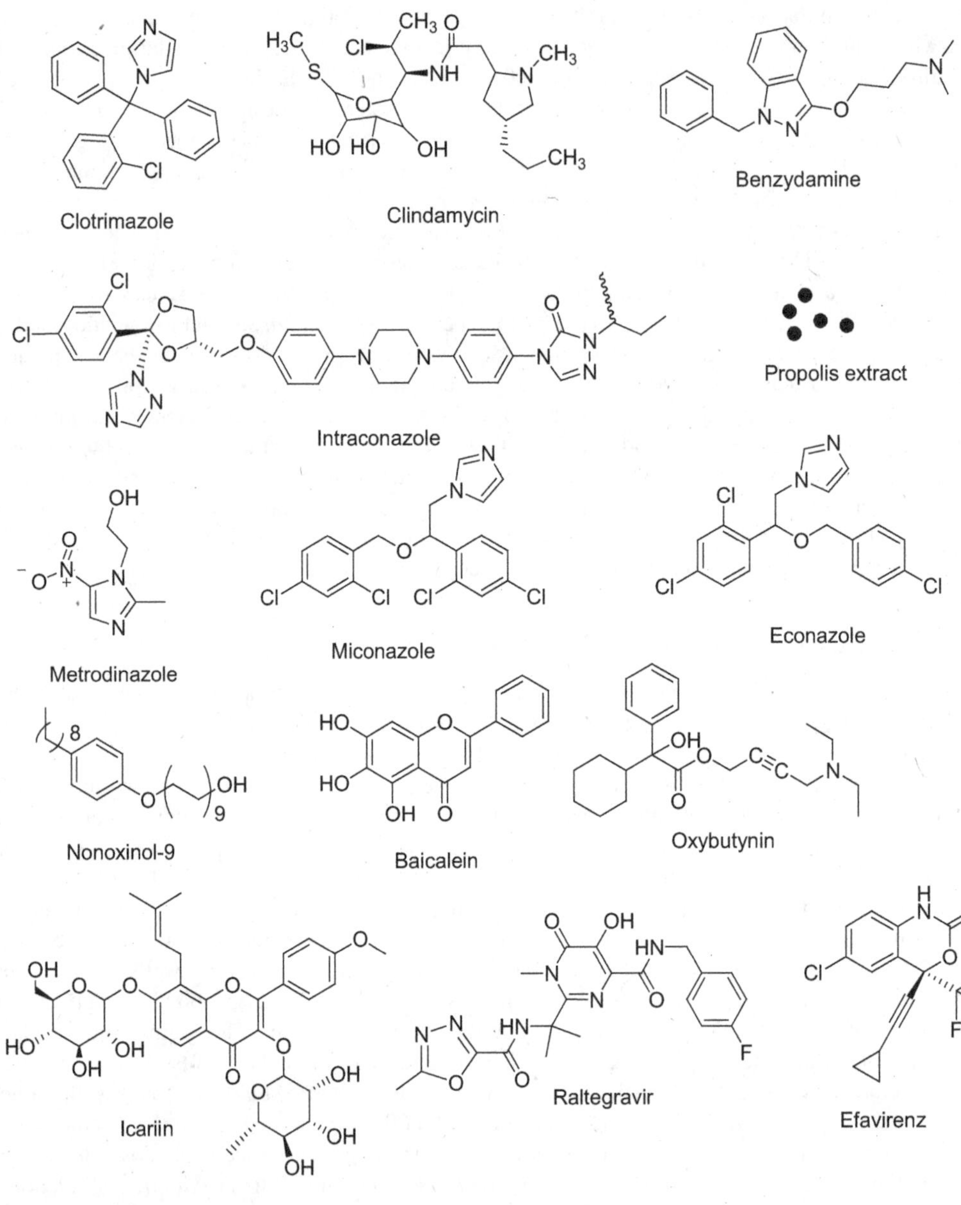

FIGURE 5.4 Selected drugs loaded into *in situ* gels for vaginal administration.

mucoadhesive properties of formulations. The pH of the *in situ* gels was in the range of 6.5–7.3, which is comparable to the commercial formulation of clotrimazole with a pH of 6.81. The formulation was characterized by high adhesiveness and low hardness with good retention at the site of application. The release of clotrimazole was rapid over a period of 6 h. The drug release decreased with an increase in the viscosity of the formulation. The formulations were effective against *Candida albicans* ATCC 90028 strain. *In vivo* distribution studies showed that the formulations were retained in the vaginal mucosa 24 h after administration [25]. Taksande et al. reported *in situ* gels prepared from Poloxamer 127 and 188, chitosan and Carbopol 974 loaded with clindamycin [26]. The loading of the drug onto the *in situ* gels enhanced the gelling capability of the gels. HPMC in the formulation enhanced the mucoadhesive property of the formulation. The formulation was retained in the vagina. The formulations were stable over a period of 3 months. *In vitro* drug release of the formulations through egg membrane revealed that the drug release was via diffusion [26].

Karavana et al. prepared a mucoadhesive *in situ* gel from Poloxamer and HPMC and loaded it with itraconazole [27]. The pH was in a range of 6.52–7.25, which is suitable for vaginal administration. The gels exhibited pseudo-plastic flow in continuous shear rheometry at 37°C, revealing their thermoresponsive property. The presence of HPMC in the formulation influenced the sol-gel transition temperature, mucoadhesive property, and mechanical and rheological properties of the formulations [27]. Hu et al. prepared pH-sensitive *in situ* gels from Carbopol and polyvinyl alcohol loaded with nystatin [28]. The formulation containing 0.66% Carbopol and 1.21% polyvinyl alcohol was suitable for vaginal administration and gelation ability *in vivo*. The release mechanism of the drug from the formulation was via erosion and diffusion [28]. Patel et al. loaded clindamycin on *in situ* gels for vaginal administration prepared from a mixture of HPMC (0.1%) and gellan gum with NaCl (0.9%) as an isotonic agent [29]. The drug content of the formulation was in the range of 98.1–101%. The cumulative drug release from the formulation was 33% over a period of 2 h followed by 67% after 6 h and 99% after 12 h. The formulation was stable over a period of 5 months with no significant changes in its physicochemical properties. An irritation test using the *hen's egg chorioallantoic membrane* (*HET-CAM*) test further showed that the formulation does not induce irritations, confirming that it is a potential therapeutic for the treatment of vaginal candidiasis [29]. Sen et al. developed *in situ* gel formulations from chitosan, gellan gum, and Poloxamer 127 and 188 for the delivery of clotrimazole [30]. The formulation released 90% of the drug in 12 h with a zero-order release profile *in vitro*. The formulations retained their gel consistency over a period of 16 h. The entrapment of drug in the gels and their ability to retain their gel consistency over a long period of time suggest that the formulations have the potential to increase drug bioavailability at the target site and reduce the side effects of the drug [30].

Tuğcu-Demiröz reported *in situ* vaginal gels loaded with benzydamine hydrochloride and prepared from Poloxamer and chitosan [31]. The pH was in the range of 3.9–5.1. The release profile of the drug from the formulations was sustained and controlled. The release of the drug was rapid over a period of 6 h, suggesting that the formulation is suitable for once-a-day administration via the vaginal route. The compressibility of the gels was influenced by the type of chitosan used. The formulation was found to be suitable for the treatment of vaginitis [31]. Glavas-Dodov et al. prepared *in situ* gels from propyleneglycolic extract of propolis, chitosan, and Poloxamer. The pH was between 4.0 and 4.5. An increase in the concentration of Poloxamer resulted in a high viscosity of the gels, thereby producing a dense gel network, which erodes at a slow rate. The prolonged release of propolis resulted from the presence of chitosan. The formulations were stable for 6 months at 5°C and for 5 months at 25°C [32]. Akhandi and Modi evaluated nystatin-loaded thermosensitive mucoadhesive *in situ* gel for vaginal drug delivery. It was prepared from HPMC and Pluronic F 68 and F127 by a cold method. Pluronic F 68 and 127 had a significant positive effect on the gelation temperature of the formulation, and HPMC influenced the mucoadhesion force and strength of the gel, which was useful in the sustained release of the drug. The formulation also exhibited good retention at the site of application. The *in situ* gels were found to be local therapy for the treatment of vaginal candidiasis [33]. Deshkar et al. prepared *in situ* gels from Poloxamer 407, HPMC, Carbopol 974, and polycarbophil by the cold dispersion method. The formulations were characterized by a quick formation of a stable gel with excellent mucoadhesion. The formulation released 74% of the drug over a period of 8 h at pH 5.4. The *in vitro* antifungal activity of the formulation against *C. albicans* was enhanced when compared with the marketed formulation. *In vivo* studies in rabbits showed the absence of vaginal irritation 10 days after the administration of the formulation [34]. Shabaan et al. compared the administration of once-daily *in situ* gel loaded with metronidazole (MTZ) with twice-daily conventional vaginal gel in the treatment of bacterial vaginosis. The cure rate of patients treated with the *in situ* gel was 75% compared with the 64% cure rate in those treated with the conventional vaginal gel. The once-daily *in situ* MTZ gel was more effective than twice-daily conventional gel over a period of 4 weeks of treatment with similar side effects [35]. Ibrahim et al. prepared *in situ* gel from the combination of Pluronic F-127 and Pluronic F-68 [36]. It was loaded with MTZ for the treatment of bacterial vaginosis. The gelation temperature of the formulations decreased with an increase in the concentration of Pluronic-127. Increasing the concentration of Pluronic decreased the rate of drug release *in vitro*, while the viscosity and mucoadhesive force were increased. *In vivo* studies on vaginas of rabbit did not reveal any abnormality in the vagina and cervix. There was no sign of inflammation. The drug release from the formulation was slow when compared with the marketed gel (Tricho® gel), suggesting that the *in situ* gel acted as a rate-controlling matrix useful for sustained drug release. It also suggests that the *in situ* gel would be retained longer in the vaginal mucosal tissue [36].

Şenyiğit et al. evaluated chitosan as a vaginal gel base for the preparation of mucoadhesive *in situ* gels loaded with the antifungal drugs miconazole nitrate and econazole nitrate [37]. The pH of the formulations was in the range of 3.79–4.82, revealing that the formulations are suitable for vaginal application and can maintain the acidic pH value of the vagina. Using different types of chitosan did not affect the pH of the gels significantly. The molecular mass and viscosity of the chitosan used influenced the mechanical properties of the formulations. Formulations prepared from high–molecular mass chitosan were characterized by a high adhesive property. However, hardness and compressibility were high and disadvantageous for their spreadability on the vagina and patient compliance. Formulations prepared from low–molecular mass chitosan were characterized by low hardness, low compressibility, and low adhesive property, suggesting that they may leak on administration and are thereby not patient compliant. The formulations prepared from medium–molecular mass chitosan exhibited good mechanical properties suitable for vaginal administration. The formulations exhibited good anticandidal activity with inhibition diameters between 26 and 34 mm. The formulations were retained in the vaginal mucosa 24 h after administration, suggesting that the rheology of vaginal gel affects its retention time [37]. Liu et al. investigated *in vivo* retention capabilities of *in situ* Poloxamer-based gel for vaginal delivery of nonoxynol-9 [38]. The *in situ* gels were composed of either 18% Poloxamer 407 and 1% Poloxamer 188 or 18% Poloxamer 407 and 6% Poloxamer 188. The formulations exhibited the same phase transition temperatures, which ranged between 27 and 32 °C. Increasing the content of Poloxamer 188 resulted in high rheological moduli in body temperature with rapid hydrogel erosion and drug release. Elimination of the gel containing 1% Poloxamer 188 was slow in rat vagina when compared with the gel containing 6% Poloxamer 188. The formulations exhibited good anti-dilution capacity, which further indicated that after administration of the formulation, the dilution by vaginal fluid did not affect its phase transition [38]. Zhou et al. prepared an *in situ* thermally sensitive gel for vaginal administration. It was prepared from the complex of baicalein and hydroxypropyl-γ-cyclodextrin for enhanced stability and solubility of baicalein [39]. It was prepared from a mixture of Poloxamers 407 and 188, sodium alginate, HPMC, and benzalkonium bromide. The drug release was via a corrosion-diffusion mechanism. *In vivo* studies revealed that the formulations restored damaged tissues and are potential therapeutics for the treatment of cervicitis, an inflammation of the cervix [39].

Tuğcu-Demiröz et al. developed mucoadhesive gel formulations of oxybutynin and compared their blood levels with orally administered oxybutynin in immediate release tablets *in vivo* [40]. The gels were prepared from chitosan, HPMC, and Poloxamer 407. The formulation exhibited a good permeation effect across the vaginal mucosa. *In vivo* studies on rabbits showed that the area under the curve (AUC) and relative bioavailability values were high for a vaginal gel formulation prepared with HPMC. The formulation was found to be useful for the treatment of overactive bladder and vaginal dryness resulting from menopause [40]. Lu et al. reported a thermosensitive *in situ* gel for local uterine administration [41]. It was prepared from Poloxamer 188 and HPMC and loaded with fertility-promoting intrauterine infusion liquid as hormone therapy. The gelation temperature was 27 °C. *In vitro* release tests showed that the release of icariin, an active compound in the hormone therapy, was slow from *in situ* forming gel. After the gel was locally administered, the serum estradiol level of the gel group was significantly higher than that of the control group ($P < 0.01$). The histological evaluation did not indicate signs of mucosal irritation in the animals that were administered the formulation. The *in situ* forming gel system was found to be a potential therapeutic for the treatment of bovine uterine diseases [41]. Date et al. developed thermosensitive gels containing raltegravir- and efavirenz-loaded PLGA NPs for pre-exposure prophylaxis of HIV [42]. The NPs were prepared by a modified emulsion–solvent evaporation method. The size and surface charge of NPs were 81.8 ± 6.4 nm and −23.18 ± 7.18 mV, respectively. The encapsulation efficiency of raltegravir and efavirenz was 56% and 98%, respectively. A thermosensitive vaginal gel was prepared using a combination of Pluronic F127 (20% w/v) and Pluronic F68 (1% w/v). The gel loaded with the NPs exhibited gelation at 33°C. The NPs and the blank gel were not cytotoxic *in vitro*. The formulation is a potential antiretroviral agent for long-term vaginal pre-exposure prophylaxis against heterosexual HIV-1 transmission [42].

5.2.3 *In Situ* Gels for Rectal Administration

The rectal route of administration of drugs is an effective route for the administration of drugs in infants who have difficulty in swallowing orally administered medicine [43]. This route is usually used in patients who are uncooperative or unconscious or when the intravenous route is compromised [44]. The length and circumference of the human rectum are in the range of 10–15 and 15–35 cm, respectively. The pH of its fluid content is in the range of 7–8 [44, 45]. The surface area is small, in the range of 200–400 cm^2. It is drained by three veins: the superior, middle, and inferior rectal veins. The superior rectal vein drains the upper part, while the middle and inferior rectal veins drain the lower part of the rectum [44, 45]. The rectal wall is composed of epithelium, and it contains cells that secrete mucus. It is usually empty, and its filling results in the defecation reflex under voluntary control [45]. Drugs are absorbed from the rectum by passive diffusion. The rate of drug absorption from the rectum is lower than from the oral route due to the small surface area of absorption [45]. The lower and middle rectal veins drain into the inferior vena cava, and the blood flows directly to the heart and to the general circulation. In contrast, the upper rectal vein drains into the portal vein, and the blood flows through the liver before reaching the heart [45]. This indicates that drug molecules from the rectum can be transported either directly to the general circulation or through the liver. Drug absorbed in the middle and lower parts of the rectum avoids the first-pass metabolism in the liver and is transported directly to the general circulation (Figure 5.5).

Few researchers have designed *in situ* gels for rectal drug delivery (Figure 5.6). Yuan et al. developed a thermosensitive, mucoadhesive *in situ* gel loaded with nimesulide for rectal delivery [46]. It was prepared from a combination of sodium alginate, Poloxamer 407, PEG, and HPMC. The gelation temperature increased significantly with an increase in the incorporation of nimesulide. Furthermore, the addition of PEG increased both the gelation temperature and the rate of drug release from the formulations. The formulation containing Poloxamer 407/nimesulide/sodium alginate/PEG 4000 in a ratio of 18/2.0/0.5/1.2% was found to exhibit an acceptable gelation temperature, a good drug release profile, and excellent retention in the rectum. No sign of mucosal irritation was observed *in vivo*, and higher initial serum concentration, C_{max}, and AUC of the loaded drug were observed when compared with a solid suppository [46].

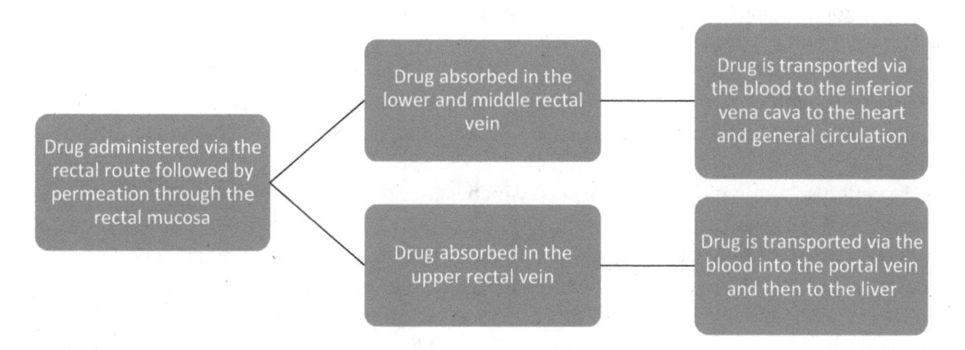

FIGURE 5.5 Mechanism of drug uptake administered rectally.

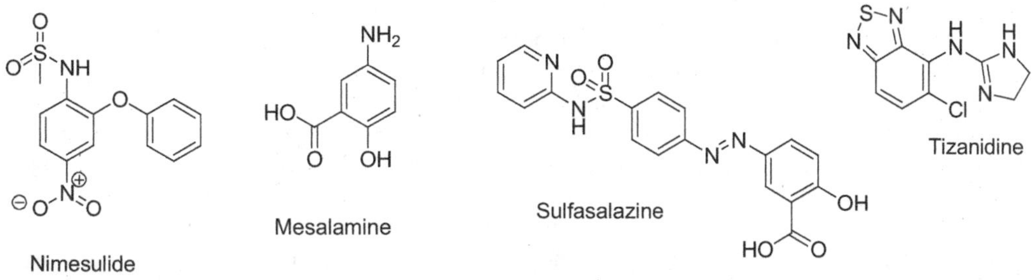

Nimesulide

Mesalamine

Sulfasalazine

Tizanidine

FIGURE 5.6 Selected drugs loaded into *in situ* gels for rectal administration.

Ramadass et al. developed a mesalamine collagen-based *in situ* gel prepared from type I collagen for the treatment of ulcerative colitis, a chronic inflammatory bowel disease [47]. The formulation was pH and temperature sensitive with the capability to undergo a sol-gel transition at physiological temperature and pH. The *in vitro* release profile of mesalamine from the formulation was sustained over a period of 12 h. *In vivo* studies in an ulcerative colitis model in BALB/c mice further indicated a significant reduction in rectal bleeding and mucosal damage score. The administration of mesalamine via the *in situ* gel enhanced the regeneration of damaged mucosa, resulting in a significant synergistic effect for the treatment of ulcerative colitis [47]. Xu et al. reported an *in situ* mucoadhesive gel to improve the efficacy of rectally administered sulfasalazine. The gel was composed of catechol–modified chitosan cross-linked by genipin. The formulation produced a lower plasma concentration of the toxic sulfapyridine, a by-product that is produced when the drug is administered orally. The delivery of sulfasalazine via the rectal route was effective and safe when compared with oral administration [48]. Moawad et al. developed tizanidine HCl–loaded nanotransfersome *in situ* gels for rectal administration [49]. The drug-loaded nanotransfersomes were prepared by a thin-film hydration method. The formulations were composed of phosphatidylcholine, Tween 80, HPMC, and Poloxamer 407. The percentage drug encapsulation efficiency was 52.39%, and the drug release was controlled over 8 h. Rectal administration of transfersome-loaded *in situ* gels in rabbits revealed that the transfersomal formulations increased the bioavailability of the drug by over 2.18-fold and increased the half-life ($t_{1/2}$) to about 10 h when compared with the oral solution. The increase in the $t_{1/2}$ of the rectal formulations suggested retardation in the drug release caused by the gel system. The encapsulation of the drug into the *in situ* gels prolonged the drug release and enhanced the drug bioavailability, which is a promising result suitable for a drug delivery system for the treatment of spasticity [49].

5.2.4 *In Situ* Gels for Nasal Administration

The nasal cavity is made up of three regions: respiratory, olfactory, and vestibule [50–53]. The anterior region of the nasal cavity, which is surrounded by cartilage with small hairs, is known as the nasal vestibule. The nasal turbinates are responsible for the turbulent airflow through the nasal passages, thereby resulting in a good contact between the inhaled air and the mucosal surface found in the respiratory region. The respiratory region contains four important cell types, which are involved in the active transport processes, in mucociliary clearance, and in trapping moisture, thereby keeping the mucosa moist; these are the basal, goblet, non-ciliated, and ciliated columnar cells present in the respiratory epithelium. The ciliated and non-ciliated cells enhance the surface area, and this is a region where drug absorption occurs [50–53].

The cilia cells also facilitate mucus transport, which is useful in mucociliary drug clearance. The olfactory region, which is located on the roof of the nasal cavity, is composed of cilia projecting down from the olfactory epithelium into a layer of mucus. The epithelial layer of the olfactory region is composed of three types of cells: the olfactory neural cells, the basal cells, and the sustentacular cells. The basal cells provide mechanical support to other cells. Small-sized therapeutics are transported via the olfactory bulb into the olfactory cortex and then to the caudal pole of the cerebral hemisphere, the cerebrum, and the cerebellum [50–53]. Drug absorption, when administered intranasally, is depicted in Figure 5.7. Some researchers have designed *in situ* gels for intranasal delivery of selected drugs (Figure 5.8).

Shah et al. designed an *in situ* gel loaded with sodium cromoglycate, an anti-asthmatic drug [54]. It was prepared from Carbopol 940 and different types of HPMC: HPMC K100, HPMC K4M, and HPMC K15M. The formulation containing 0.75% Carbopol 940 and 0.50% HPMC K4M was effective in enhancing *in vitro* permeation of the drug and the mucoadhesion, which was dependent on the residence time of the drug. The nasal mucosa can tolerate solutions within the pH range of 3–10. The pH of all the formulations was within the range of 5.6 to 6.2, which is within the physiological pH range of the nasal mucosa. The formulation did not cause any significant change in the microscopic structure of the mucosa. There was no alteration in the epithelium layer, the basal membrane, or the submucosa [54]. Gu et al. investigated *in vivo* pharmacokinetics and bioavailability of *in situ* gel loaded with risperidone, an antipsychotic drug [55].The bioavailability and absorption rate of the formulation were high when compared with the oral administration of the drug. The maximal plasma concentration was

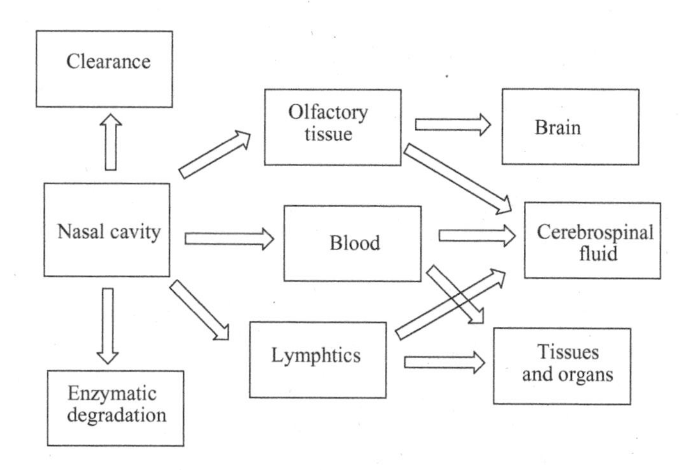

FIGURE 5.7 The mechanism of drug uptake intranasally.

15.2 μg/ml in 5 min for the formulation and 3.6 μg/ml in 30 min for oral administration of the free drug. The relative bioavailability of the drug administered via the *in situ* gel intranasally was over 1600% when compared with the oral administration of the free drug. The effect of the formulation on the nasal mucociliary movement using a toad palate model further indicated mild ciliotoxicity and adverse effects, which were temporary and reversible [55]. The distinct features of the formulation, which are rapid absorption, quick onset of action, high bioavailability, and patient compliance, are unique for the treatment of schizophrenia [55].

Galgatte et al. prepared mucoadhesive *in situ* gel formulations for prolonged residence time and enhanced drug uptake [56]. The gels were loaded with sumatriptan succinate, a drug used for the treatment of cluster headaches and migraine, which is limited by low oral bioavailability resulting from hepatic metabolism. Its transportation across the blood–brain barrier is also poor. To overcome the aforementioned limitations associated with the drug, it was loaded onto *in situ* gels prepared from deacetylated gellan gum. The *in vitro* drug release from the formulation was 98.6% within 5 h, and the *ex vivo* drug release on sheep nasal mucosa was 93% within 5 h. An *in vivo* study on Sprague-Dawley (SD) rats revealed an absolute bioavailability of 165%, and the drug targeting index for brain tissues was 1.87. The drug release from the formulation was influenced by the viscosity of the formulation. The drug targeting index suggested that sumatriptan reached the brain via the olfactory pathway. The *in situ* gel enhanced the permeation of drug molecules across the nasal mucosa through olfactory pathways [56]. Jagdale et al. loaded timolol maleate, a drug used to treat hypertension, onto *in situ* gels prepared from HPMC and Poloxamer 407 [57]. The drug suffers from extensive hepatic first-pass metabolism with a half-life of 4 h. The highest percentage drug release from both the optimized formulations via egg membrane was 81% and 84%. The drug release was influenced by the viscosity of the *in situ* gel. *Ex vivo* drug diffusion via nasal mucosa was 61% and 67% for both formulations. The gelation temperature for the formulations was in the range of 31–36°C, and it decreased with increasing concentration of Poloxamer 407 and HPMC in the formulations. Intranasal administration of the drug via *in situ* gel has the potential to improve the drug bioavailability and lower the dosing frequency [57]. Wang et al. loaded geniposide onto *in situ* mucoadhesive, thermoreversible gel to increase the residence time of the drug in the nasal cavity [58].

Geniposide is used for the treatment of brain and hepatic diseases. The *in situ* gel formulations were prepared by a cold method from Poloxamers (P407, P188) and HPMC using borneol as a permeation enhancer. The optimized amounts of Poloxamer (P407), Poloxamer (P188), and HPMC were 19.4–20.5%, 1.1–4.0%, and 0.3–0.6%, respectively. The effects of Poloxamer 407 on the gelation temperature were significant. The pH values of all the formulations were in the range of 6.3–6.5, suitable for intranasal administration. The *in vitro* drug release of geniposide was zero order, and the *ex vivo* cumulative drug release of geniposide from the formulation followed a Weibull drug release model. The release of geniposide was controlled by gel corrosion, and the permeation of geniposide was time dependent [58]. Gowda et al. developed a mucoadhesive *in situ* gel formulation loaded with diltiazem hydrochloride for

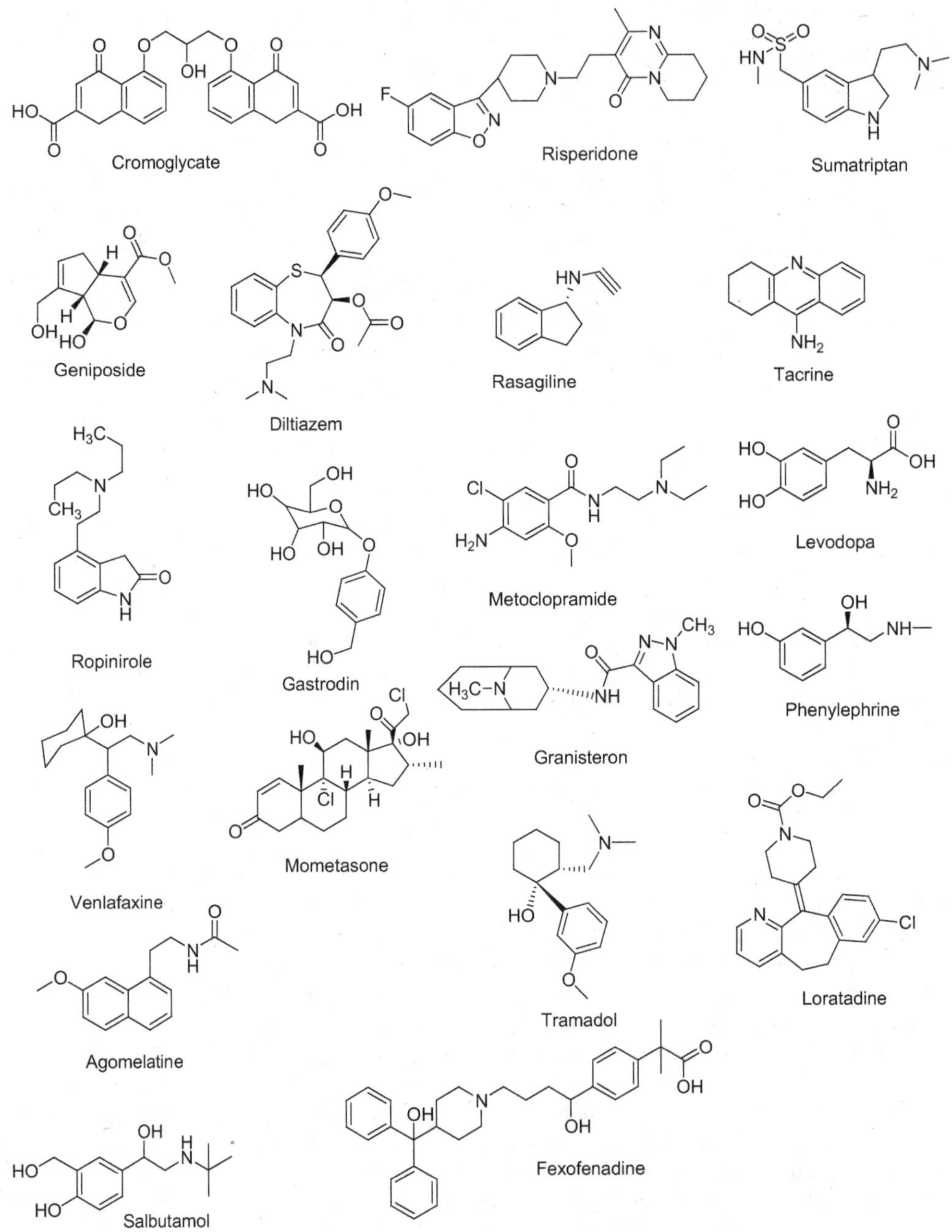

FIGURE 5.8 Selected drugs loaded into *in situ* gels for intranasal administration.

controlled drug release [59]. The formulation was prepared from Poloxamer and HPMC by a cold technique. Increasing the concentration of HPMC in the formulation decreased the gelation temperature. The drug content in the formulations was found to be in the range of 97–100%. Increasing the concentration of HPMC increased the mucoadhesive strength of the formulations. The diffusion of drug particles was enhanced by an increase in the concentration of HPMC in the formulations. The optimized formulation released 86% of the loaded drug when compared with the aqueous drug solution, in which 94% of the drug was released over a period of 5 h. The drug release was influenced significantly by the polymers used and their concentration in the formulations [59]. Barakat et al. evaluated the potential of penetration enhancer vesicles within an *in situ* forming gel network prepared from Poloxamer and hyaluronic acid for the intranasal delivery of dimenhydrinate, a drug used to treat nausea and vomiting associated with motion sickness and chemotherapy [60]. The permeation enhancer vesicles were prepared from phospholipids and a combination of labrasol, transcutol, and PEG 400. Increasing the ratio of Poloxamer P407 from 17% to 20% decreased the gelation temperature of the formulations significantly, which is attributed to the formation of a closely packed gel network, resulting in rapid gelation at lower temperatures. *In vivo* studies on a cisplatin-induced emesis model in rats indicated that an increase in the mucoadhesiveness of the formulations resulted in an increase in the drug released via the nasal mucosa [60].

Ravi et al. developed a thermosensitive gel from Poloxamer (407 and 188), Carbopol 934 P, and chitosan for the delivery of rasagiline mesylate for the treatment of Parkinson's disease. The bioavailability of the formulation was sixfold when compared with oral administration of the drug solution *in vivo*. The formulation was non-irritant and non-toxic to the rat nasal mucosa. The drug uptake in the rat brain tissue from the formulation was significant when compared with the free drug solution. The formulation extended the residence time and contact with nasal epithelium, resulting in good drug absorption from the nasal cavity. The *in vitro* drug profiles from the formulation were influenced by the mucoadhesive polymer concentration. Increasing the concentration of the mucoadhesive polymer increased the viscosity of the gel [61]. Qian et al. reported *in situ* gel formulation for intranasal delivery of tacrine, an antipsychotic drug [62]. The formulations were characterized by prolonged retention in the nasal cavity when compared with the drug solution. The formulations achieved two- to threefold higher peak plasma concentration and AUC of the drug in the plasma and brain tissue when compared with the oral solution. The extended nasal residence time, enhanced bioavailability, and increased brain uptake of the loaded drug revealed the potential of *in situ* gel in the treatment of Alzheimer disease [62]. Khan et al. prepared *in situ* gel formulations from chitosan and HPMC for intranasal delivery of the dopamine D2 agonist ropinirole to the brain [63]. *In vivo* studies in albino rats by intranasal administration of 99mTc-ropinirole *in situ* gel showed that the absolute bioavailability of ropinirole from the formulation was 82%, and the $AUC_{0-480\,min}$ in the brain after nasal administration of ropinirole was 8.5 times higher compared with the intranasal administration of the ropinirole solution. High brain uptake of the drug confirmed nose-to-brain transport from the intranasal *in situ* gel formulation [63]. Cai et al. prepared an *in situ* gel from deacetylated gellan gum loaded with gastrodin, a major bioactive of tianma used in the treatment of neurological diseases. The formulation was stable for more than 2 years. Studies using the toad palate model and the rat nasal mucociliary model confirmed the absence of ciliotoxicity [64]. Mahajan et al. loaded metoclopramide hydrochloride, an antiemetic, onto *in situ* gel formulations. The drug permeation effect of the formulations across sheep nasal mucosa was influenced by increasing the concentration of Carbopol by 0.15% or more. The bioavailability of the drug from the formulation in rabbits significantly increased to 54.61% when compared with that from the drug solution, which was 40.67%. The formulations extended the release of metoclopromide for the duration of 3 h. The mucosa treated with the formulations indicated slight degeneration of the nasal epithelium and increased vascularity in the basal membrane. No sign of a destructive effect of the formulations on the treated nasal mucosa was observed [65]. Sharma et al. prepared a thermosensitive *in situ* gel system from chitosan and poly vinyl alcohol for nasal delivery of levodopa [66]. The *in vitro* release of levodopa from the gel network decreased with an increase in the concentration of chitosan from 1% to 5% [66].

Inayat et al. loaded venlafaxine hydrochloride, an antidepressant, onto an *in situ* nasal gel prepared from gellan gum for brain delivery via the nose. One of the formulations with 0.4% w/v gellan gum exhibited 48% drug release in 4 h compared with 95% drug release from the simple solution in 2 h, indicating extended drug residence time in the nasal cavity. There was no evidence of mucosal damage [67]. Altuntaş et al. evaluated the anti-allergic effects of thermosensitive *in situ* gel loaded with mometasone furoate in an ovalbumin-induced rat model of allergic rhinitis [68]. The *in situ* gel formulation decreased nasal symptoms and ovalbumin-specific serum immunoglobulin E level when compared with the suspension of mometasone furoate. Symptoms such as mucosal edema, vascular dilatation, eosinophil infiltration, and loss of cilia were improved in the rat model treated with the *in situ* gel formulation. The nasal mucosa integrity of the rat was maintained, indicating that the formulation offers safety and efficacy advantages in long-term use [68]. Mali et al. investigated the capacity of *in situ* gel loaded with granisetron hydrochloride to prolong the residence time of the drug in the nasal cavity. The *in situ* gel was prepared from

Pluronic flake 127, moringa gum, carboxymethyl tamarind gum, PEG 6000, and sodium alginate. The optimized formulation prolonged nasal residence and improved nasal bioavailability [69]. Tejaswini and Devi prepared an *in situ* gel loaded with phenylephrine hydrochloride for sustained drug release. The formulations were prepared from Carbopol 934, Poloxamer 188, and HPMC. The drug content for the formulations was in the range of 94–99%. The drug release was high over a period of 8 h. The sustained drug release indicated the capacity of the gel to deliver the drug to the target site for the treatment of respiratory tract infections [70]. Fatouh et al. loaded agomelatine, an antidepressant drug, onto *in situ* gels. The sol-gel transition temperature was 31°C with a mucociliary transport time of 27 min. The percentage drug release from the formulations was in the range of 38–69% over a period of 8 h when compared with the free drug solution, for which it was 102%. The mucociliary transport time of the factorial design *in situ* gel formulations ranged from 4.5 to 28.5 minutes. The optimized formulation exhibited high C_{max}, $AUC_{0-360\,min}$, and absolute bioavailability of 247 ng/ml, 6677 ng min/ml, and 38%, respectively, when compared with the oral suspension of Valdoxan®. The ability of the formulation to bypass the first-pass metabolism due to the intranasal administration favored enhanced blood–brain barrier uptake of the drug [71].

Kaur et al. loaded tramadol HCl, an antidepressant drug, onto chitosan nanoparticles prepared by an ionic gelation method followed by the loading of the NPs onto *in situ* gels prepared from Pluronic and HPMC. *In vivo* studies showed a significant increase in locomotor activity, and increased glutathione and catalase levels and decreased lipid peroxidation and nitrite levels were found after intranasal administration of the formulation, revealing its potential for the treatment of depression [72]. Singh et al. developed *in situ* gels to increase the solubility of loratadine, an anti-allergic drug. The formulations were composed of β-cyclodextrin prepared by the cold method. The optimized formulation had a gelling temperature of 29°C, mucoadhesive strength of 7676.0 dyn/cm^2, and cumulative drug permeation of 98% over a period of 6 h. Histological examination revealed the absence of any significant damage to the nasal tissue [73]. Perez et al. incorporated ^{32}P-siRNA dendriplexes into *in situ* mucoadhesive gels prepared from Poloxamer (23% w/w), chitosan (1% w/w), or Carbopol (0.25% w/w). The gels retained 100% of radiolabel after 150 minutes. The gel released 35% of radiolabeled dendriplexes, and three intranasal doses of the dendriplexes in the gel did not damage the rat nasal mucosa. The administration of two intranasal doses of the formulation was required for high brain drug uptake, indicating brain delivery of the radiolabeled siRNA [74]. Salunke et al. developed a mucoadhesive *in situ* gel of salbutamol sulfate using gellan gum and HPMC for nasal administration. The mucoadhesive force was influenced by the concentration of HPMC, and the drug release was 97% in 11 h. There was no significant damage to the sheep nasal mucosa used [75]. Li et al. loaded ketorolac tromethamine, a potent drug for the treatment of moderate to severe pain, onto thermo- and ion-sensitive *in situ* hydrogels prepared from Poloxamer 407 and deacetylated gellan gum. The formulation exhibited sustained drug release, good intranasal absorption, and insignificant nasal ciliotoxicity [76]. Cho et al. loaded fexofenadine hydrochloride onto *in situ* gels prepared from Poloxamer 407, β-cyclodextrin, and chitosan. *In vitro* permeation studies in primary human nasal epithelial cell monolayers showed that an increase in chitosan content (0.1% and 0.3%, w/v) enhanced drug permeation. The addition of chitosan slightly influenced the gelation temperature and viscosity of the formulations. The plasma concentrations of the drug were significantly higher when compared with the nasal solutions. The bioavailability of the optimized thermoreversible gel containing 0.3% chitosan was 18-fold higher than that of the nasal solution, indicating the efficacy of thermosensitive gels [77].

Rao et al. studied the capability of *in situ* gels to increase the bioavailability of ropinirole for the treatment of Parkinson disease [78]. The thermoreversible nasal gels were prepared by a cold method from Pluronic F-127 and HPMC. The prolonged nasal residence time was due to the mucoadhesion and increased gel strength. The *ex vivo* drug release was between 56% and 100% in 5 h. A histological study on sheep nasal mucosa revealed moderate cellular damage. A fivefold increase in bioavailability in the brain was observed when compared with the intravenous route [78].

5.2.5 *In Situ* Gels for Injections

The modes of injection of *in situ* gels include intravenous, subcutaneous, intramuscular, intraperitoneal, etc. Several researchers have reported the efficacy of administering drug-loaded *in situ* gels via injection (Figure 5.9).

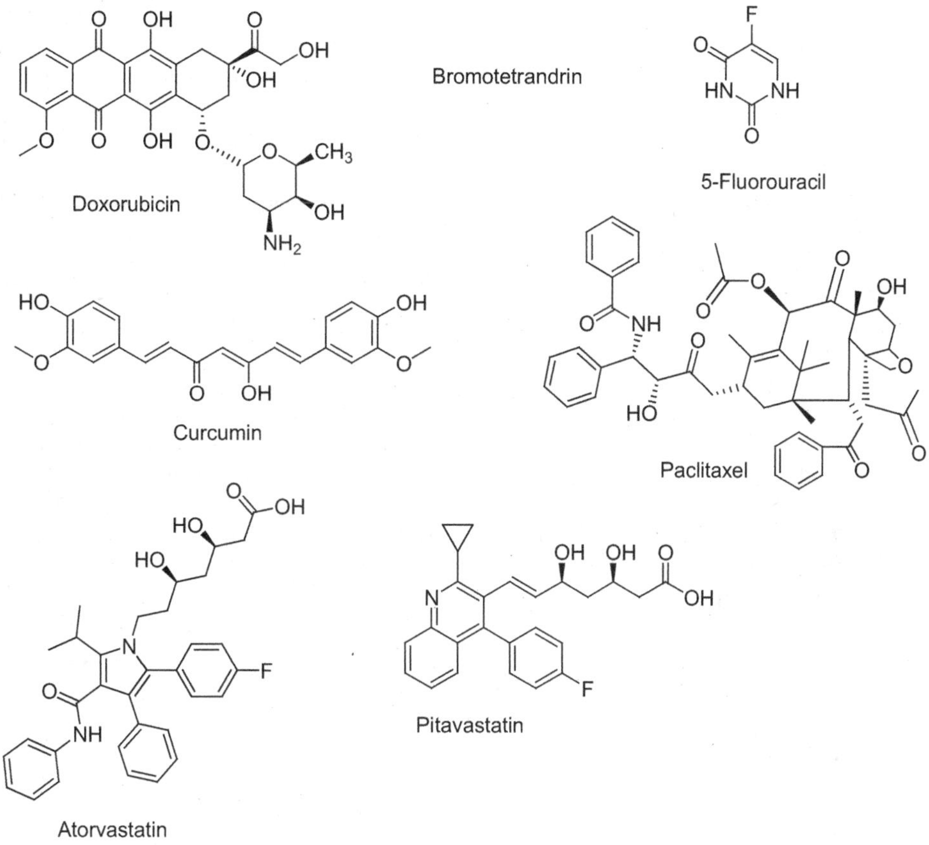

Doxorubicin
Bromotetrandrin
5-Fluorouracil
Curcumin
Paclitaxel
Atorvastatin
Pitavastatin

FIGURE 5.9 Selected drugs loaded into *in situ* gels for injections.

In situ gels have been developed for the delivery of anticancer drugs. Anticancer drugs suffer from toxicity, and the absence of lymphatic vessels in the tumors leads to high hydrostatic pressure in the tumor stroma, resulting in ineffective delivery of the drugs to the tumor tissues by systemic administration. Drug delivery systems that can deliver therapeutics to the target site can be effective for the treatment of cancer.

Shen et al. loaded doxorubicin (DOX) onto zein-based *in situ* gels for intratumoral injection for the treatment of colorectal cancer [79]. The release of DOX was rapid on the first day, followed by a sustained release over a period of 6 days. However, the release of DOX was incomplete at 7 days. Drug loading did not have any significant effect on the drug release profile ,and the protein concentration decreased burst release and prolonged drug release. *In vivo* antitumor effects of the formulation on BALB/c nude mice human colon cancer cells revealed that a single injection of the formulation produced significant inhibitory effects ($P < 0.05$) against the tumor by remarkably decreasing its propagation. The drug concentration in blood after administration of the formulation *in vivo* was reduced when compared with DOX solution. The AUC of the formulation was 9.12 μg/ml/h compared with that of the DOX solution, which was 35.94 μg/m/h. The transformation of the formulation from liquid to solid after injection resulted in the sustained release of the drug within the tumor. Drug distribution studies further indicated a higher drug concentration in the tumor for a longer time when compared with the DOX solution. The AUC of the formulation and the DOX solution in the tumor was 204.59 and 18.71 μg/ml/h, respectively, which showed that *in situ* gels are potential drug delivery systems for targeted drug delivery, thereby protecting the healthy tissues from the toxic side effects [79].

Wu et al. developed *in situ* gels from phospholipids and medium-chain triglycerides for the delivery of DOX [80]. *In vivo* studies revealed a significant antitumor effect of the formulation on S180 sarcoma tumor-bearing mice after a single administration of the formulation intratumorally. The *in vivo* biodistribution study showed a high DOX concentration in the tumors when compared with other major organs after intratumoral administration. The high concentration of DOX in the tumor and long-term retention revealed that *in situ* gels are a promising drug delivery system for chemotherapy [80]. Luo et al. developed a phospholipids-based *in situ*-forming gel for the co-delivery of DOX and bromotetrandrin. The release of DOX and bromotetrandrin from the phospholipid gel was sustained *in vitro* for over 3 weeks. The release of both drugs was prolonged for 2 weeks *in vivo* in rats that were administered the formulation via subcutaneous injection. The formulation did not induce cardiac toxicity when administered in rats via subcutaneous injection. A single administration of the formulation by intratumoral injection in the resistant MCF-7/Adr xenograft–bearing mice further showed good antitumor efficacy by reversing the multidrug resistance in breast cancers [81]. Yang et al. reported phospholipids-based *in situ* gels for intratumoral injection. which can enhance the antitumor and antimetastasis effect simultaneously [82]. The gel was co-loaded with 5-fluorouracil and magnesium oxide, and intratumoral administration in 4T1-bearing mice resulted in extended survival time and significant antitumor and antimetastasis effect when compared with the free drug [82]. Kwon et al. reported combinational chemotherapy via intratumoral injection of DOX and 5-fluorouracil to reduce the toxic effects of systemically administered DOX and 5-fluorouracil. The formulations were injected intratumorally in mice, and the formulation containing both drugs inhibited the tumor growth significantly when compared with the formulation containing a single drug. The *in vivo* biodistribution showed high concentrations of both drugs in the target tumor, suggesting that concentrations below the drug's toxic plasma concentration would not result in systemic toxicity [83].

Wan et al. loaded DOX onto temperature-sensitive *in situ* gel formulations for the treatment of liver cancer via intratumoral administration. The release of DOX was 9.4% in 24 h and 60% over a period of 10 days. The formulation exhibited an excellent antitumor effect against H22 tumor after intratumoral administration *in vivo*. *In vivo* biodistribution of DOX indicated good DOX retention in the tumor tissues when compared with the free DOX solution after intratumoral injection [84]. Chen et al. developed a phospholipid-based *in situ* gel for the intratumoral administration of paclitaxel in a brain glioma–bearing mice model [85]. An *in vivo* tolerability study showed good tolerability after the mice were treated with the drug-loaded formulation when compared with the free paclitaxel. The survival time of brain glioma–bearing mice after treatment with the formulation was extended significantly when compared with mice treated with the free drug [85]. Fong et al. prepared a folic acid–conjugated graphene oxide–based encapsulated onto hyaluronic acid-chitosan-*g*-poly(*N*-isopropylacrylamide) *in situ* gel for the targeted delivery of DOX via intratumoral administration. The release of DOX was pH dependent. Intratumoral administration of the drug-loaded *in situ* gel formulation in BALB/c nude mice subcutaneously implanted with MCF-7 cells showed that relative tumor size reduced continuously for up to 11 days with a significant tumor inhibition ratio of 52% after 21 days when compared with the free drug. These results demonstrated the capability of the formulation to extensively destroy tumor tissues, thereby enhancing the therapeutic efficiency. The enhanced intracellular uptake of the formulation resulted in the high tumor-killing ability when compared with free DOX [86]. Li et al. reported *in situ* gel formulation prepared by the reaction of a PEG derivative with α,β-polyaspartylhydrazide and loaded with DOX for cancer chemotherapy. The intratumoral administration of the formulation in mice with human fibrosarcoma indicated significant inhibition of tumor growth when compared with mice treated with the free DOX solution [87]. Ning et al. prepared *in situ* gels from thiolated chitosan and PEG diacrylate and loaded them with curcumin and lysozymes. The release of curcumin was sustained in a controlled lysozyme-responsive behavior, allowing the drug concentration to reach a therapeutic threshold quickly. The formulation delayed tumor growth and reduced side effects in tumor-bearing nude mice [88].

Zhang et al. prepared cyanoacrylate-based *in situ* gels from methoxyethyl cyanoacrylate and PEG and loaded them with 5-fluorouracil [89]. *In vivo* studies indicated 50% survival of the mice with no significant toxic effects. The formulation depot affected the flow of blood in the microvessels of tumors, thereby contributing to the synergetic anticancer effect of the loaded drug [89]. Lee et al. prepared a poly(ethylene glycol)–b-polycaprolactone (MPEG–PCL) *in situ* gel loaded with paclitaxel. Intratumoral injection of the drug-loaded formulation to mice bearing B16F10 tumor xenografts resulted in effective inhibition of the growth of B16F10 tumors over 10 days when compared with the injection of taxol.

A histological analysis revealed an increase in necrotic tissue in tumors treated with the formulation [90]. Kang et al. developed *in situ*-forming gels from PEG and polycaprolactone (PCL) loaded with DOX. The release of DOX was sustained *in vitro* over a period of 20 days. *In vivo* studies in B16F10 cancer cell xenograft–bearing mice showed that a single intratumoral injection of the *in situ* gel inhibited the tumor growth effectively and significantly when compared with the free drug. DOX released from formulation gels caused apoptosis of tumor cells. The biodistribution of DOX after a single injection of drug-loaded gel was 90% on day 1 and 13% after 15 days. Only a small amount of DOX was found in other major organ tissues [91]. *In situ* gels have also been employed for the delivery of antidiabetic drugs.

Hu et al. designed *in situ* gels for the delivery of exenatide. The *in situ* gel was prepared from phospholipid S100 and medium-chain triglyceride. *In vivo* studies in diabetic animal models revealed that a single subcutaneous injection of the formulation resulted in a hypoglycemic effect and controlled the blood glucose level comparably to exenatide solution [92]. Oh et al. prepared a Pluronic-based *in situ* gel loaded with exenatide and demonstrated the extended duration of antidiabetic effects of the drug-loaded gel. *In vivo* studies on mice in which the formulation was administered subcutaneously revealed a reduction in blood glucose level by 60.8% (from 438.46 ± 81.25 mg/dl at 0 h to 172.21 ± 26.49 mg/dl at 10 h) and 74.3% (from 382.25 ± 89.80 mg/dl at 0 h to 98.5 ± 37.68 mg/dl at 10 h) when compared with free exenatide, for which the reduction was 26.9% (from 242.34 ± 29.19 mg/dl at 0 h to 177.34 ± 25.45 mmol/l at 4 h) [93].

In situ gels are also employed to reduce cholesterol levels. Xiang et al. prepared phospholipids and soybean oil for prolonged release of pitavastatin calcium for the treatment of hyperlipidemia [94]. A single subcutaneous injection in mice released the loaded drug for 15 days. The levels of total cholesterol, total triglyceride, and low-density lipoprotein decreased significantly, and the *in vivo* therapeutic effect was sustained for a period of 20 days. The formulation showed effects that were not serious, such as mild tissue inflammatory responses [94]. Ahmed et al. loaded atorvastatin onto a biodegradable *in situ* gel prepared from poly (D,L-lactide-co-glycolide) and PEG [95]. Following the intramuscular injection of a single dose in male New Zealand White rabbits, the AUC total for the optimized formulation was 35.932 ng/ml/h, which was significantly greater compared with the marketed atorvastatin tablet, for which the AUC was 29.186 ng/ml/h. The time to reach the maximum plasma concentration was 72 h for the formulation compared with 12 h for the free drug. and the mean residence time of the formulation was 79.7 h *in vivo* [95].

5.2.6 *In Situ* Gels for Wound Dressing

In situ gels have been employed as wound dressings by incorporating metal-based nanoparticles, growth factors, peptides, and antibacterial agents for enhanced wound healing (Figures 5.10 and 5.11). The biocompatibility of *in situ* gels makes them useful in wound dressing. *In situ* gels enhance cell proliferation and wound dressing.

Lee et al. incorporated dermal fibroblasts onto gelatin-based *in situ* gels [96]. The stiffness of the wound dressings was influenced by the concentration of the hydrogen peroxide used in their preparation. The stiffness of the soft gel was 1.1 kPa, whereas that of the hard gel was 6.2 kPa. The soft gels facilitated the proliferation of dermal fibroblasts when compared with the hard gel. The subcutaneous

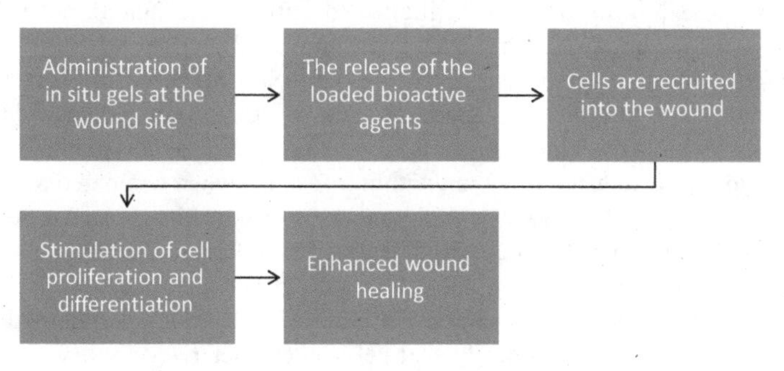

FIGURE 5.10 Mechanism of drug release from *in situ* gel at wound site.

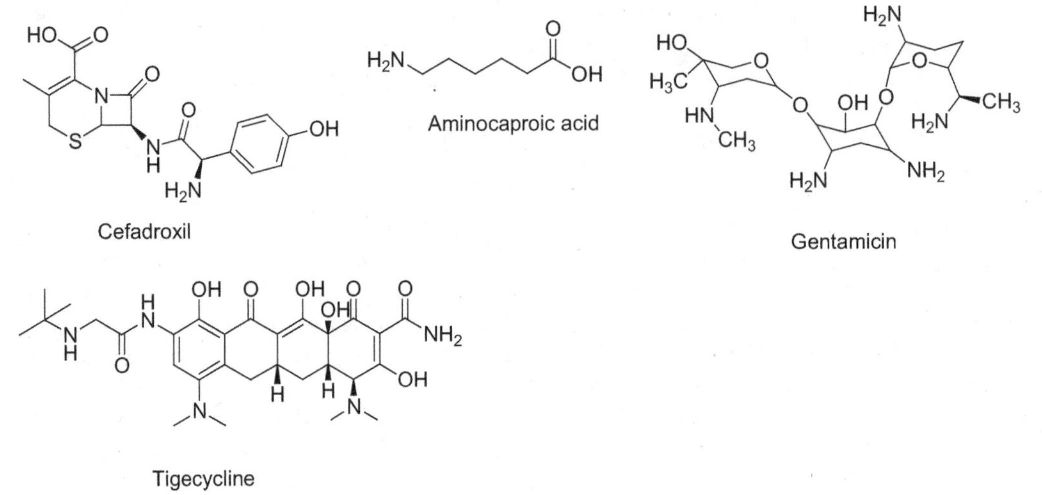

Cefadroxil

Aminocaproic acid

Gentamicin

Tigecycline

FIGURE 5.11 Selected drugs incorporated into *in situ* gels for wound dressings.

administration of the soft gel improved cell survival and retention over 14 days. *In vivo* transplantation of the formulation induced neovascularization, accelerated wound contraction, and promoted collagen deposition in incisions performed on mice skin [96]. Basha et al. loaded cefadroxil onto chitosan-based *in situ* gel. The formulation exhibited good antibacterial effects in a rat skin infection model against *S. aureus.* A remarkably accelerated wound healing process and bacterial clearance were significant when the wound dressing was administered *in vivo* [97]. Li et al. studied the role of antimicrobial peptides in the process of wound healing. Poly(L-lactic acid)-Pluronic L35-poly(L-lactic acid) (PLLA-L35-PLLA) containing human antimicrobial peptides 57 (AP-57) was developed for cutaneous wound healing. The release of the antimicrobial peptides was sustained over an extended period with low cytotoxicity and high antioxidant activity *in vitro.* An *in vivo* wound healing assay using a full-thickness dermal defect model in SD rats indicated that the *in situ* gels significantly promoted wound healing after 14 days with complete wound closure of 97%. The cutaneous wound healing enhanced granulation tissue formation, increased collagen deposition, and promoted angiogenesis in the wound tissue [98]. Du et al. developed a multifunctional *in situ*–forming gel to stop bleeding, inhibit inflammation, relieve pain, and improve healing. It was loaded with aminocaproic acid, povidone iodine, lidocaine, and chitosan. The gel enhanced the average time to hemostasis and decreased bleeding. It also exhibited good antibacterial activity and accelerated wound healing compared with gauze, indicating that it is useful for emergency situations [99].Hoque et al. reported *in situ* gels with antibacterial and hemostatic activity suitable for wound dressing. The gels were developed *in situ* from N-(2-hydroxypropyl)-3-trimethylammonium chitosan chloride and polydextran aldehyde. They were active against both Gram-positive and Gram-negative bacteria. The gels destroyed the bacteria by disrupting the membrane integrity of the pathogen. *In vivo* studies showed that gels prevented sepsis in a cecum ligation and puncture model in mice. The gels facilitated wound healing in rats with negligible toxicity toward human red blood cells and no inflammation of the surrounding tissue on subcutaneous implantation in mice [100].

Lih et al. prepared chitosan–poly(ethylene glycol)–tyramine *in situ* gels using horseradish peroxidase and hydrogen peroxide [101]. The gelation capability of the *in situ* gels was influenced by the concentration of horseradish peroxidase used in the cross-linking of the gels. The gels exhibited good hemostatic ability; the loss of blood was reduced to 59 mg within 5 seconds when compared to the control, which was 154 mg. The gels exhibited superior healing outcomes in the skin incision when compared with suture, fibrin glue, and cyanoacrylate. The wound healing was characterized by the appearance of new dermis [101]. Zhao et al. loaded curcumin onto chitosan/β-glycerophosphate *in situ* gels. The release profile of curcumin from the gel was sustained; 40% of the curcumin was released within 2 weeks. The wound healing effect of the curcumin-loaded gel was characterized by dryness, little sloughing, and no erythematous rash around the wounds, with healthy granulation tissue, which was light pink and granular in appearance. It exhibited a faster healing rate of 64% on day 7 and 95% on day 14 when compared with the formulation without curcumin and gauze. The *in situ* gels reduced bacterial load and oxidative stress in the wound site [102]. Tran et al. developed an *in situ* gel composed of rutin and tyramine-conjugated chitosan derivatives, horseradish peroxidase, and hydrogen peroxide. An *in vivo* wound healing study by injecting the gels onto rat dorsal wounds with a diameter of 8 mm for 14 days revealed the formation of neo-epithelium and a thicker granulation, similar to the original epithelial tissue, when compared with a commercial wound dressing, DuoDERM [103]. Wu et al. developed *in situ*–based gels loaded with silver nanoparticles for wound dressings with antibacterial activity. The gels exhibited slow release of silver ions with significant antibacterial activities, resulting in a 99% reduction in *E. coli*, *S. aureus*, and *Pseudomonas aeruginosa*. The gels also induced inflammation and promoted wound healing [104]. Liu et al. prepared *in situ* gels that demonstrated strong broad-spectrum antimicrobial activities against selected clinically isolated multidrug-resistant microbes [105]. The gels exhibited non-fouling properties and were found to be suitable as wound dressings to prevent infection. The gels were prepared by Michael addition from PEG and polycarbonate with quaternary ammonium groups. Gels with cationic polycarbonates exhibited high killing efficiency against clinically isolated multidrug-resistant Gram-positive and Gram-negative bacteria as well as fungi. The hydrophobic component MTC-ethyl in the cationic polycarbonate is important for antimicrobial activity [105]. De Cicco et al. prepared high–mannuronic alginate and amidated pectin blend *in situ* gels loaded with gentamicin sulfate [106]. The gels exhibited good flowability and high fluid uptake suitable for wound site filling and the prevention of bacterial proliferation. The drug release was prolonged over a period of 4–10 days, and the antimicrobial activity was high over a period of 24 days, essential to treat infected wounds [106].

Lu et al. prepared *in situ* photopolymerized gels by the ultraviolet cross-linking of a photocross-linkable azidobenzoic hydroxypropyl chitosan aqueous solution [107]. The gel maintained an extended moist environment over the wound bed for improved re-epithelialization. The gels were impermeable to bacteria but penetrable to oxygen. The gels did not cause cytotoxic effects to the skin cells [107]. Dong et al. prepared an injectable hybrid hydrogel from PEG and thiol-modified hyaluronic acid combined with adipose-derived stem cells. In *in vivo* studies in a rat dorsal full-thickness wound model. it prevented wound contraction and enhanced angiogenesis [108]. Del Gaudio et al. prepared gels containing Annexin A1 N-terminal-derived peptide for dermal wound repair application. They were prepared from a high–mannuronic content alginate and low–molecular weight chitosan. The release of the peptide was sustained over a period of 72 h. *In vivo* wound healing studies on dorsal wounds in mice showed significantly accelerated wound healing, with complete closure of the wound on day 14, indicating their potential application in dermal wound healing [109]. Nimal et al. prepared chitosan–platelet-rich plasma *in situ* gels by an ionic cross-linking method loaded with tigecycline for the treatment of *S. aureus* chronic wound infections [110]. The drug was released in a sustained manner. The gel system was an effective medium for antibiotic delivery and the treatment of skin infections caused by *S. aureus* [110].

5.2.7 *In Situ* Gels for Oral Administration

In situ gels have also been designed for the oral administration of bioactive agents. The oral route of drug administration is complex, and the gastrointestinal region extends from the mouth to the anus. There are several factors that affect the absorption of drugs administered orally (Figure 5.12). Some of the factors are gastrointestinal pH and transit times, the presence of food, age, and systemic diseases [111]. The presence of food can cause food–drug interactions, resulting in reduced absorption, delayed absorption, or improved absorption, and in some cases, the drug absorption is not affected. Food–drug interactions are influenced by the nature of the ingested foods and the administered drug. There are limitations associated with the oral administration of drugs, such as drug–drug interactions, first-pass effects, and drug–food interactions [111]. Due to the aforementioned limitations of drugs administered orally, *in situ* gels loaded with selected drugs have been designed to overcome these limitations (Figure 5.13).

Wu et al. prepared *in situ* gel systems for the oral delivery of ibuprofen [112]. The gels were prepared from gellan gum and sodium alginate. The optimized formulation was composed of 1.0% sodium alginate, 0.5% gellan gum, 0.21% sodium citrate, and 0.056% calcium chloride. A single oral dose of *in situ* gel–drug formulation to six healthy beagle dogs revealed that the T_{max} of the formulation was 1.8

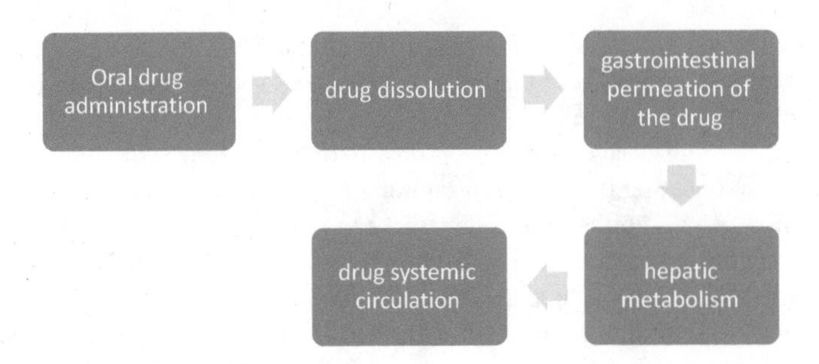

FIGURE 5.12 Mechanism of drug uptake administered orally.

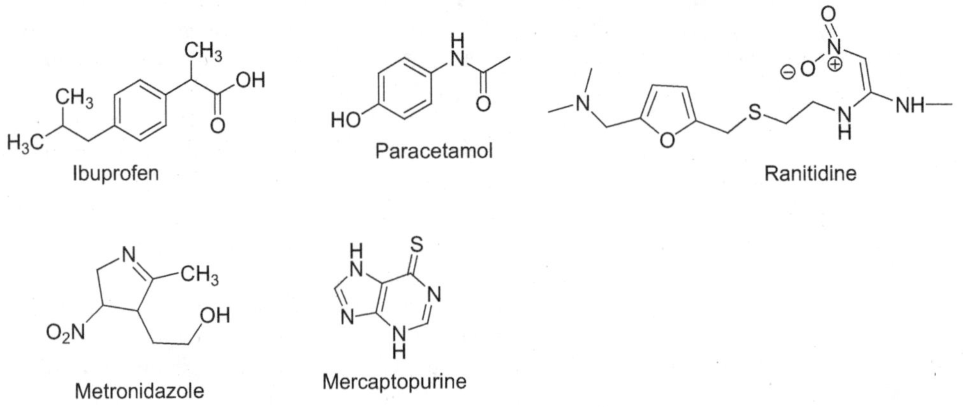

FIGURE 5.13 Selected drugs loaded into *in situ* gels for oral administration.

± 0.6 h, the C_{max} value was 29.2 ± 7.6 μg/ml, and the $T_{1/2}$ was 2.3 ± 0.5 h when compared with the free ibuprofen, for which the values were 0.4 ± 0.1 h, 37.8 ± 2.2 μg/ml, and 2.0 ± 0.9) h, respectively, for T_{max}, C_{max}, and $T_{1/2}$. The results obtained revealed enhanced drug bioavailability when loaded onto *in situ* gel [112]. Itoh et al. developed an *in situ* gel to improve the compliance of geriatric patients who have difficulties in swallowing conventional solid dosage forms. The gel was prepared from xyloglucan and sodium alginate and loaded with paracetamol [113]. The plasma levels of paracetamol after oral administration to gastric acidity–controlled rats were 1.8 and 2.5 μg/ml over 4 hours of release from the xyloglucan and alginate formulation, respectively. revealing sustained release from the gel. Visual observation of the rat stomach contents after oral administration showed reduced gel erosion, which was attributed to the sustained drug release [113]. Xu et al. developed an *in situ* gel system for the sustained delivery of ranitidine hydrochloride. It was prepared with different concentration of gellan gum, and the viscosity of the formulations increased with increased concentrations of gellan gum. The initial release of the drug was a burst effect, which was followed by moderate drug release. The animal studies indicated that the *in situ* gel formed a gel network in the stomach, followed by sustained release of ranitidine over a period of 8 h [114].

Pandya et al. loaded metronidazole onto *in situ* gelling systems to improve the residence time to eliminate *Helicobacter pylori*. The formulation was prepared from sodium alginate, xanthan gum, calcium carbonate, and sodium bicarbonate. The formulation exhibited extended drug release over a period of 10 h [115]. Mei et al. prepared *in situ* gels for the delivery of metronidazole for the treatment of periodontitis, a chronic bacterial infection. The system exhibited a sustained drug release over a period of 1 week *in vitro*. An *in vivo* study using a rabbit periodontitis model indicated that the formulation maintained metronidazole concentrations in the periodontal pockets above the minimum inhibition concentration for a period of 10 days, and the drug concentration was not detectable in the blood. These findings indicated that the formulation can release the drug over a prolonged period of time with reduced systematic side effects when compared with the metronidazole suspension, which is effective for 24 h, with metronidazole concentrations in the blood after 6 h [116]. Nagendra et al. loaded mercaptopurine, a purine antagonist, onto an *in situ* gel for sustained drug release from xanthan gum. The optimized formulation released the drug over a period of 12 h compared with the free conventional mercaptopurine tablet, which released the drug for 45 minutes. The sustained release of the drug has the capability to prolong the period of time throughout the gastrointestinal tract, thereby improving oral drug bioavailability [117].

5.3 Conclusion

In situ gels loaded with bioactive agents have been designed for administration via the nasal, ocular, oral, vaginal, and intraperitoneal routes. They are used to deliver therapeutics for the treatment of infectious and non-infectious diseases. In their design for ocular drug delivery, the particle size of the formulation influenced drug permeation, while the viscosity of the formulation influenced the sustained drug release mechanism and its easy instillation into the eye, in which there is a rapid sol to gel transition. The formulations were stable over a long period of time on storage. *In vivo* studies revealed that the drug absorption into the eyes in the aqueous humor was higher than in the systemic circulation. The efficacy of the loaded drug was maintained, and the formulations did not pose any adverse side effects. The formulations were found to be effective for the treatment of conjunctivitis, fungal eye infections, cataracts, glaucoma, etc.

In the design of *in situ* gels for vaginal administration of therapeutics, the prepared gels that were effective were characterized by sustained drug release, medium adhesiveness, and low hardness for good retention at the site of application. The dilution by the vaginal fluid did not affect the phase transition of the formulations. The *in situ* gels were suitable as prophylaxis to prevent heterosexual HIV-1 transmission and for the treatment of bovine uterine diseases, bladder and vaginal dryness resulting from menopause, cervicitis, and vaginal candidiasis. *In vivo* studies further revealed the absence of abnormality in the vagina and cervix. There has been very little research on the development of *in situ* gels for rectal drug delivery. A few reports have indicated the efficacy of gels in drug delivery via the rectal route, which is able to bypass the first-pass effects and retain the efficacy of the loaded drug.

The maximal plasma concentration was high over a short period for gel formulations administered intranasally when compared with the free drug. The nasal residence time was prolonged, with enhanced drug bioavailability and increased brain uptake revealing the potential of the formulations for the treatment of neurological disorders and respiratory diseases. *In situ* gels were found to be suitable for the delivery of anticancer and antidiabetic drugs. *In vivo* studies revealed that the formulations targeted tumor only, suggesting that concentrations below the drug's toxic plasma concentrations would not result in systemic toxicity. *In situ* gels loaded with antidiabetic drugs revealed lower blood glucose levels *in vivo* after subcutaneous injection. In wound dressing, *in situ* gels were characterized by good antibacterial activity and enhanced wound healing and skin regeneration. The sustained and prolonged release of the drug and their good stability, biodegradability, and biocompatibility characteristics make *in situ* gel formulations potential therapeutics for the treatment of a wide range of diseases.

Acknowledgment

The financial assistance of the Medical Research Council and National Research Foundation, South Africa, toward this research is hereby acknowledged.

REFERENCES

1. Peppas, N., and Langer, R. 1994. New challenges in biomaterials. *Science* 263:1715–1720.
2. Gaudana, R., Ananthula, H.K., Parenky, A., and Mitra, A.K. 2010. Ocular drug delivery. *AAPS J* 12:348–360.
3. Patel, A., Cholkar, K., Agrahari, V., and Mitra, A.K. 2013. Ocular drug delivery systems: an overview. *World J Pharmacol* 2:47–64.
4. Ahmed, T.A., and Aljaeid, B.M. 2017. A potential in-situ gel formulation loaded with novel fabricated poly (lactide-co-glycolide) nanoparticles for enhancing and sustaining the ophthalmic delivery of ketoconazole. *Int J Nanomedicine* 12:1863–1875.

5. De Logu, A., Fadda, A.M., Anchisi, C., et al. 1997. Effects of in-vitro activity of miconazole and ketoconazole in phospholipid formulations. *J Antimicrob Chemother* 40:889–893.
6. Makwana, S.B., Patel, V.A., and Parmar, S.J. 2016. Development and characterization of in-situ gel for ophthalmic formulation containing ciprofloxacin hydrochloride. *Res Pharm Sci* 6:1–6.
7. Bhowmik, M., Das, S., Chattopadhyay, D., and Ghosh, L.K. 2011. Study of thermo-sensitive in-situ gels for ocular delivery. *Sci Pharm* 79:351–358.
8. Puranik, K.M., and Tagalpallewar, A.A. 2015. Voriconazole in-situ gel for ocular drug delivery. *SOJ Pharm Pharm Sci* 2:1–10.
9. Kotreka, U.K., Davis, V.L., and Adeyeye, M.C. 2017. Development of topical ophthalmic in-situ gel-forming estradiol delivery system intended for the prevention of age-related cataracts. *PloS one* 12:e0172306.
10. Mohan, E.C., Kandukuri, J.M., and Allenki, V. 2009. Preparation and evaluation of in-situ-gels for ocular drug delivery. *J Pharm Res* 2:1089–1094.
11. Patil, S., Kadam, A., Bandgar, S., and Patil, S. 2015. Formulation and evaluation of an in-situ gel for ocular drug delivery of anticonjunctival drug. *Cellul Chem Technol* 49:35–40.
12. Cao, Y., Zhang, C., Shen, W., Cheng, Z., Yu, L.L., and Ping, Q. 2007. Poly (N-isopropylacrylamide)–chitosan as thermosensitive in-situ gel-forming system for ocular drug delivery. *J Control Release* 120:186–194.
13. Gupta, H., Velpandian, T., and Jain, S. 2010. Ion-and pH-activated novel in-situ gel system for sustained ocular drug delivery. *J Drug Target* 18:499–505.
14. Gratieri, T., Gelfuso, G.M., Rocha, E.M., Sarmento, V.H., de Freitas, O., and Lopez, R.F. 2010. A poloxamer/chitosan in-situ forming gel with prolonged retention time for ocular delivery. *Euro J Pharm Biopharm* 75:186–193.
15. Nagarwal, R.C., Kumar, R., Dhanawat, M., and Pandit, J.K. 2011. Modified PLA nano in-situ gel: a potential ophthalmic drug delivery system. *Colloids Surf B Biointerfaces* 86:28–34.
16. Gao, Y., Sun, Y., Ren, F., and Gao, S. 2010. PLGA–PEG–PLGA hydrogel for ocular drug delivery of dexamethasone acetate. *Drug Dev Ind Pharm* 36:1131–1138.
17. Liu, Y., Liu, J., Zhang, X., Zhang, R., Huang, Y., and Wu, C. 2010. In-situ gelling gelrite/alginate formulations as vehicles for ophthalmic drug delivery. *AAPS PharmSciTech* 11:610–620.
18. Mandal, S., Thimmasetty, M.K., Prabhushankar, G.L., and Geetha, M.S. 2010. Formulation and evaluation of an in-situ gel-forming ophthalmic formulation of moxifloxacin hydrochloride. *Int J Pharm Investig* 2:78–82.
19. Yu, J., Xu, X., Yao, F., et al. 2014. In-situ covalently cross-linked PEG hydrogel for ocular drug delivery applications. *Int J Pharm* 470:151–157.
20. Morsi, N., Ibrahim, M., Refai, H., and El Sorogy, H. 2017. Nanoemulsion-based electrolyte triggered in-situ gel for ocular delivery of acetazolamide. *Eur J Pharm Sci* 104:302–314.
21. Upadhayay, P., Kumar, M., and Pathak, K. 2016. Norfloxacin loaded pH triggered nanoparticulate in-situ gel for extraocular bacterial infections: optimization, ocular irritancy and corneal toxicity. *Iran J Pharm Res* 15(1):3–22.
22. Gadad, A.P., Wadklar, P.D., Dandghi, P., and Patil, A. 2016. Thermosensitive in-situ gel for ocular delivery of lomefloxacin. *Indian J Pharm Educ Res* 50:S96–S105.
23. Krishna, S.V., Ashok, V., and Chatterjee, A. 2012. A review on vaginal drug delivery systems. *Int J Biol Pharm Appl Sci* 1:152–167.
24. Hussain, A., and Ahsan, F. 2005. The vagina as a route for systemic drug delivery. *J Control Release* 103:301–313.
25. Rençber, S., Karavana, S.Y., Şenyiğit, Z.A., Eraç, B., Limoncu, M.H., and Baloğlu, E. 2017. Mucoadhesive in-situ gel formulation for vaginal delivery of clotrimazole: formulation, preparation, and in vitro/in-vivoevaluation. *Pharm Dev Technol* 22:551–561.
26. Taksande, J.B., Bhoyar, V.S., Trivedi, R.V., and Umekar, M.J. 2013. Development and evaluation in-situ gel formulation of Clindamycin HCl for vaginal application. *Pharm Lett* 5:364–369.
27. Karavana, S.Y., Rençbe, S., Şenyiğit, Z.A., and Baloğlu, E. 2012. A new in-situ gel formulation of Itraconazole for vaginal administration. *Pharmacol Pharm* 3:417–426.
28. Hu, X., Hu, R-F., and Bai, Z-W. 2013. Optimization and evaluation of a pH-sensitive in-situ gel of nystatin for vaginal delivery system. *Chinese J New Drugs* 22:1812–1818.

29. Patel, P., and Patel, P. 2015. Formulation and evaluation of clindamycin HCL in-situ gel for vaginal application. *Int J Pharma Investig* 5:50–56.
30. Sen, J., Pillai, S., Gopkumar, P., and Sridevi, G. 2014. Formulation and evaluation of muco-adhesive in-situ gels for site-specific delivery of clomatrizole. *Res Rev: J Pharm Nanotechnol* 2:8.
31. Tuğcu-Demiröz, F. 2017. Development of in-situ poloxamer-chitosan hydrogels for vaginal drug delivery of benzydamine hydrochloride: textural, mucoadhesive and in-vitrorelease properties. *Marmara Pharm J* 21:762–770.
32. Glavas-Dodov, M., Simonoska-Crcarevska, M., Raicki, R.S., Sibinovska, N., Mladenovska, K., and Zafirovska-Gapkovska, A. 2014. Bioinspired bioartifical polymer hybrid composites for propolis vaginal delivery II: formulation and characterization. *Macedonian Pharm Bull* 60:57–65.
33. Akhani, J.R., and Modi, C.D. 2014. Formulation development and evaluation of in-situ gel for vaginal drug delivery of anti-fungal drug. *Pharma Sci Monit* 5:343–364.
34. Deshkar, S.S., Patil, A.T., and Poddar, S.S. 2015. Development of a thermosensitive gel of fluconazole for vaginal candidiasis. *Int J Pharm Pharm Sci* 8:391–398.
35. Shabaan, O.M., Abbas, A.M., Fetih, G.N., et al. 2015. Once daily in-situ forming versus twice-daily conventional metronidazole vaginal gels for treatment of bacterial vaginosis: a randomized controlled trial. *J Genit Syst Disord* 4: 5.
36. Ibrahim, E.S., Ismail, S., Fetih, G., Shaaban, O., Hassanein, K., and Abdellah, N. 2012. Development and characterization of thermosensitive pluronic-based metronidazole in-situ gelling formulations for vaginal application. *Acta Pharm* 62:59–70.
37. Şenyiğit, Z.A., Karavana, S.Y., Eraç, B., Gürsel, Ö., Limoncu, M.H., and Baloğlu, E. 2014. Evaluation of chitosan based vaginal bioadhesive gel formulations for antifungal drugs. *Acta Pharm* 64:139–156.
38. Liu, Y., Yang, F., Feng, L., Yang, L., Chen, L., Wei, G., and Lu, W. 2017. In-vivo retention of poloxamer-based in-situ hydrogels for vaginal application in mouse and rat models. *Acta Pharm Sin B* 7:502–509.
39. Zhou, Q., Zhong, L., Wei, X., Dou, W., Chou, G., Wang, Z. 2013. Baicalein and hydroxypropyl-γ-cyclodextrin complex in poloxamer thermal sensitive hydrogel for vaginal administration. *Int J Pharm* 454:125–134.
40. Tuğcu-Demiröz, F., Acartürk, F., and Erdoğan, D. 2013. Development of long-acting bioadhesive vaginal gels of oxybutynin: formulation, in-vitro and in-vivo evaluations. *Int J Pharm* 457:25–39.
41. Lu, C., Liu, M., Fu, H., et al. 2015. Novel thermosensitive in-situ gel based on poloxamer for uterus delivery. *Eur J Pharm Sci* 77:24–28.
42. Date, A.A., Shibata, A., Goede, M., et al. 2012. Development and evaluation of a thermosensitive vaginal gel containing raltegravir+ efavirenz loaded nanoparticles for HIV prophylaxis. *Antiviral Res* 96(3):430–436.
43. Batchelor, H. 2014. Rectal drug delivery. In: *Pediatric Formulations*, ed. Bar-Shalom D., Rose K., 303–310. AAPS Advances in the Pharmaceutical Sciences Series, Vol 11, Springer, New York.
44. De Boer, A.G., De Leede, L.G., and Breimer, D.D. 1984. Drug absorption by sublingual and rectal routes. *Br J Anaesth* 56:69–82.
45. Garg, S., and Tukker, J.J. Rectal and vaginal drug delivery. https://clinicalgate.com/rectal-and-vaginal-drug-delivery/ (accessed 03 July 2018).
46. Yuan, Y., Cui, Y., Zhang, L., et al. 2012. Thermosensitive and mucoadhesive in-situ gel based on poloxamer as new carrier for rectal administration of nimesulide. *Int J Pharm* 430:114–119.
47. Ramadass, S.K., Perumal, S., Jabaris, S.L., and Madhan, B. 2013. Preparation and evaluation of mesalamine collagen in-situ rectal gel: a novel therapeutic approach for treating ulcerative colitis. *Eur J Pharm Sci* 48:104–110.
48. Xu, J., Tam, M., Samaei, S., et al. 2017. Mucoadhesive chitosan hydrogels as rectal drug delivery vessels to treat ulcerative colitis. *Acta Biomater* 48:247–57.
49. Moawad, F.A., Ali, A.A., and Salem, H.F. 2017. Nanotransfersomes-loaded thermosensitive in-situ gel as a rectal delivery system of tizanidine HCl: preparation, in-vitroand in-vivoperformance. *Drug Deliv* 24:252–260.
50. Aderibigbe, B.A., and Naki, T. 2018. Design and efficacy of nanogels formulations for intranasal administration. *Molecules* 23: 21.
51. Patel, Z., Patel, B., Patel, S., and Pardeshi, C. 2012. Nose to brain targeted drug delivery bypassing the blood-brain barrier: an overview. *Drug Invent Today* 4:610–615.

52. Pardeshi, C.V., and Belgamwar, V.S. 2013. Direct nose to brain drug delivery via integrated nerve pathways bypassing the blood–brain barrier: an excellent platform for brain targeting. *Expert Opin Drug Deliv* 10:957–972.
53. Ghori, M.U., Mahdi, M.H., Smith, A.M., and Conway, B.R. 2015. Nasal drug delivery systems: an overview. *Am J Pharmacol Sci* 3:110–119.
54. Shah, R.A., Mehta, M.R., Patel, D.M., and Patel, C.N. 2014. Design and optimization of mucoadhesive nasal in-situ gel containing sodium cromoglycate using factorial design. *Asian J Pharm* 5:65–74.
55. Gu, F., Ma, W., Meng, G., Wu, C., and Wang, Y. 2016. Preparation and in-vivo evaluation of gel-based nasal delivery system for risperidone. *Acta Pharm* 66:555–562.
56. Galgatte, U.C., Kumbhar, A.B., and Chaudhari, P.D. 2014. Development of in-situ gel for nasal delivery: design, optimization, in-vitro and in-vivo evaluation. *Drug Deliv* 21:62–73.
57. Jagdale, S., Shewale, N., and Kuchekar, B.S. 2016. Optimization of thermoreversible in-situ nasal gel of timolol maleate. *Scientifica* 2016:11.
58. Wang, Y., Jiang, S., Wang, H., and Bie, H. 2017. A mucoadhesive, thermoreversible in-situ nasal gel of geniposide for neurodegenerative diseases. *PloS one* 12:e0189478.
59. Gowda, D.V., Tanuja, D., Khan, M.S., Desai, J., and Shivakumar, H.G. 2011. Formulation and evaluation of in-situ gel of diltiazem hydrochloride for nasal delivery. *Pharm Lett* 3:371–381.
60. Barakat, S.S., Nasr, M., Ahmed, R.F., Badawy, S.S., and Mansour, S. 2017. Intranasally administered in-situ gelling nanocomposite system of dimenhydrinate: preparation, characterization and pharmacodynamic applicability in chemotherapy induced emesis model. *Sci Rep* 7:9910.
61. Ravi, P.R., Aditya, N., Patil, S., and Cherian, L. 2015. Nasal in-situ gels for delivery of rasagiline mesylate: improvement in bioavailability and brain localization. *Drug Deliv* 22:903–910.
62. Qian, S., Wong, Y.C., and Zuo, Z. 2014. Development, characterization and application of in-situ gel systems for intranasal delivery of tacrine. *Int J Pharm* 468:272–282.
63. Khan, S., Patil, K., Bobade, N., Yeole, P., and Gaikwad, R. 2010. Formulation of intranasal mucoadhesive temperature-mediated in-situ gel containing ropinirole and evaluation of brain targeting efficiency in rats. *J Drug Target* 18:223–234.
64. Cai, Z., Song, X, Sun, F., Yang, Z., Hou, S., and Liu, Z. 2011. Formulation and evaluation of in-situ gelling systems for intranasal administration of gastrodin. *AAPS PharmSciTech* 12:1102–1109.
65. Mahajan, H.S., and Gattani, S. 2010. In-situ gels of metoclopramide hydrochloride for intranasal delivery: in-vitro evaluation and in-vivo pharmacokinetic study in rabbits. *Drug Deliv* 17:19–27.
66. Sharma, A., Saxena, R., Dubey, A., Yadav, R., and Yadav, S.K. Development and characterization of in-situ gel system for nasal delivery of levodopa. *Rhinology and Otology* March 18–20, 2015 Dubai.
67. Pathan, I.B., Nirkhe, S.R., and Bairagi, A. 2015. In-situ gel based on gellan gum as new carrier for nasal to brain delivery of venlafaxine hydrochloride: In-vitroevaluation and in-vivostudy. *J Chem Pharm Res* 7:324–331.
68. Altuntaş, E., Yener, G., Doğan, R., Aksoy, F., Şerif Aydın, M., and Karataş, E. 2018. Effects of a thermosensitive in-situ gel containing mometasone furoate on a rat allergic rhinitis model. *Am J Rhinol Allergy* 32:132–138.
69. Mali, K.K., Dhawale, S.C., Dias, R.J., Havaldar, V.D., Ghorpade, V.S., and Salunkhe, N.H. 2015. Nasal mucoadhesive in-situ gel of granisetron hydrochloride using natural polymers. *J App Pharm Sci* 5:084–093.
70. Tejaswini, M., and Devi, A.S. 2017. Formulation and evaluation of nasal in-situ gel of phenylephrine hydrochloride. *Int J Drug Res Technol* 6:64–78.
71. Fatouh, A.M., Elshafeey, A.H., and Abdelbary, A. 2017. Agomelatine-based in-situ gels for brain targeting via the nasal route: statistical optimization, in vitro, and in-vivo evaluation. *Drug Deliv* 24:1077–1085.
72. Kaur, P., Garg, T., Vaidya, B., Prakash, A., Rath, G., and Goyal, A.K. 2015. Brain delivery of intranasal in-situ gel of nanoparticulated polymeric carriers containing antidepressant drug: behavioral and biochemical assessment. *J Drug Target* 23:275–286.
73. Singh, R.M., Kumar, A., and Pathak, K. 2013. Thermally triggered mucoadhesive in-situ gel of loratadine: β-cyclodextrin complex for nasal delivery. *AAPS PharmSciTech* 14:412–424.
74. Perez, A.P., Mundiña-Weilenmann, C., Romero, E.L., and Morilla, M.J. 2012. Increased brain radioactivity by intranasal 32P-labeled siRNA dendriplexes within in-situ-forming mucoadhesive gels. *Int J Nanomedicine* 7:1373–1385.

75. Salunke, S.R., and Patil, S.B. 2016. Ion activated in-situ gel of gellan gum containing salbutamol sulphate for nasal administration. *Int J Biol Macromol* 87:41–47.
76. Li, X., Du, L., Chen, X., et al. 2015. Nasal delivery of analgesic ketorolac tromethamine thermo-and ion-sensitive in-situ hydrogels. *Int J Pharm* 489:252–260.
77. Cho, H.J, Balakrishnan, P., Park, E.K., et al. 2011. Poloxamer/cyclodextrin/chitosan-based thermoreversible gel for intranasal delivery of fexofenadine hydrochloride. *J Pharm Sci* 100:681–691.
78. Rao, M., Agrawal, D.K., and Shirsath, C. 2017. Thermoreversible mucoadhesive in-situ nasal gel for treatment of Parkinson's disease. *Drug Dev Ind Pharm* 43:142–150.
79. Shen, N., Hu, J., Zhang, L., et al. 2012. Doxorubicin-loaded zein in-situ gel for interstitial chemotherapy of colorectal cancer. *Acta Pharma Sin B* 2:610–614.
80. Wu, W., Chen, H., Shan, F., et al. 2014. A novel doxorubicin-loaded in-situ forming gel based high concentration of phospholipid for intratumoral drug delivery. *Mol Pharm* 11:3378–3385.
81. Luo, J.W., Zhang, T., Zhang, Q., et al. 2016. A novel injectable phospholipid gel co-loaded with doxorubicin and bromotetrandrine for resistant breast cancer treatment by intratumoral injection. *Colloids Surf B Biointerfaces* 140:538–547.
82. Yang, L., Song, X., Gong, T., et al. 2018. Enhanced anti-tumor and anti-metastasis efficacy against breast cancer with an intratumoral injectable phospholipids-based phase separation gel co-loaded with 5-fluotouracil and magnesium oxide by neutralizing acidic microenvironment. *Int J Pharm* 154:181–189.
83. Kwon, D.Y., Kwon, J.S., Park, J.H., et al. 2016. Synergistic anti-tumor activity through combinational intratumoral injection of an in-situ injectable drug depot. *Biomaterials* 85:232–245.
84. Wan, J., Geng, S., Zhao, H., et al. 2016. Doxorubicin-induced co-assembling nanomedicines with temperature-sensitive acidic polymer and their in-situ-forming hydrogels for intratumoral administration. *J Control Release* 235:328–336.
85. Chen, T., Gong, T., Zhao, T., et al. 2017. Paclitaxel loaded phospholipid-based gel as a drug delivery system for local treatment of glioma. *Int J Pharm* 528:127–132.
86. Fong, Y.T., Chen, C.H., and Chen, J.P. 2017. Intratumoral delivery of doxorubicin on folate-conjugated graphene oxide by in-situ forming thermo-sensitive hydrogel for breast cancer therapy. *Nanomaterials* 7:24.
87. Li, L., Gu, J., Zhang, J., et al. 2015. Injectable and biodegradable pH-responsive hydrogels for localized and sustained treatment of human fibrosarcoma. *ACS Appl Mater Interfaces* 7:8033–8040.
88. Ning, P., Lü, S., Bai, X., et al. 2018. High encapsulation and localized delivery of curcumin from an injectable hydrogel. *Mater Sci Eng C* 83:121–129.
89. Zhang, T., Tang, Y., Zhang, W., et al. 2018. Sustained drug release and cancer treatment by an injectable and biodegradable cyanoacrylate-based local drug delivery system. *J Mater Chem B* 6:1216–1225.
90. Lee, J.Y., Kim, K.S., Kang, Y.M., et al. 2010. In-vivo efficacy of paclitaxel-loaded injectable in-situ-forming gel against subcutaneous tumor growth. *Int J Pharm* 392:51–56.
91. Kang, Y.M., Kim, G.H., Kim, J.I., et al. 2011. In-vivo efficacy of an intratumorally injected in-situ-forming doxorubicin/poly (ethylene glycol)-b-polycaprolactone diblock copolymer. *Biomaterials* 32:4556–4564.
92. Hu, M., Zhang, Y., and Xiang, N. 2016. Long-acting phospholipid gel of exenatide for long-term therapy of type II diabetes. *Pharm Res* 33:1318–1326.
93. Oh, K.S., Kim, J.Y., Yoon, B.D., et al. 2014. Sol–gel transition of nanoparticles/polymer mixtures for sustained delivery of exenatide to treat type 2 diabetes mellitus. *Eur J Pharm Biopharm* 88:664–669.
94. Xiang, N., Zhou, X., He, X., et al. 2016. An injectable gel platform for the prolonged therapeutic effect of pitavastatin in the management of hyperlipidemia. *J Pharm Sci* 105:1148–1155.
95. Ahmed, T.A., Alharby, Y.A., El-Helw, A.R., Hosny, K.M., and El-Say, K.M. 2016. Depot injectable atorvastatin biodegradable in-situ gel: development, optimization, in vitro, and in-vivo evaluation. *Drug Design Dev Ther* 10:405–415.
96. Lee, Y., Bae, J.W., Lee, J.W., Suh, W., and Park, K.D. 2014. Enzyme-catalyzed in-situ forming gelatin hydrogels as bioactive wound dressings: effects of fibroblast delivery on wound healing efficacy. *J Mater Chem B* 2:7712–7718.
97. Basha, M., AbouSamra, M.M., Awad, G.A., and Mansy, S.S. 2018. A potential antibacterial wound dressing of cefadroxil chitosan nanoparticles in-situ gel: fabrication, in-vitrooptimization and in-vivo evaluation. *Int J Pharm* 544:129–140.
98. Li, X., Fan, R., Tong, A., et al. 2015. In-situ gel-forming AP-57 peptide delivery system for cutaneous wound healing. *Int J Pharm* 495:560–571.

99. Du, L., Tong, L., Jin, Y., et al. 2012. A multifunctional in-situ–forming hydrogel for wound healing. *Wound Repair Regen* 20:904–910.
100. Hoque, J., Prakash, R.G., Paramanandham, K., Shome, B.R., and Haldar, J. 2017. Biocompatible injectable hydrogel with potent wound healing and antibacterial properties. *Mol Pharm* 14:1218–1230.
101. Lih, E., Lee, J.S., Park, K.M., and Park, K.D. 2012. Rapidly curable chitosan–PEG hydrogels as tissue adhesives for hemostasis and wound healing. *Acta Biomater* 8:3261–3269.
102. Zhao, Y., Liu, J.G., Chen, W.M., and Yu, A.X. 2018. Efficacy of thermosensitive chitosan/β-glycerophosphate hydrogel loaded with β-cyclodextrincurcumin for the treatment of cutaneous wound infection in rats. *Exp Ther Med* 15:1304–1313.
103. Tran, N.Q., Joung, Y.K., Lih, E., and Park, K.D. 2011. In-situ forming and rutin-releasing chitosan hydrogels as injectable dressings for dermal wound healing. *Biomacromolecules* 12:2872–2880.
104. Wu, J., Zheng, Y., Song, W., et al. 2014. In-situ synthesis of silver-nanoparticles/bacterial cellulose composites for slow-released antimicrobial wound dressing. *Carbohydr Polym* 102:762–771.
105. Liu, S.Q., Yang, C., Huang, Y., et al. 2012. Antimicrobial and antifouling hydrogels formed in-situ from polycarbonate and poly (ethylene glycol) via Michael addition. *Adv Mater* 24:6484–6489.
106. De Cicco, F., Reverchon, E., Adami, R., et al. 2014. In-situ forming antibacterial dextran blend hydrogel for wound dressing: SAA technology vs. spray drying. *Carbohydr Polym* 101:1216–1224.
107. Lu, G., Ling, K., Zhao, P., et al. 2010. A novel in-situ-formed hydrogel wound dressing by the photo-cross-linking of a chitosan derivative. *Wound Repair Regen* 18:70–79.
108. Dong, Y., Hassan, W.U., Kennedy, R., et al. 2014. Performance of an in-situ formed bioactive hydrogel dressing from a PEG-based hyperbranched multifunctional copolymer. *Acta Biomater* 10:2076–2085.
109. Del Gaudio, P., De Cicco, F., Aquino, R.P., et al. 2015. Evaluation of in-situ injectable hydrogels as controlled release device for ANXA1 derived peptide in wound healing. *Carbohydr Polym* 115:629–635.
110. Nimal, T.R., Baranwal, G., Bavya, M.C., Biswas, R., and Jayakumar, R. 2016. Anti-staphylococcal activity of injectable nano tigecycline/chitosan-PRP composite hydrogel using drosophila melanogaster model for infectious wounds. *ACS Appl Mater Interfaces* 8:22074–22083.
111. Abuhelwa, A.Y., Williams, D.B., Upton, R.N., and Foster, D.J. 2017. Food, gastrointestinal pH, and models of oral drug absorption. *Eur J Pharm Biopharm* 112:234–248.
112. Wu, R.L., Zhao, C.S., Xie, J.W., Yi, S.L., Song, H.T., and He, Z.G. 2008. Preparation of in-situ gel systems for the oral delivery of ibuprofen and its pharmacokinetics study in beagle dogs. *Yao Xue Xue Bao* 43:956–962.
113. Itoh, K., Tsuruya, R., Shimoyama, T., et al. 2010. In-situ gelling xyloglucan/alginate liquid formulation for oral sustained drug delivery to dysphagic patients. *Drug Dev Ind Pharm* 36:449–455.
114. Xu, H., and Shi, M. 2014. A novel in-situ gel formulation of ranitidine for oral sustained delivery. *Biomol Ther* 22:161–165.
115. Pandya, K., Aggarwal, P., Dashora, A., et al. 2013. Formulation and evaluation of oral floatable in-situ gel of ranitedine hydrochloride. *J Drug Deliv Ther* 3:90–97.
116. Mei, L., Huang, X., Xie, Y., et al. 2017. An injectable in-situ gel with cubic and hexagonal nanostructures for local treatment of chronic periodontitis. *Drug Deliv* 24:1148–1158.
117. Nagendra, R., Pai, R.S., and Singh, G. 2014. Design and optimization of novel in-situ gel of mercaptopurine for sustained drug delivery. *Braz J Pharm Sci* 50:107–119.

6

Micellar Nanoparticles for Lung Cancer Drug and Gene Delivery

Mariana Magalhães, Francisco Veiga, Ana Figueiras, and Ana Cláudia Santos

CONTENTS

6.1 Introduction

The past few years have witnessed a breakthrough in nanomedicine, in which advances in nanotechnology have allowed the development of novel therapeutic strategies and subsequently, enabled limitations associated with conventional treatments to be overcome. In this regard, nanotechnology has aroused great interest, and it is increasingly exploited due to its unique appealing features to improve drug delivery and to provide targeted delivery (Wicki et al. 2015; Shi et al. 2017). These advantages arise on the basis of the use of nanocarriers (displaying a nanoscale size), which can be composed of a wide range of materials based on organic or inorganic compounds (such as, e.g., polymers, lipids, and proteins). In this sense, developed nanoparticles (NPs) have the capability to incorporate therapeutic drugs or genetic material, improving their biodistribution, *in vivo* stability, circulation time in the bloodstream, and controlled release, as well as inducing a more specific and targeted delivery, consequently leading to a more efficient and improved treatment response with fewer side effects and off-target risks (Bertrand et al. 2014; Wicki et al. 2015; Hare et al. 2017). Currently, some nanotherapies are already in use or in the clinical trials phase for a clear majority of diseases, in which cancer is one of the main targets (Shi et al. 2017). Therefore, novel nanotherapies are emerging as a promising approach to apply in fighting lung cancer in the face of the limitations and low success rates of the available treatments, which may be due to late diagnosis (Ettinger et al. 2012; Kalemkerian et al. 2013; Hsu and Shaw 2017).

Thus, in this chapter, the authors intend to present an update on nanotechnology approaches toward lung cancer using micellar NPs. First, a short overview of lung cancer and its biological barriers will be covered, followed by a description of the available micellar NPs as well as their encapsulation methods and advantages. Finally, this chapter will discuss the available therapeutic alternatives to improve lung cancer therapy.

6.2 Lung Cancer

Nowadays, lung cancer is considered the leading cause of death from cancer in both men and women around the world. This cancer has the highest incidence and mortality rates, being the major contributor to the increase in new cancer cases per year (Dela Cruz et al., 2011). According to the World Health Organization (WHO), lung cancer is responsible for 1.69 million deaths per year, presenting a 5 year survival rate around 15%. Since 1985, the number of new lung cancer cases has increased, with a higher percentage verified in developed countries rather than in developing countries. Also, another unsettling fact is that this significant increase of new lung cancer cases is observed in women (76%) relative to men (44%). All this epidemiological evidence reflects, in general, the fact that women started to smoke two decades later than men, as well as the excessive presence of environmental and air pollution in developed countries (Jemal et al. 2008; Dela Cruz et al., 2011; Torre et al. 2015). Concerning this, there are several risk factors associated with lung cancer development, such as tobacco smoking, air pollution, and exposure to environmental carcinogens (Wender et al. 2013; Hamra et al. 2014). As an example, in China, the increased incidence and growing number of new lung cancer cases are attributed to the extreme environmental pollution (Dela Cruz et al. 2011; Torre et al. 2015). Generally, however, cigarette smoking is the main risk associated with lung cancer development, thought to be responsible for 70% of diagnosed cases (Herbst et al., 2008; Dela Cruz et al., 2011). Moreover, the probability of developing lung cancer and the risk associated with cigarette smoking are dependent on various factors, such as age, number of years of smoking, exposure to smoke, cases of cancer in the family, and genetic predisposition. In this sense, several studies have shown direct correlations between lung cancer patients and their medical history or even their age and years of smoking, which explains the higher percentage of diagnosed patients of advanced age (approximately 70 years) (Dela Cruz et al., 2011; Wender et al. 2013).

According to the American Cancer Society, lung tumors can be divided into three major subtypes: non-small cell lung cancer (NSCLC), small cell lung cancer (SCLC), and lung carcinoid tumor (American Cancer Society 2018). NSCLC is the most common type of lung cancer, with an incidence rate of 85%, being subdivided into adenocarcinoma, large-cell carcinoma, and squamous cell carcinoma. According to WHO, adenocarcinoma represents 40% of NSCLC cases, being subcategorized into adenocarcinoma *in situ* (or preinvasive lesion), minimally invasive adenocarcinoma, and invasive adenocarcinoma (Herbst et al., 2008; Ettinger et al. 2012; Inamura 2017). This division is made according to the level of invasiveness of lung adenocarcinoma and tumor size extent evaluation: adenocarcinoma *in situ* presents a diameter of 3 cm or less, while minimally invasive adenocarcinoma is characterized by an invasion size of 5 cm or less and a diameter of $\leq$3 cm (Inamura 2017). Moreover, invasive adenocarcinoma (bronchioloalveolar adenocarcinoma) has different patterns, such as lepidic growth adenocarcinoma, which happens when a proliferation of tumor cells exists along the surface of the alveolar walls but without vascular or stromal invasion (Iwata 2016; Inamura 2017). Additionally, squamous cell adenocarcinoma and large-cell carcinoma are less common types, with incidence rates around 25–30% and 10–15%, respectively (American Cancer Society 2018). SCLC is defined as very aggressive and is a less common type of lung cancer, affecting 10–15% of diagnosed patients. Relative to NSCLC, this type of lung cancer has a higher and faster growth rate and an earlier development of metastases (Kalemkerian et al. 2013). Pathologically, SCLC develops from a malignant tumor in epithelial cells, consisting of small cells with unclear cell membrane, minimal cytoplasm, and imperceptible nucleoli. The majority of SCLC cases originate in the lungs, although a small percentage can arise from extrapulmonary sites (Zakowski 2003; Kalemkerian et al. 2013). In this regard, neuroendocrine tumors include SCLC and lung carcinoid tumor, which affect 5% of diagnosed patients and have their origin in the neuroendocrine system (American Cancer Society 2018).

All lung tumors are usually asymptomatic in the early stages of the disease, which is responsible for their detection at an advanced stage and limits the therapeutic options available, leading to low survival and treatment success rates. For this reason, nanomedicine and/or gene therapy appear as refreshing alternatives to conventional treatments, improving therapeutic efficacy and specificity with fewer off-target risks and side effects (Ettinger et al. 2012; Kalemkerian et al. 2013; Hsu and Shaw 2017). Therefore, knowledge of lung cancer biology and the tumor microenvironment are a major issue to take into consideration in designing a more efficient nanotherapy. In other words, lung cancer develops through several genetic and epigenetic mutations that induce the suppression or toxic gain of function of key regulatory genes involved in pathways related to cell proliferation, invasion, migration, and apoptosis (Cooper et al. 2013; Pendharkar et al., 2013; Wood et al. 2014; Roos et al., 2015; Guo et al. 2017). Hence, these biological factors are relevant to developing targeted therapy and overcoming the biochemical barriers to drug and gene delivery.

6.2.1 Challenges and Biochemical Obstacles to Drug and Gene Delivery

The successful delivery of a therapeutic agent (drug or genetic material) is directly dependent on the capability to overcome the biological barriers and reach the target organ, tissue, or cells. Thus, in a first step, the delivery of therapeutic agents is subjected to extracellular barriers comprising unspecific interactions with serum proteins, clearance by the mononuclear phagocyte system (MPS), sequestration by the reticulum endoplasmic system (RES) organs, and passage through the cell membrane (Blanco et al., 2015; Pereira et al. 2017; Magalhães et al., 2018b).

Next, the intracellular barriers appear, which consist of transport across the cell membrane and endosomal escape. Transport across the cell membrane and cellular internalization are usually performed through endocytic pathways, namely, clathrin- and caveolae-mediated endocytosis (particles under 250 nm) or micropinocytosis (particles above 500 nm). Subsequently, cell uptake will induce the development of membrane invaginations and the formation of endosomes (intracellular vesicles) fused with lysosomes. The acidic environment of these vesicles may potentiate the degradation of the therapeutic agent. Following endosomal escape, depending on the case, there also exist additional barriers, namely, intracellular trafficking and passage through the nuclear membrane (Jones et al. 2013; Blanco et al., 2015; Magalhães et al., 2018b).

Figure 6.1 is a representative scheme of the biochemical barriers and challenges to efficient drug delivery.

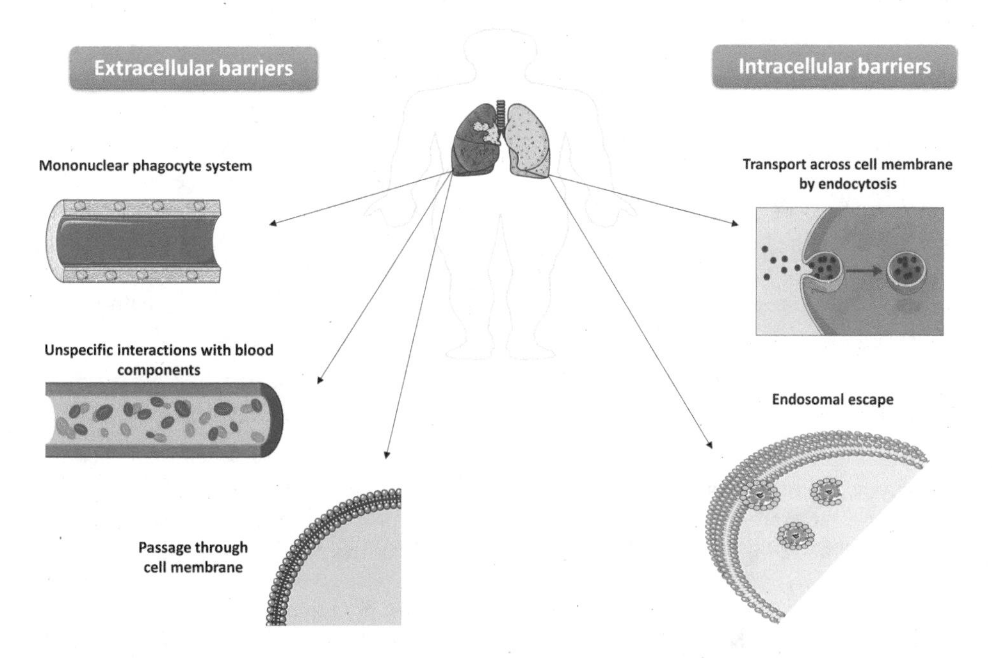

FIGURE 6.1 Framework of the various challenges and biochemical barriers (extracellular and intracellular) to an efficient drug and gene delivery.

6.3 Micellar Nanoparticles

6.3.1 Micellar Nanoparticles Developed against Lung Cancer

Micellar nanocarriers arise as an auspicious option to overcome the biochemical barriers mentioned before, increasing treatment efficacy and reducing random toxicity. Furthermore, these NPs have the capability to improve biocompatibility and biodistribution, to induce controlled drug release, and to protect the genetic material against nuclease degradation. Despite this, NPs also facilitate cell uptake and decrease the off-target risks commonly related to conventional treatments (Babu et al. 2013; Jin et al. 2014; Yin et al. 2014; Singh and Peer 2016). Polymeric micelles, solid lipid nanoparticles (SLNs), and liposomes are types of micellar NPs used in nanotechnology approaches for lung cancer therapy.

6.3.1.1 Polymeric Micelles

Polymeric micelles are small-size structures, with an optimum size between 10 and 100 nm, and usually composed of amphiphilic copolymers such as Pluronics®. These NPs, above the critical micelle concentration (CMC), have the capability, in an aqueous medium, to self-assemble into a micellar structure. These micellar systems present a hydrophobic inner core and a hydrophilic outer shell (Simões et al. 2015; Almeida et al. 2018). The hydrophobic inner core has suitable characteristics to incorporate and deliver poorly water-soluble drugs, besides the ability of these systems to be functionalized with cationic polymers and subsequently, to be able to bind and transport genetic material (Sun et al. 2011; Shen et al. 2012; Kim et al. 2014; Hao et al. 2015; Feldmann et al. 2017).

6.3.1.2 Solid Lipid Nanoparticles

SLNs are spherical nanocarriers with a solid lipid inner core composed of a solid lipid phase and a surfactant. These NPs have sizes between 40 and 1000 nm, presenting good biocompatibility and increased stability in aqueous medium (Weber et al., 2014; De Jesus, and Zuhorn 2015; Naseri et al., 2015). SLNs have competence to encapsulate a drug in the solid lipid matrix, protecting it against degradation and positively modeling the drug release profile. Besides, these spherical nanosystems also have been studied as delivery systems for genetic material through the introduction of polycations, which provide the ability to bind genetic material by electrostatic interactions (De Jesus, and Zuhorn 2015).

6.3.1.3 Liposomes

Liposomes are widely studied nanocarriers for both drug and gene delivery (Babu et al. 2013; Qu et al. 2014; Jiang et al. 2015). These bilayer phospholipid vesicles have the capacity to encapsulate drugs in the inner core or in the lipid bilayers. Additionally, the structure of cationic liposomes increases cellular uptake efficacy and genetic material incorporation through electrostatic interactions. These sphere-shaped nanocarriers present high biocompatibility and stability in physiological medium (Babu et al. 2013; Yin et al. 2014; Lee et al., 2016).

6.3.2 Methods of Encapsulation

The methods to prepare drug-loaded micellar NPs depend on each experimental goal. The physicochemical properties of the materials and the solubility of the drugs used should be taken into consideration, since these properties strongly influence efficient drug and gene delivery (Magalhães et al. 2014; Almeida et al. 2018; Magalhães et al., 2018b). Regarding this, for the preparation of drug-loaded polymeric micelles, the most commonly used methods are direct dissolution, dialysis, evaporation or film formation, microphase separation, and oil-in-water emulsion (Almeida et al. 2018). Meanwhile, to develop drug-loaded micellar lipid NPs, the conventional methods used are the thin lipid film, freezing and thawing technique, and other methods that use ethanol injections (Gubernator 2011). In the following, a brief description of the methods of encapsulation will be provided.

The direct dissolution technique relies on the dissolution of the drug and polymers in an aqueous solvent. The soluble drug is added to an excess of polymer solution, which is then subjected to moderate temperature or to mechanical agitation to induce the movement of the dissolved drug along the NP core (Sotoudegan et al. 2016; Cagel et al. 2017; Almeida et al. 2018).

In contrast to direct dissolution, dialysis is used for poorly water-soluble drugs. Therefore, for polymers with low solubility in aqueous medium, the dissolution of both polymer and drug is performed in an organic solvent. Consequently, the formation of micelles containing the drug is dependent on the solvent removal. Therefore, the chosen organic solvent should be miscible in water to enable performing dialysis of the drug and polymer mixture against deionized water. The process of drug encapsulation inside polymeric micelles is achieved while the organic phase is removed (Simões et al. 2015; Cagel et al. 2017; Almeida et al. 2018).

Additionally, to prepare poorly water-soluble drug–loaded polymeric micelles, the microphase separation method may be used. This technique involves the dissolution of the drug and the polymer in a volatile solvent miscible in water. Briefly, a mixture of drug and polymer dissolved in a volatile solvent is added dropwise into water under stirring. Subsequently, this leads to the spontaneous formation of drug-loaded polymeric micelles. Then, the volatile solvent is removed under lower pressure (Kedar et al. 2010; Mourya et al. 2011; Almeida et al. 2018).

Another relatively similar technique consists of the oil-in-water emulsion method. In this case, an organic solvent immiscible in water is gently added to an aqueous solution, containing the polymer and drug, under magnetic stirring. After this process, the formation of an emulsion occurs, in which the drug-loaded polymeric micelles are formed when the solvent is removed by evaporation (Gaucher et al. 2005; Ahmad et al. 2014; Almeida et al. 2018).

Finally, thin film formation is a method used for both drug-loaded polymeric and micellar lipid NPs. Basically, this technique involves the dissolution of the polymer or the lipid with the drug into an organic solvent. After this, the solvent is removed by evaporation, and a thin film is obtained, which is subsequently hydrated, leading to the formation of drug-loaded micellar NPs (Gubernator 2011; Monteiro et al. 2013; Simões et al. 2015; Cagel et al. 2017; Almeida et al. 2018).

Furthermore, drug-loaded lipid micelles can also be developed with the freezing and thawing technique. This consists of a solution of lipid (liposomes) and drug suspension being subjected to consecutive freezing and thawing processes. The aim of this method is to enhance the distance between the lipid bilayers to enable the transient development of holes by the ice crystals formed during the freezing process. Subsequently, this hole development provides increasing drug penetration and loading into the liposome structure (Gubernator 2011).

Lastly, there are other techniques based on ethanol injections to form lipid micelles (liposomes) containing the drug. Briefly, a solution composed of lipid and ethanol is rapidly injected into a solution containing an excess of drug. The injection is performed using a thin needle to create unilamellar liposomes, in which the diameter obtained is directly related to the composition and concentration of lipid used as well as the injection rate and temperature (Gubernator 2011).

In addition, it is important to emphasize that all the encapsulation methods described may present some problems associated with organic solvent removal, drug loading, and degree of encapsulation. Thus, it is worth mentioning once more the importance of the method chosen to achieve more stable drug-loaded NPs and consequently, increased therapeutic efficacy.

Besides, the encapsulation of genetic material in micellar NPs is usually performed by the simple method of complexation. In this method, cationic micellar nanocarriers bind the genetic material through electrostatic interactions between the negatively charged phosphate groups (P) of genes (DNA/RNA) and the positively charged amine groups (N) of polycations (Magalhães et al. 2014, 2018a, 2018b).

6.3.3 Advantages of Micellar Nanoparticles

Over the years, the constant progress in nanotechnology has brought several advantages and advances to medicine; that is, nanomedicine has arisen as a promising tool to improve treatment efficacy and reduce unspecific reactions and side effects. In this field, micellar nanocarriers, particularly liposomes and polymeric micelles, have been extensively studied and used in both drug and gene therapy approaches (Bozzuto and Molinari 2015; Landesman-Milo et al., 2015; Shi et al. 2017; Magalhães et al., 2018b).

Various advantages are attributed to NPs, such as their biocompatibility and biodegradable characteristics as well as their capability to encapsulate drugs (hydrophobic or hydrophilic) and genetic material (Bozzuto and Molinari 2015; Magalhães et al., 2018b). Moreover, NPs are able to protect from and prevent drug and gene degradation by natural processes, such as chemical inactivation or enzymatic degradation, maintaining their active and natural form. In addition, the hydrophilic outer shell of the micellar NPs provides high stability and minimizes the interaction with serum components, which increases the stability in the bloodstream (Bozzuto and Molinari 2015; Chitkara et al., 2016; Magalhães et al., 2018b). The small size of these NPs avoids recognition by the MPS, increasing the circulation time in the bloodstream (Jhaveri and Torchilin 2014; Movassaghian et al., 2015). These attributes lead to a superior biodistribution and accumulation of the therapeutic agent in the target tissue or cells (Shi et al. 2017; Magalhães et al., 2018b). Other positive outcomes of micellar NPs consist of the ability to induce a more efficient therapeutic effect, the application of lower drug doses, and the attainment of a sustained and controlled drug release profile. Thereby, these benefits are reported to promote a substantial decrease in the off-target risks and undesired side effects (to healthy tissue) (2015; Pereira et al. 2017; Magalhães et al., 2018b).

Also, the presence of cationic polymers or lipids in micellar NP formulations enhances cellular uptake because the cationic charge of the NPs is prone to interact with the negative charge of the cell membrane, facilitating drug and gene delivery into the target cells or tissue (Chitkara et al., 2016; Magalhães et al. 2018a, 2018b).

6.4 Therapeutic Approaches

6.4.1 Combined and Synergistic Therapies

The design of several therapeutic alternatives based on gene or drug delivery is presented nowadays as a promising alternative to tackle the problems associated with current treatments. However, in some cases, it was observed that a combination of types of therapy can bring more advantages and successful treatment results.

A combined approach involves the development of a therapy based on a gene therapy strategy combined with a chemotherapeutic approach targeting a specific type of cancer. In a synergistic approach, the situation is a little different; this involves the co-delivery of a drug and a therapeutic gene encapsulated inside the nanosystem or the co-delivery of two distinct drugs. This latter concept is more common (Kemp et al. 2016; Xue et al. 2016).

The representative scheme of Figure 6.2 is intended to sketch a combined and/or synergistic therapy.

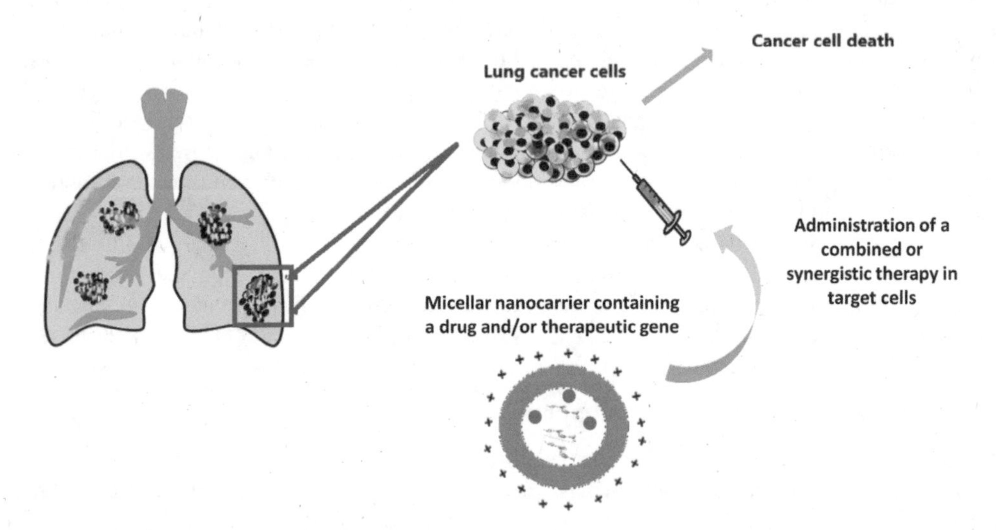

FIGURE 6.2 Schematic representation of a combined and synergistic therapeutic approach to lung cancer, namely, gene therapy combined with chemotherapy and a synergistic co-delivery of two different therapeutic agents.

Recently, some research groups have developed lung cancer therapies based on these combined and synergistic therapeutic approaches. An example is the work of Nascimento and co-workers, in which the authors combined a target RNA interference (RNAi)-based therapeutic with a chemotherapy strategy. The authors intended to develop an efficient and specific NP to deliver a small interference RNA (siRNA) encoding the mitotic arrest deficient-2 (Mad2) gene into lung cancer cells to induce the silencing of the Mad2 gene. The main purpose of this work was to knock down the Mad2 gene, which is an important gene of the mitotic checkpoint with a key role in the segregation of chromosomes, so as to promote an increase of the sensitivity of lung cancer cells to cisplatin, a platinum-based drug that acts on cell proliferation pathways by damaging DNA. In this study, a more pronounced antitumoral therapeutic effect was observed using the combined approach relative to gene therapy alone, supporting this as a promising therapeutic concept for lung cancer (Nascimento et al. 2017).

A different perspective involves the co-delivery of therapeutic agents, such as in Shen and colleagues' work. This group developed a synergistic therapy for lung cancer by using polymeric micelles with cationic characteristics to perform the co-delivery of a therapeutic gene and a chemotherapeutic drug. In this way, researchers developed a smart cationic micellar NP based on polymeric micelles to deliver, synergistically, a short hairpin RNA (shRNA) encoding the survivin gene and paclitaxel (PTX). The downregulation of the survivin gene promoted an enhancement of apoptosis-induced PTX, significantly increasing the drug's therapeutic effect in lung cancer cells (Shen et al. 2012).

A study published in 2018 also showed the success of a synergistic therapy against lung cancer. The work of Han and colleagues had as its main purpose to perform the co-delivery of two chemotherapeutic drugs. To fulfill this goal, these researchers developed polymeric micelles to encapsulate cyclosporin A and gefitinib. Thereby, after the co-delivery of both drugs, an increased therapeutic effect was revealed through a more pronounced inhibition of tumor growth when compared with the individual drug delivery (Han et al. 2018).

These examples strongly support the theory that combined and synergistic therapies are potential and auspicious options to consider in the engineering and development of novel and more efficient lung cancer therapies.

6.4.2 Strategies to Improve Anticancer Drug Effects

Lung tumors, like the majority of cancers, evidence multidrug resistance (MDR) against chemotherapeutic drugs. MDR is deeply related to lower treatment efficacy, leading to therapeutic strategies with several side effects and reduced success. In this regard, MDR is usually defined as the resistance developed by cancer cells to a distinct class of drugs. The mechanisms of drug resistance in cancer cells are mainly attributed to an overexpression of efflux drug transporters, decreasing the uptake of the drug by cancer cells.

A more promising strategy to overcome this problem and improve the antitumoral effect of drugs is based on combined therapies. Following the previous section, combined therapies enable the repair of key genes involved, for example, in cell proliferation and inhibition of apoptosis, providing a more efficient response to anticancer drugs. This approach is able to reduce MDR, avoiding the use of higher drug doses and longer treatments with reduced therapeutic effects and heavy undesired side effects (Shen et al. 2012; Chen et al. 2016; Han et al. 2018). Furthermore, the examples provided here consist of appropriate approaches to improve drug effects in combined and synergistic therapies.

Throughout, this chapter has also referred to the improvement of drug efficacy just by using micellar NPs. These nanocarriers show a remarkable ability to encapsulate a small amount of drug to transport and protect it until it reaches the target site. Arriving at the desired site, they are able to induce an improved therapeutic effect with lower doses and without promoting drug resistance, which occurs following continuous administrations of higher doses to achieve the same effect.

6.5 Conclusions

The substantial increase of new lung cancer cases worldwide and the growing death rate have raised the need to seek and develop novel therapeutic approaches to overcome the limitations associated with the available conventional treatments. In this regard, nanomedicine strategies stand out as promising and stimulating options to treat lung cancer.

Nanomedicine, precisely by the use of nanotechnology, makes use of nanocarriers, such as micellar NPs, including polymeric micelles, liposomes, and SLNs, to perform efficient drug and/or gene delivery. In the process of developing drug/gene-loaded NPs, it is very important to consider the physicochemical properties of the nanocarriers and the drug or gene as well as the method of encapsulation used, since all these factors have a vital influence on the performance of the nanosystems designed to be applied in lung cancer therapy.

Micellar NPs bring several recognized advantages, namely, the ability to protect the therapeutic agents against degradation, sequestration by RES organs, or clearance by MPS. Additionally, micellar NPs have an important role in improving drug biodistribution and bioavailability, subsequently inducing a much significantly higher therapeutic effect following the administration of lower drug doses. Therefore, these nanosystems have shown outstanding properties toward the achievement of a more efficient and specific treatment for lung cancer, avoiding undesired side effects and risks to healthy tissue. Additionally, the choice of combined therapies and the use of micellar NPs have been shown to improve drug effects and to reduce MDR in lung cancer, being, therefore, encouraging and emerging therapeutic alternatives to be applied in future clinical methods to treat lung cancer.

ACKNOWLEDGMENTS

This work received financial support from Portugal National Funds (FCT/MEC, Fundação para a Ciência e Tecnologia/Ministério da Educação e Ciência) through project UID/QUI/50006/2013, co-financed by European Union (FEDER under the Partnership Agreement PT2020). It was also supported by the grant FCT PTDC/CTM-BIO/1518/2014 from the Portuguese Foundation for Science and Technology (FCT) and the European Community Fund (FEDER) through the COMPETE2020 program. The authors acknowledge Fundação para a Ciência e a Tecnologia (FCT), Portuguese Agency for Scientific Research, for financial support through the Research Project POCI-01-0145-FEDER-016642.

REFERENCES

Ahmad, Zaheer, Afzal Shah, Muhammad Siddiq, and Heinz Bernhard Kraatz. 2014. "Polymeric Micelles as Drug Delivery Vehicles." *RSC Advances* 4 (33): 17028–38. doi:10.1039/c3ra47370h.

Almeida, M., M. Magalhães, F. Veiga, and A. Figueiras. 2018. "Poloxamers, Poloxamines and Polymeric Micelles: Definition, Structure and Therapeutic Applications in Cancer." *Journal of Polymer Research* 25 (1): 31. doi:10.1007/s10965-017-1426-x.

Amaral, Claudia, Mariana Magalhaes, Celia Cabral, Francisco Veiga, and Ana Figueiras. 2017. "Preparation and Characterization of Mixed Polymeric Micelles as a Versatile Strategy for Meloxicam Oral Administration." *Letters in Drug Design & Discovery* 14 (12): 1401–8. doi:10.2174/157018081466617 0505120728.

American Cancer Society. 2018. "American Cancer Society." Accessed September 3. https://www.cancer.org/cancer/lung-cancer.html.

Babu, Anish, Amanda K. Templeton, Anupama Munshi, and Rajagopal Ramesh. 2013. "Nanoparticle-Based Drug Delivery for Therapy of Lung Cancer: Progress and Challenges." *Hindawi* 2013: 1–12. doi:10.1155/2013/863951.

Bertrand, Nicolas, Jun Wu, Xiaoyang Xu, Nazila Kamaly, and Omid C. Farokhzad. 2014. "Cancer Nanotechnology: The Impact of Passive and Active Targeting in the Era of Modern Cancer Biology." *Advanced Drug Delivery Reviews* 66: 2–25. doi:.10.1016/j.addr.2013.11.009

Blanco, Elvin, Haifa Shen, and Mauro Ferrari. 2015. "Principles of Nanoparticle Design for Overcoming Biological Barriers to Drug Delivery." *Nature Biotechnology* 33 (9): 941–51. doi:10.1038/nbt.3330.

Bozzuto, Giuseppina, and Agnese Molinari. 2015. "Liposomes as Nanomedical Devices." *International Journal of Nanomedicine* 10: 975–99. doi:10.2147/IJN.S68861.

Cagel, Maximiliano, Fiorella C. Tesan, Ezequiel Bernabeu, Maria J. Salgueiro, Marcela B. Zubillaga, Marcela A. Moretton, and Diego A. Chiappetta. 2017. "Polymeric Mixed Micelles as Nanomedicines: Achievements and Perspectives." *European Journal of Pharmaceutics and Biopharmaceutics* 113: 211–28. doi:10.1016/j.ejpb.2016.12.019.

Chen, Zhaolin, Tianlu Shi, Lei Zhang, Pengli Zhu, and Mingying Deng. 2016. "Mammalian Drug Efflux Transporters of the ATP Binding Cassette (ABC) Family in Multidrug Resistance: A Review of the Past Decade." *Cancer Letters* 370 (1): 153–64. doi:10.1016/j.canlet.2015.10.010.

Chitkara, Deepak, Saurabh Singh, and Anupama Mittal. 2016. "Nanocarrier-Based Co-Delivery of Small Molecules and SiRNA/MiRNA for Treatment of Cancer." *Therapeutic Delivery* 7 (4): 245–55. doi:10.4155/tde-2015-0003.

Cooper, Wendy A., David C. L. Lam, Sandra A. O'Toole, and John D. Minna. 2013. "Molecular Biology of Lung Cancer." *Journal of Thoracic Disease* 5 (Suppl 5): S479–90. doi:10.3978/j.issn.2072-1439.2013.08.03.

De Jesus, Marcelo B., and Inge S. Zuhorn. 2015. "Solid Lipid Nanoparticles as Nucleic Acid Delivery System: Properties and Molecular Mechanisms." *Journal of Controlled Release* 201: 1–13. doi:10.1016/j.jconrel.2015.01.010.

Dela Cruz, Charles S., Lynn T. Tanoue, and Richard A. Matthay. 2011. "Lung Cancer: Epidemiology, Etiology, and Prevention." *Clinics in Chest Medicine* 32 (4): 605–44. doi:10.1016/j.ccm.2011.09.001.

Ettinger, David S., Wallace Akerley, Hossein Borghaei, Andrew C. Chang, Richard T. Cheney, Lucian R. Chirieac, Thomas A. D'Amico, et al. 2012. "Non–Small Cell Lung Cancer." *Journal of the National Comprehensive Cancer Network* 10 (10): 1236–71. doi:10.6004/jnccn.2012.0130.

Feldmann, Daniel P., Yuran Xie, Steven K. Jones, Dongyue Yu, Anna Moszczynska, and Olivia M. Merkel. 2017. "The Impact of Microfluidic Mixing of Triblock Micelleplexes on in Vitro/in Vivo Gene Silencing and Intracellular Trafficking." *Nanotechnology* 28: 224001.

Gaucher, Geneviève, Marie Hélène Dufresne, Vinayak P. Sant, Ning Kang, Dusica Maysinger, and Jean Christophe Leroux. 2005. "Block Copolymer Micelles: Preparation, Characterization and Application in Drug Delivery." *Journal of Controlled Release* 109 (1–3): 169–88. doi:10.1016/j.jconrel.2005.09.034.

Gubernator, Jerzy. 2011. "Active Methods of Drug Loading into Liposomes: Recent Strategies for Stable Drug Entrapment and Increased in Vivo Activity." *Expert Opinion on Drug Delivery* 8 (5): 565–80. doi:10.1517/17425247.2011.566552.

Guo, J. Y., H. S. Hsu, S. W. Tyan, F. Y. Li, J. Y. Shew, W. H. Lee, and J. Y. Chen. 2017. "Serglycin in Tumor Microenvironment Promotes Non-Small Cell Lung Cancer Aggressiveness in a CD44-Dependent Manner." *Oncogene* 36 (17): 2457–71. doi:10.1038/onc.2016.404.

Hamra, Ghassan B., Neela Guha, Aaron Cohen, Francine Laden, Ole Raaschou-Nielsen, Jonathan M. Samet, Paolo Vineis, et al. 2014. "Outdoor Particulate Matter Exposure and Lung Cancer: A Systematic Review and Meta-Analysis." *Environmental Health Perspectives* 122 (9): 906–11. doi:10.1289/ehp.1408092.

Han, Weidong, Linlin Shi, Lulu Ren, Liqian Zhou, Tongyu Li, Yiting Qiao, and Hangxiang Wang. 2018. "A Nanomedicine Approach Enables Co-Delivery of Cyclosporin A and Gefitinib to Potentiate the Therapeutic Efficacy in Drug-Resistant Lung Cancer." *Signal Transduction and Targeted Therapy* 3: 16. doi:10.1038/s41392-018-0019-4.

Hao, Shanhu, Ying Yan, Xue Ren, Ying Xu, Lanlan Chen, and Haibo Zhang. 2015. "Candesartan-Graft-Polyethyleneimine Cationic Micelles for Effective Co-Delivery of Drug and Gene in Anti-Angiogenic Lung Cancer Therapy." *Biotechnology and Bioprocess Engineering* 20 (3): 550–60. doi:10.1007/s12257-014-0858-y.

Hare, Jennifer I., Twan Lammers, Marianne B. Ashford, Sanyogitta Puri, Gert Storm, and Simon T. Barry. 2017. "Challenges and Strategies in Anti-Cancer Nanomedicine Development: An Industry Perspective." *Advanced Drug Delivery Reviews* 108: 25–38. doi:10.1016/j.addr.2016.04.025.

Herbst, Roy S., John V. Heymach, and Scott M. Lippman. 2008. "Molecular Origins of Cancer Lung Cancer." *New England Journal of Medicine* 359 (13): 1367–80. doi:10.1056/NEJMra0802714.

Hsu, Peggy P., and Alice T. Shaw. 2017. "Lung Cancer: A Wiley Genetic Opponent." *Cell* 169 (5): 777–79. doi:10.1016/j.cell.2017.05.001.

Inamura, Kentaro. 2017. "Lung Cancer: Understanding Its Molecular Pathology and the 2015 WHO Classification." *Frontiers in Oncology* 7 (August): 1–7. doi:10.3389/fonc.2017.00193.

Iwata, Hisashi. 2016. "Adenocarcinoma Containing Lepidic Growth." *Journal of Thoracic Disease* 8 (9): E1050–52. doi:10.21037/jtd.2016.08.78.

Jemal, A., R. Siegel, E. Ward, Y. Hao, J. Xu, T. Murray, and M. J. Thun. 2008. "Cancer Statistics, 2008." *CA: A Cancer Journal for Clinicians* 58 (2): 71–96. doi:10.3322/CA.2007.0010.

Jhaveri, Aditi M., and Vladimir P. Torchilin. 2014. "Multifunctional Polymeric Micelles for Delivery of Drugs and SiRNA." *Frontiers in Pharmacology* 5 (April): 1–26. doi:10.3389/fphar.2014.00077.

Jiang, Lei, Li Li, Xiaodan He, Qiangying Yi, Bin He, Jun Cao, Weisan Pan, and Zhongwei Gu. 2015. "Overcoming Drug-Resistant Lung Cancer by Paclitaxel Loaded Dual-Functional Liposomes With Mitochondria Targeting and PH-Response." *Biomaterials* 52 (1): 126–39. doi:10.1016/j.biomaterials.2015.02.004.

Jin, Lian, Xin Zeng, Ming Liu, Yan Deng, and Nongyue He. 2014. "Current Progress in Gene Delivery Technology Based on Chemical Methods and Nano-Carriers." *Theranostics* 4 (3): 240–55. doi:10.7150/thno.6914.

Jones, Charles H., Chih Kuang Chen, Anitha Ravikrishnan, Snehal Rane, and Blaine A. Pfeifer. 2013. "Overcoming Nonviral Gene Delivery Barriers: Perspective and Future." *Molecular Pharmaceutics* 10 (11): 4082–98. doi:10.1021/mp400467x.

Kalemkerian, G. P., W. Akerley, P. Bogner, H. Borghaei, L. Q. Chow, R. J. Downey, L. Gandhi, et al. 2013. "Small Cell Lung Cancer." *Journal of the National Comprehensive Cancer Network* 11: 78–98. doi:10.6004/JNCCN.2012.0130.

Kedar, Uttam, Prasanna Phutane, Supriya Shidhaye, and Vilasrao Kadam. 2010. "Advances in Polymeric Micelles for Drug Delivery and Tumor Targeting." *Nanomedicine: Nanotechnology, Biology, and Medicine* 6 (6): 714–29. doi:10.1016/j.nano.2010.05.005.

Kemp, Jessica A., Min Suk, Chan Yeong, and Young Jik. 2016. "'Combo' Nanomedicine : Co-Delivery of Multi-Modal Therapeutics for Efficient, Targeted, and Safe Cancer Therapy." *Advanced Drug Delivery Reviews* 98: 3–18. doi:10.1016/j.addr.2015.10.019.

Kim, Hyun Jin, Takehiko Ishii, Meng Zheng, Sumiyo Watanabe, Kazuko Toh, Yu Matsumoto, Nobuhiro Nishiyama, Kanjiro Miyata, and Kazunori Kataoka. 2014. "Multifunctional Polyion Complex Micelle Featuring Enhanced Stability, Targetability, and Endosome Escapability for Systemic SiRNA Delivery to Subcutaneous Model of Lung Cancer." *Drug Delivery and Translational Research* 4 (1): 50–60. doi:10.1007/s13346-013-0175-6.

Landesman-Milo, Dalit, Srinivas Ramishetti, and Dan Peer. 2015. "Nanomedicine as an Emerging Platform for Metastatic Lung Cancer Therapy." *Cancer and Metastasis Reviews* 34 (2): 291–301. doi:10.1007/s10555-015-9554-4.

Lee, Hung-Yen, Kamal A. Mohammed, and Najmunnisa Nasreen. 2016. "Nanoparticle-Based Targeted Gene Therapy for Lung Cancer." American Journal of Cancer Research 6 (5): 1118–34.

Magalhães, Mariana, Mauro Almeida, Elisiário Tavares-da-Silva, Fernanda M. F. Roleira, Carla Varela, Joana Jorge, Ana Cristina Gonçalves, et al. 2018a. "MiR-145-Loaded Micelleplexes as a Novel Therapeutic Strategy to Inhibit Proliferation and Migration of Osteosarcoma Cells." *European Journal of Pharmaceutical Sciences* 123: 28–42. doi:10.1016/j.ejps.2018.07.021.

Magalhães, Mariana, Dina Farinha, Maria Conceição Pedroso de Lima, and Henrique Faneca. 2014. "Increased Gene Delivery Efficiency and Specificity of a Lipid-Based Nanosystem Incorporating a Glycolipid." *International Journal of Nanomedicine* 9: 4979–89. doi:10.2147/IJN.S69822.

Magalhães, Mariana, Ana Figueiras, and Francisco Veiga. 2018b. "Smart Micelleplexes: An Overview of a Promising and Potential Nanocarrier for Alternative Therapies." In *Design and Development of New Nanocarriers*, edited by AM Grumezescu, 257–91. William Andrew Publishing. doi:https://doi.org/10.1016/B978-0-12-813627-0.00007-7.

Monteiro, Nelson, Albino Martins, Diana Ribeiro, Susana Faria, Nuno A. Fonseca, João N. Moreira, Rui L. Reis, and Nuno M. Neves. 2013. "On the Use of Dexamethasone-Loaded Liposomes to Induce the Osteogenic Differentiation of Human Mesenchymal Stem Cells." *Journal of Tissue Engineering and Regenerative Medicine* 9 (9): 1056–66. doi:10.1002/term.1817.

Mourya, V. K., Nazma Inamdar, R. B. Nawale, and S. S. Kulthe. 2011. "Polymeric Micelles: General Considerations and Their Applications." *Indian Journal of Pharmaceutical Education and Research* 45 (2): 128–38.

Movassaghian, Sara, Olivia M. Merkel, and Vladimir P. Torchilin. 2015. "Applications of Polymer Micelles for Imaging and Drug Delivery." *Wiley Interdisciplinary Reviews: Nanomedicine and Nanobiotechnology* 7 (5): 691–707. doi:10.1002/wnan.1332.

Nascimento, Ana Vanessa, Amit Singh, Hassan Bousbaa, Domingos Ferreira, Bruno Sarmento, and Mansoor M. Amiji. 2017. "Overcoming Cisplatin Resistance in Non-Small Cell Lung Cancer with Mad2 Silencing SiRNA Delivered Systemically Using EGFR-Targeted Chitosan Nanoparticles." *Acta Biomaterialia* 47: 71–80. doi:10.1016/j.actbio.2016.09.045.

Naseri, Neda, Hadi Valizadeh, and Parvin Zakeri-Milani. 2015. "Solid Lipid Nanoparticles and Nanostructured Lipid Carriers: Structure Preparation and Application." *Advanced Pharmaceutical Bulletin* 5 (3): 305–13. doi:10.15171/apb.2015.043.

Pendharkar, Dinesh, B. V. Ausekar, and Sumant Gupta. 2013. "Molecular Biology of Lung Cancer – A Review." *Indian Journal of Surgical Oncology* 4 (2): 120–24. doi:10.1007/s13193-013-0213-3.

Pereira, Patrícia, Maria Barreira, João A. Queiroz, Francisco Veiga, Fani Sousa, and Ana Figueiras. 2017. "Smart Micelleplexes as a New Therapeutic Approach for RNA Delivery." *Expert Opinion on Drug Delivery* 14 (3): 353–71. doi:10.1080/17425247.2016.1214567.

Qu, Mei Hua, Rui Fang Zeng, Shi Fang, Qiang Sheng Dai, He Ping Li, and Jian Ting Long. 2014. "Liposome-Based Co-Delivery of SiRNA and Docetaxel for the Synergistic Treatment of Lung Cancer." *International Journal of Pharmaceutics* 474 (1–2): 112–22. doi:10.1016/j.ijpharm.2014.08.019.

Roos, Wynand P., Adam D. Thomas, and Bernd Kaina. 2015. "DNA Damage and the Balance between Survival and Death in Cancer Biology." *Nature Reviews Cancer* 16 (December): 20. doi:10.1038/nrc.2015.2.

Shen, Jianan, Qi Yin, Lingli Chen, Zhiwen Zhang, and Yaping Li. 2012. "Co-Delivery of Paclitaxel and Survivin ShRNA by Pluronic P85-PEI/TPGS Complex Nanoparticles to Overcome Drug Resistance in Lung Cancer." *Biomaterials* 33 (33): 8613–24. doi:10.1016/j.biomaterials.2012.08.007.

Shi, Jinjun, Philip W. Kantoff, Richard Wooster, and Omid C. Farokhzad. 2017. "Cancer Nanomedicine: Progress, Challenges and Opportunities." *Nature Reviews Cancer* 17 (1): 20–37. doi:10.1038/nrc.2016.108.

Simões, Susana M. N., Ana R. Figueiras, Francisco Veiga, Angel Concheiro, and Carmen Alvarez-Lorenzo. 2015. "Polymeric Micelles for Oral Drug Administration Enabling Locoregional and Systemic Treatments." *Expert Opinion on Drug Delivery* 12 (2): 297–318. doi:10.1517/17425247.2015.960841.

Singh, Manu Smriti, and Dan Peer. 2016. "RNA Nanomedicines: The next Generation Drugs?" *Current Opinion in Biotechnology* 39: 28–34. doi:10.1016/j.copbio.2015.12.011.

Sotoudegan, Farzaneh, Mohsen Amini, Mehrdad Faizi, and Reza Aboofazeli. 2016. "Nimodipine-Loaded Pluronic® Block Copolymer Micelles: Preparation, Characterization, in-Vitro and in-Vivo Studies." *Iranian Journal of Pharmaceutical Research* 15 (4): 641–61.

Sun, Tian Meng, Jin Zhi Du, Yan Dan Yao, Cheng Qiong Mao, Shuang Dou, Song Yin Huang, Pei Zhuo Zhang, Kam W. Leong, Er Wei Song, and Jun Wang. 2011. "Simultaneous Delivery of SiRNA and Paclitaxel via a 'Two-in-One' Micelleplex Promotes Synergistic Tumor Suppression." *ACS Nano* 5 (2): 1483–94. doi:10.1021/nn103349h.

Torre, Lindsey A., Freddie Bray, Rebecca L. Siegel, Jacques Ferlay, Joannie Lortet-tieulent, and Ahmedin Jemal. 2015. "Global Cancer Statistics, 2012." *CA: A Cancer Journal for Clinicians* 65 (2): 87–108. doi:10.3322/caac.21262.

Weber, S., A. Zimmer, and J. Pardeike. 2014. "Solid Lipid Nanoparticles (SLN) and Nanostructured Lipid Carriers (NLC) for Pulmonary Application: A Review of the State of the Art." *European Journal of Pharmaceutics and Biopharmaceutics* 86 (1): 7–22. doi:10.1016/j.ejpb.2013.08.013.

Wender, Richard, Elizabeth T. H. Fontham, Ermilo Barrera, Graham A. Colditz, Timothy R. Church, David S. Ettinger, Christopher R. Flowers, et al. 2013. "American Cancer Society Lung Cancer Screening Guidelines." *CA: A Cancer Journal for Clinicians* 63 (2): 107–17. doi:10.3322/caac.21172.American.

Wicki, Andreas, Dominik Witzigmann, Vimalkumar Balasubramanian, and Jörg Huwyler. 2015. "Nanomedicine in Cancer Therapy: Challenges, Opportunities, and Clinical Applications." *Journal of Controlled Release* 200: 138–57. doi:10.1016/j.jconrel.2014.12.030.

Wood, Steven L., Maria Pernemalm, Philip A. Crosbie, and Anthony D. Whetton. 2014. "The Role of the Tumor-Microenvironment in Lung Cancer-Metastasis and Its Relationship to Potential Therapeutic Targets." *Cancer Treatment Reviews* 40 (4): 558–66. doi:10.1016/j.ctrv.2013.10.001.

Xue, Rue, Ho Lun, Hui Yi, June Young, and Xiao Yu. 2016. "Nanomedicine of Synergistic Drug Combinations for Cancer Therapy – Strategies and Perspectives." *Journal of Controlled Release* 240: 489–503. doi:10.1016/j.jconrel.2016.06.012.

Yin, Hao, Rosemary L. Kanasty, Ahmed A. Eltoukhy, Arturo J. Vegas, J. Robert Dorkin, and Daniel G. Anderson. 2014. "Non-Viral Vectors for Gene-Based Therapy." *Nature Reviews Genetics* 15 (July): 541–55.

Zakowski, Maureen F. 2003. "Pathology of Small Cell Carcinoma of the Lung." *Seminars in Oncology* 30 (1): 3–8. doi:10.1053/sonc.2003.50013.

7

Oral Delivery of Insulin by Hybrid Polymeric Hydrogel

Kalyani Prusty, Anuradha Biswal, Niladri Sarkar, and Sarat K. Swain

CONTENTS

7.1 Introduction

The discovery of insulin from the pancreas of dogs and its isolation in chemically pure crystalline form in 1923 by Nobel laureates Banting, Best, and McLeod marked a significant milestone in the evolution of biomedical research [1, 2]. The advancement of biotechnology led to the development of several forms of insulin from various sources to be used in diabetes therapeutics [3]. The conventionally available insulin formulations involve intermediate, short, and long-lasting suspensions. Subcutaneous injections have been used for insulin delivery for many decades. But, the low patient compliance for regular insulin dosage through injections has led to the hunt for oral delivery alternatives. Despite its convenience and higher patient compliance, successful oral administration of insulin is still under conflict due to the reported bioavailability of 0.1% in the harsh gastrointestinal (GI) environment [4–6]. The acidic and enzymatic degradation of insulin in the stomach and transport barriers in the small intestine have posed a great problem for effective oral delivery. Several shortcomings, such as pain, needle phobia, allergic reactions, hyperinsulinemia, inactivation by proteolytic enzymes in the GI tract, and low permeability through intestinal membranes, have made parenteral insulin administration prevalent until now [7]. However, parenteral delivery does not mimic the conventional dynamics of endogenous release of insulin and fails to achieve sustained glycemic control in diabetic patients [8]. At low concentrations (0.1 mM), insulin is a monomer and dimerizes in the pH range of 4 to 8. But, at neutral pH and in concentrations greater than 2 mM, a hexamer is formed, and the related state influences its rate of degradation [9]. This rate of degradation increases multiple times (approximately sixfold) in the presence of bile salts along with complete dissociation of insulin into its respective monomeric form [10]. The reduced internal bioavailability of insulin is further supplemented by cytochrome P450-3A4 and apically polarized efflux mediated by ATP-dependent P-glycoproteins. Currently, multiple insulin administration through daily injections is the prime mode of treatment adopted by diabetic patients [11, 12]. The oral route of insulin administration would provide an economical and beneficial alternative for these patients, as it replicates the physiological fate of insulin and imparts better glucose homeostasis [13]. But, several problems are created by the enzymatic degradation and poor absorbance of insulin in the GI tract if taken orally. Hence, an ideal drug delivery system is required to protect the drug from the harsh GI environment by prolonging its residence time in the intestine, increasing the penetration into mucosal epithelium reversibly. Furthermore, the delivery system must be safe following oral intake [14]. Moreover, the portal delivery of insulin can be attained by intestinal administration, which replicates endogenous insulin release. Evidently, from this viewpoint, designing an oral delivery system that provides substantial internal bioavailability of the drug would change the course of diabetic therapeutics. To evade the problems accompanying the parenteral administration of insulin, alternative approaches have been devised in the literature, which include polymeric micro/nanocarriers [15, 16], enzyme inhibitors/absorption enhancers [17], chemical modifications, lipid-based carriers (liposomes) [18], solid lipid nanoparticles, etc. [19]. A schematic representation of the behavior of insulin-loaded microgels during the delivery process in the GI tract is shown in [20] Scheme 1.

Previously, each of these developments encountered minimal success, and no oral insulin product is commercially available to date. But, advancements in research methods in the past decade have

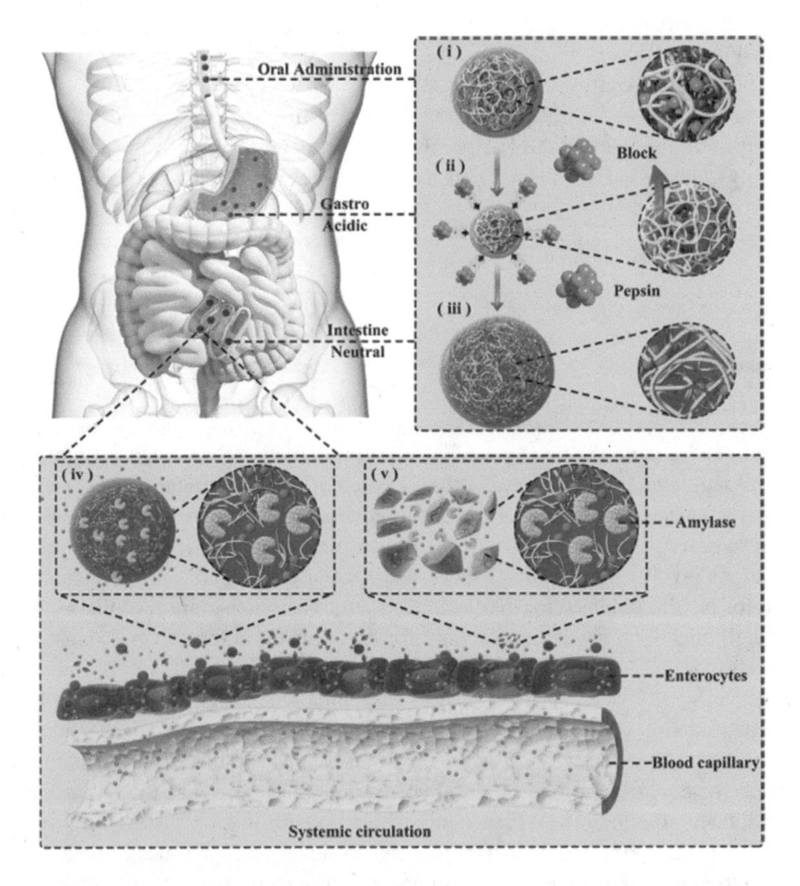

SCHEME 1 Schematic representation of the behavior of insulin-loaded microgels during the delivery process in the gastrointestinal tract. Reprinted with permission from Liu, L., Zhang, Y., Yu, S., Zhang, Z., He, C., Chen, X. pH-and amylose-responsive carboxymethyl starch/poly (2-isobutyl-acrylic acid) hybrid microgels as effective enteric carriers for oral insulin delivery. *Biomacromolecules*, 2018, 19, 2123–2136. Copyright 2018 American Chemical Society.

established the possibilities of designing polymeric drug delivery vehicles for the delivery of insulin orally. Despite the fact that some of the approaches have exhibited detrimental effects, such as damage to membrane barriers, irritation of intestinal mucosal membrane, etc., a few delivery devices, including co-polymeric hydrogels made from microparticles of poly (methacrylic acid) grafted with poly (ethylene glycol) (P(MAA-g-PEG)) have shown increased oral insulin absorption in animal experiments. These drug carriers exhibited 4.2% bioavailability of insulin along with the ability to protect the drug from enzymatic degradation and better mucoadhesive properties [21–23]. Sonaje et al. demonstrated the highest insulin bioavailability in diabetic animal models [24]. Santos and his co-workers developed an alginate–dextran sulfate (ADS) microgel using internal gelation/emulsification techniques to shield the insulin from gastrointestinal degradation and enhance its permeability through the intestinal membranes [25]. Nolan et al. designed poly (N-isopropylacrylamide) (PNIPAm) microgels and thin films of poly (N-isopropylacrylamide-co-acrylic acid) (poly (NIPAm-co-AA)) microgels and demonstrated the thermoresponsive insulin-releasing properties of these systems [26, 27]. Bai et al. used free radical polymerization techniques to design microgels from acrylate-grafted hydroxy propyl cellulose and poly (L-glutamic acid-2-hydroxyethyl methacrylate) for the oral delivery of insulin [28]. *In vitro* studies of the pH- and thermoresponsiveness of the microgels were done. Simulated environments were used to study the effective insulin-loading and releasing capacities. Insulin-loaded beads were designed by Kim and co-workers based on the terpolymers of N-isopropylacrylamide, butylmethacrylate, and acrylic acid [29]. It was observed that the release pattern of insulin depends on the molecular weight of the terpolymers. Insulin-encapsulated shells based on poly (lactic-co-glycolic acid) (PLGA) have also been examined for sustained insulin release and glycemic homeostasis [30]. An investigation of oral delivery from dextran sulfate/chitosan–based nanoparticles found enhanced pharmacological availability in diabetic rats [31]. Moreover, poly (methacrylic acid-g-ethylene glycol) and wheat germ agglutinin–functionalized poly (methacrylic acid-g-ethylene glycol)–based hydrogels, having pH-responsive releasing properties, have been analyzed for the oral delivery of insulin [32]. Cui et al. designed pH-sensitive carboxylated chitosan–grafted poly (methyl methacrylate) nanoparticles for the oral delivery of insulin [33]. A superporous hydrogel containing poly (acrylic acid-co-acrylamide)/O-carboxymethyl chitosan interpenetrating polymer networks was evaluated as an oral delivery vehicle for insulin by Yin et al. [34]. A novel type of pH- and amylose-responsive microgel as an insulin drug carrier for oral administration has been developed by Chen et al. [20].

The numbers of articles available in the literature in different years of the decade 2009–2018 (until September) are represented in Figure 7.1 with various relevant key words. In the present scenario and based on this literature, it is well understood that the oral delivery of insulin is still a challenge for researchers. Extensive literature is available that is relevant to the present topic. However, concise review reports in the form of a chapter are still rare. The present chapter reviews the oral delivery of insulin through a hybrid polymeric hydrogel. The number of works published in the preceding decade related to the present area of interest is summarized in Figure 7.1. Types of insulin, various techniques to prepare

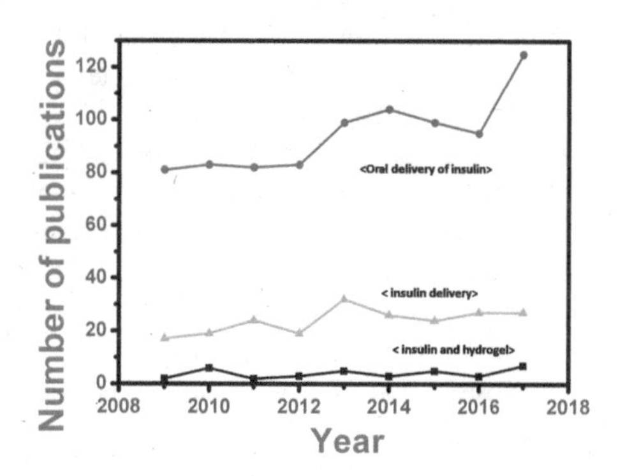

FIGURE 7.1 Number of articles available in the literature in different years of the decade 2009–2018 (until September). (Scopus key words: oral delivery of insulin; insulin delivery; insulin and hydrogels.)

hybrid polymeric hydrogels, different methods of delivery, and the properties of potential hydrogels are included in the present chapter.

7.2 Insulin

Insulin is a peptide hormone that is secreted by the beta cells present in the islets of the pancreas, an organ inside the abdomen. Type 1 diabetes is a result of the loss of these beta cells that produce insulin. Almost all diabetic people depend on human insulin or insulin analogs. The human insulin is not derived from humans; rather, it has a similar composition to that of human insulin. It is generated from yeast (Novo-Nordisk) or bacteria (Lilly) by means of genetic engineering. Insulin cannot be administrated orally due to its degradation by enzymatic attack in the stomach. Insulin is standardized as units per milliliter (cubic centimeter). The U.S. insulin, called U-100 insulin, measures 100 units per cc (ml) of the drug. Standard insulin syringes have capacities of 3/10 cc (30 units), 1/2 cc (50 units), and 1 cc (100 units). The 3/10 cc syringes are conventional to use for the measurement of small doses, as they have a higher least count value (half unit). The prescribed insulin dosage varies with body weight, blood sugar test results, food intake, especially carbohydrates, and exercise regimen. After preliminary diabetes diagnosis and treatment, patients are prescribed approximately 1/4 unit of insulin per round (1/2 unit/kg) of body weight per day. The dose is increased as per requirement up to 1/2 units per pound body weight (1 unit per kg body weight). Insulin is broadly classified into three classes: 1) "Rapid acting" (such as Humalog [H], Novolog [NL], Apidra [AP], and regular [R]); (2) "Intermediate-acting" (such as NPH [N]), and (3) "Long-acting" (such as Levemir [insulin detemir] and Lantus® [insulin glargine]). Novolog, Apidra, and Humalog insulins show rapid activity in 30–90 min and are effective for up to 3 to 4 hours. "Rapid acting insulin" could indicate any of the three insulins H, NL, or AP. The regular insulin starts functioning approximately after 30–60 min of being injected. Its peak effectiveness is shown 2 to 4 hours post injection and lasts up to 9 hours. NL, H, and AP insulins have various limitations over regular insulin. They begin their action in 10–20 min instead of 30–60 min post injection.

If insulin is administered before a meal, its activity peaks as the carbohydrates are converted into blood sugar. However, in the case of regular insulin, post-meal administration shows greater absorption and results in low blood sugar levels. The blood sugar levels are lower after 2 hours of a meal following the intake of rapid acting insulin 10–15 min before a meal. Rapid acting insulin does not last long when taken after dinner as compared with regular insulin, leading to a reduced risk of low blood sugar levels at night. Neutral protamine, Hagedorn [NPH] (N) insulin is designed using a protein that facilitates its slow absorbance in the body. Protamine is the protein that is added to the insulin to make it last longer. Human NPH shows peak activity 4 to 8 hours post injection in most people; i.e., the peak action of morning dosage is observed before supper. On average, human NPH lasts for 13 hours. The NPH insulin peak activity as well as its activity duration may be different for a few people. NPH insulin and regular insulin may be intermixed without a change in the activity of either. Lantus (insulin glargine) insulin became widely available in U.S. markets in May, 2001. It is clearly seen that the insulin lasts up to 24 hours almost without any peak (the first true basal insulin). It exhibits a similar profile to that of basal insulin secreted by the pancreas, which regulates the sugar output from the liver. Levemir (insulin detemir) formulated by Novo-Nordisk is also being used as basal insulin. It lasts up to 24 hours, but some people may need twice- daily intake.

Syringes, insulin pens, and insulin pumps are used for insulin administration for type 1 diabetes patients, who are incapable of producing insulin. To sustain them, the diabetic persons rely on multiple insulin injections per day. With the number of doses varying from person to person, insulin injections are administrated over a regular schedule during the day. Syringes or insulin pens both serve the same purposes. Insulin pens are chosen by some people in the case of requirement for a single kind of insulin; children especially are more comfortable with pen needles than with syringe needles. However, insulin self-administration age of children varies. An insulin pump, an alternative to insulin injections, is a computerized device the size of a beeper or pager that is worn on a belt or carried in the pocket. It delivers a sustained low basal dose of insulin through a flexible plastic tube called a *cannula*. It is commonly taped to a particular site, mostly the buttocks, upper arms, or other areas with fatty tissues, and the needle is removed. The person wearing the pump pushes a button on the pump while eating to deliver a surplus amount of insulin known as a *bolus*. The merits of the insulin pump are (1) better control of blood glucose levels, (2) higher adaptability to meals, exercises, and daily schedule, and (3) enhanced physical and psychological well-being. The major limitations include (1) frequent low blood sugar levels presenting as hypoglycemia, (2) increased risk of infection, (3) constant physical discomfort, (4) and ketosis and ketoacidosis, i.e., the risk of a very high blood sugar level. However, choosing an appropriate delivery method for insulin is a personal decision for a type 1 diabetic patient, generally taken by the individual, the doctor, and the family members in the case of children.

7.3 Hybrid Polymeric Hydrogel

Hydrogels are hydrophilic cross-linked macromolecular polymeric gels that are able to hold large quantities of water within their three-dimensional network structure. They have gained global recognition because of their biocompatibility, their biodegradability, and their potential applications in various industries, especially in the biomedical and pharmaceutical fields. They are generally derived from polysaccharides, peptides, and other biopolymers. The rapidly increasing demand for hydrogels is due to their rapid swelling–deswelling as a response to the variation of the surroundings [35–37]. Furthermore, the environment-triggered controlled release of the encapsulated drugs, as well as their small size and fast response, gives the hydrogels unique characteristics for the effective oral delivery of drugs (including insulin) [38–40].

Stimuli-responsive materials have become the epicenter of biomedical research and have led to numerous efforts toward preparing smart hydrogels for various purposes [41–46]. The rising interest in designing intelligent hydrogels is due to their selective response to stimuli-sensitive alterations in environmental conditions. They have gained special attention in biomedical fields because of their potential applications, especially in site-specific drug delivery systems [47–51]. Of all of them, the pH-responsive systems have acquired recognition in pharmaceutical fields due to their potential use, particularly as oral delivery systems for bioactive proteins or peptides such as insulin [52–56]. Schematic illustrations of insulin-loaded hybrid polymeric hydrogels are represented in Scheme 2. The oral delivery of insulin by different polymer-based hydrogels is represented in Table 7.1. Several polymerization techniques are employed for the fabrication of the said smart hydrogel systems for oral insulin delivery, such as solution polymerization, suspension polymerization, and emulsion polymerization.

7.3.1 Microemulsion Method

In the microemulsion method, inorganic particles with sizes in the nanometer range are obtained with minimum agglomeration [66]. Oxide and carbonate nanoparticles are generated successfully using this method [67–70]. A microemulsion is an isotropic dispersion of two immiscible liquids such as water and oil stabilized by surfactant molecules of the water/oil interface. This microemulsion is thermodynamically stable and transparent. In a water/oil microemulsion, nanosized water droplets are distributed uniformly throughout the continuous hydrocarbon phase and surrounded by the surfactant monolayer [71]. The aqueous droplets in the size range of 5–20 nm in diameter act as a microreactor or nanoreactors in which the reaction can be carried out when suitable reactants contained in the droplets collide with each other [69, 72]. Using this method, hydroxide or oxalate precursor particles are found first. The desired oxide system is achieved after subsequent drying and calcination at an appropriate temperature. Thus, the microemulsion method successfully generates dispersed nanosize organic and inorganic particles with minimum agglomeration. Despite certain limitations of this method, such as the requirement for large amounts of oil and surfactant phases and low production yield, the microemulsion method gives us an alternative technique for the synthesis of different varieties of organic as well as inorganic nanoparticles (NPs) [69,73,74].

7.3.2 Precipitation Polymerization

Thermosensitive nanogels are conventionally prepared by precipitation polymerization. When the polymerization temperature of the N-isopropyl acrylamide–based polymers exceeds the lower critical solution temperature (LCST), the growing chains of the N-isopropyl acrylamide collapse on reaching the critical length. This method results in the formation of precursor particles [75]. Three different techniques are

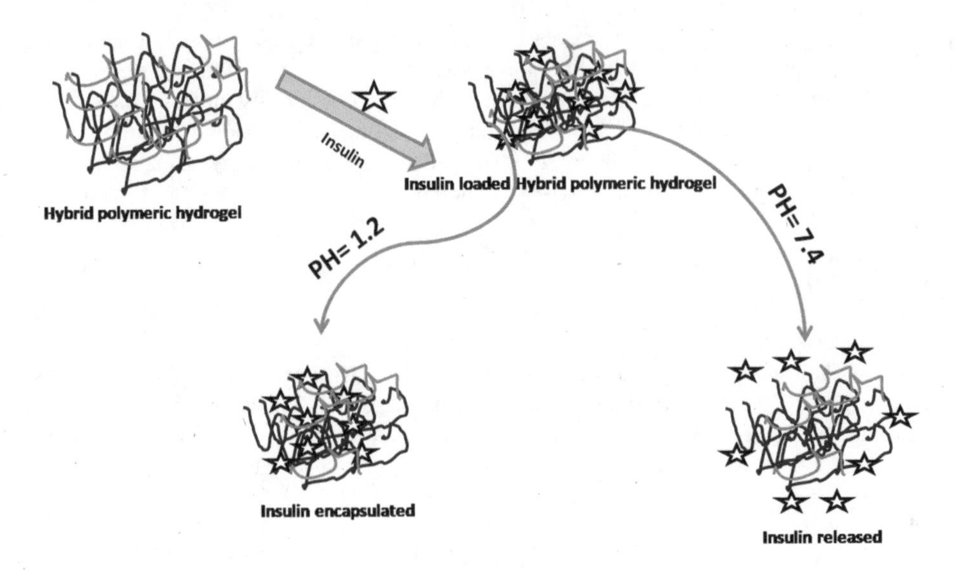

SCHEME 2 Schematic illustrations of insulin-loaded hybrid polymeric hydrogels.

employed to generate nanogels from the precursor particles: continuous growth of precursor particles by addition of monomer; deposition of a polymer on the surface of the existing polymer; and aggregation of the precursor particles to generate a large, colloidally stable polymeric gel. This technique includes the origination of charges from the initiator fragments and the incorporation of a sufficient amount of water into the collapsed chains. These methods vary from the traditional method of emulsion polymerization of water-insoluble monomers, whereby particles containing a compact structure are generated. After the execution of the complete polymerization method, the temperature of the system is reduced below the LCST, and the hydrogels produced can swell by incorporating more water. Pelton et al. established the poly(N-isopropylacrylamide) (PNIPAM)-based nanogels by the process of precipitation polymerization

TABLE 7.1
Oral Delivery of Insulin by Different Polymer-Based Hydrogels

Sl No.	Polymer	Other Components	Hydrogels	Characterization Technique	Reference
1	Poly(2-acrylamide-2-methyl-1-propane sulfonate acid)	Salecan	Salecan-g-PAMPS	FTIR, XRD, SEM	[57]
2	Chitosan	Alginate	CS/ALG	FTIR, UV, SEM, DLS	[58]
3	Chitosan	Carboxymethyl chitosan	CMCS/CS-NGs	FTIR, DLS, UV	[59]
4	Poly(acrylic acid)	Carboxymethyl cellulose	CMC-g-AA	^{1}H NMR, ESEM	[60]
5	Poly(lactic-co-glycolic acid)	N-trimethyl chitosan chloride	Insulin-LMWP	TEM, Zeta potential	[61]
6	Hyaluronic acid	$CaCO_3$	$CaCO_3$/HA NCs	SEM, TEM	[62]
7	Chitosan/albumin	Alginate/dextran sulfate	ADS/CS-ALB	TEM, Zeta potential	[63]
8	Hyaluronic acid–cysteamine	Poly vinyl alcohol	HA-Cym/PVA	SEM	[64]
9	Poly (2-isobutyl acrylic acid)	Carboxy methyl starch	CMS-g-AA	^{1}H and ^{13}C NMR, FTIR, Zeta potential	[20]
10	Polyethylene imine	Dextran sulfate	PEI/DS	TEM, DLS	[65]

Note: DLS, dynamic light scattering; ESEM, environmental scanning electron microscopy; FTIR, Fourier transform infrared; LMWP, low-molecular weight protamine; NC, nanocarrier; SEM, scanning electron microscopy; UV, ultraviolet; XRD, X-ray diffraction.

[76]. A series of core-shell nanogels with pH-responsive shell and thermosensitive core was designed by Li et al. using this method [77].

7.3.3 Surface-Initiated *In Situ* Polymerization

The "living" radical polymerization or surface-initiated controlled polymerization technique provides a potential route to prepare organic and inorganic core-shell hybrid NPs possessing the ability to control the shell structure and thickness. The nanocapsule is generated accordingly by the removal of core templates from the core-shell hybrid NPs. This technique commonly involves the use of silica NPs as templates, as they can be easily modified. Specifically, the hydroxyl groups present on silica surface templates can be conveniently modified and used as initiator for the polymerization of specific monomers. Block polymers can be prepared by the sequential addition of monomers using surface-initiated atom transfer free radical polymerization. The polymer shells, which are prepared via *in situ* polymerization of the specific surface, generally require to be stabilized by cross-linking before the removal of the core templates, which differentiates this method from the layer-by-layer self-assembly method. Mu et al. designed a pH-sensitive nanocapsule employing the surface-initiated atom transfer radical polymerization method [78]. In this method, bromoacetamide groups are conjugated onto the surface of the silica NP. The atom transfer radical polymerization (ATRP) of t-butyl acrylate and styrene is subsequently initiated on the surface of the NPs.

7.3.4 Swelling Properties

The swelling behaviors of hydrogels are studied in pH 1.2, 6.8, and 7.4 buffers, which reveal that the poly acrylic acid (PAA)/N–succinyl chitosan (PAA/S-chitosan) shows significantly less swelling as compared with the PAA–chitosan hydrogel (Figure 7.2). The remarkable swelling of the PAA–chitosan hydrogel

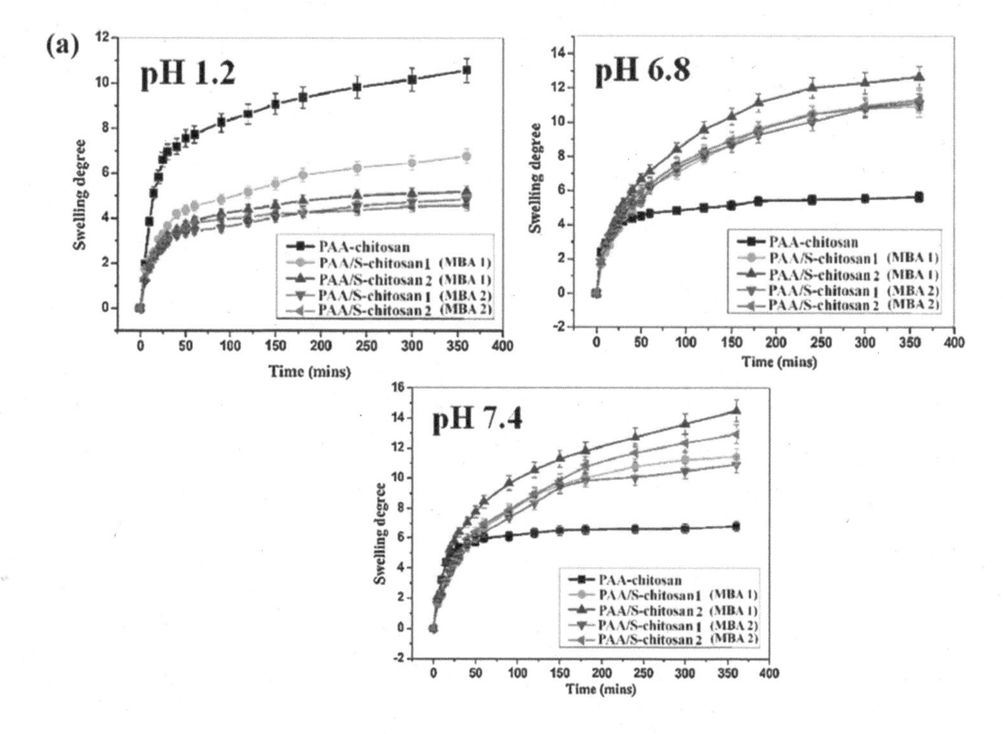

FIGURE 7.2 Swelling percentage of hydrogels in different pH media. (Reprinted from *Carbohydr. Polym.*, 112, Mukhopadhyay, P., Sarkar, K., Bhattacharya, S., Bhattacharyya, A., Mishra, R., Kundu, P.P., pH sensitive N-succinyl chitosan grafted poly acrylamide hydrogel for oral insulin delivery, 627–637, Copyright (2014), with permission from Elsevier.)

at pH 1.2 is because of the fact that the primary amino group present on chitosan becomes protonated, resulting in repulsive forces, leading to the expansion of the hydrogel network. Li et al. established, however, at pH 1.2 (~6 h incubation), the carboxyl group remains protonated, causing the hydrogel network to shrink [79]. The swelling percentage of PAA/S-chitosan decreases with an increase in the number of carboxyl groups on S-chitosan (a higher degree of succinylation), as represented in Figure 7.2. The swelling properties of PAA–chitosan and PAA/S-chitosan are altered with gradual elevation in pH from the acidic to the basic range. At 37 °C, the PAA/S-chitosan shows controlled swelling at pH 6.8 and 7.4 (intestinal) for 6 h. Swelling is initiated in all the PAA/S-chitosan hydrogels, 30 min post incubation and gradually continues for 6 h until an equilibrium swelling is achieved. However, the PAA–chitosan exhibits minimal swelling (swelling degree ~5) under the same conditions and timeframe. The carboxyl groups retain their original form at higher pH conditions, resulting in repulsive forces between the ionized COO– groups on S-chitosan, which cause swelling of the hydrogels in alkaline conditions (pH 6.8 and 7.4) in a sustained way [79].

7.3.5 Water Retention Properties

Polyacrylamide/dextran (PAM/D) with embedded silver nanoparticles and PAM/D hold water for up to 14 hours in a hot air oven at 50 °C (Figure 7.3). But, PAM/D with embedded silver nanoparticles has a higher water-holding capacity than PAM/D hydrogel, which may be due to silver nanoparticles contributing to the formation of a less interconnected pore structure, which can confer water-absorbing properties. Furthermore, cross-linking densities are also affected by water retention capacity. PAM/D network chains are delayed in their stretching at higher cross-linking density, thus suppressing the release of water. In this work, the PAM gel containing cellulose has the opposite effect; i.e., shrinkages are found. About 65% of the water is released from nanohydrogels during the first hour. That reduced swelling may be due to intermolecular bonding with water. A similar type of result is reported during an investigation into nanocomposites based on poly (acrylamide-co-acrylate) and cellulose nanowhiskers [81]. Furthermore, these results are also in accordance with the study of superabsorbent hydrogels based on cellulose for smart swelling and controllable delivery [82].

7.3.6 *In Vitro* Analysis

The insulin-releasing behavior of hydrogels is analyzed in *in vitro* conditions in artificial gastric fluid (AGF, pH $^1/_4$ 1.2) and in artificial intestinal fluid (AIF, pH $^1/_4$ 6.8); the drug loading capacities of Gel 1

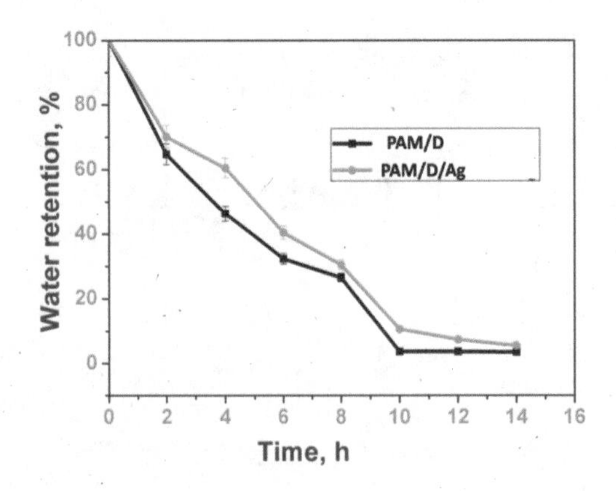

FIGURE 7.3 Water retention properties of poly acrylamide/Dextran and poly acrylamide/Dextran nanohydrogel with embedded silver nanoparticles with 1.5% concentration of nano silver. (Reprinted from *Mater. Sci. Eng. C.*, 85, Prusty, K. and Swain, S.K., Nano silver decorated polyacrylamide/dextran nanohydrogels hybrid composites for drug delivery applications, 130–141, Copyright (2018), with permission from Elsevier.)

(polygutamic acid-g-acrylic acid), Gel 2 (polygutamic acid-g-trans-crotonic acid), Gel 3 (polygutamic acid-g-trans-2-pentenoic acid), and Gel 4 (polygutamic acid-g-trans-2-hexenoic acid) were found to be 27.7%, 24.2%, 21.3%, and 18.5%, respectively. This decrease in the drug loading capacity along with the increase in the hydrophobicity is supposedly due to the lowering of the swelling ratio. As a diffusion method is used for loading drugs into the hydrogels, a higher swelling ratio would result in higher loading efficiency. Figure 7.4 shows minimal cumulative releases in AGF for all the gels due to the corresponding deswelling states in AGF. The hydrogels show remarkable release in an AIF-simulated environment because of their higher swelling ratios. Less than 10% of the loaded insulin is released in AGF, whereas in AIF, Gel 1 shows more than 90% insulin release, and for Gel 4, more than 60% drug release is observed after 6 h. Additionally, the increase in hydrophilicity in poly acrylic acid derivatives (PAAD) results in an increased release rate in either AGF or AIF. The loaded insulin is released from the hydrogels using a diffusion method. The diffusion of insulin from the hydrogels in the acidic conditions of AGF is blocked by the dense network due to the collapse of the hydrogels. It can be seen that the collapsing process of the hydrogels leads to about 5% insulin release in 30 minutes. In contrast, the neutral AIF conditions allow adequate swelling, which turns the carriers into a porous state, enabling easier diffusion of insulin. Hence, it is concluded that the insulin release is much faster in AIF as compared with AGF. Thus, intelligent hydrogels that can protect the loaded bioactive molecule (insulin) in AGF and show efficient releasing capacities in AIF can be potential candidates for the oral delivery of insulin.

7.3.7 Cytotoxicity Analysis

Cytotoxicity estimation of any drug delivery agent is one of the major challenges faced by the scientific community before animal and human trials. Since cytotoxicity is one of the most essential properties of any biomaterial, several techniques, such as MTT assays, are employed for analyzing the biocompatibility of any new material under study. MTT assays are used for *in vitro* study of the hydrogel samples to examine their biocompatibility, cytotoxicity, and residue toxicity from the loading process. The incubated HeLa cell viability with the hydrogel samples at tested concentrations up to 1 mg/ml is found to be over 90%, indicating better biocompatibility of the hydrogels with no cytotoxicity. It has been established that poly (L-glutamic acid) (PGA) and PAA are extensively used in the biomaterial and pharmaceutical fields due to their non-toxic nature. Hence, the corresponding hydrogel fabricated from PGA and PAA is expected to be non-toxic, which agrees with the results of Gel 1. It is noticeable that certain toxicity studies of PAAD have been reported earlier. The MTT assays results, represented in Figure 7.5, indicates that the hydrogels prepared from PAAD containing various hydrophobic side groups displayed low cytotoxicity, making the hydrogels suitable for *in vivo* applications.

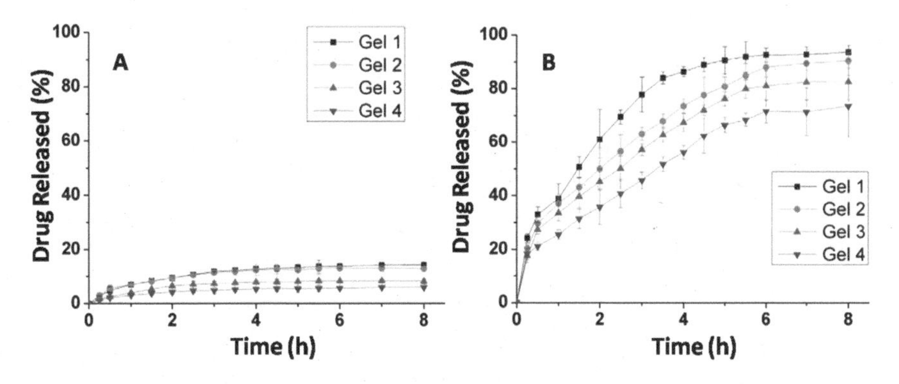

FIGURE 7.4 *In vitro* release of insulin from hydrogels as a function of time at 37 °C in (a) artificial gastric fluid (AGF) and (b) artificial intestinal fluid (AIF). (Reprinted from *Polymer*, 54, Gao, X., He, C., Xiao, C., Zhuang, X., Chen, X., Biodegradable pH-responsive poly acrylic acid derivative hydrogels with tunable swelling behaviour for oral delivery of insulin, 1786–1793, Copyright (2013), with permission from Elsevier.)

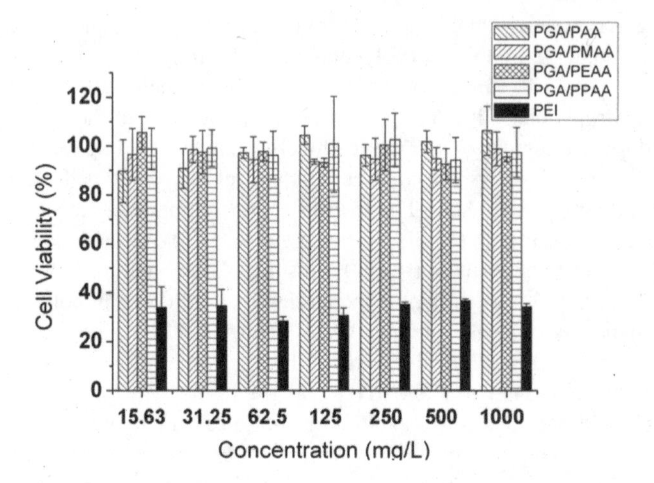

FIGURE 7.5 Evaluation of cytotoxicity against HeLa cells by the MTT assay of hydrogels ($P < 0.05$). (Reprinted from *Polymer*, 54, Gao, X., He, C., Xiao, C., Zhuang, X., Chen, X., Biodegradable pH-responsive poly acrylic acid derivative hydrogels with tunable swelling behaviour for oral delivery of insulin, 1786–1793, Copyright (2013), with permission from Elsevier.)

7.4 Polymeric Hydrogels as Vehicles for Oral Delivery of Insulin

7.4.1 Different Polymer-Based Hydrogels as Vehicles for Oral Delivery of Insulin

Release profiles of carboxylated chitosan-grafted poly (methyl methacrylate) nanoparticles (CCGN) *in vitro* at pH 2.0, 6.8, and 7.4 are represented by Figure 7.6. A slower insulin release rate is observed for pH 2.0 than for pH 6.8 and 7.4. But within the first 15 min, a burst release profile is marked for all three pH conditions. CCGN exhibits burst release because of the dissociation of the drugs from the surface of the NPs. Acidic conditions show very minimal insulin release due to protonation of the carboxyl groups of CCGN, causing agglomeration of the particles. However, the carboxyl groups of the CCGN are in the ionized state at pH 6.8 and 7.4. The increase in repulsive forces between both the negatively charged carboxyl groups on CCGN and the loaded insulin led to the comparatively faster release rate of insulin. This pH-responsive insulin release of the CCGN makes it favorable for protecting the encapsulated insulin when the NPs pass through the harsh acidic environment of the stomach.

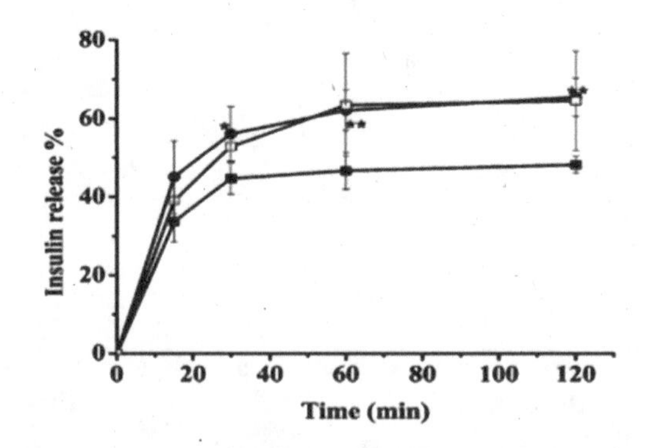

FIGURE 7.6 Insulin release profiles from insulin-loaded carboxylated chitosan-grafted poly (methyl methacrylate) nanoparticles in pH 2.0 HCl (■), pH 6.8 PBS (●), and pH 7.4 PBS (□). (Reproduced with permission from Cui, F.; Qian, F., Zhao, Z., Yin, L., Tang, C., Yin, C. Preparation, and characterization, and oral delivery of insulin loaded carboxylated chitosan grafted poly (methyl methacrylate) nanoparticles. *Biomacromolecules*. 2009, *10*, 1253–1258. Copyright 2009 American Chemical Society.)

7.4.2 Polysaccharide-Based Hydrogels as Vehicles for Oral Delivery of Insulin

The insulin release profile study *in vitro* is represented in Figure 7.7. All the PAA/S-chitosan exhibits 20–25% release of the encapsulated insulin at pH 1.2, while 73% insulin release is observed in the case of PAA–chitosan in the same conditions. Due to the protonation of carboxyl groups on S-chitosan at acidic pH 1.2, the insulin release rate may be lowered with the subsequent shrinkage of PAA/S-chitosan. In the same pH conditions, the PAA–chitosan shows greater release of insulin (73%) due to the swelling caused by protonation of the free amino groups on the chitosan structure, generating a repulsive force. The PAA/S-chitosan shows a sustained and elevated release rate of insulin in the simulated intestine environment (pH 6.8 and 7.4). Gradual swelling is observed in almost all the PAA/S-chitosan formulations 30 mins post incubation, leading to the release of nearly 98% of the encapsulated insulin, facilitated by the repulsive force between the polymer structure and the ionized carboxyl groups. Controlled release of insulin in both duodenum and ileum pH conditions is observed over a time period of 6 h, whereas; the PAA–chitosan displays 20–26% release of insulin in similar conditions. Mura et al. demonstrated the swelling behavior of the S-chitosan matrix as a response to the pH conditions of the colon in colonic drug delivery [85]. Hence, PAA/S-chitosan is proved to be more efficient in releasing the encapsulated insulin in the intestine sustainably, preserving it from the unfavorable stomach environment.

7.4.3 Protein-Based Hydrogels as Vehicles for Oral Delivery of Insulin

To enhance the functioning of microparticles (MP), insulin-loaded microparticles (ins-MP) are laminated with alginate (ALG) containing insulin. This shows no difference as compared with the un-coated MP; that is, it exhibits a fast and complete release rate at pH 1.2 and 6.8 (Figure 7.8). The coating does not offer an effective solution, in contrast to the previous study involving theophylline encapsulated with whey protein (WP)/ALG MP [86]. This may be due to steric hindrance between the insulin and the MP, making the coating unstable. The noted burst release is probably due to the repulsive forces (electrostatic)

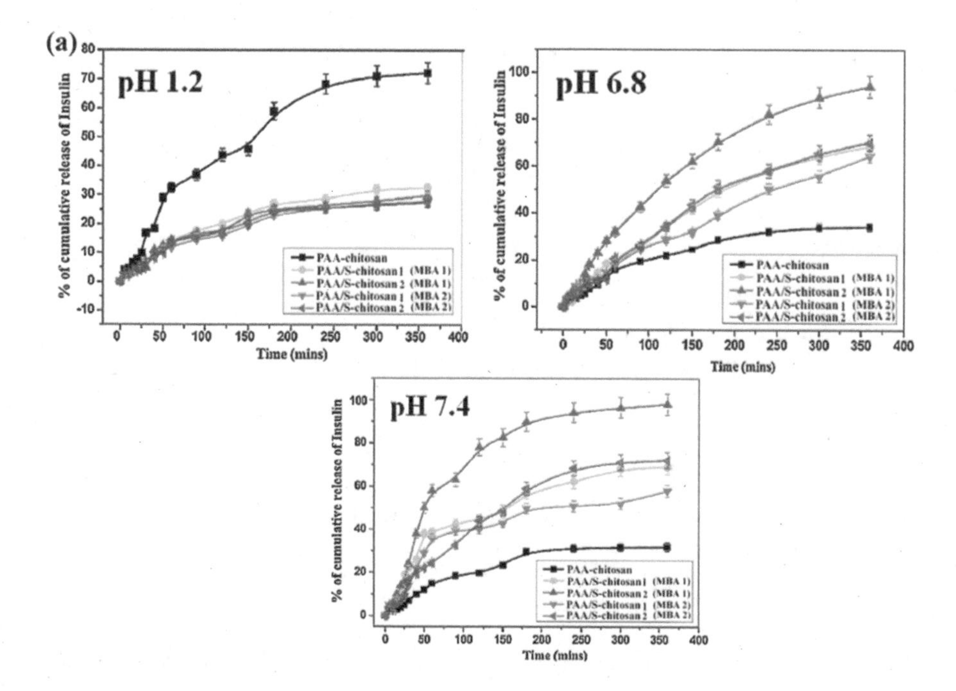

FIGURE 7.7 *In vitro* insulin release profile of hydrogels at different pH. (Reprinted from *Carbohydr. Polym.*, 112, Mukhopadhyay, P., Sarkar, K., Bhattacharya, S., Bhattacharyya, A., Mishra, R., Kundu, P.P., pH sensitive N-succinyl chitosan grafted poly acrylamide hydrogel for oral insulin delivery, 627–637, Copyright (2014), with permission from Elsevier.)

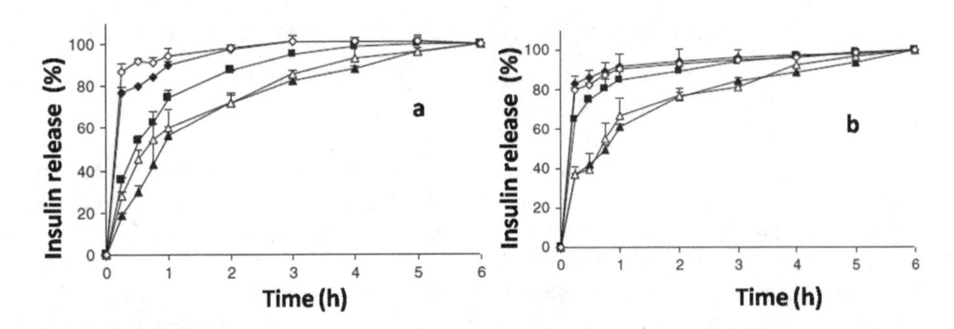

FIGURE 7.8 *In vitro* release of insulin from uncoated (black) and coated (white) insulin-loaded microparticles, composed of (a) 60/40 and (b) 80/20 WP/ALG, at 37 °C. (Reprinted with kind permission from Springer Science+Business Media: *Pharm. Res.*, Efficacy of mucoadhesive hydrogel microparticles of whey protein and alginate for oral insulin delivery, 30, 2013, 721–734, Deat-Laine, E., Hoffart, V., Garrait, G., Jarrige, J.-F., Cardot, J.-M., Subirade, M., Beyssac, E.)

between insulin and the polymers. The isoelectric point of insulin (5.4) is comparable to that of WP (pI_0 5.2), and it has the same charge as that of WP and ALG as a function of pH. The existence of repulsive interactions between the components may lead to the relaxation of the matrix [87]. An *in vitro* study of the swelling behavior of the formulation reveals that it can increase insulin absorption by the application of mechanical pressure, as established by Dorkoosh et al. [88].

7.5 Conclusion

Insulin can be orally delivered by a polymeric hydrogel vehicle, overcoming the limitations of conventional injection. From a literature review of the last decade, it is found that special attention has been paid by researchers to the oral delivery of insulin due to their biomedical and academic interest. Different types of insulin and their methods of delivery are explained in the present chapter. Hydrogels prepared from proteins, polysaccharides, and biopolymers are potential vehicles for the oral delivery of insulin. The present chapter may open a new window of understanding for scholars with motivation for the oral delivery of insulin.

Acknowledgment

The authors express their thanks to the Department of Science and Technology, Government of India, for awarding the Inspire Fellowship to Ms. Kalyani Prusty to pursue her doctoral degree.

REFERENCES

1. Banting, F.G.; Best, C.H. The internal secretion of the pancreas. *J. Lab. Clin. Med.* 1922, *7*, 251–266.
2. Bhatnagar, S.; Srivastava, D.; Jayadev, M.S.K.; Dubey, A.K. Molecular variants and derivatives of insulin for improved glycemic control in diabetes. *Prog. Biophys. Mol. Biol.* 2006, *91*, 199–228.
3. Ramesh Babu, V.; Patel, P.; Mundargi, R.C.; Rangaswamy, V.; Aminabhavi, T.M. Developments in polymeric devices for oral insulin delivery. *Expert Opin. Drug Deliv.* 2008, *5*, 403–415.
4. Yamagata, T.; Morishita, M.; Kavimandan, N.J.; Nakamura, K.; Fukuoka, Y.; Takayama, K.; Peppas, N.A. Characterization of insulin protection properties of complexation hydrogels in gastric and intestinal enzyme fluids. *J. Control. Release* 2006, *112*, 343–349.
5. Park, K.; Kwon, I.C.; Park K. Oral protein delivery: Current status and future prospect. *React. Funct. Polym.* 2011, *71*, 280–287.
6. Wong, T.W. Design of oral insulin delivery systems. *J. Drug Target.* 2010, *18*, 79–92.

7. Aoki, Y.; Morishita, M.; Asai, K.; Akikusa, B.; Hosoda, S.; Takayama, K. Regional dependent role of the mucous/glycocalyx layers in insulin permeation across rat small intestinal membrane. *Pharm. Res.* 2005, *22*, 1854–1862.
8. Agarwal, V.; Khan, M.A. Current status of the oral delivery of insulin. *Pharm. Technol.* 2001, *25*, 76–90.
9. Hansen, J.F. The self-association of zinc-free human insulin and insulin analogue B13-glutamine. *Biophys. Chem.* 1991, *39*, 107–110.
10. Li, Y.; Shao, Z.; Mitra, A.K. Dissociation of insulin oligomers by bile salt micelles and its effect on α-chymotrypsin-mediated proteolytic degradation. *Pharm. Res.* 1992, *9*, 864–869.
11. Carino, G.P.; Mathiowitz, E. Oral insulin delivery. *Adv. Drug Deliv. Rev.* 1999, *35*, 249–257.
12. Zambanini, A.; Newson, R.B.; Maisey, M.; Feher, M.D. Injection related anxiety in insulin treated diabetes. *Diabetes Res. Clin. Pract.* 1999, *46*, 239–246.
13. Owens, D.R.; Zinman, B.; Bolli, G. Alternative routes of insulin delivery. *Diabetic Med.* 2003, *20*, 886–898.
14. Khafagy, E.S.; Morishita, M.; Onuki, Y.; Takayama, K. Current challenges in noninvasive insulin delivery systems: a comparative review. *Adv. Drug Deliv. Rev.* 2007, *59*, 1521–1546.
15. Radwant, M.A.; Aboul-Enein, H.Y. The effect of oral absorption enhancers on the in vivo performance of insulin-loaded poly (ethylcyanoacrylate) nanospheres in diabetic rats. *J. Microencapsul.* 2002, *19*, 225–235.
16. Asada, H.; Douen, T.; Waki, M.; Adachi, S.; Fujita, T.; Yamamoto, A.; Muranishi, S. Absorption characteristics of chemically modified-insulin derivatives with various fatty acids in the small and large intestine. *J. Pharm. Sci.* 1995, *84*, 682–687.
17. Marschutz, M.K.; Caliceti, P.; Bernkop-Schnurch, A. Design and in vivo evaluation of an oral delivery system for insulin. *Pharm. Res.* 2000, *17*, 1468–1474.
18. Takeuchi, H.; Yamamoto, H.; Niwa, T.; Hino, T.; Kawashima, Y. Enteral absorption of insulin in rats from mucoadhesive chitosan-coated liposomes. *Pharm. Res.* 1996, *13*, 896–901.
19. Prego, C.; Garcia, M.; Torres, D.; Alonso, M.J. Transmucosal macromolecular drug delivery. *J. Control. Release* 2005, *101*, 151–162.
20. Liu, L.; Zhang, Y.; Yu, S.; Zhang, Z.; He, C.; Chen, X. pH-and amylose-responsive carboxymethyl starch/poly (2-isobutyl-acrylic acid) hybrid microgels as effective enteric carriers for oral insulin delivery. *Biomacromolecules.* 2018, *19*, 2123–2136.
21. Morishita, M.; Lowman, A.M.; Takayama, K.; Nagai, T.; Peppas, N.A. Elucidation of the mechanism of incorporation of insulin in controlled release systems based on complexation polymers. *J. Control. Release* 2002, *81*, 25–32.
22. Morishita, M.; Goto, T.; Peppas, N.A.; Joseph, J.I.; Torjman, M.C.; Munsick, C.; Nakamura, K.; Yamagata, T.; Takayama, K.; Lowman, A.M. Mucosal insulin delivery systems based on complexation polymer hydrogels: effect of particle size on insulin enteral absorption. *J. Control. Release* 2004, *97*, 115–124.
23. Foss, A.C.; Peppas, N.A. Investigation of the cytotoxicity and insulin transport of acrylic-based copolymer protein delivery systems in contact with Caco-2 cultures. *Eur. J. Pharm. Biopharm.* 2004, *57*, 447–455.
24. Sonaje, K.; Chen, Y.J.; Chen, H.L.; Wey, S.P.; Juang, J.H.; Nguyen, H.N.; Hsu, C.W.; Lin, K.J.; Sung, H.W. Enteric-coated capsules filled with freeze-dried chitosan/poly(g-glutamic acid) nanoparticles for oral insulin delivery. *Biomaterials.* 2010, *31*, 3384–3394.
25. Santos, A.C.; Cunha, J.; Veiga, F.; Cordeiro-da-Silva, A.; Ribeiro, A.J. Ultrasonication of insulin-loaded microgel particles produced by internal gelation: impact on particle's size and insulin bioactivity. *Carbohydr. Polym.* 2013, *98*, 1397–1408.
26. Nolan, C.M.; Gelbaum, L.T.; Andrew Lyon, L. 1H NMR investigation of thermally triggered insulin release from poly(Nisopropylacrylamide) microgels. *Biomacromolecules.* 2006, *7*, 2918–2922.
27. Nolan, C.M.; Serpe, M.J.; Andrew Lyon, L. Thermally modulated insulin release from microgel thin films. *Biomacromolecules.* 2004, *5*, 1940–1946.
28. Bai, Y.; Zhang, Z.; Zhang, A.; Chen, L.; He, C.; Zhuang, X.; Chen, X. Novel thermo- and pH-responsive hydroxypropyl celluloseand poly(L-glutamic acid)-based microgels for oral insulin controlled release. *Carbohydr. Polym.* 2012, *89*, 1207–1214.

29. Ramkissoon-Ganorkar, C.; Liu, F.; Baudyš, M.; Kim, S.W. Modulating insulin-release profile from pH/thermosensitive polymeric beads through polymer molecular weight. *J. Control. Release* 1999, *59*, 287–298.
30. Kim, B.S.; Oh, J.M.; Hyun, H.; Kim, K.S.; Lee, S.H.; Kim, Y.H.; Park, K.; Lee, H.B.; Kim, M.S. Insulin-loaded microparticles for in vivo delivery. *Mol. Pharm.* 2009, *6*, 353–365.
31. Sarmento, B.; Ribeiro, A.; Veiga, F.; Ferreira, D.; Neufeld, R. Oral bioavailability of insulin contained in polysaccharide nanoparticles. *Biomacromolecules.* 2007, *8*, 3054–3060.
32. Wood, K.M.; Stone, G.M.; Peppas, N.A. Wheat germ agglutinin functionalized complexation hydrogels for oral insulin delivery. *Biomacromolecules.* 2008, *9*, 1293–1298.
33. Cui, F.; Qian, F.; Zhao, Z.; Yin, L.; Tang, C.; Yin, C. Preparation, and characterization, and oral delivery of insulin loaded carboxylated chitosan grafted poly (methyl methacrylate) nanoparticles. *Biomacromolecles.* 2009, *10*, 1253–1258.
34. Yin, L.; Ding, J.; Zhang, J.; He, C.; Tang, C.; Yin, C. Polymer integrity related absorption mechanisim of superporous hydrogel containing interpenetrating polymer networks for oral delivery of insulin. *Biomaterials.* 2010, *31*, 3347–3356.
35. Bromberg, L. Intelligent hydrogels for the oral delivery of chemotherapeutics. *Expert Opin. Drug Deliv.* 2005, *2*, 1003–1013.
36. Vinogradov, S.V. Colloidal microgels in drug delivery applications. *Curr. Pharm. Des.* 2006, *12*, 4703–4712.
37. Thorne, J.B.; Vine, G.J.; Snowden, M.J. Microgel applications and commercial considerations. *Colloid Polym. Sci.* 2011, *289*, 625–646.
38. Li, J.; Mooney, D.J. Designing hydrogels for controlled drug delivery. *Nat. Rev. Mater.* 2016, *1*, 16071–16088.
39. Malmsten, M.; Bysell, H.; Hansson, P. Biomacromolecules in microgels – opportunities and challenges for drug delivery. *Curr. Opin. Colloid Interface Sci.* 2010, *15*, 435–444.
40. McClements, D.J. Designing biopolymer microgels to encapsulate, protect and deliver bioactive components: physicochemical aspects. *Adv. Colloid Interface Sci.* 2017, *240*, 31–59.
41. Wylie, R.G.; Ahsan, S.; Aizawa, Y.; Maxwell, K.L.; Morshead, C.M.; Shoichet, M.S. Spatially controlled simultaneous patterning of multiple growth factors in three-dimensional hydrogels. *Nat. Mat.* 2011, *10*, 799–806.
42. Nagy, K.J.; Giano, M.C.; Jin, A.; Pochan, D.J.; Schneider, J.P. Enhanced mechanical rigidity of hydrogels formed from enantiomeric peptide assemblies. *J. Am. Chem. Soc.* 2011, *133*, 14975–14977.
43. Li, X.M.; Kuang, Y.; Lin, H.C.; Gao, Y.; Shi, J.F.; Xu, B. Supramolecular nanofibers and hydrogels of nucleopeptides. *Angew. Chem. Int. Ed.* 2011, *50*, 9365–9369.
44. Tan, Z.Q.; Ohara, S.; Naito, M.; Abe, H. Supramolecular hydrogel of bile salts triggered by single-walled carbon nanotubes. *Adv. Mater.* 2011, *23*, 4053.
45. Lovell, J.F.; Roxin, A.; Ng, K.K.; Qi, Q.C.; McMullen, J.D.; DaCosta, R.S.; Zheng, G. Porphyrin-cross-linked hydrogel for fluorescence-guided monitoring and surgical resection. *Biomacromolecules.* 2011, *12*, 3115–3118.
46. Ouasti, S.; Donno, R.; Cellesi, F.; Sherratt, M.J.; Terenghi, G.; Tirelli N. Network connectivity, mechanical properties and cell adhesion for hyaluronic acid/PEG hydrogels. *Biomaterials.* 2011, *32*, 6456–6470.
47. Shofner, J.P.; Phillips, M.A.; Peppas, N.A. Cellular evaluation of synthesized insulin/transferrin bioconjugates for oral insulin delivery using intelligent complexation hydrogels. *Macromol. Biosci.* 2010, *10*, 299–306.
48. Lin, H.X.; Zou, Y.; Huang, Y.S.; Chen, J.; Zhang, W.Y.; Zhuang, Z.; Jenkins, G.; Yang, C.J. DNAzyme crosslinked hydrogel: a new platform for visual detection of metal ions. *Chem. Commun.* 2011, *47*, 9312–9314.
49. He, C.L.; Zhuang, X.L.; Tang, Z.H.; Tian, H.Y.; Chen, X.S. Stimuli-sensitive synthetic polypeptide-based materials for drug and gene delivery. *Adv. Healthc. Mat.* 2012, *1*, 48–78.
50. Fairbanks, B.D.; Singh, S.P.; Bowman, C.N.; Anseth, K.S. Photodegradable, photoadaptable hydrogels via radical-mediated disulfide fragmentation reaction. *Macromolecules.* 2011, *44*, 2444–2450.
51. Paquet, C.; de Haan, H.W.; Leek, D.M.; Lin, H.Y.; Xiang, B.; Tian, G.H.; Kell, A.; Simard, B. Clusters of superparamagnetic iron oxide nanoparticles encapsulated in a hydrogel: a particle architecture generating a synergistic enhancement of the T2 relaxation. *Acs Nano.* 2011, *5*, 3104–3112.

52. Huynh, C.T.; Nguyen, M.K.; Lee, D.S. Injectable block copolymer hydrogels: achievements and future challenges for biomedical applications. *Macromolecules*. 2011, *44*, 6629–6636.
53. Jeon, O.; Powell, C.; Solorio, L.D.; Krebs, M.D.; Alsberg, E. Affinity-based growth factor delivery using biodegradable, photocrosslinked heparin-alginate hydrogels. *J. Control. Release* 2011, *154*, 258–266.
54. Ito, Y.; Ochiai, Y.; Park, Y.S.; Imanishi, Y. pH-sensitive gating by conformational change of a polypeptide brush grafted onto a porous polymer membrane. *J. Am. Chem. Soc.* 1997, *119*, 1619–1623.
55. He, C.L.; Zhao, C.W.; Chen, X.S.; Guo, Z.J.; Zhuang, X.L.; Jing, X.B. Novel pH-and temperature-responsive block copolymers with tunable pH-responsive range. *Macromol. Rapid Commun.* 2008, *29*, 490–497.
56. Nguyen, H.N.; Wey, S.P.; Juang, J.H.; Sonaje, K.; Ho, Y.C.; Chuang, E.Y.; Hsu, C.-W.; Yen T.-C.; Lin, K.-J.; Sung, H.-W. The glucose-lowering potential of exendin-4 orally delivered via a pH-sensitive nanoparticle vehicle and effects on subsequent insulin secretion in vivo. *Biomaterials*. 2011, *32*, 2673–2682.
57. Qi, X.; Wei, W.; Li, J.; Zuo, G.; Pan, X.; Su, T.; Zhang, J.; Dong, W. Salecan based pH sensitive hydrogels for insulin delivery. *Mol. Pharm.* 2011, *19*, 431–440.
58. Mukhopadhyay, P.; Chakraborty, S.; Bhattacharya, S.; Mishra, R.; Kundu, P. pH sensitive chitosan/alginate core-shell nanoparticles for efficient and safe oral insulin delivery. *Int. J. Biol. Macromol.* 2015, *72*, 640–648.
59. Wang, J.; Xu, M.; Cheng, X.; Kong, M.; Liu, Y.; Feng, C.; Chen, X. Positive/negative surface charge of chitosan based nanogels and its potential influence on oral insulin delivery. *Carbohydr. Polym.* 2016, *136*, 867–874.
60. Gao, X.; Cao, Y.; Song, X.; Zhang, Z.; Zhuang, X.; He, C.; Chen, X. Biodegradable, pH-responsive crboxy methyl cellulose/poly (acrylic acid) hydrogels for oral insulin delivery. *Macromol. Biosci.* 2014, *14*, 565–575.
61. Sheng, J.; He, H.; Han, L.; Qin, J.; Chen, S.; Ru, G.; Li, R.; Yang, P.; Wang, J.; Yang, V. Enhancing insulin oral absorption by using mucoadhesive nanoparticles loaded with LMWP linked insulin conjugate. *J. Control. Release* 2016, *233*, 181–190.
62. Lin, D.; Jiang, G.; Yu, W.; Li, L.; Tong, Z.; Kong, X.; Yan, J. Oral delivery of insulin using CaCO3-based composite nanocomposite nanocarriers with hyaluronic acid coating. *Mater. Lett.* 2017, *188*, 263–266.
63. Lopes, M.; Shrestha, N.; Correja, A.; Shabazi, M.; Sarmento, B.; Hirvonn, J.; Veiga, F.; Seica, R.; Ribeiro, A.; Santos, H. Dual chitosan/albumin coated alginate/dextran sulfate nanoparticle for enhanced oral delivery of insulin. *J. Control. Release* 2016, *232*, 29–41.
64. Ding, J.; He, R.; Zhou, G.; Tang, C.; Yin, C. Multilayered mucoadhesive hydrogel films based on thiolated hyaluronic acid and poly vinylalcohol for insulin delivery. *Acta Biomater.* 2012, *8*, 3643–3651.
65. Salvioni, L.; Fiandra, L.; Gueto, M.; Mazzucchelli, S.; Allevi, R.; Truffi, M.; Sorrentino, L.; Santini, B.; Cerea, M.; Palugan, L.; Corsi, F.; Colombo, M. Oral delivery of insulin via polyethylene imine-based nanoparticles for colonic release allows glycemic control in diabetes rats. *Pharmacol. Res.* 2016, *110*, 122–130.
66. Pileni, M.P. The role of soft colloidal templates in controlling the size and shape of inorganic nanocrystals. *Nat. Mater.* 2003, *2*, 145–150.
67. Sun, Y.; Guo, G.; Tao, D.; Wang, Z. Reverse microemulsion-directed synthesis of hydroxyapatite nanoparticles under hydrothermal conditions. *J. Phys. Chem. Solids.* 2007, *68*, 373–377.
68. Singh, S.; Bhardwaj, P.; Singh, V.; Aggarwal, S.; Mandal, U.K. Synthesis of nanocrystalline calcium phosphate in microemulsion – effect of nature of surfactants. *J Colloid Interface Sci.* 2008, *319*, 322–329.
69. Lim, G.K.; Wang, J.; Ng, S.C.; Gan, LM. Processing of fine hydroxyapatite powders via an inverse microemulsion route. *Mater Lett.* 1996, *28*, 431–446.
70. Karagiozov, C.; Momchilova, D. Synthesis of nano-sized particles from metal carbonates by the method of reversed mycelles. *Chem. Eng. Process.* 2005, *44*, 115–119.
71. Arriagada, F.J. Synthesis of nanosize silica in a nonionic water-in-oil microemulsion. *J. Colloid Interface Sci.* 1999, *211*, 210–220.
72. Paul, B.K.; Moulik, S.P. Microemulsions: an overview. *J. Disper. Sci. Technol.* 1997, *18*, 301–367.
73. Lim, G.K.; Wang, J.; Ng, S.C.; Gan, L.M. Formation of nanocrystalline hydroxyapatite in nonionic surfactant emulsions. *Langmuir.* 1999, *15*, 7472–7477.
74. Bose, S.; Saha, S.K. Synthesis and characterization of Hydroxyapatite nanopowders by emulsion technique. *Chem. Mater.* 2003, *15*, 4464–4469.

75. Nunes, J.; Herlihy, K.P.; Mair, L.; Superfine, R.; De Simone, J.M. Multifunctional shape and size specific magneto-polymer composite particles. *Nano. Lett.* 2010, *10*, 1113–1119.
76. Pelton, R. Unresolved issues in the preparation and characterization of thermo responsive microgels. *Macromol. Symp.* 2004, *207*, 57–66.
77. Li, X.; Zuo, J.; Guo, Y.; Yuan, X. Preparation and characterization of narrowly distributed nanogels with temperature-responsive core and pH-responsive shell. *Macromolecules.* 2004, *37*, 10042–10046.
78. Mu, B.; Liu, P.; Tang, Z.; Du, P.; Dong, Y. Temperature and pH dual-responsive cross-linked polymeric nanocapsules with controllable structures via surface-initiated atom transfer radical polymerization from templates. *Nanomed. Nanotechnol. Biol. Med.* 2011, *7*, 789–796.
79. Li, P.; Zhang, J.; Wang, A. A novel N-succinyl chitosan-graft-poly acrylamide/attapulgite composite hydrogel prepared through inverse suspension polymerization. *Macromol. Mater. Eng.* 2007, *292*, 962–969.
80. Mukhopadhyay, P.; Sarkar, K.; Bhattacharya, S.; Bhattacharyya, A.; Mishra, R.; Kundu, P.P. pH sensitive N-succinyl chitosan grafted poly acrylamide hydrogel for oral insulin delivery. *Carbohydr. Polym.* 2014, *112*, 627–637.
81. Spagnal, C.; Rodrigues, F.H.A.; Neto, A.G.V.C.; Pereira, A.G.B.; Fajardo, A.R.; Radovanovic, E.; Rubira, A.F.; Muniz, E.C. Nanocomposites based on poly (acrylamide-co-acrylate) and cellulose nanowhiskers. *Eur. Polym. J.* 2012, *48*, 454–63.
82. Chang, C.; Duan, B.; Cai, J.; Zhang, L. Superabsorbent hydrogels based on cellulose for smart swelling and controllable delivery. *Eur. Polym. J.* 2010, *46*, 92–100.
83. Prusty, K.; Swain, S.K. Nano silver decorated polyacrylamide/dextran nanohydrogels hybrid composites for drug delivery applications. *Mater. Sci. Eng. C.* 2018, *85*, 130–141.
84. Gao, X.; He, C.; Xiao, C.; Zhuang, X.; Chen, X. Biodegradable pH-responsive poly acrylic acid derivative hydrogels with tunable swelling behaviour for oral delivery of insulin. *Polymer.* 2013, *54*, 1786–1793.
85. Mura, C.; Manconi, M.; Valenti, D.; Manca, M.L.; Diez-Sales, O.; Loy, G.; et al. *In vitro* study of N-succinyl chitosan for targeted delivery of 5-aminosalicylicacid to colon. *Carbohydr. Polym.* 2011, *85*, 578–583.
86. Hebrard, G.; Hoffart, V.; Cardot, J.M.; Subirade, M.; Alric, M.; Beyssac, E. Investigation of coated whey protein/alginate beads as sustained release dosage form in simulated gastrointestinal environment. *Drug Dev. Ind. Pharm.* 2009, *35*, 1103–1112.
87. Ouwerx, C.; Veilings, N.; Mestdagh, M.M.; Axelos, M.A.V. Physico-chemical properties and rheology of alginate beads formed with various divalent cations. *Polym. Gel. Netw.* 1998, *6*, 393–408.
88. Dorkoosh, F.A.; Borchard, G.; Rafiee-Tehrani, M.; Verhoef, J.C.; Junginger, H.E. Evaluation of superporous hydrogel (SPH) and SPH composites in porcine intestine ex vivo: assessment of drug transport, morphology effect, and mechanical fixation to intestinal wall. *Eur. J. Pharm. Biopharm.* 2002, *53*, 161–166.

8

Biodegradable Polymers in Controlled Drug Delivery Systems

Anurakshee Verma, Prabhat Kumar, and Ufana Riaz

CONTENTS

8.1 Biodegradable Polymers in Controlled Drug Delivery Systems

The design of drug delivery systems is regarded as one of the most advancing fields in which chemical engineers and chemists contribute to human health care [1–4]. Biodegradable polymers are generally used in the production of drug transport systems, as they exhibit decreased levels of toxicity and side effects besides targeted release [5]. Several biodegradable polymers have been permitted by the Food and Drug Administration (FDA), which mostly comprise polymers having amide, ester, and anhydride linkages that can be easily degraded by hydrolysis or enzymatic processes [6–10]. Drug transport systems are generally formulated to provide

- Decreased rate and intensity of side effects and toxicity
- Homogeneous distribution of drug concentration
- Reduced dosage frequency
- Prolonged therapeutic effect

8.2 Classification of Drug Release Systems Based on Biodegradable Polymers

Based on their release behavior, drug delivery systems can be classified as given in the following subsections.

8.2.1 Immediate Drug Release Formulations

In immediate drug release systems, the drug molecules break down fast and are dissolved in the dissolution media [11–13]. Mostly, immediate drug release formulations are designed in the form of oral tablets and capsules and are formulated to release the active molecules immediately through oral administration.

8.2.2 Extended Drug Release Formulations

In these systems, the drug carrier allows at least a twofold reduction in dosage frequency as compared with the drug as an immediate release dosage form. Examples of extended release dosage forms include controlled release, sustained release, and long-acting drug products. Sustained release maintains drug release over a sustained period but not at a constant rate [14]. Extended release systems allow the drug to be released over prolonged time periods. By extending the release profile of a drug, the frequency of dosing can be reduced. Extended release can be achieved using sustained or controlled release dosage forms.

8.2.3 Delayed Release Drug Formulations

Such systems release the drug at a particular time rather than rapidly after administration. An initial portion may be released promptly after administration. Enteric-coated dosage forms are common delayed release products (enteric-coated aspirin and other non-steroidal anti-inflammatory [NSAID] products). Delayed release systems can be used to protect the drug from degradation in the low-pH environment of the stomach or to protect the stomach from irritation by the drug. In this case, drug release should be postponed until the dosage has reached the small intestine. For such drug delivery systems, polymers are generally used to achieve the aim of delayed release. The tablet can be coated with a suitable polymer for release systems. The polymer dissolves at a particular pH, and the drug is transported from the low-pH fluid (stomach) to the higher-pH fluid (small intestine). When the polymer dissolves in the fluid, the drug undergoes release.

8.2.4 Pulsatile Drug Release Formulations

Pulsatile drug release systems are developed to deliver the drug at the right time, at the right site of action, and in the right amount. This method provides higher benefit as compared to the conventional dosage system along with increased patient compliance. The systems are designed according to the cardiac rhythm of the human body, and the drug is quickly as well as completely released according to the pulse of the body [15].

8.2.5 Controlled Drug Release Formulations

Controlled drug release delivers the drug at a predetermined rate, systemically, for a specific period of time. Such systems have several advantages over other drug delivery systems, including the short time span of drug release, protection of unstable drugs, and increased patient relief as well as compliance [16]. They are also significant for the release of drugs that are rapidly metabolized and eliminated from the body.

8.2.6 Modified Drug Release Formulations

A modified drug release system is a combination of immediate drug release, delayed drug release, and extended drug release systems. Through modified drug release, drug doses are delivered with a delay after administration, or for a prolonged period of time, or to a specific target in the body [17].

8.3 Biodegradable Polymers Used in Controlled Release Formulations

Recently, there has been much marketing communication on the use of biodegradable polymeric materials for controlled drug transport to overcome the burden of surgical removal of transplants [18–19]. The polymers used for designing controlled release systems must possess the following chemical characteristics:

- More hydrophilic backbone chain—ester, amide, ether, peptide linkages.
- More hydrophilic end-groups—hydroxyl, carbonyl, carboxyl—accelerate biodegradation.
- Lower degree of crystallinity.
- Enhanced porosity.
- Presence of lower–molecular weight fraction.

8.3.1 Polyesters and Their Nanocomposites

The polyester family is regarded as the most suitable for controlled release formulations due to the presence of hydrolysable ester linkages in the backbone, such as poly lactic acid (PLA), poly glycolic acid (PGA), poly (lactase-co-glycoside) (PLGA), etc. [20]. They are obtained through the ring opening polymerization of cyclic lactones [21–23]. The ring opening polymerization rate of cyclic lactones can be increased by adding an activated catalyst (Zn, Sn) with the carbonyl ester. Sn is generally used as the catalyst because it has FDA approval (food stabilizer). On the other hand, recyclable Fe(II) salts are used as initiators for polymerization temperatures above 150 C [21]. Zinc powder and CaH_2 have also been used as potential non-toxic catalysts for the production of a copolymer of PLA and poly-(ethylene oxide) [20, 24–27]. In 1967, a patent was filed that described the use of PLA as the first recyclable material in the field of drug release [28]. Astute et al. (2006) [29] synthesized PLGA copolymer by two different monomers through ring opening copolymerization. By manipulation of the copolymer/nanoparticle characteristics, the drug entrapment efficiency/release can be improved by carefully controlling all parameters. Ida et al. [30] synthesized biodegradable copolymers—poly (lactide-co-glycolide) grafted dextran (Dex-g-PGLA) microparticles—in three compositions, A (PGLA 2790), B (PGLA 3000), and C (PGLA 4330), and studied the encapsulation efficiency and *in vitro* drug release. It followed zero-order release kinetics without a significant initial burst release of drug. The rate of drug (protein) release was controllable by changing the composition of (Dex-g-PGLA). When the PGLA content was increased, the release took longer; about 50% release was observed by A for 7 days, B for 11 days, and C for 21 days [30]. Sahana et al. [31] prepared poly (lactide-co-glycolide) by an emulsion–diffusion and evaporation method using different solvents, stabilizers, and surfactants. All nanocomposites showed a similar *in vitro* release profile of estradiol for the period of 45 days in pH 7.4 phosphate buffer, following zero-order kinetics, but in the case of dichloromethane (DCM), average release was found to increase with time. Pei et al. [32] synthesized PGLA and PEGylated poly(lactic-co-glycolic acid) (PEG–PGLA) nanocomposites, which showed an initial burst release of bovine serum albumin (BSA), while only 30% release was observed from PEG–PGLA and 20% from PGLA within a 12 h time period; after 12 h, it showed sustained release. Cumulative release of BSA was observed to be 71% from PEG–PGLA and 49% from PGLA [32]. Pappas et al. [33] described the use of PLA, poly(glycolic acid) poly(ethylene glycol), and their copolymer in controlled drug release. Kranz et al. [34] prepared poly (lactide) and poly (lactide-co-glycolide) solution dispersed in the external oil phase and investigated the *in vitro* drug release of buserelin acetate and diltiazem-HCl for the period of 72 h at phosphate buffer solution. After 24 h, buserelin release was observed to be 39%, while diltiazem-HCl showed only 26% release. After 72 h, buserelin showed extended drug release, and diltiazem-HCl revealed degradation of PLA and PGLA [34]. Peng et al. [35] developed a PLA hydrogel for the release of mitomycin C and dexamethasone sodium phosphate for a period of 40 days. One milligram of mitomycin was found to be released within 35 days, and 2 mg dexamethasone sodium phosphate was released within 28 days. Dexamethasone sodium phosphate showed fast release up to 6 h, and cumulative drug release was observed to be 31%. For mitomycin, 31%

cumulative drug release was noticed in 21 days. Wang et al. [36] investigated bi- and trilayer biodegradable polymers, in which PLA was used as the supporting layer and PGLA as the drug-eluting layer, and third layer was composed of biodegradable polyester. *In vitro* release of sirolimus from these nanocomposites was detected for a period of 40 days and revealed 70% drug release, with increase in the thickness of the polymer layer [37]. Kim et al. [38] observed the release of insulin (a hydrophilic drug) from poly(lactide-co-glycolide) microcapsules; around 20% of the insulin was released within 24 h, 40% after 240 h, and around 50% after 30 days. Omayra et al. [39] developed a polyester dendritic scaffold for the *in vitro* and *in vivo* delivery of an anticancer drug. For the *in vitro* release, doxorubicin (DOX)–polymer composites decreased the number of cancer cells, and some DOX was released via these composites Figure 8.1 [39].

Sivak et al. [40–42] used PEG–PLA block copolymers for coating tablets of anhydrous theophylline [40], diprophylline [41], 7-tert-butyldi-methylsilyl-10-hydroxy-camptothecin, and metformin hydrochloride [42]. Vervaet et al. [43] synthesized thermoplastic polyurethane via hot melt extrusion, injection molding, and melt granulation for metformin hydrochloride release. The release from hot melt extrusion matrices and injection molded tablets showed sustained release for the period of 1 day (24 h); the release from the thermoplastic polyurethane–based melt granulation tablets was too fast and revealed complete release within 6 h [43]. Yang et al. (2015) [44] developed a PEG–PLA block copolymer. Rapid release of DOX was observed due to the breaking of disulfide bonds. In the presence of the reducing agent dithiothreitol, DOX was released from PEG–PLA micelles up to 64% after 14 hours. But in the absence of reducing agent, only 40% of DOX was released after 14 h [44].

8.3.2 Celluloses and Their Nanocomposites

Nanocellulose, bacterial cellulose (BC), cellulose nanocrystals (CNC), and cellulose nanofibrils (CNF) show excellent mechanical, physical, and biological properties such as biodegradability, biocompatibility, and low cytotoxicity [45]. Carboxymethyl cellulose (CMC) is the most widely used of all. CMC is biocompatible, biodegradable, hydrophilic, non-toxic in nature, and easily available [46–48]. Sideman et al. [49] synthesized hydroxyl-propyl methylcellulose for pharmaceutical devices, while Olga et al. [50] synthesized spherical cellulose nanocrystals with nanotitania and prepared nanocomposite films, which were used for the release study of triclosan. The results revealed that the CNC could be synthesized directly from filter paper with a lower concentrated acid hydrolysis time of 60 min. The drug release study was described by the Higuchi model and showed a long-term release profile for triclosan (Figure 8.2).

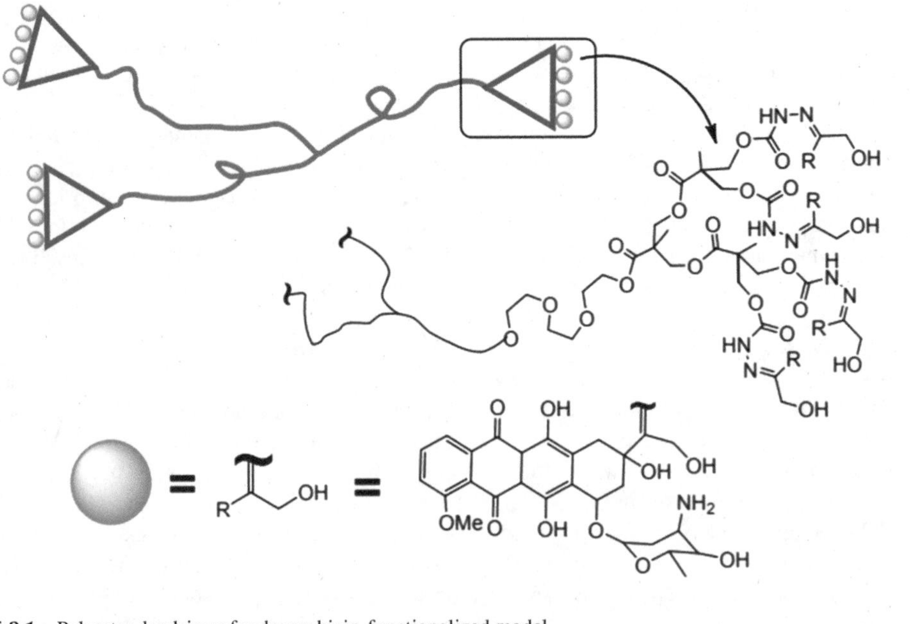

FIGURE 8.1 Polyester dendrimer for doxorubicin-functionalized model.

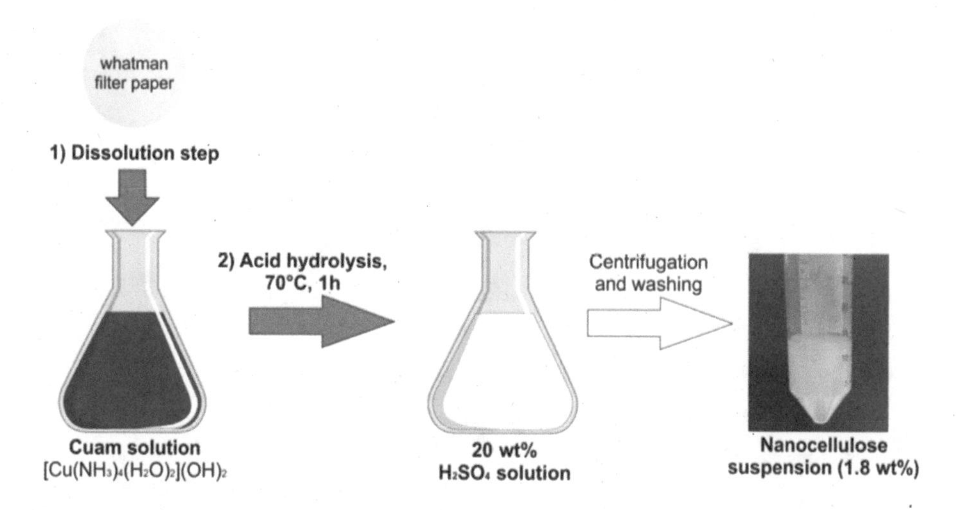

FIGURE 8.2 Synthesis route of spherical-shaped nanocellulose from filter paper.

Rasoulzadeh et al. [51] developed CMC with graphene oxide and prepared biodegradable nanohybrid hydrogel beads. The synthesized nanohybrid was used for controlled release of DOX. Graphene oxide and the anticancer drug DOX have π–π bonding interaction, and a high drug loading capacity was observed. The DOX release characteristics of the nanohybrid hydrogels were found to be more pH sensitive due to hydrogen bond interaction; they showed rapid release at pH 6.8 as compared with pH 7.4. The large amount of filler (graphene oxide) decreased the release rate due to synergistic interactions in the composites. Mahdavini et al. [52] synthesized magnetically reactive hydrogels by radical polymerization of CMC on acrylamide (Am), using potassium persulfate as a radical initiator, while N,N′methylenebis-acrylamide was used as a cross-linker component and nanoclay (montmorillonite) as a nanofiller. Under an applied magnetic field, *in vitro* drug release profiles showed cumulative drug release of approximately 79% in intestinal fluid (pH = 7.4). Wang et al. [53] reported the drug release profile of tetracycline and DOX using CNC as a vehicle, and the profile showed rapid release for 24 h. However, when the CNC were reacted with cetyl trimethyl ammonium bromide, the bound hydrophobic drugs showed a release profile for 48 h. Ntoutoume et al. [54] developed cationic β-cyclodextrin (β-CD) and cyclodextrin cellulose nanocrystals (CD-CNCs). Encapsulation efficiency and *in vitro* controlled release of an anticancer drug (curcumin) were observed for 48 h. Within 8 h, it showed only 20% drug release. Curcumin–CD/CNCx complexes revealed three to four times more effective drug release than curcumin alone or Curcumin–CD complex. Wang et al. [55] designed novel polyphosphoester-grafted CNC for binding with DOX and investigated its release into HeLa cells. The drug release was interrupted by electrostatic interaction in the acidic environment inside the tumor cells. Furong et al. [56] synthesized succinic anhydride–modified microcrystalline cellulose films (MCC) as a drug carrier for domperidone, and the release mechanism was described using the Korsmeyer–Peppas equation at pH 7.4. The developed drug delivery system was shown to improve the pharmacodynamics and bioavailability of drugs. Espinar et al. [57] prepared a film-coating material of ethyl cellulose and pectin, which was used for coating an anti-inflammatory drug (Trimcinolone) and loaded into pellets of CD. The controlled release profile of triamcinolone acetonide was investigated at pH 6 and was observed to be about 70%, while at acidic pH (1.2), it showed only 18% drug release within 5 h (Figure 8.3) [57].

Dufresne et al. [58] reported hydrogel-based CNC prepared by *in situ* host–guest inclusion. β-CD was first grafted onto CNC; then, α-CD and Pluronic polymer were attached by *in situ* inclusion to form a supramolecular hydrogel, which was used as a drug carrier for DOX. Wu et al. [59] formulated temperature- and pH-dependent polymer nanomembranes of poly (N-iso-propylacrylamide-co-methacrylic acid), used for the controlled release of lysozyme, Leuprolide, *N*-benzoyl-l-tyrosine ethyl ester HCl, momany peptide, insulin, and vitamin B_{12}.

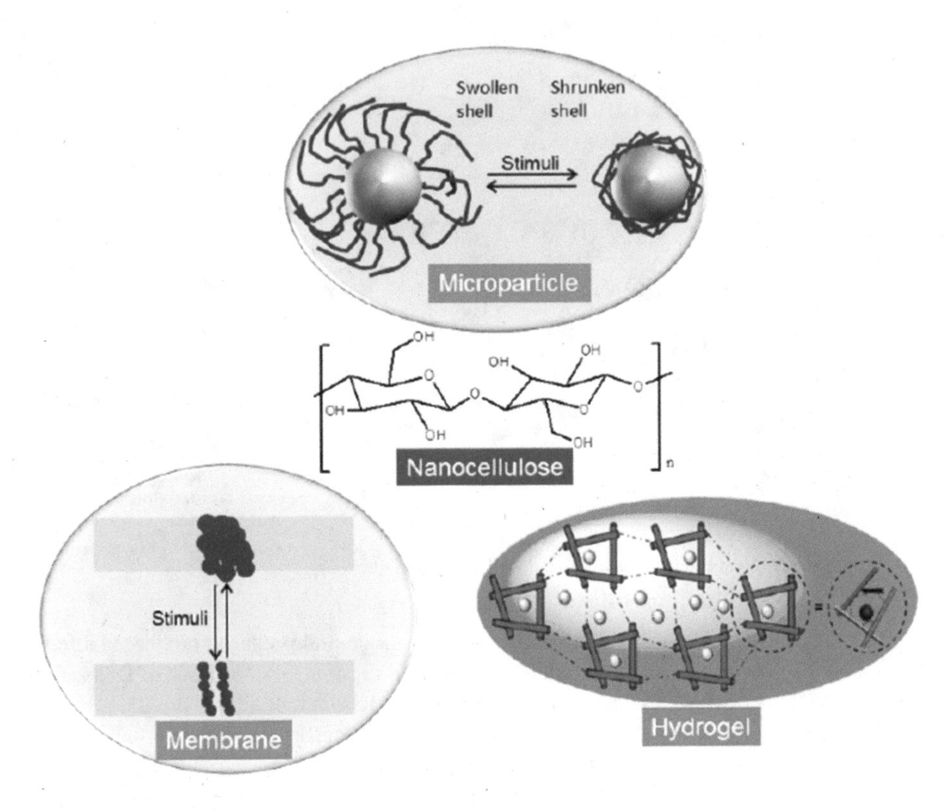

FIGURE 8.3 Nanocellulose-based smart drug delivery systems.

8.3.3 Polyethers and Their Nanocomposites

Polyethers exhibit high water solubility, good chemical stability, and low toxicity. These polymers reveal decreased interaction with blood and show high biocompatibility. Gupta et al. [60] highlighted the use of polyether-based amphiphiles in therapeutic applications with a safety profile, non-toxicity, high biocompatibility, and decreased interaction with blood components [60]. Polyethylene glycol (PEG) is the most important component in the field of drug delivery; it is regarded as a gold standard due to approval by the FDA. It possesses ion-transporting ability, non-toxicity, decreased interaction with blood components, water solubility, high chemical stability, and biocompatibility [61–63]. Zhang et al. [64] developed PEG and iron oxide nanoparticles, and methotrexate was selected as a model drug for the release studies. The drug release behavior revealed a viable solution to chemotherapeutic drug delivery for the treatment of gliomas and other cancers and demonstrated increased magnetic resonance imaging (MRI) contrast enhancement through intracellular uptake by target cells [64]. Gao et al. [65] synthesized a self-assembled copolymer of mono-methoxy polyethylene glycol and poly ε-caprolactone. An *in vitro* DOX release study showed that release from micelles at pH 5 was about 85% and at pH 7.4, only around 20%. It showed better drug release capacity in phosphate buffer as compared with acidic buffer [65]. Prabaharan et al. [66] synthesized a hyperbranched block copolymer of poly (lactide) and PEG and prepared micelles. An anticancer drug (DOX) encapsulated by these micelles provided initial rapid release up to 4 h and then showed sustained release for around 40–50 h; then, no release was observed up to 80 hours at both pH 7.4 and 5.4 [66]. Feng et al. [67] developed D-α-tocopheryl polyethylene glycol succinate nanoparticles. Natural vitamin E and Taxotere were loaded into nanoparticles. Nanoparticles extended the half-life of the drug in plasma and enhanced the cellular uptake of the drug by controlling release. Nanoparticle drug release was observed within 15 days for chemotherapy [67]. Birjand et al. [68] synthesized highly biocompatible thermoresponsive nanogels based on dendritic polyglycerol and oligo-ethylene. Drugs and dyes were evaluated based on the thermal response of the nanogel release systems. Harting et al. [69] formulated biodegradable polyether-ether-ketone (PEEK) and established the *in vitro* release bechaviour of gentamicin. It was observed that Gentamicin sulfate showed fast release, whereas gentamicin crobefat (being insoluble in water) revealed continuous release for about 10 days [69].

Zeng et al. [70] prepared a block copolymer of PEG and PLA. Paclitaxel was loaded into PEG–PLA micelles, and it was found that 17.6% of paclitaxel dissolved completely in sodium salicylate solution after 5 days, while 27.6% of paclitaxel dissolved completely after 3 days. Tziveleka et al. [71] synthesized a multifunctional hyperbranched polyether polyol. and pyrene and tamoxifen were selected as model drugs for a release study. Han et al. [72] synthesized the hydrogel via esterification with chlorinated polyethylene glycol bis-carboxymethyl ether. The swelling ratio and drug release profiles of a model drug (aspirin) were also investigated, and aspirin release profiles were observed with various cross-linking ratios at 37 °C for a period of 5 h. The release kinetics showed a good fit to the zero-order model. Ortega et al. [73] developed a polyether-block-amide copolymer film with acrylic acid, which was used for the release of vancomycin. A sustained release profile was observed for 48 h.

8.3.4 Polyamides and Their Nanocomposites

Polyamides are linear, crystalline, and thermoplastic in nature (Figure 8.4). Aliphatic polyamides such as nylons and polyglutamic acid are very useful materials regarding drug release characteristics [74]. Nathan et al. [75] described controlled drug release via amino acids and their derivatives. Anderson et al. [76] prepared γ-benzyl-L-glutamate-L-leucine copolymers in different compositions, and *in vitro* and *in vivo* tests were performed after implantation in rats for 200 days. After 14 days, a higher content of glutamic acid in the copolymer was observed, which showed faster implant degradation as compared with other compositions. Langer et al. [77] described a review focused on a controlled release system of drug, protein, and bioactive agent through biopolymeric materials. Progesterone *in vitro* release from PLA and PLA/PGA film (3 μm) were observed for a period of 9 weeks; initially, it showed very slow drug release, followed by a sudden increase in the rate of release due to mechanical deterioration and fragmentation. Schakenraad et al. [78] discussed a review of the physical, chemical, and biological properties of a poly (lactic/glycine) copolymer that was used for *in vivo* and *in vitro* drug release capacities [78, 79]. Gachard et al. [80] synthesized amino acid–based polyamide. Poly (adipoyl-L-lysine) was linked with benzocaine, and the release of benzocaine was observed for 90 h. A higher initial burst release of 26% was investigated after 15 min, while 30% of the drug was found insoluble after 24 h. But, for N-protected poly(L-cystyl-L-cystine), drug release was observed for a period of 7 weeks at pH 7.4 [80]. Khattab et al. [81] synthesized s-triazine polyamides containing glycine and thioglycolic acid, and Celecoxib

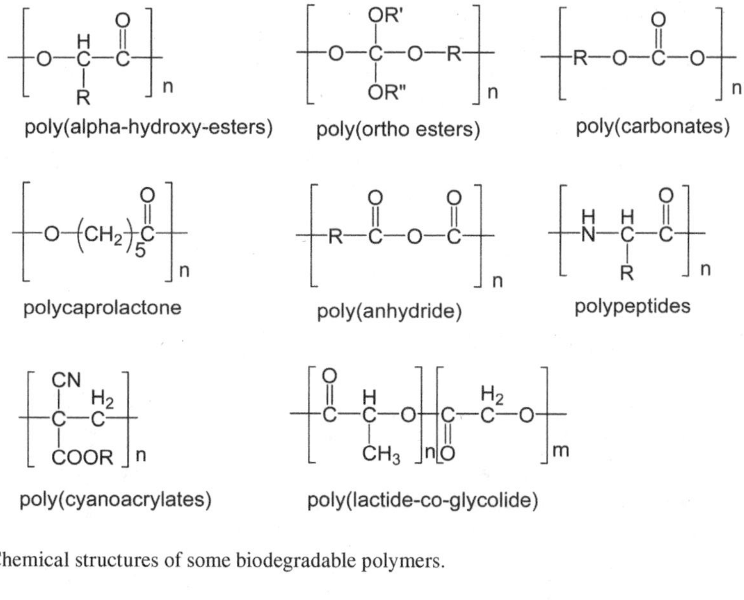

poly(alpha-hydroxy-esters) poly(ortho esters) poly(carbonates)

polycaprolactone poly(anhydride) polypeptides

poly(cyanoacrylates) poly(lactide-co-glycolide)

FIGURE 8.4 Chemical structures of some biodegradable polymers.

was selected as a model drug. The release study was analyzed for 50 h. *In vitro* cumulative release profiles of Celecoxib from different polymeric nanoparticles were investigated. After 48 h, the release was found to be 46.9%, 64.20%, 57.81%, 53.95%, and 49.43% from the polymeric nanoparticles CXB-26, CXB-43, CXB-44, CXB-45, and CXB-46, respectively. The highest drug release was observed for CXB-43 nanoparticles. Li et al. [82] prepared scaffolds from biodegradable polyurethane. Recombinant human bone morphogenetic protein (rhBMP-2) was selected for release in the skin. The release profiles showed burst release of about 35% for 21 days. Seo et al. [83] developed polyurethane by the incorporation of PEG for the controlled release of paclitaxel. The drug release from polyurethane was observed to be 8.6% higher and extended for a period of 19 days [83]. Martenelli et al. [84] prepared carboxylated polyurethane consisting of albumin nanoparticles. An antibiotic drug (cefamandole nafate) was loaded into the nanoparticles. The drug release profiles displayed a slow release of the antibiotic drug for a period of 10 days. Chen et al. [85] synthesized a copolymer of PEG with polyurethane. Different drugs (diclofenac sodium, thiamazole, and ibuprofen) were loaded into the copolymer. *In vitro* drug release profiles and kinetics depended on the solubility of the drug and drug levels. *In vitro* drug release profiles of thiamazole and diclofenac sodium showed approximately 95% release within 50 h, but for ibuprofen, the release profile showed only 75% release within the same time period. Li et al. [86] used biodegradable polyurethane scaffolds for the release of vancomycin hydrochloride, and the release profile of vancomycin showed extended release for 8 weeks [86]. Sebastian et al. [87] prepared nanoparticles of poly(glutamic acid) and copolymers of poly(glutamic acid) with ethyl, hexyl, dodecyl, and octadecyl glutamate units having 50% and 75% alkyl content. These polymers and copolymers were used for the delivery of erythromycin, and *in vitro* release of erythromycin was observed for 40 days at pH 7.4 and 37 C. The polymer showed faster release as compared with the copolymer. Brocchini et al. [88] prepared natural amino acids (polyglutamic, polyaspartic acid), which are non-toxic and biocompatible, and fabricated peptide-loaded polyarylate films. Peptide was selected as a model drug for *in vitro* release at pH 7.4 (phosphate buffer) for a period of 70 min. It was noticed that with the increase in the molecular weight of the polymer, cumulative drug release decreased. Nakanishi et al. [89] developed a polymeric micelle for a drug transport system containing PEG-conjugated DOX and poly(aspartic acid) for the delivery of DOX. NK911 has a small particle size of about 40 nm in diameter and accumulated in tumor tissue by the enhanced permeability and retention (EPR) effect, showing much stronger activity than free DOX; the drug was gradually released from the inner core by equilibrium phenomena within 8–24 h. Park et al. [90] synthesized a poly-glutamic acid (PGLA)–based hydrogel for the controlled release of clot-mixing and tissue-type plasminogen activator. Tissue-type plasminogen activator was loaded into the hydrogel for local drug delivery over a period of 10 days. The release profile showed 0% content of PGLA in the hydrogel exhibiting the highest drug release, but with an increase in the PGLA content, the drug release was decreased. Kolawole et al. [91] developed a polyamide-based monolithic drug delivery system.

Sahahuddin et al. [92] prepared microbicidal polyamides using 5 phenyl 1,3,5, oxadiazole 2 thiol 5-phenyl-1,3,5,oxadiazole-2-amine, 5-(4-chlorophenyl)–1,3,4-thiadiazole-2-thiol. Intercalation of montmorillonite was also observed for polyamides, and the nanocomposites showed slow release of drug in buffer solutions of pH 2.3, 5.8, and 7.4 for 2880 min. The release profile of nanocomposites depended on the pH, and 50% of the drug was released after 10 h, 15 h, and 19 h at pH 7.4, 5.8, and 2.3, respectively. [92]. Olkhov et al. [93] investigated drug transport properties and diffusion properties of films based on poly (3-hydroxy butyrate) and polyamide. The antiseptic drug furacillin showed constant and extended controlled release.

8.4 Proposed Mechanisms of Controlled Drug Release

A large number of scientific results based on controlled release is available, but all of them are explained by a limited number of mechanisms. The mechanisms of controlled drug release are based on release of active drug molecules from delivery systems: degradation, diffusion, swelling followed by diffusion method, and other active efflux. All these mechanisms make use of physical changes and are involved in the release system when this system is present in a biological environment. For this reason, simple drug release systems with standard active molecules (drugs) are used in the preparation of controlled drug delivery systems and also for biomedical applications (Figure 8.5) [94].

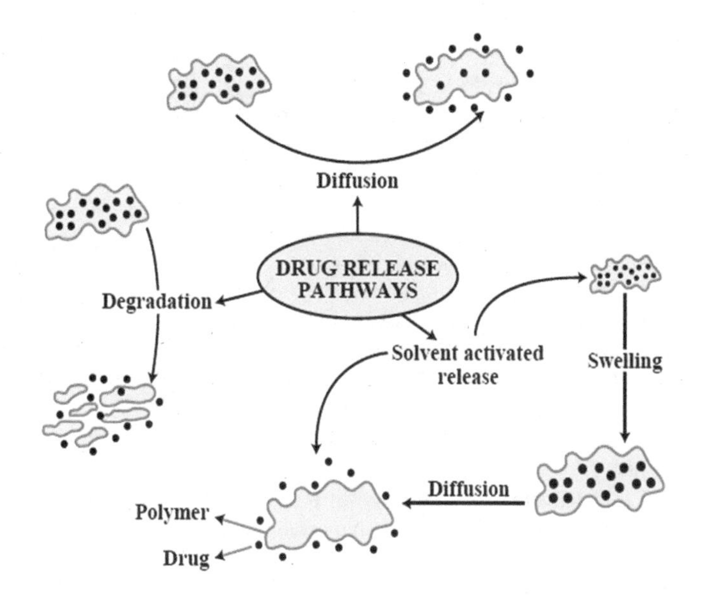

FIGURE 8.5 Mechanisms for controlled drug release.

8.4.1 Diffusion

In the diffusion process, the drug is either suspended within a polymeric matrix or encapsulated in the polymer matrix. During the planning of polymeric drug transport devices employing diffusion, parameters such as the size of the drug molecule and the porosity, degree of cross-linking, and swelling behavior of the polymer play a significant role in the rate of release [95].

8.4.2 Degradation

Polymer degradation is the most attractive way of drug release. The polymer is planned to degrade and release the drug at a required site in the animal/human body. The greatest advantage of using degradable polymers in controlled drug release systems is that they degrade into biologically acceptable agents that are metabolized and removed from the animal/human body through regular metabolic processes [96, 97].

8.4.3 Swelling Followed by Diffusion

In this mechanism, the drug is released by a pulse method only at a specific site. The delivery matrix is merged with a sensor that stimulates the drug release through the detection of the environmental parameters. The swelling followed by diffusion method is more effective for the release of insulin.

8.5 Conclusion

The processability and biodegradability of polymers can help in formulating new drug delivery systems that intensify treatment and therapy. pH-sensitive and temperature-dependent biodegradable polymers are desired for controlled drug delivery systems. The chain length, molecular weight, and polymer microstructure also affect polymer properties as well as degradation behavior, which directly influences the drug release kinetics, and thus, they should be controlled throughout the design and development of new drug delivery vehicles. The use of biodegradable polymers for pharmaceutical and medical applications therefore requires tailoring of the physicochemical properties of polymers to obtain the desired features for targeted and controlled drug delivery applications.

REFERENCES

1. Serra, L., Doménech, J., Peppas, N.A. 2006. Drug transport mechanisms and release kinetics from molecularly designed poly(acrylic acid-g-ethylene glycol) hydrogels. *Biomaterials* 27: 5440–5451.
2. Caccavo, D., Cascone, S., Lamberti, G., Barba, A.A. 2015. Controlled drug release from hydrogel-based matrices: Experiments and modeling. *Int J Pharma* 486: 144–152.
3. Gentile, P., Bellucci, D., Sola, A., Mattu, C., Cannillo, V., Ciardelli, G. 2015. Composite scaffolds for controlled drug release: Role of the polyurethane nanoparticles on the physical properties and cell behaviour. *J Mech Behav Biomed Mater* 44: 53–60.
4. Ambade, A.V., Savariar, E.N., Thayumanavan, S. 2005. Dendrimeric micelles for controlled drug release and targeted delivery. *Molecular Pharm* 2: 264–272.
5. Lewis, D.D. 1990. *Biodegradable Polymers as Drug Delivery Systems*. New York: Marcel Dekker.
6. Dorati, R., Genta, I., Colonna, C. 2007. Investigation of the degradation behavior of poly(ethylene glycolcoD,L-lactide) copolymer. *Polym Degrad Stabil* 92: 1660–1668.
7. Heller, J. 1980. Controlled release of biologically active compounds from bioerodible polymers. *Biomaterials* 1: 51–57.
8. Langer, R. 1998. Drug delivery and targeting. *Nature* 392: 5–10.
9. Shaik, M.R., Korsapati, M., Panati, D. 2012. Polymers in controlled drug delivery systems. *Int J Pharma Sci* 2: 112–116.
10. Kotwal, V.B., Saifee, M., Inamdar, N., Bhise, K. 2007. Biodegradable polymers: Which, when and why? *Indian J Pharm Sci* 69: 616–625.
11. Nyol, S., Gupta, M.M. 2013. Immediate drug release dosage form: A review. *J Drug Delivery & Ther* 3(2): 155–161.
12. Patel, U., Patel, K., Shah, D.,Shah, R. 2012. A Review on Immediate Release Drug Delivery System. *Int J Pharma Res Biosci* (5): 37–66.
13. Wale, K., Salunkhe, K., Gundecha, I., Balsane, M., Hase, S., Pande, P. 2014. Immediate Drug Release Dosage Form: A Review. *Am J Pharmtech Res* 4(1): 1–15.
14. Shargel, L., Wu-Pong, S., Yu, A.B.C. Applied Biopharmaceutics & Pharmacokinetics, 6e Pharmaceutics: Drug Delivery and Targeting, 7–13.
15. Jain, D., Raturi R., Jain, V., Bansal, P., Singh, R. 2011. Recent Technologies in Pulsatile Drug Delivery Systems. *Biomatter* 1(1): 57–65.
16. Kaplan, D.L., Wiley, B.J., Mayer, J.M., Arcidiacono, S., Keith, J., Lombardi, S.J., Ball, D., Allen, A.L. 1994. In *Biomedical Polymers*. Shalaby, S., Ed., Hansen: Cincinnati, OH.
17. Chandra, R., Rustgi, R., 1998. Polymer Science. *Prog Polym Sci* 23: 1273–1335.
18. Birnbaum, D.T., Brannon-Peppas, L. 2002. Polymers in Controlled Release. *Polym News* 27: 13.
19. Uhrich, K.E., Cannizzaro, S.M., Langer, R.S., Shakesheff, K.M., 1999. Polymeric Systems for Controlled Drug Release. *Chem Rev* 99(11): 3181–3198.
20. Lanza, R.P., Langer, R., Chick, W.L. 1997. *Principles of Tissue Engineering*. R.G. Landes Co. and Academic Press: Austin, TX.
21. Wong, W.H., Mooney, D.J. 1997. In *Synthetic Biodegradable Polymer Scaffolds*. Atata, A., Mooney D., Eds., Birkhauser: Boston, MA.
22. Griffith, L.G. 2000. Polymeric Biomaterials. *Acta Mater* 48: 263–277.
23. Schmitt, E.E., Polistina, R.A. 1967. In U.S. Patent. No. 3 297 033.
24. Hollinger, J.O. 1995. *Calvarial Bone Repair with Porous D,L-Polylactide*. CRC Press: Boca Raton, FL.
25. Shalaby, S.W. 1994. *Biomedical Polymers: Designed-to-Degrade Systems*. Shalaby, S.W., Ed., Hanser/Gardner: Cincinnati, OH.
26. Kricheldorf, H.R., Damrau, D.O. 1997. Polylactones, 38.* Polymerization of L-Lactide with Fe(II) Lactate and Other Resorbable Fe(II) Salts. *Macromol Chem Phys* 198: 1767–1774.
27. Li, S.M., Rashkov, I., Espartero, J.L., Manolova, N., Vert, M. 1996. Polymerization of Lactides and Lactones. 10. Synthesis, Characterization, and Application of Amino-Terminated Poly(ethylene glycol)-co-poly(ε-caprolactone) Block Copolymer. *Macromolecules* 29: 59.
28. Singh, L., Kumar, V., Ratner B.D. 2004. Generation of Porous Microcellular 85/15 poly (DL-Lactide-Coglycolide) Foams for Biomedical Applications. *Biomaterials* 25: 2611–2617.
29. Astete, C.E., Sabliov, C.M. 2006. Synthesis and Characterization of PLGA Nanoparticles. *J Biomater Sci Polym Ed* 17: 247–289.

30. Kakizawa, Y., Nishio, R., Hirano, T. 2009. Controlled Release of Protein Drugs From Newly Developed Amphiphilic Polymer-Based Microparticles Composed of Nanoparticles. *J Control Release* 142: 8–13.
31. Sahana, D.K., Mittal, G., Bhardwaj, V., Kumar, R.M.N.V. 2008. PLGA Nanoparticles for Oral Delivery of Hydrophobic Drugs: Influence of Organic Solvent on Nanoparticle Formation and Release Behavior In Vitro and In Vivo Using Estradiol as a Model Drug. *J Pharm Sci* 97: 1530–1542.
32. Li, Y., Pei, Y., Zhang, X., 2001. PEGylated PLGA Nanoparticles as Protein Carriers: Synthesis, Preparation and Biodistribution in Rats. *J Control Release* 71: 203–211.
33. Brannon-Peppas, L. 1995. Recent Advances on the Use of Biodegradable Microparticles and Nanoparticles Incontrolled Drug Delivery. *Int J Pharm* 116: 1–9.
34. Kranz, H., Bodmeier, R. 2007. A Novel In Situ Forming Drug Delivery System for Controlled Parenteral Drug Delivery. *Int J Pharm* 332: 107–114.
35. Li, J., Peng, L., Sun, J., Guo, H., Gup, K. 2016. Slow-Release Drug Delivery System with Polylactic Acid Hydrogels in Prevention of Tracheal Wall Fibroplasia. *Arch Clin Exp Surg* 1: 1–7.
36. Biondi, M., Ungaro, F., Quaglia, F., Netti, P.A. 2008. Controlled Drug Delivery in Tissue Engineering. *Adv Drug Deliv Rev* 60: 229–242.
37. Wang, X.T., Venkatraman, S.S., Boey, F.Y.C. 2006. Controlled Release of Sirolimus from a Multilayered PLGA Stent Matrix. *Biomaterials* 27: 5588–5595.
38. Kim, B.S., Oh, J.M., Hyun, H. 2009. Insulin-Loaded Microcapsules for In Vivo Delivery. *Mol Pharm* 6: 353–365.
39. Omayra, L., Padilla, D.J., Henrik, R.I., Lucie, G., Jean M.J. Fréchet, Francis, C.S. 2002. Polyester Dendritic Systems for Drug Delivery Applications: In Vitro and In Vivo Evaluation. *Bioconjug Chem* 13(3): 453–461.
40. Campiñez, M.D., Aguilar-De-Leyva, Á., Ferris, C., De Paz, M.V., Galbis, J.A., Caraballo, I. 2013. Study of the Properties of the New Biodegradable Polyurethane PU (TEG-HMDI) as Matrix Forming Excipient for Controlled Drug Delivery. *Drug Dev Ind Pharm* 39: 1758–1764.
41. Claeys, B., De Bruyn, S., Hansen, L., De Beer, T., Remon, J.P., Vervaet, C. 2014. Release Characteristics of Polyurethane Tablets Containing Dicarboxylic Acids as Release Modifiers: A Case Study With Diprophylline. *Int J Pharm* 477: 244–250.
42. Sivak, W.N., Pollack, I.F., Petoud, S., Zamboni, W.C., Zhang, J., Beckman, E.J. 2008. LDI-Glycerol Polyurethane Implants Exhibit Controlled Release of DB-67 and Anti-Tumor Activity In Vitro Against Malignant Gliomas. *Acta Biomater* 4: 852–862.
43. Verstraete, G., Mertens, P., Grymonpré, W., Van Bockstal, P.J., De Beer, T., Boone, M.N., Van Hoorebeke, L., Remon, J.P., Vervaet, C. 2016. A Comparative Study Between Melt Granulation/Compression and Hot Melt Extrusion/Injection Molding for the Manufacturing of Oral Sustained Release Thermoplastic Polyurethane Matrices. *Int J Pharm* 516: 602–611.
44. Yang, Q., Tan, L., He, C., Liu, B., Xu, Y., Zhu, Z., Shao, Z., Gong, B., Shen, Y.-M. 2015. Redox-Responsive Micelles Self-Assembled From Dynamic Covalent Block Copolymers for Intracellular Drug Delivery. *Acta Biomater* 17: 193–200.
45. Xie, J., Li, J. 2017. Smart Drug Delivery System Based on Nanocelluloses. *J Bioresour Bioprod* 2(1): 1–3.
46. Zhu, K., Yea, T., Liua, J., Peng, Z., Xu, S., Lei, J.H.D., Li, B. 2013. Nanogels Fabricated by Lysozyme and Sodium Carboxymethyl Cellulose for 5-Fluorouracil Controlled Release. *Int J Pharm* 441(1): 721–727.
47. Lei, H., Hongshan, L., Liufeng, L., Bin, L. 2015. Green-Step Assembly of Low Density Lipoprotein/ Sodium Carboxymethyl Cellulose Nanogels for Facile Loading and pH-Dependent Release of Doxorubicin. *Colloids Surf B Biointerfaces* (126C): 288–296.
48. Azhar, F.F., Olad, A. 2014. A Study on Sustained Release Formulations for Oral Delivery of 5-Fluorouracil Based on Alginate–Chitosan/Montmorillonite Nanocomposite Systems. *Appl Clay Sci* (101): 288–296.
49. Siepmann, J., Peppas, N.A. 2001. Modeling of Drug Release From Delivery Systems Based on Hydroxypropyl Methylcellulose (HPMC). *Adv Drug Deliv Rev* 48: 139–157.
50. Evdokimova, Olga L., Svensson, Fredric G., Agafonov, Alexander V., Håkansson, Sebastian, Seisenbaeva, Gulaim A., Kessler, Vadim G. 2018. Hybrid Drug Delivery Patches Based on Spherical Cellulose Nanocrystals and Colloid Titania-Synthesis and Antibacterial Properties. *Nanomaterials* 8: 228.

51. Rasoulzadeh, M., Namazi, H. 2017. Carboxymethyl Cellulose/Graphene Oxide Bionanocomposite Hydrogel Beads as Anticancer Drug Carrier Agent. *Carbohydr Polym* 168: 320–326.
52. Mahdavinia, G., Afzali, A., Etemadi, H., Hosseinzadeh, H. 2017. Magnetic/pH-Sensitive Nanocomposite Hydrogel Based Carboxymethyl Cellulose-g-Polyacrylamide/Montmorillonite for Colon Targeted Drug Deliver. *Nanomed Res J* 2(2): 111–122.
53. Wang, H., Roman, M. 2011. Formation and Properties of Chitosan-Cellulose Nanocrystal Polyelectrolyte–Macroion Complexes for Drug Delivery Applications. *Biomacromolecules* 12: 1585–1593.
54. Ntoutoume, G.M.A., Granet, R., Mbakidi, J.P. et al. 2016. Development of Curcumin–Cyclodextrin/Cellulose Nanocrystals Complexes: New Anticancer Drug Delivery Systems. *Bioorg Med Chem Lett* 26: 941–945.
55. Wang, H., He, J., Zhang, M., Tam, K.C., Ni, P. 2015. A New Pathway Towards Polymer Modified Cellulose Nanocrystals via a "Grafting Onto" Process for Drug Delivery. *Polym Chem* 6: 4206–4209.
56. Chengmei, S., Tao, F., Cui, Y. 2013. Cellulose-Based Film Modified by Succinic Anhydride for the Controlled Release of Domperidone. *J Biomater Sci Polym Ed* 29: 1233–1249.
57. Villar, E., Asteria, L., Álvarez, L., Blanco, J., Francisco, M., Otero-Espinar, B. 2017. Cellulose-Polysaccharide Film-Coating of Cyclodextrin Based Pellets for Controlled Drug Release. *J Drug Deliv Sci Technol* 42: 273–283.
58. Lin, N., Dufresne, A. 2013. Supramolecular Hydrogels From In Situ Host-Guest Inclusion Between Chemically Modified Cellulose Nanocrystals and Cyclodextrin. *Biomacromolecules* 14: 871–880.
59. Zhang K., Wu X.Y. 2004. Temperature and pH-Responsive Polymeric Composite Membranes for Controlled Delivery of Proteins and Peptides. *Biomaterials* 25: 5281–5291.
60. Gupta, S., Tyagi, R., Parmar, V.S., Sharma, S.K., Haag, R. 2012. Polyether Based Amphiphiles for Delivery of Active Components. *Polymer* 53: 3053–3078.
61. Knop, K., Hoogenboom, R., Fischer, D., Schubert, U.S. 2010. Poly(ethylene glycol) in Drug Delivery: Pros and Cons as Well as Potential Alternatives. *Angew Chem Int Ed* 49: 6288e308.
62. Obermeier, B., Wurm, F., Mangold, C., Frey, H. 2011. Multifunctional Poly(ethylene glycol)s. *Angew Chem Int Ed* 50: 7988–7997.
63. Rancan, F., Todorova, A., Hadam, S., Papakostas, D., Luciani, E., Graf, C., Gernert, U., Rühl, E., Verrier, B., Sterry, W., Blume-Peytavi, U., Vogt, A. 2012. Stability of Polylactic Acid Particles and Release of Fluorochromes Upon Topical Application on Human Skin Explants. *Eur J Pharm Biopharm* 80: 76–84.
64. Zang, M., Fang, C., Gunn, J., Fichtenholtz, A., Chun, C., Nathan, K. 2006. Methotrexate-Immobilized Poly(ethylene glycol) Magnetic Nanoparticles for MR Imaging and Drug Delivery. *Small* 2: 785–792.
65. Gao, J., Kim, S.J., Nasongkla, N., Hua, A., Shuai, X. 2004. Micellar Carriers Based on Block Copolymers of Poly(ε-caprolactone) and Poly(ethylene glycol) for Doxorubicin Delivery. *J Control Release* 98: 415–426.
66. Prabaharan, J., Pilla, J.G.S., Steeber, D.A., Gong, S. 2009. Folate-Conjugated Amphiphilic Hyperbranched Block Copolymers Based on Boltorn® H40, Poly(l-lactide) and Poly(ethylene glycol) for Tumor-Targeted Drug Delivery. *Biomaterials* 30(16): 3009–3019.
67. Feng, S.S., Tai, S., Zhang, Z. 2012. Vitamin E TPGS as a Molecular Biomaterial for Drug Delivery. *Biomaterials* 33(19): 4889–4906.
68. Asadian-Birjand, M., Bergueiro, J., Rancan, F., Cuggino, B.J.C., Mutihac, R.C., Achazi, A.K., Dernedde, A.J., Blume-Peytayi, C.U., Vogtb, B.A., Calderón, M. 2015. Engineering Thermoresponsive Polyether-Based Nanogels for Temperature Dependent Skin Penetration. *Polym Chem* 6: 5827–5831.
69. Rico Harting, R., Barth, M., Bührke, T., Sophia Pfefferle, R.,Petersen, S. 2017. Functionalization of Polyethetherketone for Application in Dentistry and Orthopedics. *BioNanoMaterials* 2017: 20170003.
70. Xiao, R.Z., Zeng, Z.W., Zhou, G.L., Wang, J.J., Li, F.Z., Wang, A.M. 2010. Recent Advances in PEG-PLA Block Copolymer Nanoparticles. *Int J Nanomedicine* 5: 1057–1065.
71. Tziveleka, L., Christina Kontoyianni, C., Tsiourvas, D. 2006. Novel Functional Hyperbranched Polyether Polyols as Prospective Drug Delivery Systems. *Macromol Biosci* 6(2): 161–169.
72. Jung H.H., Krochta, J.M., Kurth, M.J., Hsieh, Y.L. 2000. Lactitol-Based Poly(ether polyol) Hydrogels for Controlled Release Chemical and Drug Delivery Systems. *J Agric Food Chem* 48: 5278–5282.
73. Ortega, A., Ortiz, H.I.M., Uriostegui, L.G., Soria, G.A. 2017. Drug Delivery System Based on Poly(ether-block-amide) and Acrylic Acid for Controlled Release of Vancomycin. *J Appl Polym Sci* 135: 1–8.
74. Vyas, S.P., Khar, R.K. 2010. *Controlled Drug Delivery-Concepts and Advances*. First Edition, Vallabh Prakashan, 97–154.

75. Nathan, A., Kohn, J. 1994. *Biomedical Polymers: Designed-to-Degradable Systems*. Shalaby, S.W., Ed., Hanser/Gardner: Cincinnati, OH.
76. Anderson, J., Gibbons, D., Martin, R., Hiltner, A., Woods, R. 1974. The Potential for Poly-α-Amino Acids as Biomaterial. *J Biomed Mater Res* 5: 197–207.
77. Langer, R., Peppas, N. 1983. Chemical and Physical Structure of Polymers as Carriers for Controlled Release of Bioactive Agents: A Review. *Rev Macromol Chem Phys C* 23: 61–126.
78. Marck, K., Wildevuur, C., Sederel, W., Bantjes, A., Feijen, J. 1977. Biodegradability and Tissue Reaction of Random Copolymers of L-leucine, L-Aspartic Acid, and L-Aspartic Acid Esters. *J Biomed Mater Res* 11: 405–422.
79. Martin, E., May, P., McMahon, W. 1971. Amino Acid Polymers for Biomedical Applications. I. Permeability Properties of L-Leucine DL-Methionine Copolymers. *J Biomed Mater Res* 5: 53–62.
80. Gachard, I., Bechaouch, S., Coutin, B., Sekiguchi, H., 1997. Drug Delivery From Nonpeptidic a-Amino Acid Containing Polyamides. *Polym Bull* 38: 427–431.
81. Khattab, S.N., Naim, S.E.A., Sayed, M.E., Aly, A., Bardan, E., Ahmed, O., Elzoghby, A., Bekhitd, A., Fahama, A.E. 2013. Design and Synthesis of New s-Triazine Polymers and Its Application as Nanoparticulate Drug Delivery Systems. *NJC* 1: 1–16.
82. Li, B., Yoshii, T., Hafeman, A.E., Nyman, J.S., Wenke, J.C., Guelcher, S.A. 2009. The Effects of rhBMP-2 Released From Biodegradable Polyurethane/Microsphere Composite Scaffolds on New Bone Formation in Rat Femora. *Biomaterials* 30: 6768–6779.
83. Seo, E., Na, K. 2014. Polyurethane Membrane With Porous Surface for Controlled Drug Release in Drug Eluting Stent. *Biomater Res* 18: 15.
84. Martinelli, A., D'Ilario, L., Francolini, I., Piozzi, A. 2011. Water State Effect on Drug Release From an Antibiotic Loaded Polyurethane Matrix Containing Albumin Nanoparticles. *Int J Pharm* 407: 197–206.
85. Chen, X., Liu, W., Zhao, Y., Jiang, L., Xu, H., Yang, X. 2009. Preparation and Characterization of PEG-Modified Polyurethane Pressure-Sensitive Adhesives for Transdermal Drug Delivery. *Drug Dev Ind Pharm* 35: 704–711.
86. Li, B., Brown, K.V., Wenke, J.C., Guelcher, S.A. 2010. Sustained Release of Vancomycin From Polyurethane Scaffolds Inhibits Infection of Bone Wounds in a Rat Femoral Segmental Defect Model. *J Control Release* 145: 221–230.
87. José, A., Arias, P., Camargo, Montserrat, G.A., de-Ilarduya, A.M., Sebastián M.G. 2009. Nanoparticles Made of Microbial Poly(γ -glutamate)s for Encapsulation and Delivery of Drugs and Proteins. *J Biomater Sci* 20: 1065–1079.
88. Brocchini, S., Schachter, D.M., Kohn, J. 1997. Amino Acid Derived Polymers for Use in Controlled Delivery Systems of Peptides. *Am Chem Soc Symp Ser* 675: 154.
89. Nakanishi, T., Fukushima, S., Okamoto, K., Suzuki, M., Matsumura, Y., Yokoyama, M., Okano, T., Sakurai Y., Kataoka, K. 2001. Development of the Polymer Micelle Carrier System for Doxorubicin. *J Control Release* 74: 295.
90. Park, Y.J., Liang, J., Yang, Z., Yang, V.C. 2001. Controlled Release of Clot-Dissolving Tissue-Type Plasminogen Activator From a Poly (L-glutamic acid) Semi-Interpenetrating Polymer Network Hydrogel. *J Control Release* 75: 37–44.
91. Kolawole, O.A., Pillay, V., Choonara, Y.E. 2012. Polyamide Rate-Modulated Monolithic Drug Delivery System. US 8277841 B2.
92. Salahuddin, N., Elbarbery, A., Allam, N.G., Hasim, A.F. 2014. Polyamidemontmorillonite Nanocomposites as a Drug Delivery System: Preparation, Release of 1,3,4oxa(thia)diazoles, and Antimicrobial Activity. *J Appl Polym Sci* 131: 23.
93. Olkhov, A.A., Pankova, Y.N., Goldshtrakh, M.A. 2016. Structure and Properties of Films Based on Blends of Polyamide–Polyhydroxybutyrate. *Inorg Mater Appl Res* 7: 471–477.
94. Langer, R. 1993. Polymer-Controlled Drug Delivery Systems. *Acc Chem Res* 26: 537.
95. Siegel, R.A. 1997. *Controlled Drug Delivery Challenges and Strategies*. Park, K. Ed. Washington, DC.
96. Kost, J., Lapidot, S.A., Wise, D.L. 2000. Smart Polymers for Controlled Drug Delivery. In *Handbook of Pharmaceutical Controlled Release Technology*, 65. Brannon-Peppas, L., Klibanov, A., Langer, R., Mikos, A., Peppas, N.A., Trantolo, D.J., Wnek, G.E., Yaszemski, M.J., Eds. Marcel Dekker, Inc.
97. Manthina, M., Kalepu, S., Padavala, V. 2013. Oral Lipid-Based Drug Delivery Systems – An Overview. *Acta Pharm Sin* 3: 361–372.

9

Delivery Systems for Proteins and Peptides

Sougata Jana, Arijit Gandhi, and Kalyan Kumar Sen

CONTENTS

9.1 Introduction

Peptides and proteins have become the drugs of choice for the treatment of numerous diseases as a result of their incredible selectivity and their ability to provide effective and potent action [1]. In general, they cause fewer side effects and have great potential to cure diseases rather than merely treat their symptoms. A wide variety of peptide and protein (P/P) drugs is now produced on a commercial scale as a result of advances in the biotechnology field [2, 3]. The past decade saw an increased interest in formulating and delivering biological drugs for a range of diseases with significant unmet medical need. Unlike conventional small molecular drugs, the clinical development of these types of drug will not be possible without some sort of sophisticated pharmaceutical technology. Administering drugs orally is by far the most widely used route of administration, although it is generally not feasible for P/P drugs. The main reasons for the low oral bioavailability of biologicals are presystemic enzymatic degradation and poor penetration of the intestinal membrane [4, 5]. Much has been learned in the past few decades about macromolecular drug absorption from the gastrointestinal (GI) tract, including the barriers that restrict GI absorption. Various strategies have been pursued to overcome such barriers and to develop safe and effective oral delivery systems for proteins [3, 5]. The oral route for peptide and protein administration continues to present a significant challenge and represents a focus for many pharmaceutical researchers. However, we believe that only further research into delivery systems can make it possible for the oral route to represent a viable route of administration for P/P drugs, improving convenience for, and compliance from, patients who would benefit from these drugs.

At a cellular level, the delivery of proteins and large peptides *in vivo* can be hindered by their three-dimensional structure, spatial occupation, and hydrophilic/hydrophobic nature. Thus, the diffusion transport of these large pharmaceutical proteins is generally slower unless specific transporters are available. Protein stability, based on weak non-covalent interactions between the secondary, tertiary, and quaternary structures of proteins, is also crucial to prevent any disruptions that will destabilize the proteins [6]. These factors present proteins and peptides as highly vulnerable molecules with short *in vivo* half-lives due to degradation by enzymes and proteases, either at the administration site or en route to the site of pharmacological action, resulting in poor bioavailability. In certain cases, frequent administration at high doses might be required to obtain therapeutic effects *in vivo*, creating a risk for unexpected and unwanted side effects such as immune responses. In addition, during the preparation of P/P drugs, manufacturing processes and environmental factors may damage the proteins, reduce their biological activity, induce aggregation, render the proteins immunogenic, and lead to their precipitation [7]. These processes include sterilization and lyophilization, while the contributing environmental factors are pH, ionic strength, temperature, high pressure, non-aqueous solvents, metal ions, detergents, adsorption, agitation, and shearing.

For these reasons, a carrier or delivery system for therapeutic proteins and peptides would be ideal (Figure 9.1).

The drug delivery system (DDS) has to be able to bind the therapeutic protein/peptide in a fashion that does not affect its bioactivity but holds it sufficiently until the DDS reaches its site of action and releases the protein/peptide in a controlled, sustained manner (Figure 9.2). In this instance, DDSs formulated with biocompatible and biodegradable materials would certainly be advantageous so as to minimize any adverse host response to the DDS.

The current review aims to present ongoing research on DDSs for proteins and peptides in an informative way and the targeted delivery of such protein and peptide DDSs. Recent publications containing *in vivo* experimentation and related to delivery systems were selected to be included in this review. This review serves to update readers regarding new developments in protein and peptide delivery systems and also the future prospective.

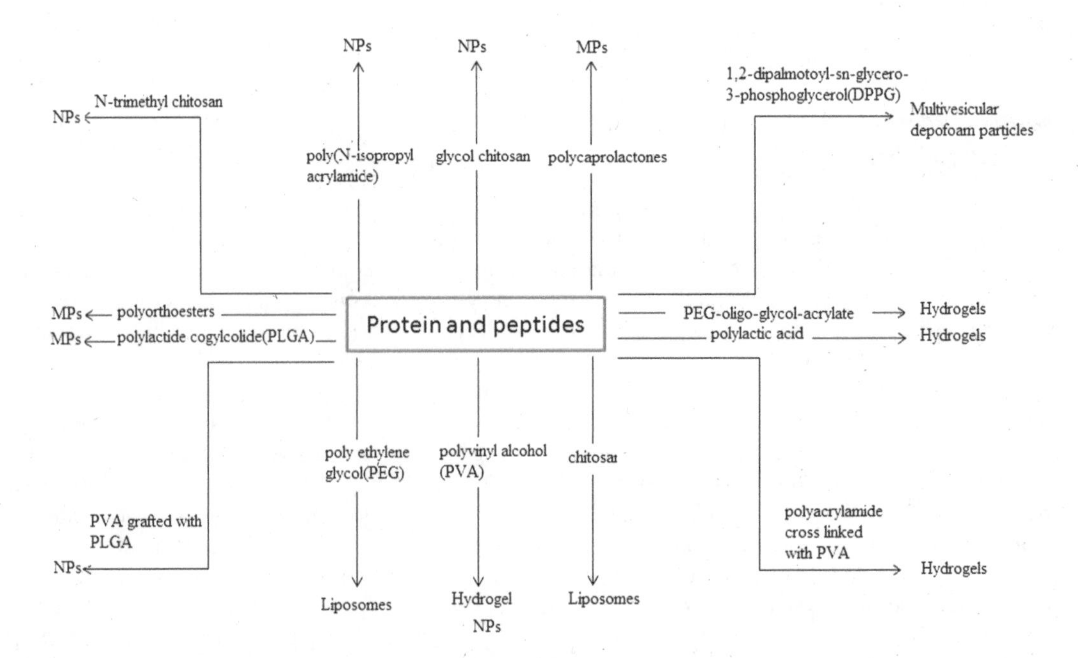

FIGURE 9.1 Approaches of different carriers for delivery of proteins and peptides.

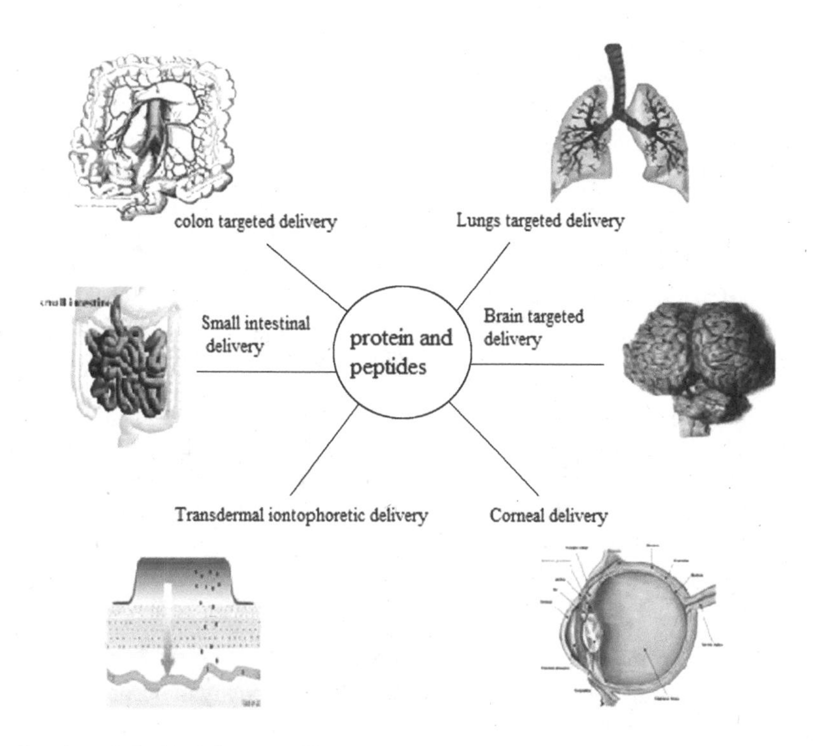

FIGURE 9.2 Protein and peptide delivery to different targeted organs.

9.2 Delivery Systems for Proteins and Peptides

9.2.1 Liposomes

A liposome is a spherical vesicle with a membrane composed of a phospholipid bilayer used to deliver a drug or genetic material into a cell. Liposomes can be composed of naturally derived phospholipids with mixed lipid chains, such as egg phosphatidylethanolamine, or of pure components such as dioleoyl phosphatidylethanolamine (DOPE). Various types of liposome formulations have been made available depending on their dimensions, composition, surface charge, and structure [8].

For the treatment of cancer, liposomal-DDSs have been developed for proteins and peptides.

A novel attempt to synthesize and deliver pro-apoptotic membrane proteins through liposomes was made in human colorectal carcinoma cells [9]. Natural liposomes derived from spinach thylacoids were prepared through the evaporation/resuspension technique together with recombinant VDAC (voltage-dependent anion channel), Bak, and BakΔBH3 protein, producing LB (Bak liposomes), LBΔBH3 (BakΔBH3 liposomes), LV (VDAC liposomes), and LVB (VDAC-Bak liposomes). On incubation of the liposomes with colorectal carcinoma cells, apoptosis was induced within 24 h with the release of cytochrome c and activation of caspases-3, -7, -9 and poly (ADP-ribose) polymerase (PARP). Thus, these results highlight the potential of pro-apoptotic proteins integrated into natural liposomes, presenting as candidates for cancer protein therapy. However, *in vivo* efficacy still needs to be determined, as the *in vitro* performance of a DDS does not necessarily predict its performance *in vivo*.

Studies were also carried out to examine the local delivery of therapeutic molecules encapsulated within liposomes as a potential method to treat ocular inflammation [10]. Using the thin-film hydration method, the authors formulated rhodamine-conjugated pegylated (PEG) liposomes composed of phosphatidylcholine (PC), phosphatidylglycerol (PG), cholesterol (Chol), 1.2-distearoyl-sn-glycero-3-phosphoethanolamine-N-[methoxypoly-(ethyleneglycol)-2000] (PEG-DSPE), and phosphatidylethanolamine-N-(lissamine rhodamine B sulfonyl) (PE-Rh). Rhodamine-conjugated liposomes loaded with vasoactive intestinal peptide (VIP), an immunosuppressive neuropeptide [11], were injected intravitreally into the eyes of rats for a study of VIP biodistribution. Twenty-four hours following the intravitreal injection, VIP-Rh-Lip fluorescent VIP-containing liposomes as well as VIP-null fluorescent liposomes were detected mainly in the posterior segment of the eye (vitreous, inner layer of the retina) and the conjunctiva and episclera. Studies also determined that the liposomes were internalized by activated retinal Muller glial cells, ocular tissue resident macrophages, and rare infiltrating activated macrophages. However, it was unclear whether both Rh-Lip and VIP-Rh-Lip were internalized in a similar manner, although some evidence pointed to the presence of VIP still contained within the conjunctiva 24 h after injection. Both Rh-Lip and VIP-Rh-Lip were distributed to the cervical lymph nodes and internalized by resident ED-3-positive macrophages adjacent to CD4- and CD8-positive T-lymphocytes. In addition, it was found that the T-lymphocytes in close contact with VIP-Rh-Lip-containing macrophages expressed VIP. The accumulation of liposomes within the subcapsular sinus of the cervical lymph nodes following intravitreal injection in the conjunctiva was possible through the conjunctival lymphatics. Thus, through this study, results showing the detection of VIP in both macrophages and T cells in cervical lymph nodes suggested that intravitreal injection of VIP-Rh-Lip may increase ocular immune privileges by modulating the locoregional immune environment. In conclusion, these observations suggested that these VIP-loaded liposomes could prove a promising therapeutic strategy to dampen ocular inflammation by modulating macrophage and T cell activation.

In yet another report using VIP, a treatment for lung diseases was described [12]. Although the administration of VIP garnered positive therapeutic potential in human trials for pulmonary artery hypertension, there is as yet no VIP agonist in clinical use [13]. This is attributed to the rapid proteolytic digestion and inactivation of VIP by endopeptidases, thus reducing the bioactivity of the peptide [14]. To enhance the local availability and therapeutic effect of the peptide, a VIP depot formulation was designed to provide a retarded release of the peptide, allowing the peptide to be protected before its uptake by the appropriate receptor [12]. The authors had previously formulated a unilamellar liposome and were now developing this liposomal formulation as an inhalable peptide carrier. VIP-loaded liposomes, composed of PEG-conjugated distearyl phosphatidylethanolamine (DSPE-PEG2000), lysosteryl-phosphatidylglycerol (lyso-PG), and palmitoyl-oleoyl-phosphatidylcholine (POPC), were synthesized by a thin-film rehydration method to incorporate an aqueous solution of VIP. PEG also formed a PEG-shell, which protected surface-attached VIP peptides from enzymatic degradation in the airway and inhibited particle aggregation. VIP-loaded liposomes were subsequently nebulized at room temperature, resulting in liposomes in the spray mist of a mean diameter below 14 mm in size. The authors reported that there were no significant changes in size before and after nebulization, indicative of the absence of degradation or aggregation of liposomes, and further supporting the suitability of an inhalable VIP liposome. When the VIP-loaded liposomes were tested biologically through relaxation experiments on rat arteries, the liposomes allowed a delay in the peptide release and produced greater vasorelaxation potential in comparison to free VIP peptides. In addition, the authors noted that there was a smooth vasorelaxation response toward effect saturation, indicative of a continuous and smooth peptide release from the liposomes. The release of the VIP peptides was also attributed to triggering by target cells, as the VIP-loaded liposomes were stable without loss of peptide during storage. This VIP-loaded liposomal formulation has shown potential for adaptation for the encapsulation of other peptide drugs. In summary, the authors have formulated a VIP-loaded liposome that increases peptide availability by (i) providing protection of peptides from peptidases and other forms of degradation and (ii) prolonging the specific activity of the peptide, indicative of sustained substance release in the lungs [12].

Liposome delivery of therapeutic proteins was also attempted as a DDS for vaccination against genital Chlamydia infection [15].

The major outer membrane protein (MOMP) is a target of both humoral and cellular immune responses during infections in humans and is a leading vaccine candidate. Despite the success of native MOMP vaccines [16, 17], recombinant MOMP, MOMP peptides, and MOMP DNA have not yielded the expected protection, as the delivery systems could not retain MOMP's strong cell-mediated immunity, resulting in a dramatic influence on the vaccine's efficacy [18, 19]. Using a recently developed cationic liposome formulation 1 (CAF01) [20], the authors compared the efficacy of MOMP/CAF01 or MOMP adjuvanted with alum (T helper cells type 2–promoting aluminum hydroxide) by mouse immunizations followed by a *Chlamydia muridarum* antigen challenge 6 weeks later. To further investigate the cell-mediated immunity mechanism involved in MOMP vaccination, the investigators carried out their studies on normal C57BL/6 mice as well as CD4+ T cells–depleted C57BL/6 mice. The results showed that mice immunized with MOMP/CAF01 experienced a clear protective effect compared with mice treated with saline or MOMP/alum, boasting significantly fewer bacteria on days 7 and 14 after infection. Mice vaccinated with MOMP/alum displayed a strong anti-MOMP humoral response with high IgG1 titers, low levels of interferon (IFN)-γ and tumor necrosis factor (TNF)-ά, and a slight reduction in Chlamydia load. On the other hand, mice vaccinated with MOMP/CAF01 displayed high titers of IgG2b and IFN-a and reduced vaginal Chlamydia load. Additionally, MOMP/CAF01 vaccinated mice showed a reduced infection-resolution time compared with saline-treated mice. Also, the results showed that vaccine-mediated protection against Chlamydia was dependent on CD4+ T cells. Having demonstrated that the MOMP/CAF01 vaccine effect was mediated primarily by CD4+ T cells, the authors also showed that similar levels of infection protection were provided by vaccination with rMOMP/CAF01, verifying that vaccination protection was not dependent on MOMP conformation. In conclusion, this study showed that a vaccine based on natively refolded as well as recombinant MOMP administered through the CAF01 delivery system induced a CD4+ T cell–dependent immune response that efficiently protects against vaginal antigen challenge with *C. muridarum*.

For diabetes mellitus, the major peptide used for treatment is insulin. Despite its importance in diabetes management, conventional insulin therapies are limited by conditions such as retinopathy, nephropathy, thickened capillary basement membranes, and cardiovascular complications [21]. In addition, effective insulin delivery could be limited by high elimination rates, resulting in a short duration of drug contact with its absorption sites and consequently, low bioavailability.

In an attempt to address these issues, the authors investigated the applicability of insulin-containing multivesicular liposomes (MVLs) with the addition of novel chitosan and Carbopol coating (CS/P-MVLs) as sustained release protein delivery systems via the nasal and ocular routes [22]. As a result, the authors obtained insulin-loaded MVLs of 26–34 mm with a high protein loading between 58% and 62%. The *in vivo* applicability of these liposomes was tested through drop-style nasal and ocular administrations in streptozotocin (STZ)-induced diabetic rats and compared with non-coated MVLs, conventional liposomes, and free insulin solution. In STZ-induced diabetic rats, CS-MVLs effectively reduced the blood glucose levels by 35% for up to 48 h after nasal administration compared with a marginal reduction of 22% by non-coated MVLs for up to 12 h after administration. A comparison between chitosan-(CS-MVLs) and Carbopol-coated MVLs (CP-MVLs) showed a maximum of 65% reduction in blood glucose levels and 55% reduction, respectively, at 8 h. On ocular administration, the CS-MVL carrier showed better blood glucose reduction levels of up to 30% at 24 h compared with the CP formulation and other tested formulations. This superior effect observed in both nasal and ocular administrations has been mainly attributed to the prolonged residence of the mucoadhesive MVLs in the mucosa as well as the protective effect of the MVLs against enzymatic attacks on the peptide drug. In conclusion, nasal administration of CS-MVLs has shown a slightly better hypoglycemic profile as compared with the ocular route.

Nonetheless, these experiments have proved to be very promising to show effective delivery of insulin to the systemic circulation via non-invasive routes and will be further investigated and developed for improved transmucosal insulin formulations.

Investigating another peptide drug, calcitonin, Werle and Takeuchi aimed to study the efficacy of liposomes coated with the polymer–protease inhibitor conjugate chitosan–aprotinin for oral peptide delivery in rats [23]. The prepared calcitonin-containing chitosan–aprotinin MVL liposome conjugate ranged from 3 to 4.5 mm in diameter and was loaded with at least 75% of the drug. Formulations of chitosan-coated MVLs, chitosan–aprotinin-coated MVLs, and free calcitonin were administered intragastrically into fasting rats, which were tested for their blood calcium levels at predetermined timepoints. The results showed a significant decrease of blood calcium levels in chitosan–aprotinin-coated MVLs after 30 min of administration as compared with chitosan-coated MVLs. The authors speculated that this effect was most likely mediated by the protease-inhibitory activity of chitosan–aprotinin, as the two formulations did not differ markedly in terms of drug loading and mucoadhesive properties. Additionally, in rats treated with chitosan-coated MVLs, calcium levels returned to the initial readings after 24 h, whereas calcium levels remained decreased after 24 h in rats treated with chitosan–aprotinin-coated MVLs. Comparing the blood calcium concentration–time curve (AUC) of both liposomes and calcitonin solution, chitosan-coated MVLs increased the AAC about 11-fold above calcitonin solution, while chitosan–aprotinin MVLs yielded a 15-fold increase. This result again suggested that the efficacy of chitosan-coated MVLs can be improved through the covalent attachment of specific protease inhibitors such as aprotinin to the polymer. Thus, it was concluded that for the liposome to achieve a more pronounced effect *in vivo*, an optimized liposomal chitosan–aprotinin delivery system that displays a maximal amount of covalently bound aprotinin without exhibiting decreased mucoadhesive properties must be designed.

When we look at the various liposomal delivery systems presented here for different proteins and peptides, there were hardly any similar formulation methods or materials. This could be attributed to the difference in protein/peptide structure, charge, stability, solubility, and other properties.

However, these varied formulations would not be helpful to researchers seeking a liposome suitable for their therapeutic of interest. A more suitable approach would be for researchers to first characterize their therapeutic of interest for properties such as structure, surface charge, and solubility before identifying the formulation method that is most appropriate. To prevent unwarranted side effects and negative interactions with the host, researchers should also opt for safer materials that are natural, biocompatible, biodegradable, and less likely to incite an adverse reaction in the host. In addition, most liposomal DDSs were formulated using complex techniques requiring specialized equipment or chemicals that might be unobtainable for other manufacturers. Thus, researchers should consider using simpler, less complex chemicals that might additionally prove to be less toxic to the host. Another concern might be due to the size of liposomes, which are typically around or above the micrometer range, as the larger the DDS, the higher the possibility of triggering an opsonization effect by the macrophages or other components of the immune system [24]. Large-sized liposomes might also encounter various barriers within the host's endothelial surfaces and various other barriers such as the blood–brain barrier (BBB).

A new acoustically active delivery vehicle was developed by Kheirolomoom et al. (2007) by conjugating liposomes and microbubbles using the high-affinity interaction between avidin and biotin. Binding between microbubbles and liposomes, each containing 5% 1,2-distearoyl-sn-glycero-3-phosphoethanolamine-N-[biotinyl (polyethylene glycol)2000] (DSPE-PEG2kBiotin), was highly dependent on avidin concentration and observed above an avidin concentration of 10 nM. With an optimized avidin and liposome concentration, they measured and calculated as many as 1000 to 10,000 liposomes with average diameters of 200 and 100 nm, respectively, attached to each microbubble. Replacing avidin with neutravidin resulted in threefold higher binding, approaching the calculated saturation level. High-speed photography of this new drug delivery vehicle demonstrated that the liposome-bearing microbubbles oscillate in response to an acoustic pulse in a manner similar to microbubble contrast agents. Additionally, microbubbles carrying liposomes could be spatially concentrated on a monolayer of PC-3 cells at the focal point of an ultrasound beam. As a result of cell–vehicle contact, the liposomes fused with the cells, and internalization of 25-[N-[(7-nitro-2-1,3-benzoxadiazol-4-yl)methyl]amino]-27-norcholesterol (25-NBD-cholesterol) occurred shortly after incubation at 37 °C, with the internalization of NBD-cholesterol substantially enhanced in the acoustic focus. This work supports the idea that the newly designed hybrid vehicle holds promise as a therapeutic delivery vehicle. These vehicles, which combine the advantages of liposomes and microbubbles, are both fusogenic and responsive to ultrasound, overcoming the limitations observed with either microbubbles or liposomes individually. The large number of liposomes attached to one bubble (1000) provides the opportunity to incorporate hydrophilic or hydrophobic therapeutics without affecting their acoustic properties. The repeated use of avidin or streptavidin could lead to an immunogenic response in some individuals. Alternate methods of microbubble–liposome conjugation, the therapeutic response, and the *in vivo* performance of the liposome-bearing microbubbles are now under evaluation [25].

9.2.2 Nanoparticles

The advancement in nanoparticles (NPs) as a drug delivery vehicle in recent years (Table 9.1) is widely expected to change the landscape of the pharmaceutical industries for the foreseeable future.

TABLE 9.1

Nanoparticle Carriers for Proteins and Peptides

Carrier	Drug	Size (nm)	Animal Model	Response	References
Poly(isobutylcyanoacrylate) NP	Insulin	220	Streptozotocin (STZ)-induced diabetic rat	Long-lasting, strong hypoglycemic response	[40]
Chitosan NP	Insulin	250–400	Alloxan-induced diabetic rat	Pharmacological availability was 14.9%	[42]
Acrylic-based copolymer NP	Insulin	200 (pH 2) – 2000 (pH 6)	STZ-induced diabetic rat	Significant reduction of serum glucose	[44]
Chitosan NP	Insulin	269, 339	STZ-induced diabetic rat	Pharmacological availability was 3.2–5.1%	[46]
Nanocubicle	Insulin	220	STZ-induced diabetic rat	Strong hypoglycemic effect	[47]
Poly(N-isopropyl acrylamide) NP	sCT	148–895	Rat	Hypocalcemic response	[sCT]
Poly(lactic-co-glycolic acid) NP	sCT	171.9–315.1	Rat	Bioavailability of sCT was 0.4%	[48]

Note: sCT, salmon calcitonin.

Nanotechnologies have become a significant priority worldwide. Several manufactured NPs—particles with one dimension less than 100 nm—are increasingly used in consumer products [26–28].

Liu et al. (2013) developed heptapeptide-conjugated active targeting NPs for the delivery of doxorubicin and small interfering RNA (siRNA) to high epidermal growth factor receptor (EGFR)–expressing breast cancer cells. The active targeting NPs were prepared using a synthesized poly(D,L-lactide-co-glycolide)–poly(ethylene glycol) (PLGA–PEG) copolymer conjugated with a heptapeptide. The particle size of the peptide-conjugated NPs was less than 200 nm with a narrow size distribution, and the surface charge was negative. The uptake of the peptide-conjugated NPs was 3.9 times more efficient in high EGFR-expressing MDA-MB-468 cells than in low EGFR-expressing HepG2 cells due to peptide specific binding to the EGF receptor followed by EGF receptor–mediated endocytosis. The NPs were used to deliver doxorubicin and siRNA, and their *in vitro* release was faster at pH 4.0 (500 U lipase) than at pH 7.4. The IC_{50} of doxorubicin-loaded peptide-conjugated NPs was 2.3 times lower than that of peptide-free NPs in MDA-MB-468 cells. Similarly, the cellular growth inhibition of siRNA/DOTAP-loaded peptide-conjugated NPs was 2.1 times higher than that of peptide-free NPs. In conclusion, the heptapeptide-conjugated PLGA-PEG NPs provided active targeting potential to high EGFR-expressing MDA-MB-468 breast cancer cells, and a synergistic cytotoxic effect was achieved by the co-delivery of doxorubicin and siRNA/DOTAP-loaded peptide-conjugated NPs [29].

In another study, peptide (insulin)-loaded NPs have been embedded into buccal chitosan films (Ch-films-NPs). These films were produced by solvent casting and involved the incorporation of NPs-insulin suspensions at three different concentrations (1, 3, and 5 mg of NPs per film) into chitosan gel (1.25% w/v) using glycerol as plasticizer. Film swelling and mucoadhesion were investigated using 0.01 M phosphate-buffered saline (PBS) at 37 C and a texture analyzer, respectively. Formulations containing 3 mg of NPs per film produced optimized films with excellent mucoadhesion and swelling properties. Dynamic laser scattering measurements showed that the erosion of the chitosan backbone controlled the release of NPs from the films preceding *in vitro* drug (insulin) release from Ch-films-NPs after 6 h. Modulated release was observed with 70% of encapsulated insulin released after 360 h. The use of chitosan films yielded a 1.8-fold enhancement of *ex vivo* insulin permeation via EpiOral™ buccal tissue construct relative to the pure drug. Flux and apparent permeation coefficient of 0.1 g/cm^2/h and 4×10^{-2} cm^2/h, respectively, were obtained for insulin released from Ch-films-NPs-3. Circular dichroism and Fourier transform infrared (FTIR) spectroscopy demonstrated that the conformational structure of the model peptide drug (insulin) released from Ch-films-NPs was preserved during the formulation process [30].

Glycol chitosan (GCS) modified with hydrophobic analogs has been studied extensively due to its ability to absorb drugs with high avidity within its hydrophobic inner cores and allow their prolonged, sustained release. GCS NPs have also been reported to display fast cellular and tissue internalization into tumors [31].

Polymeric nanocarriers (PNCs), proposed as an attractive vehicle for vascular drug delivery, remain an orphan technology for enzyme therapies due to poor loading and inactivation of protein cargoes. To unite enzyme delivery by PNCs with a clinically relevant goal of containment of vascular oxidative stress, a novel freeze–thaw encapsulation strategy was designed and provides ~20% efficiency loading of an active large antioxidant enzyme, catalase, into PNCs (200–300 nm) composed of the biodegradable block copolymer poly(ethylene glycol)-b-poly(lactic-glycolic acid). Catalase's substrate, H_2O_2, was freely diffusible in the PNC polymer. Furthermore, PNC-loaded catalase stably retained 25–30% of H_2O_2-degrading activity for at least 18 h in a proteolytic environment, while free catalase lost activity within 1 h. The delivery and protection of catalase from lysosomal degradation afforded by PNC nanotechnology may advance the effectiveness and duration of treatment of diverse disease conditions associated with vascular oxidative stress [32].

It has been established that nasal DDS is a very promising route for delivery of proteins and peptides for the reason that it can avoid degradation in the GI tract and metabolism by liver enzymes. However, the bioavailability of proteins and peptides is still low due to rapid mucociliary clearance. Here, to prolong the residence time of drugs and improve their absorption, Zheng et al. (2013) prepared the amphiphilic glycopolymer poly(2-lactobionamidoethyl methacrylate-random-3-acrylamidophenylboronic acid) (p(LAMA-*r*-AAPBA), which was able to assemble into NPs with a narrow size distribution. Insulin, as a model drug, was efficiently encapsulated within the NPs, and loading capacity was up to

12%. An *in vitro* study revealed that the insulin release could be controlled by modifying the composition of the glycopolymer. Cell viability showed that p(LAMA-*r*-AAPBA) NPs had good cytocompatibility. Moreover, the mechanism of NP internalization into Calu-3 cells was a combination mechanism of clathrin-mediated endocytosis and lipid raft/caveolae-mediated endocytosis. Importantly, there was a significant decrease in blood glucose levels after the nasal administration of p(LAMA-*r*-AAPBA) NPs to diabetic rats. Therefore, p(LAMA-*r*-AAPBA) glycopolymers have a potential application as a nasal delivery system for proteins and peptides [33].

Poly (vinyl alcohol) (PVA) hydrogel NPs were prepared by using a water-in-oil emulsion technology plus s cyclic freezing–thawing process. The PVA hydrogel NPs prepared by this method are suitable for P/P drug delivery, since the formation of the hydrogel does not require cross-linking agents or other adjuvants and does not involve any residual monomer. Particularly, there is no emulsifier involved in this new method.

Bovine serum albumin (BSA), as a model protein drug, has been incorporated into PVA hydrogel NPs. The PVA hydrogel NPs possess a skewed or log-normal size distribution. The average diameter of the PVA hydrogel NPs is 675.5 $\pm$ 42.7 nm. The protein drug loading efficiency in the PVA hydrogel NPs is 96.2 $\pm$ 3.8%. The PVA hydrogel NPs swell in aqueous solution, and the degree of swelling increases with increasing temperature. *In vitro* release studies show that the BSA release from the NPs can be prolonged to 30 h. The BSA release follows a diffusion-controlled mechanism. Both the number of freezing–thawing cycles and the release temperature influence the BSA release rate considerably. Fewer freezing–thawing cycles or a higher release temperature leads to faster drug release. The BSA is stable during preparation of the PVA hydrogel NPs [34].

The development of biotech drugs, such as peptides and proteins, that act in the central nervous system (CNS) has been significantly impeded by the difficulty of delivering them across the BBB. The surface engineering of NPs with lectins opened a novel pathway to the absorption of drugs loaded into biodegradable poly (ethylene glycol)-poly (lactic acid) NPs in the brain following intranasal administration. VIP, a neuroprotective peptide, was efficiently incorporated into poly (ethylene glycol)-poly (lactic acid) nanoparticles modified with wheat germ agglutinin, and the biodistribution, brain uptake, and neuroprotective effect of the formulation were assessed. The area under the concentration–time curve of intact ^{125}I-VIP in the brain of mice following the intranasal administration of ^{125}I-VIP carried by NPs and wheat germ agglutinin–conjugated NPs was significantly enlarged by 3.5–4.7-fold and 5.6-7.7-fold, respectively, compared with that after the intranasal application of ^{125}I-VIP solution. The same improvements in spatial memory in ethylcholine aziridium–treated rats were observed following intranasal administration of 25 μg/kg and 12.5 μg/kg of VIP loaded by unmodified NPs and wheat germ agglutinin–modified NPs, respectively. Distribution profiles of wheat germ agglutinin–conjugated NPs in the nasal cavity indicate that they have higher affinity for the olfactory mucosa than the respiratory one. An inhibition experiment with specific sugars suggested that the interaction between the nasal mucosa and the wheat germ agglutinin–functionalized NPs was due to the immobilization of carbohydrate-binding pockets on the surface of the NPs. The results clearly indicated that wheat germ agglutinin–modified NPs might serve as promising carriers, especially for biotech drugs such as peptides and proteins [35].

A novel thiomer derivative of GCS was synthesized by coupling with thioglycolic acid (TGA) and evaluated for the pulmonary delivery of peptides. NPs based on GCS and GCS–TGA were obtained by the ionic gelation method and demonstrated a particle size in the range of 0.23–0.33 μm with a positive surface charge and high calcitonin entrapment. Fluorescent GCS–TGA NPs resulted in a twofold increase in mucoadhesion to lung tissue after intratracheal administration to rats as compared with non-thiolated NPs. Evaluation of pulmonary toxicity revealed the biocompatibility of the two NP formulations with lung tissue. The efficacy of the prepared NPs to enhance the pulmonary absorption of peptides was evaluated after pulmonary administration to rats using a liquid micro-sprayer technique. Calcitonin-loaded GCS and GCS–TGA NPs resulted in a pronounced hypocalcemic effect for at least 12 and 24 h and a corresponding pharmacological availability of 27% and 40%, respectively. These findings suggest that both GCS and its thiomer derivative are promising and safe carriers for pulmonary peptide delivery [36].

In another study, the potential of *N*-trimethyl chitosan (TMC) NPs as a carrier system for the nasal delivery of proteins was investigated. TMC nanoparticles were prepared by ionic cross-linking of TMC solution (with or without ovalbumin) with tripolyphosphate at ambient temperature while stirring. The size, zeta potential, and morphology of the NPs were investigated as a function of the preparation conditions. Protein loading, protein integrity, and protein release were studied. The toxicity of the TMC NPs was tested by ciliary beat frequency measurements of chicken embryo trachea and *in vitro* cytotoxicity assays. The *in vivo* uptake of fluorescein isothiocyanate (FITC)–albumin-loaded TMC NPs by nasal epithelia tissue in rats was studied by confocal laser scanning microscopy. The NPs had an average size of about 350 nm and a positive zeta potential. The authors showed a loading efficiency up to 95% and a loading capacity up to 50% (w/w). The integrity of the entrapped ovalbumin was preserved. Release studies showed that more than 70% of the protein remained associated with the TMC NPs for at least 3 h on incubation in PBS (pH 7.4) at 37 C. Cytotoxicity tests with Calu-3 cells showed no toxic effects of the NPs, whereas a partially reversible cilio-inhibiting effect on the ciliary beat frequency of chicken trachea was observed. *In vivo* uptake studies indicated the transport of FITC–albumin-associated TMC NPs across the nasal mucosa. In conclusion, TMC NPs are a potential new delivery system for the transport of proteins through the nasal mucosa [37].

NPs are attractive carriers for vaccines. It was found that a short peptide (Hp91) activates dendritic cells (DCs), which are critical for the initiation of immune responses. In an effort to develop Hp91 as a vaccine adjuvant with NP carriers, Clawson et al. (2010) evaluated its activity when encapsulated in or conjugated to the surface of PLGA NPs. They found that Hp91, when encapsulated in or conjugated to the surface of PLGA NPs, not only activates both human and mouse DCs but is, in fact, more potent than free Hp91. Hp91 packaged within NPs was about fivefold more potent than the free peptide, and Hp91 conjugated to the surface of NPs was ~20-fold more potent than free Hp91. Because of their capacity to activate DCs, such NP–Hp91 systems are promising as delivery vehicles for subunit vaccines against infectious disease or cancer [38].

Odorranalectin (OL) was recently identified as the smallest lectin, with much less immunogenicity than other members of the lectin family. To improve nose-to-brain drug delivery and reduce the immunogenicity of a traditional lectin-modified delivery system, OL was conjugated to PEG–PLGA NPs, and its biorecognitive activity on NPs was verified by hemagglutination tests. The nose-to-brain delivery characteristics of OL-conjugated NPs (OL–NP) were investigated by an *in vivo* fluorescence imaging technique as a tracer. Furthermore, urocortin peptide (UCN), as a macromolecular model drug, was incorporated into NPs and evaluated for its therapeutic efficacy on hemiparkinsonian rats following intranasal administration by the rotation behavior test, neurotransmitter determination, and the tyrosine hydroxylase (TH) test. The results suggested that OL modification increased the brain delivery of NPs and enhanced the therapeutic effects of UCN-loaded NPs on Parkinson's disease. In summary, OL–NPs could be potentially used as carriers for nose-to-brain drug delivery, especially for macromolecular drugs, in the treatment of CNS disorders [39].

9.2.3 Microparticles

Generally, microparticles have a size range between 1 and 1000 mm. Initial experiments with microparticles started in the early 1970s, and they have since been studied extensively with many different materials and polymers.

As yet, there are very few microparticle drug delivery formulations approved for clinical use [49, 50], although more clinical trials are currently underway [51, 52]. The following articles present various microparticles designed for peptide and protein delivery (Table 9.2).

A lot of research has been carried out in the last decade to find a cure for neurodegenerative diseases, especially Parkinson's, disease but to little avail. Gujral et al. (2013) demonstrated the use of PLGA/collagen biodegradable microparticles formed using the water-in-oil-in-water (w/o/w) double emulsion method as a neurotrophic factor delivery vehicle. The microparticles were encapsulated with glial cell–derived neurotrophic factor (GDNF) fused with collagen binding peptide (CBP) immobilized to the inner collagen phase. The novelty lies in the strict regulation of release of GDNF–CBP from the microparticles as compared with a burst release from standard microparticles. The microparticles were demonstrated to be non-cytotoxic until 300 μg/2 × 10^5 cells and revealed a maximum release of 250 ng GDNF–CBP/mg microparticles in 0.3% collagenase. Differentiation of neural progenitor cells (NPCs)

TABLE 9.2

Microparticles for Delivery of Peptide and Proteins

Protein/Peptide	Carrier	Degradation Mechanism	References
Insulin	Starch	Amylase	[65]
Bovine serum albumin	Chitin	pH, enzymes	[66]
Ivermectin	Corn protein (zein)	Enzymes	[67]
Hydroxyapatite	Collagen/gelatin	Collagenase	[68]
Insulin and vasopressin	Azo-cross-linked copolymer of styrene and HEMA-coated particles	Reduction of azo bonds by microflora in large intestine	[69]
Dextran	Maleic anhydride/poly(N- isopropylacrylamide) hybrid hydrogels	Enzymes	[70]
Dopamine	Polycarbonates	Hydrolysis	[72]
Bovine serum albumin, insulin, nerve growth factor	Polycaprolactones	Hydrolysis	[73]
Virus antigen	Cross-linked albumin	Enzymes	[76]

Note: HEMA, 2-hydroxyethyl methacrylate.

into mature neurons was demonstrated by co-culturing microparticles with cells in a medium containing collagenase, which enabled the release of encapsulated GDNF–CBP, signaling the differentiation of NPCs into microtubule-associated protein 2 (MAP2)-expressing neurons. The successful ability of these microparticles to deliver neurotrophic factors and enable the differentiation of NPCs into mature neurons provides some scope for their use in the treatment of Parkinson's disease and other neurodegenerative diseases [53].

The effective microencapsulation of proteins has been limited by low encapsulation efficiencies, large required amounts of protein, and risk of protein denaturation. Sy et al. (2010) adapted a widely used immobilized metal affinity protein purification strategy to non-covalently attach proteins to the surface of microparticles. Polyketal microparticles were surface modified with nitrilotriacetic acid–nickel complexes, which have a high affinity for sequential histidine tags on proteins. The authors demonstrated that this high-affinity interaction can efficiently capture proteins from dilute solutions with little risk of protein denaturation. Proteins that bind to the Ni–NTA complex retain activity and can diffuse away from the microparticles to activate cells from a distance. In addition, this surface modification can also be used for microparticle targeting by tethering cell-specific ligands to the surface of the particles, using VE-Cadherin and endothelial cells as a model. In summary, they showed that immobilized metal affinity strategies have the potential to improve targeting and protein delivery via degradable polymer microparticles.

This work demonstrates that poly(cyclohexane 1,4-diylacetone dimethylene ketal) microparticles can be surface functionalized with Ni–NTA. Furthermore, these metal complexes can be used to non-covalently immobilize a variety of proteins in a bioactive form with little risk of denaturation compared with other processing techniques. Using His_6-Green fluorescent protein as a model protein, Ni–NTA-functionalized microparticles were able to specifically bind 40% of protein from dilute (<1 mg/ml) solutions and release these proteins with a half-life of 2 h, creating a dual delivery system with a slower release of compounds from within the particle. Bioactive vascular endothelial growth factor (VEGF) was delivered to human umbilical vein endothelial cells (HUVECs) and activated phospho-VEGF-receptor 2 (VEGFR2) in a microparticle dose–dependent manner. VE-Cadherin surface-loaded microparticles were also used to demonstrate the ability of using Ni–NTA functionalization for potentially targeting microparticles to specific cell types. Thus, functionalized polyketal microparticles (PK)-Ni–NTA particles may be a clinically useful delivery platform for diseases that require multiple factors or targeted therapy [54].

Among the different approaches to achieve protein delivery, the use of polymers, specifically biodegraded, holds great promise. A new microsphere delivery system composed of alginate microcores surrounded by a biodegradable poly-DL-lactide-poly(ethylene glycol) (PELA) was designed to improve the loading efficiency and stability of proteins. Alginate was solidified by calcium (MS-1), polylysine (MS-2), and chitosan (MS-3), respectively, to form different microcores. Human serum albumin (HSA), used as a model protein, was efficiently entrapped within the alginate microcores using a high-speed stirrer and then microencapsulated into PELA copolymer using a w/o/w solvent extraction method. Differential scanning calorimetry (DSC) analysis of the microspheres revealed the efficient encapsulation of the alginate microcores, while the microcores were dispersed in the PELA matrix. Sodium dodecyl sulfate polyacrylamide gel electrophoresis (SDS–PAGE) results showed that HSA kept its structural integrity during the encapsulation and release procedure. Microspheres were characterized in terms of morphology, size, loading efficiency, *in vitro* degradation, and protein release. The degradation profiles were characterized by measuring the loss of microsphere mass, the decrease of polymer intrinsic viscosity, and the reduction of the PEG content of the PELA coat. The release profiles were investigated by the measurement of protein present in the release medium at various intervals. The results showed that the degradation rate of these core-coated microspheres was in the order MS-2 > MS-1 > MS-3. The extent of burst release from the core-coated microspheres in the initial protein release was lower than the 27% burst release from the conventional microspheres. In conclusion, the work presents a new approach for macromolecular drug (such as P/P drugs) delivery. The core-coated microsphere system may have potential use as a carrier for drugs that are poorly absorbed after oral administration [55].

Möbus et al. (2012) prepared novel Zn^{2+}-cross-linked alginate microparticles for the controlled pulmonary delivery of protein drugs via a simple one-step spray-drying process and physicochemically characterized these systems. Microparticles were prepared by spray-drying aqueous alginate solutions containing the model protein BSA, $Zn(NH_3)_4SO_4$, and optionally, additional excipients. On ammonia evaporation, the alginate was cross-linked by Zn^{2+} ions. The microparticles were characterized by scanning electron microscopy (SEM), laser and X-ray diffraction, gel electrophoresis, aerodynamic particle size, and drug release measurements. Particles in a size range suitable for deep lung administration were obtained. Pure alginate microparticles were spherical in shape, whereas the addition of zinc led to a more collapsed geometry. Protein release depended on the (i) alginate:$ZnSO_4$ ratio (minimum release rate at 2:1); (ii) BSA content (decreasing release rate and extent with decreasing BSA content); and (iii) type of release medium (increasing release rate with increasing phosphate concentration). The emitted microparticle dose was high for all formulations (~90%). Fine particle fractions (FPF, depositing in the deep lung) up to 40% could be achieved. The FPF was affected by the BSA content, the alginate:$ZnSO_4$ ratio, and the presence/absence of poloxamer. Thus, novel Zn^{2+}-cross-linked alginate microparticles were prepared via a simple one-step process, providing an interesting potential technique for the controlled pulmonary delivery of proteins [56].

In another study, a novel composite alginate/PLGA microparticulate system was developed for protein stabilization and delivery, using bovine insulin as a model drug. Alginate particles, prepared by ionic gelation, were embedded into PLGA microparticles using the solvent diffusion–evaporation technique. The actual loading was determined by micro-BCA protein assay for total insulin and by reversed-phase high-performance liquid chromatography for soluble insulin. Insulin-loaded composite microparticles showed reproducible encapsulation efficiency with a higher soluble insulin content when compared with conventional microparticles. Bovine insulin *in vitro* release studies and adsorption behavior were investigated in 10 mM glycine buffer (pH 2.8) at 37 °C. The stability of bovine insulin, solubilized in the abovementioned buffer, was studied as well. In this case, bovine insulin was shown to be unstable in the investigated conditions, and 55% of insulin was lost after 7 days. However, composite microparticle release, characterized by a low burst effect, lasted up to 4 months. Moreover, no significant peptide adsorption on blank PLGA or blank composite microparticles was observed, while a strong interaction between alginate particles and bovine insulin was detected [57].

Moebus et al. (2012) developed novel preparation techniques for protein-loaded, controlled release alginate–poloxamer microparticles with a size range suitable for pulmonary administration. BSA-loaded microparticles were prepared by spray-drying aqueous polymer–drug solutions followed by cross-linking the particles in aqueous or ethanolic $CaCl_2$ or aqueous $ZnSO_4$ solutions. The microparticles were characterized with respect to their morphology (optical and scanning electron microscopy), particle size (laser light diffraction), calcium content (atom absorption spectroscopy), alginate content (complexation with 1,9-dimethyl methylene blue), and *in vitro* drug release (modified Franz diffusion cell). The spray-dried microparticles were spherical in shape with a size range of 4–6 μm. Aqueous cross-linking led to a significant size increase (10–15 μm), whereas ethanolic cross-linking did not. The substantial drug loss (~50%) during aqueous $CaCl_2$ cross-linking could be avoided by using aqueous $ZnSO_4$ or ethanolic

$CaCl_2$ solutions. Protein release from microparticles cross-linked with ethanolic $CaCl_2$ solutions was much faster than in the case of aqueous $CaCl_2$ solutions, probably due to the lower calcium content. The salt concentration and temperature of the cross-linking solutions also affected the composition of and drug release from the microparticles. Cross-linked alginate–poloxamer microparticles can be produced in a size range appropriate for deep lung delivery and with controlled protein release kinetics (time frame: hours to days) with these novel preparation techniques. The systems offer interesting potential for the controlled mucosal delivery of protein drugs [58].

Protein-loaded microparticles were produced from blends of PEG with poly(L-lactide) (PLA) homopolymer or poly(DL-lactide co-glycolide) copolymers (PLG) using a water-in-oil-in-oil method. The stability of ovalbumin (OVA) associated with microparticles prepared using PEG and 50:50 PLG, 75:25 PLG, and PLA, respectively, was analyzed by SDS–PAGE and quantified by scanning densitometry following incubation in PBS at 37 °C for up to 1 month. Fragmentation and aggregation of OVA was detected with all three formulations. The extent of both processes was correlated with the degradation rate of the lactide polymer used and decreased in the order PLA < 75:25 PLG < 50:50 PLG. Extensive degradation of the PLG/PEG microparticles also occurred over 4 weeks, whereas the use of PLA/PEG blends resulted in a stable microparticle morphology and greatly reduced fragmentation and aggregation of the associated protein. Following a single subcutaneous immunization, high levels of specific serum IgG antibody were elicited by OVA associated with the PLA/PEG particles. Injection of OVA associated with the 75:25 PLG/PEG microparticles resulted in very low levels of specific antibody. A higher response was induced by the 50:50 PLG/PEG formulation, but there was very large inter-animal variation in this group. Antibody levels elicited by all three formulations were significantly higher than those elicited by a single injection of soluble OVA. Analysis of antigen-specific IgG1 and IgG2a antibody subtype levels also revealed the greater efficacy of the PLA/PEG microparticles as an adjuvant system. The use of PLA/PEG microparticles shows improved protein loading and delivery capacity while maintaining a high level of stability of the associated protein. These results indicate a strong correlation between the stability of microencapsulated antigen and the magnitude of the immune response following subcutaneous immunization [59].

A microparticulate delivery system based on a thiolated chitosan conjugate was developed for the nasal application of peptides. Insulin was used as the model peptide. For the thiolation of chitosan, 2-iminothiolane was covalently linked to chitosan. The resulting chitosan–TBA (chitosan-4-thiobutylamidine) conjugate featured 304.89 ± 63.45 µmol thiol groups per gram polymer; 6.5% of these thiol groups were oxidized. A mixture of the chitosan–TBA conjugate, insulin, and the permeation mediator reduced glutathione was formulated into microparticles. Control microparticles comprised unmodified chitosan and insulin. Mannitol–insulin microparticles served as a second control. All microparticulate systems were prepared via the emulsification solvent evaporation technique. In 100 mM phosphate buffer (pH 6.8), chitosan–TBA–insulin microparticles swelled 4.39 ± 0.52-fold in size, whereas chitosan-based microparticles did not swell at all. Chitosan–TBA microparticles showed controlled release of FITC-labeled insulin over 6 h. Nasally administered chitosan–TBA–insulin microparticles led to an absolute bioavailability of $7.24 \pm 0.76\%$ (mean ± S.D.; $n = 3$) in conscious rats. In contrast, chitosan–insulin microparticles and mannitol–insulin microparticles exhibited an absolute bioavailability of $2.04 \pm 1.33\%$ and $1.04 \pm 0.27\%$, respectively (means ± S.D.; $n = 4$). Because of these results, microparticles comprising chitosan–TBA and reduced glutathione seem to represent a useful formulation for the nasal administration of peptides [60].

Chitosan microparticles (CMPs) have been developed for topical application to the eye, but their safety and efficacy in delivering proteins to the retina have not been adequately evaluated. Wassmer et al. (2013) examined the release kinetics of CMPs *in vitro* and assessed their biocompatibility and cytotoxicity to retinal cells *in vitro* and *in vivo*. Two proteins were used in the encapsulation and release studies: BSA and tat-EGFP (enhanced green fluorescent protein fused to the transactivator of transcription peptide). Not surprisingly, the *in vitro* release kinetics was dependent on the protein encapsulated, with BSA showing higher release than tat-EGFP. CMPs containing encapsulated tat-EGFP were tested for cellular toxicity in photoreceptor-derived 661 W cells. They showed no signs of *in vitro* cell toxicity at a low concentration (up to 1 mg/ml), but at a higher concentration of 10 mg/ml, they were associated with cytotoxic effects. *In vivo*, CMPs injected into the subretinal space were found beneath the photoreceptor layer of the retina and persisted for at least 8 weeks. Similarly to the *in vitro* studies, the lower concentration of CMPs was generally well tolerated, but the higher concentration resulted in cytotoxic effects and in reduced retinal function as assessed by electroretinogram amplitudes. Overall, this study suggests that CMPs are effective long-term delivery agents to the retina, but the concentration of chitosan may affect cytotoxicity [61].

The spray-congealing technique, a solvent-free drug encapsulation process, was successfully employed to obtain lipid-based particulate systems with high (10–20% w/w) protein loading. BSA was used as the model protein, and three low-melting lipids (glyceryl palmitostearate, trimyristin, and tristearin) were employed as carriers. BSA-loaded lipid microparticles were characterized in terms of particle size, morphology, and drug loading. The results showed that the microparticles exhibited a spherical shape, mean diameter in the range 150–300 μm, and an encapsulation efficiency higher than 90%. Possible changes in the protein structure as a result of the manufacturing process were then investigated for the first time using UV spectrophotometry in fourth derivative mode and FT-Raman spectroscopy. The results suggested that the structural integrity of the protein was maintained within the particles. Thermal analysis indicated that the effect of protein on the thermal properties of the carriers could be detected. Spray-congealing could thus be considered a suitable technique to produce highly BSA-loaded microparticles while preserving the structure of the protein [62].

In another study, thiol-functionalized polymethacrylic acid–polyethylene glycol–chitosan (PCP)–based hydrogel microparticles were used to develop an oral insulin delivery system. Thiol modification was achieved by grafting cysteine to the activated surface carboxyl groups of PCP hydrogels (Cys-PCP). Swelling and insulin loading/release experiments were conducted on these particles. The ability of these particles to inhibit protease enzymes was evaluated under *in vitro* experimental conditions. Insulin transport experiments were performed on Caco-2 cell monolayers and excised intestinal tissue with an Ussing chamber set-up. Finally, the efficacy of insulin-loaded particles in reducing the blood glucose level in STZ-induced diabetic rats was investigated. Thiolated hydrogel microparticles showed less swelling and had lower insulin encapsulation efficiency as compared with unmodified PCP particles. PCP and Cys-PCP microparticles were able to inhibit protease enzymes under *in vitro* conditions. Thiolation was an effective strategy to improve insulin absorption across Caco-2 cell monolayers; however, the effect was reduced in the experiments using excised rat intestinal tissue. Nevertheless, functionalized microparticles were more effective in eliciting a pharmacological response in diabetic animals as compared with unmodified PCP microparticles. From these studies, thiolation of hydrogel microparticles seems to be a promising approach to improve the oral delivery of proteins/peptides [63].

With the exception of the provision of clean water supplies, vaccination remains the most successful public health intervention strategy for the control of infectious diseases. However, the logistics of delivering at least two to three doses of vaccines to achieve protective immunity are complex, and compliance is frequently inadequate, particularly in developing countries. In addition, newly developed purified subunit and synthetic vaccines are often poorly immunogenic and need to be administered with potent vaccine adjuvants. Microparticles prepared from the biodegradable and biocompatible poly(lactide-co-glycolide) polymers (PLG) have been shown to be effective adjuvants for a number of antigens. Moreover, PLG microparticles can control the rate of release of entrapped antigens and therefore, offer potential for the development of single-dose vaccines. To prepare single-dose vaccines, microparticles with different antigen release rates may be combined as a single formulation to mimic the timing of the administration of booster doses of vaccine. If necessary, adjuvants may also be entrapped within the microparticles, or alternatively, they may be co-administered. The major problems that may restrict the development of microparticles as single-dose vaccines include the instability of vaccine antigens during microencapsulation, during storage of the microparticles, and during hydration of the microparticles following *in vivo* administration [64].

9.2.4 Miscellaneous Vehicles

9.2.4.1 Mesenchymal Stem Cells as Vehicles for Gene and Drug Delivery

Mesenchymal stem cells (MSCs) possess a set of several fairly unusual properties that make them ideally suited both for cellular therapies/regenerative medicine and as vehicles for gene and drug delivery.

These include: (1) they are relatively easy to isolate; (2) they have the ability to differentiate into a wide variety of seemingly functional cell types of both mesenchymal and non-mesenchymal origin; (3) they have the ability to be extensively expanded in culture without a loss of differentiative capacity; (4) they are not only hypoimmunogenic but produce immunosuppression on transplantation; (5) they have pronounced anti-inflammatory properties; and (6) they have the ability to home to damaged tissues, tumors, and metastases following *in vivo* administration. Studies over the last several years have now revealed that MSC have the ability to "sense" this need, migrate to the forming tumor following intravenous (i.v.) administration, and contribute to the newly forming tumor "stroma." While this may not seem ideal, since the MSCs could, in fact, provide support to the growing tumor, it has now been realized that this property presents a very powerful and unique means of selectively delivering anticancer gene products to tumor cells *in vivo* [77, 78].

9.2.4.2 Hyaluronic Acid as a Potential Delivery Vehicle for Vitronectin:Growth Factor Complexes

Novel vitronectin:growth factor (VN:GF) complexes significantly increase re-epithelialization in a porcine deep dermal partial-thickness burn model. However, the potential exists to further enhance the healing response through combination with an appropriate delivery vehicle that facilitates sustained local release and reduced doses of VN:GF complexes. Hyaluronic acid (HA), an abundant constituent of the interstitium, is known to function as a reservoir for growth factors and other bioactive species. The physicochemical properties of HA confer on it an ability to sustain elevated pericellular concentrations of these species. This has been proposed to arise via HA prolonging interactions of the bioactive species with cell surface receptors and/or protecting them from degradation. In view of this, the potential of HA to facilitate the topical delivery of VN:GF complexes was evaluated. Two-dimensional (2-D) monolayer cell cultures and three-dimensional (3-D) de-epidermized dermis (DED) human skin equivalent (HSE) models were used to test skin cell responses to HA and VN:GF complexes. 2-D studies revealed that VN:GF complexes and HA stimulate the proliferation of human fibroblasts but not keratinocytes. Experiments in 3-D models showed that VN:GF complexes, both alone and in conjunction with HA, led to enhanced development of both the proliferative and the differentiating layers in the DED–HSE models. However, there was no significant difference between the thickness of the epidermis treated with VN:GF complexes alone and with VN:GF complexes together with HA. While the addition of HA did not enhance all the cellular responses to VN:GF complexes examined, it was not inhibitory, and it may confer other advantages related to enhanced absorption and transport that could be beneficial in delivery of the VN:GF complexes to wounds [79].

9.2.4.3 Glutathione-Responsive NanoVehicles

The past couple of years have witnessed tremendous progress in the development of glutathione-responsive nanovehicles for targeted intracellular drug and gene delivery, as driven by the facts that (i) many therapeutics (e.g., anticancer drugs, photosensitizers, and antioxidants) and biotherapeutics (e.g., P/P drugs and siRNA) exert therapeutic effects only inside cells, such as in the cytosol and cell nucleus, and (ii) several intracellular compartments, such as the cytosol, mitochondria, and cell nucleus, contain a high concentration of glutathione (GSH) tripeptides (about 2–10 mm), which is 100 to 1000 times higher than that in the extracellular fluids and circulation (about 2–20 μm). Glutathione has been recognized as an ideal and ubiquitous internal stimulus for the rapid destabilization of nanocarriers inside cells to accomplish efficient intracellular drug release. GSH-responsive nanovehicles, in particular micelles, NPs, capsules, polymersomes, nanogels, dendritic and macromolecular drug conjugates, and nanosized nucleic acid complexes, are used for the controlled delivery of anticancer drugs (e.g., doxorubicin and paclitaxel), photosensitizers, antioxidants, peptides, protein drugs, and nucleic acids (e.g., DNA, siRNA, and antisense oligodeoxynucleotide). The unique disulfide chemistry has enabled novel and versatile designs of multifunctional delivery systems addressing both intracellular and extracellular barriers [80, 81].

9.2.4.4 *Polyanion-Coated Biodegradable Polymeric Micelles*

Polymeric micelles, as drug delivery vehicles, must achieve specific targeting and high stability in the body for efficient drug delivery. Arimura et al. reported the preparation of polyanion-coated biodegradable polymeric micelles by coating positively charged polymeric micelles consisting of poly (L-lysine)-*block*-poly (L-lactide) (PLys-b-PLLA) AB diblock copolymers with anionic HA by polyion complex (PIC) formation. The obtained HA-coated micelles showed significantly higher stability in aqueous solution [82].

9.2.4.5 *pH-Sensitive Drug Delivery Vehicles for Human Hepatoblastoma*

Masotti et al. (2010) prepared three novel pH-sensitive systems by derivatizing polysorbate 20 (Tween 20) with glycine, N-methylglycine, and N,N-dimethyl-glycine (TW20-GLY, TW20-MMG, and TW20-DMG). These derivatives form pH-sensitive vesicles and translocate small molecules into cells. The reported systems are efficient DDSs for human hepatoblastoma cells [83].

9.3 Targeted Delivery of Proteins and Peptides

9.3.1 Brain Targeted Delivery

P/P drugs have been identified as showing great promise for the treatment of various neurodegenerative diseases. A major challenge in this regard, however, is the delivery of P/P drugs across the BBB. Intense research over the last 25 years has enabled a better understanding of the cellular and molecular transport mechanisms at the BBB, and several strategies for enhanced P/P drug delivery over the BBB have been developed and tested in preclinical and clinical–experimental research. Among them, technology-based approaches (comprising functionalized nanocarriers and liposomes) and pharmacological strategies (such as the use of carrier systems and chimeric peptide technology) appear to be the most promising.

Strategies for enhanced P/P drug delivery over the BBB are as follows:

1. Pharmacologically based strategies
   a) Modification of P/P drugs to enhance their lipid solubility
   b) Prodrug delivery bioconversion strategies
   c) Use of colloidal drug carriers (liposomes, NPs, and nanogels)
   d) Inhibition of proteolytic enzymes and/or efflux transporters that impede P/P drug delivery over the BBB

2. Physiologically based strategies
   a) Chemical drug delivery bioconversion strategies
   b) Employing chimeric peptide technology: receptor/vector-mediated P/P drug delivery
   c) Use of transport/carrier systems

3. Strategies to increase the permeability of the BBB
   a) Osmotic opening of the BBB
   b) Biochemical opening of the BBB
   c) Ultrasound and electromagnetic radiation as modulators of the BBB

Kavitha et al. (2008) showed that NPs conjugated to the trans-activating transcriptor (TAT) peptide bypass the efflux action of P-glycoprotein and increase the transport of the encapsulated ritonavir, a protease inhibitor (PI), across the BBB to the CNS. A steady increase in the drug parenchyma/capillary ratio over time without disrupting the integrity of the BBB suggests that TAT-conjugated NPs are first immobilized in the brain vasculature prior to their transport into parenchyma. Localization of NPs in the brain parenchyma was further confirmed with histological analysis of the brain sections. The brain drug level with conjugated NPs was 800-fold higher than that with drug in solution at 2 weeks. Drug clearance was seen within 4 weeks. In conclusion, TAT-conjugated NPs enhanced the CNS bioavailability of the encapsulated PI and maintained therapeutic drug levels in the brain for a sustained period, which could be effective in reducing the viral load in the CNS, which acts as a reservoir for the replicating HIV-1 virus [84].

Glioblastoma multiforme (GBM) is the most frequent primary CNS tumor, which represents the second most common cause of cancer death in adults less than 35 years of age [85]. Since GBM differs from other cancers by its diffuse invasion of the surrounding normal tissue, it is impossible to achieve complete removal of the tumor by the conventional surgical method, and tumor recurrence from residual tumors is very possible [86].

A dual-targeting NP DDS was developed by conjugating Angiopep with poly (ethylene glycol)-co-poly (ε-caprolactone) (PEG–PCL) nanoparticles (ANG-NP) through bifunctional PEG to overcome the limitations of low transport of chemotherapeutics across the BBB and poor penetration into tumor tissue. ANG-NP can target the low-density lipoprotein receptor-related protein (LRP), which is over-expressed on the BBB and glioma cells. Compared with non-targeting nanoparticles, a significantly higher amount of rhodamine isothiocyanate-labeled dual-targeting nanoparticles were endocytosed by U87 MG cells. The antiproliferative and cell apoptosis assay of paclitaxel-loaded ANG-NP (ANG-NP-PTX) demonstrated that ANG-NP-PTX resulted in enhanced inhibitory effects to U87 MG glioma cells. The transport ratios across the BBB model *in vitro* were significantly increased, and the cell viability of U87 MG glioma cells after crossing the BBB was obviously decreased, by ANG-NP-PTX. Enhanced accumulation of ANG-NP in the glioma bed and infiltrating margin of an intracranial U87 MG glioma tumor-bearing *in vivo* model were observed by real-time fluorescence imaging. In conclusion, Angiopep-conjugated PEG–PCL NPs show prospects in dual-targeting DDS for targeting therapy of brain glioma [87].

Biomaterial vehicles that can provide sustained, site-specific molecular delivery in the CNS have potential for therapeutic and investigative applications. Song et al. (2012) presented *in vitro* and *in vivo* proof of principle tests of diblock copolypeptide hydrogels (DCHs) to serve as depots for sustained local release of protein effector molecules. They tested two DCHs, $K_{180}L_{20}$ and $E_{180}L_{20}$, previously shown to self-assemble into biocompatible, biodegradable deposits that persist 4–8 weeks after injection into mouse forebrain. *In vitro* tests demonstrated sustained release from dialysis cassettes of the representative protein lysozyme dissolved in $K_{180}L_{20}$ or $E_{180}L_{20}$ hydrogels. Release time *in vitro* varied in relation to DCH charge and mechanical properties and the ionic strength of the media. To evaluate bioactive protein delivery *in vivo*, the authors used nerve growth factor (NGF) and measured the size of mouse forebrain cholinergic neurons, which respond to NGF with cellular hypertrophy. For *in vivo* tests, the storage modulus of DCH depots was tuned to just below that of CNS tissue. In comparison with NGF injected in buffer, depots of NGF dissolved in either $K_{180}L_{20}$ or $E_{180}L_{20}$ provided significantly longer delivery of NGF bioactivity, maintaining hypertrophy of local forebrain cholinergic neurons for at least 4 weeks and inducing hypertrophy a further distance away (up to 5 mm) from injection sites. These findings showed that depots of DCH injected into the CNS can provide sustained delivery within the BBB of a bioactive protein growth factor that exerts a predicted, quantifiable effect on local cells over a prolonged subacute time [88].

Constantino et al. (2005) have reported the synthesis of the conjugate between a biodegradable poly (lactic-co-glycolic) copolymer and various peptides similar to synthetic opioid peptides to create a system able to address NPs to the CNS [89].

Recently, Kurkani et al. (2010) have shown the use of quinoline–poly(butylcyanoacrylate) (PBCA) nanoparticles intended for brain targeting for the diagnosis of Alzheimer's disease [90].

9.3.2 Colon Targeted Delivery

Day by day, there are new developments in the field of colon-specific DDS. A lot of research is being undertaken in colon-specific drug delivery, as this drug delivery route is useful not only for targeting the drugs required in the treatment of diseases associated with the colon but also as a potential site for the local and systemic delivery of P/P and other therapeutic drugs. Precise colon drug delivery requires a triggering mechanism in the delivery system that can respond only to the physiological conditions specific to the colon. The primary conventional approaches used to obtain colon-specific delivery were based on prodrugs, pH- and time-dependent systems, or microflora-activated systems and achieved limited success. However, recently, continuous efforts have been made to design colon-specific delivery systems with improved site specificity and versatile drug release kinetics to meet different therapeutic needs [91–93].

The advantages of colonic peptide delivery can be summarized as follows:

- Metabolic activity is low.
- Residence time is longer.
- It is responsive to absorption enhancers.
- The presence of colonic bacterial enzymes may offer targeting opportunities.
- The transmucosal and membrane potential differences may be of significance in the absorption of ionized or ionizable drugs.
- The bulk water absorption in this region of the intestine may provide scope for solvent drag.

Nevertheless, despite the colon's apparent attractiveness as a route for the oral delivery of peptides, only very limited bioavailability can be demonstrated for this route *in vivo*. In addition, delivery systems to achieve oral delivery to the colon remain to be resolved [94, 95].

Atchison et al. (1989) studied the colonic absorption of radiolabeled insulin using non-everted sacs of rat colon. The percentage of intraluminal insulin degradation and the transport of insulin into the surrounding media were determined. They showed that the transepithelial flux of insulin was consistently less than 0.3% of the dose. In addition, significant degradation of insulin (64%) was found within 15 min of exposure [96].

The effects of intracolonically administered human insulin, and human insulin–diethylaminoethanol (DEAE)–dextran complex entrapped in liposomes, on blood glucose in rats has been reported. In both cases, a hypoglycemic response was noted, which lasted for several hours. No indication of relative bioavailability was given, and it can only be assumed that insulin was transported across the colon [97].

Hastewell et al. (2002) used direct administration of human calcitonin into a colonic loop in anesthetized rats to examine the bioavailability of human calcitonin. This was compared with the pharmacodynamic effect, detectable in normal juvenile animals, of a reduction in plasma calcium levels in response to human calcitonin. They demonstrated the bioavailability achieved after intracolonic dosing of three different doses compared with an i.v. dose. This was 0.5% at 5.0 mg/kg, 0.9% at 10.0 mg/kg, and 0.2% at 0.1 mg/kg. In addition, intracolonically administered human calcitonin at doses of 0.1–5.0 mg/kg resulted in a dose-dependent reduction in plasma calcium levels. These doses achieved reductions in plasma calcium levels of 12% to 38%. The reference i.v. dose of 1.25 μg/kg achieved a calcium reduction of 29%. Moreover, immuno-histochemistry showed that human calcitonin transport across the rat colon was rapid, and a significant amount was via a transcellular pathway [98].

Fukunaga et al.(1984) showed that liposomally entrapped salmon calcitonin produced a hypocalcemic effect in rats when dosed orally but did not determine the mechanism or intestinal location of the absorption [99].

The impact of oral or rectal formulations of peptides is not simply an issue of patient compliance. Data on the intestinal absorption of peptides suggest that the absolute bioavailability is in the order of 1%. Even if therapeutically relevant formulations can be prepared, the implication is that approximately two orders of magnitude more drug will have to be used in GI tract formulations compared with an injection. This could have serious repercussions in two areas. First, it is likely that GI tract formulations will be considerably more expensive than current therapies because of the need for large quantities of active ingredient. Second, if the availability of the active ingredient is limited, companies providing formulations for delivery by the GI tract could be reducing the pool of patients who could be treated if injectable formulations were used.

9.3.3 Protein/Peptide Delivery by Transdermal Iontophoresis

Transdermal iontophoresis, a century-old technique, uses a small electric current to enhance the delivery of chemical compounds across the skin. It has been studied for a broad range of compounds and widely diverse clinical applications. Tedious *in vitro* work using skin fragments was extensively developed to determine whether a given peptide is a good candidate for transdermal iontophoretic delivery [100–105].

However, only a few reports have investigated the relationships between peptide physicochemical properties such as charge, size, lipophilicity, and iontophoretic transport [106–109].

Capillary zone electrophoresis (CZE) is a convenient experimental tool for mimicking the low-throughput *in vitro* skin model used to optimize the delivery of peptides by transdermal iontophoresis. Henchoz et al. (2009) demonstrated pertinent molecular parameters from CZE experiments at different pH values, the optimization of CZE experimental conditions, and the development of an *in silico* filter useful for drug design and development. The effective mobility (μ_{eff}) of ten model dipeptides was measured by CZE at different pH values, enabling their pKa values, charge, and μ_{eff} at any pH to be determined. The best linear correlation between the electromigration contributions to transdermal iontophoretic flux measured across porcine skin with donor and acceptor compartments at pH 7.4 and charge/MW ratio was obtained at pH 6.5, which seems to be the most suitable pH to mimic the *in vitro* skin model. Therefore, the experimental strategy can be considerably shortened by using a single μ_{eff} measurement at pH 6.5 as a predictor of J_{EM}. Additionally, pKa prediction software packages offer fast access to charge/MW ratio using consensual molecular charges at pH 6.5, which suggests that this simple *in silico* filter can be used as a preliminary estimation of J_{EM} [110].

Passive and iontophoretic transport of the model dipeptide tyrosine-phenylalanine (TyrPhe), which is subject to cutaneous metabolism, and the uncharged glucose derivative benzyl-2-acetamido-2-deoxy-α-d-glucopyranoside (BAd-α-Glc), used as an electroosmosis marker, through heat-separated human epidermis was investigated *in vitro*. TyrPhe and BAd-α-Glc were used separately and in combination to determine their interaction in terms of permeability and the influence of skin metabolism of TyrPhe on the permeation rate and tissue retention of itself and of BAd-α-Glc. TyrPhe was chemically and electrochemically stable but underwent considerable degradation in the epidermis under reflection boundary conditions with the generation of the degradation products tyrosine (Tyr) and phenylalanine (Phe), confirming cutaneous metabolism of TyrPhe in heat-separated human epidermis, which was more pronounced at pH 4.5 than at pH 3.0. As a result, no reproducible epidermis permeation of TyrPhe at pH 3 and no permeation at all at pH 4.5 were measured, regardless of the presence of BAd-α-Glc, accompanied by increased levels of Tyr and Phe compared with blank runs. Low temperature (4 C) at both pH values and the addition of *o*-phenanthroline at pH 3 but not at pH 4.5 yielded reproducible TyrPhe permeation and blank, i.e., endogenous levels of Tyr and Phe evidencing inhibition of degradation. Constant voltage anodal iontophoresis marginally reduced BAd-α-Glc flux at pH 3 and 4.5 compared with the passive flux. In combination with TyrPhe, the iontophoretic flux of BAd-α-Glc was increased markedly compared with the passive flux when TyrPhe was metabolized in the tissue, while no such increase was observed when TyrPhe metabolism was inhibited. The increase of BAd-α-Glc iontophoretic flux was accompanied by a considerable decrease of the BAd-α-Glc retained in the epidermis. The presence of the generated Tyr and Phe, therefore, appears to be related to a decrease of the BAd-α-Glc retained in the epidermis on the application of an electrical voltage and an enhancement of its iontophoretic flux. Thus, an interaction between the concurrent permeants at the level of tissue retention induced by metabolism can influence the apparent iontophoretic permeation [111].

The transport of proteins across the skin is highly limited due to their hydrophilic nature and large molecular size. A study was conducted to assess the skin transport abilities of a model protein across hairless rat skin during iontophoresis alone and in combination with microneedles as a function of molecular charge. The effect of microneedle pretreatment on electroosmotic flow was also investigated. Skin permeation experiments were carried out *in vitro* using daniplestim (DP) (MW, 12.76 kD; isoelectric point, 6.2) as a model protein molecule. The effect of molecular charge on protein transport was evaluated by performing studies in two different buffers—TRIS (pH 7.5) and acetate (pH 4.0). The iontophoretic transport mechanisms of DP varied with respect to molecular charge on the protein. The combination approach (iontophoresis and microneedles) gave much higher flux values compared with iontophoresis alone at both pH 4.0 and pH 7.5; however, the delivery in this case was also found to be charge dependent. The findings of this study indicate that electroosmosis persisted on microporation, thus retaining the skin's permselective properties. This enables us to explore the combination of microneedles and iontophoresis as a potential approach for the delivery of proteins [112].

9.4 Future Directions

With the advancement in protein and peptide technology, more is being understood about various proteins and their roles in human health and disease. For proteins with therapeutic value, improving technologies also mean that they could be more easily extracted, purified, or synthesized for therapeutic purposes. Unfortunately, a downside to proteins and peptides is their susceptibility to proteases and other factors. Thus, DDSs were envisioned to provide protection to proteins/peptides until they have reached their site of action. However, after a decade of protein/peptide DDS research; there is still a lack of good systems available. Thus, for more efficient protein and peptide DDSs, researchers must consider the following: safety and biocompatibility of the DDS; material and host toxicity; encapsulation efficiency; retention or enhancement of the natural therapeutic property of the protein/peptide; tissue-specific action; uncomplicated preparation procedure and administration; cost-effectiveness (not cost-prohibitive); biodegradability; and controlled, sustained release of therapeutics.

9.5 Conclusions

It is thought that a prerequisite for the successful delivery of peptides and proteins is the maximization of the absorptive uptake and stabilization of the biologicals at all stages before they reach their target. To develop and improve delivery systems with such properties, the focus should be on the development of superior materials and delivery carriers for bioactive macromolecular delivery systems. Although considerable efforts have already been made to develop delivery systems for macromolecules, extensive *in vivo* studies with these delivery systems have not been publicly reported. Therefore, the development of improved delivery devices for peptides and proteins will require continuous comparison of the *in vitro* and cellular studies with *in vivo* studies.

REFERENCES

1. Frokjaer S, D.D. Otzen. 2005. Protein drug stability: a formulation challenge. *Nat. Rev. Drug Discov.* 4: 298–306.
2. Torchilin V.P, A.N. Lukyanov. 2003. Peptide and protein drug delivery to and into tumors: challenges and solutions. *Drug Discov. Today* 8: 259–266.
3. Shah R.B. 2002. Oral delivery of proteins: progress and prognostication. *Crit. Rev. Ther. Drug Carrier Syst.* 19: 135–169.
4. Mahato R.I. 2003. Emerging trends in oral delivery of peptide and proteins. *Crit. Rev. Ther. Drug Carrier Syst.* 20: 153–214.
5. Hamman J.H. 2005. Oral delivery of peptide drugs. *BioDrugs* 19: 165–177.
6. Wang W. 2005. Protein aggregation and its inhibition in biopharmaceutics. *Int. J. Pharm.* 289: 1–30.
7. Bruno B.J, G.D. Miller, Carol S. Lim. 2013. Basics and recent advances in peptide and protein drug delivery. *Ther. Deliv.* 4: 1443–1467.
8. Dass C.R, T.L. Walker, M.A. Burton, E.E. DeCru. 1997. Enhanced anticancer therapy mediated by specialised liposomes. *J. Pharm. Pharmacol.* 49: 972–975.
9. Liguori L., B. Marques, A. Villegas-Mendez. 2008. Liposomes-mediated delivery of pro-apoptotic therapeutic membrane proteins. *J. Control. Release* 126: 21–27.
10. Camelo S., L. Lajavardi, A. Bochot. 2007. Ocular and systemic bio-distribution of rhodamine-conjugated liposomes loaded with VIP injected into the vitreous of Lewis rats. *Mol. Vis.* 13: 2263–2274.
11. Said S.I, V. Mutt. 1970. Polypeptide with broad biological activity: isolation from small intestine. *Science* 169: 1217–1218.

12. Hajos F., B. Stark, S. Hensler. 2008. Inhalable liposomal formulation for vasoactive intestinal peptide. *Int. J. Pharm.* 357: 286–294.
13. Groneberg D.A, K.F. Rabe, A. Fischer. 2006. Novel concepts of neuropeptide-based drug therapy: vasoactive intestinal polypeptide and its receptors. *Eur. J. Pharmcol.* 533: 182–194.
14. Lilly C.M, J.M. Draxen, S.A. Shore. 1993. Peptidase modulation of airway effects of neuropeptides. *Proc. Soc. Exp. Biol. Med.* 203: 388–404.
15. Hansen J., K.T. Jensen, F. Follmann. 2008. Liposome delivery of *Chlamydia muridarum* major outer membrane protein primes a Th1 response that protects against genital chlamydial infection in a mouse model. *J. Infect. Dis.* 198: 758–767.
16. Pal S., E.M. Peterson, L.M. de la Maza. 2005. Vaccination with the *Chlamydia trachomatis* major outer membrane protein can elicit an immune response as protective as that resulting from inoculation with live bacteria. *Infect. Immun.* 73: 8153–8160.
17. Pal S., I. Theodor, E.M. Peterson. 2001. Immunisation with the *Chlamydia muridarum* mouse pneumonitis major outer membrane protein can elicit a protective immune response against a genital challenge. *Infect. Immun.* 69: 6240–6247.
18. Pal S., K.M. Barnhar, Q. Wei. 1999. Vaccination of mice with DNA plasmids coding for the *Chlamydia trachomatis* major outer membrane protein elicits an immune response but fails to protect against a genital challenge. *Vaccine* 17: 459–465.
19. Shaw J., V. Grund, L. Durling. 2002. Dendritic cells pulsed with a recombinant Chlamydia major outer membrane protein antigen elicit a CD4+ type 2 rather than type 1 immune response that is not protective. *Infect. Immun.* 70: 1097–1105.
20. Christensen D., K.S. Korsholm, K.S. Rosenkrands. 2007. Cationic liposomes as vaccine adjuvants. *Expert Rev. Vaccines* 6: 785–796.
21. Davis S.N, D.K. Granner. 1996. Insulin, oral hypoglycaemic agents, and the pharmacology of endocrine pancreas. In: Hardman J.G, Limbird L.E, Molinoff P.B, Ruddeon R.W, Gilman A.G, editors. *Goodman and Gilman's the Pharmacological Basis of Therapeutics.* 9th ed., New York: Mc-Graw Hill; pp. 1487–1517.
22. Jain A.K, K.B. Chalasani, R.K. Khar. 2007. Muco-adhesive multivesicular liposomes as an effective carrier for transmucosal insulin delivery. *J. Drug Target.* 15: 417–427.
23. Werle M., H. Takeuchi. 2009. Chitosan-aprotinin coated liposomes for oral peptide delivery: development, characterisation and in vivo evaluation. *Int. J. Pharm.* 370: 26–32.
24. Moghimi S.M, J. Szebeni. 2003. Stealth liposomes and long circulating nanoparticles: critical issues in pharmacokinetics, opsonization and protein-binding properties. *Prog. Lipid Res.* 42: 463–478.
25. Kheirolomoom A., P.A. Dayton, A.F.H. Lum, E. Little, E.E. Paoli, H. Zheng, K.W. Ferrara. 2007. Acoustically-active microbubbles conjugated to liposomes: characterization of a proposed drug delivery vehicle. *J. Control. Release* 118: 275–284.
26. Chow T.S. 2003. Size-dependent adhesion of nanoparticles on rough substrates. *J. Phys. Condens. Matter* 15: L83–L87.
27. Lamprecht A., U. Schafer, C.M. Lehr. 2001. Size-dependent bioadhesion of micro- and nanoparticulate carriers to the inflamed colonic mucosa. *Pharm. Res.* 18: 788–793.
28. Kelly L., E. Coronado, L.L. Zhao, G.C. Schatz. 2003. The optical properties of metal nanoparticles: the influence of size, shape, and dielectric environment. *J. Phys. Chem. B* 107: 668–677.
29. Liu C.W., W.J. Lin. 2013. Using doxorubicin and siRNA-loaded heptapeptide conjugated nanoparticles to enhance chemosensitization in epidermal growth factor receptor high-expressed breast cancer cells. *J. Drug Target.* 21: 776–786.
30. Giovino C., I. Ayensu, J. Tetteh, J.S. Boateng. 2013. An integrated buccal delivery system combining chitosan filmsimpregnated with peptide loaded PEG-b-PLA nanoparticles. *Colloids Surf. B Biointerfaces* 112: 9–15.
31. Cho Y.W., S.A. Park, T.H. Han. 2007. In vivo tumour targeting and radionuclide imaging with self-assembled nanoparticles: mechanisms, key factors, and their implications. *Biomaterials* 28: 1236–1247.
32. Dziublaa T.D., A. Karima, V.R. Muzykantov. 2005. Polymer nanocarriers protecting active enzyme cargo against proteolysis. *J. Control. Release* 102: 427–439.
33. Zheng C., Q. Guo, Z. Wu, L. Sun, Z. Zhang, C. Li, X. Zhang. 2013. Amphiphilic glycopolymer nanoparticles as vehicles for nasal delivery of peptides and proteins. *Eur. J. Pharm. Sci.* 49: 474–482.
34. Li J.K., N. Wang, X.S. Wu. 1998. Poly(vinyl alcohol) nanoparticles prepared by freezing–thawing process for protein/peptide drug delivery. *J. Control. Release* 56: 117–126.

35. Gao X., B. Wu, Q. Zhang, J. Chen, J. Zhu, W. Zhang, Z. Rong, H. Chen, X. Jiang. 2007. Brain delivery of vasoactive intestinal peptide enhanced with the nanoparticles conjugated with wheat germ agglutinin following intranasal administration. *J. Control. Release* 121: 156–167.
36. Makhlof A., M. Werle, Y. Tozuka, H. Takeuchi. 2010. Nanoparticles of glycol chitosan and its thiolated derivative significantly improved the pulmonary delivery of calcitonin. *Int. J. Pharm.* 397: 92–95.
37. Amidi M., S.G. Romeijn, G. Borchard, H.E. Junginger, W.E. Hennink, W. Jiskoot. 2006. Preparation and characterization of protein-loaded N-trimethyl chitosan nanoparticles as nasal delivery system. *J. Control. Release* 111: 107–116.
38. Clawson C., C. Huang, D. Futalan, D.M. Seible, R. Saenz, M. Larsson, W. Ma, B. Minev , F. Zhang, M. Ozkan, C. Ozkan, S. Esener, D. Messmer. 2010. Delivery of a peptide via poly(d,l-lactic-co-glycolic) acid nanoparticles enhances its dendritic cell–stimulatory capacity. *Nanomedicine* 6: 651–661.
39. Wen Z., Z. Yan, K. Hu, Z. Pang, X. Cheng, L. Guo, Q. Zhang, X. Jiang, L. Fang, R. Lai. 2011. Odorranalectin-conjugated nanoparticles: preparation, brain delivery and pharmacodynamic study on Parkinson's disease following intranasal administration. *J. Control. Release* 151: 131–138.
40. Damge C. 1998. New approach for oral administration of insulin with polyalkylcyanoacrylate nanocapsules as drug carrier. *Diabetes* 37: 246–251.
41. Prego C. 2005. Transmucosal macromolecular drug delivery. *J. Control. Release* 101: 151–162.
42. Pan Y. 2002. Bioadhesive polysaccharide in protein delivery system: chitosan nanoparticles improve the intestinal absorption of insulin in vivo. *Int. J. Pharm.* 249: 139–147.
43. Kawashima Y. 2000. Mucoadhesive DL-lactide/glycolide copolymer nanospheres coated with chitosan to improve oral delivery of elcatonin. *Pharm. Dev. Technol.* 5: 77–85.
44. Foss A.C. 2004. Development of acrylic-based copolymers for oral insulin delivery. *Eur. J. Pharm. Biopharm.* 57: 163–169.
45. Sakuma S. 2002. Optimized chemical structure of nanoparticles as carriers for oral delivery of salmon calcitonin. *Int. J. Pharm.* 239: 185–195.
46. Ma Z. 2005. Pharmacological activity of peroral chitosan-insulin nanoparticles in diabetic rats. *Int. J. Pharm.* 293: 271–280.
47. Chung H. 2002. Self-assembled 'nanocubicle' as a carrier for peroral insulin delivery. *Diabetologia* 45: 448–451.
48. Sang H.Y., T.G. Park. 2004. Biodegradable nanoparticles containing protein fatty acid complexes for oral delivery of salmon calcitonin. *J. Pharm. Sci.* 93: 488–495.
49. Ipsen Pharmaceuticals Limited, Decapeptyl S. R.—information for the patient. URL: http://www.ipsen.ie/dkp_patient.pdf [Accessed 12/9/09].
50. Takeda UK Limited, Prostap 3 Leuprorelin acetate depot injection—summary of product characteristics. URL: http://emc.medicines.org.uk/document.aspx? DocumentId = 2237 [Accessed 12/9/09].
51. Chiappetta D.A., A.M. Carcaboso, C. Bregni. 2009. Indinavir-loaded pH-sensitive microparticles for taste masking: toward extemporaneous pediatric anti-HIV/AIDS liquid formulations with improved patient compliance. *AAPS PharmSciTech* 10: 1–6.
52. Nomura S., N. Inami, A. Shouzu. 2009. The effects of pitavastatin, eicosapentaenoic acid and combined therapy on platelet-derived microparticles and adiponectin in hyperlipidemic, diabetic patients. *Platelets* 20: 16–22.
53. Gujral C., Y. Minagawa, K. Fujimoto, H. Kitano, T. Nakaji-Hirabayashi. 2013. Biodegradable microparticles for strictly regulating the release of neurotrophic factors. *J. Control. Release* 168: 307–316.
54. Sy J.C., E.A. Phelps, A.J. García, N. Murthy, M.E. Davis. 2010. Surface functionalization of polyketal microparticles with nitrilotriacetic acid-nickel complexes for efficient protein capture and delivery. *Biomaterials* 31: 4987–4994.
55. Zhou S., X. Deng, X. Li. 2001. Investigation on a novel core-coated microspheres protein delivery system. *J. Control. Release* 75: 27–36.
56. Möbus K., J. Siepmann, R. Bodmeier. 2012. Zinc–alginate microparticles for controlled pulmonary delivery of proteins prepared by spray-drying. *Eur. J. Pharm. Biopharm.* 81: 121–130.
57. Schoubben A., P. Blasi, S. Giovagnoli, L. Perioli, C. Rossi, M. Ricci. 2009. Novel composite microparticles for protein stabilization and delivery. *Eur. J. Pharm. Sci.* 36: 226–234.
58. Moebus K., J. Siepmann, R. Bodmeier. 2012. Novel preparation techniques for alginate–poloxamer microparticles controlling protein release on mucosal surfaces. *Eur. J. Pharm. Sci.* 45: 358–366.

59. Lavelle E.C., M.-K. Yeh, A.G.A. Coombes, S.S. Davis. 1999. The stability and immunogenicity of a protein antigen encapsulated in biodegradable microparticles based on blends of lactide polymers and polyethylene glycol. *Vaccine* 17: 512–529.
60. Krauland A.H., D. Guggi, A. Bernkop-Schnürch. 2006. Thiolated chitosan microparticles: a vehicle for nasal peptide drug delivery. *Int. J. Pharm.* 307: 270–277.
61. Wassmer S., M. Rafat, W.G. Fong, A.N. Baker, C. Tsilfidis. 2013. Chitosan microparticles for delivery of proteins to the retina. *Acta Biomater.* 9: 7855–7864.
62. Sabatino M.D., B. Albertini, V.L. Kett, N. Passerini. 2012. Spray congealed lipid microparticles with high protein loading: preparation and solid state characterisation. *Eur. J. Pharm. Sci.* 46: 346–356.
63. Sajeesh S., C. Vauthier, C. Gueutin, G. Ponchel, Chandra P. Sharma. 2010. Thiol functionalized polymethacrylic acid-based hydrogel microparticles for oral insulin delivery. *Acta Biomater.* 6: 3072–3080.
64. O'Hagan D.T, M. Singh, R.K. Gupta. 1998. Poly (lactide-co-glycolide) microparticles for the development of single-dose controlled-release vaccines. *Adv. Drug Deliv. Rev.* 32: 225–246.
65. Illum L., A.N. Fisher, I. Jabbal-Gill, S.S. Davis. 2001. Bioadhesive starch microspheres and absorption enhancing agents act synergistically to enhance the nasal absorption of polypeptides. *Int. J. Pharm.* 222: 109–119.
66. Mi F.L., S.S. Shyu, Y.M. Lin, Y.B. Wu, C.K. Peng, H.Y. Tsai. 2003. Chitin/PLGA blend microspheres as a biodegradable drug delivery system: a new delivery system for protein. *Biomaterials* 24: 5023–5036.
67. Liu X., O. Sun, H. Wang, L. Zhang, J.Y. Wang. 2005: Microspheres of corn protein, zein, for an ivermectin drug delivery system. *Biomaterials* 26: 109–115.
68. Wua T.J, H.H. Huanga, C.W. Lana. 2004. Studies on the microspheres comprised of reconstituted collagen and hydroxyapatite. *Biomaterials* 25: 651–658.
69. Kompella U.B., V.H.L. Lee. 2001. Delivery systems for penetration enhancement of peptide and protein drugs: design considerations. *Adv. Drug Deliv. Rev.* 46: 211–245.
70. Zhang X., D. Wu, C.C. Chu. 2004. Synthesis and characterization of partially biodegradable, temperature and pH sensitive Dex-MA/PNIPAAm hydrogels. *Biomaterials* 25: 4719–4730.
71. Blanco-Prieto M.J., M.A. Campanero, K. Besseghir, F. Heimgatner, B. Gander. 2004. Importance of single or blended polymer types for controlled in vitro release and plasma levels of a somatostatin analogue entrapped in PLA/PLGA microspheres. *J. Control. Release* 96: 437–448.
72. Bourke S.L., J. Kohn. 2003. Polymers derived from the amino acid ltyrosine: polycarbonates, polyarylates and copolymers with poly(ethylene glycol). *Adv. Drug Deliv. Rev.* 55: 447–466.
73. Sinha V.R., K. Bansal, R. Kaushik, R. Kumria, A. Trehan. 2004. Polycaprolactone microspheres and nanospheres: an overview. *Int. J. Pharm.* 278: 1–23.
74. Lavasanifar A., J. Samuel, G.S. Kwon. 2002. Poly (ethylene oxide)-block poly (L-amino acid) micelles for drug delivery. *Adv. Drug Deliv. Rev.* 54: 169–190.
75. Berkland C., M.J. Kipper, B. Narasimhan, K. Kim, D. Pack. 2004. Microsphere size, precipitation kinetics and drug distribution control drug release from biodegradable polyanhydride microspheres. *J. Control. Release* 94: 129–141.
76. Santiago N., S. Milstein, T. Rivera, E. Garcia, T. Zaidi, H. Hong. 1993. Oral immunization of rats with proteinoid microspheres encapsulating influenza virus antigens. *Pharm Res.* 10: 1243–1247.
77. Studeny M., F.C. Marini, R.E. Champlin, C. Zompetta, I.J. Fidler, M. Andreeff. 2002. Bone marrow-derived mesenchymal stem cells as vehicles for interferon-beta delivery into tumors. *Cancer Res.* 62: 3603–3608.
78. Studeny M., F.C. Marini, J.L. Dembinski, C. Zompetta, M. Cabreira-Hansen, B.N. Bekele, R.E. Champlin, M. Andreeff. 2004. Mesenchymal stem cells: potential precursors for tumor stroma and targeted-delivery vehicles for anticancer agents. *J. Natl Cancer Inst.* 96: 1593–1603.
79. Xie Y., Z. Upton, S. Richard, S.C. Rizzi, D.I. Leavesley. 2011. Hyaluronic acid: evaluation as a potential delivery vehicle for vitronectin:growth factor complexes in wound healing applications. *J. Control. Release* 153: 225–232.
80. Soppimath K.S., T.M. Aminabhavi, A.R. Kulkarni, W.E. Rudzinski. 2001. Biodegradable polymeric nanoparticles as drug delivery devices. *J. Control. Release* 70: 1–20.
81. Torchilin V.P. 2006. Recent approaches to intracellular delivery of drugs and DNA and organelle targeting. *Annu. Rev. Biomed. Eng.* 8: 343–375.

82. Arimura H., Y. Ohya, T. Ouchi. 2004. The formation of biodegradable polymeric micelles from newly synthesized poly(aspartic acid)-block-polylactide AB-type diblock copolymer. *Macromol. Rapid Commun.* 25: 743–747.
83. Masotti A., P. Vicennati, A. Alisi, C. Marianecci, F. Rinaldi, M. Carafa, G. Ortaggi. 2010. Novel Tween 20 derivatives enable the formation of efficient pH-sensitive drug delivery vehicles for human hepatoblastoma. *Bioorg. Med. Chem. Lett.* 20: 3021–3025.
84. Rao K.S., M.K. Reddy, J.L. Horning, V. Labhasetwar. 2008. TAT-conjugated nanoparticles for the CNS delivery of anti-HIV drugs. *Biomaterials* 29: 4429–4438.
85. Allard E., C. Passirani, J.P. Benoit. 2009. Convection-enhanced delivery of nanoparticles for the treatment for brain tumors. *Biomaterials* 30: 2302–2318.
86. Ong B.Y., S.H. Ranganath, L.Y. Lee. 2009. Paclitaxel delivery from PLGA foams for controlled release in post-surgical chemotherapy against glioblastoma multiforme. *Biomaterials* 30: 3189–3196.
87. Xin H., X. Jiang, J. Gu, X. Sha, L. Chen, K. Law, Y. Chen, X. Wang, Y. Jiang, X. Fang. 2011. Angiopep-conjugated poly (ethylene glycol)-co-poly (ε-caprolactone) nanoparticles as dual-targeting drug delivery system for brain glioma. *Biomaterials* 32: 4293–4305.
88. Song B., J. Song, S. Zhang, M.A. Anderson, Y. Aoa, C. Yang, T.J. Deming, M.V. Sofroniew. 2012. Sustained local delivery of bioactive nerve growth factor in the central nervous system via tunable diblock copolypeptide hydrogel depots. *Biomaterials* 33: 9105–9116.
89. Constantino L., F. Gandolfi, G. Tosi, F. Rivasi, M.A. Vandelli, F. Forni. 2005. Peptide-derivatized biodegradable nanoparticles able to cross the blood brain barrier. *J. Control. Release* 108: 84–96.
90. Kulkarni P.V., C.A. Roney, P.P. Antich, F.J. Bonte, A.V. Raghu, T.M. Aminabhavi. 2010. Quinoline-n-butylcyanoacrylate-based nanoparticles for brain targeting for the diagnosis of Alzheimer's disease. *Wiley Interdiscip. Rev. Nanomed. Nanobiotechnol.* 2: 35–47.
91. Chourasia M.K., S.K. Jain. 2003. Pharmaceutical approaches to colon targeted drug delivery systems. *J. Pharm. Pharm. Sci.* 6: 33–66.
92. Chu Y.L., J.S. Fix. 2002. Colon-specific drug delivery: new approaches and in vitro/in vivo evaluation. *Int. J. Pharm.* 235: 1–15.
93. Gupta B.P., N. Thakur, S. Jain, P. Patel, D. Patel, N. Jain, N.P. Jain. 2010. A comprhensive review on: colon specific drug delivery system (CSDDS). *J. Pharm. Res.* 3: 1625–1629.
94. Sarasija S., A. Hota. 2000. Colon-specific drug delivery systems. *Ind. J. Pharm. Sci.* 62: 1–8.
95. Singh R. 2012. Formulation and evaluation of colon targeted drug delivery system. *Int. J. Pharm. Life Sci.* 3: 2265–2268.
96. Atchison J.A., W.E. Grizzle, D.J. Pillion. 1989. Colonic absorption of insulin; an in vitro and in vivo evaluatton. *J. Pharmacol. Exp. Ther.* 248: 567–572.
97. Manosroi A., K.H. Batter.1990. Effects of gastrointestinal administration of human insulin and a human insulin-DEAE dextran complex entrapped in different compound liposomes on blood glucose in rats. *Drug Dev. Ind. Pharm.* 16: 1521–1538.
98. Hastewell J., S. Lynch, I. Williamson, R. Fox, M. Mackay. 2002. Absorption of human calcitonin across the rat colon. *Clin. Sci.* 82: 589–594.
99. Fukunaga M., M.M. Miller, K.Y. Hostetler, L.J. Deftos. 1984. Liposome entrapment enhances the hypocalcemic action of parenterally administered calcitonin. *Endocrinology* 115: 757–761.
100. Kalia Y.N., A. Naik, J. Garrison, R.H. Guy. 2004. Iontophoretic drug delivery. *Adv. Drug Deliv. Rev.* 56: 619–658.
101. Cross S.E., M.S. Roberts. 2004. Physical enhancement of transdermal drug application: is delivery technology keeping up with pharmaceutical development? *Curr. Drug Deliv.* 1: 81–92.
102. Thomas B.J., B.C. Finzel. 2004. The transdermal revolution. *Drug Discov. Today* 9: 697–703.
103. Hirvonen J. 2005. Topical iontophoretic delivery: progress to date and therapeutic potential. *Am. J. Drug Deliv.* 3: 67–81.
104. Wang Y., R. Thakur, Q. Fan, B. Michniak. 2005. Transdermal iontophoresis: combinationstrategies to improve transdermal iontophoretic drug delivery. *Eur. J. Pharm. Biopharm.* 60: 179–191.
105. Priya B., T. Rashmi, M. Bozena. 2006. Transdermal iontophoresis. *Exp. Opin. Drug. Deliv.* 3: 127–138.
106. Schuetz Y.B., A. Naik, R.H. Guy, Y.N. Kalia. 2005. Emerging strategies for the transdermal delivery of peptides and protein drugs. *Expert Opin. Drug Deliv.* 2: 533–548.
107. Schuetz Y.B., A. Naik, R.H. Guy, Y.N. Kalia. 2005. Effect of amino acid sequence on transdermal iontophoretic peptide delivery. *Eur. J. Pharm. Sci.* 26: 429–437.

108. Mudry B., P.A. Carrupt, R.H. Guy, M.B. Delgado-Charro. 2007. Quantitative structure– permeation relationship for iontophoretic transport across the skin. *J. Control. Release* 122: 165–172.
109. Abla N., A. Naik, R.H. Guy, Y.N. Kalia. 2005. Effect of charge and molecular weight on transdermal peptide delivery by iontophoresis. *Pharm. Res.* 22: 2069–2078.
110. Henchoz Y., N. Abla, J. Veuthey, P. Carrupt. 2009. A fast screening strategy for characterizing peptide delivery by transdermal iontophoresis. *J. Control. Release* 137: 123–129.
111. Altenbach M., N. Schnyder, C. Zimmermann, G. Imanidis. 2006. Cutaneous metabolism of a dipeptide influences the iontophoretic flux of a concomitant uncharged permeant. *Int. J. Pharm.* 307: 308–317.
112. Katikanenia S., A. Badkarb, S. Nemab, A.K. Banga. 2009. Molecular charge mediated transport of a 13 kD protein across microporated skin. *Int. J. Pharm.* 378: 93–100.

10

Controlled Release of Therapeutic Proteins

Arruje Hameed, Tahir Farooq, Kanwal Rehman, and Muhammad Sajid Hamid Akash

CONTENTS

10.1 Introduction

Proteins are structurally complex macro-biomolecules that are involved in a wide range of biological pathways in different forms, such as antibodies, hormones, enzymes, and interferons. In most biological pathways and/or systems, protein–protein interactions play pivotal roles. The selectivity and specificity of proteins in biological functions originate from their complex tertiary and quaternary structures. Their involvement in a countless number of biological functions with high selectivity and specificity has highlighted them as key targets for designing agonist or antagonist peptides and proteins over the past few decades.[1–3] Based on their mode of action, therapeutic proteins have been classified into four different groups.[1] A schematic representation of therapeutic proteins into four different groups is shown in Figure 10.1.[3] The exceptional developments in the field of pharmaceutical biotechnology and genetic engineering have categorized therapeutic proteins as the leading biotherapeutics over the traditionally used therapeutic substances. Recently, therapeutic proteins have become major therapeutic agents due to their disease specificity, therapeutic potency, and minimal side effects.[4,5]

Therapeutic proteins have proved their worth as emerging alternatives to cure various types of cancers, diabetes mellitus, and other serious, life-threatening ailments incurable by traditional therapeutic agents.[6,7] On administration, therapeutic proteins play important roles at the cellular level and help to revive the normal physiological functions of the body. Therefore, the delivery of therapeutic proteins to the appropriate site of action has become a major concern for their effective response.[9] Recent developments in gene technology and pharmaceutical biotechnology have revolutionized the commercial production of therapeutic proteins, which has triggered research to develop efficient delivery strategies for their controlled release at targeted sites.[8]

Countless factors often make the delivery of therapeutic proteins a challenging task.[9] Some of them include immunogenicity, susceptibility to enzymatic degradation, poor permeability, hydrophilicity,

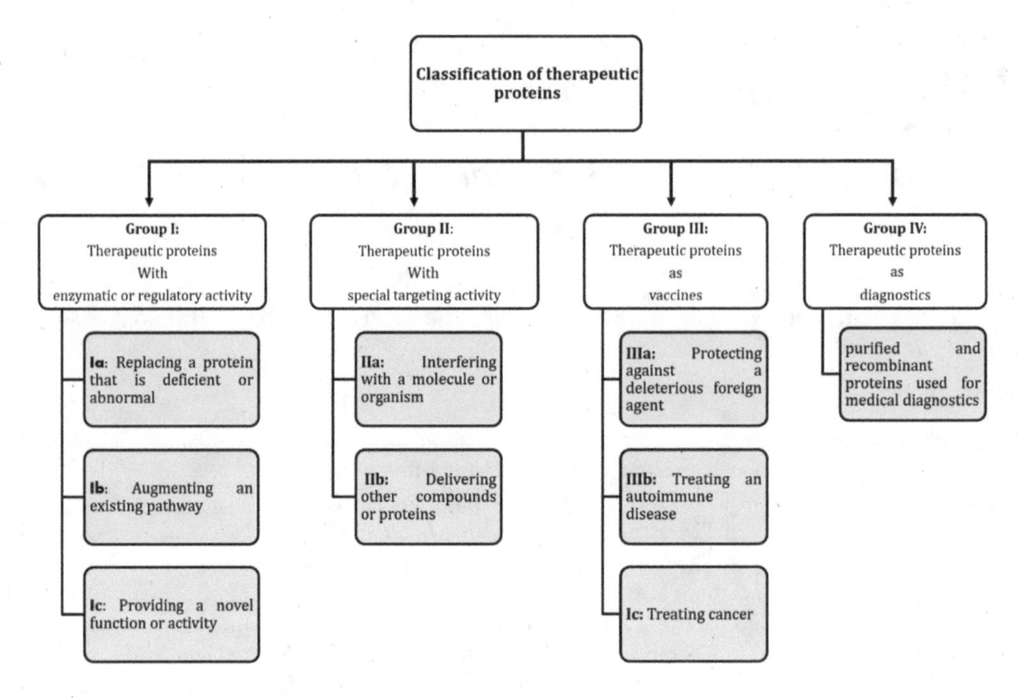

FIGURE 10.1 Classification of therapeutic proteins. (Adapted from Akash, M.S.H., Rehman, K., Tariq, M. and Chen, S., *Turk. J. Biol.*, 39, 343–358, 2015.)

and large size. The oral administration of therapeutic proteins often results in poor absorption, as their hydrophilicity and large size reduce their permeability across biological membranes.[10] Also, therapeutic proteins undergo enzymatic degradation in the GIT, reducing their bioavailability by a significant fraction. Under such circumstances, the delivery of therapeutic proteins through the oral route is considered a risky mode.[11] Considering such delivery issues, the administration of most of the available therapeutic proteins is executed through parenteral routes.

The complexity of the secondary, tertiary, and quaternary structures of therapeutic proteins renders them physically, chemically, or biologically unstable; alteration of the active conformation results in the loss of activity.[12] Consequently, the inactive proteins undergo irreversible aggregation. Therapeutic proteins administered even through the parenteral route show short half-lives due to their susceptibility to *in vivo* enzymatic degradation, so their therapeutic levels are maintained with undesirable frequent dozes. However, many complications, such as detachment, retinal hemorrhage, and cataract associated with frequent parenteral administration, result in patient non-compliance.[13] Therapeutic protein delivery using different formulations has been thoroughly investigated through non-invasive and invasive routes.[9] The frequently used non-invasive routes include inhalation, transdermal, rectal, and oral, while the invasive routes are intravenous (IV), intramuscular (IM), and subcutaneous (SC). Certain absorption enhancers, such as thiolated polymers, mucoadhesive particles, polymer inhibitor conjugates, enzyme inhibitors, and various surfactants, have been used to smooth the oral delivery of therapeutic proteins.[14–17] Further, micro- or nanoparticle-based formulations have also been used for appropriate delivery of therapeutic proteins. However, all these attempts showed side effects and drawbacks, suggesting that the oral delivery of therapeutic proteins presents a hard task to maintain the required bioavailability and therapeutic levels. The major inhibiting factors and limitations include particle aggregation and hydrophobicity, resulting in poor protein loading; formulations showing poor solubility in the GIT; and of course, the high cost of effective enzyme inhibitors.[18]

The non-invasive routes for delivery of therapeutic proteins suffer the same sort of limitations.[19] A non-invasive route, even with high patient compliance, cannot provide controlled delivery of therapeutic proteins, thus categorizing it as a failed treatment strategy. In chronic diseases, a patient's non-adherence to the treatment regimen could potentially lead to the failure of the strategy.[20,21] As per the estimates of the World Health Organization (WHO), in chronic diseases, only 50% of patients show adherence to a treatment regimen in advanced countries, and the ratio is too low in developing countries, so frequent administration is not a recommended strategy.[22] Thus, the therapeutic proteins have become appropriate candidates for controlled release formulations. Controlled delivery formulations show advantages such as fewer side effects, local delivery, and *in vivo* stability. In chronic ailments, improved patient compliance and adherence to treatment regimens reduce both dose and administration frequency. The drawbacks and limitations related to the non-invasive route could be overcome by the parenteral route, although compliance with this approach is not good. A significant improvement in patient compliance could be observed with controlled release of therapeutic proteins and lower dose and administration frequency.

The controlled release of therapeutic proteins ensures their stability and maintains the therapeutic level for a longer period, thus presenting an ideal delivery formulation. Ideally, Food and Drug Administration (FDA)-approved components of the delivery system should be non-toxic, non-immunogenic, biocompatible, and biodegradable. To avoid painful practices, narrow-gauge needles are recommended if the formulations are to be administered through the parenteral route. The controlled release delivery systems for therapeutic proteins are prepared without compromising their structure, activity, and active conformations for an efficient and controlled *in vivo* release. Some serious concerns have been observed over initial burst release patterns in cases of hydrogel- and particulate-based delivery systems. The higher surface adsorption of proteins has been considered responsible for burst releasing of therapeutic proteins in particle-based delivery systems. The burst release of therapeutic proteins poses serious threats of dose-dependent adverse effects, as it reduces the expected duration of desirable release. The formulations for controlled release of therapeutic proteins are so optimized that minimum burst release is achieved to ensure controlled release over a longer period and a lower administration frequency. Over the last few years, different polymeric-based formulations have been developed to achieve the controlled delivery of therapeutic proteins that have the ability to overcome the aforementioned challenges.[17,23] In the course of time, quite a few injectables have become commercially available, but a lot remains to be done from this perspective.[12,24] In the following subsections of this chapter, we will highlight the development of notable approaches for the controlled delivery of therapeutic proteins.

10.2 Controlled Delivery of Therapeutic Proteins Using Microparticles

Matrix-type delivery systems help to achieve the controlled delivery of therapeutic proteins from parenteral formulations. Over the past few years, the advantages of microsphere formulations have made them popular strategies for the sustainable and controlled delivery of therapeutic proteins. The development of microsphere-based biodegradable formulations to achieve the controlled delivery of therapeutic proteins has become a focused research area, as indicated by a voluminous literature in recent years. Such formulations advantageously minimize the chances of rapid *in vivo* degradation and ensure the stability of therapeutic proteins. The encapsulation of therapeutic proteins protects them from contact with esterase or protease enzymes until they are released from the microspheres. The efficient delivery of microsphere-based formulations to target sites using the oral, IV, or IM route is another added advantage. Microsphere-based formulations for the controlled delivery of therapeutic proteins have become a popular mode of delivery because of recent advancements in synthesis technology and polymer science. The polymers involved in the preparation of such formulations must not cause any alteration in the pharmacological activities of therapeutic proteins. Such polymers should be biodegradable, biocompatible, non-irritant, and non-toxic. Further, they should not undergo degradation to produce harmful products. Various types of natural and synthetic polymers have been used for the efficient delivery of therapeutic proteins.[25] Some of the most important examples of synthetic (Figure 10.2) and natural (Figure 10.3) polymers are presented here.

The desired controlled delivery of therapeutic proteins is achieved by using a suitable method of encapsulation and a polymer of biodegradable nature. Ideally, microsphere-based formulations show

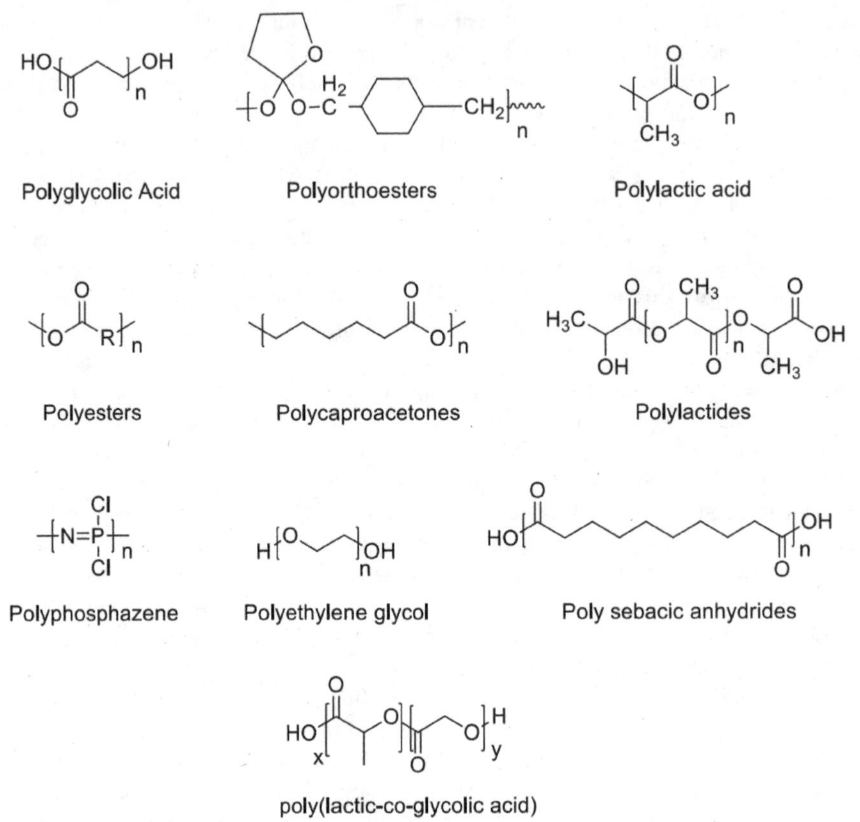

Polyglycolic Acid
Polyorthoesters
Polylactic acid
Polyesters
Polycaproacetones
Polylactides
Polyphosphazene
Polyethylene glycol
Poly sebacic anhydrides
poly(lactic-co-glycolic acid)

FIGURE 10.2 Synthetic polymers used for the preparation of controlled delivery formulations.

Starch
Chitosan
Chitin
Gelatin
Alginate

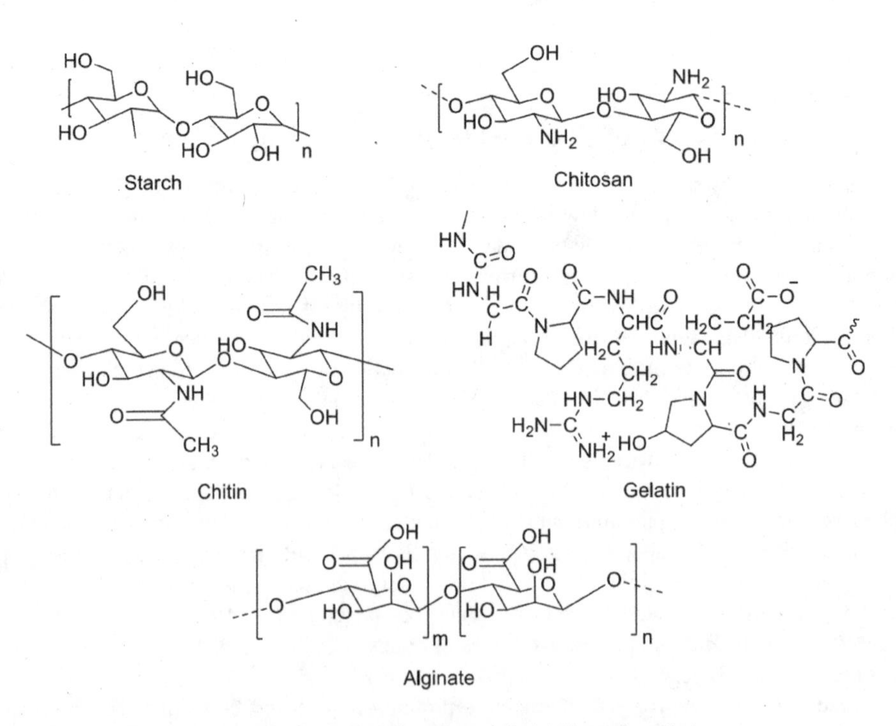

FIGURE 10.3 Natural polymers used for the preparation of controlled delivery formulations.

syringeability, which is influenced by polydispersity and particle size. The active conformations of encapsulated therapeutic proteins are not supposed to alter in microsphere production processes, and the use of denaturing solvents is avoided for such purposes. Biodegradable polymers are dissolved in organic solvents of volatile nature, such as acetone or dichloromethane. Microsphere formulations have been prepared using a spray-drying method. High-speed homogenizers are used to disperse solid drug molecules into a polymer solution. Microspheres ranging from 1 to 1000 mm in size are achieved on the evaporation of volatile organic solvent. Under a nitrogen atmosphere, the spray-drying method has been used for the successful encapsulation of human recombinant proteins such as erythropoietin. The degradation of polymers or diffusion or both mechanisms help the release of active ingredients from these microspheres. The enzyme prolidase has shown its potential involvement in the later stages of protein catabolism, and its deficiency causes chronic skin ulceration. The multiple emulsion technique has been used to prepare poly(lactic-co-glycolic acid) (PLGA) microspheres for the encapsulation of prolidase. It was observed that the polymer matrix provided the desired stability to encapsulated protein and facilitated the controlled delivery of the active ingredient. Enzyme replacement therapy could be executed by employing this technique.

Formulations of β-lactoglobulin (BLG), a major allergic protein, have also been prepared using PLGA microspheres. The oral tolerance of these proteins is improved to avoid allergic reactions, especially in newborns. For this purpose, new formulations of BLG were prepared using the water/oil/water (w/o/w) double emulsion technique whereby PLGA microspheres encapsulated the allergic protein. The use of Tween 20 in formulations was found to enhance the encapsulation efficiency and improved the controlled delivery of active protein, which consequently reduced the dose of anti-BLG IgE often required after its oral administration.

The multiple emulsion technique has also been used for the preparation of a microsphere formulation of interferon α (IFN-α) in a poly d,l-lactide-polyethylene glycol (PELA) matrix. In the formulation, PELA surrounds the calcium alginate core, and microspheres are encapsulated with IFN-α. The PELA matrix and IFN-α were stabilized by microsphere coating, resulting in the retention of bioactivity for a longer duration with high encapsulation efficiency. So, microsphere production by this method established its superiority over the conventionally practiced methods. Another study described the encapsulation of salmon calcitonin with PLGA microspheres, which demonstrated controlled release in 5 to 9 days after SC injection in rats. The same technique has also been used to prepare insulin-loaded microspheres using blends of poly(lactic acid) (PLA) and PEG homopolymer and PLGA copolymer. The resulting highly efficient microsphere formulation maintained a controlled delivery of insulin for 28 days. In rats, the glucose level remained suppressed for 9 days after the subcutaneous administration of ZnO–PLGA microspheres containing insulin, which confirmed the fast and long-lasting efficiency of the controlled delivery formulation.[26]

For the controlled delivery of therapeutic proteins, enzymatically responsive and pH-responsive hydrogel microparticles were prepared to study the influence of hydrogel degradation, cross-linking density, and particle size on system efficiency. These systems were prepared using poly(methacrylic acid-co-N-vinyl-2-pyrrolidone) (P(MAA-co-NVP)) and poly(itaconic acid-co-N-vinyl-2-pyrrolidone) (P(IA-co-NVP)) hydrogel microparticles for the controlled delivery of proteins such as rituximab, urokinase, and salmon calcitonin by the oral route. A complete and fast release of salmon calcitonin was achieved in an hour under gastric conditions without any degradation. Ideal loading and releasing behaviors were demonstrated by these hydrogels.[27] Recently, a w/o/w double emulsion method was used to develop a microstructured poly-epsilon-caprolactone particle system for an efficient controlled release of insulin *in vitro*. About 50% release was observed in the first 2 h, and complete release was achieved in 2 days. This biocompatible system showed biphasic behavior after SC administration.[28] The PLGA-based microparticles were used for the controlled co-delivery of platelet-derived growth factor (PDGF), vascular endothelial growth factor (VEGF), and bone morphogenetic protein-s (BMP-2) for bone repair.[29] In another study, the controlled delivery of pPB-HSA was made possible using subcutaneously injectable polymeric microspheres for the treatment of fibrosis. The w/o/w double emulsion technique was used to develop microspheres from multiblock copolymers of biodegradable nature. The complete release of therapeutic protein was observed in 2 weeks, while the same was achieved in 1 week when injected into mice suffering from renal fibrosis.[30]

10.3 Controlled Delivery of Therapeutic Proteins Using Nanoparticles

Nanoparticles have proved their ability for a facile delivery of a range of drugs to their target sites for longer duration. This ability has made them the carrier of choice for the controlled delivery of therapeutic proteins. Various types of nanoparticle can be employed for the delivery of therapeutic proteins (Figure 10.4).[31] The smallest capillaries in the human body show a diameter of about 5 to 6 mm, so the nanoparticles should be less than 5 μm in size to avoid any undesirable effects. The nanoparticles prepared from synthetic or natural polymers generally possess a diameter of less than 1 μm.[32] The effects of various formulation variables were evaluated on insulin-containing poly(ethyl cyanoacrylate) (PECA) nanoparticles. The nanoparticles showed a low polydispersity index, and the size, ranging from 130 to 180 nm, was found to be influenced by the mass of monomer. The mass of monomer also influenced the release rate, exhibiting zero-order controlled delivery for at least 10 days.[33] The acidic and hydrophobic nature of PLGA renders peptides and proteins unstable. Another major limitation includes the burst release of peptides/proteins from the PLGA matrix. To counter these limitations, a number of attempts were made to modify the aforementioned characteristics of the PLGA matrix to achieve an optimized formulation for the controlled delivery of therapeutic proteins. According to one study, PLGA–PVA composite nanoparticles were used for the encapsulation of bovine serum albumin (BSA).[33] The non-porous surface of the nanoparticles helped to achieve the controlled delivery of BSA for at least 2 months. The PLGA nanoparticles exhibited the controlled release of therapeutic proteins for more than 20 days, as their protein loading efficiency was improved due to the presence of a carboxyl end group on PLGA, whereas the esterification of carboxyl groups on PLGA reduced the protein loading capacity, and the nanoparticles showed a quick delivery of proteins in 14 days only.[34]

Nanoparticles with the PDGF-Receptor β (PDGFRβ) tyrphostin inhibitor AG-1295 were prepared using solvent displacement/spontaneous emulsification techniques to study release kinetics. A burst release of 25% was exhibited by nanoparticles of 120 nm, and AG-1295 (50%) was released gradually in about 30 days.[35] Another report described the preparation of PLGA particles 400 nm in diameter encapsulating active peptide using the emulsion solvent diffusion method. The pulmonary administration of an aqueous suspension of PLGA nanoparticles significantly reduced the glucose level in guinea pigs for 2 days. There was an initial burst release of 85%, and the remaining drug showed controlled release for next few hours.[36] Different nanoparticle formulations of polyanhydride and polyester were used for the

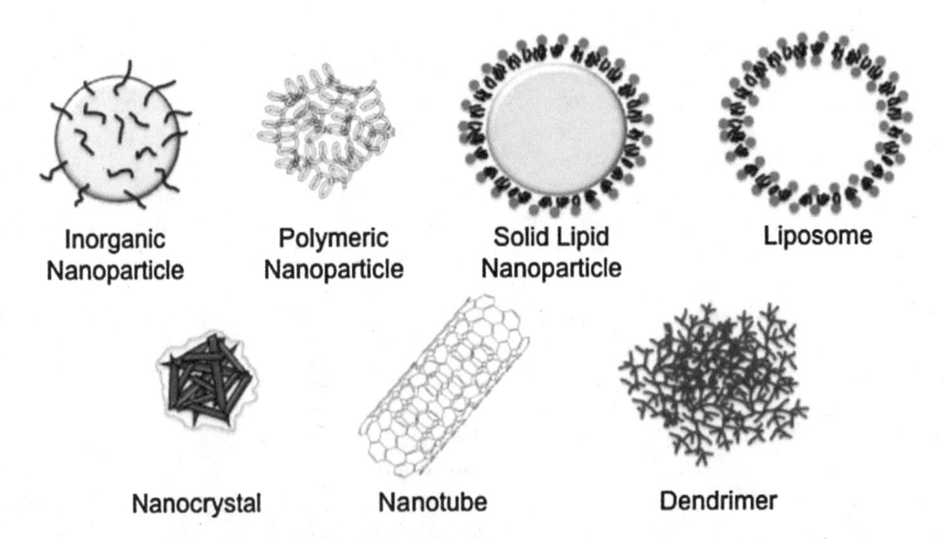

FIGURE 10.4 Schematic representation of types of nanoparticles that are being used for the delivery of therapeutic proteins. (Adapted from Faraji, A.H. and Wipf, P., *Bioorg. Med. Chem.*, 17, 2950–2962, 2009.)

encapsulation of Zn–insulin using a phase inversion nanoencapsulation technique. This controlled delivery formulation retained the bioactive insulin for 6 hours only.[37]

Among the known biodegradable and biocompatible polymers, PLGA is most frequently used for nanoparticle preparations. It has also been made available with different ratios of glycolide and lactide, but they cause instability of the encapsulated peptides/proteins. When a hot melt extrusion method was used to prepare PLGA implants, they exhibited an incomplete release of BSA, which actually was an acylated BSA.[38] The sulfated polysaccharide complexing agent dextran sulfate was successfully used to formulate a hydrophobic ion pairing (HIP) complex of lysozyme (15 kDa).[39] Nanoparticles prepared by the spontaneous emulsion solvent diffusion method displayed controlled delivery over a period of 30 days without burst release. On release, both the lysozyme–dextran sulfate complex and lysozyme were found to be fully active, so controlled delivery of antibodies could be managed by the HIP method.[39] Another report described the use of dextran sulfate for successful HIP complex formulation of BSA (66 KDa), and the nanoparticles were prepared by the solid in oil in water (s/o/w) emulsion method. Neither nanoparticle preparation nor HIP complexation influenced the secondary and tertiary structures of BSA, as evaluated by intrinsic fluorescence and circular dichroism analyses.[40] In a number of studies, mesoporous silica nanoparticles (MSN) have been found to be ideal candidates for the controlled delivery of therapeutic proteins due to their biocompatible nature and high loading capacities. Further, MSNs have been shown to be friendly to encapsulated proteins, thus minimizing the chances of their denaturation.[41] Quite often, the breaking down of nanoparticles caused the denaturation of encapsulated proteins *in vivo*. Also, sometimes, the encapsulating techniques cause the denaturation of the proteins, thus reducing their bioavailability. Recently, Molly and his co-workers have demonstrated that the controlled delivery of therapeutic proteins could be possible for several months without the requirement for an encapsulation step. This noteworthy development could be a welcome advance in the field of controlled delivery of therapeutic proteins. In this case study, a model protein was used for tissue regeneration in spinal cord injury and stroke.[42]

10.4 Controlled Delivery of Therapeutic Proteins Using *In Situ* Gel/Implants

In recent years, the developments in the field of injectable implants have revolutionized the controlled delivery of various proteins, small molecules, and other therapeutics.[43] The simplicity of these systems has made them popular and widely used. The injectable semisolid or liquid system turns to a gel/implant after injection, and a number of mechanisms are involved in this transformation. Based on the mechanism of gel/implant formation, the formulations are recognized as a cross-linking system and *in situ* phase inversion. Solvent extraction has been used for the thorough study of *in situ* implant formation, resulting in two commercial formulations. A number of factors proved the temperature-dependent sol-to-gel system to be an efficient one compared with other known systems. The implant formation has also been controlled by the development of temperature- and pH-dependent hybrid polymers. The effects of pH, phase inversion, and temperature on *in situ* gel implants have been well evaluated.[44]

10.5 Controlled Delivery of Therapeutic Proteins Using Solvent Extraction/Phase Inversion Systems

Proteins suspended in organic solvent and polymeric solutions in an appropriate water-miscible organic solvent are used for injectable implants, and the protein-loaded implants are formed after *in vivo* injection. After administration, the insoluble polymer is precipitated when organic solvent diffuses into the surrounding tissues. The therapeutic protein becomes trapped in the precipitated polymer matrix. Polymers generally used for such systems include polyesters such as polycaprolactone (PCL), poly(glycolic acid) (PGA), PLA, and PLGA. A long list of solvents with different polarities, including glycofurol, polyethylene glycol $(PEG)_{500}$-dimethylether, benzyl benzoate, ethyl acetate, triacetin, ethyl benzoate, ethanol, dimethyl sulfoxide (DMSO) and dimethyl formamide (DMF) have been used to dissolve a variety of polymers during a number of studies. In recent *in vivo* studies, glycofurol and PEG-alkyl ether were found to be biocompatible and well-tolerated solvents for PLGA implants.[45–47] Studies revealed that PLGA polymer shows minimal degradation when used in a relatively compatible solvent such as glycofurol. Such systems are employed for the controlled delivery of therapeutic proteins in active form with minimum dose frequency. Variables such as burst release, porosity of implant, and gelling rate influenced both the rate of release and the time duration for phase inversion implants/systems. Further, these factors are found to be dependent on solvent system, polymer type, crystallinity, molecular weight, concentration, and additives.[48,49]

N-Methyl-2-pyrrolidone (NMP) is a frequently used solvent for injectable implants due to its highly polar nature. It also causes rapid phase inversion for implant formation compared with other solvents such as benzyl benzoate and ethyl benzoate. The rapid phase inversion often results in the formation of porous implants, which facilitates the burst release of proteins or peptides. When NMP was employed for PLGA implant formation, the initial burst produced at least 40% lysozyme, and the rest of the dose was released at a slow rate. However, low burst releases (less than 10% of the initial dose) were observed when less porous implants were obtained with less polar solvents such as ethyl benzoate and triacetin.[50] The peptides or proteins could be solubilized or dispersed in polymer solutions, but higher release rates were observed in the dispersed state.[51] The porosity of implants increased significantly after the release of dispersed proteins, which resulted in high release rates. The burst rate and subsequent release were also influenced by the types of additives. Water-soluble additives increase the release rates even when triacetin is used for implant formation.[51] A direct proportionality was observed between the concentration of mannitol and release rates, and the effect was found to be very pronounced during the last phase. On the other hand, slow release rates were observed when linoleic triglyceride was used as a hydrophobic additive.[51] In other *in vivo* studies, similar results were observed when benzyl benzoate, ethyl benzoate, triacetin, and NMP were used for the preparation of human growth hormone (hGH)-containing PLGA implants. The controlled release of hGH was observed in plasma within a therapeutic window from a depot formed by benzyl benzoate. The physical properties of hGH were manipulated to control the burst effect and release profile. Zn-complexed and densified hGH with low aqueous solubility was prepared, which displayed negligible burst release compared with lyophilized and densified hGH.[52] Studies were conducted to establish the influence of the concentration and molecular weight of polymer and drug loading on the efficacy of leuprolide from PLGA implants prepared in NMP. In rat models, the efficacy of leuprolide implants was not influenced by the polymer concentration (40–50%) or drug loading (about 6%), whereas high–molecular weight PLGA showed a longer duration of action compared with low–molecular weight polymer. A dose of a single injection of the formulation suppressed the testosterone level for more than 3 months in a canine model.[53] In another study, surfactants such as polyvinyl alcohol (PVA) and Triton X were incorporated into the PLGA implants, thus influencing their mechanical stability to reduce the burst effect.[54] A few modified techniques have also been employed to achieve implants with better peptide loading and controlled burst release. The Durect Corporation used sucrose acetate isobutyrate (SAIB) to form a matrix as alternative injectable implant system known as SABER™ by avoiding the use of PLA or PLGA polyesters. A slow phase inversion depot was formed when SAIB was dissolved in an organic solvent such as benzyl benzoate or ethanol.

Other studies modified SAIB a bit more to deliver hGH with added advantages such as injectability through a narrow needle and high protein loading.[55] The modification involved the addition of PLA (1.0% and 10% w/w) along with SAIB in benzyl benzoate or ethanol, and lyophilized hGH was suspended in this mixture. An initial burst release of 78% of the dose was recorded in 24 h when the SABER system was used without PLA in *in vitro* studies. Interestingly, the addition of PLA (1% and 10%) to the SABER system reduced the burst release of hGH to 2% and 0.2%, respectively. But, the use of organic solvent caused the degradation of a significant amount of protein. The encapsulation of protein in crystalline form was evaluated to avoid the denaturation of proteins in the SABER system.[56] Initially, both amorphous and crystalline forms of α-amylase were suspended in PLGA/acetonitrile and SAIB/ethanol systems. The crystalline form of the protein showed high loading and retained 100% activity and stability in both systems even after long-term storage. The crystal state of α-amylase in PLGA/acetonitrile minimized the *in vitro* burst release. Grain-shaped crystals minimized the burst effect compared with rod-shaped crystals, which showed a significant high burst release. Microparticle-based *in situ* forming injectables have also been established as an alternative to injectable forming implants.[57,58] The typical emulsions are found to be less viscous than plain oil-based systems, so they could be injected using a narrow-gauge needle. Considering this fact, PLGA was dissolved in triacetin and NMP to prepare an o/o or o/w emulsion to achieve an *in situ* microparticle (ISM) forming system. In fact, *in situ* microparticles are formed via solvent extraction after injection. The effects of different process variables on the efficiency of injectable *in situ* microparticle systems were evaluated by employing myoglobin (MW 16,950 Da) and cytochrome c (MW 12,327 Da).[58] The release rate was found to be influenced by the encapsulation efficiency, the molecular weight of PLGA, and the porosity of microspheres, as was the case for PLGA injectable implants. The *in situ* microparticles with cytochrome c showed a burst release of about 40%. The addition of mannitol to myoglobin increased the burst effect. In the case of ISMs, a significant increase in burst release of 30–70% was recorded as a result of mannitol addition (5%). The encapsulated myoglobin extracted from ISMs retained its structural integrity, but the release time was only 2 weeks and was not influenced by any type of ISM modification. The release profile of leuprolide from *in situ* microspheres was evaluated under formulation parameters such as concentration of PLA and PLGA, polymer molecular weight, and polymer blending.[57] A high burst effect was observed with high–molecular weight PLGA compared with low molecular weight. PLGA with ester groups showed higher release rates compared with the polymer containing free acidic groups due to electrostatic forces between leuprolide and the polymer. An *in situ* microsphere system based on a polymer blend (polymer with 30% concentration and PLA R 202H and R 203H in 1:1 ratio) showed high loading of leuprolide (15%).

A high leuprolide loading of 15% was obtained in an ISM system containing a polymer blend (PLA R 203H and R 202H at a ratio of 1:1, polymer concentration 30%), and a minimal burst release was observed with a release period of 6 months in *in vitro* studies. The susceptibility of proteins (tertiary and quaternary structures) to degradation on exposure to oil–aqueous interfaces and organic solvents has made this controlled delivery technique applicable mainly to the delivery of peptides/proteins. NMP solvent–associated allergic dermatitis has been reported as a result of application of a commercially available injectable PLGA depot system. Indeed, the system contains leuprorelin acetate.[59] The higher protein loadings help the formulation to act successfully for the controlled delivery of therapeutic proteins. In general, the drug loading has been observed to be ~3–12%. However, use of crystalline protein and ISM-based systems could improve the drug/protein-loading capacity.

10.6 Thermosensitive Gels for Controlled Delivery of Therapeutic Proteins

In general, thermosensitive gel systems are aqueous solutions of amphiphilic block copolymers with hydrophobic segments consisting of either polyphenylene oxide (PPO), PCL, PLA, or PLGA.[60] The temperature change on injection helps the solution to undergo sol-to-gel transition and produce a solid gel structure. The excellent biocompatibility and water solubility of PEG have made it an extensively used hydrophilic block. Proteins are dissolved in aqueous polymeric solution, which becomes encapsulated on gel formation. The type, molecular weight, and concentration of the polymer that forms the hydrophobic block are important for the characteristics of gel. Thermosensitive gel systems have been found to have advantages compared with injectable implants due to the chances of *in vivo* toxicity and protein degradation with organic solvents. So, thermosensitive gels are expected to show high compatibility with therapeutic proteins and living tissues. Additionally, aqueous solution–based formulations could be sterilized, filtered, and injected with ease using narrow-gauge needles. The widely used commercially available ReGel®, an aqueous solution at room temperature, is composed of a PLGA–PEG–PLGA (~23% w/w) polymer system. It turns to gel when injected at body temperature, and this transition is reversible. Various studies have shown the significance of this thermosensitive system for the controlled delivery of therapeutic proteins, including recombinant hepatitis B surface antigen, insulin, porcine growth hormone, and other peptides.[61–64]

The dependence of the controlled release of therapeutic proteins and sol-gel transitions on the molecular weight of PEG and PLGA segments has been evaluated using PEG MW 1000 and 1450 Da for the synthesis of ReGel®-1 and ReGel®-2 with a total molecular weight of 4200. The formulations with PEG 1450 Da showed a higher critical gelling temperature compared with all formulations of ReGel® with PEG

1000 Da. Both ReGel®-1 and ReGel®-2 with 23% w/w formulations have been studied for *in vitro* controlled release of porcine growth hormone (pGH), granulocyte colony-stimulating factor, and Zn-insulin. The Zn-insulin from ReGel®-2 showed a less than 10% burst release with a very slow release rate compared with ReGel®-1, while in all other cases, a burst release of 25–40% was recorded with a controlled delivery duration of 2–3 weeks. It was suggested that the complexation of pGH and insulin caused a slow burst release. An effective injection dose of 5 mg of pGH for 14 days was successfully replaced with a single injection of ReGel®-1/Zn-pGH or ReGel®-1/pGH with a dose of 1 ml solution containing 70 mg/ml pGH in *in vivo* studies in hypophysectomized rat models. In a model study, lysozyme (30% w/v) was used to evaluate the effects of polymer concentration and composition on aspects of controlled release.[61] The dependence of reversible thermosensitive transitions on the concentration and molecular weight of block polymer was evaluated using (PLGA–PEG–PLGA) with different block lengths of PLGA (PEG MW 1000 Da). The thermosensitive transitions were found to be dependent on the concentration and molecular weight of each block. With increasing molecular weight, the upper critical gelling temperature was found to fall, and a rise was observed in the lower critical gelling temperature.[61] In the case of high–molecular weight polymers, the sol-to-gel transition was observed to be an irreversible process. The burst release decreased from 41.2 ± 5.4% to 16.1 ± 3.9% with increasing molecular weight (MW 1602 to 7859 Da) of polymer. The controlled release of lysozyme was observed in 4 weeks with all block polymers. Further, an inverse effect was recorded between polymer concentration and burst release. *In vitro* studies involving the triblock copolymer PLGA–PEG–PLGA (MW 1400–1000–1400) for the controlled delivery of pGH suggested diffusion as the main mechanistic approach responsible for the release of proteins.[65] The polymer degradation was also found to be somewhat influential in this regard. The gel containing 0.12% w/v pGH showed about 100% release, while higher concentrations displayed incomplete release, represented by a plateau phase after 4 weeks. These results were further verified when 0.12% and 0.42% w/v doses were injected in rabbit models, and the low-dose formulation provided high bioavailability (86%) compared with the high dose. The incomplete release of a high dose of pGH (0.42%) from a thermosensitive gel resulted in its low bioavailability of about 38%. In aqueous solutions, higher concentrations of proteins promote aggregation, which ultimately causes incomplete release.[66–68] In PLGA implants, histidine-HCl buffer and trehalose were used to minimize the aggregation process of a model protein and BSA.[69]

Polycaprolactone (PCL)-based triblock polymers have been used to prepare another type of thermosensitive gel system. PLGA degradation usually produces acidic byproducts such as glycolic acid or lactic acid, which has been avoided by the use of caprolactone-based polymers. In fact, proteins become denatured and lose their activity in the acidic environment created by acidic byproducts.[70,71] Comparatively better stability has been shown by protein molecules released from PCL.[72,73] Sol-to-gel transitions were also observed when PCL–PEG–PCL was used for gelling polymers with molecular weights 2600–4100 Da.[74] In *in vitro* studies, these thermosensitive gels exhibited a controlled release of horseradish peroxidase (HRP) and BSA for a period of 30 days. The aqueous solution–based systems have been used advantageously for the preparation of formulations with high loading (10% w/w) of proteins. The concentration and molecular weight of polymers were found to influence the controlled release (rate and duration) of therapeutic proteins. High–molecular weight thermosensitive gel systems exhibited a slow release rate for more than 30 days. The slow release was found to be less than 20% of the total burst release, and it decreased further with increasing polymer concentrations from 15 to 25 wt%.[74] In *in vitro* studies, BSA displayed incomplete release, but about 60% release was observed in the first 2 weeks from high–molecular weight gels. Complete delivery in 20 days was recorded with polymers of low molecular weight (MW 2600 Da). The subcutaneous administration of hydrogels in mice lasted for 45 days until the copolymers of PCL–PEG–PCL underwent degradation in *in vivo* studies. In *in vitro* studies, the controlled delivery of a model protein and BSA was evaluated using triblock copolymers with different molecular weights (2992–5046 Da) and PEG–PCL–PEG block arrangements.[75] A low burst effect and release rate for 2 weeks were displayed by hydrogels with higher protein loadings. There were no data about the stability of BSA after 2 weeks. A maximum release of 65% was shown by formulations with lower loading of therapeutic proteins. Basic fibroblastic growth factor (bFGF) was delivered using a PEG–PCL–PEG-based thermosensitive hydrogel. In BALB/c mice, strong humoral immune responses were observed for 12 weeks when bFGF-loaded hydrogels were used. Again, the initial burst and release rate were found to be dependent on protein loading. The semicrystalline and hydrophobic nature of PCL-based hydrogels avoided their degradation and acidic by-product formation for more than 30 days *in vitro*. Thus, glycolide- and lactide-based polymers could be replaced with more suitable and apparently non-hazardous PCL-based polymers. *In vitro* studies have also been performed aiming to reduce the crystalline nature of PCL segments to maintain a relatively fast process of degradation, since slow degradation leaves an empty delivery vehicle after the controlled release of therapeutic protein in a maximum duration of 3 weeks. To achieve slightly faster-degrading thermosensitive hydrogels, caprolactone and glycolide were used as hydrophobic segments for the synthesis of triblock random copolymers.[76] At neutral pH, the glycolide segment underwent degradation when a poly(ε-caprolactone-co-glycolide)–poly(ethyleneglycol)–poly(ε-caprolactone-co-glycolide)-based hydrogel was used to study the protein release pattern *in vitro* for 8 weeks.[76] BSA was loaded as 0.1% w/v in 25 wt% gel for these release pattern studies, and at least 80% release was observed in 30 days. The diffusion-controlled release mechanism was found to be responsible for this controlled delivery of BSA.

The thermosensitive hydrogel-based system offers a controlled delivery of proteins in 15–30 days, representing the advantages of simplicity, elegance, and biocompatibility of an aqueous system. The preclinical development of hydrogel-based systems should follow the following important points. The formulation should contain a high drug loading to maintain a therapeutic concentration for a longer time period. This aqueous-based thermosensitive hydrogel accommodates a higher concentration of proteins and also helps to keep intact the integrity of proteins due to the absence of an oil/water interface, which would otherwise denature therapeutic proteins. *In vivo* low bioavailability and incomplete release *in vitro* are explained by considering the degradation and aggregation of proteins as major reasons. The complications related to the handling of hydrogels during injections reduce the applications of such gel-based systems. The polymeric solutions show low viscosity and good injectability at 4 °C and turn to gel at body temperature. At room temperature, the increase in viscosity makes them hard to inject.[75]

10.7 Controlled Delivery of Therapeutic Proteins using pH-dependent *In Situ* Gels

In such cases, the pH change has been found to be the controlling factor for implant formation and the release of therapeutic proteins. Generally, the pH-dependent water-soluble polymers form such polyanionic or polycationic systems. The solution form of polymers such as chitosan and alginate is converted into a gel with the change of pH after administration. Under acidic conditions, poly(methacrylic acid) is converted into gel form.[77] Implants with controlled release of proteins have been developed using a layer-by-layer (LbL) coating approach.[78,79] The implants were coated with polyelectrolyte using chondroitin (polyanion) and poly(β-aminoester)/poly2 (polycation) to load BMP-2.[78] In *in vitro* studies at physiological pH, the BMP-2 was found to be released over a duration of 2 weeks. According to another study, the controlled delivery of BMP-264 was achieved using granules coated with hyaluronic acid/poly(L-lysine). Polyelectrolyte coating using poly(methacrylic acid) (polyanion) and poly-L-histidine (polycation) modified the surface of an anodized titanium implant. The pH-dependent release of protein was achieved due to this surface modification.[80] The controlled delivery profile was studied using fluorescently labeled poly-L-lysine (15–30 kDa). For LbL-coated implants, the coating technique and molecular weight of poly(methacrylic acid) were found to affect the duration of release and burst release. The pH-dependent sustained release was observed at pH 7–8, while maximum release was detected at pH 5–6. bFGF and BMP-2 exhibited similar release profiles for 25 days from LbL-coated implants.[81]

Chitosan (CS) is a biodegradable and biocompatible naturally occurring polysaccharide.[82] Its polyamines are ionized under acidic conditions, making it a soluble natural polymer, but it precipitates to form a gel at physiological pH. The crystalline nature of CS generates a porous gel that undergoes rapid degradation *in vivo*, thus making it impractical for the sustained release of proteins. The crystallinity of CS has been lowered by cross-linking it with poly-vinylpyrrolidone (PVP), making the gel a mechanically strong structure at physiological pH.[83] However, the toxicity of cross-linking agents or their side reactions with therapeutic proteins could lead to deleterious effects.[84] Also, such limitations have been countered by introducing new polymers for the controlled delivery of therapeutic proteins.[85,86]

The cross-linked hydrogels of CS and PVP have also been prepared using γ-radiation to minimize toxic chemical effects and were used to deliver therapeutic proteins through the oral route.[87] In acidic conditions, the cross-linked pH-dependent hydrogels showed significant swelling. The PVA content and the extent of cross-linking control the porosity of the gel. The porosity of the gel has been found to directly control the absorption of BSA. The CS and BSA show charge–charge interactions, which control the loading of BSA in cross-linked gel. In 24 h, at pH 7.4, BSA release was 40 to 90%, while at pH 1.5, the release was <10% from CS–PVP hydrogels.[87] The chemically stable cross-linking agent triethylene glycol dimethacrylate was used to develop a pH-sensitive hydrogel from poly (ethylene glycol) methacrylate (PEGMA)-grafted poly (methacrylic acid) (p(PEGMA-g-MAA)) for the controlled delivery of therapeutic proteins under acidic conditions.[88] At pH 7.4, the BSA was released completely in 12 h, while little release was observed under acidic conditions. The pH- and temperature-sensitive polymers have been developed using pH-sensitive segments and hydrophobic polymers, because pH-controlled implants do not provide protein release for a longer duration.[89–92] A pH- and temperature-sensitive SMO-PCLA–PEG–PCLA-SMO polymer was developed using thermosensitive poly(ε-caprolactone-co-lactide)–poly(ethylene glycol)–poly(ε-caprolactone-co-lactide) and pH-sensitive sulfamethazine oligomers, which existed as a solution at high temperature (70 °C) or pH (8).[93] In rats, SC injections of this gel were found to be biocompatible and underwent degradation in 6 weeks, suggesting it as a potent system for the controlled delivery of therapeutic proteins.[94] The hydrogel was found to be an effective system to encapsulate human recombinant BMP-2 and mesenchymal stem cells for at least 7 weeks in *in vivo* studies.

10.8 Controlled Delivery of Therapeutic Proteins Using Non-biodegradable Implants

Siloxanes, silicones, and polysiloxanes are organo-silicones with well-known non-reactivity, low toxicity, and biocompatibility and are thus widely used in biomedical applications. The appropriate functionalities incorporated at the silicon atom help to impart the properties of solid resins, rubbers, greases, or fluids. They have been used to prepare a number of reservoirs, matrixes, and drug eluting stents as delivery systems to achieve the controlled delivery of a range of biotherapeutics. The controlled delivery of BSA was successfully managed using implants prepared from silicon-based polymers.[95] In another study, a silicon formulation was prepared to encapsulate highly potent proteins. As a model protein, the release rate of interferon was found to be influenced by its particle size and loading. A zero-order release of 40% interferon in total was observed for a period of 28 days.[96] The release rate increased with high loading and large particle size and was enhanced further as a result of the addition of 2% glycine. The interconnected channel formation in covered-rod-type implants caused high osmotic pressure, which served to facilitate interferon release. Without any initial burst, the formulation maintained a controlled release for 1 to 3 months, exhibiting high activity in tumor model mice. The controlled release profile of BSA was also studied using the same type of formulation. The addition of sucrose was found to be an influencing factor for controlled release from silicon-based implants.[97]

Lofthouse and his colleagues compared covered-rod-type and matrix-type injectable implants for the controlled delivery of antigen—*Clostridium novyi* toxoids, *Clostridium tetani*, and avidin—in sheep.[98] The first-order delivery of antigen was managed by matrix-type implants in 30 days, while covered-rod-type implants exhibited zero-order release for a longer time. Both *in vivo* and *in vitro* studies found silicon-based implants to be reliable systems for the controlled delivery of therapeutic proteins. These biocompatible and non-toxic silicon-based systems were prepared without the use of toxic organic solvents and heat, which could cause denaturation of proteins. Further, such systems advantageously facilitated the complete release of therapeutic proteins, making them promising controlled delivery candidates.[99] It is desirable for controlled delivery formulations to follow a zero-order release pattern, ensuring that the concentration of released therapeutic proteins remains within the therapeutic window. Maintaining the therapeutic window at a constant rate for extended periods helps to avoid the establishment of toxic levels, as has been observed in a number of trial studies.[100–102] Osmotic pressure–based mechanisms operate for the controlled delivery of biotherapeutics over extended time periods. Viadur®, Alzet®, and DUROS® are well-known osmotically driven implants administered subcutaneously in humans.[103,104] Prostate cancer is treated with controlled delivery of leuprolide acetate for a period of 12 months using Viadur® as a controlled delivery implant. The DUROS® devices have been used for controlled delivery of therapeutic peptides and proteins for up to 1 year.[105] However, the applications of non-biodegradable implants for the controlled delivery of therapeutic proteins are limited, because surgery is required for their administration. Also, after the complete release of therapeutic proteins, the removal of implants requires surgery and is thus considered patient non-friendly.

10.9 Controlled Delivery of Therapeutic Proteins using Nanoparticles-in-Gel Composite Systems

Within biological systems, the denaturation and/or degradation of therapeutic proteins need to be avoided for longer biological half-lives. One such strategy involves the encapsulation of therapeutic proteins in micro- or nanoparticles. However, a few disadvantages, such as the burst effect resulting in dose-related toxicity, limit the applications of particulate-based delivery systems. The burst releasing of therapeutic proteins could be minimized by employing a nanoparticle-in-gel composite as a controlled delivery system.[106,107] A composite formulation with no burst effect was prepared by encapsulating a therapeutic protein in pentablock (PB) polymer nanoparticles and dispersing them in a thermosensitive gel.[108] This formulation with zero toxicity and high biocompatibility was used to treat posterior segment ocular diseases, which otherwise require intravitreal injections repeatedly.[107] Traditionally, therapeutic proteins such as aflibercept, ranibizumab, and bevacizumab were administered frequently to treat sight-threatening disorders, including diabetic macular edema, diabetic retinopathy, and wet age-related macular degeneration, known as posterior segment ocular diseases. The traditional practices produce severe complications, such as retinal hemorrhage, retinal detachment, endophthalmitis, etc. So, the preparation of a composite formulation with zero toxicity is a welcome advance to reduce the chances of such complications. Initially, composite formulations were prepared using bevacizumab, IgG, and BSA. Almost zero-order release with no burst effect was displayed by these biocompatible and non-toxic formulations for at least 2 months *in vitro*. The success of deficient bladder reconstruction therapy is hindered by the oxidation, diketopiperazine formation, and deamidation of administered VEGF protein. These instability factors reduce the half-life of VEGF, resulting in repeated administration of the protein, which causes progression of malignant vascular tumors as one of the known side effects.[101] To avoid such intolerable situations, initially, VEGF-nanoparticles were prepared as a controlled delivery system. However, the success of this biocompatible system was challenged by high burst release of about 40% in 48 h. Later, a composite formulation with burst effect of 15% was prepared by dispersing VEGF-encapsulated PLGA nanoparticles in Pluronic-F127 thermosensitive gel.[109] The chances of dose-related toxicity were minimized by the introduction of the controlled delivery composite system.

10.10 Controlled Delivery of Therapeutic Proteins Using Pad Technique

Type 1 diabetes patients are burdened lifelong with the repeated administration of insulin to maintain its minimum level in their blood plasma. Recently, a photo-activated depot (PAD) has been developed with the aim of minimizing the inevitable insulin administration. In the PAD formulation, through the photo-labile group, insulin is conjugated with a biodegradable but insoluble resin. *In vitro*, the release of insulin is controlled with external irradiation (light-emitting diode [LED] at 365 nm) once the PAD is subcutaneously injected.[110] This technique could be used for the controlled delivery of other proteins; depending on the size and concentration of the protein, the size of the PAD could be varied. The methods chosen for the preparation of a controlled delivery formulation for therapeutic proteins must be facile, practical, and robust. The manufacturing of a high-quality product should be ensured by the design of quality experiments, quality-by-design, and in-process quality control methods.[111,112]

Conclusion

Therapeutic proteins are expected to become the major commercially available biotherapeutics in the near future. A number of approaches for their controlled and sustained delivery have been developed to enhance poor *in vivo* half-life and improve release profiles, thus ensuring a reduced number of administrations. However, the development of controlled delivery formulations often encounter practical challenges such as stability issues of proteins during encapsulation and within formulations, release pattern, burst effect, protein–polymer interactions, protein loading, etc. Further, the susceptibility of proteins to forming irreversible aggregates at higher concentrations has also been found to hinder the development of controlled delivery formulations. Controlled delivery of therapeutic proteins has been achieved with a great deal of success by employing the aforementioned well-explained delivery techniques. However, recent advancements in the controlled delivery of therapeutic proteins for longer periods of time even without encapsulating them are expected to be a "game changer" in the field of protein therapeutics in the near future.

REFERENCES

1. Leader, B.; Baca, Q. J.; Golan, D. E. Protein therapeutics: a summary and pharmacological classification. *Nature Reviews. Drug Discovery* 2008, *7* (1), 21–39.
2. Tsomaia, N. Peptide therapeutics: targeting the undruggable space. *European Journal of Medicinal Chemistry* 2015, *94*, 459–470.
3. Akash, M. S. H.; Rehman, K.; Tariq, M.; Chen, S. Development of therapeutic proteins: advances and challenges. *Turkish Journal of Biology* 2015, *39* (3), 343–358.
4. Dimitrov, D. S. Therapeutic proteins. In V. Voynov and J. A. Caravella (Eds.), *Therapeutic Proteins*, Springer: 2012; pp 1–26.
5. Vlieghe, P.; Lisowski, V.; Martinez, J.; Khrestchatisky, M. Synthetic therapeutic peptides: science and market. *Drug Discovery Today* 2010, *15* (1–2), 40–56.
6. Akash, M. S. H.; Rehman, K.; Sun, H.; Chen, S. Interleukin-1 receptor antagonist improves normoglycemia and insulin sensitivity in diabetic Goto-Kakizaki-rats. *European Journal of Pharmacology* 2013, *701* (1–3), 87–95.
7. Akash, M. S. H.; Rehman, K.; Sun, H.; Chen, S. Sustained delivery of IL-1Ra from PF127-gel reduces hyperglycemia in diabetic GK-rats. *PloS One* 2013, *8* (2), e55925.
8. Woodnutt, G.; Violand, B.; North, M. Advances in protein therapeutics. *Current Opinion in Drug Discovery & Development* 2008, *11* (6), 754–761.
9. Rehman, K.; Akash, M. S. H.; Akhtar, B.; Tariq, M.; Mahmood, A.; Ibrahim, M. Delivery of therapeutic proteins: challenges and strategies. *Current Drug Targets* 2016, *17* (10), 1172–1188.
10. Shi, S. Biologics: an update and challenge of their pharmacokinetics. *Current Drug Metabolism* 2014, *15* (3), 271–290.
11. Weng, Z.; DeLisi, C. Protein therapeutics: promises and challenges for the 21st century. *Trends in Biotechnology* 2002, *20* (1), 29–35.
12. Carter, P. J. Introduction to current and future protein therapeutics: a protein engineering perspective. *Experimental Cell Research* 2011, *317* (9), 1261–1269.
13. Vaishya, R. D.; Khurana, V.; Patel, S.; Mitra, A. K. Controlled ocular drug delivery with nanomicelles. *Wiley Interdisciplinary Reviews: Nanomedicine and Nanobiotechnology* 2014, *6* (5), 422–437.
14. Choonara, B. F.; Choonara, Y. E.; Kumar, P.; Bijukumar, D.; du Toit, L. C.; Pillay, V. A review of advanced oral drug delivery technologies facilitating the protection and absorption of protein and peptide molecules. *Biotechnology Advances* 2014, *32* (7), 1269–1282.
15. Renukuntla, J.; Vadlapudi, A. D.; Patel, A.; Boddu, S. H.; Mitra, A. K. Approaches for enhancing oral bioavailability of peptides and proteins. *International Journal of Pharmaceutics* 2013, *447* (1–2), 75–93.
16. Gupta, S.; Jain, A.; Chakraborty, M.; Sahni, J. K.; Ali, J.; Dang, S. Oral delivery of therapeutic proteins and peptides: a review on recent developments. *Drug delivery* 2013, *20* (6), 237–246.
17. Akash, M. S. H.; Rehman, K.; Chen, S. Polymeric-based particulate systems for delivery of therapeutic proteins. *Pharmaceutical Development and Technology* 2016, *21* (3), 367–378.
18. Smart, A. L.; Gaisford, S.; Basit, A. W. Oral peptide and protein delivery: intestinal obstacles and commercial prospects. *Expert Opinion on Drug Delivery* 2014, *11* (8), 1323–1335.

19. Pisal, D. S.; Kosloski, M. P.; Balu-Iyer, S. V. Delivery of therapeutic proteins. *Journal of Pharmaceutical Sciences* 2010, *99* (6), 2557–2575.
20. Antosova, Z.; Mackova, M.; Kral, V.; Macek, T. Therapeutic application of peptides and proteins: parenteral forever? *Trends in Biotechnology* 2009, *27* (11), 628–635.
21. Bruno, B. J.; Miller, G. D.; Lim, C. S. Basics and recent advances in peptide and protein drug delivery. *Therapeutic Delivery* 2013, *4* (11), 1443–1467.
22. Sabaté, E. *Adherence to Long-Term Therapies: Evidence for Action*. World Health Organization: 2003.
23. Duncan, R. Polymer therapeutics: top 10 selling pharmaceuticals—what next? *Journal of Controlled Release* 2014, *190*, 371–380.
24. Tang, Y.; Singh, J. Biodegradable and biocompatible thermosensitive polymer based injectable implant for controlled release of protein. *International Journal of Pharmaceutics* 2009, *365* (1–2), 34–43.
25. Akash, M. S. H.; Rehman, K.; Chen, S. Natural and synthetic polymers as drug carriers for delivery of therapeutic proteins. *Polymer Reviews* 2015, *55* (3), 371–406.
26. Takenaga, M.; Yamaguchi, Y.; Kitagawa, A.; Ogawa, Y.; Mizushima, Y.; Igarashi, R. A novel sustained-release formulation of insulin with dramatic reduction in initial rapid release. *Journal of Controlled Release* 2002, *79* (1–3), 81–91.
27. Koetting, M. C.; Guido, J. F.; Gupta, M.; Zhang, A.; Peppas, N. A. pH-responsive and enzymatically-responsive hydrogel microparticles for the oral delivery of therapeutic proteins: effects of protein size, crosslinking density, and hydrogel degradation on protein delivery. *Journal of Controlled Release* 2016, *221*, 18–25.
28. Guerreiro, L. H.; Silva, D. D.; Girard-Dias, W.; Mascarenhas, C. M.; Miranda, K.; Sola-Penna, M.; Ricci Júnior, E.; Lima, L. M. T. d. R. e. Macromolecular confinement of therapeutic protein in polymeric particles for controlled release: insulin as a case study. *Brazilian Journal of Pharmaceutical Sciences* 2017, *53*(2), e16039.
29. Kirby, G. T. S.; White, L. J.; Steck, R.; Berner, A.; Bogoevski, K.; Qutachi, O.; Jones, B.; Saifzadeh, S.; Hutmacher, D. W.; Shakesheff, K. M.; Woodruff, M. A. Microparticles for sustained growth factor delivery in the regeneration of critically-sized segmental tibial bone defects. *Materials* 2016, *9* (4), 259.
30. Teekamp, N.; Van Dijk, F.; Broesder, A.; Evers, M.; Zuidema, J.; Steendam, R.; Post, E.; Hillebrands, J. L.; Frijlink, H. W.; Poelstra, K.; Beljaars, L.; Olinga, P.; Hinrichs, W. L. J. Polymeric microspheres for the sustained release of a protein-based drug carrier targeting the PDGFbeta-receptor in the fibrotic kidney. *International Journal of Pharmaceutics* 2017, *534* (1–2), 229–236.
31. Faraji, A. H.; Wipf, P. Nanoparticles in cellular drug delivery. *Bioorganic & Medicinal Chemistry* 2009, *17* (8), 2950–2962.
32. Lamprecht, A.; Ubrich, N.; Yamamoto, H.; Schäfer, U.; Takeuchi, H.; Maincent, P.; Kawashima, Y.; Lehr, C.-M. Biodegradable nanoparticles for targeted drug delivery in treatment of inflammatory bowel disease. *Journal of Pharmacology and Experimental Therapeutics* 2001, *299* (2), 775–781.
33. Watnasirichaikul, S.; Rades, T.; Tucker, I.; Davies, N. Effects of formulation variables on characteristics of poly (ethylcyanoacrylate) nanocapsules prepared from w/o microemulsions. *International Journal of Pharmaceutics* 2002, *235* (1–2), 237–246.
34. Gaspar, M. M.; Blanco, D.; Cruz, M. E. M.; Alonso, M. J. Formulation of L-asparaginase-loaded poly (lactide-co-glycolide) nanoparticles: influence of polymer properties on enzyme loading, activity and in vitro release. *Journal of Controlled Release* 1998, *52* (1–2), 53–62.
35. Fishbein, I.; Chorny, M.; Rabinovich, L.; Banai, S.; Gati, I.; Golomb, G. Nanoparticulate delivery system of a tyrphostin for the treatment of restenosis. *Journal of Controlled Release* 2000, *65* (1–2), 221–229.
36. Kawashima, Y.; Yamamoto, H.; Takeuchi, H.; Fujioka, S.; Hino, T. Pulmonary delivery of insulin with nebulized DL-lactide/glycolide copolymer (PLGA) nanospheres to prolong hypoglycemic effect. *Journal of Controlled Release* 1999, *62* (1–2), 279–287.
37. Carino, G. P.; Jacob, J. S.; Mathiowitz, E. Nanosphere based oral insulin delivery. *Journal of Controlled Release* 2000, *65* (1–2), 261–269.
38. Ghalanbor, Z.; Körber, M.; Bodmeier, R. Protein release from poly (lactide-co-glycolide) implants prepared by hot-melt extrusion: thioester formation as a reason for incomplete release. *International Journal of Pharmaceutics* 2012, *438* (1–2), 302–306.
39. Gaudana, R.; Gokulgandhi, M.; Khurana, V.; Kwatra, D.; Mitra, A. K. Design and evaluation of a novel nanoparticulate-based formulation encapsulating a HIP complex of lysozyme. *Pharmaceutical Development and Technology* 2013, *18* (3), 752–759.

40. Gaudana, R.; Parenky, A.; Vaishya, R.; Samanta, S. K.; Mitra, A. K. Development and characterization of nanoparticulate formulation of a water soluble prodrug of dexamethasone by HIP complexation. *Journal of Microencapsulation* 2011, *28* (1), 10–20.
41. Deodhar, G. V.; Adams, M. L.; Trewyn, B. G. Controlled release and intracellular protein delivery from mesoporous silica nanoparticles. *Biotechnology Journal* 2017, *12*, 1600408.
42. Pakulska, M. M.; Elliott Donaghue, I.; Obermeyer, J. M.; Tuladhar, A.; McLaughlin, C. K.; Shendruk, T. N.; Shoichet, M. S. Encapsulation-free controlled release: electrostatic adsorption eliminates the need for protein encapsulation in PLGA nanoparticles. *Science Advances* 2016, *2* (5).
43. Agarwal, P.; Rupenthal, I. D. Injectable implants for the sustained release of protein and peptide drugs. *Drug Discovery Today* 2013, *18* (7–8), 337–349.
44. Kempe, S.; Mäder, K. In situ forming implants – an attractive formulation principle for parenteral depot formulations. *Journal of Controlled Release* 2012, *161* (2), 668–679.
45. Schoenhammer, K.; Boisclair, J.; Schuetz, H.; Petersen, H.; Goepferich, A. Biocompatibility of an injectable in situ forming depot for peptide delivery. *Journal of Pharmaceutical Sciences* 2010, *99* (10), 4390–4399.
46. Schoenhammer, K.; Petersen, H.; Guethlein, F.; Goepferich, A. Poly (ethyleneglycol) 500 dimethylether as novel solvent for injectable in situ forming depots. *Pharmaceutical Research* 2009, *26* (12), 2568.
47. Schoenhammer, K.; Petersen, H.; Guethlein, F.; Göpferich, A. Injectable in situ forming depot systems: PEG-DAE as novel solvent for improved PLGA storage stability. *International Journal of Pharmaceutics* 2009, *371* (1–2), 33–39.
48. Eliaz, R. E.; Kost, J. Characterization of a polymeric PLGA-injectable implant delivery system for the controlled release of proteins. *Journal of Biomedical Materials Research* 2000, *50* (3), 388–396.
49. DesNoyer, J.; McHugh, A. Role of crystallization in the phase inversion dynamics and protein release kinetics of injectable drug delivery systems. *Journal of Controlled Release* 2001, *70* (3), 285–294.
50. Brodbeck, K.; DesNoyer, J.; McHugh, A. Phase inversion dynamics of PLGA solutions related to drug delivery: part II. The role of solution thermodynamics and bath-side mass transfer. *Journal of Controlled Release* 1999, *62* (3), 333–344.
51. Shah, N.; Railkar, A.; Chen, F.; Tarantino, R.; Kumar, S.; Murjani, M.; Palmer, D.; Infeld, M.; Malick, A. A biodegradable injectable implant for delivering micro and macromolecules using poly (lactic-co-glycolic) acid (PLGA) copolymers. *Journal of Controlled Release* 1993, *27* (2), 139–147.
52. Brodbeck, K. J.; Pushpala, S.; McHugh, A. J. Sustained release of human growth hormone from PLGA solution depots. *Pharmaceutical Research* 1999, *16* (12), 1825–1829.
53. Ravivarapu, H. B.; Moyer, K. L.; Dunn, R. L. Parameters affecting the efficacy of a sustained release polymeric implant of leuprolide. *International Journal of Pharmaceutics* 2000, *194* (2), 181–191.
54. Bouissou, C.; Rouse, J.; Price, R.; Van der Walle, C. The influence of surfactant on PLGA microsphere glass transition and water sorption: remodeling the surface morphology to attenuate the burst release. *Pharmaceutical Research* 2006, *23* (6), 1295–1305.
55. Okumu, F. W.; Dao, L. N.; Fielder, P. J.; Dybdal, N.; Brooks, D.; Sane, S.; Cleland, J. L. Sustained delivery of human growth hormone from a novel gel system: SABERTM. *Biomaterials* 2002, *23* (22), 4353–4358.
56. Pechenov, S.; Shenoy, B.; Yang, M. X.; Basu, S. K.; Margolin, A. L. Injectable controlled release formulations incorporating protein crystals. *Journal of Controlled Release* 2004, *96* (1), 149–158.
57. Luan, X.; Bodmeier, R. Influence of the poly (lactide-co-glycolide) type on the leuprolide release from in situ forming microparticle systems. *Journal of Controlled Release* 2006, *110* (2), 266–272.
58. Jain, R. A.; Rhodes, C. T.; Railkar, A. M.; Malick, A. W.; Shah, N. H. Controlled release of drugs from injectable in situ formed biodegradable PLGA microspheres: effect of various formulation variables. *European Journal of Pharmaceutics and Biopharmaceutics* 2000, *50* (2), 257–262.
59. Ruiz-Hornillos, J.; Henríquez-Santana, A.; Moreno-Fernández, A.; González, I. G.; Sánchez, S. R. Systemic allergic dermatitis caused by the solvent of Eligard®. *Contact Dermatitis* 2009, *61* (6), 355–356.
60. Akash, M. S. H.; Rehman, K.; Chen, S. Pluronic F127-based thermosensitive gels for delivery of therapeutic proteins and peptides. *Polymer Reviews* 2014, *54* (4), 573–597.
61. Chen, S.; Pieper, R.; Webster, D. C.; Singh, J. Triblock copolymers: synthesis, characterization, and delivery of a model protein. *International Journal of Pharmaceutics* 2005, *288* (2), 207–218.

62. Jeong, B.; Lee, K. M.; Gutowska, A.; An, Y. H. Thermogelling biodegradable copolymer aqueous solutions for injectable protein delivery and tissue engineering. *Biomacromolecules* 2002, *3* (4), 865–868.
63. Kim, Y. J.; Choi, S.; Koh, J. J.; Lee, M.; Ko, K. S.; Kim, S. W. Controlled release of insulin from injectable biodegradable triblock copolymer. *Pharmaceutical Research* 2001, *18* (4), 548–550.
64. Zentner, G. M.; Rathi, R.; Shih, C.; McRea, J. C.; Seo, M.-H.; Oh, H.; Rhee, B.; Mestecky, J.; Moldoveanu, Z.; Morgan, M. Biodegradable block copolymers for delivery of proteins and water-insoluble drugs. *Journal of Controlled Release* 2001, *72* (1–3), 203–215.
65. Chen, S.; Singh, J. Controlled release of growth hormone from thermosensitive triblock copolymer systems: in vitro and in vivo evaluation. *International Journal of Pharmaceutics* 2008, *352* (1–2), 58–65.
66. Simpson, R. J. Stabilization of proteins for storage. *Cold Spring Harbor Protocols* 2010, doi:10.1101/pdb.top79
67. Szenczi, A.; Kardos, J.; Medgyesi, G. A.; Zavodszky, P. The effect of solvent environment on the conformation and stability of human polyclonal IgG in solution. *Biologicals* 2006, *34* (1), 5–14.
68. Chi, E. Y.; Krishnan, S.; Randolph, T. W.; Carpenter, J. F. Physical stability of proteins in aqueous solution: mechanism and driving forces in nonnative protein aggregation. *Pharmaceutical Research* 2003, *20* (9), 1325–1336.
69. Rajagopal, K.; Wood, J.; Tran, B.; Patapoff, T. W.; Nivaggioli, T. Trehalose limits BSA aggregation in spray-dried formulations at high temperatures: implications in preparing polymer implants for long-term protein delivery. *Journal of Pharmaceutical Sciences* 2013, *102* (8), 2655–2666.
70. van de Weert, M.; Hennink, W. E.; Jiskoot, W. Protein instability in poly (lactic-co-glycolic acid) microparticles. *Pharmaceutical Research* 2000, *17* (10), 1159–1167.
71. Schwendeman, S. P. Recent advances in the stabilization of proteins encapsulated in injectable PLGA delivery systems. *Critical Reviews™ in Therapeutic Drug Carrier Systems* 2002, *19* (1), 73–98.
72. Stanković, M.; Tomar, J.; Hiemstra, C.; Steendam, R.; Frijlink, H. W.; Hinrichs, W. L. Tailored protein release from biodegradable poly (ε-caprolactone-PEG)-b-poly (ε-caprolactone) multiblock-copolymer implants. *European Journal of Pharmaceutics and Biopharmaceutics* 2014, *87* (2), 329–337.
73. Stanković, M.; de Waard, H.; Steendam, R.; Hiemstra, C.; Zuidema, J.; Frijlink, H. W.; Hinrichs, W. L. Low temperature extruded implants based on novel hydrophilic multiblock copolymer for long-term protein delivery. *European Journal of Pharmaceutical Sciences* 2013, *49* (4), 578–587.
74. Ma, G.; Miao, B.; Song, C. Thermosensitive PCL-PEG-PCL hydrogels: synthesis, characterization, and delivery of proteins. *Journal of Applied Polymer Science* 2010, *116* (4), 1985–1993.
75. Gong, C. Y.; Dong, P. W.; Shi, S.; Fu, S. Z.; Yang, J. L.; Guo, G.; Zhao, X.; Wei, Y. Q.; Qian, Z. Y. Thermosensitive PEG–PCL–PEG hydrogel controlled drug delivery system: sol–gel–sol transition and in vitro drug release study. *Journal of Pharmaceutical Sciences* 2009, *98* (10), 3707–3717.
76. Jiang, Z.; Hao, J.; You, Y.; Gu, Q.; Cao, W.; Deng, X. Biodegradable thermogelling hydrogel of P (CL–GL)-PEG-P (CL–GL) triblock copolymer: degradation and drug release behavior. *Journal of Pharmaceutical Sciences* 2009, *98* (8), 2603–2610.
77. Joshi, R.; Robinson, D. H.; Himmelstein, K. J. In vitro properties of an in situ forming gel for the parenteral delivery of macromolecular drugs. *Pharmaceutical Development and Technology* 1999, *4* (4), 515–522.
78. Macdonald, M. L.; Samuel, R. E.; Shah, N. J.; Padera, R. F.; Beben, Y. M.; Hammond, P. T. Tissue integration of growth factor-eluting layer-by-layer polyelectrolyte multilayer coated implants. *Biomaterials* 2011, *32* (5), 1446–1453.
79. La, W. G.; Park, S.; Yoon, H. H.; Jeong, G. J.; Lee, T. J.; Bhang, S. H.; Han, J. Y.; Char, K.; Kim, B. S. Delivery of a therapeutic protein for bone regeneration from a substrate coated with graphene oxide. *Small* 2013, *9* (23), 4051–4060.
80. Peterson, A. M.; Möhwald, H.; Shchukin, D. G. pH-controlled release of proteins from polyelectrolyte-modified anodized titanium surfaces for implant applications. *Biomacromolecules* 2012, *13* (10), 3120–3126.
81. Peterson, A. M.; Pilz-Allen, C.; Kolesnikova, T.; Möhwald, H.; Shchukin, D. Growth factor release from polyelectrolyte-coated titanium for implant applications. *ACS Applied Materials & Interfaces* 2013, *6* (3), 1866–1871.
82. Kumar, M. R.; Muzzarelli, R. A.; Muzzarelli, C.; Sashiwa, H.; Domb, A. Chitosan chemistry and pharmaceutical perspectives. *Chemical Reviews* 2004, *104* (12), 6017–6084.

83. Vaghani, S. S.; M. M. Patel. Hydrogels based on interpenetrating network of chitosan and polyvinyl pyrrolidone for pH-sensitive delivery of repaglinide. *Current Drug Discovery Technologies* 2011, *8* (2), 126–135.
84. Mi, F. L.; Sung, H. W.; Shyu, S. S. Synthesis and characterization of a novel chitosan-based network prepared using naturally occurring crosslinker. *Journal of Polymer Science Part A: Polymer Chemistry* 2000, *38* (15), 2804–2814.
85. Chan, A. W.; Neufeld, R. J. Tuneable semi-synthetic network alginate for absorptive encapsulation and controlled release of protein therapeutics. *Biomaterials* 2010, *31* (34), 9040–9047.
86. Koetting, M. C.; Peppas, N. A. pH-Responsive poly (itaconic acid-co-N-vinylpyrrolidone) hydrogels with reduced ionic strength loading solutions offer improved oral delivery potential for high isoelectric point-exhibiting therapeutic proteins. *International Journal of Pharmaceutics* 2014, *471* (1–2), 83–91.
87. Dergunov, S. A.; Mun, G. A. γ-irradiated chitosan-polyvinyl pyrrolidone hydrogels as pH-sensitive protein delivery system. *Radiation Physics and Chemistry* 2009, *78* (1), 65–68.
88. Zhang, Y.-H.; Shang, Q.; Zheng, T.; Liang, Y.-Y.; Chen, T. Preparation of pH-sensitive hydrogels for oral delivery of protein. *Biomedical Engineering and Biotechnology (iCBEB)*, 2012 International Conference on, IEEE: 2012; pp 437–440.
89. Determan, M. D.; Cox, J. P.; Mallapragada, S. K. Drug release from pH-responsive thermogelling pentablock copolymers. *Journal of Biomedical Materials Research Part A* 2007, *81* (2), 326–333.
90. Milašinović, N.; Krušić, M. K.; Knežević-Jugović, Z.; Filipović, J. Hydrogels of N-isopropylacrylamide copolymers with controlled release of a model protein. *International Journal of Pharmaceutics* 2010, *383* (1–2), 53–61.
91. Milašinović, N.; Knežević-Jugović, Z.; Milosavljević, N.; Filipović, J.; Krušić, M. K. Controlled release of lipase from *Candida rugosa* loaded into hydrogels of N-isopropylacrylamide and itaconic acid. *International Journal of Pharmaceutics* 2012, *436* (1–2), 332–340.
92. Manokruang, K.; Lee, D. S. Albumin-conjugated pH/thermo responsive poly (amino urethane) multiblock copolymer as an injectable hydrogel for protein delivery. *Macromolecular Bioscience* 2013, *13* (9), 1195–1203.
93. Shim, W. S.; Kim, J.-H.; Park, H.; Kim, K.; Kwon, I. C.; Lee, D. S. Biodegradability and biocompatibility of a pH-and thermo-sensitive hydrogel formed from a sulfonamide-modified poly (ε-caprolactone-co-lactide)–poly (ethylene glycol)–poly (ε-caprolactone-co-lactide) block copolymer. *Biomaterials* 2006, *27* (30), 5178–5185.
94. Kim, H. K.; Shim, W. S.; Kim, S. E.; Lee, K.-H.; Kang, E.; Kim, J.-H.; Kim, K.; Kwon, I. C.; Lee, D. S. Injectable in situ–forming pH/thermo-sensitive hydrogel for bone tissue engineering. *Tissue Engineering Part A* 2008, *15* (4), 923–933.
95. Hsieh, D. S.; Rhine, W. D.; Langer, R. Zero-order controlled-release polymer matrices for micro-and macromolecules. *Journal of Pharmaceutical Sciences* 1983, *72* (1), 17–22.
96. Kajihara, M.; Sugie, T.; Mizuno, M.; Tamura, N.; Sano, A.; Fujioka, K.; Kashiwazaki, Y.; Yamaoka, T.; Sugawara, S.; Urabe, Y. Development of new drug delivery system for protein drugs using silicone (I). *Journal of Controlled Release* 2000, *66* (1), 49–61.
97. Maeda, H.; Ohashi, E.; Sano, A.; Kawasaki, H.; Kurosaki, Y. Investigation of the release behavior of a covered-rod-type formulation using silicone. *Journal of Controlled Release* 2003, *90* (1), 59–70.
98. Lofthouse, S.; Kajihara, M.; Nagahara, S.; Nash, A.; Barcham, G.; Sedgmen, B.; Brandon, M.; Sano, A. Injectable silicone implants as vaccine delivery vehicles. *Vaccine* 2002, *20* (13–14), 1725–1732.
99. Kajihara, M.; Sugie, T.; Hojo, T.; Maeda, H.; Sano, A.; Fujioka, K.; Sugawara, S.; Urabe, Y. Development of a new drug delivery system for protein drugs using silicone (II). *Journal of Controlled Release* 2001, *73* (2–3), 279–291.
100. Cukierski, M. J.; Johnson, P. A.; Beck, J. C. Chronic (60-week) toxicity study of DUROS leuprolide implants in dogs. *International Journal of Toxicology* 2001, *20* (6), 369–381.
101. Fowler, J. E.; Gottesman, J. E.; Reid, C. F.; Andriole, G. L.; Soloway, M. S. Safety and efficacy of an implantable leuprolide delivery system in patients with advanced prostate cancer. *The Journal of Urology* 2000, *164* (3), 730–734.
102. Wright, J. C. Critical variables associated with nonbiodegradable osmotically controlled implants. *The AAPS Journal* 2010, *12* (3), 437–442.

103. Fowler, J. E.; Flanagan, M.; Gleason, D. M.; Klimberg, I. W.; Gottesman, J. E.; Sharifi, R. Evaluation of an implant that delivers leuprolide for 1 year for the palliative treatment of prostate cancer. *Urology* 2000, *55* (5), 639–642.
104. Wright, J. C.; Leonard, S. T.; Stevenson, C. L.; Beck, J. C.; Chen, G.; Jao, R. M.; Johnson, P. A.; Leonard, J.; Skowronski, R. J. An in vivo/in vitro comparison with a leuprolide osmotic implant for the treatment of prostate cancer. *Journal of Controlled Release* 2001, *75* (1–2), 1–10.
105. Rohloff, C. M.; Alessi, T. R.; Yang, B.; Dahms, J.; Carr, J. P.; Lautenbach, S. D. DUROS® Technology delivers peptides and proteins at consistent rate continuously for 3 to 12 months. *Journal of Diabetes Science and Technology* 2008, *2* (3), 461–467.
106. Kang, C. E.; Baumann, M. D.; Tator, C. H.; Shoichet, M. S. Localized and sustained delivery of fibroblast growth factor-2 from a nanoparticle-hydrogel composite for treatment of spinal cord injury. *Cells Tissues Organs* 2013, *197* (1), 55–63.
107. Patel, S. P.; Vaishya, R.; Mishra, G. P.; Tamboli, V.; Pal, D.; Mitra, A. K. Tailor-made pentablock copolymer based formulation for sustained ocular delivery of protein therapeutics. *Journal of Drug Delivery* 2014, *2014*, 15.
108. Mitra, A. K.; Mishra, G. P. Pentablock polymers. Google Patents: 2013.
109. Geng, H.; Song, H.; Qi, J.; Cui, D. Sustained release of VEGF from PLGA nanoparticles embedded thermo-sensitive hydrogel in full-thickness porcine bladder acellular matrix. *Nanoscale Research Letters* 2011, *6* (1), 312.
110. Jain, P. K.; Karunakaran, D.; Friedman, S. H. Construction of a photoactivated insulin depot. *Angewandte Chemie* 2013, *125* (5), 1444–1449.
111. Eon-Duval, A.; Broly, H.; Gleixner, R. Quality attributes of recombinant therapeutic proteins: an assessment of impact on safety and efficacy as part of a quality by design development approach. *Biotechnology Progress* 2012, *28* (3), 608–622.
112. Rathore, A.; Bhambure, R.; Ghare, V. Process analytical technology (PAT) for biopharmaceutical products. *Analytical and Bioanalytical Chemistry* 2010, *398* (1), 137–154.

11

Polyhydroxybutyrate-Based Nanoparticles for Controlled Drug Delivery

K.V. Radha and S. Saranya

CONTENTS

11.1 Introduction

Biodegradable polymers are a rapidly emerging field that holds great promise for revolutionizing drug delivery systems, tissue engineering applications, and biomedical sensors. The first reported biomedical application of a polymer was in nylon sutures (Blaine 1946). Recently, the pharmaceutical and biotechnology industries have developed a wide array of drug candidates based on proteins and nucleic acids, which cannot be delivered by conventional methods because of low solubility, fast metabolism, and site specificity. The successful clinical application of these new drugs is providing the driving force for the development of new biomaterials for controlled drug delivery and gene therapy applications. Advances in new synthetic methods and polymer designs have resulted in a large expansion of biomedical polymer applications.

Polymers are desirable carriers for clinical applications due to the ease of tailoring their chemical, physical, and biological properties. Synthetic and natural polymers, classifications of polymers based on their physical and chemical properties, have equal importance in drug delivery systems.

Natural polymers, proteins and polysaccharides, have advantages over their synthetic counterparts and are used for various biomedical applications due to their biocompatibility and non-immunogenic properties. Chan and Leong (2008) studied implants for bovine discs made from collagen. In *in vivo* studies, it was observed that the disc height was restored by the collagen implant. Alginate, a polysaccharide derived from seaweed, has been used for biomedical applications such as tissue engineering (Sotome et al. 2004). However, the applications are limited to soft tissue engineering applications, as they do not possess the mechanical strength for preparing bone implants. Biopolymers synthesized from microorganisms have gained more interest in the biomedical and pharmaceutical industry, as they are highly biodegradable and biocompatible. Polyhydroxy alkanoates (PHA) have gained major importance in biopolymers due to their mechanical properties, biocompatibility, and biodegradability (D-ong et al. 2006).

Polyhydroxy butyrate (PHB) is a PHA, a polymer belonging to the polyester class that area of interest as bioderived and biodegradable plastics. There has been a focus on the use of PHB and related copolymers as carriers or implants for drug delivery. When compared with chemically produced polymers such as polyglycolate, polylactate, and poly (lactide-co-glycolide), which are well known as biologically degradable drug carriers with good retarding characteristics, PHB has the inherent advantage that it is an easily processed, biologically degradable, and compatible polymer (Pouton & Akhtar 1996). Additionally, the controllable retarding properties of PHB can be manipulated by variations in processing and the molecular weight of the polymer (Kassab et al. 1997). Thus, it could be anticipated to be an ideal basic material for matrix composition.

PHB is produced by different bacterial species as inclusion bodies to serve as carbon sources in nutritional stress conditions. PHB is synthesized from acetyl-CoA, produced by bacteria by the sequential action of three enzymes:

1. 3-Ketothiolase (PHB A gene) catalyzes the formation of a carbon–carbon bond by the condensation of two acetyl-CoA molecules (Mercan & Beyatli 2001).
2. NADPH-dependent acetoacetyl-CoA reductase (PHB B gene) catalyzes the stereoselective reduction of acetoacetyl-CoA formed in the first reaction to R-3 hydroxybutyryl-CoA.

3. The third reaction of this pathway is catalyzed by PHB synthase (PHB C gene), which catalyzes the polymerization of R-3 hydroxybutyryl-CoA to form PHB (Steinbuchel & Schlegel 1991).

PHB is a partially crystalline polymer, which has material properties similar to those of polypropylene (PP) and polyethylene (PE) (Holmes 1988).

PHB has been considered as a most promising biopolymer, since its biocompatibility and biodegradable property make it a substitute for synthetic polymers (Albertsson et al. 1985). Biopolymers are widely used as biomaterials due to their favorable properties, such as good biocompatibility and easy design and preparation, and can deliver therapeutic agents directly into the intended site of action with superior efficacy. They play a significant role in the field of smart drug delivery applications (Shariati & Peters 2003). The ideal requirements for designing a nanoparticulate delivery system are particle size, surface character, enhanced permeation, flexibility, solubility, and release of therapeutically active agents to attain the target and specific activity at a predetermined rate and time (Bazzo et al. 2008).

Advances in biopolymer science in the nanotechnology field have seen success in smart drug delivery systems. Recently, these advances have been found in various medical applications for nanoscale structures in smart drug delivery. Smart delivery systems should possess important features such as a prescheduled rate, self-control, targeting, predetermined time, and monitoring delivery. They are drug carriers of natural, semisynthetic, and synthetic polymeric nature at the nano to micro scale (Gultekin & Değim 2013). Most polymeric nanoparticles with surfactants offer stability of various forms of active drugs and have useful smart release properties.

Numerous biological applications, such as tissue engineering, wound healing, etc., have been reported for nano- and micro-sized particles (Wang et al. 2004). Moreover, polymeric particles have proved their effectiveness in stabilizing and protecting drug molecules, proteins, peptides, or DNA molecules from degradation by various environmental hazards (Liu et al. 2016). Various applications such as cancer targeting have been shown to be promising in the near future (Saito et al. 1996). This chapter discusses the encapsulation of the model drug dipyridamole with PHB.

11.2 Biosynthesis of Polymers

Bacteria can synthesize a wide range of biopolymers that serve diverse biological functions and have material properties suitable for numerous industrial and medical applications. The most widely produced microbial bioplastics are PHAs and their derivatives (Madison & Huisman 1999). PHAs are stored as intracellular granules by microorganisms, especially by bacteria. In 1926, Lemoigne discovered the presence of PHA in *Bacillus megaterium* as an intracellular reserve material.

PHB is the first discovered and most intensively studied PHA. To date, over 150 types of hydroxy carboxylic acid have been identified as components of PHAs (Steinbuchel & Valentin 1991). The PHA/PHBs that are present intracellularly in an amorphous state have a particular surface layer consisting of proteins and phospholipids. This surface layer is sensitive to physical chemical stress, and these are called native or intracellular PHA/PHB granules. PHA/PHBs are the only 100% biodegradable polymers. Synthesis by numerous microorganisms under unbalanced growth conditions, such as limitation of an essential nutrient such as nitrogen, oxygen, magnesium, or phosphorus, and the presence of excess carbon source creates polyesters of various butyric acids. They possess properties like those of various synthetic thermoplastics, such as PP, and can be recycled. They are also completely degraded to water and carbon dioxide under aerobic conditions and to methane under anaerobic conditions by microorganisms in soil, sea, lake water, and sewage.

11.3 PHB-Producing Microorganisms

PHBs are produced by different bacterial cultures. *Cupriavidus necator* (formerly known as *Ralstonia eutropha* or *Alcaligenes eutrophus*) is the one of the most extensively studied microbes. At present, bacterial fermentation of *C. necator* seems to be the most cost-effective process, and even if production switches to other bacteria or agricultural crops, these processes are likely to use *C. necator* genes. Other important strains that have been recently studied include *Bacillus* sp., *Alcaligenes* sp., *Pseudomonas* sp., *Aeromonas hydrophila*, *Rhodopseudomonas palustris*, and recombinant *Escherichia coli*.

11.3.1 PHB Production by Cupriavidus necator

Studies on *C. necator* are most extensive due to its ability to accumulate large amounts of PHB from simple carbon sources: for example, glucose, fructose, and acetic acid. Imperial Chemical Industries (ICI) in the United Kingdom has been producing PHB on a large scale from glucose, and poly(3-hydroxybutyra te-co-3-hydroxyvalerate) (P(3HB-co-3HV)) from a mixture of glucose and propionic acid by fed-batch culture of *A. eutrophus*. Initially, the cells were grown on a glucose-containing medium containing only calculated amounts of phosphate to support a desired amount of cell growth. Cells encounter phosphate limitation after about 60 h and accumulate PHB during the next 40–60 h.

11.3.2 PHB Production by Alcaligens latus

A. latus produces PHB from carbon sources such as glucose and sucrose at a rate comparable to that of *A. eutrophus* H16. *A. latus* DSM 1123 can grow and produce PHB at 35°C. This strain requires only 5 h to reach an intracellular PHB concentration of 80% of dry biomass, and the synthesis of PHB is growth associated. The PHB concentration in this strain has been increased from 50% to 87% in fed-batch cultivation by the induction of limited nitrogen in the medium.

11.3.3 PHB Production by Recombinant E. coli

E. coli has proved to be a versatile performer not only in expediting the molecular analysis and biosynthesis but also in (a) synthesizing the biopolymer to extremely high intracellular levels and (b) being amenable to specific genetic strategies such as genetically mediated lysis and (c) the use of mutants to metabolically engineer a strain that produces P(3HB-co-3HV) copolymer. The synthesis of PHB by recombinant *E. coli* does not require limitation of a specific nutrient but is dependent on acetyl-CoA availability. The advantages of employing recombinant *E. coli* for production include fast growth, high cell density, the ability to use several inexpensive carbon sources, and easy purification.

11.3.4 Bacillus—a Potential PHB Producer

Bacillus spp. have been reported to produce PHB, which varies from 11% to 69% of yield. Some of the *Bacillus* spp. producing PHB are *B. amyloliquefaciens* DSM7, *B. laterosporus*, *B. licheniformis*, *B. macerans*, *B. cereus*, *B. circulans*, *B. firmus* G2, *B. subtilis K*8, *B. sphaericus X*3, *B. megaterium* Y6, *B. coagulans*, *B. brevis*, *B. sphaericus ATCC*14577, *B. thuringiensis*, *B. mycoides* RLJ B-017, and *Bacillus* spp. JMa5. Among these, *B. subtilis* offers the advantage of lack of lipopolysaccharide (LPS) and excretes proteins at a high rate into the medium (Morimoto et al. 2008). It thus stands a better chance as a PHB producer for biomedical applications (Valappil et al. 2007). *B. subtilis* is a generally regarded as safe (GRAS) organism by the Food and Drug Administration (FDA) (Harwood & Wipat 1996; Apetroaie-Constantin et al 2009) and thus offers additional benefits.

11.4 Factors Affecting Large-Scale PHB Production

The choice of the substrate used to produce PHB is determined by the physiological-biochemical properties of PHB-producing microorganisms, the economic efficiency of the preferred strategy, and the field of application. The quality and cost of PHB for different applications vary depending on the substrates, degree of reduction, and energy content.

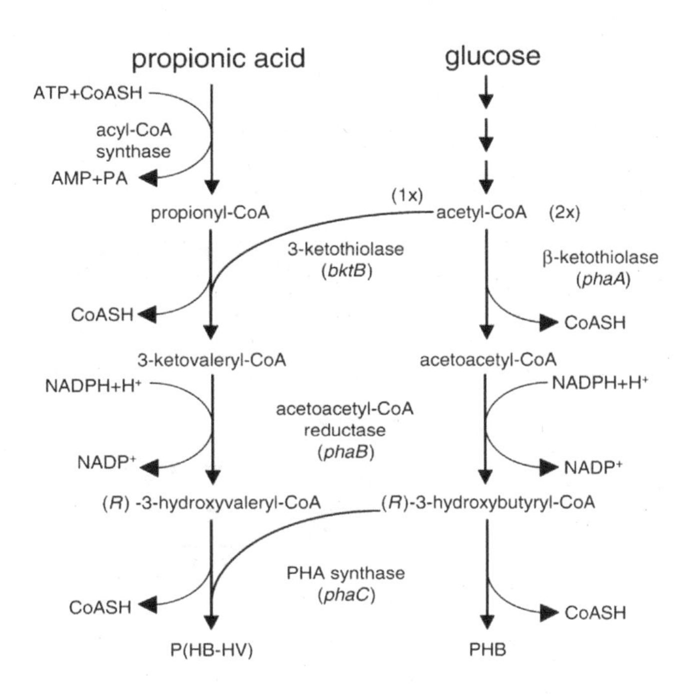

FIGURE 11.1 PHB and P(HB–HV) biosynthetic pathways.

11.5 Biosynthetic Pathway of PHB

The biosynthetic pathway of P(3HB) involves three enzymatic reactions catalyzed by three different enzymes (Figure 11.1). The condensation of two acetyl-CoA molecules forms the first reaction, which is converted to acetoacetyl-CoA by β-ketothiolase (encoded by phbA). The second reaction, facilitating the reduction of acetoacetyl-CoA to (R)-3-hydroxybutyryl-CoA, is catalyzed by an NADPH-dependent acetoacetyl-CoA dehydrogenase (encoded by phbB). Finally, the (R)-3-hydroxybutyryl-CoA monomers are polymerized into PHB by P(3HB) polymerase, encoded by phbC (Huisman et al. 1989).

11.6 Dipyridamole–a Model Drug

Dipyridamole is used as a model drug for studying controlled release formulations with a PHB nanoparticle matrix. Dipyridamole (molecular formula: $C_{24}H_{40}N_8O_4$ with molar mass 504.625 KDa) is an odorless, yellow crystalline powder having a melting point in the range of 164–168°C and is soluble in dilute acids, methanol, ethanol, and chloroform. It is a phosphodiesterase inhibitor that blocks the uptake and metabolism of adenosine by erythrocytes and endothelial cells. Dipyridamole inhibits the formation of blood clots (Qiu et al. 2014). The drug can be administered in oral or intravenous form for the treatment of cardiovascular, hypertension, and stroke patients. At high doses over a short time, it inhibits the phosphodiesterase enzymes that normally break down cAMP by increasing cellular cAMP levels and blocks the platelet aggregation response to ADP and cGMP.

The presence of a higher level of related substances or impurities may have harmful effects on the body. Hence, controlled release becomes a necessity, and it has been achieved by encapsulating the drug into a PHB matrix. Nanoparticles based on PHB are prepared and the processing conditions varied to evaluate their impact on morphology, size, and drug encapsulation capabilities. From the viewpoint of a real-world application in the pharmaceutical field, the biocompatibility of PHB and related nanoparticles are tested by *in vitro* cytotoxicity assays, and the development of a final pharmaceutical form is discussed.

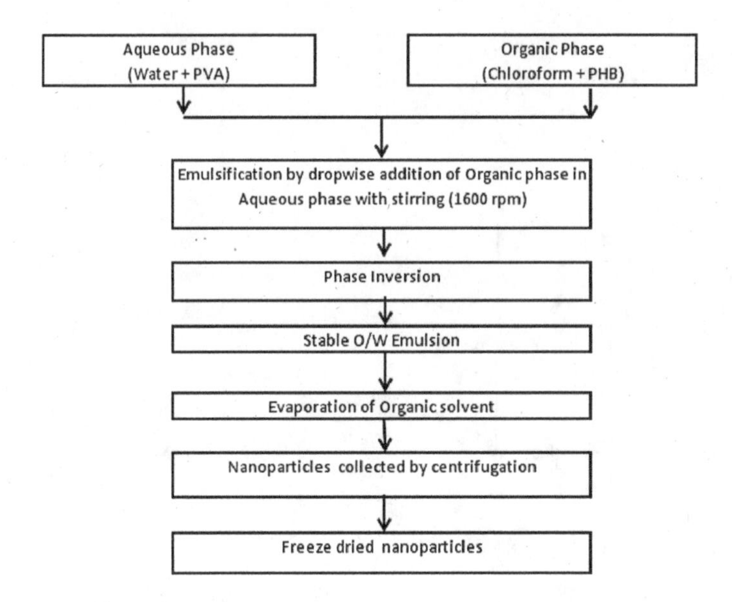

FIGURE 11.2 Schematic representation of nanoparticle preparation.

11.6.1 Preparation of Dipyridamole-Loaded PHB

Solubility is one of the key factors for this polymer; hence, the drug must be loaded. The oil water emulsion solvent evaporation method is normally adopted for the preparation of dipyridamole-loaded PHB, as both the polymer and the drug were found to be soluble in the organic phase. Figure 11.2 shows a schematic representation of the steps involved in nanoparticle formation. The emulsion formed is converted to a nanoparticle suspension on evaporation. The obtained nanoparticles will be freeze-dried for further use. Emami et al. (2014) reported the preparation of poly(lactic-co-glycolic acid) (PLGA) nanoparticles of about 200 nm in size by using 1.0% (w/v) dichloromethane as solvent and polyvinyl alcohol (PVA) as stabilizing agent. Bohrey et al. (2016) reported on PLGA nanoparticles prepared by the solvent evaporation method using chloroform as solvent and achieved nanoparticles in the size range of 250 nm.

11.7 Drug–Polymer Interaction

11.7.1 Optimization of Process Parameters

The effects of various process parameters, such as concentration of polymer, drug, stabilizer, and stirring rate, are analyzed for high nanoparticle yield, smaller particle size, and maximum entrapment efficiency for obtaining dipyridamole-loaded PHB. The concentration of polymer is one of the decisive formulation variables, which plays a crucial role in modulating the particle size, shape, and entrapment efficiency of the nanoparticles and also the prolonged release of this polymeric system. Although polymeric nanoparticles have difficulty of scaling up, low drug loading capacity and wide size distribution, they have attracted increasing attention, because they provide the possibility of transporting bioactive compounds to specific tissues, cells, and cell compartments When drug-loaded nanoparticles are injected intravenously, they cross epithelial barriers and circulate in the blood vessels before reaching the target site. Escape of nanoparticles from the vascular circulation then occurs.

Some of the factors that disturb the interaction of nanoparticles with the drug are particle size, viscosity of the polymer solution, concentration of the polymer, mechanism introduced for interaction, etc. The interactions are described briefly as follows:

Particle Size: Size is analyzed using a particle size analyzer, as the particle size increases with the amount of PHB.

Viscosity: Due to the increasing viscosity of the dispersed phase (polymer solution), poorer dispersibility of PHB was obtained in the aqueous phases (Quintanar et al. 1996).

Polymer concentration: The polymer concentration in the internal phase of the emulsion is an important factor in determining the size of the nanoparticles. Coarse emulsions are obtained at a higher polymer concentration, which leads to the buildup of bigger particles during the diffusion process. There is a greater probability for the desolvated macromolecules to coalesce in a more viscous solution, thereby forming larger particles.

Technique applied: Kwon et al. (2001) produced PLA nanodispersions with an average size of 200–300 nm by the solvent evaporation method under optimized conditions. There is high viscous resistance against the shear forces during this type of emulsification.

11.7.1.1 Effect of Concentration of Drug

The impact of biopolymer particles on rheology depends mainly on their concentration, composition, interactions, shape, and size. The effect of varying drug concentration is studied to observe the effect of increase in the amount of drug. The yield of nanoparticle increases with maximum entrapment efficiency. The encapsulation efficiency was determined by using the following formula:

$$\text{Encapsulation efficiency } (\%) = \left[1 - (\text{Drug in supernatant liquid / Total drug added})\right] \times 100$$

When smaller amounts of drug are taken, smaller particles are obtained after the evaporation of the organic solvent, whereas if the amount of drug were increased, the particle size would increase because of the high solid content after evaporation. The amount of drug loading is determined using the following formula:

$$\%\text{ Drug loading } = (\text{Mass of drug in NP / Mass of NP recovered}) \times 100$$

Win and Feng (2005), showed that the particle size of PLA increased from 367 to 475 nm as the concentration of drug increased to 2.5–15 mg/ml. This could be due to higher viscosity of the drug solution in aqueous phase, which leads to a lower diffusion rate of the solvent and thus results in a larger particle size with low entrapment efficiency. The entrapment efficiency decreases with increased polymer concentration because of the inadequate amount of polymer in the system. This makes it insufficient to entrap the drug.

11.7.1.2 Effect of Emulsifier Concentration

A high concentration of emulsifier leads to the production of nanoparticles of reduced size. This phenomenon can be expected because of the stabilizing function of emulsifiers. The higher amount of emulsifier would fail to stabilize nanoparticles and would tend to form aggregates, leading to the production of larger-sized nanoparticles. Different emulsifiers with different polymers were used to prevent aggregation of droplets during emulsification and solvent evaporation by reducing the interfacial tension between the droplets and the external phase.

It was observed by Feng and Huang (2001) that with decrease in the nanoparticle size and increase in the PVA concentration, the PVA transfer rate from the oil/water interface onto the newly formed curved interface of the nanoparticle increases. This observation agreed with the results obtained in this study. In the emulsion solvent evaporation method, the emulsification and stabilization of the nanoparticles are crucial factors. Murakami et al. (1999) studied the role of surfactant in the emulsification process and in the protection of droplets, because it can avoid the coalescence of globules determining the size of the nanoparticles.

Yang et al. (2001) observed that at lower PVA concentrations. such as 0.025% w/v, the water droplets inside the nanoparticles coalesced with each other, thereby forming interconnecting water channels that increased the release rate of BSA. A higher concentration of PVA, on the other hand, increases the viscosity of the external phase, thus making it increasingly difficult to break the emulsion droplets into smaller sizes. Sahoo et al. (2002) observed that the prepared nanoparticles were washed three times with water to remove any residual PVA that was present on the surface of the nanoparticles. Residual PVA on the surface of the nanoparticles can have a significant effect on their physical properties, such as surface charge and hydrophobicity. This method was established to be the most common, effective, and robust in the encapsulation of drugs with different properties. Furthermore, the insolubility of the drug and polymers in PVA makes it a good candidate as an external phase in the emulsification process

11.7.1.3 Effect of Stirring Rate

Emulsification can be considered one of the most important steps in the process, because an insufficient dispersion of phases results in larger particles with a wide size distribution. The high energy released in the stirring process leads to a rapid dispersion of the polymeric organic phase as nanodroplets of smaller size. The final size of the nanoparticles depends on the globule size throughout the emulsification process. Table 11.1 shows the parameters that affect the particle size.

A reduction of the emulsion globule size allows the formation of smaller nanoparticles. Jung et al. (2000) observed the formation of larger PLA nanoparticles when the stirrer speed was maintained at 500 rpm in comparison with the smaller nanoparticles produced when the stirrer speed was maintained at 800 rpm. Similar observations were made in this study, where at a higher stirrer speed of 1600 rpm, nanoparticles with an average size of 241 nm were produced. Martin et al. (2000) observed the formation of larger PHB microparticles of 100 μm when the stirrer speed was maintained at 500 rpm when compared with the smaller microparticles of 5 μm produced when the stirrer speed was increased to 800 rpm.

11.7.1.4 Screening of Lyoprotectants for Freeze-Drying

The dipyridamole-loaded PHBs are freeze-dried using various lyoprotectants (glucose, sucrose, fructose, and dextran) and screened for successful freeze-drying. The particle sizes of nanoparticles before and after freeze-drying are analyzed using a zetasizer. In addition, different properties of freeze-dried nanoparticles, such as their physical appearance and reconstitution nature, are studied. The critical analysis of freeze-dried products normally includes the size after reconstitution, entrapment efficiency, and appearance of the cake after freeze-drying. Redispersion of the freeze-dried product is achieved by manual shaking, sonication, and vortexing. Based on observations such as size and ease of redispersibility, glucose and dextran were found to be compatible. In a study carried out by Saez et al. (2000), 20% glucose and sucrose were found to be the best lyoprotectant for the stabilization of cyclosporine A–loaded PLGA nanoparticles.

11.7.1.4.1 Effect of Freeze- Drying on Entrapment Efficiency and Morphology of Drug-Loaded PHB

During freeze-drying, drug may leach out from the nanoparticles and cause a reduction in the entrapment efficiency. The morphology of freeze-dried nanoparticles needs to be intact and agglomerated after freeze-drying. Muller et al. (1986) have reported that the solubility of poloxomers is higher in cold water than in hot water due to hydrogen bond formation between the water molecules and the oxygen bonds of the poloxomers, ultimately leading to the aggregation of nanoparticles. Based on different observations such as size and ease of redispersibility, glucose and dextran were found to be compatible (Figure 11.3).

TABLE 11.1

Parameters That Control Drug with Particle Size

Parameter	Larger Size	Smaller Size
Molecular. weight of polymer	Decreases	Increases
Stirring rate	Decreases	Increases
Polymer concentration	Increases	Decreases

FIGURE 11.3 Morphology of freeze-dried drug-loaded PHB.

11.8 Characterization of Drug with Polymer

11.8.1 Surface Morphology of Nanoparticles

The influence of the concentration of polymer on the particle morphology and shape is studied using scanning electron microscopy (SEM) analysis. Yang et al. (2010) observed a rough surface morphology with increased surface porosity when a low volume of methylene chloride was used as the solvent for fabrication of polycaprolactone (PCL) and PLGA. A decrease in the volume of the solvent is also known to affect the surface morphology of the nanoparticles, thereby increasing the number of surface pores at a lower volume of the solvent; the water droplets entrapped within the nanoparticles are known to evaporate, leaving behind empty spaces both on the surface as well as within the nanospheres, thus contributing to a rough, porous surface morphology. Chee et al. (2008) studied mitragynine in PHB solvent cast films (Figure 11.4).

Lu et al. (1999) have also made similar observations with the emergence of rougher surface morphology in the presence of increased concentration of drug. Bazzo et al. (2008) also confirmed that the presence of drug contributes to rough morphology.

11.8.2 Zeta Potential Measurement

The zeta potential value depends on the charges present on the particle surface. An increase or a decrease in the surface charged groups will result in a higher or lower zeta potential. The COOH groups on the nanoparticles exhibit a negative charge at neutral pH due to deprotonation.

At a lower pH, a charge reversal from negative to positive is observed. This phenomenon is observed for all nanoparticles prepared using different concentrations of PVA. Similarly, a decrease in positive zeta potential values was observed when the concentration of residual PVA on the surface of nanoparticles was increased. Often, non-ionic surfactants such as PVA are known to strongly adhere to the nanoparticle surface by anchoring the hydrophobic tail in the polymer, leaving the polar head protruding on the surface (Figure 11.5).

During preparation of the nanoparticles, the hydrophobic ends penetrate into the organic phase and interact with PHB nanoparticles, due to which an irreversible binding of PVA occurs on the surface of

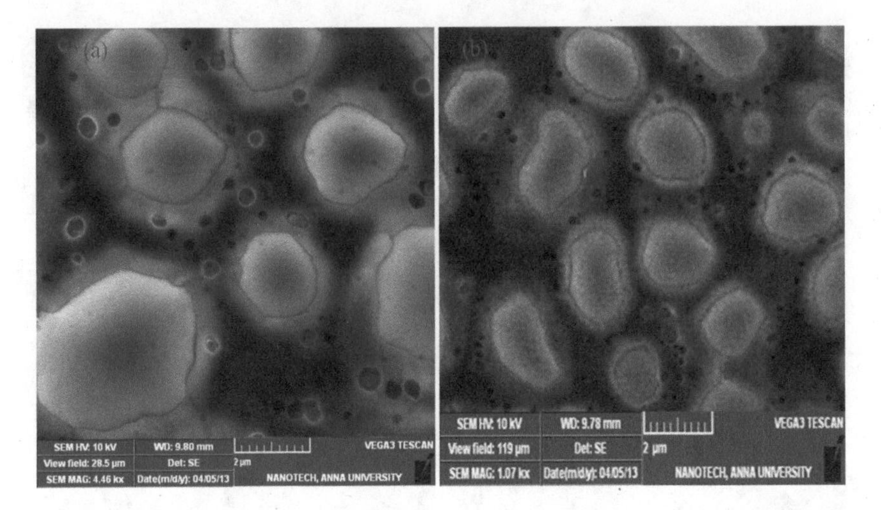

FIGURE 11.4 SEM image of PHB particles in normal and magnified view.

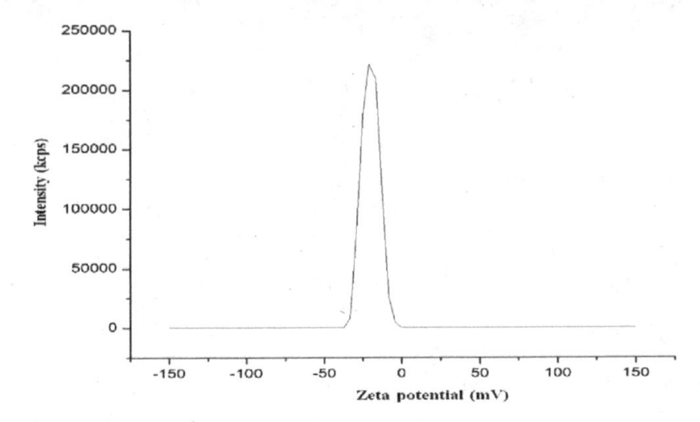

FIGURE 11.5 Zeta potential analysis of PHB.

nanoparticles when the solvent evaporates. Lee (1995) showed that with an increase in the PVA concentration from 0.5% to 5%, the zeta potential values were decreased. In acidic conditions, the zeta potential values showed a high positive charge when compared with the lower surface charge of nanoparticles prepared with a high PVA concentration. Poletto et al. (2007) observed an increase in the specific surface area of poly(3-hydroxybutyrate-co-3-hydroxyvalerate) (PHBV) particles.

11.9 Degradation and Controlled Release of Drug

11.9.1 *In Vitro* Degradation, Release, and Stability Studies of Dipyridamole-Loaded PHB

The bioavailability of drug from the polymer matrix can be altered by different parameters, such as formulation and the manufacturing process. Successful drug availability from the polymer matrix can be studied by the *in vitro* degradation of polymer, which is like *in vivo* body conditions. The release of drug should be studied to evaluate its release pattern in simulated body fluid (SBF).

11.9.1.1 In Vitro Degradation Studies

The *in vitro* degradation studies of the drug (dipyridamole)-loaded PHB (DLPHB) and unloaded PHB (ULPHB) are carried out in SBF. The water uptake and weight loss of the polymer matrix are evaluated.

The water adsorption in the drug-loaded nanoparticles increases due to the solubility of drug, as the drug is water soluble. This creates pores, leading to water uptake and resulting in weight loss when compared with unloaded PHB. Bazzo et al. (2008) studied piroxicam-loaded PHB microparticles, where the desorption of drug crystals occurred on the surface of the microparticles, leading to increased water uptake within the pores. The studies of Chee et al. (2008) showed that PHB films exhibited a higher pore size when the concentration of drug increased. Also, it was observed that before drug loading, the polymer pores were present close to each other; however, after drug loading, the drug became entrapped within the pores, thus increasing their size. As a result, the larger amount of drug present on the surface and entrapped within the pores underwent rapid dissolution when immersed in the release buffer, thereby increasing the water channels in the films. Due to this, the polymer degradation rate by hydrolytic scission was accelerated.

11.10 Nanoparticle Property Assessment—after Degradation

11.10.1 Change in pH

The study on change in pH is carried out to observe the change in the pH of SBF with an increase in immersion time to adjust the ion concentration close to that of human blood plasma. Due to the polymer degradation and release of the drug in SBF, there is an increase in the pH of the SBF solution. The alpha half-life of the drug is usually a few minutes, but the terminal life of the drug concentration is nearly 10 h in SBF. After 30 days of degradation, the drug and the polymer matrix are totally degraded, and the pH is decreased. At the initial stage, the polymer tends to degrade slowly, which causes a pH increase, followed by rapid degradation of the polymer with the drug causing a reduction in pH. The initial dissolution of the drug on immersion in the SBF leads to the formation of aqueous channels on the surfaces that are filled with the aqueous medium.

Water penetration into the polymer matrix results in hydrolytic scission of the ester linkages in the polymer backbone, thus accelerating the dissolution of the polymer and drug, which tends to reduce pH. Mustafa et al. (2000) studied the biodegradation of PCL scaffolds in SBF over 28 days, which showed a reduction in pH from 8.0 to 6.8 due to the slow biodegradation rate.

11.10.2 Change in Molecular Weight

The molecular weight of the drug-loaded PHB matrix is found to decrease with time. Nanoparticles show increased water uptake within the pores of the polymer matrix, which results in weight loss and leads to a gradual decrease in the molecular weight of the polymer from 390 to 270 kDa over time.

The decrease in molecular weight is attributed to the degraded end products, such as monomers and the carrier drug. Sulheim et al. (2016) demonstrated the decrease in the molecular weight of polyalkyl cyanoacrylate nanoparticles by the degraded end products and the type of polymerization of the monomers.

11.11 Characterization of Degraded Nanoparticles

11.11.1 Morphological Changes of Degraded PHB Nanoparticles

The surface of drug-loaded PHB nanoparticles after degradation is studied using SEM (Figure 11.6). Drug release occurs through the porous surface of the degraded polymer matrix. The increased porosity is due to the increased water uptake and weight loss of drug within the polymer matrix. During the formation of nanoparticles, after solvent is removed, the water droplets within the nanoparticles are quickly trapped, and these later dry out, leaving pores.

The outer wall of the nanoparticles prevents a further influx of water and the formation of larger pores. Kamoun et al. (2015) observed the formation of pores in the presence of a high concentration (0.5% w/v) of PVA. The water droplets are stabilized, preventing them from coalescing and leading to porosity; thus, the optimal PVA concentration contributes an even pore size distribution throughout the polymer matrix.

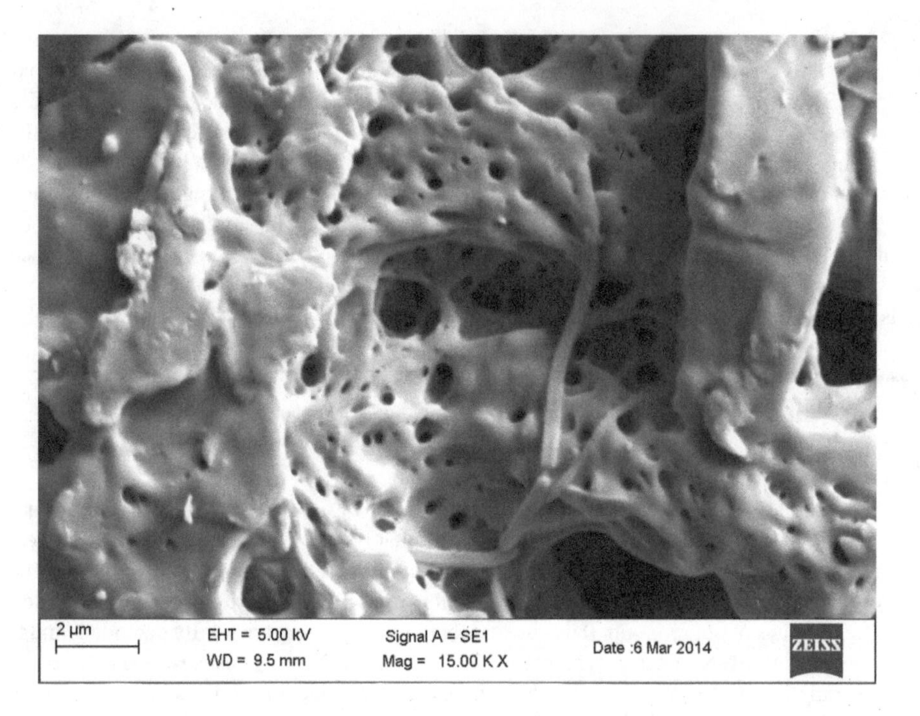

FIGURE 11.6 Morphology of degraded drug-loaded PHB.

11.11.2 Effect of Water Adsorption on Thermal Properties

The differential scanning calorimetry (DSC) thermogram gives the values of the thermal properties. The crystallinity of a polymer contributes to changes in Tg values. Smith et al. (2009) studied the decrease of Tg values of a polymer with an increase in the uptake of water. With an increase in the uptake of water by the polymer, the Tg value is reduced. Frank et al. (2005) showed that the Tg values of iodosalt-loaded PLGA microparticles started to decrease after 14 days of degradation in buffer.

The degree of crystallinity (%Xc) and Tm value of drug-loaded nanoparticles exhibit significant reduction due to degradation in the amorphous region followed by degradation in the crystalline domains, thus decreasing the crystalline nature of the polymer matrix.

11.11.3 *In Vitro* Release Studies in Simulated Body Fluid (SBF) and Phosphate-Buffered Saline (PBS)

The drug release rate is dependent on the size of nanoparticles (Edlund & Albertson 2002). The cumulative release of dipyridamole from a PHB matrix in SBF and PBS was seen to be biphasic, whereby an initial burst release was observed, followed by controlled release. Measurement of dipyridamole release from nanoparticles in SBF and PBS showed initial burst release (0 min to 2 h), during which the drug released 52% of the total encapsulated drug, followed by a controlled release period from 5 to 12 h, in which drug slowly diffused from the nanoparticles into the buffer. After 20 h, a total release of 95.15% of drug had occurred, while the drug release in PBS was slow and followed a controlled release pattern.

The initial release from 0 to 2 h was 45% of total encapsulated drug, followed by controlled release, in which 53–73% of drug was released into the medium. A total drug release of 83.3% in PBS was found. Nanoparticles of larger size tend to release the drug slowly for a longer period when compared with small nanoparticles (Berkland et al. 2002). Reduced drug diffusion and the low specific area of larger nanoparticles are the reasons for slow release of drug. In this study, drug release within a period of 20 h was investigated, which may be due to the small size of nanoparticles.

Distribution of the drug is also an important factor of drug release. The initial burst is high is due to surface-associated drug (Kassab et al. 1997). Other parameters, such as the type of drug and polymer, also determine the rate of drug release (Poletto et al. 2007). Drug diffuses slowly from the nanoparticles into the buffer. The controlled release is dependent on the type of polymer (Naraharisetti et al. 2005) and the degradation process. Polymer degradation can take place by surface erosion or bulk degradation, which increase water uptake into the polymer, and surface erosion occurs, followed by degradation.

11.11.4 Accelerated Stability Studies

Freeze-dried nanoparticles containing lyoprotectant were used for accelerated stability studies. Accelerated stability studies were carried out to evaluate the change in mean particle size, percentage entrapment efficiency, and drug content over a period of months at various temperatures. At low temperatures, there were no significant changes in particle size up to 3 months. At higher temperatures, the particle size of nanoparticles in the freeze-dried state increased. The particle size was found to have decreased at the end of the second and the third month. Gasper et al. (1998) reported that the particle size increases over time due to the aggregation of nanoparticles at high temperature, resulting in increased polymer size. The drug content of dipyridamole in freeze-dried PHB nanoparticles had decreased at the end of the second and the third month. At 40°C, the nanoparticles were unstable, and a significant change in the mean particle size and drug content were observed, due to the aggregation of the polymer and degradation at high temperature. The extent of aggregation of nanoparticles increased with an increase in the storage temperature from 4°C to 40°C. The best storage temperature in the freeze-dried state is 4°C, where the nanoparticles remain intact and stable.

11.11.5 Determination of Drug–Polymer Interaction

To study the effect of drug–polymer interaction on the *in vitro* release of drug, analyses such as scanning electron microscopy coupled with energy dispersive X-ray spectroscopy (SEM-EDX), Fourier transform infrared spectroscopy (FTIR), DSC, and X-ray diffraction (XRD) were carried out.

11.11.5.1 Elemental Analysis–SEM-EDX

SEM-EDX analysis of dipyridamole-loaded PHB nanoparticles was done to characterize the presence of carbon and oxygen content on the surface of nanoparticles. The elemental analyses of drug-loaded nanoparticles are shown in Figure 11.7.

The presence of 55% carbon and 42% oxygen on the surface of the polymer was observed, which indicates the presence of the elements in the polymer and drug composition. Naraharisetti et al. (2005) evaluated the surface bound elements on PLLA microparticles; microparticles of high particle size indicate high drug interaction with more surface elements.

11.11.5.2 Fourier Transform Infrared Studies (FTIR)

The drug–polymer interaction plays an important role in drug release characteristics. The chemical interaction between the drug and the polymer affects the release pattern of the drug (Hussain et al. 2013). FTIR was used to study the interaction between drug and PHB.

Studies using nuclear magnetic resonance (NMR) and mass spectroscopic analysis were performed to study the drug–polymer interaction. The concentration of the drug present within the nanoparticles would be too low to detect any interaction with the polymer with FTIR.

11.11.5.3 Thermal Analysis

The analysis of thermal properties was used to detect the existence of a possible interaction between the drug and the polymer. The increase in the Tg value (4.9°C) of loaded PHB nanoparticles is due to the

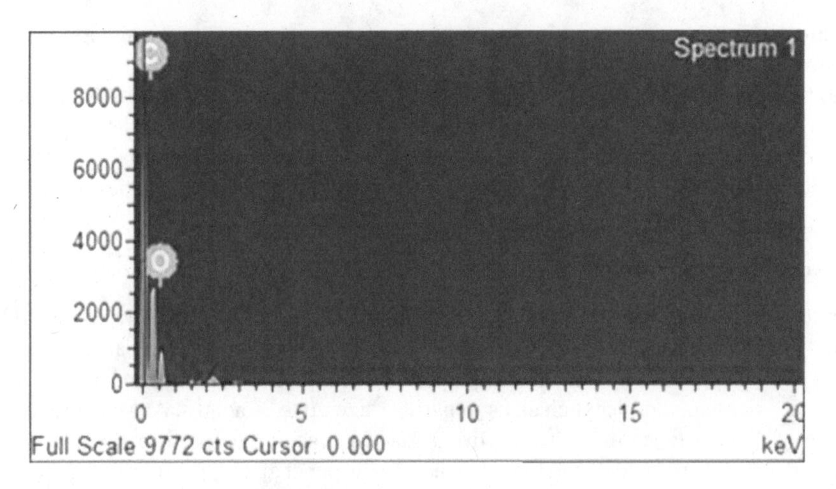

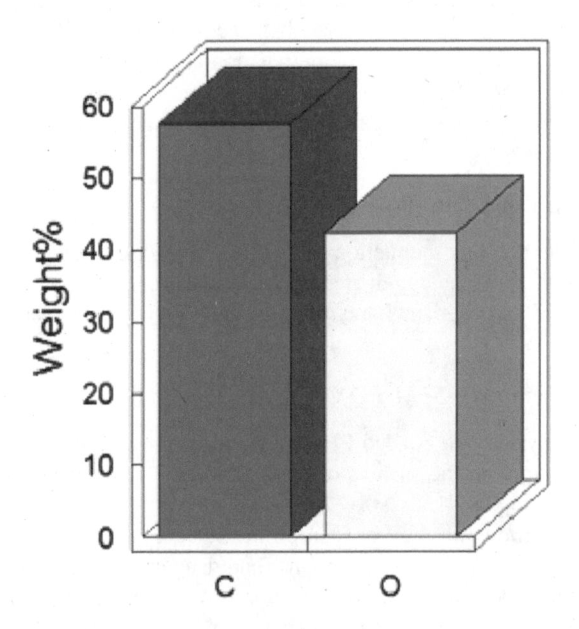

FIGURE 11.7 SEM-EDX analysis of drug-loaded PHB nanoparticles.

increase in the concentration of the emulsion in the presence of the drug and the mobility of the polymer chains in reducing the Tg values of the polymer (Chang et al. 2006). The Tm of the drug-loaded nanoparticles, at 162.35°C, was also lower than that of the unloaded PHB nanoparticles, indicating that the crystalline properties of the drug-loaded PHB nanoparticles decreased in the presence of drug, probably due to the incorporation of drug within the crystalline structure of the polymer disrupting the crystal nature.

Martin et al. (2000) reported that the processing conditions of a polymer increase the crystallinity of the polymer. Chang et al. (2006) studied the increase in the concentration of drug and mobility of polymer that results in a reduction of Tg values. Bidone et al. (2009) have also observed a decrease of 10°C in the Tg of ibuprofen-loaded PHB microparticles compared with unloaded polymer due to molecular dispersion of the drug within the polymer matrix affecting the crystalline property, which in turn, affects the thermal property. Bidone et al. (2009) reported a decrease in the melting temperature of ibuprofen-loaded PHB when compared with unloaded PHB. It was found that ibuprofen had coexisted as both a crystalline and a molecular dispersion of the polymer in the microparticles.

11.11.5.4 X-ray Diffraction (XRD) Analysis

The crystalline nature of the polymer is usually hindered by the presence of drug, and this can be analyzed using XRD. The XRD spectrum of dipyridamole-loaded nanoparticles confirms the dispersion of dipyridamole within PHB nanoparticles. The diffraction patterns of dipyridamole-loaded PHB exhibited an intense peak, confirming their high crystallinity. A well-defined diffraction pattern with intense peaks confirms the crystalline nature of the polymer, not affected by drug loading. Palau et al. (2008) used small angle diffraction analysis (SAXD) to get additional information on lamellar structures between the domains of drug and polymer. Shah et al. (2010) also conducted a study in which PHB and chitosan were synthesized using a single and double emulsion technique and analyzed by XRD to identify possible interactions between PVA in the emulsion phase. From the results, it can be confirmed that the solvent evaporation technique had an influence on the physical properties of the polymer.

11.12 Biocompatibility of the Drug-Loaded Polymer and Its Application

11.12.1 Evaluation of Biocompatibility, Bioactivity, and Toxicity by *In Vitro* and *In Vivo* Assay

Biocompatibility is the ability of a biomaterial to perform its specific function with respect to therapy without eliciting any immune response by the host cells (Li et al. 2004). Nanoparticles are a new class of biomedical products that has potential applications in drug delivery devices. The safety of nanomaterials is of major concern (Shen et al. 2013). Nanoparticles for pharmaceutical application should be subjected to biocompatibility testing before regulatory approval for administration to a host (Dobrovolskaia et al. 2008). *In vitro* and *in vivo* assays for dipyridamole-loaded PHB are necessary for the evaluation of its biocompatibility, bioactivity, and toxicity.

11.12.2 *In Vitro* Cytocompatibility Study

Biologically synthesized PHB nanoparticles show significant antiproliferation activity against the Vero cell line; thus, they are considered fully biocompatible. The concentration required for inhibition (IC_{50}) was studied against the Vero cell line treated with PHB nanoparticles, and it was noted that early apoptotic cells with nuclear shrinkage and chromatin condensation occurred, which was observed by changes in the morphology of the cell structure at higher concentrations, while mild csswellular damage was observed at lower concentration. The Vero cells remain normal when treated with a very low concentration, say 62.5 µg/ml, over a period of 72 h. Errico et al. (2009) reported on PHB in the mouse fibroblast 3T3 cell line, which exhibited a high level of cytocompatibility and higher IC_{50} values.

11.12.2.1 Hemocompatibility Assay

Hemolysis assays are performed to analyze the hemocompatibility of drug–PHB by studying the effect of nanoparticles on erythrocytes. Chan and Leong (2008) evaluated the hemocompatibility of PHB–PEG–PHB nanoparticles at a concentration of 120 µg/ml. They observed no hemolysis, and the nanoparticles were regarded as hemocompatible for drug delivery applications.

11.12.3 *In Vitro* Bioactivity Study of Dipyridamole Clot Lysis Assay

It is necessary to study the thrombolytic activity of the drug (dipyridamole) encapsulated in a PHB matrix. Various methods have been developed to measure the thrombolytic (clot lysis) activity of antithrombotic drugs. The best way to study the thrombolytic drugs is through an *in vitro* clot lysis assay.

Figure 11.8 shows experimental results of the clear, visual representation of clot lysis. Tube no. 1 is a control clot; no clot lysis is observed in tube no. 1. In tubes no. 2–5 (positive control), the intact clot was lysed by four different concentrations of dipyridamole-loaded nanoparticles in decreasing order

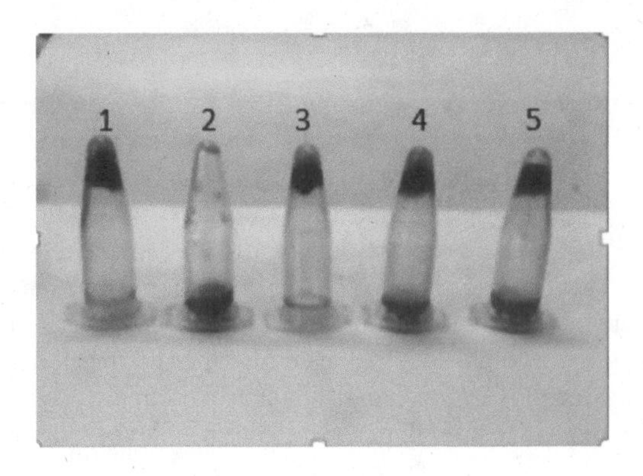

FIGURE 11.8 Visual representation of clot lysis of drug-loaded PHB.

of concentration. After the dissolution of the clots, the tubes were inverted, and fluid along with the remnants of clots could be clearly seen. The percentage clot lysis was found to increase with high concentrations of nanoparticles, whereas the lysis of clots gradually decreased with low concentrations of nanoparticles. Basta et al. (2005) worked on artificial clots and used ultrasound methods to measure the thrombolytic activity of the enzyme streptokinase. Several other models have been reported, but these are very costly and not affordable; these problems demand a simple and cost-effective clot lytic model for the measurement of the clot lysis activity of thrombolytic drugs. An attempt has been made to develop an *in vitro* clot lytic model using a known thrombolytic drug, dipyridamole.

11.12.4 *In Vivo* Toxicity Study of Dipyridamole-Loaded PHB Nanoparticles in Zebrafish Embryo

The analysis of off-target effects of drug-loaded nanoparticles requires systemic analysis, as they often involve the complete organism rather than specific tissue or cell targets. Off-target effects represent a major concern in the development of new drug candidates and require animal toxicity testing. The most reliable, fastest, and cheapest assay is based on the zebrafish (*Danio rerio*) embryo. The zebrafish is an appropriate model system, used in developmental biology and the pharmaceutical sciences, including toxicology, for evaluating material toxicity (Teraoka et al. 2003). The zebrafish embryo has a transparent body, which makes it easy to collect data, and there is a close similarity of genetic characters between zebrafish and mammals in the early stages of development (Arora et al. 2012) The hatching rate of PHB nanoparticle–treated embryos was studied over 96 h post fertilization. The hatching rate was increased to 85% with a low concentration of nanoparticles. The optimum concentration of nanoparticles shows no significant impact on the hatching and mortality rate of zebrafish embryos. No significant changes were observed in the morphology of embryos; thus, PHB nanoparticles did not produce any major toxicity in the development of zebrafish.

Langer and Tirrell (2004) studied *in vivo* toxicity testing using the zebrafish assay. The *in vivo* zebrafish embryo assay seems to be more sensitive for picking up potential adverse effects than *in vitro* experiments in cells. Consequently, the zebrafish embryo assay might prove to be an interesting and important intermediate tool for facilitating efficient and ethical *in vivo* nanotoxicity testing. Owens and Peppas (2006) presented a review of the effect of nanoparticles on zebrafish hatching. They reported complete inhibition of hatching and embryo death within the chorion on nanoparticle exposure. They also concluded that nanoparticles interact with the hatching enzymes and are thus responsible for toxicity rather than in its ionic forms. Xing et al. (2007) studied the influence of silver nanoparticles (AgNPs) on neurological development; these resulted in small heads along with the presence of hypoplastic hind brain, little eye, and cardiac defects.

Another study performed by Lu et al. (1999) revealed carbon nanotubes (CNT)-induced biochemical alterations in zebrafish embryos. Further studies showed that CNT exposure can stimulate the brain and cause gonadal alterations. Chang et al. (2006) studied the toxicity of TiO_2 nanoparticles in zebrafish and revealed that these nanoparticles are distributed and accumulated in different parts of zebrafish, such as gill, liver, and brain. In comparison to other nanoparticles studied in above discussed literature, PHB-based nanoparticles do not cause any developmental changes in the zebrafish embryo; thus, this study suggested that PHB nanoparticles could be a highly biocompatible material for biomedical and drug delivery applications.

11.13 Conclusion

The biodegradable PHB-based nanoparticle system developed has potential biomedical and drug delivery applications. It is used as a biocompatible drug carrier system for controlled drug delivery. *In vitro*, PHB nanoparticles were found to be biocompatible; they did not inhibit the growth or metabolic activity of the mouse fibroblast 3T3 cell line. The optimized formulation shows that preparation with a low concentration of polymer and drug and high PVA concentration with a high stirring rate produces smaller-size, spherical, porous-surface nanoparticles with optimum zeta potential values. This chapter has discussed the suitability of the nanoparticles for application in the controlled delivery of dipyridamole with probable enhancement in absorption, bioavailability, and therapeutic efficiency. The drug release from the PHB matrix is biphasic, whereby an initial burst is followed by controlled release. The interaction between drug and polymer was characterized using SEM-EDX, FTIR, DSC, and XRD, showing the possible interaction between the drug and the polymer. The hemocompatibility assay confirms the compatibility of the drug-loaded nanoparticles with blood cells, and the *in vitro* clot lysis assay confirms the thrombolytic activity of drug embedded in the polymer matrix. The toxicity of PHB nanoparticles was studied in *in vivo* conditions in zebrafish embryos, and no significant changes were observed in the morphology and development of the embryos; thus, PHB nanoparticles did not produce any major toxicity in the development of zebrafish, and hence, PHB nanoparticles could be highly biocompatible as well as an effective carrier of drugs. These properties of PHB nanoparticles allow more diverse applications in biomedical fields. The future scope of the present work will emphasize developing PHB into an active scaffold formulation and studying its tissue engineering applications, and developing PHB and copolyesters of 3-hydroxy butyrate and evaluating them as a carrier system for intracellular sustained release with model drugs.

REFERENCES

Albertsson, AC, Donaruma, LG & Vogl, O 1985. Synthetic polymers as drugs. *Annals of the New York Academy of Sciences*. 446: 105–115.

Apetroaie-Constantin, C, Mikkola, R, Andersson, MA, Teplova, V, Suominen, I, Johansson, T & Salkinoja-Salonen, M 2009. *Bacillus subtilis* and *Bacillus mojavensis* strains connected to food poisoning produce the heat stable toxin amylopsin. *Journal of Applied Microbiology*. 106: 1976–1985.

Arora, S, Rajwade, JM & Paknikar, KM 2012. Nanotoxicology and invitro studies. *Toxicology and Applied Pharmacology*. 258: 151–165.

Basta, HA, Buzaz, AJ & Mcclure, MA 2005. Identification of novel retroid agents in *Danio rerio*. *Evolution Bioinformation*. 3: 179–195.

Bazzo, GC, Lemo, E & Goncalvos, MC 2008. Effect of preparation conditions on morphology, drug content and release profiles of polyhydroxy butyrate microparticles containing piroxicam. *Journal of Brazilian Chemical Society*. 19: 914–921.

Berkland, C, Kim, K & Pack, DW 2002. PLG microsphere size controls drug release rate through several competing factors. *Pharmaceutical Research*. 20: 1055–1062.

Bidone, J, Paula, APM, Bazzo, GC, Carmignan, F, Soldi, MS, Pires, ATN & Lemos sennaa, E 2009. Preparation and characterization of ibuprofen loaded microparticles consisting of poly (3-hydroxybutyrate) and methoxy poly (ethylene glycol) blends or poly hydroxybutyrate and gelatin composites for controlled drug release. *Material Science and Engineering*. 29: 588–593.

Blaine, G 1946. The uses of plastics in surgery. *The Lancet*. 248(6424): 525–528.

Bohrey, S, Chourasiya, V & Pandey, A 2016. Polymeric nanoparticles containing diazepam: preparation, optimization, characterization, invitro drug release and release kinetic study. *Nano Convergence*. 3: 1–7.

Cesare, E, Cristina, B, Federica, C & Emo, C 2009. PHA based polymeric nanoparticles for drug delivery. *Journal of Biomedicine and Biotechnology*. 2: 1–10.
Chan, BP & Leong, KW 2008. Scaffolding in tissue engineering: general approaches and tissue specific considerations. *European Spine Journal*. 17: 467–479.
Chang, H, Perrie, IY, Coombes, AGA 2006. Delivery of the antibiotic gentamycin sulphate from precipitation cast matrices of polycaprolactone. *Journal of Controlled Release*. 110: 414–421.
Chee, JW, Amirul, AA, Majid, MIA & Mansor, SM 2008. Factors influencing the release of mitragyna speciosa crude extracts from biodegradable P(3HB–co–4HB). *International Journal of Pharmacy*. 361: 1–6.
Cheng, J, Flahaut, E & Cheng, SH 2007. Effect of carbon nanotubes on developing zebrafish (DanioRerio) embryos. *Environmental Toxicology and Chemistry*. 26(4): 708–716.
Dobrovolskaia, MA, Aggarwal, P, Hall, JB & Mc Neil, SE 2008. Preclinical studies to understand nanoparticle interaction with the immune system and its potential effects on nanoparticle biodistribution. *Molecular Pharmaceutics*. 5: 487–495.
Edlund, U & Albertson, AC 2002. Degradable polymer microparticles for controlled drug delivery. *Advances in Polymer Science*. 157: 67–112.
Emami, J, Pourmashhadi, A, Sadeghi, H, Varshosaz, J & Hamishehkar, H 2014. Formulation and optimization of celecoxib-loaded PLGA nanoparticles by the Taguchi design and their *in vitro* cytotoxicity for lung cancer therapy. *Pharmaceutical Development and Technology*. 20(7): 791–800.
Feng, SS & Huang, G 2001. Effects of emulsifier on the controlled release of paclitaxel (Taxol) from nanospheres of biodegradable polymers. *Journal of Controlled Release*. 71: 53–69.
Frank, A, Rath, SK & Venkataraman, SS 2005. Controlled release from bioerodible polymers: Effect of drug type and polymer composition. *Journal of Controlled Release*. 102: 333–344.
Gasper, RMM, Blanco, D, Cruz, ME & Alonso, MJ 1998. Formulation of L-asparginase loaded poly (lactide-co-glycoside) nanoparticles; influence of polymer properties on enzyme loading, activity and invitro release. *Journal of Controlled Release*. 52: 53–62.
GültekinHE & Değim, Z 2013. Biodegradable polymeric nanoparticles are effective systems for controlled drug delivery. *FABAD Journal of Pharmaceutical Sciences*. 38(2): 107–118.
Harwood, CR & Wipat, A 1996. Sequencing and functional analysis of the genome of Bacillus subtilis strain 168. *FEBS Letters*. 389: 84–87.
Holmes, PA 1988. Biologically produced PHA polymers and copolymers. In: Bassett DC & Bassett DC (eds) *Developments in Crystalline Polymers*, Vol. 2. London: Elsevier, pp. 1–65.
Huisman, GW, Leeuwde, O, Eggink, G & Witholt, B 1989. Synthesis of poly-3-hydroxyalkanoates is a common feature of fluorescent pseudomonads. *Applied Environmental Microbiology*. 55: 1949–1954.
Hussain T, Saeed T, Mumtaz AM, Javaid Z, Abbas K, Awais A & Idrees HA 2013. Effect of two hydrophobic polymers on the release of gliclazide from their matrix tablets. *Acta Poloniae Pharmaceutica*. 70(4): 749–757.
Jung, T, Breitenbach, A & Kissel, T 2000. Sulfobutylated poly (vinyl alcohol) grafted PLGA facilitated the preparation of small negatively charged biodegradable nanospheres for protein delivery. *Journal of Controlled Release*. 6: 157–169.
Kamoun, EA, Kenawy, E-RS, Tamer, TM, El-Meligy, MA, Mohy Eldin, MS 2015. Poly (vinyl alcohol)-alginate physically crosslinked hydrogel membranes for wound dressing applications: Characterization and bio-evaluation. *Arabian Journal of Chemistry*. 8(1): 38–47.
Kassab, AC, Xu, K, Denkbas, EB, Dou, Y, Zhao, S & Piskin, E 1997. Rifampicin carrying polyhydroxybutyrate micro particles as potential chemoembolization agent. *Journal of Biomaterial Science Polymer Edition*. 8: 947–961.
Kwon, HY, Lee, JY, Choi, SW, Jang, Y & Kim, JH 2001. Preparation of PLGA nanoparticles containing estrogen by emulsification diffusion method. *Colloids and Surfaces A*. 182: 123–130.
Langer, R & Tirrell, DA 2004. Designing materials for biology and medicine. *Nature*. 428: 487–492.
Lee, SY 1995. Review bacterial polyhydroxy alkanoates. *Biotechnology Bioengineering*. 49: 1–14.
Lemoigne, M 1926. Products of dehydration and of polymerization of poly-hydroxybutyric acid. *Bulletin de la Société Chimique de France (Paris)*. 8: 770–782.
Li, H & Chang, J 2004. Fabrication and characterization of bioactive wollastonite/PHBV composite scaffolds. *Biomaterials*. 25: 5473–5480.
Liu, D, Yang, F, Xiong, F & Gu, N 2016. The smart drug delivery system and its clinical potential. *Theranostics*. 6(9): 1306–1323.

Madison, LL & Huisman, GW 1999. Metabolic engineering of poly(3-hydroxyalkanoates): from DNA to plastic. *Microbiology and Molecular Biology Reviews*. 63(1): 21–53.

Martin, MA, Miguens, FC, Rieumont, J & Sanchez, R 2000. Tailoring of the external and internal morphology of poly 3 hydroxy butyrate microparticles. *Colloids and Surfaces B Interfaces*. 17: 111–116.

Mercan, N & Beyatli, Y 2001. Production of poly-β-hydroxybutyrate (PHB) by *Bacillus sphaericus* strains. *Journal of Biotechnology*. 25(2): 1–7.

Morimoto, T, Kadoya, R, Endo, K, Tohata, M, Sawada, K, Liu, S, Ozawa, T, Kodama, T, Kakeshita, H, Kageyama, Y, Manabe, K, Kanaya, S, Ara, K, Ozaki, K & Ogasawara, N 2008. Enhanced recombinant protein productivity by genome reduction in *Bacillus subtilis*. *DNA Research*. 15: 73–81.

Muller, RH, Davis, SS & Illum, L 1986. Particle charge and surface hydrophobicity of colloidal drug carriers. *Targeting of Drugs with Synthetic Systems*. 113: 239–263.

Murakami, H, Kobayashi, M, Takeuchi, H & Kawashima, Y 1999. Preparation of PLGA nanoparticles by modified spontaneous emulsification solvent diffusion method. *International Journal of Pharmacy*. 187: 143–152.

Mustafa, K, Helen Billman, J & Don, MN 2000. Quantitative determination of the biodegradable polymer poly hydroxyl butyrate in a recombinant *E.coli* strain by use of mid infrared spectroscopy and multivariate statistics. *Applied and Environmental Microbiology*. 68(8): 3415–3420.

Naraharisetti, PK, Lewa, MDN & Fub, YC 2005. Gentamycin loaded discs and microparticles and their modifications. *Biomaterials*. 77: 329–337.

Owens, DE & Peppas, NA 2006. Opsonization, biodistribution and pharmacokinetics of polymeric nanoparticles. *International Journal of Pharmaceutics*. 307: 93–102.

Palau, MM, Franco, L & Puiggali, J 2008. Microparticles of new alternating copolyesters derived from glycolic acid units for controlled drug release. *Journal of Applied Polymer Science*. 110: 2127–2138.

Poletto, FS, Jager, E, Re, MI, Guterress, SS & Pohlmann, AR 2007. Rate modulating PHBHV/PCL microparticles containing weak acid model drugs. *International Journal of Pharmacy*. 345: 70–80.

Pouton, CW, & Akhtar, S 1996. Biosynthetic polyhydroxy alkanoates and their potential in drug delivery. *Advances in Drug Delivery Reviews*. 18: 133–162.

Qiu, Y, Brown, AC, Myers, DR, Sakurai, Y, Mannino, RG, Tran, R, Ahn, B, Hardy, ET, Kee, MF, Kumar, S, Bao, G, Barker, TH & Lam, WA 2014. Platelet mechanosensing of substrate stiffness during clot formation mediates adhesion, spreading, and activation. *PNAS*. 111(40): 14430–14435.

Quintanar, GD, Fessi, H, Allemann, E & Doelker, E 1996. Influence of stabilizing agents and preparative variables on the formation of poly (D,L- lactic acid) nanoparticles by an emulsification diffusion technique. *International Journal of Pharmacy*. 143: 133–141.

Saez, A, Guzman, M, Molpeceres, J & Aberturas, MR 2000. Freeze drying of polycaprolactone and PLGA nanoparticles induces minor particle size changes affecting the oral pharmacokinetics of loaded drugs. *European Journal of Pharmacy and Biopharmacy*. 50: 379–387.

Sahoo, SK, Panyam, J, Prabha, S & Labhasetwar, V 2002. Residual Poly vinyl alcohol associated with PLGA nanoparticles and cellular uptake. *Journal of Control Release*. 82: 105–114.

Saito, Y, Nakamura, S, Hiramitsu, M & Doi, Y 1996. Microbial synthesis and properties of poly hydroxyl butyrate. *Poly International*. 39: 169–171.

Shah, M, Naseer, MI & Choi, MH 2010. Amphilic PHA-m PEG copolymeric nanocontainers for drug delivery. *International Journal of Pharmacy*. 400: 165–175.

Shariati, A & Peters, CJ 2003. Recent developments in particle design using supercritical fluids. *Current Opinion in Solid State Material Science*. 7: 371–383.

Shen, JM, Yin, T & Tian, X 2013. Surface charge switchable polymeric magnetic nanoparticles for the controlled release of anticancer drug. *Applied Materials & Interfaces*. 5: 7014–7024.

Smith, KE, Sawicki, S, Hyjek, MA, Downey, S & Gall, K 2009. The effect of the glass transition temperature on the toughness of photopolymerizable (meth)acrylate networks under physiological conditions. *Polymer*. 50(21): 5112–5123 (Guildf 9).

Sotome, S, Uemura, T, Kikuchi, M, Chen, J, Itoh, S, Tanaka, J, Tateishi, T & Shinomiya, K 2004. Synthesis and In vivo evaluation of a novel hydroxyapatite/collagen- alginate as a bone filler and a drug delivery carrier of bone morphogenetic protein. *Material Science and Engineering*. 24: 341–347.

Steinbuchel, A & Schlegel, HG 1991. Physiology and molecular genetics of poly beta hydroxyl alkanoic acid synthesis in *Alcaligenes eutrophus*. *Molecular Microbiology*. 5(3): 535–542.

Sulheim, E, Baghirov, H, von Haartman, E, Bøe, A, Åslund, AKO, Mørch, Y & de Lange Davies, C 2016. Cellular uptake and intracellular degradation of poly (alkyl cyanoacrylate) nanoparticles. *Journal of Nanobiotechnology*. 14: 1.

Teraoka, H, Dong, W & Hiraga, T 2003. Zebrafish as a novel experimental model for developmental toxicology. *Journal of Neurophysiology*. 43: 123–132.

Valappil, SP, Peiris, D, Langley, GJ, Herniman, JM, Boccaccini, AR, Bucke, C & Roy, I 2007. Polyhydroxyalkanoate (PHA) biosynthesis from structurally unrelated carbon sources by a newly characterized *Bacillus* sp. *Journal of Biotechnology*. 127: 475–487.

Wang, YW, Wu, Q & Chen, GQ 2004. Attachment, proliferation, and differentiation of osteoblast on random biopolysester scaffolds. *Biomaterials*. 25: 669–675.

Win, KY & Feng, SS 2005. Effects of particle size and surface coating on cellular uptake of polymeric nanoparticles for oral delivery of anticancer drugs. *Biomaterials*. 26: 2713–272.

Xing, J, Zhang, D & Tan, T 2007. Studies on the oridonin-loaded poly (d,l- lactic acid) nanoparticles in vitro and in vivo. *International Journal of Biological Macromolecules*. 40: 153–158.

Yang, A, Liu, W, Li, Z & Jiang, L 2010. Influence of polyethylene glycol modification on phagocytic uptake of polymeric nanoparticles mediated by immunoglobulin G and complement activation. *Journal of Nanoscience and Nanotechnology*. 10: 622–628.

Yang, Y, Chung, TS & Ng, NP 2001. Morphology, drug distribution and in vitro release profiles of biodegradable polymeric microparticles containing protein fabricated by double emulsion solvent evaporation method. *Biomaterials*. 22: 231–241.

12

Smart Delivery Systems for Personal Care and Cosmetic Products

Fanwen Zeng and Nilesh Shah

CONTENTS

12.1 Introduction

Personal care and cosmetic markets cover a diverse range of products that consumers rely on every day for personal hygiene, beautification, and protection. Among them are makeup, lotion, cream, powder, lipstick, lip balm, perfume, color cosmetics, mascara, face mask, body wash, shampoo, conditioner, bath oil, antiperspirant, deodorant, hair color, hair spray, hair gel, shaving gel, nail polish, toothpaste, and sunscreen.[1] New forms of cosmetic products are launched periodically in this innovative industry. This is partially attributed to the innovations from the ingredient suppliers.[2] Among them, skin care is one of the most dynamic segments and commands a significant market share. The desire to look younger continues to create the need for high-performance cosmetic products that deliver aesthetic and anti-aging benefits. Increasingly, active ingredients are formulated into the personal care products to offer these benefits. These products are commonly referred to as *cosmeceuticals*, which have both a cosmetic and a "therapeutic" function.

Smart delivery systems for active ingredients will be a key growth driver for companies that are striving to differentiate themselves from others in the personal care and cosmetic markets.[3] Due to the increasing complexity of formulations and actives involved, various innovative delivery systems are finding increasing use in these markets. In addition to protecting and delivering cosmetic active ingredients, a smart delivery system can also enhance formulation stability, improve formulation compatibility by separating incompatible ingredients, maximize the aesthetics of the formulation, provide desired visual effects, and increase the safety profiles of the consumer products. This chapter will review the development and application of smart delivery systems for cosmetic active ingredients from both patent and academic literature.

12.2 Encapsulation Materials and Processes

Encapsulation processes, both chemical and physical/mechanical, have been adapted for the encapsulation of cosmetic active ingredients. Tables 12.1 and 12.2 exemplify both of these processes as well as the typical materials used in the process and the resulting particle sizes. Most of the resulting delivery systems tend to have well-defined core-shell morphology, where the active or payload resides inside the core and is surrounded by the shell or wall material. For some, the shell can be a highly cross-linked network, which can be called a hard shell, while for others, the shell, which can be called a soft shell, is composed of non-cross-linked polymer or simply a lipid bilayer. Furthermore, some delivery systems may not have a completely defined core-shell morphology. The active is simply absorbed in a solid matrix, which acts as a reservoir.

Table 12.3 lists common cosmetic active ingredients and sensorial and aesthetic benefit agents that are broadly used in the personal care and cosmetic markets.

TABLE 12.1

Encapsulation Processes by Chemical Methods

Processes	Typical Materials	Typical Size
Simple coacervation	Gelatin and others	2 μm and above
Complex coacervation	Gelatin, gum arabic, glutaraldehyde	2–2000 μm
Interfacial and *in situ* polymerization	Polyureas, polyurethanes, polyamides, aminoplasts	0.5–1000 μm
Layer by layer deposition	Polycation and polyanion	0.02–20 μm
Calcium alginate process	Alginate, $CaCl_2$	~500 μm
Emulsion and mini-emulsion polymerization	Monomers	0.1–1 μm
Sol-gel	Silane	2–20 μm

TABLE 12.2

Encapsulation Processes by Physical/Mechanical Methods

Processes	Typical Materials	Typical Size
Spray-drying/chilling	Various	10–5000 μm
Pan coating	Various	>800 μm
Fluidized bed coating	Various	20–1500 μm
Extrusion	Melt-sensitive materials	10–2000 μm
Spinning disc coating	Various	>100 μm
Solvent evaporation	Polymeric material	Variable
Microsponge, porous particles	Porous materials from suspension polymers	50–200 μm
Cyclodextrin cages	Cyclodextrin	Variable
Emulsions (microemulsion, liquid crystals, multiple emulsion, Pickering emulsion)	Emulsifiers with various hydrophilic–lipophilic balance, solid particles for Pickering emulsion	Variable
Vesicular systems	Phospholipids (lecithin) and co-surfactant, amphiphilic block or graft copolymer	~1 μm and below
Liquid particulate systems	Solid and liquid lipids	Variable

TABLE 12.3

Common Cosmetic Active Ingredients and Sensorial, Aesthetic Benefit Agents

Active ingredients	Vitamins, peptides/proteins, enzymes, fatty oils, UV absorbers, fruit acids, oxidants/reductants, natural extracts
Sensorial and aesthetic agents	Pigments, colorants, fragrances, phase change materials, menthol, silicones and petrolatum

12.3 Desired Release Kinetics and Physicochemical Challenges

Depending on the cosmetic active and the end use application, the desired release kinetics can range from controlled release to even zero release. For controlled release, the idealized release profile can range from triggered release, to constant release, to targeted release. Various triggered release mechanisms can be employed in the design and are listed here:

1. Abrasion, pressure
2. Osmotic pressure
3. Moisture concentration
4. Temperature
5. Ionic strength
6. pH
7. Enzymatic shell decomposition
8. Oxidation/reduction

However, significant physicochemical challenges arise in the design of a commercially viable delivery system.[4] These challenges can come from either the manufacturing of the system itself or the downstream application. In the manufacture of the smart delivery system, challenges are generated by either the cosmetic active or the delivery system. Physical characteristics of the cosmetic actives, such as hydrophobicity, crystallinity, and their partition coefficient between oil and aqueous phases, can dictate the delivery system that can be employed. The delivery system itself plays a key role in the release profile and cost/performance balance. In the downstream application, the delivery system has to survive the formulation process, storage, and aging conditions, while the actives need to be available at an appropriate time point to deliver the benefits.

12.4 Applications

The diversity of chemical and physical/mechanical approaches characteristic of this field is aptly demonstrated by the examples shown in this section.

12.4.1 Encapsulation of UV Absorbers

Sunscreens are important skin care products that are designed to block ultraviolet (UV) radiation, which causes photoaging and potentially, skin cancer. The UV radiation that reaches the surface of the earth is composed of UVB (290 to 320 nm) and UVA (320 to 400 nm). While both UVA and UVB cause skin damage, they do not have identical modes of action. UVB is about 50–100 times more energetic than UVA. UVB can cause skin reddening and sunburn, but the damage is limited to the skin's superficial epidermal layers. It plays a key role in the development of skin cancer. UVA is 30 to 50 times more prevalent than UVB and penetrates the skin more deeply, which makes it a major contributor to skin aging and wrinkling. The active ingredients in sunscreen formulations are UV absorbers. The UV absorbers that are currently approved for use by the Food and Drug Administration (FDA) can be either organic molecules or inorganic metal oxide particles, with each having its own unique UV absorption spectrum. To meet the broad spectrum coverage claim of a critical wavelength of >370 nm as defined by the FDA, formulators will have to incorporate various types of UV absorbers into the sunscreens.

Incompatibility of UV absorbers in the sunscreen formulations and during the end use application can be a major concern. For example, the presence of octyl methoxycinnamate (OMC) can accelerate the degradation of avobenzone through a Norrish type addition reaction pathway. Encapsulation technology will be one of the key approaches to resolve these problems.[5] The sol-gel encapsulation process developed and patented by Avnir and coworkers at Sol-Gel Technologies was designed for precisely this purpose.[6] OMC and avobenzone were encapsulated separately by the sol-gel method. In this sol-gel method, the hydrophobic solution containing the UV absorber and sol-gel precursors was emulsified in an aqueous solution under high shear. The formation of inorganic silica spheres, as shown in Figure 12.1, was carefully controlled by the appropriate reaction conditions, such as the selected pH change, so that the leaching rate of UV absorbers from the encapsulated product was very low. Examples of sol-gel precursors include tetraethoxysilane, methyltriethoxysilane, and poly (diethoxysiloxane). When these separate UV absorber microcapsules were formulated into the sunscreen formulation, good photostability was demonstrated for the OMC/avobenzone pair due to their physical separation inside their individual capsule walls. In addition to the improvement of photostability, the sol-gel UV absorber microcapsules were shown to be safer than their unencapsulated counterparts in terms of phototoxicity, plasmid DNA nicking, and amount of extractable UV absorber.[7] The last property suggests that the encapsulation of UV absorber can prevent the penetration of actives into the skin, which is not desirable in this application, as shown in Figure 12.2.[8] Furthermore, the encapsulation of UV absorbers can improve aesthetic feel without the oily and sticky feel encountered from the unencapsulated counterpart. Finally, the technology can provide formulation flexibility, with the product supplied as either a water-based slurry or isolated dry powder.[9]

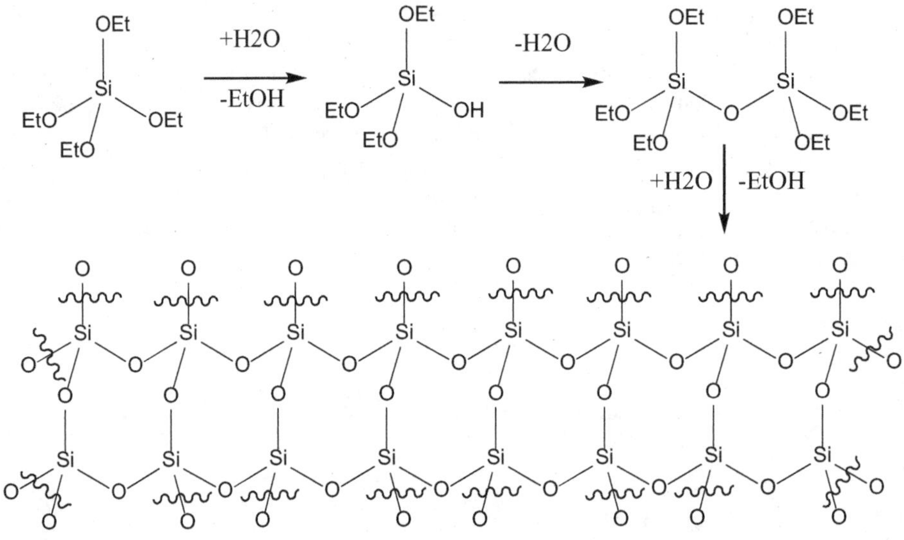

FIGURE 12.1 Sol-gel chemistry.

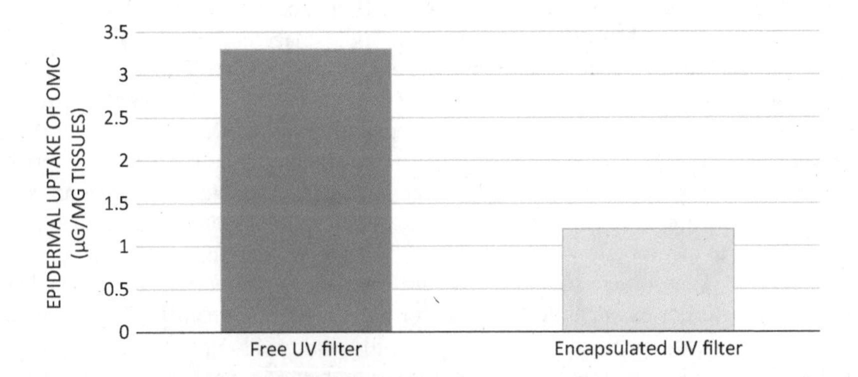

FIGURE 12.2 Comparison of epidermal uptake of OMC for free and encapsulated UV filter.

Similar UV absorber microcapsules were described by the team from Seiwa Kasei Co., except that the wall membrane was formed by the polycondensation of an organic silane molecule with an amphiphilic molecule from the continuous and/or dispersion phase.[10] The amphiphilic molecule can be a silylated peptide or protein. Examples of the amphiphilic molecule are hydrolyzed proteins modified with N-[2-hydroxy-3-(3'-trihydroxy silyl) propoxyl] propyl groups. Examples of organic silanes are methyl triethoxysilane, phenyl triethoxysilane, and tetraethoxysilane. It is noted that the presence of organic moieties in the wall membrane results in a soft and lighter after-feel for consumers.

To completely prevent the interaction between UV absorbers and the penetration of UV absorbers into the skin, Berg-Schultz[11] from DSM patented a process to make microcapsules of UV absorbers where the UV absorbers are covalently linked to the cross-linkable functional group. The process further includes at least one cross-linkable monomer but excludes any non-cross-linkable UV absorber. The cross-linkable functional group described in the patent is a triethoxysilyl group, while the cross-linkable monomer is tetraethoxysilane. While the actual morphology was not disclosed, one can postulate that the UV absorber chromophore is completely immobilized into the silica, which resides inside the core of the microcapsule, while silica without the UV absorber chromophore resides in the shell. Interestingly, sufficient UV absorption activity is retained in this microcapsule design.

Teams from Aquea Scientific Corporation[12] patented a bodywash formulation containing the sol-gel UV absorber microcapsule coated with a cationic component. This bodywash formulation imparts sun protection to skin with an SPF of at least 2 after a single application of the bodywash to skin followed by rinsing. Examples of the cationic component in the patent include mostly cationic polymers such as polyquaternium-4. For the encapsulation of low-solubility organic UV filters, Andre et al. from Merck GmbH[13] reported the use of an emollient that can help to dissolve more than 40% of the active at room temperature in the microencapsulation process. Examples of low-solubility organic UV filters are a triazine derivative, a diarylbutadiene derivative, a hydroxybenzophenone derivative, and/or a methyleneb isbenzotriazolyl-tetramethylbutylphenol derivative. The emollient disclosed in the patent application was N,N-dimethyldecanamide.

Polymer chemistry plays a vital role in the design of innovative microencapsulation technology. While interfacial and *in situ* polymerizations are widely used as chemical routes to manufacture microcapsules, other heterophase polymerization technologies, including emulsion, mini-emulsion,[14] dispersion, and suspension polymerization processes, can be modified and adapted to the encapsulation of liquid or low–melting point bioactive agents. While particle sizes obtained from a suspension process are in the micron range, the particle size from a mini-emulsion process is in the submicron range.[15] In a typical mini-emulsion process, oil phase ingredients, including bioactive agents, monomers, initiator, and ultrahydrophobes, are thoroughly mixed and then emulsified in an aqueous phase containing the appropriate amount and type of surfactant. The ultrahydrophobe is used as an osmotic pressure agent to prevent Ostwald ripening.[16] The particle size of the oil-in-water emulsion is controlled by applying a high-shear homogenization step (either a high-pressure or an ultrasonic device). Polymerization occurs in these droplets, also termed *nanoreactors*, to form submicron particles as shown in Figure 12.3.

Mini-emulsion encapsulation can be achieved by choosing monomers that are miscible with the active materials, while the resulting polymers are not miscible with the actives. Before the start of polymerization, the active material and monomers are dispersed in the oil phase. After the polymerization starts, the newly formed polymer is no longer compatible with the active material, and phase separation

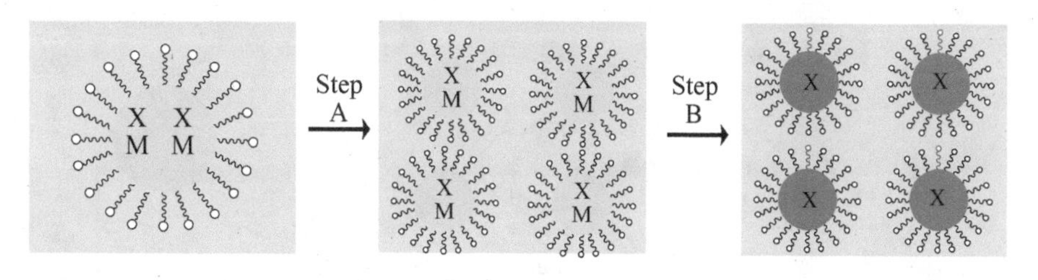

FIGURE 12.3 Mini-emulsion encapsulation. X: active material; M: monomers; Step A: high shear homogenization; Step B: polymerization.

occurs, resulting in the formation of a capsule containing the active material. A variety of polymer chemistries have been used in this process, including free radical polymerization and polycondensation.[17] Hereuchokeu et al. from BASF[18] reported the mini-emulsion encapsulation of UV absorbers with average particle size of less than 1 μm. The process consists of dissolving, emulsifying, and homogenizing the oil-soluble organic UV absorber in at least one ethylenically unsaturated monomer followed by polymerization with a redox initiator pair. In one example, ethylenically unsaturated monomers include methyl methacrylate (MMA), stearyl methacrylate (SMA), and a small amount of butanedioldiacrylate. The UV absorbers in one example are the mixture of ethylhexyl methoxycinnnamate/bis-ethylhexyloxy phenol methoxyphenyl triazine/benzotriazolyl dodecyl p-cresol in a 65/10/25 ratio. The redox initiator pair in one example is ascorbic acid and hydrogen peroxide. The *in vitro* SPF studies show that sunscreen formulations containing encapsulated organic UV absorbers have higher *in vitro* SPF than counterparts containing unencapsulated organic UV absorbers for various combinations of UV absorbers.

Independently, Jones and Wills from Dow Chemical Company (formerly Rohm and Haas)[19] reported the encapsulation of UV absorbers with even smaller average particle size range via mini-emulsion polymerization. The shell is highly cross-linked and prepared by free radical polymerization of a monomer mixture containing 10–100 wt% cross-linkers. Examples of cross-linkers include allyl methacrylate, allyl acrylate, divinylbenzene, and trimethylolpropane trimethacrylate. In one example, the encapsulated particle containing homosalate and avobenzone was shown to provide a significant *in vitro* SPF and UVA absorption boost in comparison to the control sample without encapsulation.

Encapsulation processes from interfacial and *in situ* polymerizations have been the incumbent technology for various active ingredients, but they tend to provide capsules in the micron size range.[20] One way to reduce the particle size is through a high–internal phase emulsion process developed by the Dow Chemical Company, as reported by Fletcher et al.[21] In this process, hydrophobic oil, water, and surfactant are fed through a rotor stator under heated conditions. Subsequently, the obtained emulsion is diluted with water to give a final emulsion with reasonable viscosity, as shown in Figure 12.4. Particle sizes close to 1 micron can be achieved with this novel process. In this process, wall-forming materials can be introduced at both the oil tank and the dilution water tank to form a shell encircling the hydrophobic oil. The wall-forming material in the example is a polyurea shell formed by interfacial reaction of a multifunctional isocyanate and a multifunctional amine. The UV absorber capsule described in this patent application, when incorporated in a personal care composition, provides multiple benefits to consumers, such as improving aesthetics, ease of spread, improving adsorption onto skin, reduced tackiness, and reduced greasiness.

A different approach to manufacture submicron encapsulation particles through interfacial polymerization was reported by Zeng et al at the Dow Chemical Company.[22] As described in the patent application, the aqueous phase contains partially hydrolyzed polyvinyl alcohol and a cosmetically acceptable non-ionic surfactant, while the personal care active includes UV absorbers such as avobenzone, homosalate, octocrylene, octyl methoxycinnamate, etc. Examples of the non-ionic surfactant include fatty alcohol ethoxylates and alkoxylates of fatty acid esters. A polyurea shell formed by interfacial reaction of a multifunctional isocyanate and a multifunctional amine was used in the example. It is also worthy of mention that the homogenizing step can be carried out under ambient pressure. In comparison with the

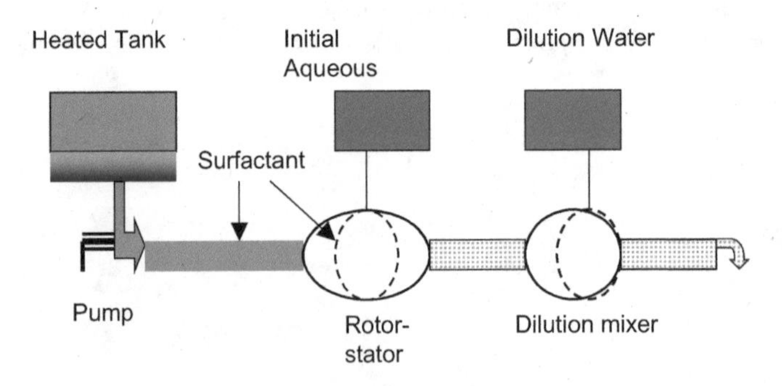

FIGURE 12.4 Encapsulation through high–internal phase emulsion.

micron-sized UV absorber capsule, the submicron UV absorber capsule provides an even better *in vitro* SPF boost in sunscreen formulation.

The manufacture of micron-size polymer composite particles containing UV absorbers through suspension polymerization was reported by Lee et al from Sunjin Chemical.[23] The process included the following steps: (1) dissolving and dispersing a sunscreen agent in a free radical polymerizable monomer or a multifunctional cross-linking monomer; (2) emulsifying the mixture from step 1 with homogenizer in a solvent under the presence of a dispersion stabilizer to form droplets; (3) polymerizing the droplets from step 2 with initiator to manufacture polymer composite particles; and (4) cleaning and drying the manufactured polymer composite particles. In one example, the monomer is methylmethacrylate, the cross-linking monomer is ethyleneglycol dimethacrylate, the initiator is 2,2′-azobis-2,4-dimethylvaleronitrile, the dispersion stabilizer is polyvinyl alcohol, and the UV absorber is avobenzone. Table 12.4 lists encapsulation products of UV absorbers commercially available in the market with company and trade names, the active ingredients, and shell chemistry.

The encapsulation of UV absorbers also receives significant attention from the academic community. Weiss-Angeli et al.[24] reported the preparation of nanocapsules of octyl methoxycinnamate and quercetin (3,3',4',5,7-pentahydroxyflavone dehydrate) through the interfacial deposition of polymer approach. Quercetin acts as a reactive oxygen species (ROS scavenger) and helps to further stabilize the UV absorber. Under UVA irradiation, co-nanoencapsulation provides better stabilization than the nanoencapsulation of the individual components. do Nascimento et al.[25] reported the preparation of poly(epsilon-caprolactone) nanoparticles of UV absorbers by the solvent emulsification and evaporation method. The UV absorbers in this study included 2-ethylhexyl-p-methoxycinnamate, benzophenone-3, and octocrylene. The nanoparticles were characterized extensively for their particle size, polydispersity index, morphology, zeta potential, encapsulation efficiency, and UV absorber content. Also, the safety, photostability, and performance of these nanoparticles were evaluated.

12.4.2 Microcapsules and Microparticles for Exfoliation, Aesthetics, and Consumer Benefits

Microcapsules and microparticles, as shown in Figure 12.5, are used in a wide range of personal care cleansing products, such as body wash, facial wash, facial scrubs, and hand soap formulations. Their functions include exfoliation, creating attractive visual effects in products, and active delivery. The active ingredients either inside the core or in the particle can include a mixture of natural oils, synthetic oils, vitamins, pigments, and dyes. The shell or the carrier materials can be either natural, naturally derived, or synthetic materials. Examples of natural and naturally derived materials include cellulose, hydroxypropyl methylcellulose, gelatin, gum, alginate, agar, carrageenan, chitosan, and jojoba esters (wax). Pigments and dyes are usually added along with active ingredients. Therefore, these microcapsules and microparticles can be supplied in a wide range of colors. Also, most of these products have particles that are large enough to be visible. Therefore, they provide aesthetic appeal to consumers. Furthermore, these particles are hard enough that they can function as abrasive/exfoliation agents when

TABLE 12.4

Selected Commercial Encapsulated UV Absorbers

Company	Trade Name	Active	Shell
EMD Chemicals	Eusolex UV-Pearls™	Ethylhexyl methoxycinnamate , ethylhexyl salicylate	Silica
BASF	Tinosorb® S Aqua	Bis-Ethylhexyloxyphenol methoxyphenyl triazine	pMMA
Seiwa Kasei Co., Ltd	Silasoma	Ethylhexyl methoxycinnamate, butyl methoxydibenzoylmethane	Silica/peptide
Tagra Biotechnologies Ltd	SunCaps™	Avobenzone, octyl methoxycinnamate, octisalate, homosalate	N/A
Sunjin Chemical	Hybrid PMMA Series	Avobenzone or bis-ethylhexyloxyphenol methoxyphenyl triazine	pMMA

FIGURE 12.5 Microparticles for cleansing applications. (Adapted from www.cosmeticsdesign-europe.com/.)

they contact the skin. Yet, these particles tend to be soft enough to be crushable. Gentle scrubbing and rubbing can deform these particles. This results in the release of actives that provide biological and/or sensorial benefits to consumers. Table 12.5 lists a selection of microcapsules and microparticles available in the markets.

The manufacture of cellulose microparticles was recently disclosed by Schweikert et al. from Induchem.[26] The process involves the following steps: (1) mixing the majority of the cellulose with a small amount of hydroxypropyl methylcellulose and lipo-soluble active ingredients; (2) adding water to this blend to form a wet granulate; (3) drying the wet granulate in a fluid bed dryer; (4) sieving the final product to the desirable particle size range. It is worth noting that the resulting cellulose microparticles exhibit and maintain exfoliating properties in body care compositions.

Novel microbeads based on partially hydrogenated triglyceride oils were reported by Kleiman et al. from International Flora Technologies, Ltd.[27] When more than 50% partially hydrogenated triglyceride oils, which contain more than 15% by weight of >C18 fatty acid moieties, are used in the composition, the resulting microbeads are hard particles as supplied but can transform into soft particles in personal care formulations. Potentially, the particle hardness is dictated by the melting point of the triglyceride, which itself is dependent on the carbon chain length distribution. Cosmetic additives such as pigments can be entrapped into the particles during manufacturing, which could involve melting the wax/pigment mixture followed by cooling and granulation. During the topical application, the soft particles can easily deform and release the cosmetic additive, while the triglyceride itself delivers an emollient benefit to the skin and hair.

Due to the recent scrutiny of microbeads in personal care products, there is a strong desire for the development of biodegradable alternatives. For example, Havens et al. from Virginia Institute of Marine Science[28] disclosed personal care formulations containing biodegradable polyhydroxyalkanoate (PHA)

TABLE 12.5

Selected Microcapsules and Microparticles for Exfoliation, Aesthetics, and Consumer Benefits

Supplier	Trade Name	Active	Shell	Size (μm)
Induchem	Unispheres® XS	Vitamin E (1%)	Microspheres of mannitol, cellulose, hydroxypropyl methylcellulose	200–500
Floratech	Florabeads	Pigment	Microspheres of jojoba esters (wax)	150–600
Ashland	Captivates HC encapsulates series	Blends of natural oils, synthetic oils, and many oil-soluble actives, range of pigments and dyes	Complex coacervate based on naturally derived polymers such as gelatin and gum	10–2000
Ashland	Captivates GL encapsulates series	Pigments, clays, oils, insoluble vitamins, proteins	Alginate, agar, carrageenan, or chitosan	250–3000
Vantage	Lipospheres	Customized products	N/A	N/A

microbeads. The selection of PHA polymers includes poly-3-hydroxybutyrate, poly-4-hydroxybutyrate, polyhydroxyvalerate, poly-3-hydroxyhexanoate, and their copolymers.

12.4.3 Color-Changing Pigment through Microencapsulation

Recently, microcapsules containing pigments[29] have been adopted in cosmetic and skin care formulations, especially in BB (blemish balm) and CC (color correction) creams. The pigment, for example, iron oxide, is effectively encapsulated inside the polymeric shell, and its original color is concealed. The resulting microcapsules tend to have a white or light looking color. Therefore, ultra-light color cosmetic products containing iron oxide can be obtained. On rubbing in as topical application by consumers, the controlled release of the pigments provides uniform color development to correct, even, and perfect the skin tone. Figure 12.6 illustrates the pigment microcapsules as supplied, as formulated in personal care products, and their release on application. Furthermore, other pigment-incompatible cosmetic ingredients can be introduced to formulate all-in-one facial cosmetic products that offer benefits such as serum replacement, moisturization, primer, foundation, and SPF protection. The pigment microcapsules also help to simplify the formulation process, as the products can be introduced to the formulation at the end of the manufacturing process without the need of homogenization and/or adding heat. They also minimize the risk of pigment aggregation even in the absence of a stabilizing agent. Table 12.6 lists selected commercially available microencapsulation products containing pigments.

Kvitnitsky et al. from Tagra Biotechnologies Ltd. have disclosed their process for making pigment microcapsules in their patent application.[30] The process is adapted from the solvent removal method that has been used for other microencapsulations.[31] This method is applicable for the encapsulation of oil-soluble and oil-dispersible substances that have limited water solubility. The process involves the following steps: (1) preparing an organic solution or dispersion containing the pigment, wall-forming polymer, and an organic solvent that is partially miscible with water and can dissolve or disperse the pigment and wall-forming polymer; (2) preparing an aqueous continuous phase saturated with the organic solvent and an emulsifier; (3) adding 1 to 2 with mixing to form an emulsion; (4) adding an excess amount of water to the emulsion from step 3 to initiate extraction of the organic solvent from the emulsion, and

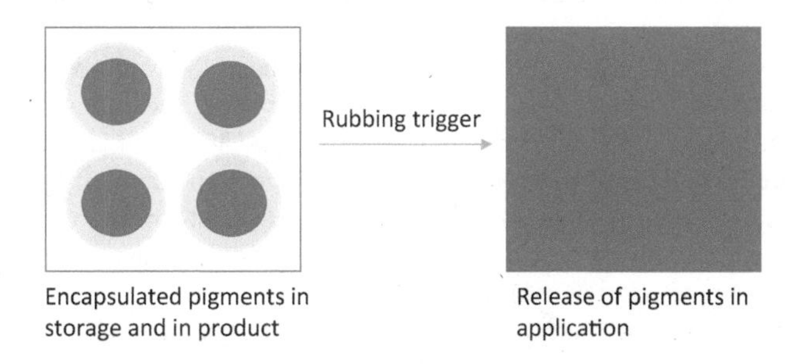

FIGURE 12.6 (a) Pigment microcapsules in as-produced commercial product and in a personal care product. (b) Release on application.

TABLE 12.6

Selected Microencapsulation Products Containing Pigments

Supplier	Trade Name	Active	Shell
Tagra Biotechnologies Ltd	TagraCaps™ series	Pigment combination such as iron oxide, titanium dioxide	Acrylates/ammonium methacrylate copolymer
Biogenics	Magicolor™ series	Iron oxide	Polyester, polyvinyl pyrrolidone, hydroxypropyl methylcellulose, shellac
Induchem	Unisphere® bicolor series	Matt white (coated), pigmented core bead (up to 50% pigment)	Mannitol, cellulose, hydroxypropyl methylcellulose

continuing the extraction by incubating the solvent, thus promoting the formation of solid single-layer microcapsules; (5) isolating microcapsules, following by washing with water and drying them, to form single-layer microcapsules. Multilayer microcapsules can be obtained by repeating step 4 or by carrying out further surface modifications. The organic solvent used in the procedure is ethyl acetate. The wall-forming polymer can be selected from a polyacrylate, a polymethacrylate, a low–molecular weight poly(methyl methacrylate)-co-(methacrylic acid), poly(ethyl acrylate)-co-(methyl methacrylate)-co-(trimethylammmonium-ethyl methacrylate chloride), poly(butyl methacrylate)-co-(2-dimethylaminoethyl methacrylate)-co-(methyl methacrylate), poly(styrene)-co-(maleic anhydride), copolymer of octylacrylamide, cellulose ethers, cellulose esters, and poly(ethylene glycol)-block-poly(propylene glycol)-block-poly(ethylene glycol).

Independently, Kim et al. from BioGenic Inc. disclosed an alternative process to make pigment microcapsules.[32] These pigment microcapsules have onion-type morphology. The core particles contain color pigment residing in the core, with the first polymer shell encapsulating the color pigment. The core particles are further encapsulated with an outer shell, which itself contains an inner shell containing the functional pigment and outer polymer layer. The color pigment includes mostly iron oxide with different colors, but other organic or natural pigments can be used. The functional pigment can be selected from titanium dioxide, zinc oxide, boron nitride, talc, mica, and their mixtures. The polymer used in the examples includes polyester, a polyester emulsion, polyaminomethacrylate, polyvinylpyrrolidone, hydroxypropylmethylcellulose, and shellac. Plasticizers such as 1,3-butanediol, polyethylene glycol, dipropylene glycol, or their mixtures are present in the core particles. This helps to easily break the microcapsules in the consumer end use application. Once the capsule is rubbed into the skin, the color of the iron oxide appears and develops a match to the user's natural skin tone, while the functional pigment also provides gloss, lubricity, and sensorial feel. It is very interesting to note that spray-drying is the main process step besides dissolution, mixing, and sieving. This helps to eliminate the waste water generation that is encountered in the solvent removal method.

A similar double-layer pigment microcapsule was reported by Yayoi.[33] A combination of materials that have different hardness and solubility in water is used as a shell component, so that the time required for pigment release can be optimized to deliver the target color formation and gradation pattern. These various shell materials include mannitol, hydrogenated lecithin, poly(methyl methacrylate), cellulose, and shellac.

Benaltabet et al. from Tagra Biotechnologies Ltd have extended their microencapsulation of pigment to carbon black.[34] Since carbon black is lighter and has more color effect at low use concentrations, the resulting encapsulated product may find use particularly in mascaras with less clumping, less beading, less buildup, better lash separation, and better lash curling. Since the end use application requires that the black color needs to be manifested without the rubbing release, the research team was able to meet these criteria by eliminating the use of plasticizer in the process. The plasticizer can potentially soften the shell and cause the pigment to release prematurely.

Innovative cosmetic formulations containing color-changing microcapsules have been disclosed extensively by researchers from L'Oreal. For example, Chai et al. disclosed a color-changing composition in the form of an oil/water emulsion.[35] Lemoine and Chu disclosed color-changing compositions that also contain a UV absorber.[36] Lemoine et al. disclosed a cosmetic gel product containing color-changing microcapsules, a hydrophilic gelling agent, and additional cosmetic ingredients such as volatile and non-volatile silicon or hydrocarbon oils, film-forming agents, silicone elastomer, and self-tanning agents.[37] Kawamoto disclosed a color-changing cosmetic product in a foam form where a propellant can be used in one compartment to pressurize and deliver the product as a foam.[38] Furthermore, Shimizu and Yu[39] from L'Oreal expand color-changing microcapsule technology to the microencapsulation of reflective particles and its use in cosmetic products. Examples of the reflective particles include inorganic particles coated with metallic (poly)oxide(s), such as mica coated with titanium dioxide or iron oxide.

Besides skin care applications, color-changing microcapsules also find use in other personal care products. One example is to incorporate color-changing microcapsules into cleansing products, as reported by Lachmann et al. from Clariant.[40] During the washing process, the shell decomposes, and the colorants are released into the foam, inducing a color change of the foam. This color change can be used as an indicator to the consumers that sufficient scrubbing or cleansing has been achieved to meet the target hygiene standard, while a desirable amount of antibacterial agent has been delivered to the hands of consumers. The microencapsulated colorant granules as described have a core/shell morphology. The core comprises a colorant, microcrystalline cellulose, and a polyol, while the shell comprises a polymer selected from the group consisting of polycarboxylic acids, copolymers of polycarboxylic acids, vinyl polymers, cellulose, and cellulose derivatives as well as a white pigment that can conceal the colorant in the core.

12.4.4 Hollow Sphere Polymers

Hollow sphere particles have been conventionally used in various industries as light-weight fillers.[41] However, when uniform hollow sphere polymers (HSP) are used, where an air void core is encapsulated inside the polymer shell, unique optical properties result from scattering of radiation from the air–polymer interface. Dow Chemical Company is a pioneer in the development of HSP technology for a variety of industrial applications, ranging from architectural coatings for enhanced hiding, to paper coating for high gloss, to sunscreen formulations with enhanced UV protection. HSP with a narrow particle size distribution, with the size range from 50 nm to more than 1 μm, can be prepared through multiple steps of emulsion polymerization as described by Chang et al.[42] Due to the refractive index difference between the polymer shell and air, UV radiation, passing through HSP, will be refracted or scattered. Therefore, the presence of HSP in the sunscreen formulation increases the scattering efficiency of UV radiation. This increases the effective path length for UV rays as they pass through the sunscreen layer on the skin. Also, the likelihood that the radiation will be absorbed by the UV active is also increased. Both effects result in boosting the SPF value for sunscreen formulation, as predicted by Beer's Law. Figure 12.7 shows the refractive index contrast of shell and air for HSP and the mechanism of SPF boost. Figure 12.8 shows the SPF boost for sunscreen formulations containing HSP in either the aqueous or the

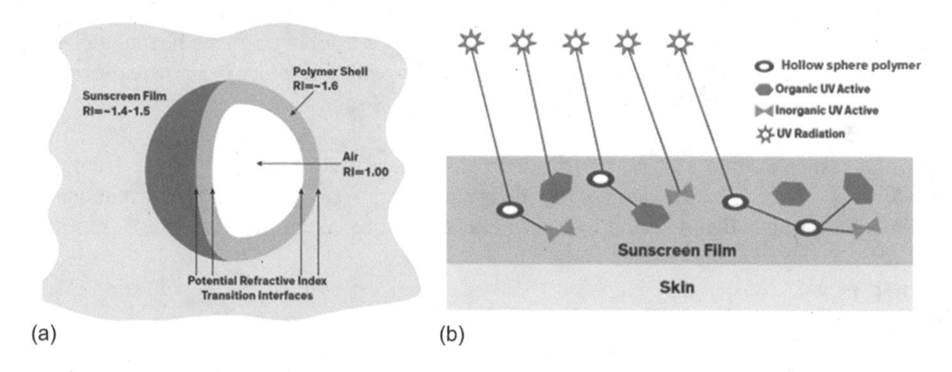

FIGURE 12.7 (a) Structure of HSP particle and (b) mechanism of SPF boost by HSP.

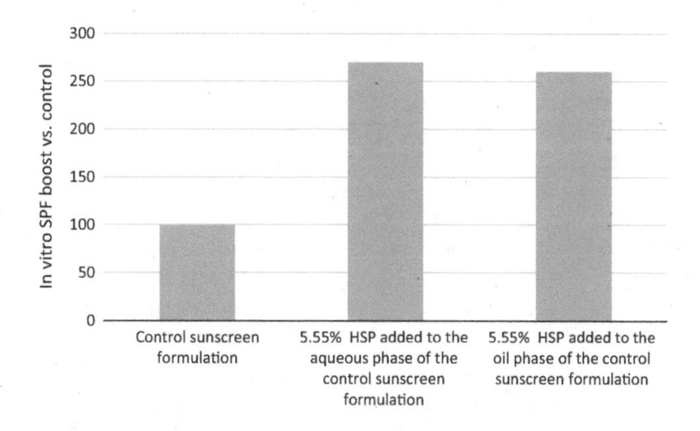

FIGURE 12.8 *In vitro* SPF boost of sunscreen formulations containing HSP versus control.

oil phase versus the control formulation without HSP. Alternatively, the addition of HSP to an existing formulation allows the formulator to use significantly less UV active to deliver the same level of protection. As an extra benefit, the ability to reduce UV active levels in the formulation can also help to reduce skin irritation potential. Table 12.7 lists the HSP products for enhancing SPF performance in sunscreen formulations.

12.4.5 Microspheres and Microabsorbents

A wide range of microsphere particles have been used in cosmetic formulations.[43] They are usually supplied as a free-flowing powder with a well-defined and narrow particle size distribution. Due to their micron particle size range and shape, microspheres offer enhanced sensory properties for a wide range of personal care products. They can scatter visible light to diminish the appearance of fine lines and wrinkles on the skin, creating a soft focus or optical blurring effect. Furthermore, they provide a "ball-bearing" effect, which imparts to personal care products an elegant silky texture, enhanced slip, and the ability to promote a more natural skin tone. Finally, some microspheres also absorb sebum, which helps to reduce the perceived skin oiliness and leaves a dry feel on skin. Table 12.8 lists selected microsphere products used in cosmetic formulations from various suppliers with a wide range of chemistry.

One special class of microparticles can be classified as polymeric microabsorbents. These microabsorbents are highly cross-linked polymer particles that contain a continuous non-collapsible network of pores from the inside of the particle to the surface. The particle size of these microabsorbents ranges from close to 1 micron to up to 100 microns. Due to their unit structure, they have high pore volume, high surface area, and low apparent bulk density. They function as a reservoir to absorb, retain, and deliver actives for both pharmaceutical and cosmetic applications. The release of active ingredients occurs through a combination of friction and diffusion, providing a substantially slowed release of the active. The slowing of release of actives helps minimize irritation while providing actives available for extended effectiveness. Table 12.9 lists selected microabsorbent particles used in cosmetic formulations, all from Amcol Health and Beauty Solutions. They are commercially available either in a powder form or as pre-loaded forms with a range of cosmetic actives such as retinol, salicylic acid, glycerin, cyclomethicone, and benzoyl peroxide.

One of the earlier examples of polymeric microabsorbents was disclosed by Katz from Advanced Polymer Systems, Inc.[44] In this example, the microabsorbents were prepared by conventional suspension polymerization in a liquid–liquid system. First, a mixture containing monomers, cross-linker, a

TABLE 12.7

HSP for Enhancing SPF Performance in Sunscreen Formulation

Supplier	Trade Name	Core	Shell
Dow Chemical Company	SunSpheres™ PGL, LCG, Powder	Void	Styrene/acrylic copolymer

TABLE 12.8

Selected Microsphere Products Used in Cosmetic Formulations

Supplier	Trade Name	INCI[a] Name
Dow Corning	Dow Corning® 9701 Cosmetic Powder	Dimethicone/vinyl dimethicone crosspolymer (and) silica
Momentive Performance Materials	Tospearl Microspheres	Polymethylsilsesquioxane
Presperse	Luxsil Cosmetic Microspheres	Calcium aluminum borosilicate
Akzo Nobel	DRY-FLO® TS	Tapioca starch polymethylsilsesquioxane
Kobo Products	MSP-825	Methyl methacrylate crosspolymer
	SP-10	Nylon-12
	CARESS® series	Boron nitride with various particle size (PS)

[a] INCI: International Nomenclature of Cosmetic Ingredients.

TABLE 12.9

Selected Microabsorbent Particles Used in Cosmetic Formulations

Supplier	Trade Name	Active	Polymer
AMCOL Health & Beauty Solutions	Poly-Pore® series	Retinol, salicylic acid, glycerin, cyclomethicone, benzoyl peroxide	Allyl methacrylates crosspolymer
	Microsponge®		Methyl methacrylate/glycol dimethacrylate crosspolymer
	Polytrap®		Lauryl methacrylate/glycol dimethacrylate crosspolymer

polymerization catalyst, and a suitable porogen, which is fully miscible with monomers but immiscible with water, was prepared. Then, the solution was suspended in an aqueous solution, which contained surfactants and dispersants to aid the suspension. Once the suspension was established with suitable mixing to achieve discrete droplets of the desired size, polymerization was initiated. On the completion of polymerization, the resulting rigid but porous microbeads were recovered from the suspension. Then, the porogen was removed through washing and drying. Finally, the cosmetic active was added and retained inside the microabsorbents through capillary force. Examples of polymer materials include cross-linked copolymers of styrene and divinylbenzene, methyl methacrylate and ethylene glycol dimethacrylate, and 4-vinylpyridine and ethylene glycol dimethacrylate.

Microabsorbents prepared by the suspension polymerization process were also disclosed by Sojka from Amcol International Corp.[45] Examples of monomer mixtures to make these microabsorbents include butyl methacrylate, allyl methacrylate, and/or ethylene glycol dimethacrylate. One example cosmetic active disclosed is benzoyl peroxide, which is used in the treatment of acne. In comparison with conventional formulations, the advantages of this microabsorbent delivery system include sustained release of active with a continuous fresh supply, protection of the active from oxidation decomposition, reduced irritancy, and reduction of frequency of applying the skin formulation to the affected area. Finally, as the benzoyl peroxide is released from the carrier particles, the natural skin oils such as sebaceous fluids are absorbed into microabsorbents. This exchange of benzoyl peroxide for natural skin oils further improves the effectiveness of this delivery system for acne treatment, since the presence of skin oils is a strong contributing factor to acne production, and provides a better sensory and dry feel to consumers.

Sojka from Amcol International Corp.[46] further disclosed a precipitation polymerization process to make polymeric microabsorbents with high oil adsorbency. The precipitation process comprises the following steps: (1) dissolving monomer, cross-linkers, and organic polymerization initiator in an inert water-immiscible organic solvent such as a silicone to form a monomer mixture; (2) carefully mixing this mixture to prepare suspended microdroplets, which can be converted via heat-triggered radical polymerization to produce microporous particles, and their aggregates; (3) separating the microporous particle/aggregates from the organic solvent to produce a product in the solid form on drying in a vacuum oven. The unit microparticles have a diameter in the range from 0.1 to about 80 microns, while the aggregates, or assemblies of microparticles, have diameters from about 1 to about 500 μm. The product has an extremely low bulk density at about 0.03 to 0.06 g/cm^3. Examples of cross-linkers include monoethylene glycol dimethacrylate, diethylene glycol dimethacrylate, triethylene glycol dimethacrylate, allyl methacrylate, and their mixtures. The oil adsorption capacity was measured by the addition of incremental amounts of hydrophobic liquid to a known amount of powder, with gentle mixing, until the powder was no longer free flowing and yet adsorbing liquid. Examples of hydrophobic liquids include mineral oil, artificial sebum, glycerin, cyclomethicone, dimethicone, vitamin E acetate, benzophenone-3, fragrance, and others. In each case, the microabsorbent can absorb at least 7.5 times the weight of oil per weight of microabsorbent.

Finally, an improved controlled release system of this microabsorbent was disclosed by Sojka from Amcol International Corp.[47] Besides the microabsorbent particles and the oil-soluble topically active compound, the system also contains a water-soluble release retardant, which is coated and adsorbed onto the microadsorbent and active compound complex. In one example, the controlled release profile of salicylic acid was substantially improved by adding an increasing amount of stearyl alcohol as a release retardant to the microabsorbent–salicylic acid combination. This improvement is attributed to the desirable properties of stearyl alcohol, which is a solid at room temperature and has a good balance of hydrophobic/hydrophilic nature.

12.4.6 Lipid-Based Delivery Systems for Cosmetic Applications

Lipid-based delivery systems have been adapted for cosmetic applications to deliver actives such as antioxidants. These can be classified into three major systems:

1) Emulsion systems such as oil-in-water (o/w) emulsions, multiple (w/o/w) emulsions, microemulsions, and liquid crystal-forming emulsions
2) Vesicular systems such as liposomes, niosomes, and nanotopes
3) Lipid particulate systems

While extensive reviews are available for emulsion[48] and vesicular systems,[49] this section focuses on lipid particulate systems, particularly on solid lipid nanoparticles (SLN) and nanostructured lipid carriers (NLC) as pioneered by R.H. Müller and his co-workers.[50] SLN are similar to emulsion systems except that lipids used in these systems remain solid at both room and body temperature. In a typical procedure, SLN can be prepared by (1) dissolving or dispersing actives in the melted lipid; (2) dispersing the mixture from step 1 in a hot surfactant aqueous solution to form a pre-emulsion; (3) passing the pre-emulsion from step 2 through a high-pressure homogenizer, followed by cooling to form SLN. Depending on the formulation compositions and process conditions, there are three ways that the actives can be incorporated and distributed into SLN: homogeneously throughout the solid lipid matrix, in an active-enriched shell, or in an active-enriched core. The degree of stabilization of actives can depend on the optimal selection of lipid and surfactant as well as particle size of SLN. In one study, retinol-containing SLN had the highest retinol stabilization effect in the smallest particles, at 0.9 μm. The typical release profile of active from SLN is shown to be biphasic—an initial burst release followed by extended release. But, the relative ratio of burst versus extended release is shown to correlate mostly with the process temperature and amount of the surfactant. In one study, both low process temperature and low surfactant level suppressed the water solubility of the active and enhanced active incorporation inside the lipid matrix. This led to little or no burst release. Despite the advantages of SLN for active stabilization and prolonged release, one limitation is the potential exclusion of active during storage, where high-energy, imperfect solid lipid morphology transitions into a highly ordered low-energy liquid crystal state. To overcome this limitation, a blend of solid and liquid lipids was used to form NLC. In comparison with the relatively ordered structure of SLN, the lipid matrix in NLC can be designed to exist as either an imperfect crystal form, an amorphous form, or a multiple form with both solid and liquid compartments. All these irregular morphologies help NLC to minimize active exclusion. SLN and NLC are commercially available from PharmaSol with the trade name of Nanopearls. Products can be supplied as a cream-like dispersion, with solids up to 35–45%, or a gel form by the addition of a gelling agent. They can be incorporated into creams or lotions, preferably at the end of production, especially if the actives are temperature sensitive.

12.5 The Future

One of the mega-trends in personal care is the demand for user-friendly products that can carry out multiple functions in a single application. This is well demonstrated in all-in-one BB products that include functions such as serum replacement, moisturization, primer, foundation, and SPF protection. Novel encapsulation technologies will continue to provide solutions for addressing the challenges raised by the formulation complexity and potential incompatibility of active materials in such products. Furthermore, smart delivery systems for active ingredients will be a key growth driver for companies aiming to innovate and differentiate themselves in the personal care market. However, success will be predicated on intimate collaborations between innovative raw material suppliers and cosmetic product companies. Two immediate innovation interests in this area are novel delivery systems that meet the increasingly strict regulatory requirements and have more user-friendly release triggers. For example, when toxic materials such as formaldehyde[51] or isocyanate[52] are involved in encapsulation processes, the residual levels in the final product must be at very low or undetectable levels. Some of these issues are being addressed by both academic and industrial innovators. For example, novel formaldehyde-free core/shell microcapsules have been developed based on di-aldehyde and polyhydroxy compounds.[53] One example of a novel release mechanism[54] involves the use of skin's natural pH to trigger the controlled release of sensitive active ingredients when a pH-sensitive methacrylate copolymer[55] was employed for encapsulation. Another example for active release involves the shell degradation triggered by enzymes available from the skin itself.[56] Overall, one can envision that more robust and innovative controlled release technologies that can meet tougher success criteria will be commercially available in the near future. This in turn will allow formulators to launch innovative personal care products that deliver health benefits along with indulgence to the consumers.

REFERENCES

1. Schlossman ML (2002) *Chemistry and Manufacture of Cosmetics* (3rd edn, Volume I–III). Allured Publishing Corp., Carol Stream, IL.
2. Srinivasa R (2013) Global markets for chemicals for cosmetics & toiletries. BCC Research, January 2013.
3. Rosen MR (ed.) (2005) *Delivery System Handbook for Personal Care and Cosmetic Products – Technology, Applications and Formulations*. William Andrew Publishing, New York.
4. Zeng F, Shah N (2014) Physico-chemical challenges of designing delivery systems for consumer and industrial applications. Controlled Release Society Annul Meeting, July 2014.
5. Hewitt JP (2009) Clinical guide to sunscreens and photoprotection. *Basic and Clinical Dermatology* 43:155–168.
6. Avnir D et al (2002) Method for obtaining photostable sunscreen compositions. US6436375B1, 20 Aug 2002.
7. Avnir D et al (2002) Sunscreen composition containing sol-gel microcapsules. US6468509B2, 22 Oct 2002.
8. Frank P et al (2007), Sunscreen composition. US7264795, 4 Sep 2007.
9. Avnir D et al (2001) Sunscreen composition containing sol-gel microcapsules. US6238650B1, 29 May 2001.
10. Yoshioka M et al (2010) Cosmetics with blended micro capsule. JP04521792B2, 11 Aug 2010.
11. Berg-Schultz K (2014) Microcapsules with UV filter activity and process for producing them. US 8765967 B2, 1 Jul 2014.
12. Compton DL et al (2006) Sunscreen compositions and methods of use. US7001592B1, 21 Feb 2006.
13. Andre V et al (2010) UV filter capsule. US20100209463A1, 19 Aug 2010.
14. Asua JM (2002) *Progress in Polymer Science* 27:1283–1346;
    Shork F (2005) *Advances in Polymer Science* 175:129;
    Landfester K (2006) *Annual Review of Materials Research* 36:231;
    Landfester K (2009) *Angewandte Chemie International Edition* 48:4488.
15. Zeng F et al (2012) Advances of sub-micron particle technologies for consumer and industrial applications. Controlled Release Society Annul Meeting, July 2012.
16. Ugelstad J et al (1974) *Makromolekulare Chemie* 175:507;
    Davies SS (1976) In: Smith AL (ed.) *Theory and Practice of Emulsion Technology* (p. 325). Academic Press, New York.
17. Landfester, K (2006). In: Ghosh SK (ed.) *Functional Coatings by Polymer Microencapsulation* (p. 29). Wiley-VCH, Weinheim.
18. Hereuchokeu B et al (2010) Water based concentrated product forms of oil-soluble organic UV absorbers. US20100284950A1, 11 Nov 2010.
19. Jones CE, Wills MC (2009) Particle containing ultraviolet absorber. US20090311336A1, 17 Dec 2009.

20. Simon B (1996) *Microencapsulation: Methods and Industrial Application.* Marcel Dekker Inc. New York.
21. Fletcher R et al (2009) Encapsulated hydrophobic actives via interfacial polymerization. WO2009091726A1, 23 Jul 2009.
22. Zeng F et al (2013) Encapsulation of personal care actives. WO2013059167A1, 30 May 2014.
23. Lee S-H et al (2009) Polymer composite particles containing sunscreen agent and manufacturing method thereof. US20090053153A1, 26 Feb 2009.
24. Weiss-Angeli V et al (2008) Nanocapsules of octyl methoxycinnamate containing delayed the photodegradation of both components under ultraviolet A radiation. *Journal of Biomedical Nanotechnology* 4(1):80–89.
25. do Nascimento DF et al (2012) Characterization and evaluation of poly(epsilon-caprolactone) nanoparticles containing 2-ethylhexyl-p-methoxycinnamate, octocrylene, and benzophenone-3 in anti-solar preparations. *Journal of Nanoscience and Nanotechnology* 12(9):7155–7166.
26. Schweikert K et al (2015) Exfoliating cellulose beads and cosmetic uses thereof. EP 2907498 A1, 19 Aug 2015.
27. Kleiman R et al (2013) Cosmetic particles that transform from hard to soft particles comprising hydrogenated long-chain triglyceride oils. US 8613956, 24 Dec 2013.
28. Havens KJ et al (2014) Method for reducing marine pollution using polyhydroxyalkanoate microbeads. US20140026916 A1, 30 Jan 2014.
29. Jones, SR et al (2009) The development of colour-encapsulated microspheres for novel colour cosmetics. *Journal of Microencapsulation* 26(4):325–333 (CODEN: JOMIEF; ISSN: 0265–2048).
30. Kvitnitsky E et al (2010) Compositions for topical application comprising microencapsulated colorants. WO 2009138978 A2, 15 July 2010.
31. Babtsov V et al (2005) Method of microencapsulation. US 6932984 B1, 23 Aug 2005.
32. Kim C-H et al (2013) Color capsule composition for cosmetics, preparation method thereof and cosmetic formulation comprising the same. US 8834906 B2, 16 Sep 2014.
33. Yayoi O (2011) Pigment-encapsulating microcapsule and cosmetic obtained by formulating the same. JP2011079804A.
34. Benaltabet L et al (2012) Microcapsules comprising black pigments. WO 2012156965 A2, 22 Nov 2012.
35. Chai Y et al (2013) Colour changing composition in o/w emulsion form. WO 2013106998 A1, 25 Jul 2013.
36. Lemoine C, Chu J (2013) Colour changing composition with UV filter(s). WO 2013107001 A1, 25 Jul 2013.
37. Lemoine C et al (2013) Colour changing composition. WO 2013107350 A1, 25 Jul 2013.
38. Kawamoto M (2013) Changing-color composition in a foam form. WO 2013108410 25 Jul 2013.
39. Shimizu M, Yu Q (2015) Composition comprising microcapsules containing reflective particles. WO 2015166454 A1, 5 Nov 2015.
40. Lachmann A et al (2014) Color changing cleaning composition. US 8680032 B2, 25 Mar 2014.
41. Lipovetskaya Y (2013) Microspheres: Technologies and global markets. BCC Research, Wellesley, MA.
42. Chang CJ et al (2007) Method for preparing ultraviolet radiation absorbing compositions. EP1092421, 21 Mar 2007.
43. Shao Y, Pekarek D (2002) Microspheres. In: Schlossman ML (ed.) *Chemistry and Manufacture of Cosmetics* (3rd edn, Volume 3, pp. 601–617). Allured Publishing Corp., Carol Stream, IL.
44. Katz MA et al (1999) Methods and compositions for topical delivery of benzoyl peroxide. US 5879716A, 9 Mar 1999.
45. Sojka MF (1997) Process for producing an oil sorbent polymer and the product thereof. US 5677407 A, 14 Oct 1997.
    Sojka MF (1998) Process for producing an oil sorbent copolymer and the product thereof. US 5712358 A, 27 Jan 1998.
    Sojka MF (1998) Process for producing an oil sorbent polymer and the product thereof. US 5777054 A, 7 Jul 1998.
    Sojka MF (1998) Process for producing an oil and water adsorbent polymer capable of entrapping solid particles and liquids and the product thereof. US 5830967 A, 3 Nov 1998.
    Sojka MF (1998) Process for producing an oil sorbent copolymer and the product thereof. US 5834577 A, 10 Nov 1998.

46. Sojka MF (1998) Precipitation Polymerization process for producing an oil adsorbent polymer capable of entrapping solid particles and liquids and the product thereof. US 5830960 A, 3 Nov 1998 and US 5837790 A 17 Nov 1998.
47. Sojka MF, Spindler R (2002) Controlled release compositions and method. US 6491953 B1 10 Dec 2002.
48. Mufti J et al (2005) Skin care delivery systems. HAPPI, 9 Nov 2005.
49. Magdassi S, Touitou E (eds) (1999) *Novel Cosmetic Delivery Systems.* Marcel Dekker, New York.
50. Souto EB, Müller RH (2008) Cosmetic features and applications of lipid nanoparticles (SLN®, NLC®). *International Journal of Cosmetic Science* 30(3):157–165.
    Müller RH et al (2002) Solid lipid nanoparticles (SLN) and nanostructured lipid carriers (NLC) in cosmetic and dermatological preparations. *Advanced Drug Delivery Reviews* 54(S): S131–S155.
51. Long Y et al (2009) Microcapsules with low content of formaldehyde: preparation and characterization. *Journal of Materials Chemistry* 19(37):6882–6887.
52. Jacquemond M et al (2009) Perfume-containing polyurea microcapsules with undetectable levels of free isocyanates. *Journal of Applied Polymer Science* 114(5):3074–3080.
53. Last, K (2012) New formaldehyde-free stable-core/shell microcapsules for industrial use. *SOFW Journal* 138(4):52–64.
54. Esser-Kahn AP et al (2011) Triggered release from polymer capsules. *Macromolecules* 44(14):5539–5553.
55. Klee SK et al (2009) Triggered release of sensitive active ingredients upon response to the skin's natural pH. *Colloids and Surfaces, A: Physicochemical and Engineering Aspects* 338(1–3):162–166.
56. Perrier E et al (2005) Enzymatically activated encapsulation technologies. An ultimate delivery system for active ingredient. *Fragrance Journal* 33(11):29–37.

13

Controlled Release of Pesticide Formulations

Dhruba Jyoti Sarkar, S. Majumder, P. Kaushik, Najam Akhtar Shakil, and Jitendra Kumar

CONTENTS

13.1 Introduction

Pesticides are considered as one of the critical inputs for securing global food production. Unlike fertilizers, which are used in tonne quantities in agricultural fields, pesticides are used in gram quantities, ranging from a few grams to several hundred grams per hectare. So, to solve the problem of uniform distribution of such a small quantity of pesticide over a vast agricultural field, the pesticide technical materials are diluted with a suitable solid or liquid material for spraying or applying easily. While developing formulations of pesticides, a few points such as optimum biological efficacy, ease of handling, storage stability, and safety must be taken into consideration. However, the use of pesticides in their conventional formulation approaches brings several health and environmental concerns. Most of the problems caused by pesticidal products are associated with their mode of applications. In the environment, when a pesticidal product is applied, the full content does not reach the target pest. This situation arises due to the use of poor formulation or delivery approaches while preparing pesticidal products, whereby the active ingredient (a.i.) is loosely entrapped in an adjuvant or carrier, and on application, most of the a.i. is released instantaneously. The released a.i. eventually loses activity due to adsorption into soil clay and organic matter or due to dissemination through leaching, runoff, volatilization, etc., reaching various compartments of ecosystems (Parmar and Tomar, 2004). To get a more pronounced effect, sometimes, users apply higher than recommended doses without realizing that most of the applied chemical will not reach the target and will be lost to the environment, causing pollution and other dreaded health hazards. To compensate for these losses and to obtain effective pest control, further application of pesticidal products is required to obtain the desired effect. This excess not only increases the cost of pesticide application but also poses potential hazards to the environment and non-target organisms. To address these issues, researchers all over the world have developed newer pesticidal products with advanced controlled release (CR) formulation technology. CR pesticidal formulation technology has been accepted globally as a potent and environmentally benign tool for combating pests. A CR formulation is defined as a combination of a biologically active agent and a polymer arranged to allow the delivery of the agent to the target at controlled rates over a specified period (Lewis and Cowser, 1977).

13.2 Controlled Release Principles

CR technology offers the opportunity to improve the efficacy of a.i. by protecting them against environmental losses. The amount lost to the environment is replaced by the amount released from the CR formulation until a balance is achieved. The objective is to deliver the a.i. to the target site at a constant effective level over a predetermined period of time (Figure 13.1). A CR formulation with a constant output rate has a zero-order rate. If the total amount of active agent is M_∞, and the amount that will be released in a unit of time is M_t, then the release rate dM_t/dt is zero order; dM_t/dt is constant. A CR formulation with a zero-order release is an ideal formulation, proving a release duration that has no deleterious effect on non-target biota. CR formulations generally refer to the use of polymers containing a.i., which are released into the target site at relatively constant rates over an extended period of time to avoid the risk of the a.i. being lost by environmental factors. The release of a.i. from CRs depends on the physical and chemical properties of the polymer, such as permeability or hydrolysis and degradation. Polymers have been employed extensively either as encapsulating agents for a.i. or as a backbone for chemically attaching active agrochemical groups.

13.3 Controlled Release Strategies

Structurally CR formulations of pesticides can be divided into four main types (Figure 13.2a). These are polymer membrane reservoir, monolithic matrix, covalently bound a.i., and coated active ingredient granule systems.

13.3.1 Polymer Membrane Reservoir

This approach includes microcapsules and microstrips. Microcapsules are sprayable and are generally in the 5–50 micron particle diameter range. They are composed of a core of a.i. surrounded by a polymer shell or membrane (Figure 13.2a I). There may be one or several cores and one or many shells. The

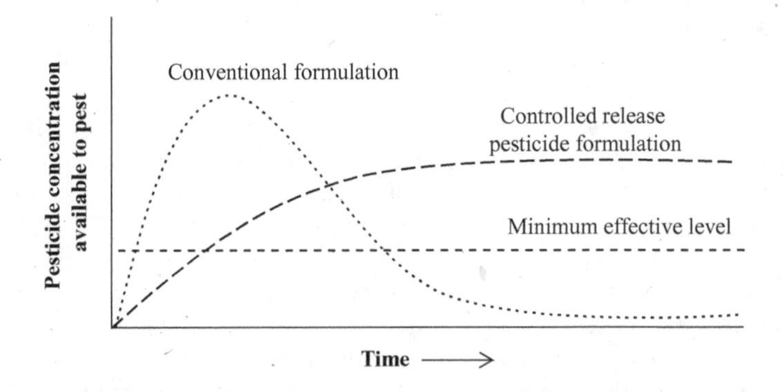

FIGURE 13.1 Release of bioactive molecules from conventional and CR formulations.

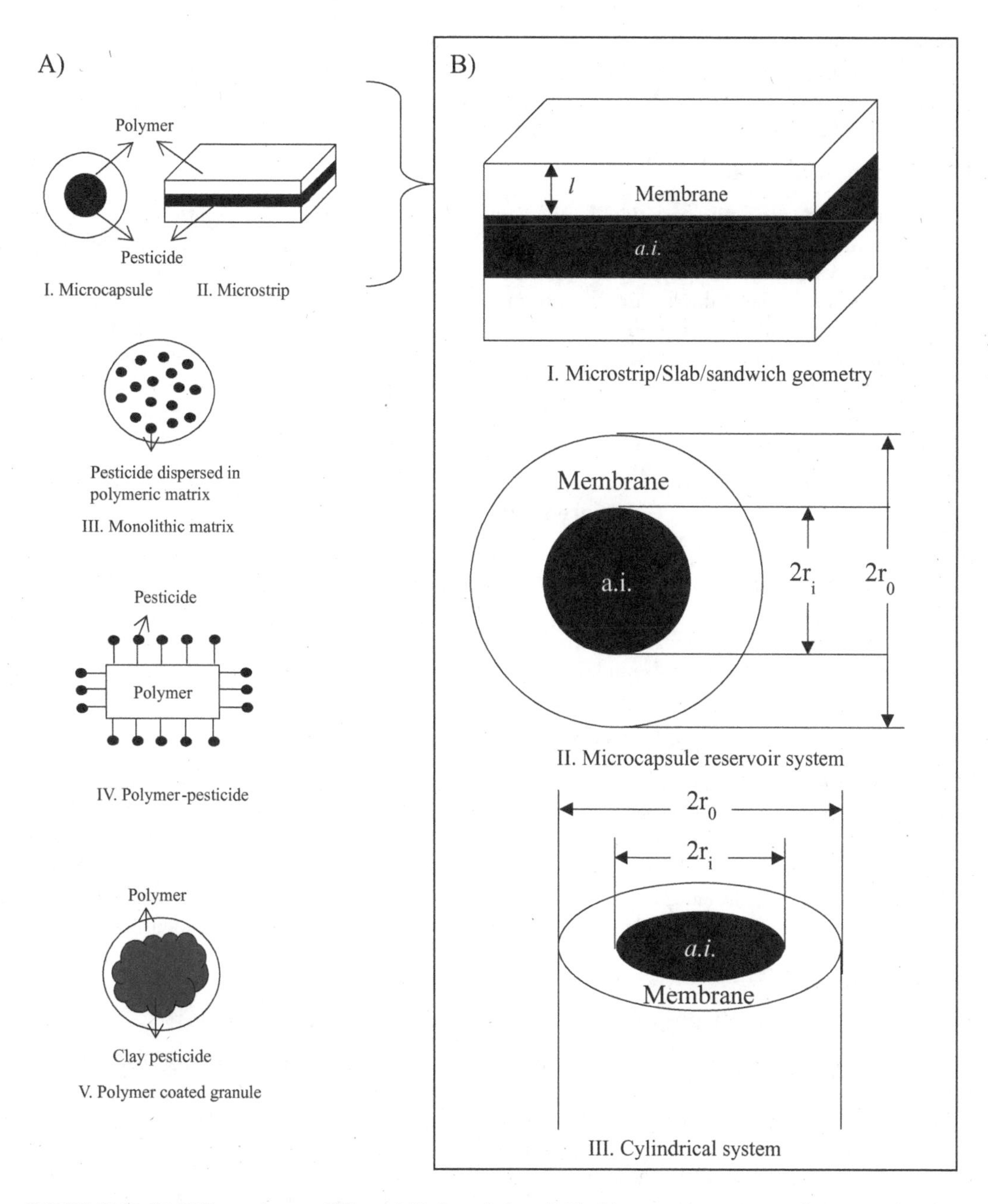

FIGURE 13.2 (a) Different classes of CR pesticide formulations (I–V); (b) geometric parameters that govern the release of entrapped a.i. from membrane reservoir formulations (I–III).

a.i. can be a solid or a liquid or even a gas. The outer polymeric layer(s) protect(s) the a.i. from adverse reactions and volatilization and restrict direct exposure to the outer environment. Here, the release rate depends on the physiochemical nature of the a.i. and the polymer and the geometry of the release device. A microcapsule formulation of a pesticide can be produced by interfacial polycondensation, *in situ* polymerization, or coacervation. However, the most suitable approach for industrial application is interfacial polycondensation, whereby the shell thickness and particle size can be easily controlled by changing the monomer content and other process conditions. The shell material or polymer used for the preparation of a microcapsule formulation should satisfy the following conditions: the appropriate molecular weight, glass transition temperature, and molecular structure to achieve the optimum release rate; non-reactiveness with pesticidal technical materials; environmentally benign nature; low cost; high storage stability; etc. Other than microcapsules, pesticide can be entrapped in a sandwich-type system, in which the a.i. is contained in a central reservoir layer surrounded above and below by two protective layers (Figure 13.2a II).

13.3.2 Monolithic Matrix

In the monolithic matrix, the pesticidal a.i. is dispersed or dissolved in a rate-controlling polymeric matrix. Similarly to the previously described polymer membrane reservoir, here also the release of entrapped a.i. depends on the physiochemical nature of the a.i. and the polymer and the geometry of the release device. Initially, the release starts from the a.i. dispersed in the surface layer of the device, followed by subsequent a.i. dispersed in the inner core. It was observed that when the loading of a.i. is low, slower release of a.i. was observed due to a more diffusional pathway of the polymer network. The polymeric matrixes may include rubber, polyvinyl chloride (PVC), a gypsum–wax mixture, polyester and acrylic resins, polyvinyl acetate (PVA), cellulose, starch, and gels such as alginate and lignin (Figure 3.2a III).

13.3.3 Covalently Bound Active Ingredient

In this system, pesticide molecules are covalently bound to a polymer or bound to the monomer units followed by polymerization (Figure 13.2a IV). Pesticide is released into the environment from the polymeric matrix by cleavage of bonds through a chemical reaction such as hydrolysis. Pesticidal molecules that contain very reactive functional groups, such as carboxyl, hydroxyl, amino, etc., can be reacted with polymer or monomer to develop this kind of CR formulation.

13.3.4 Coated Active Ingredient Granule Systems

With a slight difference from the covalently bound system, this approach consists of clay or other mineral granules of about 1 mm diameter, pesticidal molecules, and polymer. Clay minerals, having high sorptivity, are impregnated with pesticide, followed by coating with polymer film (Figure 13.2a, V). In this system, due to the tortuosity of the clay mineral and high binding energy between clay and the pesticidal molecules, the release of the latter into the outer environment is further extended. Several clay minerals, such as montmorillonite, talc, attapulgite, etc., are being used to develop such formulations.

13.4 Mechanism of Release from CR Formulations of Pesticides

Based on the release mechanism of the a.i., the polymeric CR systems can be classified into diffusion release systems, erosion or chemical reaction release systems, swelling CR systems, and osmosis pumping release systems. The classification of CR systems into these categories is governed by the kind of interaction between the polymer matrix and the surrounding environment. If during interaction, the polymer remains unchanged and release occurs only through diffusion, then the formulation system falls into diffusion CR systems. If the formulation releases a.i. through erosion of the polymeric wall or matrix or breaking of the chemical bonds between the a.i. and the polymer, the system may be categorized as an erosion or chemical reaction CR system. If the release occurs through swelling of polymeric matrix driven by absorption of environmental fluid, the system may be termed a swelling CR system. There are some CR systems in which release is triggered by osmosis, using osmotic pressure as the driving force. In a practical scenario, it must be noted that two or more of these release mechanisms may be simultaneously active in CR systems.

13.4.1 Diffusion CR Systems

A CR formulation is considered to be diffusion controlled when the diffusion of a.i. through the polymeric matrix or membrane controls the release rate. There are two major types of diffusion-controlled system: membrane reservoir and monolithic matrix. In a membrane reservoir, the a.i. is encapsulated within a rate-controlling membrane, while a.i. is homogeneously dispersed or dissolved throughout a monolithic matrix.

The main steps in the release of a.i. from a membrane reservoir formulation are depicted in Figure 13.3a. The steps include diffusion of a.i. within the reservoir, dissolution or partition between the reservoir and the membrane, diffusion through the membrane, partition between the membrane and the environmental fluid, and finally, release to the bulk environment (elution medium) through diffusion from the formulation surface. If the membrane reservoir formulation contains excess a.i., then a steady-state concentration profile as shown in Figure 13.3a I can be obtained, however, with time, the reservoir will be depleted, and the concentration profile will change in either pseudo-steady-state (Figure 13.3a II) or unsteady fashion.

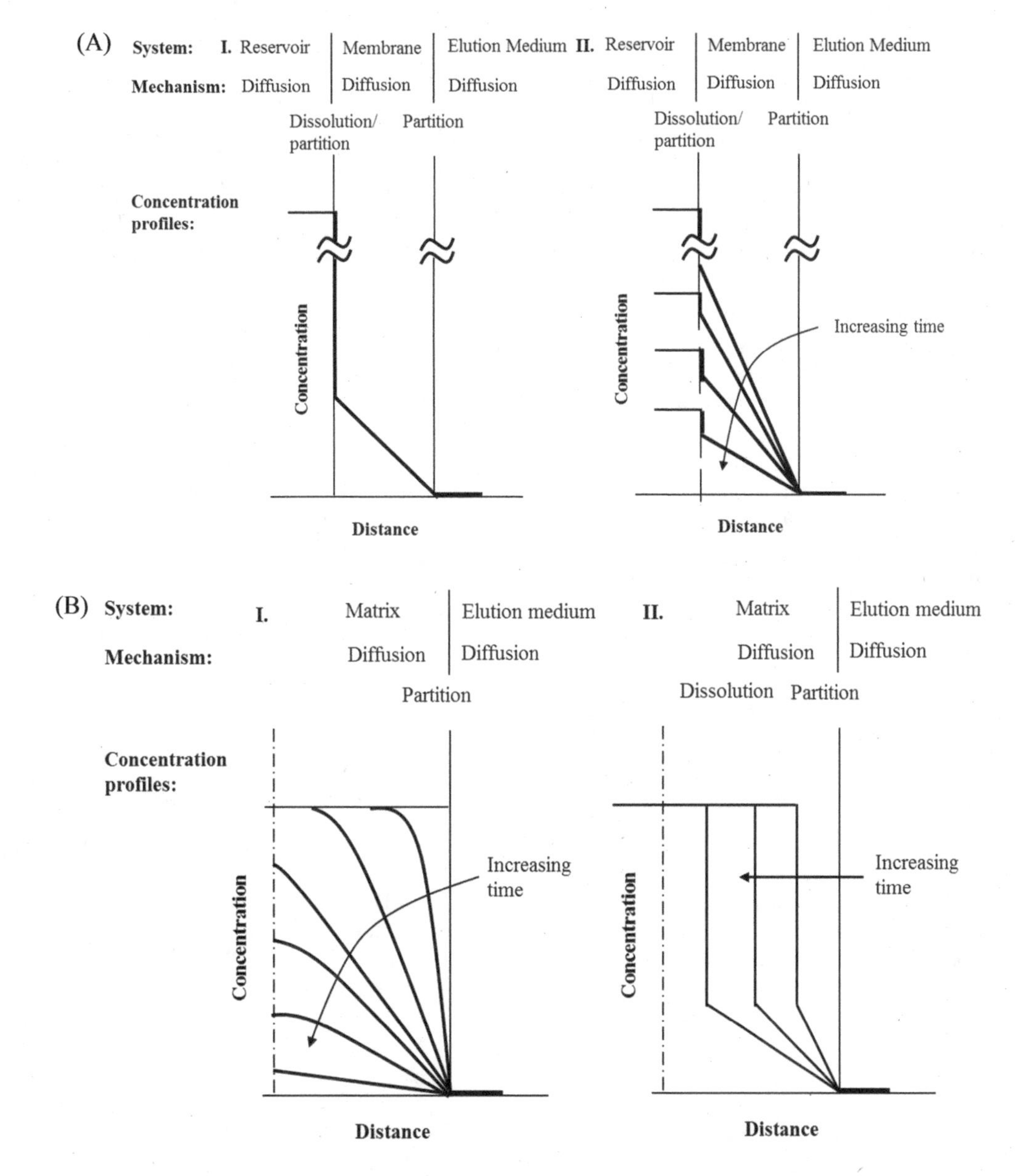

FIGURE 13.3 (a) Schematic representations of membrane reservoir formulation function: reservoir device with excess active agent (I), reservoir device without excess active agent (II); (b) schematic representation of the functioning of a monolithic matrix formulation with dissolved a.i. (I) and dispersed a.i. (II). (From Baker, R.W., *Controlled Release of Biologically Active Agents*. John Wiley and Sons, New York, 1987. With permission.)

The principal steps of a.i. release from a monolithic matrix are exhibited in Figure 13.3b. When a.i. is dissolved in the monolithic matrix, the steps include diffusion to the matrix surface, partition between the matrix and the environment fluid (elution medium), and transport away from the formulation surface (Figure 13.3b I). If the a.i. is dispersed as a separated phase in the matrix, then it will first dissolve in the matrix prior to diffusion to the surface (Figure 13.3b II).

Diffusion CR of pesticides from a polymeric matrix, either a monolithic or a reservoir system, has been known for a long time. In the 1960s, a number of polymeric systems were tested for controlled delivery of aquatic herbicides. Natural rubber was used for diffusion CR of butoxyethanol ester of 2,4-D (2,4-D BEE), which showed membrane-moderated diffusion (Langer and Peppas, 1983). Later, more sophisticated microencapsulation techniques were used to prepare a spherical-shaped pesticide reservoir device. The frequently used microencapsulation techniques were phase separation, coacervation, spray coating, solvent evaporation, interfacial polymerization, etc., depending on the properties of the pesticide and the type of release rate–controlling membrane. Some of the common membrane materials include polyamide, polyesters, polyurea, cellulose, and gelatin. A CR formulation of chloridazon using lignin and polyethylene glycol as core material followed by coating with ethyl cellulose and dibutyl sebacate using Wurster-type fluidized-bed equipment was reported (Fernández-Pérez et al., 2011). The developed formulation of chloridazon showed high encapsulation efficiency (>90%), and the release was found to follow first-order release kinetics, with exponential decline of chloridazon in the granules (Table 13.1). Diffusion-controlled release of alachlor was reported previously by encapsulating the same in different biopolymeric matrixes involving alginates and pectin (Gerstl et al., 1998). It was observed that among the tested biopolymers and their combinations, the CR formulation prepared with alginate showed the slowest release properties, with a rate constant (k) of 0.0309, followed by pectin (k 0.0570) < alginate–pectin (k 0.0755). However, if a clay mineral such as montmorillonite is added as a filler material, it introduces more barrier properties into the polymer matrix, the rate constant value is further reduced, and eventually, the $T_{50\%}$ is increased (Gerstl et al., 1998) (Table 13.1). Similarly, CR formulations of the herbicides Cloridazon and Metribuzin were prepared using Na alginates as the base polymer, followed by the addition of filler materials, bentonite, anthracite, and charcoal, alone or in combination (Céspedes et al., 2007). It was found that the use of these filler materials imparted more slow-release properties

TABLE 13.1

Diffusion Controlled Release of Alachlor and Imidacloprid from CR Formulations

			Equations			
			$\left(1-\frac{M_t}{M_o}\right)^{\frac{1}{3}} = 1-kt^{\frac{1}{2}}$			
Pesticides	**Polymer**	**Filler material**	***K*** **(h^{-1})**		**r^2**	**$t_{50\%}$ (h)**
Alachlor	Aliginate	–	0.0309		0.997	44
Alachlor	Pectin	–	0.057		0.982	13
Alachlor	Alginate-pectin	–	0.0755		0.959	7
Alachlor	Alginate	Montmorillonite	0.0109		0.997	357
Alachlor	Pectin	Montmorillonite	0.0341		0.996	37
Alachlor	Alginate-Pectin	Montmorillonite	0.0327		0.993	40
			$\frac{M_t}{M_\infty} = kt^n + c$			
		Granule size (mm)	***K*** **(day^{-1})**	***n***	**r**	**$t_{50\%}$ (day)**
Imidacloprid	Pine kraft lignin	<0.25	59.67	0.32	0.9785	0.94
Imidacloprid	Pine kraft lignin	0.25–0.50	37.95	0.48	0.9932	2.08
Imidacloprid	Pine kraft lignin	0.50–0.71	19.5	0.65	0.9988	4.26
Imidacloprid	Pine kraft lignin	0.71–1.00	13.58	0.53	0.9977	11.88

Source: Gerstl et al., 1998; Fernández-Pérez et al., 1998.

TABLE 13.2

In-vitro Release Kinetics Parameters of Lignin Based Granules of Chloridazon Herbicide after Curve Fitting Water Release Data in to Korsmeyer-Peppas and First Order Kinetics Equation

		Korsmeyer-Peppas: $\frac{M_t}{M_\infty} = k_1 t^n$			
	Size Distribution of Granules	K_1 **(×10² h⁻¹)**	n	**r**	T_{50} **(h)**
Lignin	d < 0.2 mm	67.47	0.28	0.93	0.34
	0.2 mm < d < 0.5 mm	42.85	0.32	0.99	1.62
	0.5 mm < d < 1 mm	22.96	0.40	0.98	7.00
	1 mm < d < 2 mm	12.80	0.51	0.99	14.46
	2 mm < d < 3 mm	13.86	0.35	0.94	39.08
	0.2 mm < d < 1 mm	41.82	0.32	0.98	1.74
	Coating with Ethyl cellulose & Dibutyl sebacate	**First order:** $\frac{M_t}{M_\infty} = 1 - e^{k_2 t}$			
Lignin, Ethyl cellulose, Dibutylsebacate		K_2 **(×10² h⁻¹)**	**r**	T_{50} **(h)**	
	Ethyl cellulose 10%	4.45	0.96	15.58	
	Ethyl cellulose 20%	2.21	0.92	31.34	
	Ethyl cellulose 20% + Dibutylsebacate 2.25%	1.06	0.92	65.39	

Source: Fernández-Pérez et al., 2011.

K_1& K_2, *rate constant; r,* correlation coefficient; T_{50}, the time (hour) taken for 50% of the active ingredient released into water.

to these pesticidal formulations. The diffusional release of imidacloprid from a CR formulation, using pine kraft lignin, was also reported, whereby greater size of the granules led to slower release of the a.i. (Fernández-Pérez et al., 1998) (Table 13.2).

13.4.2 Chemical Reaction CR Systems

Unlike the polymers used in diffusional release of a.i., there are some polymers that play an active role in the release process by undergoing chemical reaction or erosion at the target site. These chemically responsive polymeric systems can be categorized as physically entrapped systems and chemically immobilized systems (Baker, 1987; Langer and Peppas, 1983).

13.4.2.1 Physically Entrapped Systems

These are degradable or erodible polymeric systems (both reservoir and monolithic) in which a.i. is physically immobilized. An erodible monolithic device releases the a.i. through erosion, diffusion, or a combination of both, but the release rate is determined by the rate-determining release mechanism. If the erosion of polymer is slow, than diffusion will govern the release rate, and vice versa. The release rate will be purely governed by an erosion mechanism where the a.i. is extensively physically immobilized in the polymer. A combination of both mechanisms is possible if their contribution to the rate determination is comparable.

Erosion of a polymeric matrix can occur either from the surface (heterogeneous erosion) or homogeneously throughout the device. A polymeric matrix that erodes from the surface can lead to a zero-order release rate if the surface area remains constant (Langer and Peppas, 1983). Properties such as crystallinity, hydrophilicity, etc. determine the erosion movement of the polymeric matrix. Hydrophobic polymers will try to exclude water from the interior and are more likely to erode from the surface, in contrast to homogeneous erosion by hydrophilic polymers. Similarly, hydrophobicity induced by the crystallinity of a polymer favors a surface erosion mechanism more than homogeneous erosion by amorphous

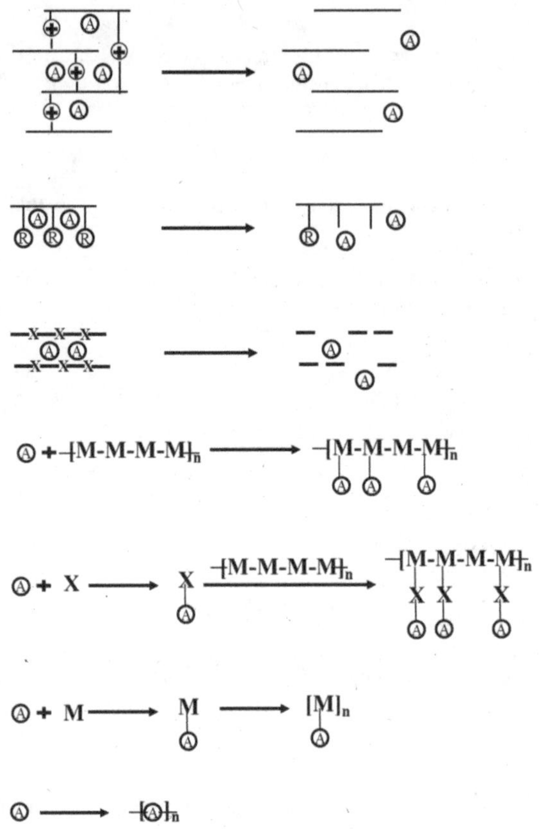

FIGURE 13.4 Schematic representation of degradation mechanisms of a.i. physically entrapped in a polymeric matrix (Type I to Type III) and chemically immobilized a.i. (Type IV A to Type V B). (From Heller and Baker, 1980; Pitt and Schindler, 1983.) A, active ingredient; X, labile bond; M, polymerizable monomer.

polymers. The preferred erosion or degradation mechanism for this kind of polymeric system involves hydrolytic and enzymatic cleavage of labile bonds such as polycarbonates, polyesters, polyurethanes, poly(orthoesters), and polyamides (Langer and Peppas, 1983).

Polymers can be classified into three categories based on the mechanism of erosion or degradation (Figure 13.4) (Heller and Baker, 1980; Pitt and Schindler, 1983). The first category (Type I) involves water-soluble polymers insolubilized by labile chemical bonding or cross-linking. Cleavage of chemical bonds causes the polymer matrix to solubilize in water and thus release a.i. These types of polymer matrixes are useful for the delivery of poorly water-soluble a.i., since highly soluble a.i. will be released immediately. The second category (Type II) involves polymers, initially water insoluble, becoming soluble due to the hydrolysis or ionization of substituent hydrophobic pendant groups without much change of molecular structure or molecular weight. The third category (Type III) of polymers consists of water-insoluble polymers that become soluble due to hydrolytic cleavage into small molecules. This polymeric matrix undergoes a combination of surface and homogeneous erosion mechanisms.

Erosion of the polylactic acid (PLA) matrix was found to be the governing mechanism to release the loaded sodium salt of 2,4-D (Sinclair, 1973). A combination of diffusion and erosion release mechanisms was reported from oxidatively cross-linked starch xanthate microencapsulated pesticide formulations with or without incorporation of latex rubber (Sasha et al., 1976). Polymer relaxation CR of triazine herbicides (ametryn, atrazine, and simazine) was reported from nanocapsules based on polymeric poly(ε-caprolactone) (Grillo et al., 2012). λ-Cyhalothrin release from a PLA matrix was reported to be initiated by diffusion followed by hydrolytic degradation (Liu et al., 2016). Macromolecular relaxation followed by diffusion was reported to be the major release mechanism of avermectin from starch microcapsules (Li et al., 2016). A coupling of Fickian diffusion and polymer relaxation was found to be the release mechanism of thiamethoxam insecticide from amphiphilic polymer–based formulations (Table 13.1) (Sarkar et al., 2012b). Polymer chain relaxation of carboxymethyl starch-based microparticles was found to be the major release mechanism of entrapped isoproturon (Wilpiszewska et al., 2016).

13.4.2.2 Chemically Immobilized Systems

A polymeric matrix with erodible bonds can be chemically bonded to a.i. either within the backbone or on side groups. There are examples in the literature where the polymer itself is biologically active (Qi et al., 2004; Hernandez-Munoz et al., 2008; Aziz et al., 2006). In the case of a chemically erodible matrix where the a.i. is chemically bonded to polymer chains, the principal release rate–controlling factor is the rate of erosion of the labile bonds. This kind of carrier system with chemically bonded a.i. has the advantage of higher a.i. loading than the physically entrapped carrier systems. Based on the mode of attachment of a.i. in the polymeric matrix, such systems can be classified into two major categories (Figure 13.4, Types IV and V) (Baker, 1987; Pitt and Scgindler, 1983). The first class involves chemical bonding of a.i. to the polymer backbone, and the second class involves co- or homo-polymerization of a.i. or their derivatives themselves. These two categories were further fine tuned into subgroups (A and B) based on the actual chemical reaction of polymerization.

The classes of chemical bonds involved in the formation of chemically immobilized systems of Type IV A are esters, amide, anhydrides, urethanes, hydrazines, thioethers, etc. If the a.i. and polymer do not have suitable reactive groups, either or both of them can be derivatized to form a linkage (Figure 13.4, Type IV B). Type V A and B involve the derivatization of the a.i. followed by co- or homo-polymerization. One such herbicide–polymer combination, a polymer containing pendant herbicides, was first reported in 1971 (Allan et al, 1971). Similarly, 2,4-D and 2-(2,4,5-trichlorophenoxy) propionic acid as pendant substituents on vinyl and acryloyloxyethyl ester polymers have been reported with a hydrolysis-induced release mechanism (Harris and Post, 1975). It was observed that the rate of hydrolysis of ester linkages can be enhanced by the incorporation of aminimide or carboxylic groups along the polymer chain. Linear or cross-linked functionalized polymers containing pendant pesticide-active (pentachlorophenol) groups using a free radical polymerization technique were reported, in which pentachlorophenol was linked to the polymer chain through oxyethylene as a spacer (Akelah et al., 1987). Release was initiated through the hydrolysis of the polymer chain, and the degree of cross-linking, hydrophilicity of the spacer group, pH, and temperature influence the release rate.

13.4.3 Swelling CR Systems

A polymer matrix in which there is a change of polymer morphology (glass to gel) on interaction with the release medium follows the swelling CR phenomenon. If the release medium is thermodynamically compatible with the polymer, and the temperature is above the glass transition temperature (T_g) of the polymer, then a glassy polymer begins to undergo a glass-to-gel transition. In the gel state, the polymer chain is more mobile, which allows the diffusion of a.i. more rapidly out of the matrix. Such a swelling CR matrix can be classified into two categories: the first involves homogeneous or continuous swelling of the entire polymer matrix, and the second involves spatially discontinuous swelling of the polymer matrix (Peppas et al., 1984). During swelling, an interface separates the gel region of polymer from the glassy core, and this interface moves as the swelling advances (Figure 13.5). The release rate of the a.i. depends on this advancement of the interface or transition process (glass to gel), which is governed by the time-dependent macromolecular relaxation of polymer chains.

Swelling CR polymeric formulations can be prepared by dissolving or dispersing the a.i. in a polymer solution. The solvent can then be evaporated to produce a glassy polymeric matrix (Hopfenberg et al., 1978). The a.i. can also be loaded into a polymer matrix by soaking the gel in a solution or emulsion containing a.i. and allowing them to equilibrate. After loading, the penetrant (water/solvent) in the polymeric gel can be removed by evaporation, leaving a dry, glassy polymeric matrix (Lee, 1984).

Recently, natural polymer–based superabsorbent polymers have been observed as a material of great interest to encapsulate pesticides with CR properties due to their biodegradable character and high swelling capacity (Kenawy, 1998; Rudzinski et al., 2002). A fungicide, carbandazim, was loaded in

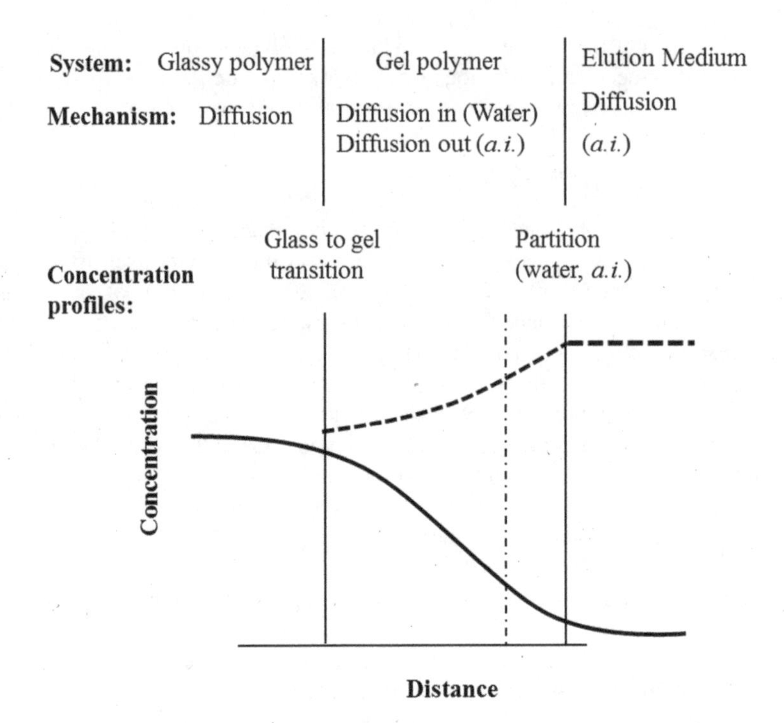

FIGURE 13.5 Schematic representation of the functioning of a swelling-controlled formulation. Solid line indicates release of a.i. and dashed line indicates penetration of elution medium. (Reprinted with kind permission from Fan, L.T. and Singh, S.K., Swelling-controlled release, *Controlled Release: A Quantitative Treatment*, Springer Science+Business Media: 1989, 110–156.)

starch-g-(acrylic acid-co-methyl methacrylate) hydrogel beads, and a burst release phenomenon (8–22% of loaded carbandazim) was observed due to loosely entrapped carbandazim on the hydrogel surface, followed by swelling-controlled diffusional release (Bai et al., 2015). Swelling and erosion CR of chlorpyrifos was reported from starch and sodium alginate–based hydrogel microspheres prepared by cross-linking with $CaCl_2$ (Roy et al., 2009). Polyacrylamide–methyl cellulose–based hydrogels were used as a controlled delivery device for paraquat (Aouada et al., 2010). Analysis of the release of paraquat from the hydrogels using a semi-empirical Ritger–Peppas model revealed initial fast release followed by sustained release for up to 45 days. Citric acid–cross-linked carboxymethyl cellulose hydrogel bentonite composites were reported as a triggered delivery device for thiamethoxam (Sarkar and Singh, 2017). The delivery device showed higher water absorbency in basic pH conditions. The release kinetics study of thiamethoxam from developed hydrogel-based formulations followed the Gallagher–Corrigan equation, with an immediate burst release phenomenon, and a higher release rate of thiamethoxam was observed at alkaline pH than in neutral conditions (pH 7.0). The sensitivity of hydrogels toward external stimuli such as pH arises due to the presence of hydrophilic functional groups, such as amino, carboxyl, and hydroxyl groups, in the polymer chains (Gupta et al., 2002). Hydrogels with anionic functional groups ($-COO^-$, $-OH$) swell more rapidly in basic pH conditions due to the generation of a larger ion swelling pressure than hydrogels with cationic functional groups ($-NH_2$), which swell more rapidly in acidic pH conditions. This differential swelling of hydrogels affected by external stimuli such as pH makes these polymers a stimuli-responsive delivery device for biologically active substances (Ninni et al., 2013).

13.5 CR of Pesticide from Nanoreservoir

Polyethylene glycol (PEG) ($HO-(CH_2CH_2O)_n-H$) is available in a wide range of molecular weights (from 200 to tens of thousands). It is a highly water-soluble hygroscopic polymer used extensively in the pesticide formulation industry as a non-ionic surfactant and a base polymer for seed coat formulation (Kumar et al., 2007a; Sarkar et al., 2012). PEG (MW 6000) was reported to be used as a matrix to develop a CR formulation of phorate (Rao et al., 1989). PEG-based nanomicelles have recently emerged as a novel promising colloidal carrier for formulating non-polar and hydrophobic a.i. Polymeric micelles are found to be considerably more stable than surfactant micelles and have the loading capacity to carry hydrophobic compounds in their inner core. Copolymers of PEG and various dimethyl esters of aromatic di-acids, which self-assemble into nanomicellar aggregates in aqueous media, have been reported (Figure 13.6) (Shakil et al., 2010; Sarkar et al., 2012). The effects of different molecular weights of PEG and the type of hydrophobic group in the block copolymer were reported to alter the release of entrapped pesticides in water media (Table 13.3). It was found that in the case of entrapped imidacloprid, there was no effect of PEG molecular weight on the release rate, whereas the release rate of thiamethoxam was reduced when the molecular weight of PEG was increased (Table 13.3) (Sarkar et al., 2012; Adak et al., 2012). These nanomicelles were also used for the encapsulation of carbofuran, a systemic insecticide-nematicide, for the development of a CR formulation (Shakil et al., 2010). CR formulations based on amphiphilic nanopolymers of carbofuran were reported to reduce nematode penetration and the number of galls in tomatoes under field conditions (Pankaj et al., 2012). Other than these pesticides, nanoformulations of azadirachtin-A, carbofuran, β-cyfluthrin, and thiram based on amphiphilic polymers, synthesized from aromatic diesters and PEGs, have also been reported, and their bioefficacy has been evaluated against different pests (Shakil et al., 2010; Kumar et al., 2010; Sarkar et al., 2012; Loha et al., 2011, 2012; Kaushik et al., 2013; Adak et al., 2012). The use of nanotechnology in pesticide delivery systems promises to reduce pesticide use and to enable the development of a range of inexpensive nanotech applications for slow release and efficient pest control at lower environmental cost (Liu et al., 2008). Nanoformulations of pesticides can result in enhanced activity against pests yet reduce the side effects of the pesticide because of the low dose used. Due to the small size of the nanoformulations, they can easily deposit on the leaves

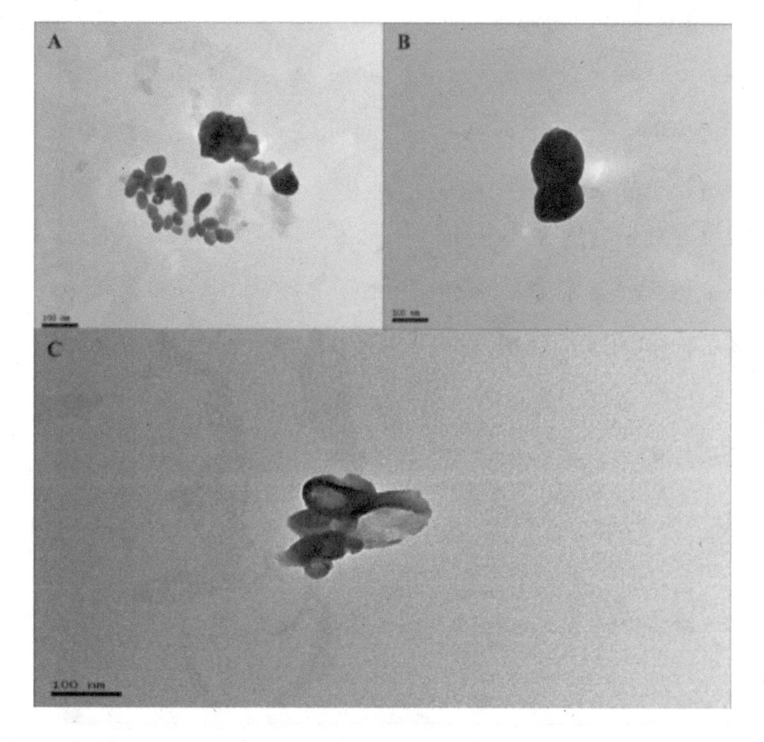

FIGURE 13.6 Transmission electron micrograph of amphiphilic block co polymer (a), encapsulated thiamethoxam in amphiphilic polymer (b), and ruptured nanocapsules of thiamethoxam (c). (From Sarkar et al., 2012b.)

TABLE 13.3

Diblock Amphiphilic Co-Polymers Used for Preparation of CRFs of Imidacloprid and Thiamethoxam; and Their *In-vitro* Release Kinetics Parameters after Curve Fitting in Korsmeyer-Peppas Equation

Diblock Amphiphilic Co-Polymers	Diacids & Diesters	Avg. MW of PEG	Pesticides	$M_t/M_\infty = kt^n$		
				k	n	R^2
Poly[poly-(oxyethylene-300)-oxyglut aroyl]	Glutaric acid	300	Imidacloprid	0.338	0.282	0.93
Poly [poly (oxyethylene-600)-oxy glutaroyl]	Glutaric acid	600		0.415	0.222	0.91
Poly [poly (oxyethylene-1000)-oxy glutaroyl]	Glutaric acid	1000		0.239	0.370	0.93
Poly [poly (oxyethylene-300)-oxy adipoyl]	Adipic acid	300		0.364	0.241	0.84
Poly [poly (oxyethylene-600)-oxy adipoyl]	Adipic acid	600		0.346	0.261	0.94
Poly [poly (oxyethylene-1000)-oxy adipoyl]	Adipic acid	1000		0.333	0.264	0.94
Poly [poly (oxyethylene-300)-oxy pimeloyl]	Pimelic acid	300		0.340	0.279	0.92
Poly [poly (oxyethylene-600)-oxy pimeloyl]	Pimelic acid	600		0.391	0.215	0.91
Poly [poly (oxyethylene-1000)-oxy pimeloyl]	Pimelic acid	1000		0.314	0.290	0.92
Poly [poly (oxyethylene-300)-oxy suberoyl]	Suberic acid	300		0.317	0.320	0.95
Poly [poly (oxyethylene-600)-oxy suberoyl]	Suberic acid	600		0.267	0.345	0.92
Poly [poly (oxyethylene-1000)-oxy suberoyl]	Suberic acid	1000		0.273	0.272	0.94
Poly[poly-(oxyethylene-1000)-o xyazelaioyl]	Azelaic acid	1000	Thiamethoxam	0.254	0.532	0.90
Poly[poly-(oxyethylene-2000)-o xyazelaoyl]	Azelaic acid	2000		0.177	0.666	0.93
Poly[poly-(oxyethylene-4000)-o xyazelaoyl]	Azelaic acid	4000		0.126	0.806	0.95
Poly[poly-(Oxyethylene-1000)-o xysebacoyl]	Sebacic acid	1000		0.227	0.599	0.93
Poly[poly-(oxyethylene-2000)-o xysebacoyl]	Sebacic acid	2000		0.160	0.731	0.95
Poly[poly-(oxyethylene-4000)-o xysebacoyl]	Sebacic acid	4000		0.117	0.846	0.98
Poly[poly-(oxyethylene-1000)-oxyiso phthaloyl]	Dimethyl Isophthalate	1000		0.205	0.610	0.90
Poly[poly-(oxyethylene-2000)-oxyiso phthaloyl]	Dimethyl Isophthalate	2000		0.187	0.636	0.99
Poly[poly-(oxyethylene-4000)-oxyiso phthaloyl]	Dimethyl Isophthalate	4000		0.166	0.676	0.99
Poly[poly-(oxyethylene-1000)-oxyter ephthaloyl]	Dimethyl Terephthalate	1000		0.183	0.662	0.92
Poly[poly-(oxyethylene-2000)-oxyter ephthaloyl]	Dimethyl Terephthalate	2000		0.129	0.793	0.95
Poly[poly-(oxyethylene-4000)-oxyter ephthaloyl]	Dimethyl Terephthalate	4000		0.102	0.881	0.95

Source: Sarkar et al., 2012; Adak et al., 2012.

R^2, Coefficient of determination; n, Difussion exponent and k, release rate constant.

of the plants, which helps to reduce waste of the pesticide (Yang et al., 2009). Stabilized polymeric nanoparticles have been used for the controlled and efficient release of the synthetic pyrethroid bifenthrin (Liu et al., 2011).

13.6 Mathematical Models Used for Describing CR Release Systems

13.6.1 Diffusion CR System

There are several factors that govern the *in vitro* performance of a diffusion CR system. Some of these factors are the diffusivity and solubility of the a.i. in the polymer, the partition coefficient between the environmental elution medium and the polymer, the loading of the a.i., matrix geometry, barrier or membrane thickness, etc. It has been proved that the diffusion of a.i. from CR systems follows Fick's law of diffusion:

$$J = -D\frac{dC_m}{dx} \tag{13.1}$$

where

J is the flux in g/cm²/s
C_m is the concentration of a.i. in the membrane in g/cm³
dC_m/dx is the concentration gradient
D is the diffusion coefficient of the a.i. in the membrane in cm²/s

Equation 13.1 can be restated as

$$J = \frac{DK\Delta C}{l} \tag{13.2}$$

where

l is the thickness of the membrane
ΔC is the difference in concentration of a.i. on either side of the membrane

In the case of a membrane reservoir system (Figure 13.2a I), if a.i. is enclosed within an inert membrane, according to Fick's law, a steady state zero-order release rate will be established if the concentration of a.i. is maintained constant within the reservoir. For a microstrip or slab or sandwich geometry (Figure 13.2b I), Fick's law can be redesigned as

$$\frac{dM_t}{dt} = \frac{2ADK\Delta C}{l} \tag{13.3}$$

where

dMt/dt is the steady-state release rate at time t
A is the surface area of the membrane
l is the thickness of the membrane
DK is the membrane permeability

For a microcapsule reservoir system (Figure 13.2b II), the rate of diffusion at a given time can be given as

$$\frac{dM_t}{dt} = 4\pi DK\Delta C\frac{r_o r_i}{r_o - r_i} \tag{13.4}$$

For a cylinder reservoir (Figure 13.2b III), the steady-state release rate is given by

$$\frac{dM_t}{dt} = \frac{2\pi DK\Delta C}{\ln\left(\frac{r_o}{r_i}\right)} \tag{13.5}$$

For a monolithic matrix, the a.i. is either dispersed or dissolved in the rate-controlling polymer matrix, and the release rate does not follow zero-order kinetics. The a.i. is first released from the surface layer of the monolithic matrix, and with time, the diffusion distance to reach the surface increases, thus showing a slowly declining rate of release. If an a.i. is dissolved in a spherical monolithic matrix (radius r_o), then the rate of release is expressed by two equations:

$$\frac{dM_t / M_\infty}{dt} = 3\left(\frac{D}{r_o \pi t}\right)^{\frac{1}{2}} - \frac{3D}{r_o^2} \tag{13.6}$$

$$\frac{dM_t / M_\infty}{dt} = \frac{6Dt}{r_o^2} \exp\left(\frac{-\pi^2 Dt}{r^2}\right) \tag{13.7}$$

where

M_∞ is the total amount of a.i. loaded in the matrix
M_t is the a.i. released at time t

Equation 13.6 is called the *early time approximation*, which is valid for $M_t/M_\infty < 0.4$, and Equation 13.7 is called the *late time approximation*, which is valid for $M_t/M_\infty > 0.6$. These equations reveal that the first 40% of the loaded a.i. is released at a rate that decreases as the square root of time, and the release of the remaining a.i. (60%) follows exponential first-order kinetics.

If the a.i. is dispersed in a spherical monolithic matrix (radius r_0), the equation changes, and the rate of release is expressed by (Higuchi, 1961)

$$\frac{dM_t / M_\infty}{dt} = \frac{3DC_s}{r_o^2 C_o}\left(\frac{\left[1 - M_t / M_\infty\right]^{\frac{1}{3}}}{1 - \left[1 - M_t / M_\infty\right]^{\frac{1}{3}}}\right) \tag{13.8}$$

where

C_o is the total content of a.i. (dispersed plus dissolved) in the spherical monolithic matrix
C_s is the solubility of a.i.in the polymer matrix

13.6.2 Swelling CR System

Release of the a.i. from swellable polymer matrixes as a function of time can be expressed as

$$\frac{M_t}{M_\infty} = kt^n \tag{13.9}$$

where

M_t is the amount of a.i. released by time t
M_∞ is the amount of a.i. released at infinite time, theoretically corresponding to the total amount of a.i. incorporated into the matrix
k, n are system parameters that depend on the nature of the polymer/penetrating elution medium/a.i. interaction

According to Peppas (1984), the release mechanisms corresponding to the n values are Fickian diffusion ($n = 0.5$), pseudo-case II transport ($n = 1.0$), pseudo-super-case II transport ($n > 1.0$), and anomalous transport ($0.5 < n < 1.0$). Sometimes, on placement in the release medium, a CR system

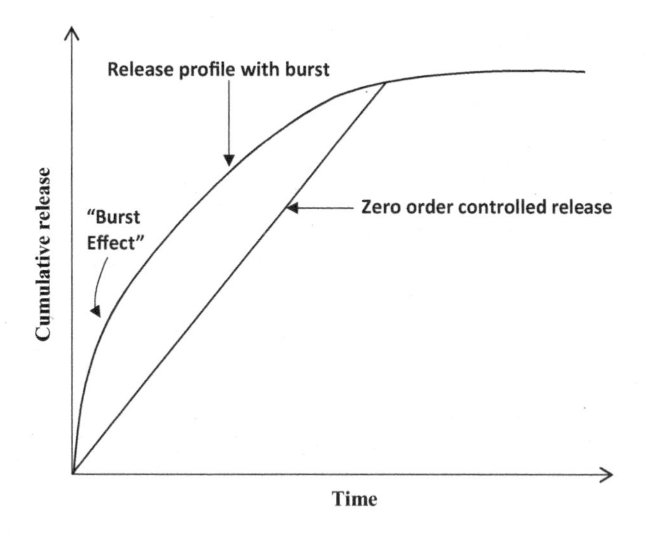

FIGURE 13.7 Schematic representation of the initial burst effect in a controlled release delivery system.

will initially release a large amount of a.i. before its release rate achieves a stable profile. Such an event is referred to as *burst release* (Huang and Brazel, 2001). As depicted in Figure 13.7, the initial burst release leads to higher initial a.i. release, and it has been found that release through burst happens for a very short period of time as compared with the total release period. The initial burst effect in the release kinetic equation was accounted for by Gallagher and Corrigan (2000) through the following model:

$$\frac{M_t}{M_\infty} = \left(\frac{M_t}{M_\infty}\right)_B \left[1 - e^{-k_1 t}\right] + \left[1 - \left(\frac{M_t}{M_\infty}\right)_B\right]\left[\frac{e^{k_2 t - k_2 t_{2\max}}}{1 + e^{k_2 t - k_2 t_{2\max}}}\right]$$

Where

$\frac{M_t}{M_\infty}$ is the fraction of thiamethoxam released at a time t

k_1 and k_2 are the kinetic constants for the I and II stage, respectively

$\left(\frac{M_t}{M_\infty}\right)_B$ indicates the fraction released in the first release stage I

$t_{2\max}$ indicates the time at which the maximal rate of the a.i. release is accomplished in the II stage

13.7 Advantages of CR Formulation of Pesticides

The regulated release of plant protection chemicals not only combats the pest and the related biotic stress but also protects the environment through an economical, efficient, and safe delivery of the xenobiotics. Besides, it also mitigates the problem of phytotoxicity, active ingredient (a.i.) wastage due to environmental degradation, leaching, etc. and enables convenient handling and distribution and an extended release period of the chemical. CR formulation products increase pest control efficiency through the use of a reduced quantity of toxicant, reduced toxicity to non-target organisms, reduced leaching, and extended residual activity (Scher, 1977). The fundamental properties of CR formulations, i.e., interaction of pest and pesticide, pesticide mobility kinetics, optimization of quantities for bioactivity, ease of handling, cost economics, safety and environment impact–related issues, and how to improve the overall effectiveness of the pesticide have been studied (Mcfarlane and Pedley, 1979). Due to various advantages, numerous examples are available in the literature wherein such products have been effectively employed to combat pests (Fernandez-Perez et al., 1999; Kumar et al., 2007b; Kumar et al., 2010a, 2010b). The advantages of CR formulation of pesticides can be summarized as follows:

- The effective duration of non-persistent pesticides can be prolonged.
- The CR formulation allows much less pesticide to be used for the same period of activity, resulting in less wastage and fewer applications.
- Losses due to environmental factors (evaporation, photolysis, leaching with water, and degradation due to chemical and microbiological factors) are reduced. This results in savings in the cost of the active agent.
- Environmental contamination is reduced.
- Toxicity to non-target species of plants, mammals, birds, fish, and other organisms is reduced.
- Pesticides can be better targeted to desired areas, and their efficacy can be improved.
- CR formulations are safer to applicators, handlers, and others coming into contact with the pesticides.

13.8 Summary

Pesticidal products with improved safety and environment-friendly features are currently in demand for agriculture. To fulfill this demand, CR technology provides the opportunity to develop pesticidal formulations that are safe, more efficient, low cost and labor saving. The CR mechanism involves slow release of the a.i. to the target site, which can be regulated by employing different release mechanisms such as diffusion, erosion, etc. Polymers being an integral part of the CR system, several synthetic and natural polymers are being reported for the development of CR formulations of pesticides. Traditional approaches to the CR formulation of pesticides involved entrapping the a.i. in a polymeric membrane followed by slow release through a diffusion mechanism. More recently, other approaches, such as entrapping pesticide molecules in hydrogel/superabsorbent-like materials, are being developed extensively, in which the release of a.i. is governed by the swelling-controlled diffusion mechanism and the stimuli-response release mechanism. At present, more emphasis is being placed on the development of nanoformulations of pesticides with CR properties. This approach is suited to maximizing bioefficacy along with environment-friendly features. To make pesticide application in agriculture safer and more efficient, in future, more emphasis will be given to developing various CR formulations involving novel devices such as nanoreservoirs, hydrogels, and others.

REFERENCES

Adak, T.; Kumar, J.; Shakil, N.A.; Walia, S. Development of controlled release formulations of imidacloprid employing novel nano-ranged amphiphilic polymers. *J. Environ. Sci. Health Part B*. 2012, *47*(3), 217–225.

Akelah, A.; Hassaneien, M.; Selim, A.; Rehab, A. Preparation and study of functionalized polymers containing pendent pesticide-active groups. *J. Chem. Technol. Biotechnol.* 1987, *37*(3), 169–181.

Allan, G.G.; Chopra, C.S.; Neogi, A.N.; Wilkins, R.M. Design and synthesis of controlled release pesticide-polymer combinations. *Nature*. 1971, *234*, 349–351.

Aouada, F.A.; de Moura, M.R.; Orts, W.J.; Mattoso, L.H.C. Polyacrylamide and methylcellulose hydrogel as delivery vehicle for the controlled release of paraquat pesticide. *J. Mater. Sci.* 2010, *45*, 4977–4985.

Aziz, A.; Trotel-Aziz, P.; Dhuicq, L.; Jeandet, P.; Couderchet, M.; Vernet, G. Chitosan oligomers and copper sulphate induce grapevine defence reactions and resistance to gray mold and downy mildew. *Phytopathology*. 2006, *96*, 1188–1194.

Bai, C.; Zhang, S.; Huang, L.; Wang, H.; Wang, W.; Ye, Q. Starch-based hydrogel loading with carbendazim for controlled-release and water absorption. *Carbohyd. Polym.* 2015, *125*, 376–383.

Baker, R.W. *Controlled Release of Biologically Active Agents*. John Wiley and Sons, New York, 1987, 1–83.

Céspedes, F.F.; Sánchez, M.V.; Garcia, S.P.; Pérez, M.F. (2007). Modifying sorbents in controlled release formulations to prevent herbicides pollution. *Chemosphere. 69*(5), 785–794.

Fan, L.T.; Singh, S.K.Swelling-controlled release. In *Controlled Release: A Quantitative Treatment*, vol. *13*; Cantow, H.J.; Harwood, H.J.; Kennedy, J.P.; Ledwith, A.; Meiβner, J.; Okamura, S.; Henrici-Olive, G.; Olivé, S., Eds.; Springer, 1989, 110–156.

Fernández-Pérez, M.; Gonzalez-Pradas, E.; Urena-Amate, M.D.; Wilkins, R.M.; Lindup, I. Controlled release of imidacloprid from a lignin matrix: water release kinetics and soil mobility study. *J. Agric. Food Chem.* 1998, *46*(9), 3828–3834.

Fernández-Pérez, M.; Villafranca-Sánchez, M.; Flores-Céspedes, F.; Daza-Fernández, I. Ethylcellulose and lignin as bearer polymers in controlled release formulations of chloridazon. *Carbohyd. Polym.* 2011, *83*(4), 1672–1679.

Fernandez-Perez, M.; Villafranca-Sanchez, M.; Gonzalez-Pradas, E.; Flores-Cespedes, F. Controlled release of diuron from an alginate–bentonite formulation: water release kinetics and soil mobility study. *J. Agric. Food Chem.* 1999, *47*, 791–798.

Gallagher, K.M.; Corrigan, O.I. Mechanistic aspects of the release of levamisole hydrochloride from biodegradable polymers. *J. Control. Release.* 2000, *69*(2), 261–272.

Gerstl, Z.; Nasser, A.; Mingelgrin, U. Controlled release of pesticides into soils from clay– polymer formulations. *J. Agric. Food Chem.* 1998, *46*(9), 3797–3802.

Grillo, R.; dos Santos, N.Z.P.; Maruyama, C.R.; Rosa, A.H.; de Lima, R.; Fraceto, L.F. Poly(εcaprolactone) nanocapsules as carrier systems for herbicides: physico-chemical characterization and genotoxicity evaluation. *J. Hazard. Mater.* 2012, *231*, 1–9.

Gupta, P.; Vermani, K.; Garg, S. Hydrogels: from controlled release to pH-responsive drug delivery. *Drug Discov. Today.* 2002, *7*(10), 569–579.

Harris, F.W.; Post, L.K. Synthesis and polymerization of the vinyl and acryloyloxyethyl esters of 2, 4-dichlorophenoxyacetic acid and 2-(2, 4, 5-trichlorophenoxy) propionic acid. *J. Polym. Sci. Polym. Lett. Ed.* 1975, *13*(4), 225–229.

Heller, J.; Baker, R.W. Theory and practice of controlled drug delivery from bioerodible polymers. In *Controlled Release of Bioactive Materials*; Baker, R.W., Ed.; Academic Press, New York, 1980, 1–17.

Hernandez-Munoz, P.; Almenar, E.; Del Valle, V.; Velez, D.; Gavara, R. Effect of chitosan coating combined with postharvest calcium treatment on strawberry (Fragaria× ananassa) quality during refrigerated storage. *Food Chem.* 2008, *110*(2), 428–435.

Higuchi, T. Rate of release of medicaments from ointment bases containing drugs in suspension. *J. Pharm. Sci.* 1961, *50*(10), 874–875.

Hopfenberg, H.B.; Hsu, K.C. Swelling-controlled, constant rate delivery systems. *Polym. Eng. Sci.* 1978, *18*, 1186–1191.

Huang, X.; Brazel, C.S. On the importance and mechanisms of burst release in matrix-controlled drug delivery systems. *J. Control. Release.* 2001, *73*(2), 121–136.

Kaushik, P.; Shakil, N.A.; Kumar, J.; Singh, M.K.; Singh, M.K.; Yadav, S.K. Development of controlled release formulations of thiram employing amphiphilic polymers and their bioefficacy evaluation in seed quality enhancement studies. *J. Environ. Sci. Health Part B.* 2013, *48*(8), 677–685.

Kenawy, E.R. Recent advances in controlled release of agrochemicals. *J. Macromol. Sci. C. Polym. Rev.* 1998, *38*, 365–390.

Kumar, J.; Nisar, K.; Arun Kumar, M.B.; Walia, S.; Shakil, N.A.; Prasad, R.; Parmar, B.S. Development of polymeric seed coats for seed quality enhancement of soybean (Glycine max). *Indian J. Agric. Sci.* 2007a, *77*(11), 738–743.

Kumar, J.; Nisar, K.; Shakil, N.A.; Walia, S.; Parsad, R. Controlled release formulations of metribuzin: release kinetics in water and soil. *J. Environ. Sci. Health Part B.* 2010a, *45*, 330–335.

Kumar, J.; Shakil, N.A.; Singh, M.K.; Pankaj; Singh, M.K.; Pandey, A.; Pandey, R.P. Development of controlled release formulations of Azadirachtin – A employing poly(ethylene glycol) based amphiphilic copolymers. *J. Environ. Sci. Health Part B.* 2010b, *45*(4), 310–314.

Kumar, J.; Walia, S.; Shakil, N.A.; Parmar, B.S. Water and soil release kinetics of Cartap hydrochloride from controlled release formulations. *Pesticide Res. J.* 2007b, *19*(1), 122–127.

Langer, R.; Peppas, N. Chemical and physical structure of polymers as carriers for controlled release of bioactive agents: a review. *J. Macromol. Sci. Rev. Macromol. Chem. Phys.* 1983, *E23*, 61–126.

Lee, P.I. Novel approach to zero-order drug delivery via immobilized nonuniform drug distribution in glassy hydrogels. *J. Pharm. Sci.* 1984, *73*, 1344–1347.

Lewis, D.H.; Cowsar, D.R. Principles of controlled release pesticide. In *Controlled Release Pesticides*; Scher, H.B., Ed.; ACS Symposium Series: Washington, DC, 1977, *53*, 1–16.

Li, D.; Liu, B.; Yang, F.; Wang, X.; Shen, H.; Wu, D. Preparation of uniform starch microcapsules by premix membrane emulsion for controlled release of avermectin. *Carbohyd. Polym.* 2016, *136*, 341–349.

Liu, B.; Wang, Y.; Yang, F.; Wang, X.; Shen, H.; Cui, H.; Wu, D. Construction of a controlled-release delivery system for pesticides using biodegradable PLA-based microcapsules. *Colloid. Surf. B.* 2016, *144*, 38–45.

Liu, S.; Yuan, L.; Yue, X.; Zheng, Z.; Tang, Z. Recent advances in nanosensors for organophosphate pesticide detection. *Adv. Powder Technol.* 2008, *19*, 419e441.

Liu, Y.; Wei, F.; Wang, Y.; Zhu, G. Studies on the formation of bifenthrin oil-in-water nano-emulsions prepared with mixed surfactants. *Colloid. Surf. A.* 2011, *389*(1), 90–96.

Loha, K.M.; Shakil, N.A.; Kumar, J.; Singh, M.K.; Adak, T.; Jain, S. Release kinetics of *β*-Cyfluthrin from its encapsulated formulations in water. *J. Environ. Sci. Health Part B.* 2011, *46*, 201–206.

Loha, K.M.; Shakil, N.A.; Kumar, J.; Singh, M.K.; Srivastava, C. Bio-efficacy evaluation of nanoformulations of β-cyfluthrin against *Callosobruchusmaculatus* (Coleoptera: Bruchidae). *J. Environ. Sci. Health B.* 2012, *47*(7), 687–691.

McFarlane, R.N.; Pedley, B.J. Some fundamental considerations of controlled release. *Weed Sci.* 1978, *9*, 411–424.

Ninni, L.; Ermatchkov, V.; Hasse, H.; Maurer, G. Swelling equilibrium of hydrogels of (N-isopropyl acrylamide+ anionic and cationic comonomers) in aqueous solutions of sodium chloride: experimental results and modeling. *Fluid Phase Equilibr.* 2013, *337*, 137–149.

Pankaj; Shakil, N.A.; Kumar, J.; Singh, M.K.; Singh, K. Bioefficacy evaluation of controlled release formulations based on amphiphilicnano-polymer of carbofuran against *Meloidogyne incognita* infecting tomato. *J. Environ. Sci. Health Part B.* 2012, *47*(6), 520–528.

Parmar, B.S.; Tomar, S.S. Preparation and performance. In *Pesticide Formulation: Theory & Practice*, 1st edn; Parmar, B.S.; Tomar, S.S., Eds.; CBS Publishers &Distributers, New Delhi, 2004, 8–142.

Peppas, N.A. Mathematical modeling of diffusion processes in drug delivery polymeric systems. In *Controlled Drug Bioavailability, Vol. I, Drug Product Design and Performance*; Smolen, V.F.; Ball, L.A., Eds.; Wiley, New York, 1984, 206–237.

Pitt, C.G.; Schindler, A. Biodegradation of polymers. In *Controlled Drug Delivery, Vol. I, Basic Concepts*; Bruck, S.D., Ed.; CRC Press, Boca Raton, FL, 1983, 53–80.

Qi, L.F.; Xu, Z.R.; Jiang, X.; Hu, C.H.; Zou, X.F. Preparation and antibacterial activity of chitosan nanoparticles. *Carbohyd. Res.* 2004, *339*, 2693–2700.

Rao, K.N.; Srivastava, K.P.; Parmar, B.S. Development of controlled release phorate formulations and their evaluation for pest control and grain yield on sorghum. *Pesticide Res. J.* 1989, *1*(1), 7–11.

Roy, A.; Bajpai, J.; Bajpai, A.K. Dynamics of controlled release of chlorpyrifos from swelling and eroding biopolymeric microspheres of calcium alginate and starch. *Carbohyd. Polym.* 2009, *76*(2), 222–231.

Rudzinski, W.E.; Dave, A.M.; Vaishnav, U.H.; Kumbar, S.G.; Kulkarni, A.R.; Aminabhavi, T.M. Hydrogels as controlled release devices in agriculture. *Des. Monomers Polym.* 2002, *5*, 39–65.

Sarkar, D.J.; Kumar, J.; Shakil, N.A.; Adak, T.; Watterson, A.C. Synthesis and characterization of amphiphilic PEG based aliphatic and aromatic polymers and their self-assembling behavior. *J. Macromol. Sci. Pure & Appl. Chem. A.* 2012, *49*(6), 455–465.

Sarkar, D.J.; Kumar, J.; Shakil, N.A.; Walia, S. Quality enhancement of soybean seed coated with nanoformulated thiamethoxam and its retention study. *Pesticide Res. J.* 2012a, *24*(1), 55–64.

Sarkar, D.J.; Kumar, J.; Shakil, N.A.; Walia, S. Release kinetics of controlled release formulations of thiamethoxam employing nano-ranged amphiphilic PEG and diacid based block polymers in soil. *J. Environ. Sci. Health Part A.* 2012b, *47*(11), 1701–1712.

Sarkar, D.J.; Singh, A. Base triggered release of insecticide from bentonite reinforced citric acid crosslinked carboxymethyl cellulose hydrogel composites. *Carbohyd. Polym.* 2017, *156*, 303–311.

Scher, H.B. Microencapsulated pesticide. In *Controlled Release Pesticides*; Scher, H.B., Ed.; ACS Symposium Series: Washington, DC, 1977, 126–144.

Shakil, N.A.; Singh, M.K.; Pandey, A.; Kumar, J.; Pankaj, B.; Parmar, V.S.; Pandey, R.P.; Watterson, A.C. Development of poly (ethylene glycol) based amphiphilic copolymers for controlled release delivery of carbofuran. *J. Macromol. Sci. Pure & Appl. Chem. A.* 2010, *47*, 241–247.

Shasha, B.S.; Doane, W.M.; Russell, C.R. Starch-encapsulated pesticides for slow release. *J. Polym. Sci. Polym. Lett. Ed.* 1976, *14*(7), 417–420.

Sinclair, R.G. Slow-release pesticide system. Polymers of lactic and glycolic acids as ecologically beneficial, cost-effective encapsulating materials. *Environ. Sci. Technol.* 1973, *7*(10), 955–956.

Wilpiszewska, K.; Spychaj, T.; Paździoch, W. Carboxymethyl starch/montmorillonite composite microparticles: properties and controlled release of isoproturon. *Carbohyd. Polym.* 2016, *136*, 101–106.

Yang, F.L.; Li, X.G.; Zhu, F.; Lei, C.L. Structural characterization of nanoparticles loaded with garlic essential oil and their insecticidal activity against Triboliumcastaneum (Herbst) (Coleoptera: Tenebrionidae). *J. Agric. Food Chem.* 2009, *57*, 10156–10162.

14

Microcapsule-Assisted Smart Coatings

Mahendra S. Mahajan and Vikas V. Gite

CONTENTS

14.1 Conventional and Smart Coatings

These days, everyone wants a product with an aesthetic look along with performance. Such products can be cars, televisions, computers, household items, etc., and their aesthetic look is normally achieved by the application of decorative coatings.

The performance of articles can also be measured in terms of anticorrosive properties, because the corrosion of metals is a serious issue in all metal industries due to many problems such as leakages, explosions, and other consequences. All over the world, approximately 4% of gross domestic product is spent on account of the replacement or repair of rusted materials.[1,2] Hence, the protection of metal against corrosion has emerged as a major economical investment for industrial sectors. To avoid such losses, polymeric coatings or paints are playing a significant role. In general, a coating is a type of covering applied to the surface of an object as a continuous film used for protection and decoration purposes. The main constituents used for coatings are polymeric binders, pigments, solvents, and additives. In this case, binders or resins are responsible for the formation of a protective layer and its strong adherence to the substrate. Pigment mostly provides color and improves some mechanical properties of the paint film. Solvents or thinners are used to control the viscosity of the coatings,[3] whereas additives are used in small quantities for specific purposes.

During day-to-day use, coated articles come into contact with various environmental factors such as light, moisture, manmade chemicals, mechanical pressure, etc., which sometimes create cracks in the coatings. Hence, at the point of damage, the metal surface interacts with corrosive factors such as moisture and O_2, which lead to the start of corrosion. Thus, conventional coatings are able to control the corrosion process to a certain extent but unable to defend metal against corrosion after mechanical damage. In such cases, corrosion protection can be enhanced by adding corrosion inhibitors to the coating formulation. The common conventional anticorrosive compounds used are phosphates, nitrites, chromates, etc., which can inhibit the corrosive process, but the damage cannot be repaired automatically. Likewise, they have own merits and demerits; they are widely toxic in nature, damaging to the environment, and hazardous to human health. Moreover, their direct addition to a coating formulation may demonstrate a negative impact on the performance of the coatings. The exposure of the protected substrate leads to the beginning of wet and dry corrosion processes. Therefore, cracks or damage must be repaired manually, but this process is insufficient and not accurately identifiable. Recently, self-healing coatings have been developed, which have extra ability to protect metal surfaces, far better than conventional systems.

Self-healing coatings repair cracks automatically without any human intervention. One of the types of self-healing coating contains an encapsulated healing agent in the form of microcapsules. The healing materials are entrapped in the microcapsules as a storage material and possess the capability to polymerize after the breakdown of the capsules. These microcapsules are incorporated into coatings to be applied to metal substrates. When cracks occur on the coated surface, the loaded microcapsules break, and encapsulated healing material oozes out and spreads on the exposed substrate. This agent can cure by several mechanisms, such as oxidative curing, moisture curing, ultraviolet (UV) light curing, or other processes, resulting in a new passive layer being formed on the cracked surface. Thus, it repairs or heals the cracks and protects the metal from further deterioration. Such coatings are also formulated to show two types of functions: healing of cracks and protection from corrosion.

14.2 Types of Smart Self-Healing Coatings

Smart materials change their properties in response to external stimuli.[4] A smart self-healing coating works in a similar way; it responds to various external factors and is responsible for the healing of cracks without human intervention. The phenomenon is similar to the natural healing processes of wounds in living things. Nature provides self-defense against damage for the recovery of stable operation. Similarly, a self-healing material possesses the functionality to recover from damage or cuts resulting from mechanical or long-time use. These coatings can be classified into two types, as shown in Figure 14.1.

14.2.1 Extrinsic Self-Healing

The mechanism of extrinsic self-healing involves the addition of an entrapped self-healing agent in external materials such as microcapsules or vascular materials possessing a self-healing agent. The microcapsule comprises a polymeric shell and healing agent (the core). At the time of coating formulation, microcapsules are added to the coating matrix. When the coated surface is damaged and scratches form, the added polymeric microcapsules break, releasing healing material, which spreads onto the scratches. The healing agent forms a liquid layer on the damaged surface and is

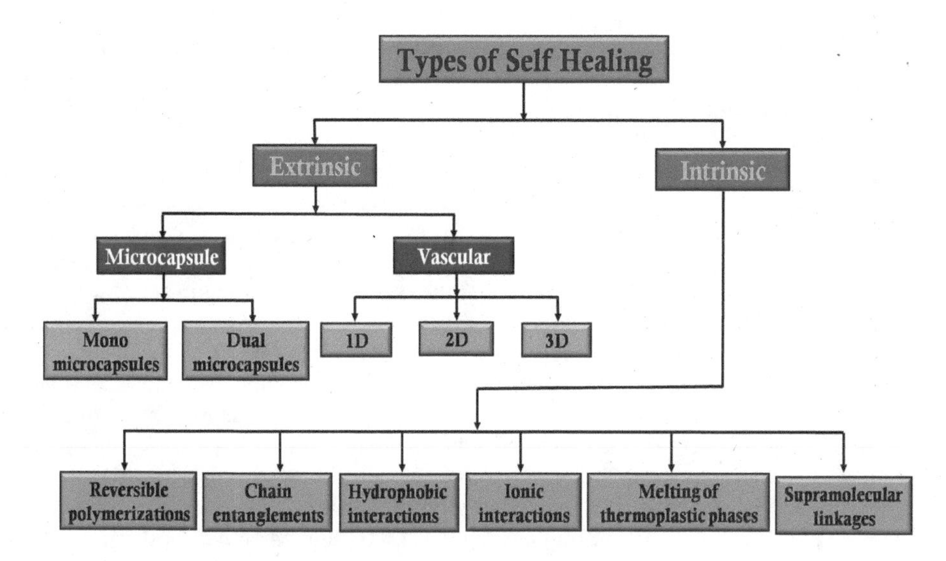

FIGURE 14.1 Types of self-healing system.

converted into a new polymeric solid protective layer with the help of atmospheric moisture, light, or other factors.

The development of a microcapsule-based self-healing coating was first reported by White et al.[5] In their system, dicyclopentadiene (DCPD) and transition metal (Grubbs' catalyst) were used as healing agent and catalyst, respectively, in a urea formaldehyde (UF) polymeric shell. Further, using the same approach, a number of researchers modified microcapsule-based self-healing coatings.[6–8]

14.2.2 Intrinsic Self-Healing Coatings

Intrinsic self-healing systems have repeated healing ability toward external scratches. The repair of damage is based on different mechanisms: reversible polymerizations,[9–13] hydrophobic interactions,[14,15] ionic interactions,[16] chain entanglements,[17,18] supramolecular linkages,[19,20] melting of thermoplastic phases, etc.[21,22]

14.3 Classification of Extrinsic Self-Healing Coatings

In this type of coating, the healing is achieved by the breaking of the container and releasing and curing the preloaded healing agent on the scratched surface of a coating matrix. Based on the type of container, extrinsic self-healing can be subdivided into capsule- and vascular-based healing.

14.3.1 Capsule-Based Self-Healing

In capsule-based self-healing, capsules containing the active healing agent are uniformly dispersed in a variety of coating formulations, and at the time of scratching, healing is triggered over the damaged surface by curing of the healing agent that is released from the broken microcapsules. This kind of self-healing mechanism can be further classified into two types: mono-capsules and dual capsules.[23]

A mono-capsule-based system consists of polymeric microcapsules with only healing agent, which does not require any catalyst or hardener for curing or repair. Instead, it is cured by atmospheric oxygen or moisture or other factors. Several healing agents have been used to prepare this type of healing system, which include linseed,[6,24,25] tung,[26,27] and soybean oils.[28] In addition to oils, diisocyanate-containing capsules are also prepared for their use in mono-capsule-based self-healing coatings.[29,30] Even the anticorrosive performance of coatings formulated with a mono-capsule-based system can also be improved using a corrosion inhibitor (quinoline) as a core in a polyurea protective shell.[31] Similarly, epoxy coatings formulated with this system have achieved dual functional properties, that is, corrosion and self-healing, by encapsulating mercaptobenzothiazole as a corrosion inhibitor and linseed oil as an active healing agent in a polyurea formaldehyde shell.[32]

In some cases, healing materials are encapsulated, and catalyst is added directly into the polymeric coating. When the damage occurs on the coating, the capsule breaks and releases the healing agent, which forms a new passive layer on the scratched surface after contacting the catalyst present in the binder. Reported capsule and catalyst–based systems include DCPD as a healing agent and tungsten (VI) chloride as a catalyst[33] and polydimethoxysiloxane (PDMS) as a healing material with dibutyltin dilaurate (DBTDL) and platinum as catalysts.[34] Further, epoxy resin is used as a healing agent along with boron trifluoride diethyl etherate ($(C_2H_5)_2OBF_3$)) as an active catalyst.[35]

A dual capsule–based system includes two capsules, one with a healing agent and the other with a hardener. At the time of damage to the coating substrate, both kinds of capsule are broken, and the healing agent and the hardener are released separately from the capsules. After their contact with each other, they form a new layer on the damaged surface. Reported dual capsule–based self-healing systems include epoxy as a repairing resin along with aliphatic polyamide,[36] pentaerythritol tetrakis(3-mercaptopropionate),[37] BF_3–amine complex,[38] mercaptan,[39] tetrathiol,[40] etc. as hardeners. Figure 14.2 represents a capsule-based self-healing system.

14.3.2 Vascular-Based Self-Healing Mechanism

In the case of vascular-based self-healing, a vascular network of hollow glass fibers is used to encapsulate the healing agent, and the healing mechanism is similar to that of the capsule-based formulation except for different synthesis and addition methods. The healing agent is isolated in the vascular network and properly dispersed in a coating. When the coating cracks, the vascular network breaks, and healing agent is released to form a fresh polymer film after curing, as mentioned for the capsule-based system. The vascular-based system has some advantages over the capsule-based system; that is, a higher loading capacity and external refilling possibility of the healing agent after its first delivery. Thus, a vascular-based system has the ability to heal cracks repeatedly and more efficiently, and it provides reinforcement to the coating matrix.[41]

On the basis of arrangements of vascular networks, they may be divided into three types: one- (1-D), two- (2-D), and three-dimensional (3-D). A 1-D network is developed by integrating hollow glass fibers or a pipeline in a 1-D arrangement filled with suitable types of healing agent.[42,43] An epoxy healing agent loaded in hollow glass fibers and hardener infiltrated into different types of glass fibers have been used for formulating 1-D healing systems.[44] After the breaking of the hollow glass fibers present in the coating, the healing agent is released over the damaged area through capillary action and polymerizes to form a new passive layer to complete its self-healing ability. In 2-D and 3-D self-healing, the interconnected microvascular networks are arranged in a 2-D and 3-D manner, respectively, keeping the rest of the mechanism similar to the 1-D system. Such arrangements are suitable for repairing multiple harmful actions by a continuous flow of the healing materials present in the network structure.[45,46]

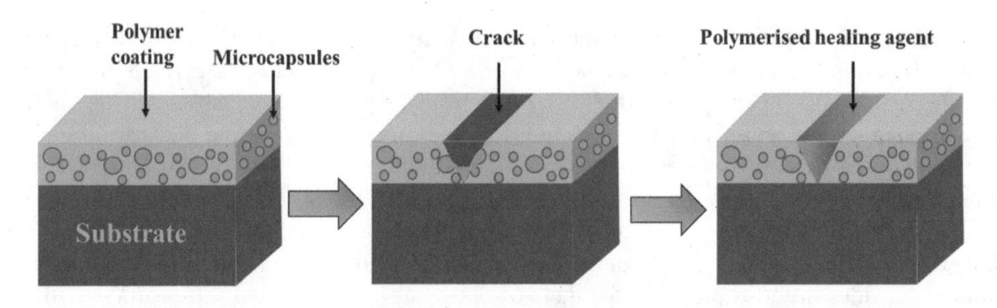

FIGURE 14.2 Representation of capsule-based self-healing system.

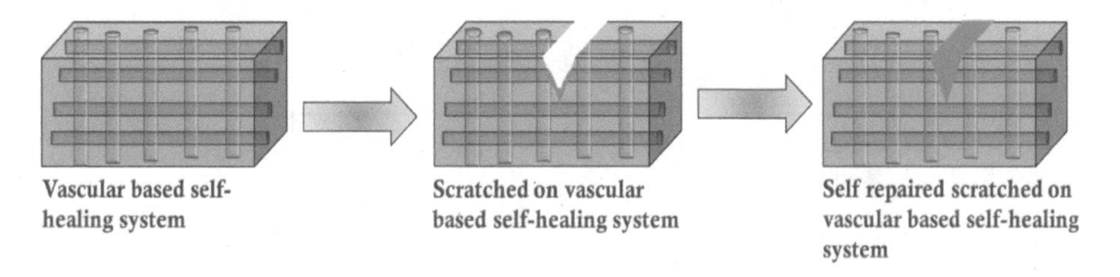

FIGURE 14.3 Representation of vascular-based self-healing system.

The major difficulties in enhancing the efficiency of the vascular self-healing system are the isolation of the healing agent and its proper arrangement into polymeric substrates. Further, a number of healing agents possess high viscosities and low wetting properties that directly affect the filling of the vascular network and chemical incompatibility. Figure 14.3 represents a vascular-based self-repairing mechanism.

14.4 Concept of Microencapsulation

Microencapsulation is a technique used to isolate solid, liquid, or gaseous active substances from the surrounding environment. Encapsulation of materials is needed for a number of reasons, such as increasing the stability and shelf-life of materials, easy handling of hazardous materials, masking of test substances, handling of liquid and gaseous materials in the solid state, separating incompatible components, and controlling the release of materials in a suitable time and intervals. The encapsulated materials are called the *core*, whereas the surrounding polymeric layer is referred to as the *shell*, which gives protection to the core component, and finally, the formed structure is termed a *microcapsule*. This technique has been used in various fields, such as self-healing and anticorrosive coatings, pharmaceuticals,[47] printing,[48] controlled release of pesticides,[49] chemicals, cosmetics,[50] etc. The microencapsulation method has gained importance with the opening up of new applications, which are shown in Figure 14.4.

14.5 Components of Microcapsules

The important components of microcapsules are the core and the polymeric shell. The core is the substance that is present inside the polymeric shell wall, which protects it from the surroundings. The components of microcapsules are represented in detail in Figure 14.5 and will be described further.

14.5.1 Shell Materials

The shell of a microcapsule is an outer layer used to protect the inner core material from atmospheric conditions. The selection of the shell is crucial in the preparation of microcapsules, as it decides most of

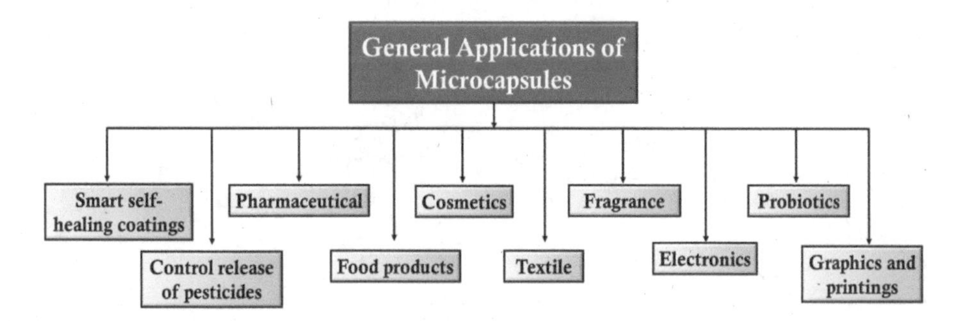

FIGURE 14.4 General applications of microcapsules.

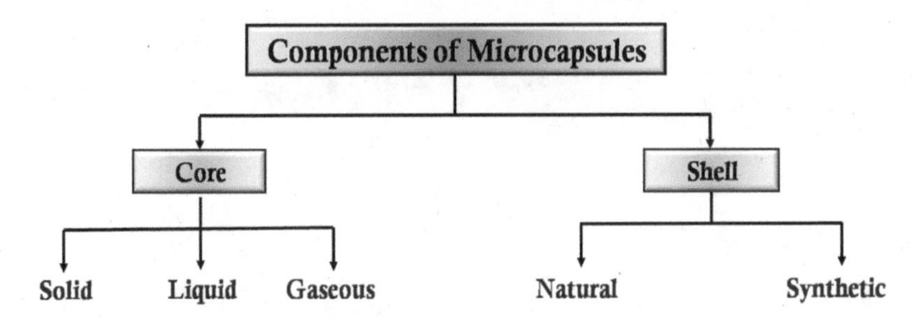

FIGURE 14.5 Important components of microcapsules.

the properties of capsules, that is, thermal stability, release behavior, protection of the core, and stability against several environmental factors. General features of shells include chemical inertness, non-reactive nature against the core moiety, and the formation of solid layer. Additionally, they provide strength, stability, permeability, flexibility, etc. to microcapsules.[51] The following are the common shell materials used in the preparation of microcapsules for the development of self-healing coatings.

14.5.1.1 Phenol Formaldehyde

The phenol formaldehyde (PF) type of shell is the most used material in the preparation of microcapsules and even for controlled release formulations of pesticides.[102] Under basic conditions, phenol and formaldehyde are reacted to form mono-, di-, and tri-methylol phenols. Simultaneously, the methylol groups react with each other to form methyl ether or methylene bridges. Further, the cross-linking process is carried out in either the presence or the absence of a cross-linker under acidic conditions. In the entire process, PF being organic in nature, the healing agent concentrates inside the sphere as a core, and simultaneously, a shell forms around it. The final structure of the capsule totally depends on the reaction conditions, including temperature, quantity of emulsifier, stirring speed, amount of core material, and mole ratio of phenol to formaldehyde. So far, various core materials such as linseed oil, tung oil, etc. have been encapsulated into PF shells.[6,52,53] The representative reaction of PF shell formation is shown in Figure 14.6 with resorcinol as a cross-linker.

14.5.1.2 Urea Formaldehyde

The UF shell is another widely used material for the encapsulation of self-healing agents. It is formed by the condensation of urea with formaldehyde under acidic or basic conditions. The final nature of the capsules is spherical, and they are obtained as a free-flowing powder after drying. There are two general methods of UF capsule preparation. The basic method goes through two steps: formation of a prepolymer and cross-linking after the addition of cross-linking agent. The first step of the reaction involves major formation of mono-, di-, and tri-methylol ureas, whereas in the second step, the methylol groups condense to form a cross-linked UF shell. The final structure of the capsule depends on the urea to formaldehyde ratio, reaction temperature and pH, stirring speed, and time.[54] If the pH of the reaction

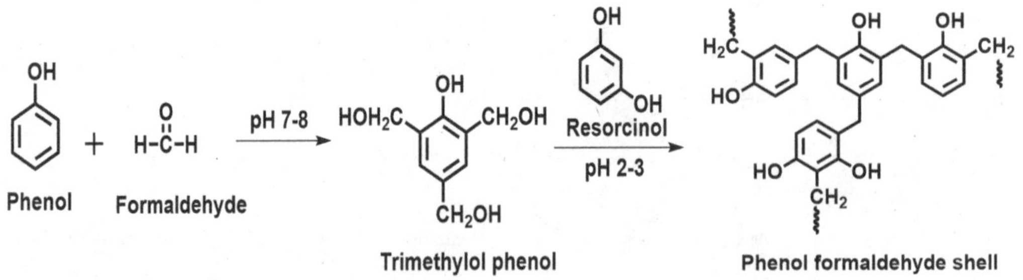

Phenol + Formaldehyde —pH 7-8→ Trimethylol phenol —Resorcinol, pH 2-3→ Phenol formaldehyde shell

FIGURE 14.6 Representative reaction of PF shell formation using resorcinol.

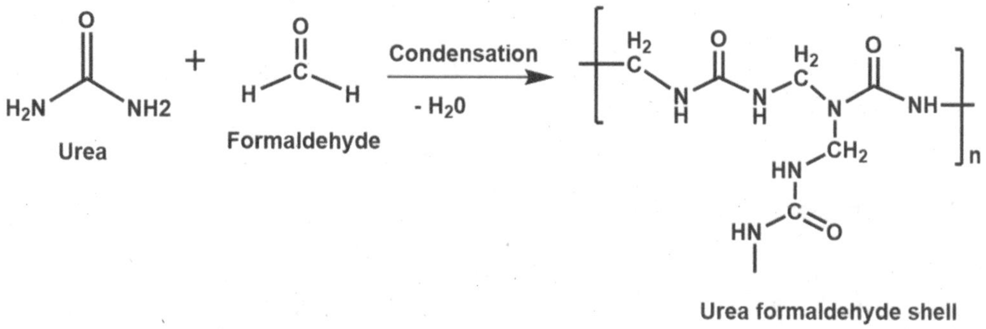

FIGURE 14.7 Representative reaction of UF shell formation.

mixture is too high, the polymerization of methylol urea leads to the formation of a cured prepolymer unsuitable for a shell.

In the acidic method, shell formation does not require precondensation, and the entire process is completed in a single step. On comparison of both the processes, the acidic method is time saving and simple, and it is easy to adjust the pH during the reaction.

The diameter of the capsules can be optimized by maintaining the proper stirring speed, because the higher the agitation speed, the lower will be the diameter of the capsules, and vice versa. Commonly used healing materials encapsulated into UF-based shells are DCPD,[55] linseed oil,[56] tung oil,[27] epoxy resin,[57] etc. The representative reaction for the formation of a UF shell is given in Figure 14.7.

14.5.1.3 Melamine Formaldehyde

The condensation of melamine with excess of formaldehyde can be used to prepare a melamine formaldehyde (MF) shell of microcapsules in acid or basic media. In the preparation process, first, an MF prepolymer is formed containing one to six methylol groups, while the second step involves polycondensation of these groups to methylene bridges. The polycondensation occurs on the surface of core droplets to form a water-insoluble cross-linked MF shell. The chemical composition and structural arrangements of the shell are totally dependent on the melamine to formaldehyde ratio, pH of the reaction medium, and reaction time. Due to the higher functionality of melamine, MF shells possess better hardness and higher heat resistance than UF and PF shells due to the presence of a heterocyclic ring. MF shells have been used for encapsulating different types of core materials including linseed oil,[58] fragrant oils,[59,60,61] epoxy resin,[62] DCPD,[63] etc. The representative MF shell formation reaction is given in Figure 14.8.

14.5.1.4 Polyurea

Polyurea shell material is normally synthesized by the condensation reaction of different types of diisocyanates and diamines. The structure, size, thickness, and morphology of shell walls are dependent on the raw materials used and reaction conditions such as temperature and agitation speed. A comparative

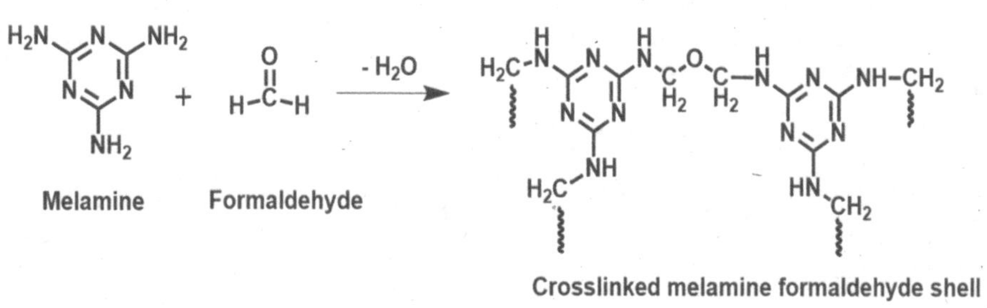

FIGURE 14.8 Representative reaction of MF shell formation.

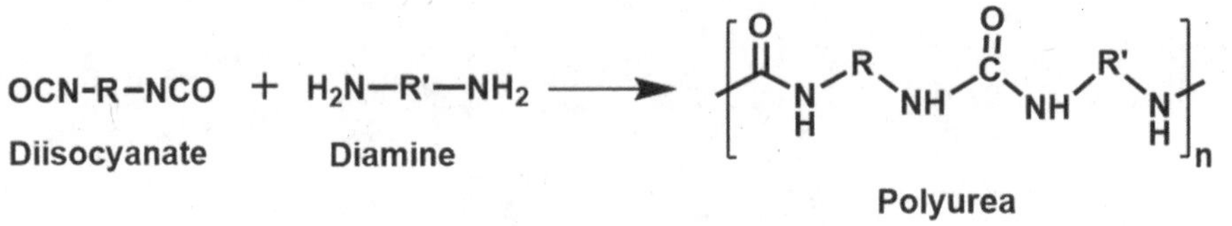

OCN-R-NCO + H_2N—R'—NH_2 → Polyurea

Diisocyanate Diamine

FIGURE 14.9 Representative reaction of polyurea shell formation.

study investigating the morphologies and release behaviors of isophorone diisocyanate (IPDI) and toluene diisocyanate (TDI)[64] indicated that the wall thickness and smoothness of a capsule increase with increasing percentage of IPDI due to its lower reactivity than TDI. Polyurea microcapsules are developed using poly(amidoamine) (PAMAM) dendrimers as the amine part by reacting it with methylene diphenyl diisocyanate (MDI) and hexamethylene diisocyanate (HDI) for formulating self-healing anticorrosive coatings containing linseed and tung oils as healing agents.[65–67] The thermal stability of PAMAM-based polyurea microcapsules is higher than that of PF- and UF-based shells, even these microcapsules found to have elastic nature. The representative reaction for the preparation of polyurea shell material is given in Figure 14.9.

14.5.1.5 Polyurethane

Polyurethane (PU) is a group of polymers containing urethane [-NH-COO-] groups in their backbone chains. They have many applications as polymer binders due to their excellent property profile and the possibility of tailoring their structure.[68] PU has been used as a shell material for microcapsule preparation by the addition reaction of diisocyanates with dihydroxy compounds. Different types of polyols used in the preparation of PU shells include 1,4-butandiol, 1,6-hexanediol, glycerol,[69] etc. Renewable dimer acid–based polyols are also used in combination with TDI as a diisocyanate compound to prepare PU shells.[70] Likewise, the formulation of self-healing coatings using PU-based microcapsules has been developed from glycerol and TDI.[71] The representative reaction of PU shell formation is given in Figure 14.10.

14.5.2 Core Materials

Core materials used in the preparation of microcapsules can be in solid, liquid, or gaseous states. Their selection depends on the type of shell and the final applications of the microcapsules. Core materials used in the preparation of microcapsules for self-healing coatings are in liquid form and include linseed oil, tung oil, DCPD, epoxy resin, IPDI, HDI, corrosion inhibitors, etc. As the core oozes out from the microcapsules, it involves different healing mechanisms, which normally depend on the selection of healing agent. Even the performance of self-healing coatings is mainly decided by the types of healing agent used, and hence, their selection is crucial. The general characteristics required for the core in its use for a self-healing mechanism are the presence of inertness within the shell material, liquid state, free-flowing nature, ability to seal the crack without any difficulty, etc. If the core is reactive to the shell, the properties of the material might change and directly affect its self-repairing and release mechanisms over a damaged surface. The general healing mechanisms represented by healing materials are described in the following section.

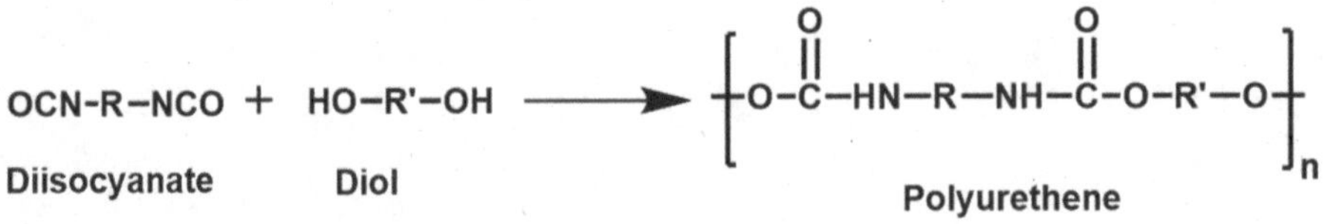

OCN-R-NCO + HO-R'-OH → $\left[\text{O-C(=O)-HN-R-NH-C(=O)-O-R'-O} \right]_n$

Diisocyanate Diol Polyurethene

FIGURE 14.10 Representative reaction of PU shell formation.

14.6 Healing Mechanisms

14.6.1 Moisture Cured

In this mechanism, a healing agent or core polymerizes in the presence of atmospheric moisture and forms a thin film. Diisocyanates are good candidates for moisture-cured self-healing coatings[72] because of their inherent property to react with moisture. The reaction of diisocyanate with moisture forms polyurea, and the properties of the resultant polymer are decided by the type of diisocyanate. However, diisocyanates have some disadvantages, such as difficulty of storage, lower stability due to their high reactivity, presence of toxicity, and high vapor pressure. Therefore, sometimes their prepolymers are used in encapsulation as a core for self-healing coatings. This is a catalyst-free, one-pack, and mono-capsule-based healing system.

In the case of a moisture-cured self-healing mechanism based on microcapsules, the formation of a scratch is responsible for the breaking of microcapsules, release of diisocyanate, and formation of polyurea film on the reaction of diisocyanate with moisture.

Several attempts have been made to use IPDI as a healing agent in different types of shells for designing a variety of self-repairing coating formulations, including PU capsules[29] and UF capsules.[73,74] In a similar manner, monomeric HDI is also used as a healing agent in a MDI prepolymer–based PU shell for formulating self-healing coatings using epoxy binders.[75] The prepared microcapsules are normally found with 5–350 μm diameter and 1–15 μm shell thickness. Moreover, derivatives of diisocyanates, such as a trimer of HDI, have also been used as a core material with polyurea as a shell material.[76]

Even water-borne PU paint has been successfully encapsulated within PU microcapsules. The synthesized microcapsules showed an average diameter of 39–72 mm with 3.8-5.5 mm thickness of shell and repairing efficiency in the range of 47–100% by loading various levels of healing agent.[77]

The main challenge in the development of moisture-cured self-healing is the encapsulation of diisocyanates or their derivatives into a polymeric shell due to the possibility of their undesired reaction with entrapped water during capsule formation. The moisture-cured self-repairing mechanism of diisocyanate healing agent is represented in Figure 14.11.

14.6.2 Oxidative Polymerization

In oxidative polymerization, healing agents have the ability to polymerize in the presence of atmospheric oxygen and form a new thin film. General healing agents used for oxidative curing systems are linseed, tung, and dehydrated castor oils. If the drying oil is added into the coating directly, it starts polymerization after having contact with oxygen and will not be available for healing damage as and when required. Hence, encapsulation in a suitable container avoids its contact with oxygen and unwanted interactions with the coating matrix and achieves self-healing ability over the damaged surface.

In the oxidative polymerization mechanism, the initially encapsulated drying oil releases from the broken microcapsules and fills or covers the scratch. Then, it starts to oxidatively polymerize in the presence of oxygen and creates a new passive layer on the cracked surface that increases the life span of the substrate against attack by a corrosive environment.

Linseed oil is the most used in self-healing coatings because of low cost, easy availability, and the presence of substantial unsaturation. The beauty of oxidative polymerization is that it does not need an additional catalyst and heals the crack simply when it comes into contact with atmospheric oxygen. In the

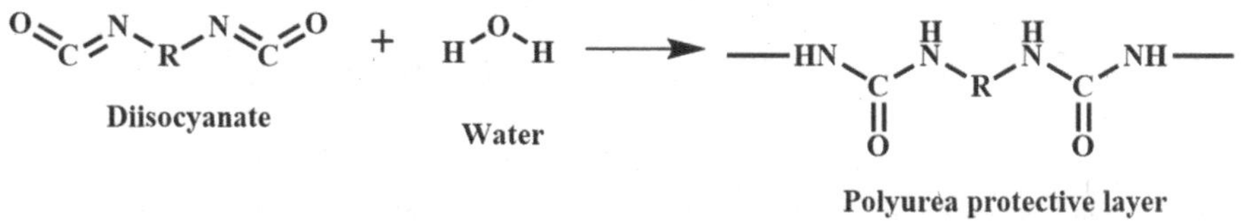

Diisocyanate
Water
Polyurea protective layer

FIGURE 14.11 General reaction of diisocyanate healing agent with atmospheric moisture.

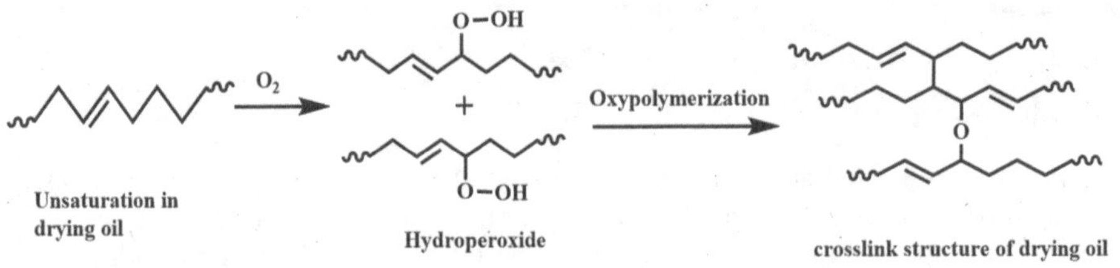

FIGURE 14.12 Representative self-healing reaction of drying oil.

same way, tung oil can also be used as a self-healing agent for developing anticorrosive smart self-healing coatings.[67] As compared with tung oil, the curing time of linseed oil is lower due to the presence of non-conjugated double bonds in linoleic and linolenic acids. However, the double bonds present in tung oil are in conjugation through the presence of eleostearic acid, which leads to faster oxidation and polymerization.[25,67,78] The representation of the healing mechanism of drying oil is given in Figure 14.12.

14.6.3 Hardener Cured

In this type of curing system, the polymeric liquid resin and hardener are used as self-healing agents in combination. For this, the resin and the hardener are encapsulated in separate microcapsules due to their high reactivity toward each other, and the system falls into the category of a dual capsule–based system. However, in some cases, only the self-healing agent is encapsulated, while the hardener is added directly to the coating matrix. At the time of scratch or breakdown, capsules are broken, and the resin comes into contact with the hardener, forming a new polymeric layer. Hardeners used for curing epoxy resins include mercaptans, amines, anhydrides, phenols, etc. In such cases, the epoxy matrix possesses the added advantage of high adhesion. Amine curing agents are more reactive and show amphoteric nature.[37] However, they should not be entrapped in a UF-based shell, as in an acidic medium, they may be easily deactivated.

A hardener-cured self-healing system has been developed that comprises an epoxy resin encapsulated in a UF shell and $CuBr_2(2\text{-MeIm})_4$, a complex of $CuBr_2$ and 2-methylimidazole, dissolved in the coating matrix[79] as a hardener. When the damage occurs, the dispersed capsules break and release isolated healing agent, which is cured after coming into contact with $CuBr_2(2\text{-MeIm})_4$ present in the coating. However, the system requires a high temperature, 130–170 °C, for healing damage.

The use of epoxy resins as a healing agent in epoxy matrix is advantageous due to their structural symmetry.[80] Figure 14.13 represents the general reaction of a hardener-cured self-repairing system.

14.6.4 Catalyst Cured

The healing agents used in the catalyst-cured system require a suitable catalyst to cure or repair the cracks. Important monomers and polymers used in catalyst-cured self-healing systems include DCPD, ethylidene norbornene (ENB), poly(dimethylsiloxane), etc. In the common mechanism, healing agent is released from the capsules as in the previous case and comes into contact with the catalyst, which

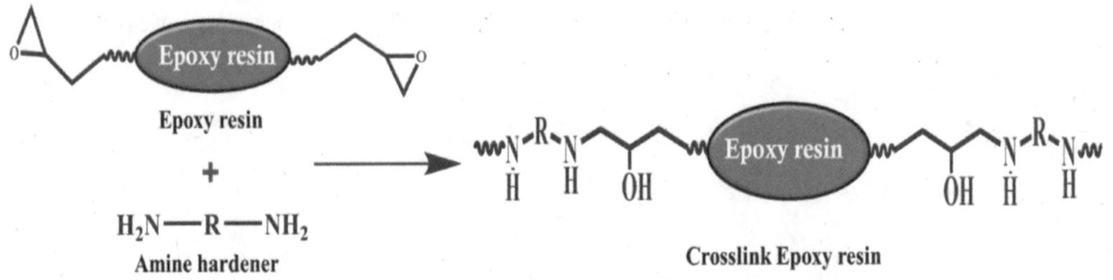

FIGURE 14.13 Representative reaction of hardener cured self-healing system.

is present in the polymer matrix and starts to heal the crack by curing. The preferred healing agent in catalyst-cured systems is DCPD, which was first used by White et al. for developing a catalyst-cured self-healing system. They used PF as a shell for encapsulation with DCPD monomer as a self-healing agent and preloaded Grubbs' catalyst in the epoxy matrix. The healing mechanism goes through the ring opening metathesis polymerization of DCPD in the presence of Grubbs' catalyst. The prepared repairing system recovers about 75% average fracture load. The main reasons for the popularity of DCPD are that it has a longer shelf-life, a high rate of polymerization even at room temperature, low viscosity, and low volatility. Commercially, DCPD is available in endo and exo isomeric forms. Due to the presence of different steric interactions, the exo form of DCPD is more reactive and faster in curing than its endo form.[81] Due to high reactivity, exo-DCPD shows very fast gelation and requires a very short time (5–10 min) for catalyst dissolution. Consequently, released healing agent on a scratched surface may be completely cured with poor self-healing performance.[82]

Despite many advantages, the DCPD/Grubbs' catalyst system has some disadvantages, such as compatibility of DCPD and Grubbs' catalyst in the coating matrix, agglomeration, improper dissolution of catalyst, and inadequate interactions of healing agent with the catalyst.[83,84]

This problem is addressed by blending DCPD and ENB in a 1:3 ratio, which increases the rigidity of the cured film, requires a small amount of catalyst, and decreases the curing time. In this case, ENB is responsible for increasing the curing rate, whereas DCPD contributes to increasing rigidity.[80,85] DCPD is commonly used as a healing agent along with first-, second-, and third-generation Grubbs' catalysts.[86,87] Other catalysts, such as WCl_2, tungsten (VI) chloride[33], and scandium (III) triflate[88] have also been investigated due to their cost-effectiveness, higher thermal stability, and high availability. The representative reaction of catalyst-cured self-healing is given in Figure 14.14.

14.6.5 UV and Photochemical

The healing agents can also be cured in the presence of UV or sunlight. The repairing of scratches involves rearrangement, reversible bond breaking and forming, shape memory, or maybe other mechanisms. The system is eco-friendly, as it is very fast and does not require any catalyst, heat treatment, hardener, or additive.

Photo-initiated cycloaddition reactions occurring at definite wavelengths of light are also used for bond formation or self-healing. Light-induced reactions are simple and mostly used in the development of self-healing coatings.

The initially reported light-induced self-healing materials are synthesized by particular photochemical cycloaddition of cinnamoyl groups.[89] In this case, the self-healing ability is due to the photo-cycloaddition and re-cycloaddition rearrangements of cinnamoyl groups.

Particular types of unsaturated compounds with possibilities of [2+2] cycloaddition reactions are used to form a cyclobutene functionality by photo-radiation.[90] In these reactions, covalent bonds are formed through the presence of a short irradiation wavelength of light to give the starting olefins. Photosensitive groups are used to make polymeric materials to be used in self-healing coatings because of their reversible cross-linking reaction. The compounds showing reversible rearrangements of bonds under UV light are photosensitive molecules such as coumarin, anthracene, maleimide, cinnamic acid, and butadiene. Taking advantage of the reversible rearrangement of coumarin, researchers developed a photo-assisted self-healing system containing PU-bonded dihydroxy coumarin derivatives.[91] The prepared healing

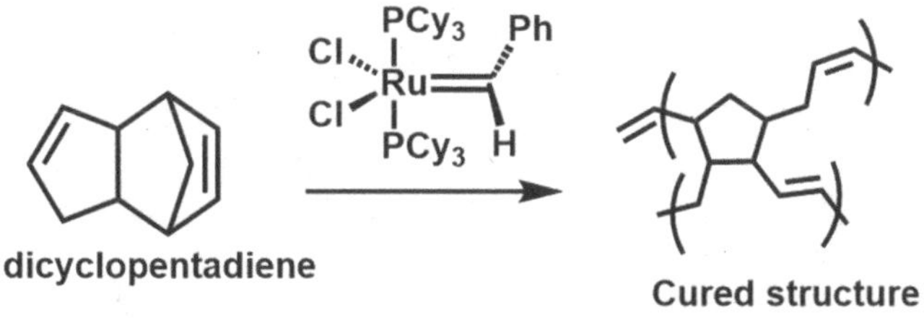

PCy3 Ph
Cl
Ru
Cl
PCy3 H
dicyclopentadiene
Cured structure

FIGURE 14.14 Representative reaction of catalyst cured self-healing system.

system has the ability to heal damage repeatedly by photochemical cross-linking and de-cross-linking reactions in the presence of UV light at wavelengths of 350 and 254 nm.

Moreover, considering the disadvantages of petroleum-based feedstocks, such as limited availability, risk to the environment, decline in the near future, etc., a renewable source–based light-triggered self-repairing cross-linked polymer network has also been developed using poly(2,5-furandimethylene succinate) in the presence of bismaleimide as a cross-linker. The prepared network healed a broken surface at room temperature within 1–10 days.[92]

Generally, most UV- or photo-assisted self-repairing materials are based on intrinsic self-healing systems. However, researchers also developed an acrylate-based UV-triggered microcapsule-based extrinsic self-healing system using polyurethane acrylate[93] and methacryloxypropyl terminated dimethyl silicone oligomers as self-healing agents in UF capsules.[94] Moreover, a sunlight-induced microcapsule-based healing system has also been developed from methacryloxypropyl terminated polydimethylsiloxane as a healing agent and benzoin isobutyl ether as the photo-initiator in UF as a shell-forming material.

14.6.6 Thermally Reversible

A thermally reversible system shows repair of internal damage through a thermoreversible covalent bonding process. This reaction has provided the idea of making a new class of thermally reversible polymers through Diels–Alder (DA) reactions. These reactions involve [4+2] cycloaddition of a diene and a dienophile in the presence of light with a particular range of wavelengths. Based on DA reactions, several thermally reversible polymers have been reported, which include furan and maleimide,[95–97] polyimide,[98] and modified maleimide with PU.[99,100] The DA reaction chemistry of dienes and dienophiles has a number of advantages, such as high reaction yield, absence of side products, water solubility, and no need for a catalyst.[101]

These types of system do not require microcapsules or preloading with any type of material to obtain their self-repairing properties. Due to the presence of the inherent self-healing ability of the polymer network, the system shows intrinsic-type self-healing and cures substrates repeatedly after damage.

14.7 Important Chemical Methods of Encapsulation

Several types of preparation methods are available in the literature for encapsulation of core materials. In common methods, healing agents are mostly encapsulated by chemical methods. Among these, various subtechniques are available to form healing agent–based microcapsules, such as interfacial/*in situ*, emulsion, suspension, and precipitation or dispersion polymerization.

14.7.1 Interfacial Polymerization

In the interfacial polymerization technique, small droplets of the oil phase are dispersed in the reaction medium, and the polycondensation reaction of monomers is carried out at the interface of the oil and aqueous phases. Thus, a shell starts to form around the surface of the droplets. For encapsulation of the core, it must have the ability to disperse properly, which may require an appropriate amount of stabilizer to avoid coagulation of droplet particles during the polycondensation reaction or microcapsule formation. After the cross-linking, solid polymeric microcapsules are obtained with a permanent shape, and these can be filtered to separate them from the medium. Various polymeric microcapsules are prepared by interfacial polymerization, such as PU,[68] polyurea[65,102,103], PF,[104,105] UF[105], etc.

14.7.2 *In Situ* Polymerization

In situ polymerization is the most versatile chemical technique of encapsulation, and it has similarities to interfacial polymerization. In general, the oil phase is emulsified with water using a suitable emulsifying agent with high-speed stirring to obtain a stable emulsion. The size of droplets or core can be controlled by adjusting the stirring rate. Initially, a low–molecular weight prepolymer forms, in which cross-linking occurs to increase molecular weight and finally converts the surface of the droplets into a solid form as microcapsules. There are several examples of polymeric shells that are prepared by *in situ* polymerization, such as UF, PF, and MF.[62,106,107]

14.7.3 Emulsion Polymerization

Generally, water-soluble monomer and water-insoluble core are most appropriate for encapsulation through the emulsion polymerization method.[108] Here, an emulsion is prepared by dropwise addition of monomer for shell under stirring in an aqueous medium that contains core and suitable emulsifying agent. The polymerization process starts and primarily develops a precipitate of polymer molecules in the form of microcapsules in an aqueous system. Further, as the rate of polymerization increases, the polymeric shell develops slowly on a surface of core, and the entire core becomes covered with the newly formed polymeric shell. Polymer shells prepared by emulsion polymerization include polyurea, polystyrene, etc.

14.8 Future Scope

For the last few years, self-healing coatings have been used to enhance the lifespan of substrates against corrosion by repairing the micro-cracks caused by damage. Thus, the overall outlook of such a new field is most interesting and advantageous for researchers and industries. The concept of self-healing was proved properly with the completion of synthesis and evaluation protocols. Today, microcapsule-based self-healing coatings are rapidly improving with respect to capsule types, synthesis processes, mechanical properties of capsules, different types of healing agents with diverse repairing mechanisms, etc. Due to the advantages of capsule-based self-healing, the system is mainly focused on the automobile industries. In future, the system has vast scope to extend applications in different fields, such as biotechnology, material science, solar cells, satellites, the defense sector, marine vessels, aircraft, electronic devices, etc. Despite many advantages, the system suffers from disadvantages, such as a single healing event, obtaining microcapsules with uniform size and shape, agglomeration of microcapsules during wall formation, and deterioration in the properties of the polymer matrix. So, future research can be concentrated on tackling such problems for addressing the real practical applications or the commercial viability of microcapsule-based self-healing coatings.

REFERENCES

1. Makhlouf, Abdel Salam Hamdy, ed. *Handbook of Smart Coatings for Materials Protection*, vol. 64. Elsevier, 2014.
2. Marathe, Ravindra J., Ashok B. Chaudhari, Rahul K. Hedaoo, Daewon Sohn, Vijay R. Chaudhari, and Vikas V. Gite. "Urea formaldehyde (UF) microcapsules loaded with corrosion inhibitor for enhancing the anti-corrosive properties of acrylic-based multi-functional PU coatings." *RSC Advances* 5, no. 20 (2015): 15539–15546.
3. Jones, Frank N., Mark E. Nichols, and Socrates Peter Pappas. *Organic Coatings: Science and Technology*. John Wiley & Sons, 2017.
4. Hager, Martin D., Peter Greil, Christoph Leyens, Sybrand van der Zwaag, and Ulrich S. Schubert. "Self-healing materials." *Advanced Materials* 22, no. 47 (2010): 5424–5430.
5. White, Scott R., N. R. Sottos, P. H. Geubelle, J. S. Moore, M. R. Kessler, S. R. Sriram, E. N. Brown, and S. Viswanathan. "Autonomic healing of polymer composites." *Nature* 409, no. 6822 (2001): 794.
6. Jadhav, Rajendra S., Dilip G. Hundiwale, and Pramod P. Mahulikar. "Synthesis and characterization of phenol–formaldehyde microcapsules containing linseed oil and its use in epoxy for self-healing and anticorrosive coating." *Journal of Applied Polymer Science* 119, no. 5 (2011): 2911–2916.
7. Zhang, Cheng, Haoran Wang, and Qixin Zhou. "Preparation and characterization of microcapsules based self-healing coatings containing epoxy ester as healing agent." *Progress in Organic Coatings* 125 (2018): 403–410.

8. Alizadegan, Farhad, S. Mojtaba Mirabedini, Shahla Pazokifard, Saba Goharshenas Moghadam, and Ramin Farnood. "Improving self-healing performance of polyurethane coatings using PU microcapsules containing bulky-IPDI-BA and nano-clay." *Progress in Organic Coatings* 123 (2018): 350–361.
9. Deng, Guohua, Chuanmei Tang, Fuya Li, Huanfeng Jiang, and Yongming Chen. "Covalent cross-linked polymer gels with reversible sol– gel transition and self-healing properties." *Macromolecules* 43, no. 3 (2010): 1191–1194.
10. Chen, Xiangxu, Fred Wudl, Ajit K. Mal, Hongbin Shen, and Steven R. Nutt. "New thermally remendable highly cross-linked polymeric materials." *Macromolecules* 36, no. 6 (2003): 1802–1807.
11. Murphy, Erin B., Ed Bolanos, Christine Schaffner-Hamann, Fred Wudl, Steven R. Nutt, and Maria L. Auad. "Synthesis and characterization of a single-component thermally remendable polymer network: Staudinger and Stille revisited." *Macromolecules* 41, no. 14 (2008): 5203–5209.
12. Zechel, Stefan, Robert Geitner, Marcus Abend, Michael Siegmann, Marcel Enke, Natascha Kuhl, Moritz Klein et al. "Intrinsic self-healing polymers with a high E-modulus based on dynamic reversible urea bonds." *NPG Asia Materials* 9, no. 8 (2017): e420.
13. Jo, Young Yeol, Albert S. Lee, Kyung-Youl Baek, Heon Lee, and Seung Sang Hwang. "Thermally reversible self-healing polysilsesquioxane structure-property relationships based on Diels-Alder chemistry." *Polymer* 108 (2017): 58–65.
14. Tuncaboylu, Deniz C., Murat Sari, Wilhelm Oppermann, and Oguz Okay. "Tough and self-healing hydrogels formed via hydrophobic interactions." *Macromolecules* 44, no. 12 (2011): 4997–5005.
15. Xia, Nan Nan, Xiao Min Xiong, Min Zhi Rong, Ming Qiu Zhang, and Fangong Kong. "Self-healing of polymer in acidic water toward strength restoration through the synergistic effect of hydrophilic and hydrophobic interactions." *ACS Applied Materials & Interfaces* 9, no. 42 (2017): 37300–37309.
16. Xiao, Ye, Huihua Huang, and Xiaohong Peng. "Synthesis of self-healing waterborne polyurethanes containing sulphonate groups." *RSC Advances* 7, no. 33 (2017): 20093–20100.
17. Andreeva, Daria V., Dmitri Fix, Helmuth Möhwald, and Dmitry G. Shchukin. "Self-healing anticorrosion coatings based on pH-sensitive polyelectrolyte/inhibitor sandwichlike nanostructures." *Advanced Materials* 20, no. 14 (2008): 2789–2794.
18. Yamaguchi, Masayuki, Susumu Ono, and Kenzo Okamoto. "Interdiffusion of dangling chains in weak gel and its application to self-repairing material." *Materials Science and Engineering: B* 162, no. 3 (2009): 189–194.
19. Burattini, Stefano, Barnaby W. Greenland, Daniel Hermida Merino, Wengui Weng, Jonathan Seppala, Howard M. Colquhoun, Wayne Hayes, Michael E. Mackay, Ian W. Hamley, and Stuart J. Rowan. "A healable supramolecular polymer blend based on aromatic π–π stacking and hydrogen-bonding interactions." *Journal of the American Chemical Society* 132, no. 34 (2010): 12051–12058.
20. Li, Guifei, Jie Wu, Bo Wang, Shifeng Yan, Kunxi Zhang, Jianxun Ding, and Jingbo Yin. "Self-healing supramolecular self-assembled hydrogels based on poly (L-glutamic acid)." *Biomacromolecules* 16, no. 11 (2015): 3508–3518.
21. Chen, Yulin, Aaron M. Kushner, Gregory A. Williams, and Zhibin Guan. "Multiphase design of autonomic self-healing thermoplastic elastomers." *Nature Chemistry* 4, no. 6 (2012): 467–472.
22. Grande, Antonio M., A. Rahman, Luca Di Landro, Maurizio Penco, and Gloria Spagnoli. "Self healing of blends based on Sodium Salt of Poly (Ethylene-co-Methacrylic Acid)/poly (ethylene-co-vinyl alcohol) and epoxidized natural rubber following high energy impact." In *Proceeding of 3rd International Conference on Self-Healing Materials*. 2011.
23. Hia, Iee Lee, Vahdat Vahedi, and Pooria Pasbakhsh. "Self-healing polymer composites: prospects, challenges, and applications." *Polymer Reviews* 56, no. 2 (2016): 225–261.
24. Wang, Haoran, and Qixin Zhou. "Evaluation and failure analysis of linseed oil encapsulated self-healing anticorrosive coating." *Progress in Organic Coatings* 118 (2018): 108–115.
25. Leal, Debora Abrantes, Izabel Cristina Riegel-Vidotti, Mario Guerreiro Silva Ferreira, and Claudia Eliana Bruno Marino. "Smart coating based on double stimuli-responsive microcapsules containing linseed oil and benzotriazole for active corrosion protection." *Corrosion Science* 130 (2018): 56–63.
26. Samadzadeh, M., S. Hatami Boura, M. Peikari, A. Ashrafi, and M. Kasiriha. "Tung oil: An autonomous repairing agent for self-healing epoxy coatings." *Progress in Organic Coatings* 70, no. 4 (2011): 383–387.
27. Li, Haiyan, Yexiang Cui, Zhike Li, Yanji Zhu, and Huaiyuan Wang. "Fabrication of microcapsules containing dual-functional tung oil and properties suitable for self-healing and self-lubricating coatings." *Progress in Organic Coatings* 115 (2018): 164–171.

28. Shisode, Priyanka S., Chetan B. Patil, and Pramod P. Mahulikar. "Preparation and characterization of microcapsules containing soybean oil and their application in self-healing anticorrosive coatings." *Polymer-Plastics Technology and Engineering* 57, no. 13 (2018): 1334–1343.
29. Yang, Jinglei, Michael W. Keller, Jeffery S. Moore, Scott R. White, and Nancy R. Sottos. "Microencapsulation of isocyanates for self-healing polymers." *Macromolecules* 41, no. 24 (2008): 9650–9655.
30. Kong, Fanhou, Weichang Xu, Xuelong Zhang, Xin Wang, Yu Zhang, and Jinglong Wu. "High-efficiency self-repairing anticorrosion coatings with controlled assembly microcapsules." *Journal of Materials Science* 53, no. 18 (2018): 12850–12859.
31. Gite, Vikas V., Pyus D. Tatiya, Ranindra J. Marathe, Pramod P. Mahulikar, and Dilip G. Hundiwale. "Microencapsulation of quinoline as a corrosion inhibitor in polyurea microcapsules for application in anticorrosive PU coatings." *Progress in Organic Coatings* 83 (2015): 11–18.
32. Siva, T., and S. Sathiyanarayanan. "Self healing coatings containing dual active agent loaded urea formaldehyde (UF) microcapsules." *Progress in Organic Coatings* 82 (2015): 57–67.
33. Kamphaus, Jason M., Joseph D. Rule, Jeffrey S. Moore, Nancy R. Sottos, and Scott R. White. "A new self-healing epoxy with tungsten (VI) chloride catalyst." *Journal of the Royal Society Interface* 5, no. 18 (2008): 95–103.
34. Cho, Soo Hyoun, H. Magnus Andersson, Scott R. White, Nancy R. Sottos, and Paul V. Braun. "Polydimethylsiloxane-based self-healing materials." *Advanced Materials* 18, no. 8 (2006): 997–1000.
35. Xiao, Ding Shu, Yan Chao Yuan, Min Zhi Rong, and Ming Qiu Zhang. "A facile strategy for preparing self-healing polymer composites by incorporation of cationic catalyst-loaded vegetable fibers." *Advanced Functional Materials* 19, no. 14 (2009): 2289–2296.
36. Jin, Henghua, Chris L. Mangun, Dylan S. Stradley, Jeffrey S. Moore, Nancy R. Sottos, and Scott R. White. "Self-healing thermoset using encapsulated epoxy-amine healing chemistry." *Polymer* 53, no. 2 (2012): 581–587.
37. Yuan, Yan Chao, Min Zhi Rong, Ming Qiu Zhang, Jian Chen, Gui Cheng Yang, and Xue Mei Li. "Self-healing polymeric materials using epoxy/mercaptan as the healant." *Macromolecules* 41, no. 14 (2008): 5197–5202.
38. Song, Yi Xi, Xiao Ji Ye, Min Zhi Rong, and Ming Qiu Zhang. "Self-healing epoxy with a fast and stable extrinsic healing system based on BF 3–amine complex." *RSC Advances* 6, no. 103 (2016): 100796–100803.
39. Lee, Jim, Mingqiu Zhang, Debes Bhattacharyya, Yan Chao Yuan, Krishnan Jayaraman, and Yiu Wing Mai. "Micromechanical behavior of self-healing epoxy and hardener-loaded microcapsules by nanoindentation." *Materials Letters* 76 (2012): 62–65.
40. Hillewaere, Xander KD, Roberto FA Teixeira, Le-Thu T. Nguyen, José A. Ramos, Hubert Rahier, and Filip E. Du Prez. "Autonomous self-healing of epoxy thermosets with thiol-isocyanate chemistry." *Advanced Functional Materials* 24, no. 35 (2014): 5575–5583.
41. Trask, R. S., and I. P. Bond. "Biomimetic self-healing of advanced composite structures using hollow glass fibres." *Smart Materials and Structures* 15, no. 3 (2006): 704.
42. Dry, Carolyn M., and Nancy R. Sottos. "Passive smart self-repair in polymer matrix composite materials." In *Smart Structures and Materials 1993: Smart Materials*, vol. 1916, pp. 438–445. International Society for Optics and Photonics, 1993.
43. Dry, Carolyn. "Procedures developed for self-repair of polymer matrix composite materials." *Composite Structures* 35, no. 3 (1996): 263–269.
44. Bleay, S. M., C. B. Loader, V. J. Hawyes, L. Humberstone, and P. T. Curtis. "A smart repair system for polymer matrix composites." *Composites Part A: Applied Science and Manufacturing* 32, no. 12 (2001): 1767–1776.
45. Huang, C-Y., R. S. Trask, and I. P. Bond. "Characterization and analysis of carbon fibre-reinforced polymer composite laminates with embedded circular vasculature." *Journal of the Royal Society Interface* 7, no. 49 (2010): 1229–1241.
46. Trask, R. S., and I. P. Bond. "Bioinspired engineering study of Plantae vascules for self-healing composite structures." *Journal of the Royal Society Interface* 7, no. 47 (2010): 921–931.
47. Paulo, Filipa, and Lucia Santos. "Design of experiments for microencapsulation applications: A review." *Materials Science and Engineering: C* 77 (2017): 1327–1340.
48. Jacobson, Joseph M. "Electronically addressable microencapsulated ink and display thereof." U.S. Patent 6,652,075, issued November 25, 2003.

49. Hedaoo, Rahul Kishore, and Vikas Vitthal Gite. "Renewable resource-based polymeric microencapsulation of natural pesticide and its release study: an alternative green approach." *RSC Advances* 4, no. 36 (2014): 18637–18644.
50. Miyazawa, K., I. Yajima, I. Kaneda, and T. Yanaki. "Preparation of a new soft capsule for cosmetics." *Journal of Cosmetic Science* 51, no. 4 (2000): 239–252.
51. Wei, Huige, Yiran Wang, Jiang Guo, Nancy Z. Shen, Dawei Jiang, Xi Zhang, Xingru Yan et al. "Advanced micro/nanocapsules for self-healing smart anticorrosion coatings." *Journal of Materials Chemistry A* 3, no. 2 (2015): 469–480.
52. Jadhav, Rajendra S., Vishal Mane, Avinash V. Bagle, Dilip G. Hundiwale, Pramod P. Mahulikar, and Gulzar Waghoo. "Synthesis of multicore phenol formaldehyde microcapsules and their application in polyurethane paint formulation for self-healing anticorrosive coating." *International Journal of Industrial Chemistry* 4, no. 1 (2013): 31.
53. Bagle, Avinash V., Rajendra S. Jadhav, Vikas V. Gite, Dilip G. Hundiwale, and Pramod P. Mahulikar. "Controlled release study of phenol formaldehyde microcapsules containing neem oil as an insecticide." *International Journal of Polymeric Materials and Polymeric Biomaterials* 62, no. 8 (2013): 421–425.
54. Katoueizadeh, Elham, Seyed Mojtaba Zebarjad, and Kamal Janghorban. "Investigating the effect of synthesis conditions on the formation of urea–formaldehyde microcapsules." *Journal of Materials Research and Technology* 8, no. 1 (2019): 541–552.
55. Brown, Eric N., Michael R. Kessler, Nancy R. Sottos, and Scott R. White. "In situ poly (urea-formaldehyde) microencapsulation of dicyclopentadiene." *Journal of Microencapsulation* 20, no. 6 (2003): 719–730.
56. Suryanarayana, C., K. Chowdoji Rao, and Dhirendra Kumar. "Preparation and characterization of microcapsules containing linseed oil and its use in self-healing coatings." *Progress in Organic Coatings* 63, no. 1 (2008): 72–78.
57. Liu, Qi, Jiupeng Zhang, Wolong Liu, Fucheng Guo, Jianzhong Pei, Cunzhen Zhu, and Wenwu Zhang. "Preparation and characterization of self-healing microcapsules embedding waterborne epoxy resin and curing agent for asphalt materials." *Construction and Building Materials* 183 (2018): 384–394.
58. Asadi, Amir Khalaj, Morteza Ebrahimi, and Mohsen Mohseni. "Preparation and characterisation of melamine-urea-formaldehyde microcapsules containing linseed oil in the presence of polyvinylpyrrolidone as emulsifier." *Pigment & Resin Technology* 46, no. 4 (2017): 318–326.
59. Hwang, Jun-Seok, Jin-Nam Kim, Young-Jung Wee, Jong-Sun Yun, Hong-Gi Jang, Sun-Ho Kim, and Hwa-Won Ryu. "Preparation and characterization of melamine-formaldehyde resin microcapsules containing fragrant oil." *Biotechnology and Bioprocess Engineering* 11, no. 4 (2006): 332–336.
60. Bone, Stéphane, Claire Vautrin, Virginie Barbesant, Stéphane Truchon, Ian Harrison, and Cédric Geffroy. "Microencapsulated fragrances in melamine formaldehyde resins." *CHIMIA International Journal for Chemistry* 65, no. 3 (2011): 177–181.
61. Hong, K., and S. Park. "Melamine resin microcapsules containing fragrant oil: synthesis and characterization." *Materials Chemistry and Physics* 58, no. 2 (1999): 128–131.
62. Sharma, Shilpi, and Veena Choudhary. "Poly (melamine-formaldehyde) microcapsules filled with epoxy resin: effect of M/F ratio on the shell wall stability." *Materials Research Express* 4, no. 7 (2017): 075307.
63. Hu, Jianfeng, Huan-Qin Chen, and Zhibing Zhang. "Mechanical properties of melamine formaldehyde microcapsules for self-healing materials." *Materials Chemistry and Physics* 118, no. 1 (2009): 63–70.
64. Hong, K., and S. Park. "Morphologies and release behavior of polyurea microcapsules from different polyisocyanates." *Journal of Materials Science* 34, no. 13 (1999): 3161–3164.
65. Tatiya, Pyus D., Rahul K. Hedaoo, Pramod P. Mahulikar, and Vikas V. Gite. "Novel polyurea microcapsules using dendritic functional monomer: synthesis, characterization, and its use in self-healing and anticorrosive polyurethane coatings." *Industrial & Engineering Chemistry Research* 52, no. 4 (2013): 1562–1570.
66. Chaudhari, Ashok B., Pyus D. Tatiya, Rahul K. Hedaoo, Ravindra D. Kulkarni, and Vikas V. Gite. "Polyurethane prepared from neem oil polyesteramides for self-healing anticorrosive coatings." *Industrial & Engineering Chemistry Research* 52, no. 30 (2013): 10189–10197.
67. Tatiya, Pyus D., Pramod P. Mahulikar, and Vikas V. Gite. "Designing of polyamidoamine-based polyurea microcapsules containing tung oil for anticorrosive coating applications." *Journal of Coatings Technology and Research* 13, no. 4 (2016): 715–726.

68. Gite, V. V., P. P. Mahulikar, D. G. Hundiwale, and U. R. Kapadi. "Polyurethane coatings using trimer of isophorone diisocyanate." 63 (2004): 348–354.
69. Kardar, Pooneh. "Preparation of polyurethane microcapsules with different polyols component for encapsulation of isophorone diisocyanate healing agent." *Progress in Organic Coatings* 89 (2015): 271–276.
70. Koh, Eunjoo, and Sooyeoul Park. "Self-anticorrosion performance efficiency of renewable dimer-acid-based polyol microcapsules containing corrosion inhibitors with two triazole groups." *Progress in Organic Coatings* 109 (2017): 61–69.
71. Sondari, Dewi, Athanasia Amanda Septevani, Ahmad Randy, and Evi Triwulandari. "Polyurethane microcapsule with glycerol as the polyol component for encapsulated self healing agent." *Journal of Engineering and Technology* 2, no. 6 (2010): 466–471.
72. Billiet, Stijn, Xander KD Hillewaere, Roberto FA Teixeira, and Filip E. Du Prez. "Chemistry of crosslinking processes for self-healing polymers." *Macromolecular Rapid Communications* 34, no. 4 (2013): 290–309.
73. Wang, Wei, Likun Xu, Feng Liu, Xiangbo Li, and Lukuo Xing. "Synthesis of isocyanate microcapsules and micromechanical behavior improvement of microcapsule shells by oxygen plasma treated carbon nanotubes." *Journal of Materials Chemistry A* 1, no. 3 (2013): 776–782.
74. Di Credico, Barbara, Marinella Levi, and Stefano Turri. "An efficient method for the output of new self-repairing materials through a reactive isocyanate encapsulation." *European Polymer Journal* 49, no. 9 (2013): 2467–2476.
75. Huang, Mingxing, and Jinglei Yang. "Facile microencapsulation of HDI for self-healing anticorrosion coatings." *Journal of Materials Chemistry* 21, no. 30 (2011): 11123–11130.
76. Nguyen, Le-Thu T., Xander KD Hillewaere, Roberto FA Teixeira, Otto van den Berg, and Filip E. Du Prez. "Efficient microencapsulation of a liquid isocyanate with in situ shell functionalization." *Polymer Chemistry* 6, no. 7 (2015): 1159–1170.
77. Koh, Eunjoo, Nam-Kyun Kim, Jihoon Shin, and Young-Wun Kim. "Polyurethane microcapsules for self-healing paint coatings." *RSC Advances* 4, no. 31 (2014): 16214–16223.
78. Bradley, Thedore F. "DRYING OILS AND RESINS polymeric functionality with relation to the addition polymerization of drying oils." *Industrial & Engineering Chemistry* 30, no. 6 (1938): 689–696.
79. Yin, Tao, Min Zhi Rong, Ming Qiu Zhang, and Gui Cheng Yang. "Self-healing epoxy composites–preparation and effect of the healant consisting of microencapsulated epoxy and latent curing agent." *Composites Science and Technology* 67, no. 2 (2007): 201–212.
80. Murphy, Erin B., and Fred Wudl. "The world of smart healable materials." *Progress in Polymer Science* 35, no. 1–2 (2010): 223–251.
81. Rule, Joseph D., and Jeffrey S. Moore. "ROMP reactivity of endo-and exo-dicyclopentadiene." *Macromolecules* 35, no. 21 (2002): 7878–7882.
82. Patel, Amit J., Nancy R. Sottos, Eric D. Wetzel, and Scott R. White. "Autonomic healing of low-velocity impact damage in fiber-reinforced composites." *Composites Part A: Applied Science and Manufacturing* 41, no. 3 (2010): 360–368.
83. Mauldin, Timothy C., Joseph D. Rule, Nancy R. Sottos, Scott R. White, and Jeffrey S. Moore. "Self-healing kinetics and the stereoisomers of dicyclopentadiene." *Journal of the Royal Society Interface* 4, no. 13 (2007): 389–393.
84. Lee, Jong Keun, Sun Ji Hong, Xing Liu, and Sung Ho Yoon. "Characterization of dicyclopentadiene and 5-ethylidene-2-norbornene as self-healing agents for polymer composite and its microcapsules." *Macromolecular Research* 12, no. 5 (2004): 478–483.
85. Liu, Xing, Jong Keun Lee, Sung Ho Yoon, and Michael R. Kessler. "Characterization of diene monomers as healing agents for autonomic damage repair." *Journal of Applied Polymer Science* 101, no. 3 (2006): 1266–1272.
86. Wilson, Gerald O., Mary M. Caruso, Neil T. Reimer, Scott R. White, Nancy R. Sottos, and Jeffrey S. Moore. "Evaluation of ruthenium catalysts for ring-opening metathesis polymerization-based self-healing applications." *Chemistry of Materials* 20, no. 10 (2008): 3288–3297.
87. Wilson, Gerald O., Keith A. Porter, Haim Weissman, Scott R. White, Nancy R. Sottos, and Jeffrey S. Moore. "Stability of second generation Grubbs' alkylidenes to primary amines: formation of novel ruthenium-amine complexes." *Advanced Synthesis & Catalysis* 351, no. 11–12 (2009): 1817–1825.
88. Coope, Tim S., Ulrich FJ Mayer, Duncan F. Wass, Richard S. Trask, and Ian P. Bond. "Self-healing of an epoxy resin using scandium (iii) triflate as a catalytic curing agent." *Advanced Functional Materials* 21, no. 24 (2011): 4624–4631.

89. Chung, Chan-Moon, Young-Suk Roh, Sung-Youl Cho, and Joong-Gon Kim. "Crack healing in polymeric materials via photochemical [2+ 2] cycloaddition." *Chemistry of Materials* 16, no. 21 (2004): 3982–3984.
90. Wu, Dong Yang, Sam Meure, and David Solomon. "Self-healing polymeric materials: a review of recent developments." *Progress in Polymer Science* 33, no. 5 (2008): 479–522.
91. Ling, Jun, Min Zhi Rong, and Ming Qiu Zhang. "Photo-stimulated self-healing polyurethane containing dihydroxyl coumarin derivatives." *Polymer* 53, no. 13 (2012): 2691–2698.
92. Zeng, Chao, Hidetake Seino, Jie Ren, Kenichi Hatanaka, and Naoko Yoshie. "Bio-based furan polymers with self-healing ability." *Macromolecules* 46, no. 5 (2013): 1794–1802.
93. Zhao, Dong, Mo-zhen Wang, Qi-chao Wu, Xiao Zhou, and Xue-wu Ge. "Microencapsulation of UV-curable self-healing agent for smart anticorrosive coating." *Chinese Journal of Chemical Physics* 27, no. 5 (2014): 607–615.
94. Xing, Rui-ying, Qiu-yu Zhang, and Jiu-li Sun. "Preparation and properties of self-healing microcapsules containing an UV-curable oligomer of silicone." *Polymers & Polymer Composites* 20, no. 1/2 (2012): 77.
95. Gaina, C., O. Ursache, V. Gaina, E. Buruiana, and D. Ionita. "Investigation on the thermal properties of new thermo-reversible networks based on poly (vinyl furfural) and multifunctional maleimide compounds." *Express Polymer Letters* 6, no. 2 (2012): 129–141.
96. Toncelli, Claudio, Dennis C. De Reus, Francesco Picchioni, and Antonius A. Broekhuis. "Properties of reversible Diels–Alder furan/maleimide polymer networks as function of crosslink density." *Macromolecular Chemistry and Physics* 213, no. 2 (2012): 157–165.
97. Fan, Mengjin, Jialin Liu, Xiangyuan Li, Junying Zhang, and Jue Cheng. "Recyclable Diels–Alder Furan/Maleimide polymer networks with shape memory effect." *Industrial & Engineering Chemistry Research* 53, no. 42 (2014): 16156–16163.
98. Patel, Yogesh S., and Hasmukh S. Patel. "Thermoplastic-thermosetting merged polyimides via furan-maleimide Diels–Alder polymerization." *Arabian Journal of Chemistry* 10 (2017): S1373–S1380.
99. Li, Jinhui, Guoping Zhang, Libo Deng, Kun Jiang, Songfang Zhao, Yongju Gao, Rong Sun, and Chingping Wong. "Thermally reversible and self-healing novolac epoxy resins based on Diels–Alder chemistry." *Journal of Applied Polymer Science* 132, no. 26 (2015).
100. Rivero, Guadalupe, Le-Thu T. Nguyen, Xander KD Hillewaere, and Filip E. Du Prez. "One-pot thermo-remendable shape memory polyurethanes." *Macromolecules* 47, no. 6 (2014): 2010–2018.
101. Tasdelen, Mehmet Atilla. "Diels–Alder "click" reactions: recent applications in polymer and material science." *Polymer Chemistry* 2, no. 10 (2011): 2133–2145.
102. Hedaoo, Rahul K., Pyus D. Tatiya, Pramod P. Mahulikar, and Vikas V. Gite. "Fabrication of dendritic 0 G PAMAM-based novel polyurea microcapsules for encapsulation of herbicide and release rate from polymer shell in different environment." *Designed Monomers and Polymers* 17, no. 2 (2014): 111–125.
103. Hedaoo, Rahul K., Pramod P. Mahulikar, Ashok B. Chaudhari, Sandip D. Rajput, and Vikas V. Gite. "Fabrication of core–shell novel polyurea microcapsules using isophorone diisocyanate (IPDI) trimer for release system." *International Journal of Polymeric Materials and Polymeric Biomaterials* 63, no. 7 (2014): 352–360.
104. Hedaoo, R. K., P. P. Mahulikar, and V. V. Gite. "Synthesis and characterization of resorcinol-based cross linked phenol formaldehyde microcapsules for encapsulation of pendimethalin." *Polymer-Plastics Technology and Engineering* 52, no. 3 (2013): 243–249.
105. Marathe, R. J., and V. V. Gite. "Encapsulation of 8-HQ as a corrosion inhibitor in PF and UF shells for enhanced anticorrosive properties of renewable source based smart PU coatings." *RSC Advances* 6, no. 115 (2016): 114436–114446.
106. Marathe, Ravindra, Pyus Tatiya, Ashok Chaudhari, Jeongwook Lee, Pramod Mahulikar, Daewon Sohn, and Vikas Gite. "Neem acetylated polyester polyol—Renewable source based smart PU coatings containing quinoline (corrosion inhibitor) encapsulated polyurea microcapsules for enhance anticorrosive property." *Industrial Crops and Products* 77 (2015): 239–250.
107. Karandikar, Pravin S., Jamatsing D. Rajput, Suresh D. Bagul, Vikas V. Gite, and Ratnamala S. Bendre. "Controlled release study of phenol formaldehyde based microcapsules containing various loading percentage of core cypermethrin at different agitation rates." *Polymer Bulletin* (2018): 1–18.
108. Dubey, Rama. "Microencapsulation technology and applications." *Defence Science Journal* 59, no. 1 (2009): 82.

15

Encapsulation Technologies for Modifying Food Performance

Maria Inês Ré, Maria Helena Andrade Santana, and Marcos Akira d'Ávila

CONTENTS

15.1 Introduction

The food industry faces serious challenges in the twenty-first century. Consumers are demanding more from the foods they eat. Now, foods must not only taste good and aid immediate nutrition but also assist in mitigating disease, provide clear health benefits, and help to reduce health care costs. To meet these demands, food manufacturers must prepare safe, healthy, and convenient foods that are of good value and of great taste. The global functional foods market size is estimated at US$ 160 billion in 2019. This number is expected to climb up to about US$ 250 billion in the next 5 to 6 years (Rattanachaikunsopon and Phumkhachorn, 2018).

To date, a number of national authorities, academic bodies, and the industry have proposed definitions for functional foods. Although the term *functional food* has already been defined several times, there is no universally accepted definition for this emerging food category. Definitions range from the very simple to the more complex (Siró et al., 2008): "foods that, by virtue of physiologically active components, provide health benefits beyond basic nutrition" (International Life Sciences Institute, 1999); "food similar in appearance to a conventional food, consumed as part of the usual diet, with demonstrated physiological benefits, and/or to reduce the risk of chronic disease beyond basic nutritional functions" (Health Canada, 1998).

Examples of functional foods that have potential benefits for health and whose market has grown tremendously include baby food, ready meals, snacks, soft drinks such as energy and sport drinks, meat products, and spreads.

A functional benefit is usually obtained by fortification with an active (functional) ingredient. Examples of functional ingredients are flavors, vitamins, minerals, enzymes, peptides, bioactive lipids, antioxidants, and probiotic microorganisms. Functional ingredients come in a variety of molecular and physical forms, with different polarities (polar, non-polar, amphiphilic), molecular weights (low to high), and physical states (solid, liquid). They are rarely used directly in their pure form. Instead, they are often incorporated into some form of delivery system generated by encapsulation technologies.

A delivery system must perform a number of different roles. First, it serves as a vehicle for carrying the functional ingredient to the desired site of action. Second, it may have to protect the functional ingredient from chemical or biological degradation (e.g., oxidation) during processing, storage, and use; this maintains the functional ingredient in its active state. Third, it may have to be capable of controlling the release of the functional ingredient, either responding to specific environmental conditions that trigger release (e.g., pH, ionic strength, or temperature) or varying the release rate. Fourth, the delivery system has to be food-grade and also compatible with the physicochemical and qualitative attributes (i.e., appearance, texture, taste, and shelf life) of the final product (Weiss et al., 2006).

The characteristics of the delivery systems are among the most important factors influencing the efficacy of functional ingredients in many industrial products. The delivery systems can be used to deliver a host of ingredients in a range of food formulations. The demand for encapsulation technologies is growing at around 10% annually, driven by both increasing fortification with healthy ingredients and the consumer demand for novel products (Brownlie, 2007). Encapsulation technologies are attracting growing interest because they can decrease costs for food makers, particularly those using sensitive ingredients such as probiotics, by reducing the need for preservatives. Encapsulation also allows manufacturers of food and beverages, as well as other consumer products, to add into their formulations ingredients that would be normally used in traditional processing.

15.2 Encapsulation Technologies

Encapsulation is a topic of interest in a wide range of scientific and industrial areas, varying from pharmaceutics to agriculture and from pesticides to enzymes. The development of these technologies is characterized by strong fundamental research and several industrial applications, demonstrated by the growing number of scientific papers and patent applications. The European network of patent databases (esp@cenet) has approximately 600 patent documents worldwide containing "microencapsulation" in the title or in the abstract, whereas the U.S. Patent and Trademark Office has approximately 160 patents and 140 applications with at least one claim containing the term "microencapsulation."* In the past 5 years, there have been 200–300 peer-reviewed articles per year on microencapsulation in the *Web of Science*®† database (Figure 15.1), in contrast to 1600 papers in the previous 20 years.

Encapsulation involves the coating or entrapment of a desired component (active or core material) within a secondary material (encapsulating, carrier, coating, shell, or wall material) to prevent or delay the release of the active or core ingredient until a certain time or a set of conditions is achieved. Encapsulation can potentially offer numerous benefits to the materials being encapsulated. Various properties of active ingredients may be changed by encapsulation. For example, handling and flow properties can be changed by converting a liquid to a solid encapsulated form, and hygroscopic materials can be protected from moisture.

* Data related to a search performed on the U.S. Patent and Trademark Office: 1976–Jan. 2009 (patents) and 2001–Jan. 2009 (patent applications); and on the espac@cenet: 1970–Jan. 2009.

† Registered trademark of Thomson Reuters, New York.

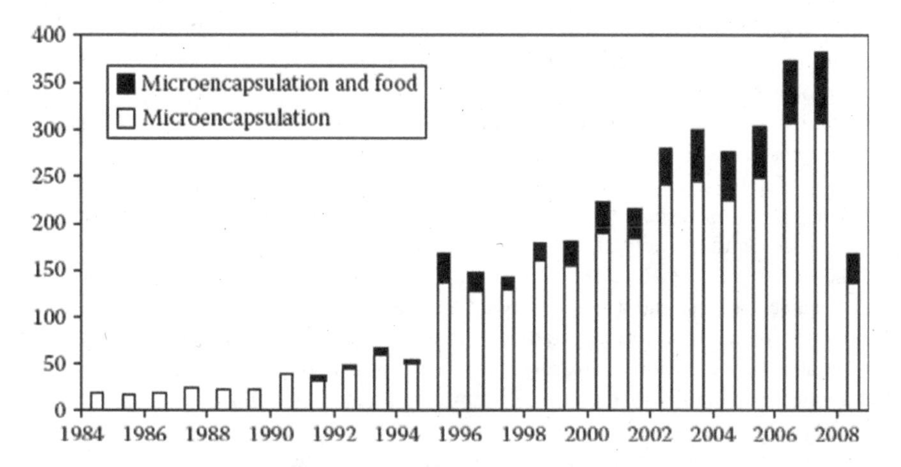

FIGURE 15.1 Evolution of the number of scientific publications in the Web of Science database (up to July 2008) on "microencapsulation" and "microencapsulation of food ingredients." Microencapsulation applied to food represents approximately 20% of the papers published in the microencapsulation area.

Encapsulation systems in food applications are typically used with at least one of the following purposes:

- To solve formulation problems arising from a limited chemical or physical stability of the active ingredient
- To overcome incompatibility between the active ingredient and the food matrix
- To control the release of a sensorial active compound
- To assist or enhance the absorption of a nutrient

Examples of food additives that may benefit from encapsulation and controlled release are flavors, minerals, and lipids, among others.

There are several items to consider when choosing or developing an encapsulated food ingredient. Each of these is very important to the success of the product. The molecular characteristics of the ingredient, its desired function, the type of coating, the required protection rate, the particle size, and the processing conditions must all be clearly defined. In this regard, there are some important questions that must be answered:

- What are the molecular structure, size, and charge of the ingredient?
- What is it that you are trying to achieve with the encapsulation?
- When or how do you want the core ingredient to be released?
- Do you want the coating to melt and release the core ingredient at a certain temperature?
- Do you want the coating to break away by mechanical action or by dissolving in water?
- Is a change in pH, such as in the gastrointestinal (GI) tract, going to release the core? Or is there another way your core ingredient will be released?

A critical step in developing encapsulated food products is to determine the encapsulant formulation that meets the desired stability and release criteria. The GRAS (generally recognized as safe) encapsulating material must stabilize the core material, must not react with the active ingredient or cause it to deteriorate, and should release it under the specific conditions based on product application. In addition, delivery systems should be developed from inexpensive ingredients, since the additional costs associated with the encapsulation of the active ingredient should be overcome by its benefits: for example, improved shelf life, better bioavailability, or enhanced marketability.

The variety of encapsulating materials allows food producers to select compounds that work for water- or fat-soluble food ingredients; dissolve, melt, or rupture to release core material; and provide textural characteristics to satisfy consumer palates. Commonly used encapsulating materials are carbohydrates (due to their ability to absorb and retain flavors), cellulose (based on its permeability), gums (which offer good gelling properties and heat resistance), lipids (based on their hydrophobicity), and usually gelatin as a protein, which is non-toxic, inexpensive, and commercially available. Some combinations of these encapsulating agents are also commonly used.

New encapsulating materials for foods have not really emerged in recent years. However, a great effort has been made with respect to food proteins. Food proteins have been engineered as a range of new GRAS matrices with the potential to incorporate nutraceutical compounds and provide controlled release via the oral route. The advantages of food protein matrices include their high nutritional value, abundant renewable sources, and acceptability as naturally occurring food components degradable by digestive enzymes. In addition, food proteins can be used to prepare a wide range of matrices and multicomponent matrices in the form of hydrogels or micro- or nanoparticles, all of which can be tailored for specific applications in the development of innovative functional food products, as recently reviewed (Chen et al., 2006).

Encapsulation and microencapsulation are often used interchangeably when discussing the process technology. Microencapsulation is encapsulation at the microscale, producing delivery devices ranging from 1 to 1000 μm in size and generally less than 200 μm.

Delivery devices can have many morphologies, depending on the materials and methods used in their preparation. In general, one can distinguish between two main groups of device architecture, depending on the way the core (solid or liquid) is distributed within the system (Figure 15.2):

1. A reservoir system, in which the core is largely concentrated near the center and enveloped by a continuous film (wall) of the encapsulating material
2. A matrix system, in which the core is finely dispersed throughout a continuous matrix of the encapsulating material

The active constituent/encapsulating material ratio is usually high in reservoir systems (between 0.70 and 0.95), whereas for matrix systems, this ratio is generally lower than 1.5 (more commonly between 0.2 and 0.35). The delivery devices defined in 1 are often referred to as *microcapsules*, and those described in 2 are called *microspheres*.

Delivery devices do not necessarily have a spherical shape, as illustrated in Figure 15.2e. A great variety of shapes can be obtained when a solid core material is encapsulated by a shell. Particle size is an important characteristic of these structures, because it is one of the many parameters that can be tailored to control the release rates of encapsulated ingredients. However, the production of microcapsules often gives a certain particle size polydispersity. The active ingredient release kinetics depends on the particle size distribution. It is thus necessary to determine both the mean particle size and the size distribution for the targeted delivery.

The release of the core material from a delivery device may be programmed to be immediate, delayed, pulsatile, or prolonged over an extended period of time. In general, the release depends on the architecture and the physical structure of the device as well as on the barrier properties of the encapsulating material used to form the system.

Delivery devices can be formulated to release the core material for food applications through a variety of mechanisms to meet product performance requirements, which may be based on temperature or solvent effects, diffusion, degradation, or particle fractures. For example, the encapsulating material can be fractured by external or internal forces such as chewing (Figure 15.3). Melting of the core or the encapsulating material by means of an appropriate solvent or thermally is another way of controlling the release of the active ingredient. Solvent release is based on the solubilization of the encapsulant (water is typically the solvent) followed by subsequent release of the encapsulated ingredient. Release may be regulated by controlling the dissolution rate of the encapsulant and pH effects. For example, coating materials can be selected to dissolve on consumption, slowly or quickly in the acidic gastric medium, or only when a certain pH is reached. Thermal release is commonly used for fat capsules and occurs during baking. The release of the core material may be delayed until the proper temperature is reached, delaying a chemical reaction. For example, sodium bicarbonate is a baking ingredient that reacts with food acids

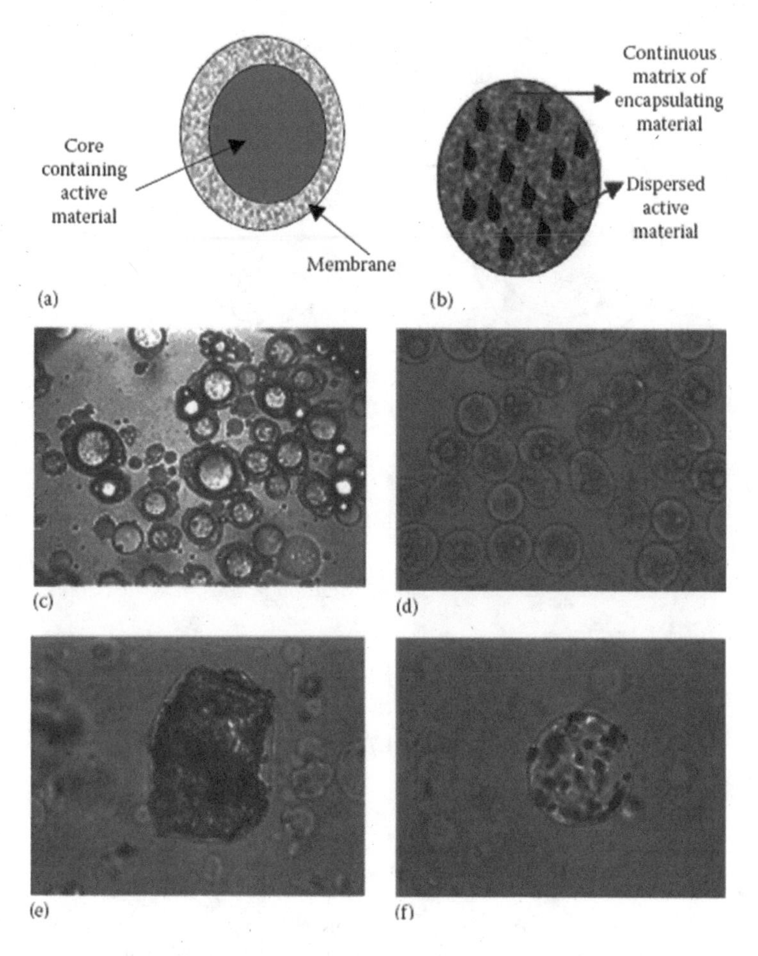

FIGURE 15.2 Schematic diagram of microcapsule morphology: (a) reservoir system (simple wall); (b) matrix system; micrographs: (c) simple wall (liquid core); (d) multicore; (e) simple wall (solid and irregular core); and (f) matrix (solid core dispersed into the polymeric matrix).

to produce leavening agents, which give baked goods their volume and lightness of texture. To delay and control the leavening process, the sodium bicarbonate is encapsulated in a fat, which is solid at room temperature but melts at a temperature of about 50–52 °C. Release of the active ingredient from microcapsules can be accomplished through biodegradation processes if the encapsulating agent is sensitive to enzymatic actions. For example, lipid coatings may be degraded by the action of lipases.

Diffusion is another important mechanism in release into foods, because it is dominant in controlled release from matrix systems (microspheres). Diffusion occurs when the active ingredient passes through the encapsulating material. This mechanism can occur on a macroscopic scale (such as through pores in the matrix) or on a molecular level by permeation through the structuring material. Examples of diffusion-release systems are shown in Figure 15.4.

The typical release profiles shown in Figure 15.4 for reservoir and matrix delivery systems may present variations: a burst effect due to the presence of some core material too close to the external device surface for matrix devices, or a delayed time to start diffusion due to the diffusion of the core through the encapsulating layer of the reservoir device. Also, the physical state of the core material (dissolved or dispersed) defines the release kinetics. For example, a reservoir system in which the active core is not dissolved results in zero-order kinetics (constant flow), whereas it results in first-order kinetics (exponentially decreasing flow) if the core is dissolved in the encapsulated material.

The encapsulation of food ingredients can be achieved by physical, physicochemical, or chemical techniques (Shahidi and Han, 1993; Desai and Park, 2005a; Champagne and Fustier, 2007). The various

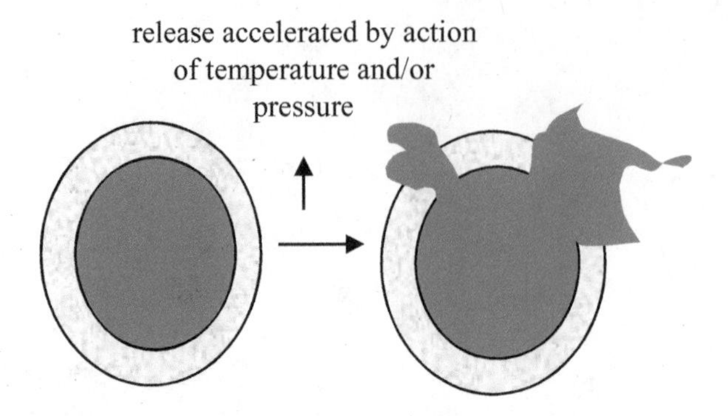

FIGURE 15.3 Release from reservoir device (microcapsule) fractured by mechanical forces.

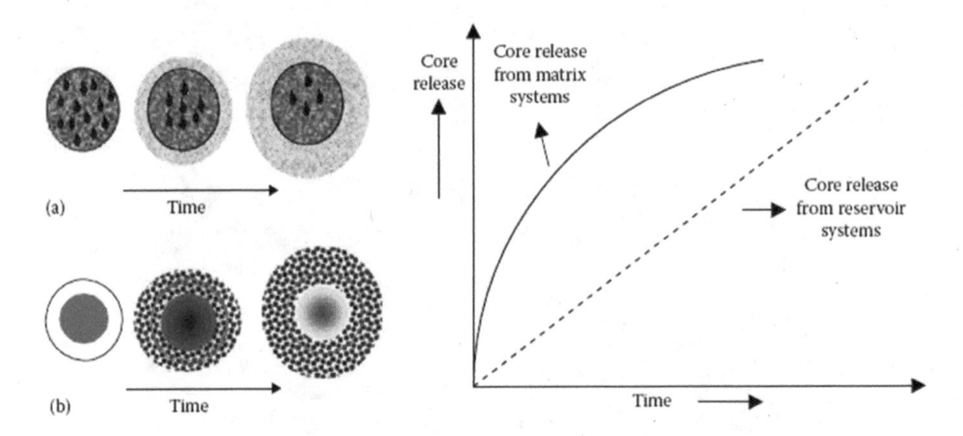

FIGURE 15.4 Examples of diffusion-release systems. (a) Diffusion from a typical matrix delivery system: diffusion occurs when the active ingredient passes from the structuring matrix into the external environment. As the release continues, the rate normally declines with time, since the active agent has a progressively longer distance to travel and therefore, requires a longer diffusion time to release. (b) Diffusion from a reservoir system, whether the active core is surrounded by a film or a membrane of an encapsulating material. The only structure effectively limiting the release of the core material is the encapsulating layer surrounding the core. Since this layer is uniform and has a constant thickness, the diffusion rate of the core can be kept fairly stable throughout the lifetime of the delivery system.

encapsulation technologies allow product formulators to make delivery devices from less than one micrometer to several thousand micrometers in size. Each technology offers specific attributes, such as high production rates, large production volume, high product yield, and different capital and operating costs. Other process variables include degree of flexibility in the selection of the encapsulating material and differences in the device size and morphology. The selection of an encapsulation technique is governed by the properties (physical and chemical) of the core and encapsulating materials and the intended application of the food ingredients.

The material to be encapsulated can be solid or liquid. The principle of most encapsulation technologies is quite simple, combining three consecutive steps (Poncelet and Dreffier, 2007):

1. The active ingredient is mixed within an encapsulating material, in most cases a polymer solution.
2. The resultant formulation, in a liquid form, is dispersed into fine droplets by dripping (drop-by-drop), spraying, or emulsification. This step increases the liquid surface available for further transformations (evaporation, cooling, gelation, chemical reactions) and favors the generation of dispersed (liquid or solid) delivery structures.

3. The oil or aqueous droplets are stabilized in a third step. Oil droplets can be made of a molten material and can be transformed into solid particles by cooling. Depending on their formulation, aqueous droplets can be subjected to a number of solidification processes such as gelation, polymerization, or crystallization.

When the active ingredient is in a solid form (solid particles), encapsulation can also be achieved by spraying a coating solution onto the particle surface by many different processes that will keep the particles in motion (e.g., agitation in a fluid bed or a pan rotating bed), followed by a consecutive step of stabilization of the coating by solidification or membrane formation.

15.3 Encapsulating Systems

Delivery systems can be solid or liquid depending on the food matrix where they are introduced. Some examples of *solid systems* are spray-dried and gel microparticles, whereas *liquid systems* include liposomal and emulsified systems. Each of these delivery systems has its own specific advantages and disadvantages for the encapsulation, protection, and delivery of food ingredients. These aspects are briefly discussed in the following, together with a description of the basic principles of each technique, its physicochemical characteristics, and the current challenges for its application in foods.

15.3.1 Spray-Dried Microparticles

15.3.1.1 Concept, Structure, and Properties

Spray-drying is a preservation technique commonly used in the food industry, mainly for dairy products. By decreasing water content and water activity when converting liquids into powders, this technique increases the storage stability of products, minimizes the risk of chemical or biological degradation, and also reduces the storage and transport costs.

Spray-drying is a unique drying process, since it involves both particle formation and drying. From a microstructural viewpoint, the formation of spray-dried powders involves droplet formation from a

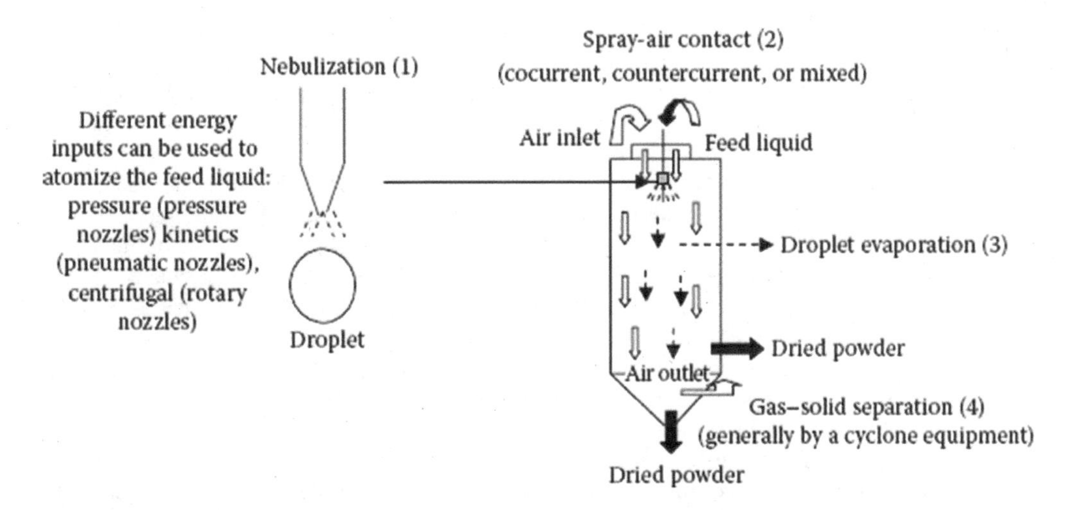

FIGURE 15.5 The main stages involved in the spray-drying process: (1) nebulization of feed liquid into small droplets (spray formation); (2) spray–air contact (mixing and flow), which can be made by several modes: cocurrent (the liquid is nebulized in the same direction as the airflow, as illustrated), countercurrent (liquid droplets and hot air flowing in opposite directions), or mixed flow (the liquid is sprayed upward and only remains in the hot zone for a short time); (3) droplet evaporation; and (4) separation of dried product from the air.

liquid state followed by a solidification operation driven by solvent evaporation, as schematically represented in Figure 15.5.

Liquid atomization is a decisive stage in spray-drying, defining the evaporation surface. It covers the process of liquid bulk breakup into millions of individual droplets forming a spray. To illustrate, consider the division of one liquid droplet with an initial diameter of 1 cm into N droplets of an equal final diameter of 100 μm. For the same liquid volume, this disintegration mechanism generates 10^6 droplets of 100 μm. The superficial area of the liquid, for example, the available surface for heat and mass transfer between the liquid and the drying air, is thereby increased 100 times.

Liquid nebulization into small droplets can be carried out by kinetic pressure or centrifugal energy. Conventional atomizer or nebulizer devices include centrifugal or pneumatic nozzles and rotary discs. Selection of the atomizer configuration is one of the most important choices in the spray-drying design. It depends on the nature and viscosity of the liquid to be sprayed and has a significant effect on the size distribution of the final dry particles. As a general rule, an increase in the energy available for nebulization (i.e., rotary nozzle speed, nozzle pressure, or air-to-liquid ratio in a pneumatic nozzle) reduces the particle size (Masters, 1985). However, for pneumatic nozzles, there is an optimum air-to-liquid ratio, above which an increase in energy input does not increase the efficiency of the nozzle to disintegrate the liquid and represents a waste of energy (Ré et al., 2004). By modifying the atomization and the drying conditions as well as the liquid formulation, it is possible to alter and control properties of spray-dried powders such as particle size, solubility, dispersibility, moisture content, hygroscopicity, flowability, and bulk density.

Spray-dried particles are produced as a very fine powder. Mean particle sizes range from a few microns to several tens of microns, with a relatively narrow size distribution, which actually depends on the spray–air contact mode. Particles in the range of 1–50 μm in diameter are typically produced in the cocurrent mode, the most usual contact mode to dry thermosensitive products. Larger particles, ranging from 50 to 200 μm, can be obtained in a countercurrent mode due to their agglomeration inside the drying chamber. Spray-dried particles may be spherical or present superficial indentations formed by shrinkage of the droplets during the early stages of the drying process due to the effect of a surface tension–driven viscous flow (Masters, 1985). The interested reader is referred to Chapter 9 for further details on the principles of spray-drying and its use as a preservation method in the food industry.

Spray-drying is used as a microencapsulation method, because structured microparticles can be created, starting from complex liquid mixtures comprising an active ingredient dissolved, dispersed, or emulsified with an encapsulating material in an aqueous or an organic solvent (water is used in most cases). An appropriate combination of the active ingredient with a range of encapsulating materials and other formulating ingredients, such as surfactants, can give the in-use properties of the encapsulated product. In a single operation, after nebulization of the liquid feed into the spray-dryer, the active ingredient can be entrapped into the main component of the structure: for example, the encapsulating material. This process permits the isolation of microspheres or microcapsules, depending on the initial formulation (Figure 15.6), as discussed by Ré (2006).

An important step in developing spray-dried microcapsules for food application is the choice of the encapsulating material. This greatly affects the encapsulation efficiency, appropriate thermal or dissolution release, mechanical strength, microcapsule stability, and compatibility with the food product. The criteria for this choice are mainly based on the solubility in water at an acceptable level and the film-forming and emulsifying properties of the limited number of food-grade materials available. Carbohydrates such as starches, maltodextrins, corn syrup solids, and gum arabic are used extensively in spray-dried encapsulations of food ingredients as the encapsulating material (Ré, 1998). Proteins such as whey proteins, soy proteins, and sodium caseinate have also been used (Gharsallaoui et al., 2007). However, the selection of the encapsulating material for spray-dried microcapsule production has traditionally involved trial-and-error procedures. First, a material is chosen, and microcapsules are produced using different mass proportions of encapsulating material to active ingredient and various drying conditions. The spray-dried powders are then evaluated for encapsulation efficiency, stability under different storage conditions, and degree of protection provided to the core material, among other physical characterizations (particle size, powder density, powder flowability, etc.). This procedure is costly and time consuming. Some efforts have been made to develop quantitative methods for selecting the most suitable wall materials for spray-dried microcapsules, mainly for lipid encapsulation.

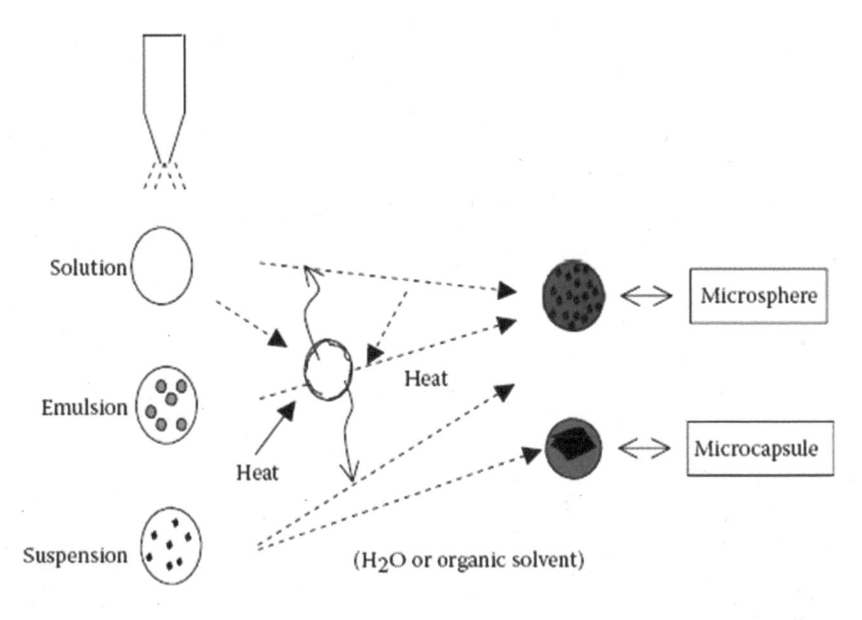

FIGURE 15.6 Architecture of spray-dried particles, depending on the initial formulation (solution, suspension, or emulsion). (From Ré, M.I., *Drying Technol.*, 24, 433, 2006. With permission.)

A method proposed by Matsuno and Adachi (1993) for screening the most suitable wall materials for lipid encapsulation is based on measurements of the drying rate of an emulsion as a function of its moisture content. A suitable material for this application should possess high emulsifying activity, high stability, and a tendency to form a fine and dense network during drying, and at the same time, should not permit lipid separation from the emulsion during dehydration. Because the isothermal drying rate is governed by the water diffusion rate, the drying rate may reflect the sample matrix characteristics; that is, the finer and denser the matrix, the lower the drying rate. According to this method, a characteristic drying curve as a function of moisture content for a suitable group of encapsulating materials is presented in Figure 15.7 (Type 1 curve). This curve has been interpreted in terms of the ability of the wall material to form a dense network. The drying rate decreases rapidly as the water content decreases, reflecting a rapid formation of a dense skin and good protection of the core ingredient against oxygen transfer and possible deterioration. Some materials presenting this type of drying curve are maltodextrin and gum arabic, which are considered as the most suitable for microencapsulation by spray-drying. According to this method, materials that do not form dense skins at the early stages of drying are unsuitable for efficient lipid encapsulation (Types 2, 3, and 4 curves). Characteristic Type 2 materials are caseinate and albumin, Type 3 materials are low–molecular weight saccharides that do not crystallize readily, such as glucose, and Type 4 materials are those that easily crystallize on dehydration, such as mannitol (Matsuno and Adachi, 1993). However, not all the materials that show an early decreasing rate, in which water evaporation is controlled by diffusion mechanisms, are suitable for lipid encapsulation when used alone. Therefore, it is desirable to determine an optimal combination of materials that will provide excellent emulsifying capacity and very low oxygen diffusion. In this respect, as analyzed by Pérez-Alonso et al. (2003), the Adachi and Masuno method does not allow effective discrimination between materials showing similarly shaped drying curves.

Another method, proposed by Pérez-Alonso et al. (2003), uses the activation energy of carbohydrate polymer blends dried isothermally as a discriminating parameter for selecting the most suitable mixture as wall material for spray-dried microcapsules. The activation energy provides a measure of the necessary energy required for evaporating a mass of water from the material to be dried. This method requires the knowledge of the drop volume shrinkage of every conceivable blend, which can be achieved as follows. A drop of a blend constituent aqueous solution is put on a glass slide, and micrographs of the *X–Y*,

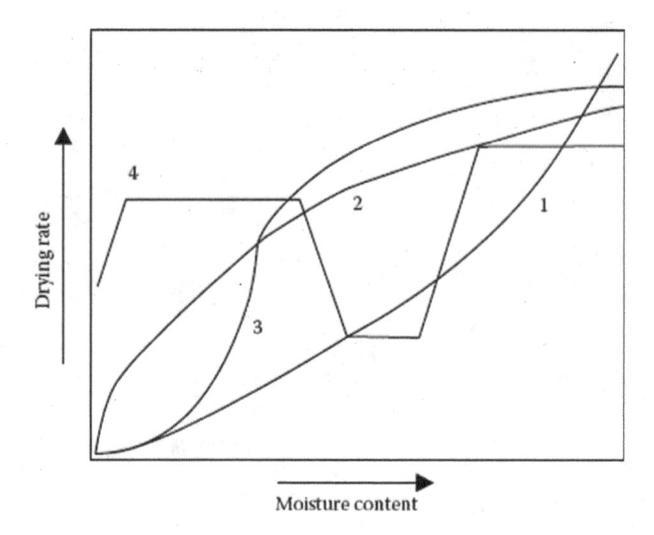

FIGURE 15.7 Schematic representation of isothermal drying curve for the selection of encapsulating materials for spray-dried microcapsules (Matsuno and Adachi method). Type 1 curve corresponds to materials that form fine, dense, two-dimensional skins immediately on drying. Types 2, 3, and 4 curves correspond to materials that do not form dense skins at an early stage of drying. (From Pérez-Alonso, C. et al., *Carbohydr. Polym.*, 53, 197, 2003. With permission.)

X–Z, and *Y–Z* planes of the drop are taken. The area of each plane is calculated and approximated to that of a sphere. These steps are repeated as the drops are dried isothermally at intervals of approximately 10% moisture content decrease (determined by drop mass loss). The experimental points are then fitted with a polynomial reported for each blend constituent and assumed additive volumes of the blend constituents to determine the drop shrinkage of studied blends.

Despite these efforts, screening for new wall materials, such as milled citrus fruit fibers as a potential replacement for maltodextrin-type carriers (Chiou and Langrish, 2007), is still mainly done by trial and error.

When selecting spray-drying to produce an encapsulated food ingredient, one is generally looking for high production in a short time and for a product in a powder form. Spray-dried microparticles are commonly used to encapsulate flavors or lipids, and the release mechanism is generally linked to the dissolution of the encapsulating agents. The encapsulated ingredients are, in general, rapidly released due to the dissolution of the spray-dried structures at the time of consumption by dispersing the powder in a wet formulation (e.g., instant beverages). The challenge is how to encapsulate thermosensitive and volatile compounds by a drying process operation, avoiding thermal degradation and volatile losses during drying, and generating an encapsulated product in a powder form with good flowability, stability, and acceptable shelf life after drying.

15.3.2 Gel Microparticles

15.3.2.1 Concept, Structure, and Properties

Gel microparticles are formed from the concept of gelation as an encapsulation technology. Gelation is based on the formation of a solution, a dispersion, or the emulsification of the core material in an aqueous solution containing a hydrophilic polymer (hydrocolloid) capable of forming a gel under an external action, either physical or chemical.

There are many techniques for physical gelation, whose use depends on whether the hydrocolloid can gel in water without additives (thermal gelation) or whether ions are required to aid gelation (ionotropic gelation). In thermal gelation, typically, a solution is made by dissolving a hydrocolloid in powder form in water at high temperature and then cooling to room temperature. As the solution cools, enthalpically stabilized chain helices may form from segments of individual chains, leading to a three-dimensional network (Burey et al., 2008). Examples of such systems are gelatin and agar.

Ionotropic gelation occurs via the cross-linking of hydrocolloid chains with ions, generally cation-mediated gelation of negatively charged polysaccharides. Examples of such systems are alginate, carrageenan, and pectin. A typical example is the formation of alginate beads by dropping an alginate solution into a bath containing calcium chloride to form the insoluble calcium alginate. There are two main methods by which ionotropic gelation can be done: external and internal gelation. External gelation involves the introduction of a hydrocolloid solution into an ionic solution, with gelation occurring via diffusion of ions from outside into the hydrocolloid solution. This is the easiest and most often used method for encapsulation by ionotropic gelation. However, it can often cause inhomogeneous gelation of gel particles due to its diffusion-based mechanism, which constitutes a drawback. Surface gelation often occurs prior to core gelation, and the former can inhibit the latter, leading to gel particles with firm outer surfaces and soft cores (Chan et al., 2006). Internal gelation overcomes the main disadvantage of the external gelation method, as it requires the dispersion of ions prior to their activation to cause gelation of hydrocolloid particles. This usually involves the addition of an inactive form of the ion that will cause cross-linking of the hydrocolloid, which is then activated, for example, by a change in pH after the ion dispersion is sufficiently complete. Internal gelation is particularly useful in alginate systems, which can gel rapidly and may become inhomogeneous if gelation occurs before adequate ion dispersion has occurred (Poncelet et al., 1992; Chan et al., 2002). For example, in the production of alginate particles by external gelation, the alginate solution is extruded as droplets into a solution of a calcium salt. For internal gelation, an insoluble calcium salt is added to the alginate–drug solution, and the mixture is extruded into oil (Liu et al., 2002). The latter is acidified to bring about the release of Ca^{2+} from the insoluble salt for cross-linking with the alginate. Despite their homogeneity, internal gelated matrices may be more permeable, resulting in lower encapsulation efficiencies and faster release rates (Vanderberg and De La Noüe, 2001), which may also be overcome by manipulating the pH of the medium and the amount of calcium salt used.

Whatever the technique used (thermal or ionotropic gelation), gel particles are generally formulated in a two-step procedure involving droplet formation and hardening. The droplet formation step determines the mean size and the size distribution of the resulting gel particles. In the following, the main procedures used for droplet formation—droplet extrusion, nebulization (spray), and emulsification—are described.

15.3.2.1.1 Droplet Extrusion

Extrusion denotes feeding the hydrogel solution, typically containing the active material to be encapsulated, through a single pathway or multiple pathways directly into the continuous gelation bath. Henceforth, for simplicity, the hydrogel solution containing the active material to be encapsulated (generally an aqueous dispersion) is referred to only as *hydrogel* solution.

In the droplet extrusion technique, also referred to as the *drop method*, hydrogel solutions are extruded through a small tube or needle (Figure 15.8a), permitting the formed droplets to fall freely into a gelation

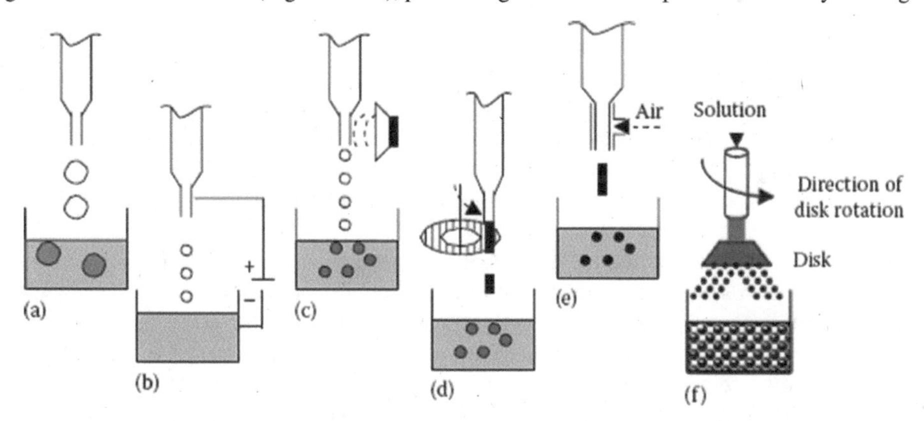

FIGURE 15.8 Schematic representation of the different techniques for drop formation employed in gelation: (a) droplet extrusion; (b) electrostatic dripping; (c) laminar jet breakup; (d) jet-cutting; (e) jet nebulizer; and (f) disk nebulizer.

bath. The droplets may be cross-linked by the addition of an appropriate cross-linker to the receiving solution. The size of the droplets, and thus, the size of the subsequent gel particles, depends on the diameter of the needle, the flow rate of the solution, its viscosity, and the concentration of the ionic solution. Typical gel particle sizes obtained using the conventional syringe-drop method are 0.5–6 mm, and on the scale of hundreds of microns if modified techniques suitable for large-scale processes are used to disperse the hydrocolloid solution into droplets (Burey et al., 2008). The main difference among the reported techniques for mass production of small narrowly dispersed or monosized hydrocolloid gel particles lies in the way the drops are formed, that is, electrostatic dripping (Figure 15.8b), jet breakup through mechanical vibrations (Figure 15.8c), jet-cutting (Figure 15.8d), and jet and rotating disc atomizers (Figure 15.8e and f, respectively).

The electrostatic technique uses an electric potential difference to pull the droplets from a needle tip (Figure 15.8b). The electrostatic potential difference is established between the needle feeding the solution and the gelling bath. In the absence of an electric field, a droplet forming on a needle tip will grow until its mass is large enough to escape the surface tension at the needle–droplet interface. With the introduction of an electric field, a charge is induced on the droplet surface. Mutual charge repulsion results in an outwardly directed force acting downward on the forming droplet. The additional electrostatic force pulls the droplet from the needle tip at a much lower mass and hence, size. Capsule size may be controlled by adjusting the magnitude of the voltage. The higher the voltage, the higher the electrical force pulling the droplet, and therefore, the smaller the obtained droplet and consequently, the capsule. The literature reports different capsule ranges that can be achieved using this method; for example, 40–2500 μm according to Burey et al. (2008) or 50–800 μm according to Poncelet and Dreffier (2007).

In the laminar jet breakup technique, a laminar flow of the hydrocolloid solution is converted into a succession of identical droplets by the action of an ultrasound vibrating nozzle (Figure 15.8c). Jet-cutting is another method, in which the fluid is pressed through a nozzle in the form of a liquid jet. This jet is cut into uniform cylindrical segments by a means of a rotating cutting tool (Figure 15.8d). Due to surface tension effects, these segments form spherical beads while falling down. The diameter of the resulting bead is determined by the number of cutting wires, the number of rotations of the cutting tool, and the mass flow through the nozzle, which in turn, depends on both the nozzle diameter and the fluid velocity.

15.3.2.1.2 Nebulization

The dispersion of the hydrocolloid solution may also be achieved by jet nebulizers using, for example, a coaxial air stream that pulls droplets from a needle tip into a gelling bath (Figure 15.8e). Small quantities of gel particles ranging in size down to around 400 μm are achieved by this method (Herrero et al., 2006). Droplet formation is aided by the shear energy of gas flow used to overcome the viscous and surface tension forces of the fluid. The viscosity and surface tension of a liquid being nebulized can thus alter the properties of the aerosol generated (Figure 15.8f).

15.3.2.1.3 Emulsification

In the emulsion technique, solutions are mixed and dispersed into a non-miscible phase. For food applications, vegetable oils are used as the continuous phase. In some cases, emulsifiers are added to form a better emulsion, since such chemicals lower the surface tension, resulting in smaller droplets. After emulsion formation, gelation and/or membrane formation is initiated by cooling and/or addition of a gelling agent to the emulsion or by introducing a cross-linker. In a last step, the gel particles formed are washed to remove oil (Chan et al., 2002).

Stirring is the most straightforward method to generate droplets of a dispersed phase in a continuous phase to produce an emulsion. In the simplest approach, the continuous phase is poured into a vessel and stirred by an impeller (Figure 15.9a). The dispersed phase is then added, dropwise or all at once, under agitation at a sufficient speed to reach the desired droplet size. The final droplet size of the liquid–liquid dispersion in stirred vessels depends on parameters such as the physicochemical characteristics of the two phases (e.g., viscosity, interfacial tension, and stabilizer concentration), the preparation conditions of the emulsion (e.g., temperature, addition order of the components), and the stirring system (e.g., shear rate, design of the stirrer, and containing vessel). Hydrocolloid solution droplet sizes normally range from 0.2 to 80 μm, although they can be as large as 5000 μm, and gel particles can range from 10 to 3000 μm, as summarized in a recent review (Burey et al., 2008).

A number of approaches have been proposed for continuous emulsification and improved control of product size distribution: static mixers, membrane emulsification, and microchannel emulsification. For the last two, emulsions are produced by extruding a liquid through many individual pores or microchannels.

Static mixers consist of a series of geometric mixing elements fixed within a pipe. The particular arrangement of these mixing elements repeatedly splits and recombines the stream of fluid passing through the tube (Figure 15.9b). Recombination occurs through impingement of the substreams, creating turbulence and inducing back mixing. Static mixers installed in tubes allow continuous production and have already been used to produce gel particles (Belyaeva et al., 2004).

Membrane emulsification is a relatively new method for the preparation of spherical particles with a highly uniform size distribution, and it can be used to extrude a hydrocolloid solution into a gelation bath (Oh et al., 2008). The method involves the use of a porous membrane with a highly uniform pore size. The dispersed phase is pressed through the membrane pores, whereas the continuous phase flows along the membrane surface. Droplets grow at pore outlets until they detach on reaching a certain size (Figure 15.9c). A low-pressure drop is applied to force the dispersed phase to permeate through the microporous membrane into the continuous phase. Details on such methods using membrane emulsification are summarized in several review papers (Joscelyne and Trägardh, 2000; Vladisavljevic and Williams, 2005). The distinguishing feature is the fact that the resulting droplet size is primarily controlled by the choice of the membrane and not by the generation of turbulent droplet breakup. The technique is highly attractive given its simplicity, potentially lower energy demand, need for less surfactant, and the resulting narrow droplet size distributions. It is applicable to both oil-in-water (O/W) and water-in-oil (W/O) emulsions, and it has been recently applied to prepare microgel particles with a uniform size distribution (Wang et al., 2005; Zhou et al., 2007).

Microfluidic methods have recently been mentioned as an emerging technology for the production of monosized gel particles (Amici et al., 2008). Typically, the term *microfluidics* refers to the manipulation of fluids in systems or devices having a network of chambers and reservoirs connected by channels, whose typical cross-sectional dimensions range from 1.0 to 500 μm, the so-called *microchannels.*

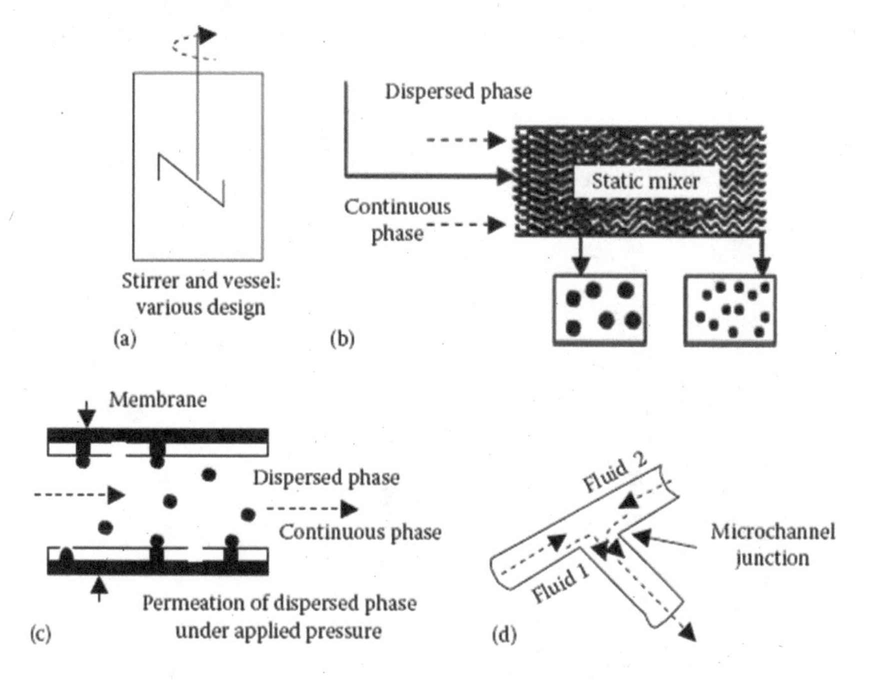

FIGURE 15.9 Schematic diagram of different emulsification processes to produce gel particles: (a) mechanical stirring; (b) static mixing; (c) membrane emulsification; and (d) microchannel emulsification.

Materials such as silicon, cofired ceramic, glass, quartz, and polymers have been explored for the fabrication of microfluidic systems (Nguyen and Zhigang, 2005). Due to the large surface-to-volume ratio and low inertial forces encountered at the microscale, highly precise and specific flow manipulation and control can be achieved by appropriate microfluidic design.

Emulsions are produced in microfluidic devices when two immiscible fluids, flowing in two separate microchannels, are forced through a microchannel junction, and the flow of one fluid (usually the fluid that wets the channel surface) breaks the flow of the other to form microdroplets (Figure 15.9d). Drop formation is reproducible, and the drop size can be regulated by operating on factors such as flow rates, fluid viscosities, and surfactant concentration. Gel formation in microfluidic systems has been reported, including, for instance, the thermal gelation of κ-carrageenan (Walther et al., 2005) and agarose (Xu et al., 2005). Ionotropic gelation can also be achieved in microchannels through both the internal and the external approaches (Huang et al., 2007; Amici et al., 2008).

Alginate microspheres have been extensively used as delivery systems, because they are very easy to prepare, the process is very mild, and virtually any ingredient can be encapsulated (Vladisavljevic and Williams, 2005). This is the reason why they are chosen to exemplify the use of the membrane and channel emulsification processes to produce gel particles.

Figure 15.10 illustrates the preparation of Ca alginate gel particles by membrane emulsification by two different approaches. In the first one, an aqueous Na alginate solution is extruded through a hydrophobic membrane into an oil phase to form a W/O emulsion. The droplets are then gelled by adding a $CaCl_2$ solution, and the gelled particles can be collected by filtration. This process was used by Weiss et al. (2004), enabling the formation of uniform gel particles with an adjustable mean diameter (20–300 μm). In the second approach, Ca alginate gel particles can be produced by internal gelation; in brief, a dispersion of an aqueous Na alginate solution containing $CaCO_3$ is extruded through a microporous membrane into an oil phase. The gelation is initiated by the addition of acetic acid into the emulsion, dissolving $CaCO_3$ to release Ca^{2+} and form Ca alginate. Using the second approach, Liu et al. (2003) obtained Ca alginate gel particles with a mean size of 55 μm (with a coefficient of variation of 27%) using a nickel membrane with a pore size of 2.9 μm.

Figure 15.11 illustrates the preparation of Ca alginate gel particles by microchannel emulsification. To produce alginate gel particles by external gelation, an alginate aqueous solution flows through a flow-focusing channel, and an alginate droplet is formed from the balance of interfacial and viscous drag forces resulting from the continuous (oil) phase flowing past the alginate solution. It immediately

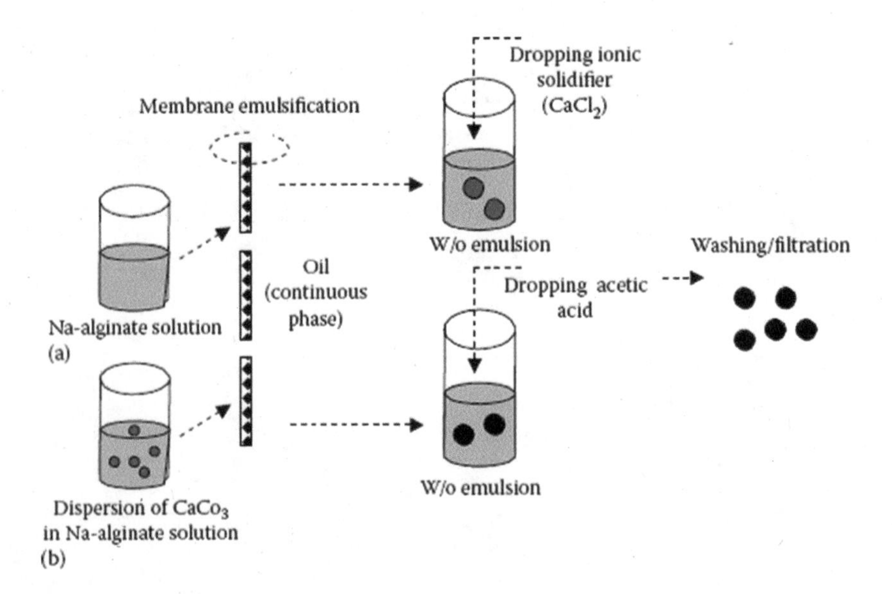

FIGURE 15.10 Illustration of the production of alginate microgels by membrane emulsification: (a) external gelation and (b) internal gelation.

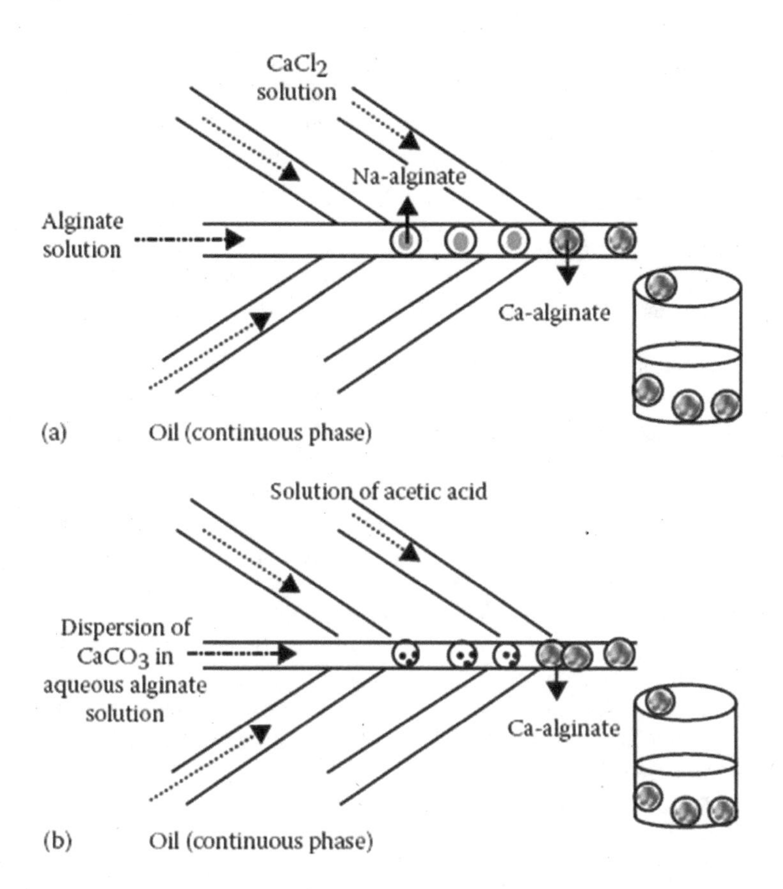

FIGURE 15.11 Illustration of the microfluidic production of alginate microgels by (a) internal gelation and (b) external gelation.

reacts with an adjacent $CaCl_2$ drop that is extruded into the main flow channel by another flow-focusing channel located downstream in relation to the site of the alginate drop creation. This procedure has been used in the literature, generating monosized alginate beads within a range of 50–200 μm depending on flow conditions (Hong et al., 2007). To produce alginate drops in a microfluidic system by internal gelation (Amici et al., 2008), two aqueous streams, one acidic and one containing alginate and calcium carbonate, can merge immediately prior to entering a channel where a continuous flow of an oil phase breaks the flow of the aqueous phase to form microdroplets (Figure 15.11b). Several variations of both approaches have been developed in the literature by using new microfluidic systems (Liu et al., 2006; Choi et al., 2007).

15.3.3 Liposomal Systems

15.3.3.1 Basic Principles

Liposomes are conceptually biomimetic model systems. They allow studies of the lipid matrix of biomembranes, as well as the investigation of membrane-embedded proteins and certain fundamental aspects of organelles in a biomimicking environment, outside the living cell. Structurally, liposomes are self-assembled colloidal particles in which an aqueous nucleus is enclosed by one or several concentric phospholipid bilayers. Phospholipids are amphiphilic molecules, which in aqueous solution form energy-favorable structures as a result of hydrophilic and hydrophobic interactions. Depending on the size and the number of bilayers, liposomes are classified as multilamellar vesicles (MLVs) or large and small unilamellar vesicles (LUVs and SUVs). Other classifications for larger liposomes include plurilamellar vesicles (PLVs), in which non-concentric bilayers enclose various aqueous compartments, and oligolamellar vesicles (OLVs), in which small vesicles are included in the structure of a large vesicle. The size of unilamellar liposomes may vary from 20 to 500 nm approximately, and the thickness of one lipid bilayer is about 4 nm. Due to the presence of hydrophilic and hydrophobic domains in the structure, liposomes are able to encapsulate water-soluble molecules in the aqueous nucleus, water-insoluble molecules within the lipid membrane, or amphiphiles between the two domains. Therefore, liposomes are a powerful solubilizing system for a wide range of compounds.

Figure 15.12 shows schematically the phospholipid aggregation in a bilayer, and the vesiculation, which occurs in the presence of excess water, forming the liposomal structure. Due to their structure, chemical composition, and size, all of which can be well controlled by preparation processes, liposomes exhibit several properties that are useful in a large range of applications. The most important properties are colloidal size, bilayer phase behavior, mechanical properties, permeability, and charge density. Liposomes also allow surface modifications through the attachment of ligands and bonded or grafted polymers. Figure 15.13 shows the various modifications of liposome surfaces.

For historical reasons, when no physical or chemical surface modification is introduced, liposomes are called *conventional* to distinguish them from surface-modified liposomes.

Liposomes can be made entirely from naturally occurring substances and are therefore biodegradable and non-immunogenic. In some cases, the introduction of synthetic lipids is useful for specific characteristics, such as stability and charge. The lipid composition, colloidal characteristics, and surface modifications also allow liposomes to have functional properties, such as stability, controlled release of the encapsulated molecules, specific targeting, and controlled pH and temperature sensitivity.

From the thermodynamic point of view, liposomes are not stable structures, and so, they cannot form themselves spontaneously. To produce liposomes, some energy from extrusion, homogenization, or sonication must be dissipated into the system. Subsequently, after formation, liposomes can aggregate, fuse, and form larger structures that eventually settle out of the liquid. Thermodynamically stable systems, such as micelles, stay at the same phase forever. The precursor phase of liposomes is the symmetric bilayer composed of self-assembled phospholipid molecules with comparable areas of the polar and non-polar portions. Liposomes are formed when, at high concentration, the self-assembled molecules form a long-range ordered liquid-crystalline lamellar phase, which when diluted in excess water, is dispersed in stable colloidal particles. These particles retain the short-range symmetry of their original parent phase.

An extension of Tanford's (1980) treatment of the shapes of micelles, based on the molecular shape analysis and the concept of the shape parameter (ratio between the non-polar and polar cross-section areas), was introduced by Israelachvili (1991) for a qualitative understanding of the topology of lipid aggregates with different lipid compositions. However, the liposome models are approximate, and a

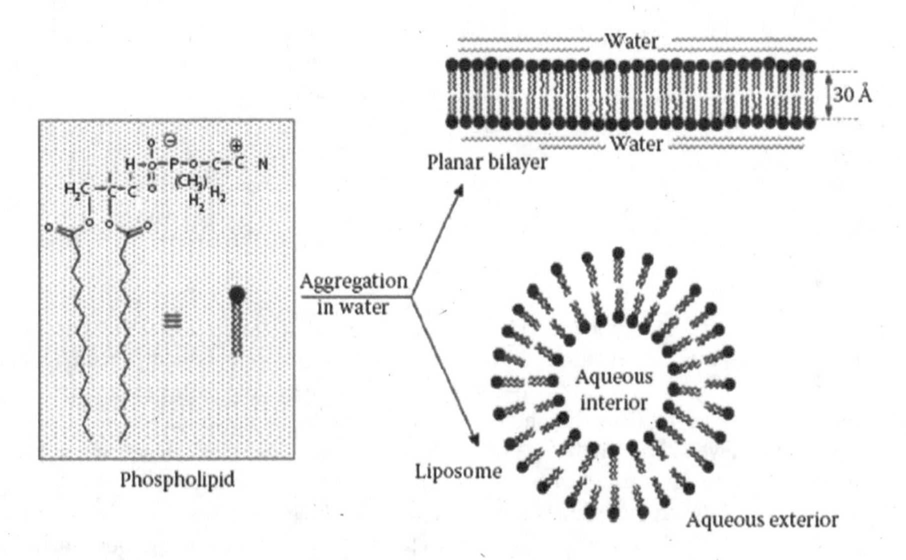

FIGURE 15.12 Phospholipid aggregation in planar bilayer and vesiculation in liposomal structure.

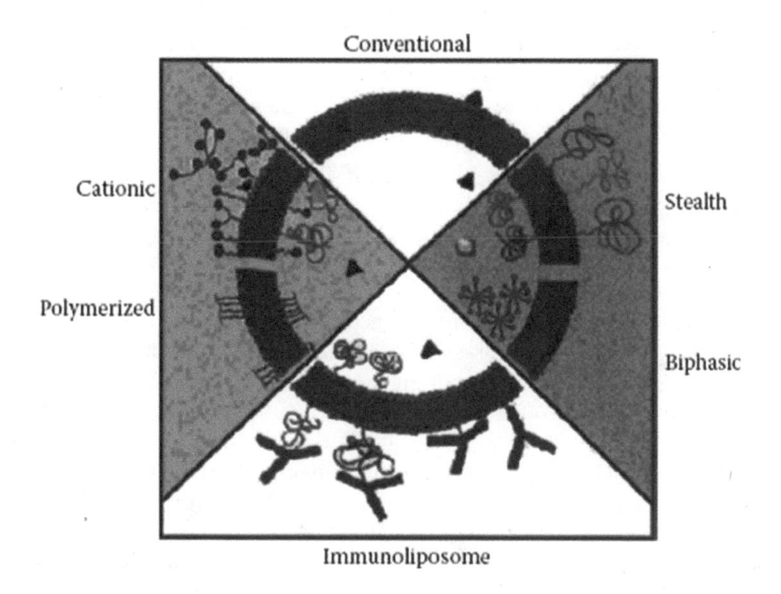

FIGURE 15.13 Surface modifications in liposomes.

rigorous thermodynamic analysis fails, because liposomes are not at thermodynamic equilibrium. Rigorous analysis would yield a very narrow size distribution, which is never observed in practice, as well as spontaneous formation, while a high-energy process is typically needed to produce liposomes.

Helfrich (1973) explained the instability of liposomes by the bending elasticity. Symmetric membranes prefer to be flat (spontaneous curvature equal to zero), and energy is required to curve them. Various factors, such as asymmetric changes by ionization or the insertion of molecules such as surfactants, change the spontaneous curvature to non-zero values, but the spontaneous formation does not produce stable liposomes for drug encapsulation. Stable liposomes for drug delivery must be in a kinetically trapped and thermodynamically unstable state. As a consequence of that, liposomes maintain their integrity despite changes in the environment such as dilution or *in vivo* administration. Micelles and microemulsions, which are thermodynamically stable systems, disintegrate, aggregate, or change phase under perturbations of the medium. The vesiculation accumulates an excess of free energy around 10^{-50} kT (1 kT = 4.11 $\times 10^{-21}$ J at T = 298 K) in the curvature of the liposomes, which generates instabilities, promoting fusion and disintegration, or may become bioavailable on vesicle fusion with cell membranes.

Phase transition is one of the most important properties of liposomes. It happens when lipid bilayers change from a solid-ordered phase at low temperature to a fluid-disordered phase above the phase transition temperature. Bilayers can also undergo transitions into different liquid-crystalline phases, such as hexagonal or micellar. The phase transitions can be triggered by physical or chemical factors, resulting in different effects on the liposomes. Thus, transitions from the gel to the liquid-crystalline phase caused by changes in temperature or ionic strength cause liposome leakage.

Physicochemical stability determines the shelf-life stability of liposomes. By optimizing the size distribution, pH, and ionic strength, as well as by the addition of antioxidants and chelant agents, the stability of liposomes can be preserved for years. They can be stored in frozen or dried form, but cryoprotectants have to be added to prevent fusion. Electrostatic and steric stabilization also reduce fusion and disintegration by freezing. Biological *in vivo* stabilization is obtained by reducing the interactions of liposomes with macromolecules, blood protein, and disintegrating enzymes as well as adverse pH conditions. Therefore, the *in vivo* stability of liposomes depends on the route of administration. Steric stabilization is generally provided by a protective coating made by grafting the liposome surface with inert hydrophilic polymers such as polyethylene glycol (Lasic, 1995) or hyaluronic acid (Eliaz and Szoka, 2001).

Since their discovery in the 1960s and the first studies in which they were used as a drug delivery system in the 1970s, liposomes have been used as an encapsulation system in a myriad of applications ranging from material science to analytical chemistry, food, and medicine.

Numerous books and reviews describe the various physicochemical and biological aspects of liposomes, as well as their construction and characterization. This literature has been written by scientists who developed the scientific basis of the design and construction of liposomes for various applications (Gregoriadis, 1988; Lasic, 1993; Lasic and Papahadjopoulus, 1998).

15.3.3.2 Applications in Foods

Gomez-Hens and Fernandez-Romero (2006) depict the number of articles and patents published in the 1990–2004 period regarding the use of liposomal systems in five areas: drug and gene delivery systems, biochemical and biotechnological applications, cosmetics, nutrition, and foods. Figure 15.14 shows the evolution of articles and patents from 2004 to 2008. In both cases, the tendency is the same: there have been significant advances in the applications of liposomes in the biomedical and pharmaceutical industries related to therapeutic drugs, opposed to their applications in foods, which are presently at an early stage of development.

The applications of liposomes in foods fulfill similar requirements to their applications in pharmaceuticals and have been focused on the following categories: formulation aids, processing, preservation, stabilizers, nutritional supplements, and nutraceutical carriers.

Liposomes aid in formulation because they are powerful in solubilizing non-water-soluble compounds, enhancing their bioavailability. Furthermore, liposomes entrap hydrophilic molecules into their interior and hydrophobic molecules into their lipophilic membrane, encapsulating various nutritional molecules. The first example of liposomes in foods is human milk, which has been studied for years. Electron microscopic studies show the presence of liposomes along with emulsion droplets and casein micelles (Roger and Anderson, 1998). Liposomes, as a microstructural component of breast milk, may play an important role in enhanced nutrient absorption, colloidal stability, and immunogenicity (Keller, 2001).

In the processing of foods, liposomes accelerate cheese ripening and increase the yield in bioconversion through uniform distribution of hydrophilic enzymes in a hydrophobic medium.

Applications of liposomes in cheese ripening were developed by the 1980s (El Soda, 1986). The enhancement of proteolysis by encapsulated cyprosins was evident 24 h after the manufacture of Manchego cheese. The addition of encapsulated cyprosins to milk perceptibly accelerated the development of flavor intensity in experimental cheese through 15 days of age without enhancing bitterness (Picon et al., 1996). The capability of neutral and charged liposomes to entrap the proteolytic enzyme

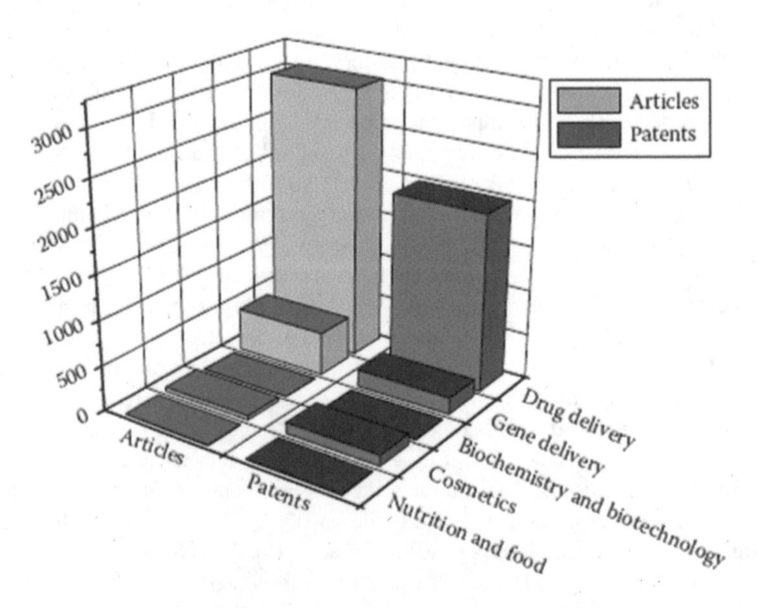

FIGURE 15.14 Articles and patents published on liposomal delivery systems in the 2004–2008 period (last year incomplete). (From SciFinder Scholar®, American Chemical Society, Columbus, OH.)

neutrase, and the stability of the preparation, were evaluated in the ripening of Saint-Paulin cheese milk (Alkhalaf et al., 1989).

Liposomes also promote sustained release of antimicrobial peptides, ensuring protection of the formulation. Liposome-entrapped nisin retained higher activity against *Listeria innocua* and improved stability in cheese production, proving to be a powerful inhibitor in the growth of *L. innocua* in cheese, but did not prevent the detrimental effect of nisin on the actual cheese-ripening process. The coencapsulation of calcein and nisin, and calcein and lysozyme, demonstrated that the production and optimization of stable nanoparticulate aqueous dispersions of polypeptide antimicrobials for microbiological stabilization of food products depends on the selection of suitable lipid–antimicrobial combinations (Benech et al., 2002; Were et al., 2003).

Besides solubilization, the encapsulation capability of liposomes protects labile compounds from chemical degradation, light oxidation during storage, and harmful compounds from the environment. The stabilizing capability of liposomes also preserves the taste and flavors of foods during processing and storage. Liposomes improve the nutritional effects of foodstuffs by entrapping nutritionally important compounds, such as vitamins, polyunsaturated fatty acids, minerals, and antioxidants.

Liposome encapsulation enhances the health benefits of nutrients by changing their kinetics of release and increasing their bioavailability. Recently, immunoliposomes, which are liposomes containing antibodies for site-specific targeting, have been studied for nutrient targeting regulation. A useful model involving the leptin protein and immunoliposomes was used to illustrate the nutrient regulation of the endocrine system (Xianghua and Zirong, 2006).

Liposomes as a carrier matrix in foods have become an attractive system, because they can be constructed entirely from acceptable edible compounds (food-grade ingredients), such as proteins and carbohydrates. Lecithin is the main natural phospholipid, routinely extracted from nutrients such as egg yolks and soybeans. Additionally, the phospholipids in the liposome matrix are also versatile nutraceuticals for functional foods. The benefits are for the brain, liver, and blood circulation. Phosphatidylcholine is a highly effective nutraceutical for recovery of the liver following toxic or chronic viral damage. It has exceptional emulsifying properties, which the liver draws on to produce the digestive bile fluid. The lung and intestinal lining cells use phosphatidylcholine to make the surfactant coating essential for their gas and fluid exchange functions. Phosphatidylcholine exhibits potentially lifesaving benefits against pharmaceutical and death cap mushroom poisoning, alcohol-damaged liver, and chronic hepatitis B. Phosphatidylserine has established benefits for higher brain functions such as memory, learning and word recall, mood elevation, and action against stress. Phosphatidylserine also has a salutary revitalizing effect on the aging brain and may also be helpful to children with cognitive and mood problems. The fast access of glycerophosphocholines to the human brain and their capacity to sharpen mental performance also make them well suited for drink formulations. The nutraceutical properties of phospholipids are described extensively by Kidd (2002). Therefore, the product value comes from the health benefits of the phospholipids associated with the benefits of the selected nutrient. This combined phospholipid–nutrient approach is suited to producing chewable tablets, confectionery products, cookies, granulates, spreads, bars, and emulsified or purely aqueous-phase beverages.

Although liposomes carrying nutrients are ingested via the GI system, the oral route also offers a way through the sublingual mucosal membranes. In the first case, the adverse conditions of the environment (the low pH of the stomach, the surfactant action of bile salts, and the presence of lipases) destabilize conventional liposome formulations. The sublingual route avoids the first-pass liver clearance and metabolism, offering direct uptake of nutrients into the bloodstream through the mucosal membranes. Additionally, the sublingual administration avoids swallowing difficulties from the ingestion of tablets or large capsules by old people or children, in addition to being an alternative for personal preference.

The performance of the sublingual administration of nutrients has been demonstrated using CoEnzyme Q10 in a spray formulation compared with the powder formulation in hard gelatin capsules. The results showed increased bioavailability of 100% for CoQ10 over endogenous levels with the sublingual spray compared with 50% increase over baseline levels with a two-piece gelatin capsule as measured by the area under the curve in 24 h (Gibaldi, 1991). Additionally, the time of onset with the spray formulation administration was shorter than with the capsule, adding benefit to treatments that require immediate onset, such as the cardiogenic supplement CoQ10, diet aids, and the treatment of pain, fever, or insomnia

(Rowland and Towzer, 1995). Keller (2001) listed some products that have been formulated using this novel oral liposomal delivery system.

The main issue in liposome encapsulation for food industry is the scaling up of the processes at an acceptable cost. Methods of liposome formation now exist that do not make use of sonication (Batzri and Korn, 1973; Kirby and Gregoriadis, 1984; Zhang et al., 1997) or of any organic solvents (Frederiksen et al., 1997; Zheng et al., 1999) and allow the continuous production of microcapsules on a large scale. Nowadays, liposome encapsulation has also become a routine process in the food industry (Gregoriadis, 1987; Kirby and Law, 1987; Kim and Baianu, 1991; Gouin, 2004).

The great advantage of liposomes over other encapsulation technologies is the stability imparted by liposomes to water-soluble materials in high–water activity applications: spray-dried, extruded, and fluidized beds impart great stability to food ingredients in the dry state but release their content readily in high–water activity applications, giving up all protective properties.

Microfluidization techniques have been shown to be an effective and solvent-free continuous method for the production of liposomes with high encapsulation efficiency. The method can process a few hundred liters per hour of aqueous liposomes on a continuous basis. The process has been reported in the literature (Vuillemard, 1991; Maa and Hsu, 1999; Zheng et al., 1999).

Multitubular systems represent a scalable version of the Bangham method. This method is adequate to prepare liposomes for food applications due to its simplicity and ease of scaling up (Tournier et al., 1999; Carneiro and Santana, 2004; Latorre et al., 2007).

Dry liposomes circumvent the drawback of liposome stability in the large-scale production, storage, and shipping of encapsulated food ingredients. Freeze-drying the liposome suspension can only be carried out using high-price encapsulated ingredients in a niche market due to the considerable cost of large-scale freeze-drying processes. Moreover, not all liposome formulations can be freeze-dried, and the reconstitution of the wet formulation is not always straightforward and usually requires complex steps and processes (Lasic, 1993).

These problems are reduced when the bioactive ingredient is incorporated in a lipid matrix by spray-drying, and subsequently, liposomes are made through mechanical stirring (Alves and Santana, 2004). The operational conditions modulate the crystallinity of the lipid matrix and the efficiency of incorporation of the bioactive compound. Conventional or special mechanical stirrers can be used to adjust the size and distribution of liposomes.

Supercritical fluid offers another attempt to avoid the use of organic solvent in the production of liposomes. Basically, the process involves the solubilization of the phospholipids under supercritical conditions followed by the release of the supercritical mixture into an aqueous phase containing the dissolved active ingredient, which results in the formation of liposomes containing the active ingredient in their aqueous cores. Although this method is scientifically interesting, the encapsulation efficiency reported so far is limited to 15% (Frederiksen et al., 1997), which would have to be dramatically increased for this technique to become interesting from an industrial point of view.

Liposomes can also be associated with other technologies. Many hydrophilic and hydrophobic compounds, including various vitamins and antioxidants, can be dispersed in other matrices by liposome encapsulation.

15.3.4 Emulsified Systems

15.3.4.1 Concept, Structure, and Properties

Emulsions are defined as mixtures of two immiscible liquids wherein one phase is dispersed in the other in the form of droplets. Food emulsions usually contain an aqueous phase (polar), an oil phase (apolar), and surfactants that are added to stabilize the system by reducing the interfacial tension between the dispersed and continuous phases. When oil droplets are present in the emulsion, they are called O/W emulsions. On the other hand, when the aqueous phase is dispersed, the emulsion is called W/O, and sometimes, it is referred to as an *inverse emulsion*. Figure 15.15 shows a schematic picture of O/W and W/O emulsions stabilized by surfactant molecules. These molecules are adsorbed on the interface and oriented according to the type of emulsion. In the case of O/W emulsions, the hydrophilic head is

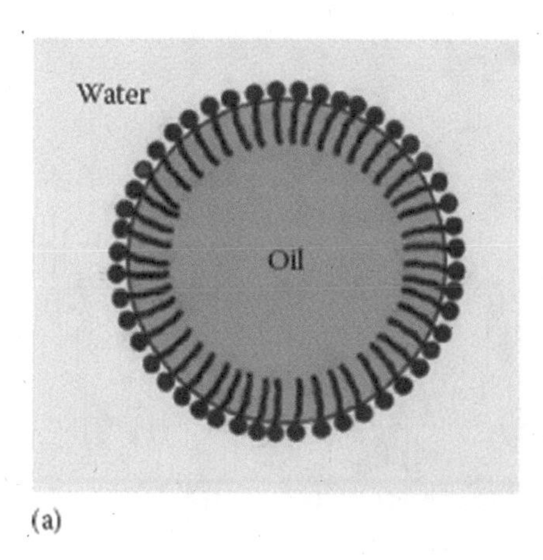

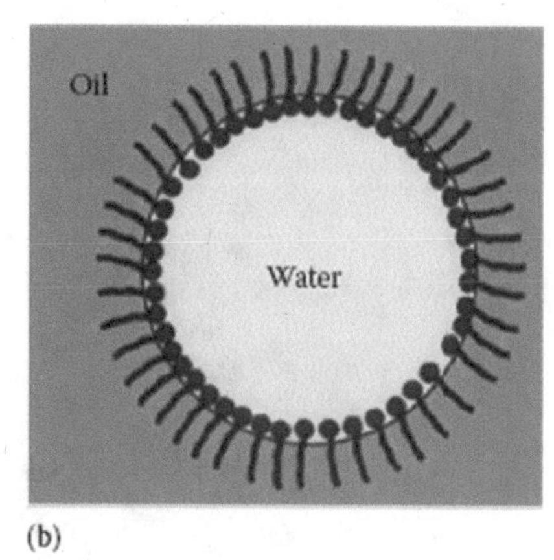

(a) (b)

FIGURE 15.15 Schematic picture of (a) O/W and (b) W/O emulsions stabilized by surfactant molecules.

"dissolved" in the continuous phase, whereas in W/O emulsions, it is in the dispersed phase. The formation of O/W or W/O emulsions depends on the migration of surfactant molecules to the interface, which stabilizes the droplets, and coalescence, leading to the destruction of the droplets. Thus, this is a competing process, and the phase that presents higher coalescence rates will become the continuous phase (Evans and Wennerstrom, 1994). The rates of migration and coalescence depend on the components' chemical structure, leading to specific surfactant conformations when adsorbed on the oil–water interface. One practical parameter is the hydrophilic–lipophilic balance (HLB) of the surfactant, which is a number that relates the number of hydrophobic and hydrophilic groups in a surfactant molecule. In general, surfactants with HLB from 3 to 9 tend to stabilize W/O emulsions, whereas molecules with HLB higher than 9 tend to stabilize O/W emulsions. Therefore, it is possible to prepare emulsions with droplet volume fractions up to 90%, provided that an adequate surfactant or a mixture of surfactants is used. Sometimes, emulsions are also stabilized by adding high–molecular weight components, such as long-chain polymers and proteins, which adsorb on the interface, acting as a surfactant. Small solid particles also tend to adsorb on the oil–water interface and can act as surfactant in an emulsion.

Another type of emulsion that has gained interest in food applications is the so-called *multiple emulsion*, which is basically an emulsion contained in a droplet. For example, a water-in-oil-in-water (W/O/W) emulsion means a multiple emulsion of water droplets inside an oil droplet that is dispersed in a continuous water phase. Recently, potential industrial applications in encapsulating active food components were recognized. One of the main difficulties in applying multiple emulsions is their low stability, which limits the applicability when prolonged stability and release are necessary (Muschiolik, 2007). Figure 15.16 shows a schematic representation of a W/O/W emulsion.

The fact that the dispersed phase can be composed of either oil or water shows that emulsions can be used to encapsulate both lipophilic and hydrophilic bioactives (Flanagan and Singh, 2006). Food systems based on emulsions are recognized to have great potential in delivering functional components such as omega-3, β-carotene, fatty acids, phytosterols, and antioxidants, among others (McClements et al., 2007). The main preparation methods are based on addition and stirring components, mechanical mixing, homogenization, and heating, which make these systems suited for industrial scale-up.

Emulsified systems can be classified according to their thermodynamic stability and their droplet size. Macroemulsions (or simply emulsions) are metastable systems; that is, the system is not in thermodynamic equilibrium, and it will break down into two distinct phases if sufficient time is allowed. However, emulsions that keep their kinetic stability for periods of months or years can be prepared by using appropriate components and amounts (McClements et al., 2007). This is the most common type of emulsion, and it is found in many food systems, such as milk and salad dressing. Macroemulsions are usually polydisperse, with droplet sizes in the range of 1–100 μm. The main destabilization mechanisms in macroemulsions are droplet creaming, flocculation, and coalescence.

Microemulsions, on the other hand, are thermodynamically stable emulsions. Therefore, at a given temperature, pressure, and composition, these systems keep their morphological characteristics and are

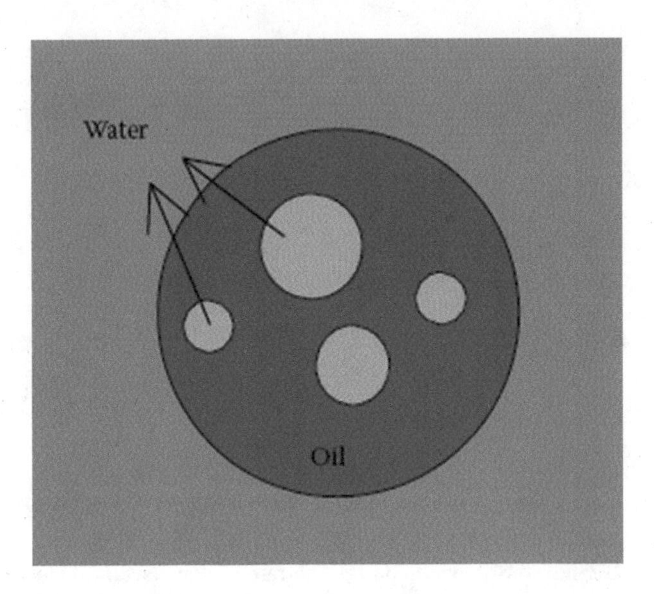

FIGURE 15.16 Schematic representation of a W/O/W emulsion.

not affected by the destabilization mechanisms cited in the previous paragraph. Microemulsions are usually monodisperse, with droplets in a nanoscale range (10–100 nm) (Flanagan and Singh, 2006). Thermodynamic stability is achieved by the proper choice of the components, as well as their proportions, leading to a negative overall free energy of mixing (Evans and Wennerstrom, 1994). Usually, large amounts of surfactants are required, and different surfactants and cosurfactants are generally used. In microemulsions, surfactants are important, because they not only decrease the interfacial tension between the oil and water phases but also affect the energy balance of the system through the formation of self-assembled structures in the continuous phase, such as micelles. Furthermore, the chemical structure of the surfactant affects the spontaneous curvature of the interface, which is an important factor in determining the droplet size as well as the type of emulsion.

Microemulsions exhibit different phase behavior under equilibrium, which is classified using the Winsor classification system. A Winsor I system means that the microemulsion coexists with an oil-rich region. When there is a water-rich region present in the system, the microemulsion is said to be a Winsor II system. In the case where there are both oil-rich and water-rich regions coexisting with the microemulsion, one speaks of a Winsor III system. Finally, a Winsor IV system means that there is no phase coexistence, and only the microemulsified phase is observed. This phase behavior is desired for food delivery systems and, as said earlier, depends on the proper choice of system components as well as temperature conditions (Flanagan and Singh, 2006). Figure 15.17 shows a schematic representation of the different microemulsion regimes.

Depending on its composition, a single-phase microemulsified system can also exhibit different morphologies. The three main structures are O/W, W/O, and bicontinuous. The last is a structure in which both oil and water exist as a continuous phase, but all three structures have a surfactant monolayer in the interface separating both phases. These structures are shown in Figure 15.18. Usually, an O/W microemulsion is formed when oil concentration is low, and a W/O microemulsion is formed when water concentration is low. Bicontinuous systems are formed when the amounts of water and oil are similar (Lawrence and Rees, 2000).

Microemulsions as food systems have great potential, which can be attested by patented products (Bauer et al., 2002; Allgaier et al., 2004; Chanamai, 2007). The incorporation of proteins in microemulsions might also have an impact on food applications in the future (Rohloff et al., 2003). Studies on microemulsions applied to the pharmaceutical field might also be of interest when seeking food applications, since biocompatible components are used in this field. Studies in this area have been summarized in review articles (Lawrence and Rees, 2000; Rane and Anderson, 2008).

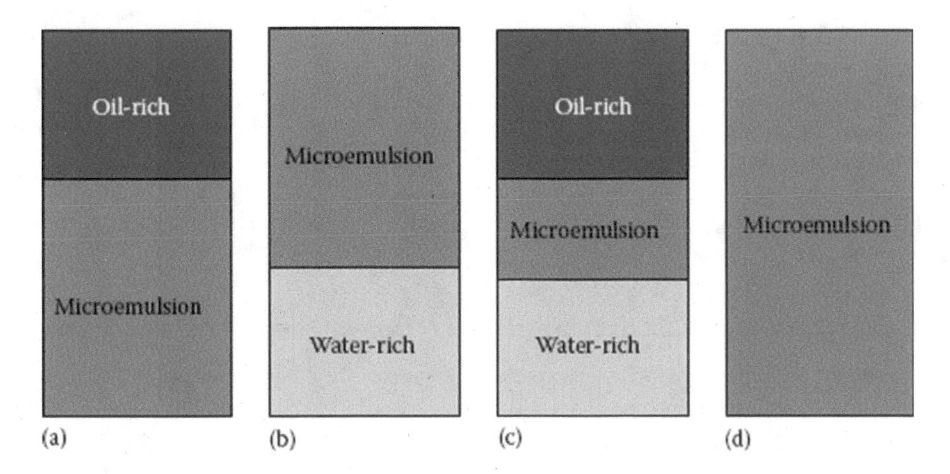

FIGURE 15.17 Schematic representation of the microemulsion regimes: (a) Winsor I, (b) Winsor II, (c) Winsor III, and (d) Winsor IV.

Recently, metastable emulsions with nanosized droplets have started to receive attention due to their technological potential in the pharmaceutical and food industries (Solans et al., 2005), and their fundamental properties have been studied (Mason et al., 2006b). Such systems are called *nanoemulsions* or *mini-emulsions*. Basically, they are emulsions with droplet sizes in the range of 50–200 nm. Nanoemulsions have the same physical appearance as a microemulsion; that is, they have droplets in the nanoscale range and usually exhibit transparency and low viscosity. Although they can lose stability through coalescence and flocculation, the main destabilization mechanism is Ostwald ripening due to the high Laplace pressure of the droplets (Porras et al., 2004).

Nanoscale emulsions have gained technological interest, because the transport efficiency of functional components in emulsion food systems is increased when droplets are in the nano scale (Spernath and Aserin, 2006). In addition, these emulsions are transparent, and they have lower viscosity when compared with conventional emulsions, which makes them suitable for use in beverages, for example. In recent years, considerable research effort has been made to understand the physical properties, preparation, phase behavior, and stability of micro- and nanoemulsions.

Nanoemulsions are metastable structures. This fact confers on nanoemulsions advantages and disadvantages for functional food applications in comparison with microemulsions. The main advantages are that nanoemulsions do not require the use of large amounts of surfactants, and there is a wider range of possible combinations of different components for a given system (Solans et al., 2005). Furthermore, concentrated systems can be prepared, and their rheological properties can be explored for different food applications (Mason et al., 2006b). The main disadvantages are the limited kinetic stability of the system, which has to be monitored to keep the desired properties for a sufficient period of time for a

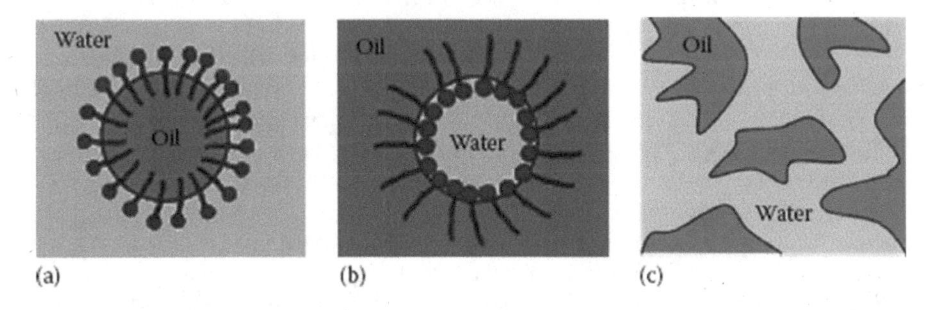

FIGURE 15.18 Schematic representation of the main single-phase microemulsion systems: (a) O/W, (b) W/O, and (c) bicontinuous.

given application. This is an important factor for determining the shelf life of food products based on nanoemulsions.

Although the system is metastable, its creaming stability is highly enhanced due to the small droplet size, which leads to homogeneous systems even for low-viscosity continuous phases. It has been reported that nanoemulsions that are kinetically stable from flocculation and coalescence can be prepared (Solans et al., 2005; Mason et al., 2006a). In fact, the main destabilization mechanism in nanoemulsions is Ostwald ripening due to high Laplace pressures of droplets, which is significantly higher when compared with conventional emulsions.

15.3.4.2 Preparation and Characterization of Microemulsions

Microemulsions are prepared by adding the proper amounts of the components, which form the microemulsion after a given period of time. This is the great advantage of microemulsion preparation when compared with conventional emulsification methods, since the preparation does not require the input of high amounts of energy into the system. However, microemulsion systems can lose their characteristic morphology with variations in temperature and composition. Therefore, the range of the parameters that maintain the microemulsion characteristics has to be determined to define the applicability range of a given system.

Preparation methods for microemulsions consist essentially in adding and mixing the components to the system in different ways and conditions to form a microemulsion. Usually, a single surfactant is not sufficient to decrease the interfacial tension up to the point at which spontaneous emulsification occurs. Thus, one or more cosurfactants are used, which are usually amphiphilic molecules with a different HLB from the main surfactant, or alcohols, but their applicability as food systems can be limited due to toxicity issues. Although the system to be formed is thermodynamically favorable, i.e., the microemulsion is formed spontaneously, it is usually necessary to overcome kinetic energy barriers. The main preparation methods are the low-energy emulsification and the phase inversion temperature (PIT) method. The first method can be achieved by (1) adding water in a mixture of oil and surfactant; (2) adding oil in a mixture of water and surfactant; and (3) mixing all components together at once. It has been reported in the literature that the order of ingredient addition can play a significant role in the formation of the microemulsion (Flanagan and Singh, 2006).

In the PIT method, an initial emulsion, for example, a W/O emulsion, is heated up to a temperature, called the PIT, at which the interfacial tension between the oil and water phases reaches a minimum. At this point, there is an inversion, and an O/W emulsion is formed. The system is then cooled while stirring, and a stable microemulsion is formed. Sometimes, high-pressure homogenization is used to prepare microemulsions, but this method is limited due to the high heat dissipation involved.

The characterization of microemulsions requires the construction of phase diagrams, which can be done by titration methods (Lawrence and Rees, 2000). Ternary or pseudoternary phase diagrams are usually built by varying the amounts of the components and observing the phase behavior of the system to identify the region where a clear and isotropic emulsion is formed, that is, a Winsor IV system. In general, pseudoternary diagrams are found, since microemulsion systems usually contain cosurfactants. Figure 15.19 shows a schematic pseudoternary phase diagram of a microemulsion system, indicating the phase behavior for a given composition. Each axis corresponds to the volume or mass fraction of each component or group of components, which are usually water, oil, and surfactant/cosurfactant.

The microstructure (or nanostructure) characterization of microemulsions can be performed by using dynamic light scattering (DLS) to determine droplet sizes. Scanning electron microscopy (SEM), small-angle x-ray scattering (SAXS), small-angle neutron scattering, and nuclear magnetic resonance (NMR) can be used to determine other structural features, such as the presence of wormlike reverse micelles and other liquid-crystalline phases.

The microstructure and dynamic behavior of microemulsions strongly affect their macroscopic properties. Therefore, it is important to characterize the macroscopic properties of a determined system. In food applications, it is very important to characterize the rheological properties of microemulsions, which can be performed using conventional rheometers. Techniques such as measurements of conductive and dielectric properties can be used to determine the type of microemulsion formed (W/O or

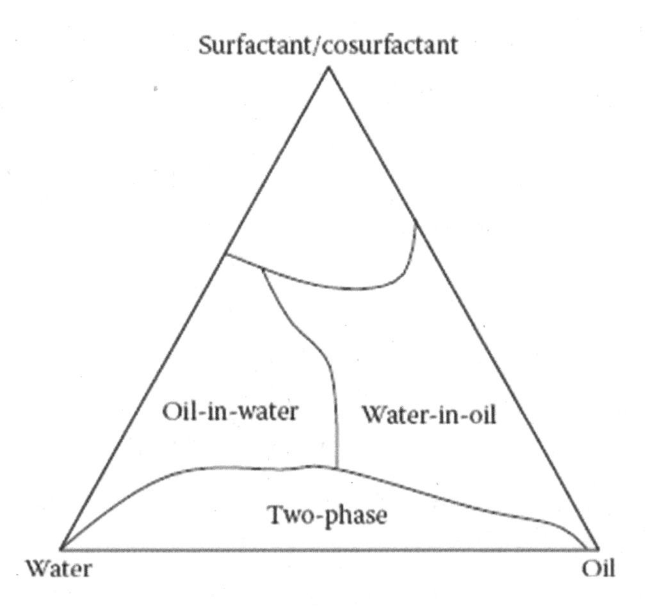

FIGURE 15.19 Schematic pseudoternary diagram showing regions of two-phase and microemulsion regimes.

O/W) and monitor percolation, phase inversion, and other structural and dynamic features. The optical features of microemulsions are important for food applications, since it is necessary for the system to be visually appealing when one is considering a commercial functional food system (Flanagan and Singh, 2006).

15.3.4.3 Preparation and Characterization of Nanoemulsions

Nanoemulsions can be prepared by low-energy methods based on PIT, which are similar to the one described in Section 15.3.4.2, but in this case, the system, when cooled, keeps its morphology in a non-equilibrium state (Förster et al., 1995; Morales et al., 2003; Izquierdo et al., 2004, 2005). Other low-energy methods resulting in nanoemulsions are the phase inversion composition and autoemulsification methods. The former is similar to the PIT method, but the phase inversion occurs by modifying the composition of the system, leading to a kinetically stable nanoemulsion (Forgiarini et al., 2001; Porras et al., 2004). The latter is based on the dilution of an initial stable microemulsion, usually at the bicontinuous phase, resulting in a nanoemulsion (Pons et al., 2003; Wang et al., 2007, 2008).

Preparation using low-energy methods results in nanoemulsions lying in a region close to thermodynamic equilibrium. Therefore, their properties are very similar to a Winsor IV microemulsion system, and a distinction between this type of nanoemulsion and a thermodynamic stable microemulsion has been contested by some research groups (Mason et al., 2006b). However, the stability of nanoemulsions prepared by low-energy methods, showing their non-equilibrium character, has been studied and summarized in review articles, and such emulsions have gained wide acceptance due to their potential applications in food systems and pharmaceuticals (Tadros et al., 2004; Solans et al., 2005; Gutiérrez et al., 2008).

Another method that has gained importance is preparation under high shear. In the past, this method was limited due to the lack of high-pressure homogenizers that were able to generate shear rates high enough to break droplets into nanometer sizes. Therefore, most nanoemulsions prepared using this method were limited to laboratory research using home-built homogenizers. Recently, affordable high-pressure homogenizers with the capacity to generate high shear, such as those based on microchannel flows, have appeared on the market. Consequently, the industrial interest in nanoemulsions is expected to increase. This method is particularly interesting for food applications, since most of the conventional food emulsions are prepared in this way, and high-pressure homogenization operations are suited for scale-up when industrial production rates are desired.

Nanoemulsions prepared by high-shear methods form metastable systems far from equilibrium, and despite the comparable droplet sizes, their characteristics are not similar to those of microemulsions. The main advantage of these nanoemulsions is the fact that kinetically stable systems can be prepared using considerably lower amounts of surfactants while exhibiting physical properties similar to their counterpart microemulsions, such as low viscosity and transparency (Mason et al., 2006b). Therefore, nanoemulsions prepared under high shear can be advantageous if surfactant cost is an important issue when developing food delivery systems. Also, it is possible to develop concentrated systems, which can exhibit unique rheological properties that could be explored in food applications.

The main morphological parameters in nanoemulsions are their average droplet size, droplet size distribution, and droplet volume fraction. Usually, droplet sizes are measured using DLS, and the stability of the system can be monitored by size measurements in a time interval. NMR can be used to measure droplet sizes and droplet interactions in concentrated systems, and it has been used in stability studies of emulsified systems. It can also be used when *in situ* measurements are required, and it is able to characterize concentrated systems, the emulsion type (W/O or O/W), and flow and mixing properties; although this technique has more widespread use in characterizing conventional emulsions, it is feasible for applications in nanoemulsions. Morphological aspects of concentrated nanoemulsions have been studied using small-angle neutron scattering (Mason et al., 2006a). Rheological and optical properties are important when seeking food applications. Therefore, the dependence of viscosity and other viscoelastic properties on droplet sizes and droplet concentrations is important in nanoemulsion characterization.

Reports of nanoemulsions applied as drug delivery systems can be found in the literature (Solans et al., 2005). However, few studies are found concerning applications in food systems, but this number can be expected to grow due to the increase in the availability of homogenization systems of ultrahigh shear based on microchannel flow and other recently developed preparation methods (Kentish et al., 2008; Yuan et al., 2008). Furthermore, fundamental research on nanoemulsion formation through high shear has recently appeared (Meleson et al., 2004). Reviews on the fundamentals and potential applications of nanoemulsions can be found in the literature (Solans et al., 2005; Mason et al., 2006b; Gutiérrez et al., 2008). Recently, a patent was filed claiming the production of a W/O nanoemulsion for food applications (Del Gaudio et al., 2007).

15.4 Applications of Encapsulated Active Ingredients in Foods

Encapsulated ingredients are used in many food applications. They can be incorporated into beverages, dairy products, baked products, and manufactured meats, including infant and other specialized formulations. Examples of food products in which encapsulated ingredients can be incorporated are ultraheat-treated (UHT) milk, cheese, ice cream, margarines, muesli bars, yogurt, infant foods, dietetic food supplements, spreads, health drinks, mayonnaise, baked products, and breakfast cereals.

Encapsulation may be used to deliver traditional active ingredients, such as flavors, vitamins, minerals, sweeteners, and antioxidants, or relatively novel ones, such as probiotic microorganisms. Spray-dried and gel microparticles, liposomes, and emulsified systems, which are under focus in this chapter, have been used for some of these applications. Their functionality as delivery systems is discussed in Sections 15.4.1 through 15.4.5.

15.4.1 Flavors

One of the largest food applications is the encapsulation of flavors. Flavors can be among the most valuable ingredients in any food formula. Even small amounts of some aromatic substances can be expensive, and because they are usually sensitive and volatile, preserving them is often a top concern of food manufacturers. Encapsulating these high-cost materials can result in cost savings, as the loss through storage and processing is limited.

The encapsulation of flavors has been attempted and commercialized using many different methods, such as spray-drying, spray-chilling or spray-cooling, extrusion, coacervation, and molecular inclusion. Among them, spray-drying is the most widely used commercial process in the large-scale production of encapsulated flavors and volatiles. One of the reasons is the large-scale capacity of production in a continuous mode. The microencapsulation of flavors by spray-drying presents the challenge of removing water by evaporation while retaining substances that are much more volatile than water, which is the case for most organic compounds. However, spray-dried microcapsules with high retention of aromatic compounds can be obtained due to the phenomenon known as *selective diffusion* (Thijssen, 1971). This concept is based on the fact that the diffusion of water in concentrated solutions behaves differently from the diffusion of other substances. According to this concept, favorable conditions for obtaining high retention of volatiles can be created in spray-drying due to the rapid decrease of the water concentration at the drying droplet surface in contact with the hot drying air. Once the droplet surfaces have dried sufficiently, selective diffusion comes into effect, because the diffusion coefficients of organic compounds in the surface region become much lower than that of water. When volatile substances are encapsulated, successful microencapsulation relies on achieving high retention of the core material during processing and storage. Many studies have been carried out on the influence of encapsulating material compositions and operating conditions on the encapsulation efficiency and controlled release of encapsulated flavors. The most frequently used carriers include carbohydrates, gums, and food proteins. Each group of materials has certain advantages and disadvantages. For this reason, many coatings are actually composite formulations. Thorough reviews of the technology have been published (Madene et al., 2006); some of them, with special emphasis (Ré, 1998), are dedicated to the encapsulation of flavors by spray-drying (Gharsallaoui et al., 2007).

A subject of increasing interest in this area concerns the development of alternative and inexpensive polymers that may be considered as natural, such as gum arabic (good emulsifying properties), and could encapsulate flavors with good efficiency. For example, sugar beet pectin (Drusch, 2007) has been regarded as an alternative emulsifying encapsulating material for flavors.

Another area of interest to optimize the encapsulation efficiency of food flavors and oils by spray-drying is the submicronization of the oil droplets of the emulsion. It has been well documented that the emulsion droplet size has a pronounced effect on the encapsulation efficiency of different core materials by spray-drying (Jafari et al., 2008). The findings clearly show that reducing the emulsion size can result in encapsulated powders with higher retention of volatiles and lower content of unencapsulated oil at the surface of powder particles. The presence of oil on the surface of the powder particles is the most undesirable property of encapsulated powders, and it has been pointed out as a frequent problem with the quality of spray-dried products. This surface oil not only causes the wettability and dispersibility of the powder to deteriorate, but it is also readily susceptible to oxidation and to the development of rancidity.

Much of the work in this area has been done in emulsions having a droplet size of more than 1 μm, and the application of submicron (nano) emulsions in the encapsulation of oils and flavors is relatively new in the literature. Some work has been carried out to determine the influence of submicron emulsions produced by different emulsification methods on encapsulation efficiency and to investigate the encapsulated powder properties after spray-drying for different emulsion droplet sizes and surfactants. The process has been referred to as *nanoparticle encapsulation*, since a core material in the nanosize range is encapsulated into a matrix of micron-sized powder particles (Jafari et al., 2008). This area of research is developing. Some patents were filed in the past describing microemulsion formulations applied to flavor protection (Chung et al., 1994; Chmiel et al., 1997) and applications in flavored carbonated beverages (Wolf and Havekotte, 1989). However, there is no clear evidence on how submicron emulsions or nanoemulsions can improve the encapsulation efficiency and stability of food flavors and oils into spray-dried powders.

Spray-drying of emulsions has been traditionally used for the encapsulation of flavors; however, novel encapsulation and delivery properties can be achieved by encapsulating flavors into liposomes. Because liposomes have the ability to carry hydrophilic and fat-based flavors, they protect them from degradation during processing and storage and also increase the longevity of the flavor in the system where they are being used. Therefore, their use in the beverage industry has been widespread (Reineccius, 1995).

The bioflavor compounds of blue cheese, obtained from fermentation of *Aspergillus* spp., were encapsulated in soy lecithin liposomes and spray-dried to obtain the powder form by Santana et al. (2005). A sensory evaluation was performed by adding the liposome–bioflavor powder in a base of light cream cheese, which was spread on toast. Flavor intensity, acceptance by consumers, and purchasing intention were tested in the sensory evaluation. The results showed that the encapsulation maintained the characteristic flavor of blue cheese, and the product was classified by the consumers as acceptable. The dried liposome-stabilized flavor was useful to add to foods and to be kept in storage.

15.4.2 Antioxidants

There is a growing demand for delivery of antioxidants through functional foods with the concomitant challenge of protecting their bioactivity during food processing and subsequent passage through the GI tract. Antioxidants such as lycopene and β-carotene can be encapsulated by spray-drying. Blends of sucrose and gelatin have been successfully used in the encapsulation of lycopene (Shu et al., 2006), and blends of maltodextrin and starches can be used to encapsulate β-carotene (Loksuwan, 2007).

The product quality has been analyzed with respect to the retention of the antioxidant activity of the spray-dried powder. Polyphenolic compounds present in several extracts (grape seed, apple polyphenolic extract, or olive leaf) were also encapsulated by spray-drying in protein–lipid emulsions (Kosaraju et al., 2008) and chitosan (Kosaraju et al., 2006).

The common carotenoids in fruits and vegetables (lycopene, lutein, zeaxanthin, astaxanthin, and β-cryptoxanthin) are used as ingredients in foods. They substitute for artificial colorants and are also functional ingredients due to their pro-vitamin A activity, apart from the fact that they act as antioxidants (Fernandez-Garcia et al., 2007). The potential market for water-dispersible carotenoids is broad, including ice creams, soups, desserts, meat products, and animal foods (Delgado-Vargas et al., 2000). Nevertheless, carotenoids lose color under oxidation, as they suffer isomerization easily under heat, acidic pH, or exposure to light. The necessity of protection, as well as the lipophilic or amphiphilic nature of carotenoids, makes them attractive for liposome encapsulation. However, the approach of the studies of carotenoids in liposomes focuses only on their oxidant activity and interactions with the lipid bilayer (Socaciu et al., 2000, 2002; Kostecka-Gugala et al., 2003; Gruszecki and Strzalka, 2005; Jemiola-Rzeminka et al., 2005; McNulty et al., 2007; Sujak et al., 2007). Applications of carotenoids in liposomes are intended to increase their longevity in foods as well as to increase their oral bioavailability due to the presence of lipids as coadjuvants in the formulation (Fernandez-Garcia et al., 2007; Parada and Aguilera, 2007).

Reports of conventional O/W emulsions used to encapsulate lycopene and β-carotene are found in the literature (Ribeiro et al., 2006; Santipanichwong and Suphantharika, 2007). The stability of emulsified lycopene was evaluated after its incorporation in liquid food matrices (Ribeiro et al., 2003). Microemulsions have been applied to increase the efficiency of antioxidants such as ascorbic acid (Moberger et al., 1987). Furthermore, microemulsions have successfully been used for lycopene solubilization (Spernath et al., 2002). Applications of microemulsions to increase the stability of antioxidants have been reported (Moberger et al., 1987; Yi et al., 1991).

15.4.3 Omega-3 Fatty Acids

Omega-3 acids are considered essential to human health but cannot be manufactured by the human body and must therefore be obtained from food. These acids are naturally present in most fishes and certain plant oils such as soybean and canola, which are foods that people rarely consume in large quantities. Moreover, the direct addition of omega-3 fatty acids to many foods is impractical due to some characteristics (fishy flavors, readily oxidized), which together reduce the sensory acceptability of foods containing fatty acids, limit shelf life, and potentially reduce the bioavailability of the acids. Encapsulation responds to the challenges of omega-3 fatty acid delivery and extends the reach of its health benefits.

These acids can be encapsulated by spray-drying. The success of the encapsulation is based on the ability of the spray-dried powder to provide first, good retention of these compounds within the structures and second, good oxidative stability. The goal is to find the appropriate carriers to create good barrier properties as shown in the literature. For example, glucose syrup was used in combination with proteins such as whey protein isolate or soy protein isolate (Rusli et al., 2006) or with sugar beet pectin (Drusch et al., 2007) to encapsulate fish oil, leading to a product that was more stable to oxidation than bulk oils. Other formulation compositions, such as a blend of modified celluloses (methylcellulose and hydroxypropylmethylcellulose) with good emulsifying properties and maltodextrin (Kolanowski et al., 2004), have also been tested to encapsulate fish oil. Another recently developed strategy was based on the production of multilayer membranes of lecithin and chitosan around the oil droplets of the O/W emulsion before spray-drying (Klinkesorn et al., 2006). Despite research being in progress, the influence of various process variables on oil oxidation during the emulsifying and drying stages is still not well known.

The encapsulation of omega-3 fatty acids using O/W emulsions was recently reported in the literature (Lee et al., 2006). The potential for emulsion-based delivery systems of omega-3 molecules in different types of foods such as yogurts, milk, and ice cream has been recognized (McClements et al., 2007). However, the emulsion technology to encapsulate these molecules is difficult, requiring the development of antioxidant technologies to stabilize the system due to the complex oxidation reaction of omega-3 molecules (McClements and Decker, 2000).

Omega-3 fatty acids can also be protected against oxidation when encapsulated within liposomes (Haynes et al., 1991; Wallach and Mathur, 1992).

15.4.4 Vitamins and Minerals

Encapsulating vitamins and minerals offers several benefits. It increases their stability when exposed to air, heat, or moisture. Many of these micronutrients are often destroyed in the baking process. Loss through processing and storage is prevented, resulting in the ability to use less of these products and thus save cost. Finally, many of these micronutrients have undesirable flavors or odors, which can be masked, keeping the micronutrients available to be absorbed in the GI tract.

Fortifying foods with minerals and vitamins is becoming more and more common. Mineral deficiency is one of the most important nutritional problems in the world. The best method to overcome this problem is to make use of an external supply, which may be nutritional or supplementary, such as the fortification of foods with highly bioavailable mineral sources. The major interests of mineral encapsulation are linked to the fact that this technique enables mineral reactions with other ingredients to be reduced when they are added to dry mixes to fortify a variety of foods, and it can also incorporate time-release mechanisms of the minerals into the formulations. For example, iron is the most difficult mineral to add to foods and ensure adequate absorption, and iron bioavailability is severely affected by interactions with food ingredients (e.g., tannins, phytates, and polyphenols). Additionally, iron catalyzes the oxidative degradation of fatty acids and vitamins (Schrooyen et al., 2001).

Liposomes have been used to encapsulate bioactive vitamins and minerals. Milk enriched with ferrous sulfate encapsulated in liposomes enabled an increase in the iron concentration compared with free iron. The encapsulated ferrous sulfate was stable to heat sterilization (100 °C, 30 min) and storage at 4 °C for 1 week.

Furthermore, liposomes provided the same bioavailability as the free sulfate, adding the advantage of being coated with a phospholipid membrane, which kept the iron from contact with the other components of food, thus preventing undesirable interactions (Boccio et al., 1997; Uicich et al., 1999; Lysionek et al., 2000, 2002; Shuqin and Shiying, 2005).

Orange juice, cereals, and even candies are fortified with vitamins and minerals such as vitamin C and calcium. Vitamin C, also known as ascorbic acid or ascorbate, is added extensively to many types of foods for two quite different purposes: as a vitamin supplement, to reinforce dietary intake of vitamin C, and as an antioxidant, to protect the sensory and nutritive quality of the food itself. The encapsulation of vitamin C improves and broadens its applications in the food industry. Spray-dried structures encapsulating vitamin C can be produced by using several carriers as encapsulating materials, including methacrylate copolymers named Eudragit®.* The resulting delivery systems were able to offer controlled release at different pH values due to the characteristics of Eudragit (Esposito et al., 2002). Chitosan, a hydrophilic polysaccharide also used as a dietary food additive, was used to encapsulate vitamin C by spray-drying (Desai and Park, 2005b). Chitosan was cross-linked with a non-toxic cross-linking agent, tripolyphosphate.

Spray-drying was also used to formulate calcium microparticles using cellulose derivatives and polymethacrylic acid as encapsulating materials (Oneda and Ré, 2003) to modify the dissolution rate of

* Registered trademark of Rohm GmbH & Co. KG, Darmstadt, Germany.

calcium from calcium citrate and calcium lactate. Microparticulate systems with an incorporated time-release mechanism were obtained to modify the calcium release from these commercial salts used in fortification of the diet.

Liposomes composed of phosphatidylcholine, cholesterol, and DL-(-tocopherol improved the shelf life of vitamin C from a few days up to 2 months, especially in the presence of common food components that normally speed up decomposition, such as copper ions, ascorbate oxidase, and lysine (Kirby et al., 1991). Calcium lactate was also encapsulated in lecithin liposomes, in this case to prevent undesirable calcium–protein interactions (Champagne and Fustier, 2007). The liposomal calcium levels of fortified soymilk were equivalent to those found in cow's milk. A synergistic effect of coencapsulation of vitamins A and D in liposomes promoted calcium absorption in the GI tract (Champagne and Fustier, 2007).

W/O emulsions based on olive oils were also used to encapsulate vitamin C (Mosca et al., 2008). Solubilization of vitamin E in microemulsions based on polyoxyethylene (POE) surfactants was reported in the past (Chiu and Yang, 1992). Phase behavior studies were conducted on microemulsions based on different oil phases, such as limonene, medium-chain triglycerides (MCT), short-chain alcohols, polyols, and different surfactants (Garti et al., 2001; Papadimitriou et al., 2008; Zhang et al., 2008). In addition, the phase behavior of microemulsions prepared with food-grade components based on lecithin has been investigated (Patel et al., 2006), showing potential for applications in encapsulating vitamins and minerals.

15.4.5 Probiotics

According to the Food and Agriculture Organization (FAO) of the United States and the World Health Organization (WHO), a probiotic is a live microorganism that when administered in adequate amounts, confers a health benefit to the host. The FAO/WHO Expert Consultation lists benefits that had substantial support from peer-reviewed publication of human studies (FAO/WHO, 2001).

Probiotics may consist of a single strain or a mixture of several strains. Most common are lactic acid bacteria from the *Lactobacillus* and *Bifidobacterium* genera. Species of bacteria and yeasts used as probiotics include *Bifidobacterium bifidum*, *B. breve*, *Lactobacillus casei*, *L. acidophilus*, *Saccharomyces boulardii*, and *Bacillus coagulans*, among others (Champagne et al., 2005).

Dietary supplements containing viable probiotic microorganisms (referred to herein as *probiotics*) are increasing in popularity in the marketplace as their health benefits become recognized. Probiotics are sensitive to various environmental conditions such as pH, moisture, temperature, and light. When these conditions are not properly controlled, the product's viability (measured in colony forming units [CFU]) and therefore, its efficacy can be substantially reduced. The viability and stability of probiotics have been both a marketing and a technological challenge for industrial producers.

Losses of microorganisms occur during manufacture and during the product's shelf life. In addition, probiotics with good characteristics for effectiveness against disease and other conditions may not have good survival characteristics during transit through the GI tract. The probiotic cultures encounter gastric juices in the stomach ranging from pH 1.2 (on an empty stomach) through pH 5.0. These cultures stay in the stomach from around 40 min up to 5 h. In the stomach and the small intestine, these probiotics also encounter bile salts and hydrolytic and proteolytic enzymes, which are able to kill them. These cultures are able to grow or survive only when they reach the higher-pH regions of the GI tract. During this transit, probiotics also have to compete with resident bacteria for space and nutrients. In addition, they have to avoid being flushed out of the tract by normal peristaltic action and being killed by antimicrobials produced by other microorganisms. The ability to adhere to surfaces, such as the intestinal mucosal layer and the epithelial cell walls of the gut, is an important characteristic of a probiotic. The term *colonization* is used, meaning that the microorganism has mechanisms that enable it to survive in a region of the GI tract on an ongoing basis.

Intensive research efforts have been focused on protecting the viability of probiotic cultures both during product manufacture and storage and through the gastric transit until the target site is reached. Protection may be achieved in several ways, including encapsulation.

At least five encapsulation methods have been investigated to protect probiotics: spray-coating, spray-drying, extrusion droplets, emulsion, and gel particle technologies, including spray-chilling. Several reviews in the literature are dedicated to probiotics (Mattila-Sandholm et al., 2002; Champagne et al.,

2005) or more specifically, to probiotic encapsulation (Kailasapathy, 2002; Krasaekoopt et al., 2003; Anal and Singh, 2007). In the specific case of fermented dairy products, a thorough discussion of both probiotics and prebiotics is given in Chapter 20.

Spray-drying is rarely considered for cell immobilization because of the high mortality resulting from simultaneous dehydration and thermal inactivation of microorganisms. Despite this limitation, several studies have evaluated spray-drying as a process for encapsulating probiotics. Technical alternatives have been proposed to increase the thermal resistance of the microorganisms during the dehydration process, such as the proper adjustment and control of the inlet and outlet drying temperatures (O'Riordan et al., 2001), the use of complex thermoprotector (prebiotic) carbohydrates as encapsulating materials (Rodriguez-Huezo et al., 2007), or even a previous encapsulation of the microorganisms by another technique, as proposed by Oliveira et al. (2007). These authors encapsulated probiotics (*B. lactis* and *L. acidophilus*) in a casein–pectin complex formed by complex coacervation, and the wet encapsulated microorganisms were dried by spray-drying.

It has been demonstrated that a variety of probiotic cultures can be protected via encapsulation by spray-drying in a variety of carriers, including whey protein (Picot and Lacroix, 2004), a matrix of gelatin, soluble starch, skim milk, or gum arabic (Lian et al., 2003) and cellulose phthalate, which is an enteric release pharmaceutical compound (Favaro-Trindade and Grosso, 2002), and a matrix of protective colloids (whey protein isolate, mesquite gum, and maltodextrin in a 17:17:66 ratio) associated with aguamiel as a prebiotic thermoprotector (Rodriguez-Huezo et al., 2007). In fact, the protective effect exerted by spray-drying encapsulation against stressful conditions (the environment in the food product, during storage, and during the passage through the stomach or intestinal tract) may vary with the carriers or encapsulating materials and the microorganisms, but in all cases, the thermal resistance of strains is a critical parameter that should always be taken into consideration if spray-drying is the intended method for encapsulation (Picot and Lacroix, 2004; Su et al., 2007).

Most of the literature reported on the encapsulation of probiotics has investigated the use of gel particles for improving their viability in food products and the intestinal tract. The bacterial cells are dispersed into the hydrocolloid solution before gelation.

Entrapping probiotic bacteria in gels with ionic cross-linking is typically achieved with polysaccharides (alginate, pectin, and carrageenan). By far, the most commonly used material for this purpose is alginate, and the most commonly reported encapsulation procedure is based on calcium alginate gel formation. The droplets form gel spheres instantaneously (sodium alginate in calcium chloride), entrapping the cells in a three-dimensional lattice of ionically cross-linked alginate. The success of this method is due to the gentle environment it provides for the entrapped material, its cheapness, its simplicity, and its biocompatibility (Anal and Singh, 2007). In the past 10 years, there have been 93 peer-reviewed articles on "probiotics encapsulation" (Web of Science database), 47 of them especially on "encapsulation using alginate."

Various researchers have studied factors affecting the gel particles' characteristics and their influence on the encapsulation of probiotics, such as concentrations of alginate and $CaCl_2$, timing of hardening of the gel particles, and cell concentrations (Chandramouli et al., 2004). Most of them have shown that probiotics can be protected in calcium alginate beads, which is generally demonstrated by an increase in the survival of bacteria under different harsh conditions, compared with free microorganisms. Alginates also demonstrate easy release of the encapsulated bacteria when suspended in an alkaline buffer. However, the degree of protection might depend on the gel particle size, suggesting that these microorganisms should be encapsulated within a specific gel particle size range. For example, very large calcium alginate beads (>1 mm) can negatively affect the textural and sensorial properties of food products in which they are added (Hansen et al., 2002), whereas reduction of the sphere size to less than 100 μm would be advantageous for texture considerations, allowing the direct addition of encapsulated probiotics to a large number of foods. However, it has been demonstrated that particles smaller than 100 μm do not significantly protect the probiotics in simulated gastric fluid compared with free cells (Hansen et al., 2002). One limitation for cell loading in small particles is also the large size of microbial cells, typically 1–4 μm, or particles of freeze-dried culture (more than 100 μm). On the other hand, there is evidence in the literature that calcium alginate gel particles with mean diameters of 450 (Chandramouli et al., 2004) and 640 μm (Shah and Ravula, 2000) could protect probiotics from adverse gastric conditions. In the latter case, the particles, after being freeze-dried, also protected the viability of the microorganisms in fermented frozen dairy desserts. In fact, Chandramouli et al. (2004) found an optimal particle size of 450 μm for the calcium alginate gel particles to protect the cells (*L. acidophilus*) when testing gel particles of different sizes (200, 450, and 1000 μm).

The composition of the alginate also influences bead size (Martinsen et al., 1989). Alginates are heterogeneous groups of polymers with a wide range of functional properties. Alginates with a high content of guluronic acid blocks (G blocks) are preferable for capsule formation because of their higher mechanical stability and better tolerance to salts and chelating agents.

In addition to the reports of benefits of encapsulation in protecting probiotics against the stressful conditions of the GI tract, there is increasing evidence that the procedure is helpful in protecting the probiotic cultures destined to be added to foods. For example, encapsulation technologies have been used satisfactorily to increase the survival of probiotics in high-acid fermented products such as yogurts (Krasaekoopt et al., 2003), including Ca alginate gel particles. Other reported food vehicles for the delivery of encapsulated probiotic bacteria are cheese, ice cream, and mayonnaise (Kailasapathy, 2002).

Despite the suitability of alginate as an entrapment matrix material, this system has some limitations due to its low stability in the presence of chelating agents such as phosphate, lactate, and citrate. The chelating agents share an affinity for calcium and destabilize the gel (Kailasapathy, 2002). Special treatments, such as coating the alginate particles, can be applied to improve the properties of encapsulated gel particles. Coated beads not only prevent cell release but also increase mechanical and chemical stability. It has been reported that cross-linking with cationic polymers, coating with other polymers, mixing with starch, and incorporating additives can improve the stability of beads (Krasaekoopt et al., 2003). For example, alginate can be coated with chitosan, a positively charged polyamine. Chitosan forms a semipermeable membrane around a negatively charged polymer such as alginate. This membrane, like alginate, does not dissolve in the presence of Ca^{2+} chelators or antigelling agents and thus, enhances the stability of the gel and provides a barrier to cell release (Krasaekoopt et al., 2004, 2006; Urbanska et al., 2007).

Various other polymer systems have been used to encapsulate probiotic microorganisms. Kappa-carrageenan (Adhikari et al., 2003), gellan gum, gelatin, starch, and whey proteins (Reid et al., 2007) have also been used as gel encapsulating systems for probiotics. An increasing interest in developing new compositions of gel particles to improve the viability of the probiotic microorganisms to harsh conditions (thermotolerance, acid tolerance, etc.) is marked by the more recent research reported in the literature. Some of these systems include alginate plus starch (Sultana et al., 2000), alginate plus methylcellulose (Kim et al., 2006), alginate plus gellan (Chen et al., 2007), alginate–chitosan–enteric polymers (Liserre, 2007), alginate-coated gelatin (Annan et al., 2008), gellan plus xanthan (McMaster et al., 2005), |-carrageenan with locust bean gum (Muthukumarasamy et al., 2006), and alginate plus pectin plus whey proteins (Guerin et al., 2003). In some cases, systems have been developed not only to provide better probiotic viability but also to deliver a prebiotic synergy (Iyer and Kailasapathy, 2005; Crittenden et al., 2006).

Improving the number of possibilities to encapsulate probiotics is an important tool, because in recent years, the consumer demand for non-dairy-based probiotic products has increased (Prado et al., 2008), and the application of probiotic cultures in non-dairy products represents a great challenge, because they may represent a new, hostile environment for probiotics (heat-processed foods, storage at room temperature, more acid foods such as fruit juices, etc.).

Emulsified systems have also been investigated to protect probiotics. The incorporation of *L. acidophilus* in a W/O/W emulsion was recently reported, and the protective effect of the probiotic in a low-pH environment was evaluated (Shima et al., 2006). Lactic acid bacteria were encapsulated in sesame oil emulsions, and when they were subjected to simulated high gastric or bile salt conditions, a significant increase in survival rate was observed (Hou et al., 2003).

15.5 Encapsulation Challenges

The challenges in developing a commercially viable encapsulated food ingredient depend on selecting appropriate and food-grade (GRAS) encapsulating materials, selecting the most appropriate process to provide the desired size, morphology, stability, and release mechanism, and the economic feasibility of large-scale production, including capital, operating, and other miscellaneous expenses, such as transportation and regulatory costs.

However, the development of any encapsulation technique must not be treated as an isolated operation but as part of an overall process starting with ingredient production and followed by processes including encapsulation, right through to the liberation and use of the ingredient. Furthermore, a selection has to be made between batch, semicontinuous, and continuous encapsulation processes, resulting in a difficult choice for process designers. Cost is often the main barrier to the implementation of encapsulation, and multiple benefits are generally required to justify the cost of encapsulation. Indeed, in the food industry, regulations with respect to ingredients, processing methods, and storage conditions are tight, and the price margin is much lower than in, for example, the pharmaceutical industry.

This procedure is something of an art, as Asajo Kondo asserts in *Microcapsule Processing and Technology* (Kondo, 1979):

> Microencapsulation is like the work of a clothing designer. He selects the pattern, cuts the cloth, and sews the garment in due consideration of the desires and age of his customer, plus the locale and climate where the garment is to be worn. By analogy, in microencapsulation, capsules are designed and prepared to meet all the requirements in due consideration of the properties of the core material, intended use of the product, and the environment of storage.

Encapsulation technology remains something of an art, although firmly grounded in science. Combining the right encapsulating materials with the most efficient production process for any given core material and its intended use requires extensive scientific knowledge of all the materials and processes involved and a good feel for how materials behave under various conditions.

Continuing research is clearly necessary to improve and extend the technology to the encapsulation of a wide variety of beneficial ingredients. Researchers are investigating the next generation of encapsulation technologies, including

- The development of *new, natural food materials and encapsulated products* that can be used by food manufacturers, among them non-proteinaceous materials to eliminate allergens, that protect the encapsulated ingredients while they travel through the body to a targeted site in the GI tract
- The increase in the range of *processing techniques*, with special interest for processes producing in continuous mode with high productivity
- The potential use of *coencapsulation methodologies*, where two or more bioactive ingredients can be combined to have a synergistic effect
- The *targeted delivery* of bioactives to various parts of the GI tract
- The trial of new ways of *incorporating bioactives into foods* with minimal loss of bioactivity and without compromising the quality of the food that is used as a delivery vehicle
- The understanding of the self-assembly and stabilization of *nanoemulsions* during food processing

These developments will give food manufacturers new opportunities to produce a greater variety of innovative functional foods that promote the health and well-being of consumers.

REFERENCES

Adhikari, K., Mustapha, A., and Grun, I.U. 2003. Survival and metabolic activity of microencapsulated *Bifidobacterium longum* in stirred yogurt. *J. Food Sci.* 68:275–280.

Alkhalaf, W., El Soda, M., Gripon, J.-C., and Vassal, L. 1989. Acceleration of cheese ripening with liposomes-entrapped proteinase: Influence of liposomes net charge. *J. Dairy Sci.* 72:2233–2238.

Allgaier, J., Willner, L., Richter, D., Jakobs, B., Sottmann, T., and Strey, R. 2004. Method for increasing the efficiency of surfactants with simultaneous suppression of lamellar mesophases and surfactants with an additive added thereto. U.S. Patent 2004054064-A1, filed Aug. 19, 2003, and issued Mar. 18, 2004.

Alves, G.P. and Santana, M.H.A. 2004. Phospholipid dry powders produced by spray drying processing: Structural, thermodynamic and physical properties. *Powder Technol.* 145:141–150.

Amici, E., Tetradis-Meris, G., Pulido de Torres, C., and Jousse, F. 2008. Alginate gelation in microfluidic channels. *Food Hydrocolloids* 22:97–104.

Anal, A.K. and Singh, H. 2007. Recent advances in microencapsulation of probiotics for industrial applications and targeted delivery. *Trends Food Sci. Technol.* 18:240–251.

Annan, N.T., Borza, A.D., and Hansen, L.T. 2008. Encapsulation in alginate-coated gelatin microspheres improves survival of the probiotic *Bifidobacterium adolescentis* 15703T during exposure to simulated gastro-intestinal conditions. *Food Res. Int.* 41:184–193.

Batzri, S. and Korn, E.D. 1973. Single bilayer liposomes prepared without sonication. *Biochim. Biophys. Acta.* 298:1015–1019.

Bauer, K., Neuber, C., Schmid, A., and Voelker, K.M. 2002. Oil in water microemulsion. U.S. Patent 6426078-B1, filed Feb. 26, 1998, and issued July 30, 2002.

Belyaeva, E., Della Valle, D., Neufeld, R.J., and Poncelet, D. 2004. New approach to the formulation of hydrogel beads by emulsification/thermal gelation using a static mixer. *Chem. Eng. Sci.* 59: 2913–2920.

Benech, R.-O., Kheadr, E.E., Lacroix, C., and Fliss, I. 2002. Antibacterial activities of nisin Z encapsulated in liposomes or produced in situ by mixed culture during cheddar cheese ripening. *Appl. Environ. Microbiol.* 68:5607–5619.

Boccio, J.R., Zubillaga, M.B., Caro, R.A., Gotelli, C.A., and Weill, R. 1997. A new procedure to fortify fluid milk and dairy products with high bioavailable ferrous sulfate. *Nutr. Rev.* 55:240–246.

Brownlie, K. 2007. Marketing perspective of encapsulation technologies in food applications. In *Encapsulation and Controlled Release Technologies in Food Systems*, ed. J.M. Lakkis. Ames, IA: Blackwell Publishing Professional, pp. 213–233.

Burey, P., Bhandari, B.R., Howes, T., and Gidley, M.J. 2008. Hydrocolloid gel particles: Formation, characterization, and application. *Crit. Rev. Food Sci. Nutr.* 48:361–377.

Carneiro, A.L. and Santana, M.H.A. 2004. Production of liposomes in a multitubular system useful for scaling- up of processes. *Prog. Colloid Polym. Sci.* 128:273–277.

Champagne, C.P. and Fustier, P. 2007. Microencapsulation for the improved delivery of bioactive compounds into foods. *Curr. Opin. Biotechnol.* 18:184–190.

Champagne, C.P., Gardner, N.J., and Roy, D. 2005. Challenges in the addition of probiotic cultures to foods. *Crit. Rev. Food Sci. Nutr.* 45:61–84.

Chan, L., Lee, H., and Heng, P. 2002. Production of alginate microspheres by internal gelation using an emulsification method. *Int. J. Pharm.* 241:259–262.

Chan, L.W., Lee, H.Y., and Heng, P.W.S. 2006. Mechanisms of external and internal gelation and their impact on the functions of alginate as a coat and delivery system. *Carbohydr. Polym.* 63:176–187.

Chanamai, R. 2007. Microemulsions for use in food and beverage products. U.S. Patent 087104-A1, filed Oct. 6, 2006, and issued Apr. 19, 2007.

Chandramouli, V., Kailasapathy, K., Peiris, P., and Jones, M. 2004. An improved method of microencapsulation and its evaluation to protect *Lactobacillus* spp. in simulated gastric conditions. *J. Microbiol. Methods* 56:27–35.

Chen, L., Remondetto, G.E., and Subirade, M. 2006. Food protein-based materials as nutraceutical delivery systems. *Trends Food Sci. Technol.* 17:262–283.

Chen, M.J., Chen, K.N., and Kuo, Y.T. 2007. Optimal thermotolerance of *Bifidobacterium bifidum* in gellan-alginate microparticles. *Biotechnol. Bioeng.* 98:411–419.

Chiou, D. and Langrish, T.A.G. 2007. Development and characterisation of novel nutraceuticals with spray drying technology. *J. Food Eng.* 82:84–91.

Chiu, Y.C. and Yang, W.L. 1992. Preparation of vitamin E microemulsion possessing high resistance to oxidation. *Colloids. Surf.* 63:311–322.

Chmiel, O., Traitler, H., and Vopel, K. 1997. Food microemulsion formulations. WO Patent 96/23425, filed Jan. 24, 1996, and issued Aug. 8, 1996.

Choi, C.H., Jung, J.H., Rhee, Y.W., Kim, D.P., Shim, S.E., and Lee, C.S. 2007. Generation of monodisperse alginate microbeads and in situ encapsulation of cell in microfluidic device. *Biomed. Microdevices* 6:855–862.

Chung, S.L., Tan, C.-T., Tuhill, I.M., and Scharpf, L.G. 1994. Transparent oil-in-water microemulsion flavor or fragrance concentrate, process for preparing same, mouthwash or perfume composition containing said transparent microemulsion concentrate, and process for preparing same. U.S. Patent 5283056, filed July 1, 1993, and issued Feb. 1, 1994.

Crittenden, R., Weerakkody, R., Sanguansri, L., and Augustin, M. 2006. Symbiotic microcapsules that enhance microbial viability during nonrefrigerated storage and gastrointestinal transit. *Appl. Environ. Microbiol.* 72:2280–2282.

Del Gaudio, L., Lockhart, T.P., Belloni, A., Bortolo, R., and Tassinari, R. 2007. Process for the preparation of water-in-oil and oil-in-water nanoemulsions. WO Patent 2007/112967-A1, filed Mar. 28, 2007, and issued Oct. 11, 2007.

Delgado-Vargas, F., Jimenez, A.R., and Paredes-Lopez, O. 2000. Natural pigments: Carotenoids, anthocyanins and betalains—Characteristics, biosynthesis, processing and stability. *Crit. Rev. Food Sci. Nutr.* 40:173–189.

Desai, K.G. and Park, H.J. 2005a. Recent developments in microencapsulation of food ingredients. *Drying Technol.* 23:1361–1394.

Desai, K.G. and Park, H.J. 2005b. Encapsulation of vitamin C in tripolyphosphate cross-linked chitosan microspheres by spray drying. *J. Microencapsul.* 22:179–192.

Drusch, S. 2007. Sugar beet pectin: A novel emulsifying wall component for microencapsulation of lipophilic food ingredients by spray-drying. *Food Hydrocolloids* 21:1223–1228.

Drusch, S., Serfert, Y., Scampicchio, M., Schmidt-Hansberg, B., and Schwarz, K. 2007. Impact of physicochemical characteristics on the oxidative stability of fish oil microencapsulated by spray-drying. *J. Agric. Food Chem.* 55:11044–11051.

El Soda, M. 1986. Acceleration of cheese ripening: Recent advances. *J. Food Prot.* 49:395–399.

Eliaz, R.E. and Szoka, F.C. Jr., 2001. Liposome-encapsulated doxorubicin targeted to CD44: A strategy to kill CD44-overexpressing tumor cells. *Cancer. Res.* 61:2592–2601.

Esposito, E., Cervellati, F., Menegatti, E., Nastruzzi, C., and Cortesi, R. 2002. Spray dried Eudragit microparticles as encapsulation devices for vitamin C. *Int. J. Pharm.* 242:329–334.

Evans, D.F. and Wennerstrom, H. 1994. *The Colloidal Domain—Where Physics, Chemistry, Biology and Technology Meet.* New York: Wiley-VCH.

FAO/WHO. 2001. Evaluation of health and nutritional properties of powder milk and live lactic acid bacteria. Food and Agriculture Organization of the United Nations and World Health Organization Expert Consultation Report. Cordoba, Argentina. Available at http://www.who.int/foodsafety/publications/fs_management/en/probiotics.pdf, accessed Feb. 16, 2009.

Favaro-Trindade, C.S. and Grosso, C.R.F. 2002. Microencapsulation of L-acidophilus (La-05) and B-lactis (Bb-12) and evaluation of their survival at the pH values of the stomach and in bile. *J. Microencapsul.* 19:485–494.

Fernandez-Garcia, E., Minguez-Mosquera, M.I., and Perez-Galvez, A. 2007. Changes in composition of the lipid matrix produce a differential incorporation of carotenoids in micelles. Interaction effect of cholesterol and oil. *Innov. Food Sci. Emerg. Technol.* 8:379–384.

Flanagan, J. and Singh, H. 2006. Microemulsions: A potential delivery system for bioactives in food. *Crit. Rev. Food Sci. Nutr.* 46:221–237.

Forgiarini, A., Esquena, J., Gozales, C., and Solans, C. 2001. Formation of nanoemulsions by low-energy emulsification methods at constant temperature. *Langmuir* 17:2076–2083.

Förster, T., Rybinski, W.V., and Wadle, A. 1995. Influence of microemulsion phases on the preparation of fine-disperse emulsions. *Adv. Coll. Int. Sci.* 58:119–149.

Frederiksen, L., Anton, K., van Hoogevest, P., Keller, H.R., and Leuenberger, H. 1997. Preparation of liposomes encapsulating water-soluble compounds using supercritical CO2. *J. Pharm. Sci.* 86:921–928.

Garti, N., Yaghnur, A., Leser, M.E., Clement, V., and Watzke, H.J. 2001. Improved oil solubilization in oil/water food grade microemulsions in the presence of polyols and ethanol. *J. Agric. Food Chem.* 49:2552–2562.

Gharsallaoui, A., Roudaut, G., Chambin, O., Voilley, A., and Saurel, R. 2007. Applications of spray-drying in microencapsulation of food ingredients: An overview. *Food Res. Int.* 40:1107–1121.

Gibaldi, M. 1991. *Biopharmaceutics Clinical Pharmacokinetics*, 4th edn. Philadelphia, PA: Lea & Febiger.

Gomez-Hens, A. and Fernandez-Romero, J.M. 2006. Analytical methods for the control of liposomal delivery systems. *Trends Anal. Chem.* 25:167–177.

Gouin, S. 2004. Microencapsulation: Industrial appraisal of existing technologies and trends. *Trends Food Sci. Technol.* 15:330–347.

Gregoriadis, G. 1987. Encapsulation of enzymes and other agents liposomes. In *Chemical Aspects in Food Enzymes*, ed. A.J. Andrews. London: Royal Society of Chemistry.

Gregoriadis, G., ed. 1988. *Liposomes as Drug Carriers.* Chichester: John Wiley & Sons Ltd.

Gruszecki, W.I. and Strzalka, K. 2005. Carotenoids as modulators of lipid membrane physical properties. *Biochim. Biophys. Acta* 1740:108–115.

Guerin, D., Vuillemard, J.C., and Subirade, M. 2003. Protection of bifidobacteria encapsulated in polysaccharide-protein gel beads against gastric juice and bile. *J. Food Prot.* 66:2076–2084.

Gutiérrez, J.M., Gonzalez, C., Maestro, A., Solè, I., Pey, C.M., and Nolla, J. 2008. Nanoemulsions: New applications and optimization of their preparation. *Curr. Opin. Colloid Interface Sci.* 13:245–251.

Hansen, L.T., Allan-Wojtas, P.M., Jin, Y.L., and Paulson, A.T. 2002. Survival of Ca-alginate microencapsulated *Bifidobacterium* spp. in milk and simulated gastrointestinal conditions. *Food Microbiol.* 19:35–45.

Haynes, L.C., Levine, H., and Finley, J.W. 1991. Liposome composition for stabilization of oxidizable substances. U.S. Patent 5015483, filed Sep. 02, 1989, and issued May 14, 1991.

Health Canada. 1998. Policy Paper—Nutraceuticals/functional foods and health claims on foods. Available at http://www.hc-sc.gc.ca/fn-an/label-etiquet/claims-reclam/nutrafunct_foods-nutra-fonct_aliment-eng.php, accessed Feb. 16, 2009.

Helfrich, W. 1973. Elastic properties of lipid bilayers: Theory and possible experiments. *Z. Naturforsch.* 28C:693–703.

Herrero, E.P., Martin Del Valle, E.M., and Galan, M.A. 2006. Development of a new technology for the production of microcapsules based in atomization processes. *Chem. Eng. J.* 117:137–142.

Hong, J.S., Shin, S.J., Lee, S.H., Wong, E., and Cooper-White, J. 2007. Spherical and cylindrical microencapsulation of living cells using microfluidic devices. *Korea-Aust. Rheol. J.* 19:157–164.

Hou, R.C.W., Lin, M.Y., Wang, M.M.C., and Tzen, J.T.C. 2003. Increase of viability of entrapped cells of *Lactobacillus delbrueckii* ssp bulgaricus in artificial sesame oil emulsions. *J. Dairy Sci.* 86:424–428.

Huang, K.S., Lai, T.H., and Lin, Y.C. 2007. Using a microfluidic chip and internal gelation reaction for monodisperse calcium alginate microparticles generation. *Front. Biosci.* 12:3061–3067.

International Life Sciences Institute. 1999. Safety assessment and potential health benefits of food components based on selected scientific criteria. ILSI North America Technical Committee on Food Components for Health Promotion. *Crit. Rev. Food Sci. Nutr.* 39:203–316.

Israelachvili, J.N. 1991. *Intramolecular and Surface Forces.* New York: Academic Press.

Iyer, C. and Kailasapathy, K. 2005. Effect of co-encapsulation of probiotics with prebiotics on increasing the viability of encapsulated bacteria under in vitro acidic and bile salt conditions and in yogurt. *J. Food Sci.* 70:18–23.

Izquierdo, P., Esquena, J., Tadros, T.F. et al. 2004. Phase behavior and nano-emulsion formation by the phase inversion temperature method. *Langmuir* 20:6594–6598.

Izquierdo, P., Feng, J., Esquena, J. et al. 2005. The influence of surfactant mixing ratio on nano emulsion formation by the pit method. *J. Colloid Interface Sci.* 285:388–394.

Jafari, S.M., Assadpoor, E., Bhandari, B., and He, Y. 2008. Nano-particle encapsulation of fish oil by spray drying. *Food Res. Int.* 41:172–183.

Jemiola-Rzeminka, M., Pasenkiewicz-Gierula, M., and Strzalka, K. 2005. The behaviour of ®-carotene in the phosphatidylcholine bilayer as revealed by a molecular simulation study. *Chem. Phys. Lipids* 135:27–37.

Joscelyne, S.M. and Trägardh, G. 2000. Membrane emulsification—A literature review. *J. Membr. Sci.* 169:107–117.

Kailasapathy, K. 2002. Microencapsulation of probiotic bacteria: Technology and potential applications. *Curr. Issues Intest. Microbiol.* 3:39–48.

Keller, B.C. 2001. Liposomes in nutrition. *Trends Food Sci. Technol.* 12:25–31.

Kentish, S., Wooster, T.J., Ashokkumar, M., Balachandran, S., Mawson, R., and Simons, L. 2008. The use of ultrasonics for nanoemulsion preparation*Innovat. Food Sci. Emerg. Technol.* 9:170–175.

Kidd, P.M. 2002. Phospholipids: Versatile Nutraceuticals for Functional Foods. In *Functional Foods and Nutraceuticals.* January:1–11.

Kim, C.J., Jun, S.A., and Lee, N.K. 2006. Encapsulation of *Bacillus polyfermenticus* SCD with alginatemethylcellulose and evaluation of survival in artificial conditions of large intestine. *J. Microbiol. Biotechnol.* 16:443–449.

Kim, H.H.Y. and Baianu, I.C. 1991. Novel liposome microencapsulation techniques for food applications. *Trends Food Sci. Technol.* 2:55–61.

Kirby, C.J. and Gregoriadis, G. 1984. Dehydration-rehydration vesicles: A simple method for high yield drug entrapment in liposomes. *Biotechnology* 2:979–984.

Kirby, C.J. and Law, B. 1987. Development in the microencapsulation of enzymes in food technology. In *Chemical Aspect of Food Enzymes*, ed. A.T. Andrews. London: Royal Society of Chemistry.

Kirby, C.J., Whittle, C.J., Rigby, N., Coxon, D.T., and Law, B.A. 1991. Stabilization of ascorbic acid by microencapsulation in liposomes. *Int. J. Food Sci. Technol.* 26:437–449.

Klinkesorn, U., Sophanodora, P., Chinachoti, P., Decker, E.A., and McClements, J. 2006. Characterization of spray-dried tuna oil emulsified in two-layered interfacial membranes prepared using electrostatic layer-by-layer deposition. *Food Res. Int.* 39:449–457.

Kolanowski, W., Laufenberg, G., and Kunz, B. 2004. Fish oil stabilisation by microencapsulation with modified cellulose. *Int. J. Food Sci. Nutr.* 55:333–343.

Kondo, A. 1979. *Microcapsule Processing and Technology*. New York: Marcel Dekker, Inc.

Kosaraju, S.L., D'ath, L., and Lawrence, A. 2006. Preparation and characterisation of chitosan microspheres for antioxidant delivery. *Carbohydr. Polym.* 64:163–167.

Kosaraju, S.L., Labbett, D., Emin, M., Konczak, L., and Lundin, L. 2008. Delivering polyphenols for healthy ageing. *Nutr. Diet.* 65:48–52.

Kostecka-Gugala, A., Latowski, D., and Strzalka, K. 2003. Thermotropic phase behaviour of alpha-dipalmitoylphosphatidylcholine multibilayers is influenced to various extents by carotenoids containing different structural features—Evidence from differential scanning calorimetry. *Biochim. Biophys. Acta* 1609:193–202.

Krasaekoopt, W., Bhandari, B., and Deeth, H. 2003. Evaluation of encapsulation techniques of probiotics for yoghurt. *Int. Dairy J.* 13:3–13.

Krasaekoopt, W., Bhandari, B., and Deeth, H. 2004. The influence of coating materials on some properties of alginate beads and survivability of microencapsulated probiotic bacteria. *Int. Dairy J.* 14:737–743.

Krasaekoopt, W., Bhandari, B., and Deeth, H.C. 2006. Survival of probiotics encapsulated in chitosan-coated alginate beads in yoghurt from UHT and conventionally treated milk during storage. *LWT—Food Sci. Technol.* 39:177–183.

Lasic, D.D. 1993. *Liposomes: From Physics to Applications*. New York: Elsevier.

Lasic, D.D. 1995. Pharmacokinetics and antitumor activity of anthracyclines precipitated in sterically (stealth) liposomes. In *Stealth Liposomes*, eds. D.D. Lasic and F.J. Martin. Boca Raton, FL: CRC Press.

Lasic, D.D. and Papahadjopoulos, D., eds. 1998. *Medical Applications of Liposomes*. New York: Elsevier.

Latorre, L.G., Carneiro, A.L., Rosada, R.S., Silva, C.L., and Santana, M.H.A. 2007. A mathematical model describing the kinetic of cationic liposome production from dried lipid films adsorbed in a multitubular system. *Braz. J. Chem. Eng.* 24:1–10.

Lawrence, M.J. and Rees, G. 2000. Microemulsion-based media as novel drug delivery systems. *Adv. Drug Deliv. Rev.* 45:89–121.

Lee, S., Hernandez, P., Djordjevic, D. et al. 2006. Effect of antioxidants and cooking on stability of n-3 fatty acids in fortified meat products. *J. Food Sci.* 71:C233–C238.

Lian, W.C., Hsiao, H.C., and Chou, C.C. 2003. Viability of microencapsulated bifidobacteria in simulated gastric juice and bile solution. *Int. J. Food Microbiol.* 86:293–301.

Liserre, A.M., Ré, M.I., and Franco, B.D.G.M. 2007. Microencapsulation of *Bifidobacterium animalis* subsp. lactis in modified alginate-chitosan beads and evaluation of survival in simulated gastrointestinal conditions. *Food Biotechnol.* 21:1–16.

Liu, K., Ding, H.J., Chen, Y., and Zhao, X.Z. 2006. Shape-controlled production of biodegradable calcium alginate gel microparticles using a novel microfluidic device. *Langmuir* 22:9453–9457.

Liu, X.D., Bao, D.C., Xue, W. et al. 2003. Preparation of uniform calcium alginate gel beads by membrane emulsification coupled with internal gelation. *J. Appl. Polym. Sci.* 87:848–852.

Liu, X.D., Yu, X.W., Zhang, Y., Xue, W.M. et al. 2002. Characterization of structure and diffusion behaviour of Ca-alginate beads prepared with external or internal calcium sources. *J. Microencapsul.* 10:775–782.

Loksuwan, J. 2007. Characteristics of microencapsulated beta-carotene formed by spray drying with modified tapioca starch, native tapioca starch and maltodextrin. *Food Hydrocolloids* 21:928–935.

Lysionek, A.E., Zubillaga, M.B., Sarabia, M.I. et al. 2000. Study of industrial microencapsulated ferrous sulfate by means of the prophylactic-preventive method to determine its bioavailability. *J. Nutr. Sci. Vitaminol.* 6:125–129.

Lysionek, A.E., Zubillaga, M.B., Salgueiro, M.J., Pineiro, A., Caro, R.A., Weill, R., and Boccio, J.R. 2002. Bioavailability of microencapsulated ferrous sulfate in powdered milk produced from fortified fluid milk: A prophylactic study in rats. *Nutrition* 18:279–281.

Maa, Y.F. and Hsu, C. 1999. Performance of sonication and microfluidization for liquid-liquid emulsification. *Pharm. Dev. Technol.* 4:233–240.

Madene, A., Jacquot, M., Scher, J., and Desobry, S. 2006. Flavour encapsulation and controlled release—A review. *Int. J. Food Sci. Technol.* 41:1–21.

Market Research. 2004. Global market review of functional foods—Forecasts to 2010. Available at http://www.marketresearch.com, accessed July 20, 2008.

Martinsen, A., Skjak-Braek, C., and Smidsrod, O. 1989. Alginate as immobilization material. I. Correlation between chemical and physical properties of alginate gel beads. *Biotechnol. Bioeng.* 33:79–89.

Mason, T.G., Graves, S.M., Wilking, J.N., and Lin, M.Y. 2006a. Extreme emulsification: Formation and structure of nanoemulsions. *Condens. Matter Phys.* 9:193–199.

Mason, T.G., Wilking, J.N., Meleson, K., Chang, C.B., and Graves, S.M. 2006b. Nanoemulsions: Formation, structure, and physical properties. *J. Phys. Condens. Matter* 18:635–666.

Masters, K. 1985. *Spray Drying Handbook.* New York: Halsted Press.

Matsuno, R. and Adachi, S. 1993. Lipid encapsulation technology—Techniques and applications to food. *Trends Food Sci. Technol.* 4:256–261.

Mattila-Sandholm, T., Myllarinen, P., Crittenden, R., Mogensen, G., Fondén, R., and Saarela, M. 2002. Technological challenges for future probiotic foods. *Int. Dairy J.* 12:173–182.

McClements, D.J. and Decker, E.A. 2000. Lipid oxidation in oil-in-water emulsions: Impact of molecular environment on chemical reactions in heterogeneous food systems. *J. Food Sci.* 65:1270–1282.

McClements, D.J., Decker, E.A., and Weiss, J. 2007. Emulsion-based delivery systems for lipophylic bioactive components. *J. Food Sci.* 72:109–124.

McMaster, L.D., Kokott, S.A., Reid, S.J., and Abratt, V. 2005. Use of traditional African fermented beverages as delivery vehicles for *Bifidobacterium lactis* DSM 10140. *Int. J. Food Microbiol.* 102:231–237.

McNulty, H.P., Byun, J., Lockwood, S.F., Jacob, R.F., and Mason, R.P. 2007. Differential effects of carotenoids on lipid peroxidation due to membrane interactions: X-ray diffraction analysis. *Biochim. Biophys. Acta* 1768:167–174.

Meleson, K., Graves, S., and Mason, T.G. 2004. Formation of concentrated nanoemulsions by extreme shear. *Soft Matter* 2:109–123.

Moberger, L., Larsson, K., Buchheim, W., and Timmen, H. 1987. A study on fat oxidation in a microemulsion system. *J. Disper. Sci. Technol.* 8:207–215.

Morales, D., Gutierrez, J.M., Garcia-Celma, M.J., and Solans, Y.C. 2003. A study of the relation between bicontinuous micro emulsions and oil/water nano-emulsion formation. *Langmuir* 19:7196–7200.

Mosca, M., Ceglie, A., and Ambrosone, L. 2008. Antioxidant dispersions in emulsified olive oils. *Food Res. Int.* 41:201–207.

Muschiolik, G. 2007. Multiple emulsions for food use. *Curr. Opin. Colloid Interface Sci.* 12:213–220.

Muthukumarasamy, P., Allan-Wojtas, P., and Holley, R.A. 2006. Stability of *Lactobacillus reuteri* in different types of microcapsules. *J. Food Sci.* 71:20–24.

Nguyen, N.T. and Zhigang, W. 2005Micromixers—A review. *J. Micromech. Microeng.* 15:R1–R6.

Oh, J.K., Drumright, R., Siegwart, D.J., and Matyjaszewski, K. 2008. The development of microgels/nanogels for drug delivery applications. *Prog. Polym. Sci.* 33:448–477.

Oliveira, A.C., Moretti, T.S., Boschini, C., Baliero, J.C.C., Freitas, O., and Favaro-Trindade, C.S. 2007. Stability of microencapsulated B lactis (Bl 01) and L acidophilus (LAC 4) by complex coacervation followed by spray drying. *J. Microencapsul.* 24:685–693.

Oneda, F. and Ré, M.I. 2003. The effect of formulation variables on the dissolution and physical properties of spray-dried microspheres containing organic salts. *Powder Technol.* 130:377–384.

O'Riordan, K., Andrews, D., Buckle, K., and Conway, P. 2001. Evaluation of microencapsulation of a *Bifidobacterium* strain with starch as an approach to prolonging viability during storage. *J. Appl. Microbiol.* 91:1059–1066.

Papadimitriou, V., Pispas, V., Syriou, S. et al. 2008. Biocompatible microemulsions based on limonene: Formulation, structure, and applications. *Langmuir* 24:3380–3386.

Parada, J. and Aguilera, J.M. 2007Food microstructure affects the bioavailability of several nutrients. *J. Food Sci R Conc. Rev. Hypoth. Food Sci.* 72:21–32.

Patel, N., Schmid, U., and Lawrence, M.J. 2006. Phospholipid-based microemulsions suitable for use in foods. *J. Agric. Food Chem.* 54:7817–7824.

Pérez-Alonso, C., Báez-González, J.G., Beristain, C.I., Vernon-Carter, E.J., and Vizcarra-Mendonza, M.G. 2003. Estimation of the activation energy of carbohydrate polymers blends as selection criteria for their use as wall material for spray-dried microcapsules. *Carbohydr. Polym.* 53:197–203.

Picon, A., Serrano, C., Gaya, P., Medina, M., and Nunhez, M. 1996. The effect of liposome-encapsulated cyprosins on manchego cheese ripening. *J. Dairy Sci.* 79:1694–1705.

Picot, A. and Lacroix, C. 2004. Encapsulation of bifidobacteria in whey protein-based microcapsules and survival in simulated gastrointestinal conditions and in yoghurt. *Int. Dairy J.* 14:505–515.

Poncelet, D. and Dreffier, C. 2007. Les methods de microencapsulation de A à Z (ou presque). In *Microencapsulation: Des sciences aux technologies*, eds. T. Vandamme, D. Poncelet, and P. Subra-Paternault. Paris: Tec&Doc (Editions), pp. 23–33.

Poncelet, D., Lencki, R., Beaulieu, C., Halle, J.P., Neufeld, R.J., and Fournier, A. 1992. Production of alginate beads by emulsification internal gelation. I. Methodology. *Appl. Microbiol. Biotechnol.* 38:39–45.

Pons, R., Carrera, I., Caelles, J., Rouch, J., and Panizza, P. 2003. Formation and properties of miniemulsions formed by microemulsions dilution. *Adv. Colloid Interface Sci.* 106:129–146.

Porras, M., Solans, C., Gonzalez, C., Martinez, A., Guinart, A., and Gutierrez, J.M. 2004. Studies of formation of W/O nano-emulsions. *Colloids Surf. A* 249:115–118.

Prado, F.C., Parada, J.L., Pandey, A., and Soccol, C.R. 2008. Trends in non-dairy probiotic beverages. *Food Res. Int.* 41:111–123.

Rane, S.S. and Anderson, B.D. 2008. What determines drug solubility in lipid vehicles: Is it predictable? *Adv. Drug Deliv. Rev.* 60:638–656.

Rattanachaikunsopon, P. and Phumkhachorn P. 2018. Functional Food: What Are They? and Why Are They So Popular? *Acta Scientific Nutritional Health.* 2:26–27.

Ré, M.I. 1998. Microencapsulation by spray drying. *Drying Technol.* 16:1195–1236.

Ré, M.I. 2006. Formulating drug delivery systems by spray drying. *Drying Technol.* 24:433–446.

Ré, M.I., Messias, L.S., and Schettini, H. 2004. The influence of the liquid properties and the atomizing conditions on the physical characteristics of the spray-dried ferrous sulfate microparticles. Paper presented at the Annual International Drying Symposium, Campinas, Brazil, Aug. 2004, 1174–1181.

Reid, A.A., Champagne, C.P., Gardner, N., Fustier, P., and Vuillemard, J.C. 2007. Survival in food systems of *Lactobacillus rhamnosus* R011 microentrapped in whey protein gel particles. *J. Food Sci.* 72:31–37.

Reineccius, G.A. 1995. Liposomes for controlled release in the food industry. In *Encapsulation and Controlled Release of Food Ingredients* (American Chemical Society Symposium Series 590), eds. S.J.Risch and G.A.Reineccius. Washington, DC: American Chemical Society, pp. 113–131.

Ribeiro, H.S., Ax, K., and Schubert, H. 2003. Stability of lycopene emulsions in food systems. *J. Food Sci.* 68:2730–2734.

Ribeiro, H.S., Guerrero, J.M.M., Briviba, K., Rechkemmer, G., Schuchmann, H.P., and Schubert, H. 2006. Cellular uptake of carotenoid-loaded oil-in-water emulsion in colon carcinoma cell in vitro. *J. Agric. Food Chem.* 54:9366–9369.

Rodriguez-Huezo, M.E., Duran-Lugo, R., Prado-Barragan, L.A. et al. 2007. Pre-selection of protective colloids for enhanced viability of *Bifidobacterium bifidum* following spray-drying and storage, and evaluation of aguamiel as thermoprotective prebiotic. *Food Res. Int.* 40:1299–1306.

Roger, J.A. and Anderson, K.E. 1998. The potential of liposomes in oral drug delivery. *Crit. Rev. Ther Drug. Carrier Syst.* 15:421–481.

Rohloff, C.M., Shimek, J.W., and Dungan, S.R. 2003. Effect of added alpha-lactalbumin protein on the phase behavior of AOT-brine-isooctane systems. *J. Colloid Interface Sci.* 261:514–523.

Rowland, M. and Towzer, T.N. 1995. *Clinical Pharmacokinetics: Concepts and Applications*, 3rd edn. Baltimore, MD: Williams & Williams.

Rusli, J.K., Sanguansri, L., and Augustin, M.A. 2006. Stabilization of oils by microencapsulation with heated protein-glucose syrup mixtures. *J. Am. Oil Chem. Soc.* 83:965–997.

Santana, M.H.A., Martins, F., and Pastore, G.M. 2005. Processes for stabilization of bioflavors through encapsulation in cyclodextrins and liposomes. Brazilian Patent Application, PI 0403279–9 A, filed Dec. 2005 (in Portuguese).

Santipanichwong, R. and Suphantharika, M. 2007. Carotenoids as colorants in reduced-fat mayonnaise containing spent brewer's yeast beta-glucan as fat replacer. *Food Hydrocolloids* 21:565–574.
Schrooyen, P.M.M., van der Meer, R., and De Krif, C.G. 2001. Microencapsulation: Its application in nutrition. *Proc. Nutr. Soc.* 60:475–479.
Shah, N.P. and Ravula, R. 2000. Microencapsulation of probiotic bacteria and their survival in frozen fermented dairy desserts. *Aust. J. Dairy Technol.* 55:139–144.
Shahidi, F. and Han, X.Q. 1993. Encapsulation of food ingredients. *Crit. Rev. Food Sci. Nutr.* 33:501–547.
Shima, M., Morita, Y., Yamashita, M., and Adachi, S. 2006. Protection of *Lactobacillus acidophilus* from the low pH of a model gastric juice by incorporation in a W/O/W emulsion. *Food Hydrocolloids* 20:1164–1169.
Shu, B., Yu, W., Zhao, Y., and Liu, X. 2006. Study on microencapsulation of lycopene by spray-drying. *J. Food Eng.* 76:664–669.
Shuqin, X. and Shiying, X. 2005. Ferrous sulfate liposomes: Preparation, stability and application in fluid milk. *Food Res. Int.* 38:289–296.
Siró, I., Kálpona, E., Kálpona, B., and Lugasi, A. 2008. Functional food. Product development, marketing and consumer acceptance—A Review. *Appetite* 51:456–467.
Socaciu, C., Bojarski, P., Aberle, L., and Diehl, H.A. 2002. Different ways to insert carotenoids into liposomes affect structure and dynamics of bilayer differently. *Biophys. Chem.* 99:1–15.
Socaciu, C., Jessel, R., and Diehl, H.A. 2000. Competitive carotenoid and cholesterol incorporation into liposomes: Effects on membrane phase transition, fluidity, polarity and anisotropy. *Chem. Phys. Lipids* 106:79–88.
Solans, C., Izquierdo, P., Nolla, J., Azemar, N., and Garcia-Celma, M.J. 2005. Nano-emulsions. *Curr. Opin. Colloid Interface Sci.* 10:102–110.
Spernath, A. and Aserin, A. 2006. Microemulsions as carriers for drugs and nutraceuticals. *Adv. Colloid Interface Sci.* 128–130:47–64.
Spernath, A., Yaghmur, A., Aserin, A., Hoffman, R.E., and Garti, N. 2002. Food-grade microemulsions based on nonionic emulsifiers: Media to enhance lycopene solubilization. *J. Agric. Food Chem.* 50:6917–6922.
Su, L.C., Lin, C.W., and Chen, M.J. 2007. Development of an oriental-style dairy product coagulated by microcapsules containing probiotics and filtrates from fermented rice. *Int. J. Dairy Technol.* 60:49–54.
Sujak, A., Strzalka, K., and Gruszecki, W.I. 2007. Thermotropic phase behaviour of lipid bilayers containing carotenoid pigment canthaxanthin: A differential scanning calorimetry study. *Chem. Phys. Lipids* 145:1–12.
Sultana, K., Godward, G., Reynolds, N., Arumugaswamy, R., Peiris, P., and Kailasapathy, K. 2000. Encapsulation of probiotic bacteria with alginate-starch and evaluation of survival in simulated gastrointestinal conditions and in yoghurt. *Int. J. Food Microbiol.* 62:47–55.
Tadros, T., Izquierdo, R., Esquena, J., and Solans, C. 2004. Formation and stability of nanoemulsions. *Adv. Colloid Interface Sci.* 108:303–318.
Tanford, C. 1980. *The Hydrophobic Effect: Formation of Micelles and Biological Membranes.* New York: Wiley-Interscience.
Thijssen, H.A.C. 1971. Flavour retention in drying preconcentrated food liquids, *J. Appl. Chem. Biotechnol.* 21:372–376.
Tournier, H., Schneider, M., and Guillot, C. 1999. Liposomes with enhanced entrapment capacity and their use in imaging. U.S. Patent 5,980,937, filed Aug. 12, 1997 and issued Nov. 9, 1999.
Uicich, R., Pizarro, F., Almeida, C. et al. 1999. Bioavailability of microencapsulated ferrous sulfate in fluid cow milk: Studies in human beings. *Nutr. Rev.* 19:893–897.
Urbanska, A.M., Bhathena, J., and Prakash, S. 2007. Live encapsulated *Lactobacillus acidophilus* cells in yogurt for therapeutic oral delivery: Preparation and in vitro analysis of alginate-chitosan microcapsules. *Can. J. Physiol. Pharmacol.* 85:884–893.
Vanderberg, G.W. and De La Noüe, J. 2001Evaluation of protein release from chitosan-alginate microcapsules produced using external or internal gelation. *J. Microencapsul.* 18:433–441.
Vladisavljevic, G.T. and Williams, R.A. 2005. Recent developments in manufacturing emulsions and particulate products using membranes. *Adv. Colloid Interface Sci.* 113:1–20.
Vuillemard, J.-C. 1991. Recent advances in the large-scale production of lipid vesicles for use in food products: Microfluidization. *J. Microencapsul.* 8:547–562.

Wallach, D.F.H. and Mathur, R. 1992. Method of making oil filled paucilamellar lipid vesicles. U.S. Patent 5160669, filed Oct. 16, 1990, and issued Nov. 03, 1992.
Walther, B., Cramer, C., Tiemeyer, A. et al. 2005. Drop deformation dynamics and gel kinetics in a co-flowing water-in-oil system. *J. Colloid Interface Sci.* 286:378–386.
Wang, L., Li, X., Zhang, G., Dong, J., and Eastoe, J. 2007. Oil-in-water nanoemulsions for pesticide formulations. *J. Colloid Interface Sci.* 314:230–235.
Wang, L., Mutch, K.J., Eastoe, J., Heenan, R.K., and Dong, J. 2008. Nanoemulsions prepared by a two-step low-energy process. *Langmuir* 24:6092–6099.
Wang, L.Y., Ma, G.H., and Su, Z.G. 2005. Preparation of uniform sized chitosan microspheres by membrane emulsification technique and application as a carrier of protein drug. *J. Control. Release* 106:62–75.
Weiss, J., Kobow, K., and Muschiolik, G. 2004. Preparation of microgel particles using membrane emulsification. Abstracts of Food Colloids, Harrogate, U.K., B–34.
Weiss, J., Takhistov, P., and McClements, D.J. 2006. Functional materials in food nanotechnology. *J. Food Sci.* 71:107–116.
Were, L.M., Bruce, B.D., Davidson, P.M., and Weiss, J. 2003. Size, stability, and entrapment efficiency of phospholipid nanocapsules containing polypeptide antimicrobials. *J. Agric. Food Chem.* 51:8073–8079.
Wolf, P.A. and Havekotte, M.J. 1989. Microemulsions of oil in water and alcohol. U.S. Patent 4,835,002, filed July 10, 1987, and issued May 30, 1989.
Xianghua, Y. and Zirong, X. 2006. The use of immunoliposome for nutrient target regulation (a review). *Crit. Rev. Food Sci. Nutr.* 46:629–638.
Xu, S., Nie, Z., Seo, M. et al. 2005. Generation of monodisperse particles by using microfluidics: Control over size, shape, and composition. *Angew. Chem. Int. Ed.* 44:724–728.
Yi, O.S., Han, D., and Shin, H.K. 1991. Synergistic antioxidative effects of tocopherol and ascorbic-acid in fish oil lecithin water-system. *J. Am. Oil Chem. Soc.* 68:881–883.
Yuan, Y., Gao, Y., Zhao, J., and Mao, L. 2008. Characterization and stability evaluation of beta-carotene nanoemulsions prepared by high pressure homogenization under various emulsifying conditions. *Food. Res. Int.* 41:61–68.
Zhang, H., Feng, F., Li, J. et al. 2008. Formulation of food-grade microemulsions with glycerol monolaurate: Effects of short-chain alcohols, polyols, salts and nonionic surfactants. *Eur. Food Res. Technol.* 226:613–619.
Zhang, L., Liu, J., Lu, Z., and Hu, J. 1997. Procedure for preparation of vesicles with no leakage from water-in-oil emulsion. *Chem. Lett.* 8:691–692.
Zheng, S., Alkan-Onyuksel, H., Beissinger, R.L., and Wasan, D.T. 1999. Liposome microencapsulations without using any organic solvent. *J. Disper. Sci. Technol.* 20:1189–1203.
Zhou, Q.Z., Wang, L.Y., Ma, G.H., and Su, Z.G. 2007. Preparation of uniform-sized agarose beads by microporous membrane emulsification technique. *J. Colloid Interface Sci.* 311:118–127.

16

Encapsulation of Aroma

Christelle Turchiuli and Elisabeth Dumoulin

CONTENTS

16.1 Introduction

Aromas are used in the food, pharmaceutical, chemical, and cosmetic industries to create or to modify the sensory perception (taste, odor) of food products, objects, or atmosphere, to finally reach human beings or animals. They are used with different concentrations and quantities, either for direct consumption (a dish, a drink, or a medicine) or as a component at an industrial level.

Aroma compositions (called *aroma*) are complex mixtures of different organic molecules of different size, molecular weight (MW), structure, and polarity (i.e., hydrophilic and hydrophobic groups) with various physicochemical properties. Depending on these properties, they are more or less sensitive to heat, light, and oxygen, and more or less soluble in solvents (water, alcohol), and are characterized by their diffusivity in various media, relative volatility, and polarity. They give a global complex characteristic, the *aroma profile*, which may be modified because of volatility (balance between volatile and non-volatile components), oxidation (off-flavors), and diffusivity of some of the molecules (Richard and Multon, 1992).

Very often, aroma molecules are hydrophobic oils. Their characteristic taste or smell may be pleasant, strong (if pure) or not, and may need to be modified, masked, and diluted. The aroma will be used in association with other substances, in an environment which may be a liquid, a paste, a solid, or a gas, and subjected to different use conditions (temperature, pH) according to the final objectives.

Some examples of products containing aroma are given to show the variety of environmental conditions to deliver its characteristics:

- In food products: liquid yogurts (a_w, pH), beverages (a_w, pasteurization), aromatized tea (hot water), spiced pizzas (cooking), drinks in vending machines (dosage, rapid dissolution), chewing gum with menthol (flavor and gum), coated products (meat/fish, frying), enriched bars, animal feed, etc.
- In chemicals, cosmetics, medicine, agriculture: washing products (solid, liquid), coated seeds, fertilizers (protection against insects), controlled release medicines to mask a bad taste (saliva), perfumes (liquid, solid), perfumed objects (sponges, scarves), scratch and sniff, aerosols (liquid, powder, gas), antimicrobial agents (essential oils), cleaning agents (i.e., dissolution of oil in limonene), etc.

Due to the aroma composition and use in various media, the stability, the intensity, the risk of losses, and the conservation of the profile will represent important parameters during preparation (dosage, formulation, pH) and processing operations such as heating, cooking, freezing, scratching, dissolution, etc., including storage and delivery conditions. This means that efficient protection is necessary to maintain the aroma's properties until use: to prevent or reduce the action of atmosphere, oxygen, water, or heat; interactions with other components; modifications of structure; and dosage. Then, the products with protected aroma will have to deliver the right balance of flavor/smell (no off-flavor) under specific stress conditions of temperature, humidity, and mechanical stress, often with the desired odor on a long-term basis.

There is not one unique problem and solution, but many, all specific.

In nature, we may observe many products (flowers, fruits, and vegetables) existing with "protection" (Richard and Multon, 1992):

- Vegetable cells are good encapsulation tools to release active molecules at a given time.
- The membranes, skin, and shell of fruits, nuts, and flowers control the transfer of water and gas, give protection against oxygen/air and heat, and offer a physically resistant structure.

The microcapsules of aroma will imitate nature with a film, a skin, to be a *barrier* for the active aroma composition (in the core). The aroma core material will be finally dispersed homogeneously (multicore) in a liquid or solid support material (~20–30% of total weight): if soluble in water, as a solution; if not soluble in water, as an emulsion or in a solid/paste/powder. In some cases, a high core loading is obtained (90%) with specific techniques (coacervation, coating, and coextrusion). Or, multicore capsules are prepared initially by emulsion and then transformed into powder by spray-drying, extrusion, etc.

The aroma support material (or wall) chosen as the barrier (skin) will often be a film-forming and surface-active polymer with several roles (in liquids or solids); the main benefits for the aroma are protection, dosage, stability, and possible controlled release in different media and end use conditions.

The different steps to consider must be well defined: the preparation of microcapsules (composition, process), their storage, and their final use.

The total encapsulation process will be chosen and achieved with the close collaboration of several partners:

- The aroma specialist for knowledge on aroma properties and behavior
- The process engineer to propose different processes according to the final use and shape and aroma behavior, with the necessary technical help of support providers
- The final user, industry, and/or consumer to define their needs in terms of use, concentration, medium, etc.

In this chapter, different aspects of aroma encapsulation will be considered: the choice of support material depending on the searched properties of microcapsules and the production of microcapsules, and examples of aroma encapsulation to produce complex structured products.

16.2 Properties of Microcapsules and Choice of Support Material

The general definitions of encapsulation (micro- or nanoencapsulation) may be applied to any active component (AC) to be encapsulated: it is placed in a medium where its mobility and reactivity are reduced and controlled. AC is protected inside the microcapsule with a minimal fraction of "unprotected AC" on the surface.

16.2.1 Properties of Microcapsules

The core, or the AC, is usually dispersed/placed in the encapsulating agent, called a support, wall, carrier, shell, matrix, skin, or film. In the case of aroma, the protection is applied to a mixture of organic components, some of them possibly volatile at normal pressure and temperature, with chemical functions more or less sensitive to oxidation and hydrolysis; often concentrated; oil or water soluble. The final capsules with different sizes will exist in products as

- A liquid, as an emulsion or a gel (nanometer or micrometer size)
- A solid, as powder particles (micrometer or millimeter size) in the shape of spheres, sticks, or crystals
- A gas, as particles to reach the lungs

Microcapsules must allow, according to Shahidi and Han (1993),

1. Reducing and controlling the reactivity of the aroma with its outside environment: light, oxygen, water, and other components
2. Decreasing the evaporation or transfer rate of the aroma components to the environment (modifying composition, profile, and intensity) during all processing operations and storage, depending on temperature/heat, water content (a_w, atmosphere), and duration
3. Promoting defined handling of the aroma:
    a. Dilution in a support (if necessary)
    b. Achievement of uniform dispersion in the host/support material at the right concentration

c. Change into a shape, with size and outside surface, allowing mixing (in liquid, paste, or solid) with other ingredients; sometimes incompatible if no encapsulation
d. Conversion of the liquid aroma into a solid form easy to associate with other constituents (i.e., in a support)

4. Controlling the final release of the aroma: under specific operating conditions, with uniform distribution, masking other tastes and smells, with a lasting effect (proper delay)

The desired final release of aroma from the microcapsule structure may be rapid or progressive (in water, solvent, saliva, washing, or atmosphere), by mechanical breaking (scratching), physical or chemical action, diffusion, dissolution, or melting, according to the final use (Figure 16.1).

The parameters of release will be related to

- The structure and support composition (including coating) and properties (solubility, permeability, hygroscopicity [Tg])
- The water activity ($a_{w)}$, water content, relative humidity (RH) of atmosphere/medium in contact
- The temperature, pH, agitation (or not)
- The time in relation to reaction kinetics
- The preserved storage conditions (time (t), Temperature (R), RH, O_2, etc.) between the preparation of the encapsulated aroma and its final use

For example, during the storage of aroma powders, the unwanted release of encapsulated molecules (especially components with different volatilities) will depend on atmosphere, RH, and the composition of supports/emulsifiers: this may be related to water adsorption by wall materials, the possible modification of structure (collapse, glass transition), and the appearance (or not) of cracks.

16.2.2 Choice of Support Material

16.2.2.1 Needed Properties for Supports

The choice of the aroma-encapsulating supports (matrix) is the basis for the success of encapsulation, throughout the different steps: from the formulation through the encapsulation process, followed by storage and finally, use/release. For example, a good support for the encapsulation process may show bad properties during storage!

The properties of supports to consider for the composition depend on the final use:

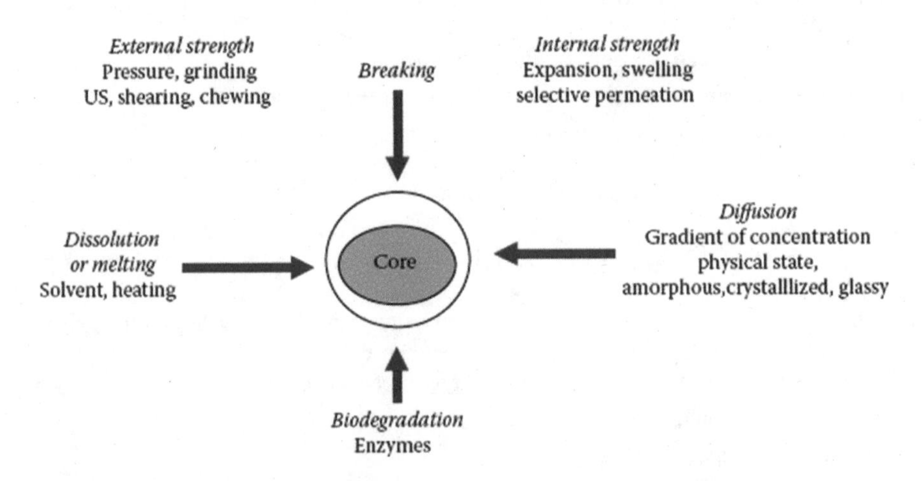

FIGURE 16.1 Modes of core release from capsule (including time, pH, and temperature).

- No reaction with aroma during process and storage; natural components (increasing demand), food/pharmaceutical grade
- Capacity to retain AC during process and storage to limit diffusion and evaporation of aroma throughout the needed shelf life
- Good protection against the environment (heat, light, humidity, and oxygen) to avoid physical changes (collapse, crystallization, reactions, etc.)
- Capacity to release AC in the end (operating conditions)
- Compatibility for integration into the final formula

Then, depending on the *processes* of encapsulation and on the desired final physical structure, other properties to consider for supports are

- *Solubility* in solvents used in the industry concerned: food, pharmacy, cosmetics, chemistry
- Providing easy *dispersion* or emulsion of aroma with the desired concentration
- Capacity to *eliminate solvent* (or other constituents) used in the process (i.e., water, alcohol) if necessary
- *Rheological properties* in concentrated solutions for easy use (i.e., pumping)

And finally, but importantly from an industrial point of view, the final decision will take care of costs (i.e., raw materials, process, and storage conditions), *security*, and *legislation*.

The main idea is to build a protective network for the aroma, made of long chains, with possible polar, hydrophobic, and hydrophilic properties, with a defined physical state (solid, liquid, crystallized, or amorphous) in the working conditions (i.e., temperature, a_w). The necessary good knowledge of the aroma's constituent properties provided by the aroma suppliers will complete the data given by the support suppliers.

16.2.2.2 Examples of Supports

The following examples are commonly applied, especially for food products (Risch and Reineccius, 1988, 1995; Karbowiak et al., 2007; Augustin and Hemar, 2009; De Vos et al., 2010).

16.2.2.2.1 Polysaccharides

Maltodextrins (equivalent dextrose [DE] <20) are cornstarch derivates with high solubility in water. They are neutral (white and tasteless), they provide a reducing environment, and they have poor emulsifying properties. If the DE increases, the viscosity decreases, and the hygroscopicity increases. They are used as support for the retention of volatile substances, mainly in spray-drying and extrusion.

Modified starch is a good emulsifier because of the presence of both hydrophilic and hydrophobic groups and is able to retain oil and hydrophilic components. It is used in spray-drying for the retention of volatile substances such as aroma.

Cyclodextrins α, β, and γ have a molecular structure with a hydrophobic cavity. Their solubility in water is low. They lead to a very stable inclusion complex (in process, storage, and use).

Dextrose is used as a texture and storage agent, with high water solubility (sweet taste), good heat stability, and low hygroscopicity. It is used in extrusion and cocrystallization processes.

Lactose is commonly used in pharmaceutical components. In powders, its ability to crystallize in high RH must be considered for storage.

Celluloses from plants (hydroxymethyl, methyl, hydroxypropyl, ethyl, etc.) may be added to fatty acids.

Chitosan (from shellfish and mushrooms) and inulin (oligofructose, chicory, agave) are natural supports soluble in water.

Soybean soluble polysaccharide (SSPS) is used as an emulsifier in spray-drying.

16.2.2.2.2 Gums

These are polymers with long chains, able to form gels and films and to control crystallization. They comprise algae extracts (alginate, agar, and carrageenan) and plant exudates (acacia and mesquite gums).

Acacia gum (polysaccharide, proteins [5%], and fibers) brings emulsifying, film-forming, and antioxidant properties, with low viscosity in water solution.

Their respective viscosity in water (1%) is: gum guar 3.5, locust bean 3, tragacanth 0.7, carrageenan 0.5, and acacia 5×10^{-3} Pas.

They are used in spray-drying and are associated, for example, with maltodextrins.

16.2.2.2.3 Lipids

These are used as a barrier to water transfer (in a film or coating).

Wax is used for the coating of components soluble in water.

Acetoglycerides have a water permeability that is a function of the degree of acetylation.

Lecithin (lipid mix) is a good emulsifier, providing a low encapsulation temperature.

Liposomes (in soya, egg, and milk fat) are also used in cosmetics. They are single- or multilayered vesicles with an aqueous phase enclosed within a phospholipid-based membrane.

16.2.2.2.4 Proteins

Milk proteins, caseinates, egg albumin, gluten, zein, and soy are examples.

Proteins have the ability to assemble at interfaces. Whey proteins serve as emulsifying and film-forming agents against oxygen and aroma diffusion. Sodium caseinate is reported to be superior to whey proteins with regard to emulsifying properties and resistance to heat-induced denaturation. It is associated with lactose in spray-drying (Vega and Roos, 2006).

Gelatin is soluble in water and gives thermally reversible gels that are not easy to dry.

They are used in spray-drying, in coacervation (with gum), and for coating (with sugar).

16.2.3 Summary

The encapsulation support will be chosen according to the final use conditions (temperature, pH, delayed release, controlled water content, etc.), the mode of preparation, and the solubility properties (in water and solvent). It may need to be natural, edible (i.e., food or medicine), neutral or with color, taste, and odor; with hydrophilic properties (–OH groups) and/or hydrophobic, more or less polar (Figure 16.2).

The active aroma may be entrapped in

- An amorphous matrix (glassy state) (fast cooling or drying), the glass being less permeable to organic components and gases (i.e., O_2) for $T < Tg$
- In a medium with fat or crystals (sugar)
- Into linked polymers (coacervation) or into liposomes

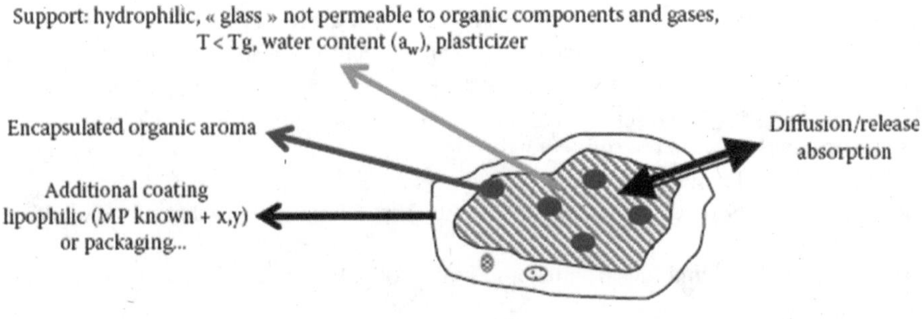

FIGURE 16.2 Schematic aroma encapsulation in powder particle.

Usually, several supports will be mixed in specific proportions to combine their complementary properties (and cost!) with the help—if necessary—of other necessary agents for emulsification, reticulation, and antioxidant action. For example, acacia gum can be associated with a less expensive carbohydrate provided an oil/gum ratio of 1 is maintained to keep good encapsulation efficiency (McNamee et al., 2001). And it also has to be taken into consideration that the whole composition (supports + aroma) will be integrated as a part of the final product composition, including regulatory limits.

16.3 Some Encapsulation Processes

16.3.1 Introductory Remarks

For aroma, the decision to invest in a delivery technology is driven by the balance between the cost and the added value as compared with a standard liquid flavor.

Several works may be consulted according to the objectives: Dziezak (1988); Marion and Audrin (1988); Risch and Reineccius (1988); Richard and Multon (1992); Risch and Reineccius (1995); De Roos (2003); Gouin (2004); Onwulata (2005); Madene et al. (2006); Lakkis (2007); Bouquerand et al. (2012).

Research into the best aroma encapsulation process very often needs studies at different scales, from laboratory via pilot scale to industry scale, with the usual but not easy engineering problem of scaling up. The choice of the process depends on the needed aroma protection and the possible supports for the desired final use and release.

Again, the choice and the study will be done in close cooperation between the aroma and support suppliers, the process engineer, and the final producer/user. Very often, preliminary trials will be done using aroma model molecules (i.e., one or two, or oils) at a pilot scale with different supports to study. Then, the proper scale and conditions will be studied and chosen.

The choice will also depend on the desired production scale: either limited production in batches for a high-value product (reproducibility and traceability) or a continuous controlled process with high capacity. That means also finding equipment and conditions adapted to the different scales, and often, the need for specific lines (cleaning and contamination).

The process may lead to a liquid (emulsion, suspension, or slurry) or a solid (paste or powder) or a gas, by which the aroma will be protected until release in the desired conditions. Solid forms represent 95% of delivery systems, and of these, 78% are obtained by spray-drying (Bouquerand et al., 2012).

The first encapsulation process to examine will be the liquid emulsion, used either as it is or as a first step toward other encapsulation forms. Then, several processes use an initial formulated liquid to prepare, in the end, a dry product such as a powder. Starting with a liquid state provides a good initial homogeneous aroma dispersion, which will be more or less maintained in the powdered form.

Why produce aroma powders?

- To facilitate long-term storage at ambient temperature by decreasing a_w
- To reduce weight and volume so as to facilitate worldwide trade and handling
- To give a physical structure (i.e., porosity and surface properties) to facilitate agglomeration, granulation, and mixing with other powders and to improve flowability and rehydration

The final product quality must satisfy the demands of the initial objectives. This means that through all the different operations/processes, the transformations will be controlled: not only to define, measure, and optimize the operating conditions but also to be able to analyze the constituents; to describe the history of evolution of their physical and chemical structures. As usual, the analyses adopted are a very important complementary part of processing.

16.3.2 Emulsification

Emulsions represent a *liquid encapsulation form*, a macroscopic dispersion of two non-miscible liquids. For aroma encapsulation, the *continuous phase* is usually the solution of supports, and the *dispersed phase* is the core aroma material (or aroma solubilized/dispersed in an oil base). They are often *oil-in-water* emulsions:

→polar hydrophilic liquid phase + lipophilic nonpolar liquid phase←

Aroma emulsions have specific characteristics and needs:

- Size and size distribution (narrow) of dispersed oil drops (microcapsules/nanocapsules) (100–0.1 μm)
- Stability/protection of microcapsules during emulsion preparation and storage

Microcapsules containing fish oil with proteins as well were prepared using oil-in-water-in oil (O/W/O) double emulsification followed by gelation method (heat or enzyme) (Cho et al., 2003). The oil phase may be the base to include oil-soluble ingredients.

When the emulsion is the first step before further processing (i.e., spray-drying for powders with pumping, pulverization, and drying), stability properties of the emulsion will have to be adapted.

16.3.2.1 Preparation/Composition

Several components are associated with obtaining the dispersion and stability of small regular oil drops in a continuous phase by acting more or less rapidly and permanently at the interface:

- Emulsifiers: to facilitate the formation/dispersion of oil drops (glycerides, proteins, lecithin, etc.). They are adsorbed on the periphery of oil drops (the oil/aqueous phase interface) to decrease the surface tension of the drops and to form a barrier to prevent their coalescence. They are amphiphilic molecules including both hydrophilic and lipophilic groups. They may have a role in protection against oxidation.
- Stabilizers: to maintain the dispersion by different phenomena such as steric repulsion, electrostatic interactions, viscosity, and gelling effects (polysaccharides, pectins, modified starches, gums, etc.).

The rate v of separation (rising or falling) of drops (mean radius r) is predicted by the Stokes law:

$$v = 2r^2 g(d_d - d_c)/9\mu$$

where

g is the acceleration due to gravity

d_d, d_c are the densities of the two phases, dispersed and continuous

μ is the viscosity of the continuous phase

Decreasing the drop radius represents one way to stabilize an emulsion.

- Thickeners: weighing agents to improve the stability by decreasing the difference in density between the dispersed and continuous phases. They are used to increase the density of oil in which they are soluble (usually $d_d < d_c$).

These different components may sometimes interact together in a *positive or negative* way on the emulsion size because of their structure or concentration.

16.3.2.2 Equipment

Equipment must bring mechanical energy (shear forces) to break the oil aroma phase into small regular drops (initial coarse emulsion) and then, decrease more or less the dispersed drop size (fine emulsion) to improve the stability of the emulsion, which is directly linked to the diameter of the dispersed drops.

Different techniques such as ultrasound treatment, mixers (agitator, Ultra Turrax), homogenizers (with pressure), and membranes (Microfluidizer®) are used in relation to the desired final emulsion size, the composition of the emulsion, and the volume to produce (100 ml or 10 l), and with energy consumption linked to the energy density concept (Schubert et al., 2009).

During the emulsion preparation, some air may be incorporated, which is able to participate in aroma oxidation over time (or air may be included in a spray-dried powder). So, some sort of deaeration (i.e., under vacuum) and preservatives may be necessary.

Also, part of the energy used to disrupt oil into small drops is transformed into heat, which is transferred to the emulsion, increasing the temperature, which must be controlled (i.e., refrigeration) to prevent losses of volatile molecules.

The preparation of high volumes of emulsions for industrial capacity production (different from laboratory scale) needs special study and adapted equipment (Turchiuli et al., 2013b).

16.3.2.3 Conclusion

Emulsions represent a liquid aroma encapsulation form that is inexpensive and suitable for a stable concentrated liquid with low volume (drinks, sauces, or cosmetics) (Figure 16.3). But it may be also the preliminary step before another transformation to give a dried emulsion.

Nanoemulsions (<100 nm) are used for "clear, transparent" drinks/liquids; however, there is a possibility of coalescence and loss of clarity with time.

16.3.3 Spray-Drying

Spray-drying is a convective drying technique used to transform a feed in liquid or slurry form into a dry, free-flowing powder. The aroma encapsulated and dispersed in a liquid form will be transformed into an aroma encapsulated in solid powder particles with minimal modification of the aroma properties and dosage. A dry powder well dosed with aroma (~20–25% total solid [TS]) will be easy to handle, to incorporate into a dry system, or to dissolve again.

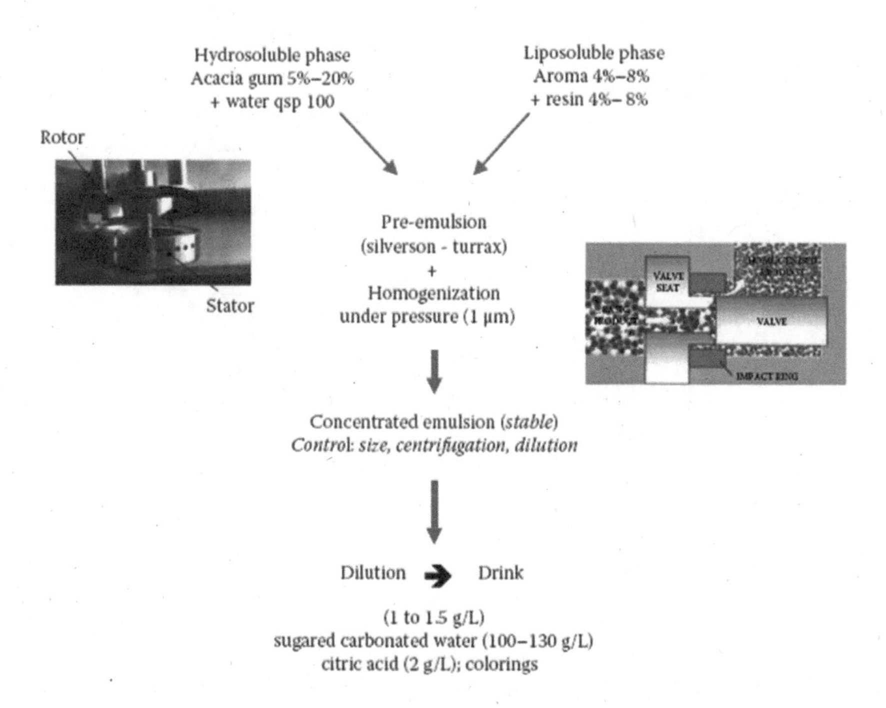

FIGURE 16.3 Preparation of concentrated emulsion for drinks.

Safety considerations will appear with flavor formulations containing high levels of low-boiling components with explosion risks.

Many research works have been published concerning aroma encapsulation by spray-drying. Some of the works worthy of mention are the following: Kerkhof and Thijssen (1974); King et al. (1984); Masters (1985); Risch and Reineccius (1988); Rosenberg et al. (1990); King (1995); Risch and Reineccius (1995); Dumoulin and Bimbenet (1998); Re (1998); Gibbs et al. (1999); Finney et al. (2002); Gouin (2004); Reineccius (2004); Vega and Roos (2006); Vandamme et al. (2007); Jafari et al. (2008); Jafari (2009); Drusch et al. (2012).

16.3.3.1 Principle of Spray-Drying

The formulated liquid feed containing the aroma (solution, emulsion, or slurry) is dispersed with an atomizer (a bifluid or high-pressure nozzle, or a centrifugal wheel) into spherical drops into a hot air flow (or inert nitrogen if the absence of O_2 is required for oxidation/safety reasons). The contact configuration between the hot air and the sprayed drops is usually cocurrent. The hot air brings the necessary energy to evaporate water (solvent) from liquid drops through their surface. The mass transfer of heat and water between the drops and the air is favored by the large contact surface, and the air flow is used to transport the resulting dry drops/particles of aroma powder along trajectories with time–temperature histories. Spray-drying is a *continuous process* with possible air treatment (dehydration, fines) and partial outlet air recycling.

When aqueous solution/emulsion drops are formed with atomizers followed by solvent (water) evaporation in air, the risk of modification of the structure and composition of the solution/emulsion in the drop is limited for different reasons.

Spray-drying permits the removal of water by evaporation through the drop surface while retaining the volatile aroma (low concentration). This is based on the relative diffusivity of water and aroma in liquid droplets, which varies with the dry matter content (which changes during drying) and temperature (Rulkens and Thijssen, 1972; Coumans et al., 1994). When the water concentration in drops decreases (on drying), the diffusion coefficients of both water and aroma decrease. From a certain TS concentration, , however, the aroma diffusivity decreases faster than the water diffusivity (Figure 16.4). So, by fast drying, the drop surface layer becomes selective, which is favorable to volatile aroma retention.

During drying, the temperature of the drops remains below the wet bulb temperature of the drying gas. However, even if the drying time is short (some seconds) and the drop temperature low because of a high

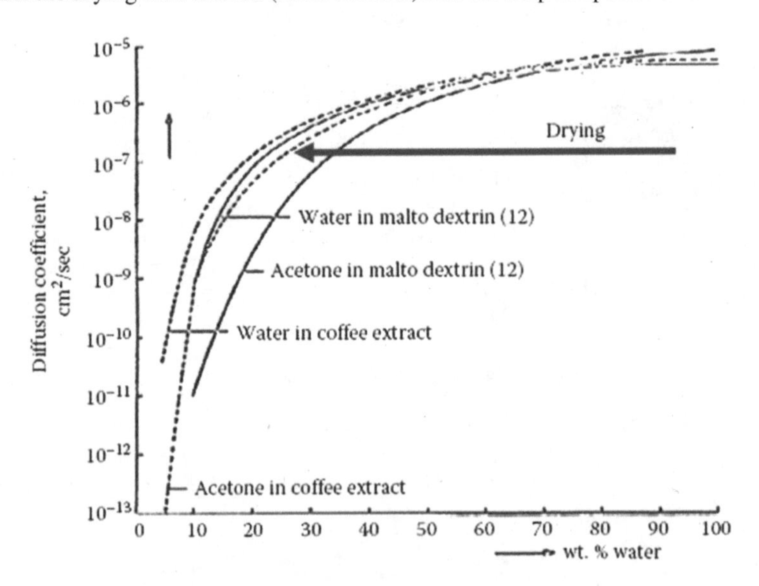

FIGURE 16.4 Thijssen theory of selective diffusion. Diffusion coefficients of water and acetone in coffee extracts and maltodextrin solutions (0.1% w/w acetone, 25°C) (acetone: volatile molecule).

water evaporation rate, some molecules of volatile aroma may migrate together with water molecules to the drop surface, where they may either disappear and/or stay unprotected at the surface.

The fast drying of the drop surface layer depends on good conditions for air/drop mixing, which are closely connected with the viscosity, surface tension, and the size of liquid drops, and optimized operating conditions without stopping further fast drying. Also, it is necessary to have a narrow size distribution of drops (spraying system/feed properties) so as to have homogeneous drop drying behavior. The formation of spherical drops with the desired size and narrow size distribution will depend on the physical properties of the sprayed liquid (viscosity and surface tension) associated with the chosen spraying system (sheets or ligaments broken into drops), and control of liquid movement/mixing inside the drops before solid particle formation.

16.3.4 Influence of Operating Conditions

1. *A fine emulsion* (1 μm) (aroma in an aqueous solution of support) leads to better aroma retention and a lower fraction of aroma on the surface of powder particles compared with coarser emulsions.
2. The rapid formation of *a dry layer at the drop surface* (decreased a_w) must limit the diffusion of volatile components (selective diffusion compared with water). That will happen with
   a. High temperature of drying air (high evaporation rate) limited by the temperature reached by drying drops (volatile evaporation)
   b. Optimal high TS concentration/viscosity of feed liquid to dry (limited by pumping and spraying need!) fixing the drop structure rapidly with optimal drop size
3. A low, controlled temperature for particles throughout the drying process (i.e., no sticking) and an optimal short drying time will limit the losses of volatile components.

For example, during drop drying, the main losses concern the more volatile constituents of the aroma. To compensate for these losses, the initial mix may be enriched in these constituents. This is also linked to the size of the aroma molecule, the aroma concentration in the liquid to be dried, and the possible migration of aroma into the drop.

Operating conditions depend on the chosen equipment (spray-dryer geometry and atomizer type), airflow rate and temperature levels (inlet/outlet difference), and liquid feed flow rate, fixing the drying rate. For a given inlet air temperature, a higher outlet temperature corresponds to faster drying but higher powder temperature. And according to the chosen supports, drying may be accompanied by surface cracking and modified aroma retention.

16.3.4.1 Characteristics of the Resulting Aroma Powder

The powder characteristics will be defined as

- Size and size distribution of particles (i.e., 20–100 μm).
- Water content, a_w (i.e., <0.2).
- Aroma fraction in 100 g of powder.
- Fraction of aroma on particle surface (non-protected); surface state, smooth or with asperities (microscopy).
- Aroma well dispersed in the particles (possible to reconstitute the initial emulsion by dissolving the powder).
- Physicochemical and end use properties of the bulk powder, such as rehydration, density (bulk and packed), flowability, hygroscopicity, mechanical resistance, porosity, etc. Depending on the emulsion composition, the gas content of the drop, and the drying conditions, the dry particles may be compact or hollow, with a different distribution of aroma.

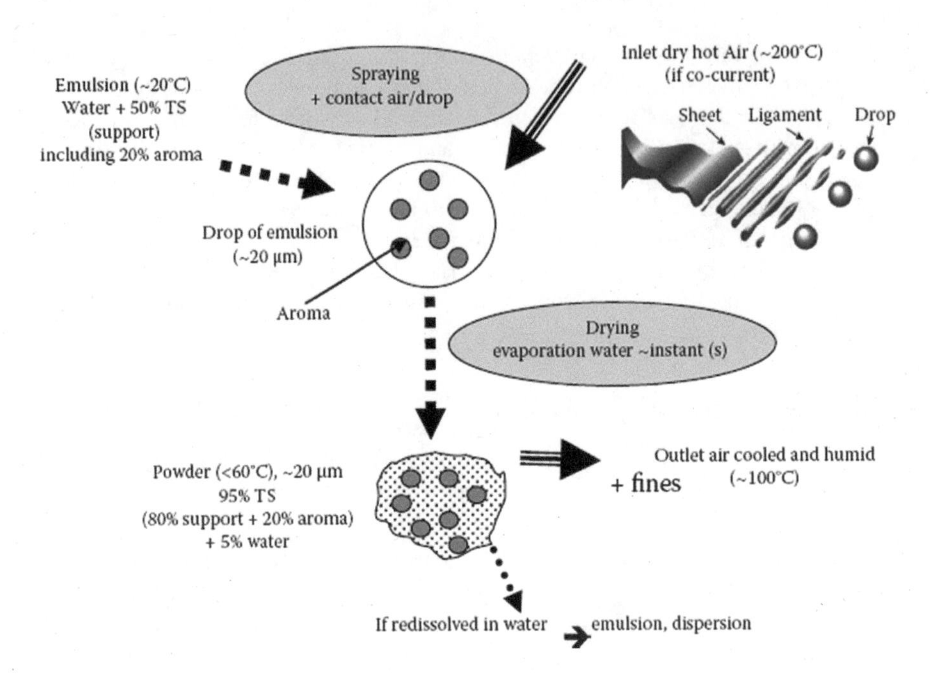

FIGURE 16.5 Schematic spray-drying process for dried emulsion.

Figure 16.5 shows typical spray-drying conditions, and the different operations with corresponding variables are summarized in Table 16.1. Some schemes for atomizers (nozzles or wheel) and spray-dryers (one or two steps) are given in Figures 16.6 through 16.8, adapted from Pisecky (1997). The choice of spray system, spray-dryer, and operating conditions will impact the resulting powder properties.

Spray-dried powders have a small particle size (10–100 μm) with poor handling properties. Very often in the food industry, at the outlet of the spray-drying chamber, the spray-dried particles are modified by additional treatment such as agglomeration, allowing the modification/increase of powder size/porosity and improvement of solubility/dispersibility properties (i.e., instant powder; decreased proportion of fine particles) (Buffo et al., 2002).

Computational fluid dynamic simulations have been used to model the drop/particle trajectories in spray-drying (Gianfrancesco et al., 2010), but optimization is still complex.

16.3.5 Summary

With the spray-drying process, the main objectives are to obtain an aroma powder with the "desired quality"

- By easy elimination of solvent from a liquid, an emulsion, or a slurry (drying properties)
- In a short time (i.e., a few seconds)
- By building solid particles with homogeneous size (i.e., 100 μm) and shape
- Maintaining the integrity of constituents (keeping the aroma profile) with minimal aroma loss
- Giving the final product properties adapted to the use

This will determine the choice of the spraying mode and drying parameters, including the liquid feed composition.

16.3.5.1 Spray-Drying of Aroma Emulsions

Aromas in emulsions are dried with a "support" with the following properties:

- Edible, non-toxic, and soluble in water (usual solvent, or ethanol) leading to a viscosity (optimal concentration) suitable for pumping and spraying

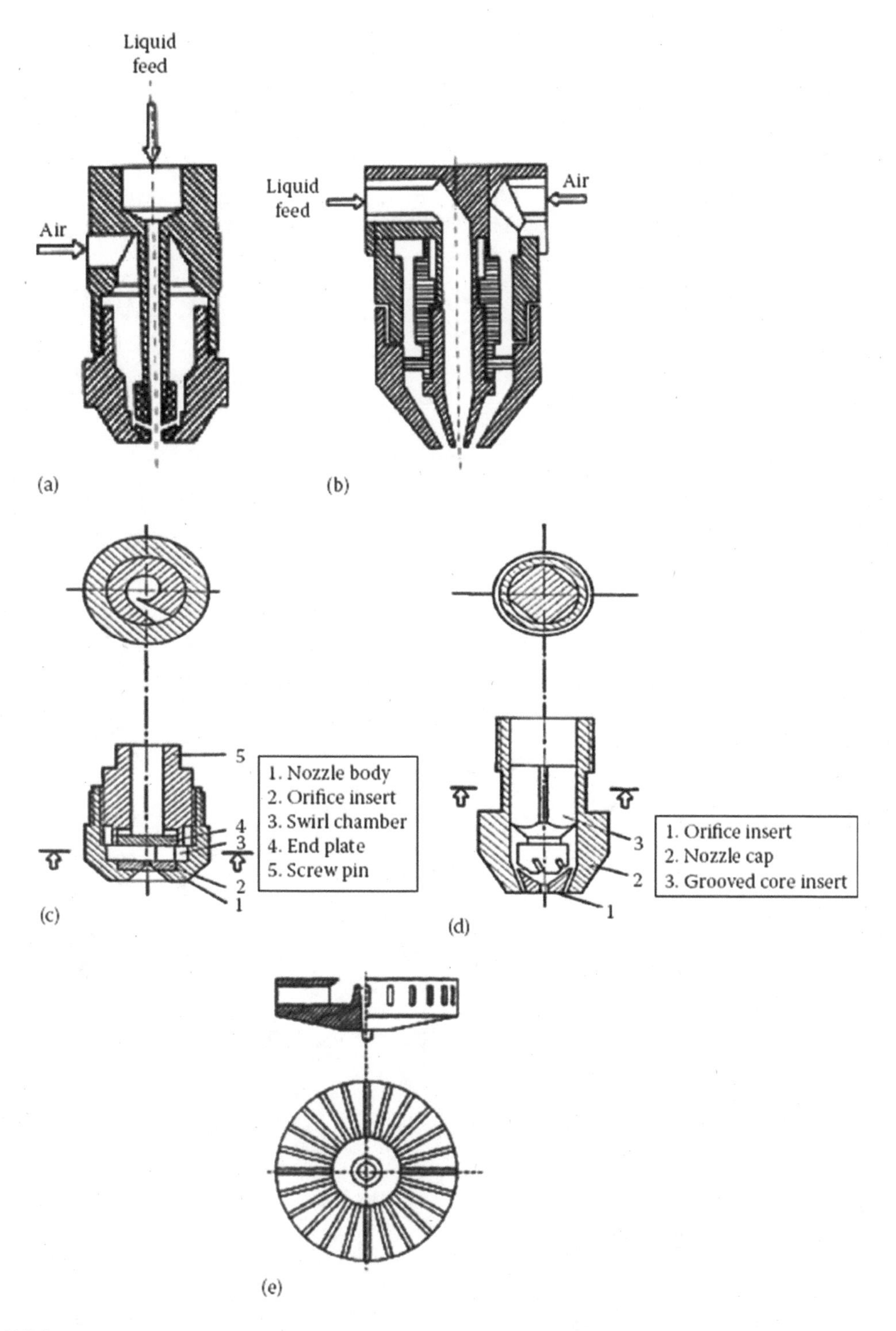

FIGURE 16.6 Types of atomizers in spray-drying for structured liquid drops. Two-fluid nozzle with (a) internal air/liquid mixing and (b) external mixing; pressure nozzle with (c) internal, (d) external air/liquid mixing, (e) rotary/centrifugal atomizer.

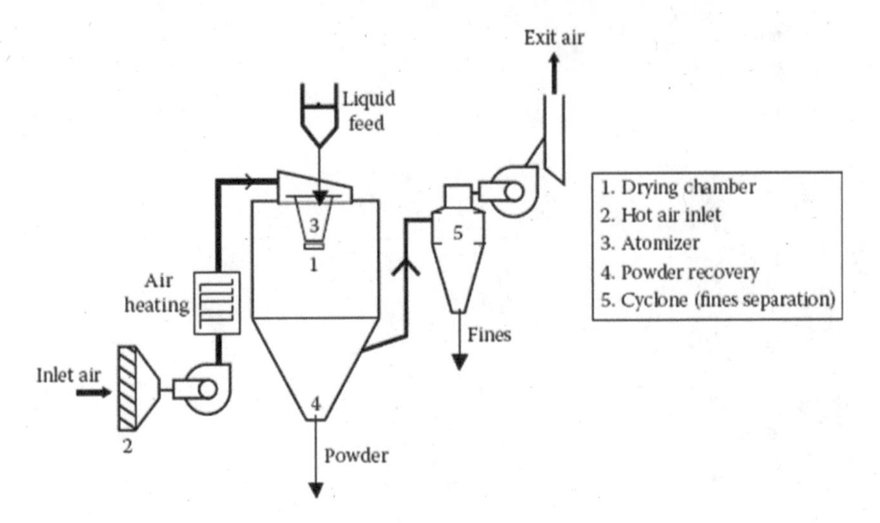

FIGURE 16.7 Spray-drying process scheme (one stage). Hot air drying/drops cocurrent = no high T for dry product. Residence time ~10 s, continuous process.

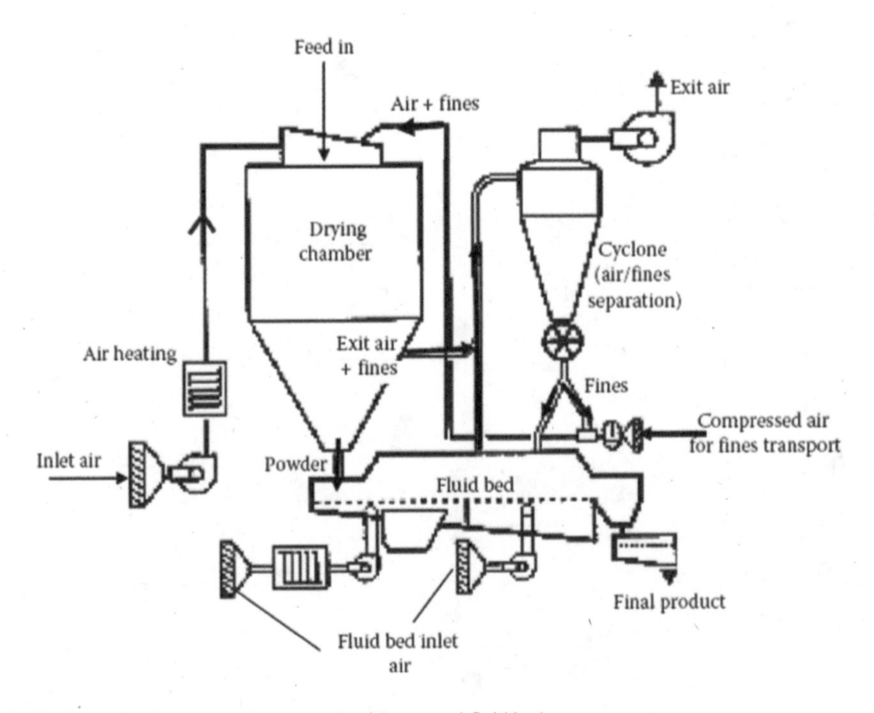

FIGURE 16.8 Spray-drying process (two stages) with external fluid bed.

- Without action on aroma; neutral (i.e., odor, color, and taste)
- Good ability to form a crust/layer at the surface of drops (easy to dry), avoiding sticking problems (Tg)
- Low hygroscopicity for further storage (maintains capsule intact)
- Able to release aromas when further used in controlled release conditions
- Particle size and size distribution of particles linked to spraying in drops (~20 μm)
- Cost studied in relation to final use

TABLE 16.1

Spray-Drying Process and Main Parameters

Operations	Variables
Dispersion, support + water	Total solid (TS) content, composition
+ emulsifier, + stabilizer	Surface tension, viscosity, pH, temperature
+ aroma	% support/aroma (<20%)
Homogenization emulsion O/W	Equipment (agitator, pressure, membrane)
	Emulsion size, size distribution
	Viscosity, surface tension
Spray drying	Equipment (chamber, geometry, etc.)
Spraying into drops	Atomizers: nozzle (air, pressure; position, number); rotary, etc.
+	Shape/size, size distribution of drops
	Trajectory of drops
Drying hot air	Flow rate/pressure drying air
	Temperature air inlet/outlet
	Flow rate/temperature product
	Powder yield (TS), sticking
>> Aroma encapsulated in powder	Water content, a_w; composition, retention, degradation,% surface; structure (amorphous, crystallized)
(+ agglomeration in fluid bed)	Size, density, wettability, flowability, surface state
	Temperature, Tg, MP

In food, the supports (alone or mixed, more or less soluble in water) are the following:

- Glucides, maltodextrins, modified starch, cyclodextrins, sucrose, celluloses
- Lipids, wax, lecithins
- Milk proteins, gelatin
- Gums, algae extracts

The load of extract or aroma in the support (in powder) is usually 10–30% w/w of TS. Lipids, proteins, gums, and modified starch act as emulsifying and film-forming agents. The carbohydrates act as a matrix material, which may be protective against oxidation during storage (high DE) (Sheu and Rosenberg, 1995; Re, 1998; Baranauskiene et al., 2006; Baranauskiene et al., 2007).

Some examples of encapsulation studies are given, using simple model aroma molecules that are more or less volatile and soluble in water.

- Different food supports were tested (maltodextrin, inulin, and acacia gum with emulsifying properties) for the encapsulation by spray-drying (air T 180–90°C) of a model lipophilic component (α-tocopherol dispersed in olive oil, 1/4) to prepare dry emulsions in powders. The emulsions (2 μm) contained 40% w/w TS, which contained 8% w/w olive oil, 2% w/w α-tocopherol, and 30% w/w support. No significant influence of the nature of these carriers on the physical properties of the powders (size, size distribution, densities, or flowability) or on the spray-drying efficiency was observed; 73% of the initial oil was recovered in powder, with 5% of the oil phase at the particle surface (non-encapsulated). But due to its emulsifying and film-forming properties, the use of acacia gum, in combination with maltodextrin and/or inulin, helped to obtain more stable initial emulsions with a monodispersed size distribution (~2 μm) and higher powder yield for spray-drying. When redissolved in water, dry emulsions led to reconstituted emulsions with a size distribution similar to that of the initial emulsion, indicating that spraying of the emulsion into drops did not modify its structure (Turchiuli et al., 2014).

- The presence of low-MW carbohydrates in supports may contribute, via their amorphous state ($T < Tg$), to possible modification/deterioration of the quality and functionality of powders linked with temperature and RH, especially during the spray-drying process and storage: crystallization, stickiness, or caking (Vega and Roos, 2006). The determination of Tg will be important in predicting the onset of deteriorative reactions. Also, the choice of supports (protein adsorption) may help to modify the powder surface (Jayasundera et al., 2009).

One example study used aroma molecules as models with different MW and volatility (diacetyl [DA] and vanillin [VA], citral, and linalyl acetate) and different supports, such as maltodextrin, sucrose, and skim and whole milk (Senoussi et al., 1994, 1995).

The (partial) replacement of maltodextrin by sucrose led to more regular particles and better powder wettability. The flowability and regular shape of skim milk powder were better than those of whole milk due to the absence/presence of fat.

Higher retention of VA was observed compared with DA (MW: VA > DA; volatility: VA < DA), and retention was better with whole milk (due to the fat phase) compared with skim milk. The partial replacement of maltodextrin by sucrose (lower MW) is favorable to the departure of aroma molecules with water during spray-drying.

The study of volatile DA retention (boiling point [BP] 88°C) in various constituents of milk (lactose, fat, and proteins) during spray-drying showed the importance for aroma retention of the amorphous/crystalline ratio in lactose, which is in turn, linked to glass transition phenomena, temperature, and water/RH. The presence of microcrystals in drops (due to concentration or fast drying) may be a barrier to the diffusion of volatile molecules, therefore increasing the retention in the final spray-dried powder.

During spray-drying and storage, the milk proteins showed a high retention capacity. During storage of powders in different humid atmospheres, volatile DA losses followed lactose crystallization, which was higher with high RH. Changes were related to the difference $T - Tg$. This shows the importance of controlled atmosphere humidity between 22 and 53 % RH at 25°C during powder storage to avoid crystallization (i.e., lactose/skim milk) and aroma loss.

Several studies have shown (see Table 16.2) the importance of retention efficiency in spray-drying encapsulation:

- *The properties of aroma molecules*, such as solubility in water, size and shape of molecules, volatility, MW, and polarity, and the importance of matrix choice, especially in relation to glass transition temperature/crystallization, which can be the source of possible modification of the powder structure. Also, comparing volatiles and non-volatiles (D-limonene and fish oil), the surface oil content of non-volatile encapsulated powders was much higher: volatile compounds can be evaporated and removed during spray-drying (Jafari et al., 2007b).
- *The size of the feed emulsion transformed into emulsion drops with spraying systems.* The forces applied to change a film/sheet of emulsion into drops are able to break or to associate the aroma oil drops of the emulsions inside the drops and consequently, to modify the aroma dispersion. Some research (Table 16.2) (Risch and Reineccius, 1988; Soottitantawat et al., 2005a,b) shows that an emulsion of ~1 μm (compared with 4 μm) is more stable, adapted to good aroma retention in a spray-dried powder, with a low fraction on the dried particle surface. The use of smaller emulsions needing more energy is not required for high retention in these studies. Also, the presence of smaller drops in capsules within particles may provide a greater surface for oxidation in the case of oxygen penetration.
- *The emulsifiers and emulsifying techniques* to produce nanoparticle encapsulated powders (Jafari et al., 2007a). In the case of fish oil encapsulated in maltodextrin combined with modified starch or whey protein concentrate, spray-dried powders were obtained from nanoemulsions (210–280 nm) prepared by microfluidization with good efficiency (Jafari, 2009). Also, the saturation of the carrier solution of wall materials influenced the flavor retention and surface oil content (Penbundiktul et al., 2012).

TABLE 16.2

Examples of Studies on Aroma Encapsulation by Spray-drying, with Some Important Parameters

Aroma		Carriers/Emulsifiers		Feed/Emulsion			
Molecule	**Properties (MW, BP, S/W)**	**Product**	**Properties, Role**	**Composition % TS, Viscosity (mPa-s), % Aroma**	**Flavor Size in Emulsion →Powder**	**Spray-drying Air Temperature In/Out, Flow Rate, Feed Flow Rate, Powder Size Range**	**References, Main Objectives**
D-limonene (linalool)	Low solubility (1589 mg/l water)	Modified starch (HI-CAP 100)		–10% to 40% TS (viscosity × 60)	Polytron + homogenizer, 8000 rpm	3–9 ml/min feed, 180–120°C	Penbundiktul et al. (2012)
Linalyl acetate (= bergamot oil)	30 mg/l	Acacia gum	Carriers	–10% w/w TS, flavor/solid 1:4 w/w	High-pressure homogenizer, 1 (GA) to 5 µm, non-saturated or saturated overnight	Two fluid nozzle atomizer	Effect of saturation of carrier solution on high retention and low surface oil
Cardamom spice oleoresin (cooking, acid beverages)	Balance volatile (terpenoids)/ non-volatile, chemical changes for T > 149°C; pH <4.	Acacia gum, modified starch (HICAP), maltodextrin + Tween 80	Wall + emulsifier + for good emulsion	Oleoresin: 5% of carrier	Shear homogenizer 5 min, 3000 rpm, + 2 drops Tween	Buchi 190, 178 /120°C, nozzle air 5 bar, 300 g/h liquid feed	Krishnan et al. (2005), content and stability (6 w) (volatile/ non-volatile); GA = good protection
D-limonene, ethyl butyrate	Terpene, MW 136; 176°C	MD 15–20, acacia gum	Carrier, emulsifier	40% w/w wet basis (10% E, 30% MD), GA-MD ~180 mPa-s, SSPS-MD ~2600	25% TS (mass), 0.8–4.09 µm	200/110 ± 10°C, air 110 kg/h	Soottitantawat et al. (2003)
Ethyl propionate	Low sol. ester, MW 116; 116°C; 6.7 × 10^{-3} v/v	Modified starch (HI-CAP 100)	Emulsifier		→ 1 to 3 µm	Centrifugal 30,000 rev/ min; feed 45 ml/min; powder 42 to 55 µm	Role of size of aroma feed emulsion, with different carriers/ aromas, for powders
		SSPS (soybean soluble polysaccharide); SAIB (sucrose acetoisobutyrate)	Emulsifier; weighing agent for esters (density)				
	Ester			HI-CAP-MD			% aroma retention
	MW 102; 102°C			~60–90			% aroma on surface
	1.7 × 10^{-2} v/v						

(*Continued*)

TABLE 16.2 (CONTINUED)

Examples of Studies on Aroma Encapsulation by Spray-drying, with Some Important Parameters

Aroma		Carriers/Emulsifiers		Feed/Emulsion			
Molecule	Properties (MW, BP, S/W)	Product	Properties, Role	Composition % TS, Viscosity (mPa-s), % Aroma	Flavor Size in Emulsion →Powder	Spray-drying Air Temperature In/Out, Flow Rate, Feed Flow Rate, Powder Size Range	References, Main Objectives
d-limonene Ethylbutyrate	Low solubility MW 136 BP 176°C 6 g/L, 20°C MW 116 BP 121°C	Maltodextrin +acacia gum (emulsion) Or + Soybean polys. SSPS (emulsion) SAIB sucrose acetoisobutyrate + Triglycerides MCT	Carrier Emulsifier 10% w/w in liquid feed To regulate density of ethyl butyrate (stable emulsion)	10% to 30% TS (w/w) Flavor/emulsifier 0.25–1	Polytron homogenizer, 3 min + Microfluidizer (82.8 MPa)	45 ml/min feed, 150/74°C–100°C Centrifugal atomizer	Liu et al. (2001) Very good retention for D-limonene (GA) For ethyl butyrate, good with SSPS or adjusting density (stable emulsion)
Citral, linalyl acetate	BP 228°C, 1.34 g/l, 37°C; BP 220°C, 0.9 g/ml, 25°C; proportion 80/20	Acacia gum, maltodextrin 17	Emulsifier, carrier, MD/GA 4:1; 3:2; 2:3; 0:1	30%–40%–50% TS, flavor/ support 1:4	Shear homogenizer 20 min or + pressure homogenizer 1–5 µm	Leaflash 350/100°C, nozzle 70 kg/h liquid feed, powder 50–150 µm	Bhandari et al. (1992), optimal MD/GA ratio (3/2) for 82% encapsulation
Orange oil	0.83 mg/l	Acacia gum, modified starch, glyceral abiebate, glycerol tribenzoate, brominated vegetable oil	Carriers, possible weighing agents to increase oil density (solubility)	Carrier:flavor ratio 4:1, 37.5% TS	Coarse (whisk) 4 µm; medium coarse (high-shear mixer) 2.5 µm Medium fine (+ high speed) 1.8 µm; fine, homogenizer 1.7 µm; microfluidizer 0.9 µm	200°C/100°C, Niro	Risch and Reineccius ( ) Emulsion size decrease related to better flavor retention and less surface oil for powder

BP, boiling point (°C); E, emulsifier; AG, acacia gum; MD, maltodextrin; MW, molecular weight (g); S/W, solubility in water.

- *Spray-drying of multiple emulsions* results in a double-layered microcapsule, providing better protection. For example, orange oil-in-water emulsion is further encapsulated in another oil phase to form a double emulsion (O-W-O). Spray-drying this double emulsion mixed with an aqueous solution of lactose and caseinate provides an efficient secondary coating (Edris and Bergnstähl, 2001).

16.3.6 Spray-Cooling

16.3.6.1 Principle

In the spray-cooling process, aroma encapsulated (uniform dispersion) in molten wax or fat with surfactant is sprayed in drops at a certain airflow rate at ambient or cold temperature. The objective is to rapidly solidify the fat, avoiding crystallization.

The spray-dryer is equipped with a heated nozzle (or a centrifugal atomizer) to spray the liquid.

The main supports for food products are vegetable fat and stearin (melting point [MP] 45–122°C); monoglycerides and diglycerides (MP 45–67°C); and vegetable hydrogenated oils (MP 32–42°C). The final products can be used in bakery and in soups. Other supports, solid at working temperature, may be chosen as a function of domain of use, viscosity, and crystallization properties.

Usually, capsules have a large size (0.5–2 mm), and they are non-porous and not soluble in water. The AC may be partly found at the particle surface. The main advantages are the absence of solvent, low consumption of energy, and no degradation of heat-sensitive components. But, possible sticking onto walls may lead to losses. This process may be used for a double encapsulation of spray-dried particles.

Another possibility could be the atomization of emulsions (simple or double) with a twin-fluid nozzle, in a cold atmosphere such as –30°C, to produce solid particles. This process is used to study the atomization process (drop size distribution and shape) through the microstructure of the solid particles.

16.3.7 Freeze-Drying

16.3.7.1 Principle

The aqueous aroma composition (solid, liquid, or paste) includes components/supports that are able to form by freezing an amorphous matrix or a gel to immobilize aroma compounds. This will be transformed into an aroma powder, without modification of the aroma, for easy storage. After freezing of the aroma composition, there is drying by sublimation of ice, usually under vacuum. The driving force for sublimation is the difference in pressure between the water vapor pressure at the ice interface and the partial pressure of water vapor in the chamber. Because of the low processing temperatures, thermal degradation reactions are excluded. Atmospheric pressure may be used if the water vapor pressure of gas is inferior to the vapor pressure of ice.

16.3.7.2 Main Characteristics

Usually, the structure of the matrix is preserved during the whole process, which is a long one. Temperature and pressure are low (Coumans et al., 1994; Buffo and Reineccius, 2001).

The main parameters are

- High initial concentration in dissolved solids to reduce the volume fraction of dispersed aroma, but giving a very dense product
- Freezing rate not too high (pure ice, aroma components concentrated in matrix phase)
- Sample size not too big (low thickness)
- Nature of volatile component: retention increases with MW

Volatile retention occurs through entrapment in the amorphous matrix of hydrogen-bonded carbohydrate molecules, and the volatile retention regions are quite small.

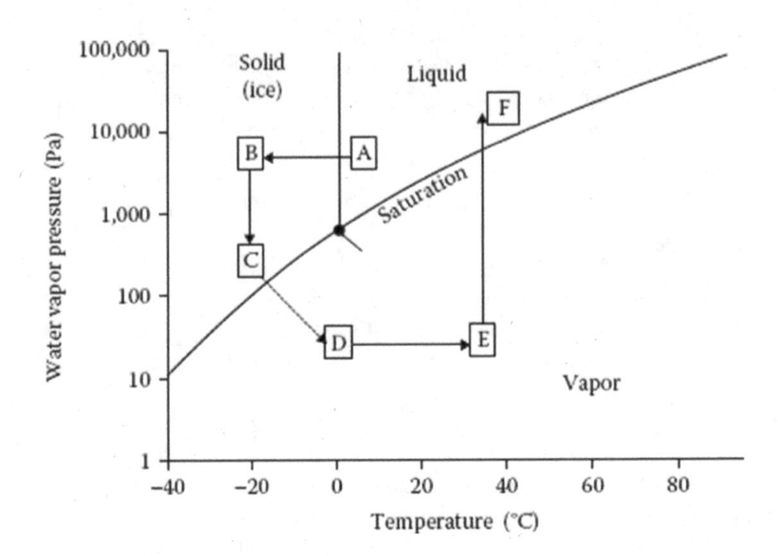

FIGURE 16.9 Freeze-drying principle (AB freezing; BC vacuum; CD heating/ice sublimation; DE water desorption; EF ambient pressure).

The equipment consists of a chamber with plates, which can be either refrigerated or heated, under vacuum, with a cold trap for water vapor (sublimation).

Freeze-drying consists of several steps (Figure 16.9):

- Freezing (AB), whereby most of the water is frozen (~ −20°C, ice crystals) (internal or external), leaving a concentrated matrix phase (with aroma)
- Vacuum (BC), then heating for ice sublimation (CD, leaving open pores) and desorption (last % water…) (DE)
- Ambient pressure (in dry inert gas) (EF)

The final residual moisture content (low) is determined by the secondary drying phase.

It is also possible to prepare a spray–freeze-dried powder. Atomized drops are immediately frozen in liquid nitrogen, and the frozen solvent is removed by sublimation. This technique is long and costly, requiring specific storage and transport.

16.3.8 Cocrystallization

16.3.8.1 Principle

A solution of support able to crystallize (i.e., sucrose/lactose in water) is concentrated until it reaches supersaturation. Then, the aroma is added into the supersaturated sugar syrup under vigorous agitation. After nucleation, crystallization occurs, with the emission of a substantial amount of heat. Then, solid crystals (3–30 μm) with entrapped flavor and agglomerates are formed and separated, dried, and sieved (Chen et al., 1988; Beristain et al., 1996; Kim et al., 2001).

16.3.8.2 Main Characteristics

The aroma is trapped between microcrystals.

In the case of sucrose and water, the aroma must be soluble or dispersible in water. The encapsulated aroma is easy to dissolve, but the solution has a sweet taste! The addition of an antioxidant may be recommended (porous structure).

The principle may be adapted to other supports that are able to crystallize.

16.3.9 Coacervation

16.3.9.1 Principle

The process is based on the phase separation of one or several hydrocolloids from an initial solution and the subsequent deposition of this newly formed coacervation phase around the active ingredient (aroma) suspended or emulsified in the same reaction medium. Then, the hydrocolloid shell may be cross-linked with appropriate cross-linkers. The phase separation will be provoked by reaction between two different colloids by varying the temperature, pH, and medium composition (solvent and salts). This process was tested in the 1950s for carbonless copy paper (dye in particles ~20 μm) (Risch and Reineccius, 1988; Gouin, 2004; Leclercq et al., 2009).

This is used for the encapsulation of oils or lipohydro-soluble products. The capsules are very rich in internal phase (oil/aroma 50–90%), with a very thin wall ("core/shell" system). If a final dry product is needed, the drying of capsules (by a fluid bed or freeze-drying) is delicate, with the possible loss of molecules with low MW.

16.3.9.2 Methods of Coacervation

Several methods exist; not all of them are applicable to food products, which use the evaporation of solvents, or non-compatible polymers, or interfacial polymerization.

In simple coacervation or gelification (500 μm to 2 mm), an emulsion of oil in an aqueous solution of a polymer/substance able to form a gel is prepared. Changing the pH or temperature, or adding salts, causes the substance to precipitate around the drops (alginate/$CaCl_2$; gelatin hot/cooled oil). Then, the particles are separated and dried. Nanospherical particles (100 nm of essential oils in zein (a protein) were prepared by phase separation and then lyophilized (Parris et al., 2005).

In complex coacervation (20 μm to 1 mm), for example, aqueous solutions of the AC, a polyanion (–), and a polycation (+) are mixed. The two polymers with opposite charges (electrostatic interactions) will interact to form a deposit of coacervate at the surface of the AC (i.e., acacia gum, alginate; carboxymethylcellulose [CMC] with gelatin; proteins and anionic polysaccharides) (De Kruif et al., 2004). Reticulation may be provoked by dilution and modification of pH or temperature (Figure 16.10).

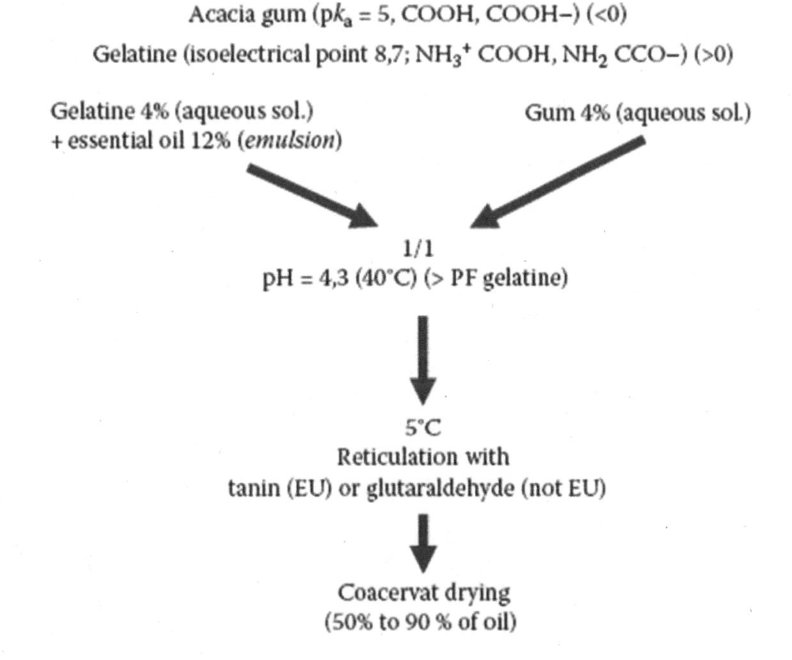

FIGURE 16.10 Example of complex coacervation.

Gelatin and acacia gum (opposite charges at low pH) were used to encapsulate a lipophilic flavor oil to be used in frozen foods and released on heating (Yeo et al., 2005) with a liquid or solid core (Leclercq et al., 2009).

16.3.10 Extrusion

16.3.10.1 Principle

The aroma to be encapsulated is dispersed in the melted or dissolved support (i.e., melted sucrose at 110–130°C) in a premix, in the screw, or injected at the end. Then, the dispersion is extruded (forced to pass through a hole with a selected size and shape, a needle tip) into a medium where the matrix is rapidly transformed into a gel, a solid precipitate, with the surface cleaned/dried by the medium (i.e., isopropyl alcohol for sugar). The main parameters are good mixing and controlled operating conditions such as mechanical forces, temperature (relatively low), pressure, and residence time, usually with no expansion at the outlet (Figure 16.11).

The shape of extruded solid products may be monodisperse microspheres or small sticks (~300 µm to 1 mm).

If the mix contains water (matrix, water, and aroma), extractive drying may be achieved by removing water from the formed rigid drops by means of contact with a water-absorbing second phase such as polyethylene glycol (PEG)400 (Kerkhof and Thijssen, 1974). In another example, an aqueous emulsion of core (volatile esters and paprika oleoresin)/wall (acacia gum and CMC) materials was injected/sprayed into a cold, dehydrating liquid such as absolute ethanol. The resulting slurry was filtered and then dried under vacuum (50°C). Retention was improved with a high solid concentration and a low core-to-shell ratio (Zilberboim et al., 1986).

The main supports in food products are glassy carbohydrates, alone or mixed: dextrose, corn syrup, glycerin, maltodextrins, sucrose with added emulsifiers, monoglycerides, gum, pectin, and lecithin. Cyclodextrin-complexed flavors may be prepared prior to extrusion (Bhandari et al., 2001).

Extrusion products typically contain 8–10% of flavor oil, but the final washed surface makes the product very stable (Rish and Reineccius, 1988; Qi and Xu, 1999). Carbohydrate matrices in the glassy state have very good barrier properties (i.e., against atmospheric gases). Dripping and jet breakup are the

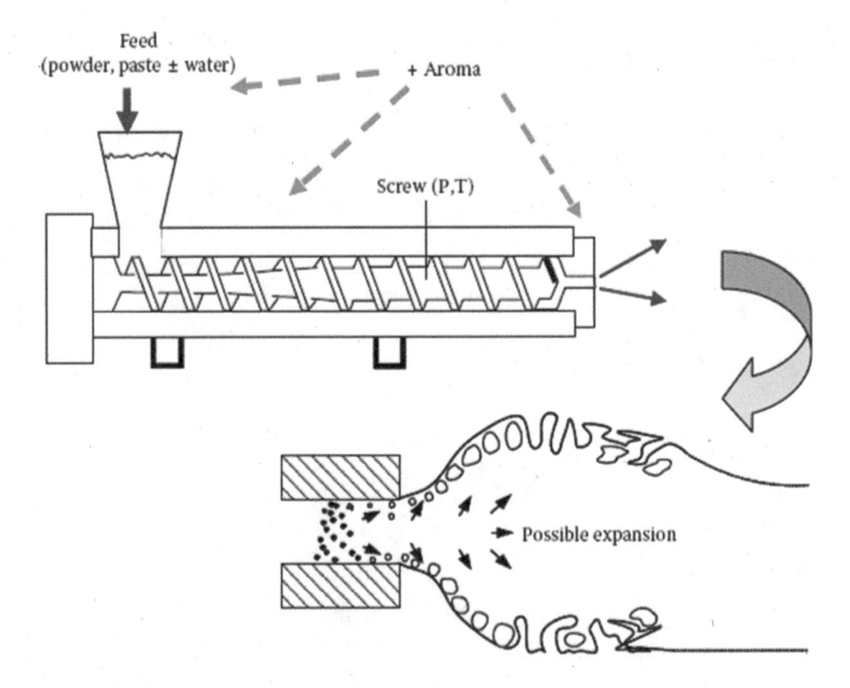

FIGURE 16.11 Extrusion process scheme.

methods for forming microcapsules (Whelehan and Marison, 2011). The parameters are screw temperature and speed, and residence time distribution (Yuliani et al., 2006).

One example of AC encapsulation (embedding) is described as a continuous process (several screw sections) with a matrix composition made of a hydrophobic component for controlling the release (i.e., fat, wax, paraffin, etc.) and a material plasticizable at low shear (i.e., starch or cyclodextrin). The AC is added (5–20% by weight) to the melt matrix at a low temperature with reduced postextrusion drying and expansion. Particles are extruded through a die with multiple apertures (i.e., 0.5–7 mm). They may be covered with an additional film-forming substance (i.e., wax, fat, etc.) (Van Lengerich, 2003).

In coextrusion, two concentric jets are used (i.e., a double fluid nozzle) (Schlameus, 1995; Gouin, 2004):

- An internal jet with the product to be encapsulated
- An external jet made of the wall product: liquid, melted (wax or fat), or in solution in water (hydrocolloid)

At the edge of the nozzle, Rayleigh instabilities lead to the formation of round beads. The wall solidifies by cooling (wax) in a cooled organic solvent (gelatin) or in $CaCl_2$ solution if it is made of alginate. The obtained capsules are spherical, with uniform size (>500 μm), and the internal phase may be important, making up 90% of the capsule mass.

16.3.11 Molecular Inclusion

16.3.11.1 Principle

The cyclodextrin (CD) molecule (usually β) is a truncated cone made of seven glucopyranose units; the external part is hydrophilic, and the internal part is hydrophobic (MW 1145; solubility 1.85 g/100 ml water (25°C) (α-CD 12.7 g/100 ml; γ-CD 25.6 g/100 ml). The internal cavity (diameter 5–8 Å) is non-polar and may accept a non-polar "guest" molecule of a similar size to aroma molecules (6–15% w/w). This is a selection by the host molecule configuration. If formed, the aroma/cyclodextrin complexes are very stable to evaporation, oxidation, light, and heat, with good protection of the aroma profile.

16.3.11.2 Production

Several techniques may be used:

- In the liquid state, the aroma, dissolved in a solvent soluble in water (i.e., ethanol), is added drop by drop to an aqueous solution (i.e., + ethanol) of CD. Then, the final CD/oil complex will precipitate in a kind of cocrystallization (for lemon oil, see Bhandari et al., 1998).
- In the solid state, the CD in a paste (powder, water) is mixed by kneading with the product to encapsulate, with just the right quantity of water, some of the complex. The paste is vacuum- or spray-dried (for lemon oil, see Bhandari et al., 1999).
- Aroma vapor may flow through an aqueous solution of CD with precipitation of the complex.

In all cases, the product may be filtered and dried to obtain a final solid. The stable final product is interesting for strong flavors or tastes.

CDs may be used in combination with carbohydrates (by spray-drying or extrusion) to ensure both encapsulation and further protection (Qi and Xu, 1999).

16.3.12 Coating: Impregnation

16.3.12.1 Principle of Impregnation

The aroma is dispersed at the surface of a support in powder (sugar, salt, modified starch, or silica gel) in a mixer (i.e., sucrose with vanilla).

The process is simple and economical but with bad protection of aroma. The product needs additional protection, such as packaging, against temperature and light.

Microcapsules for textile finishing or for incorporation (fragrance or essential oil) into the textile fibers have to be resistant to mechanical and thermal stress, with impermeable and pressure-sensitive walls. *In situ* polymerization of aminoaldehyde polymers may be chosen with good results (Boh et al., 2011).

16.3.12.2 Principle of Coating in Turbine/Drum/Screw (Endless) or in Fluid Bed

A coating can be used with solid particles containing aroma as an additional protection (i.e., a smooth or hard structure). An additional possibility may be to include aroma (the same or different) in the coating layer (where it would be well protected).

This is realized by the agitation/fluidization of individual solid particles at a certain flow rate of hot or cold air. Then, the coating product is sprayed (by a nozzle) onto the moving particles, progressively forming a homogeneous coating layer.

For air-fluid-bed particle coating, spraying may be done from either above or below the fluidized bed of particles (100 µm to mm). The added layer is usually thin (~no size change) and may need final drying (Figure 16.12). Several layers may be formed successively. Usually, the particle size is superior to 100 µm for good fluidization, which implies the spraying of small coating drops in relation to this size to achieve a uniform coating and avoid agglomeration.

The coating products are melted (without solvent; hot-melt coating) or dissolved in a volatile solvent (usually water): they may be derivatives of cellulose, dextrins, emulsifiers, lipids, or derivatives of proteins or starch. Among the polysaccharides, cellulose derivatives are often used due to their good barrier properties against gases such as oxygen; they have good mechanical properties but are often permeable to water. A wide variety of possible coatings offer many possibilities for controlled release (Clark and Shen, 2004).

In a fluid bed, many parameters will be considered (Guignon et al., 2002, 2003; Karbowiak et al., 2007) (Table 16.3):

- The particles to coat (composition, size, density, surface state, and load) and the coating (composition, adhesion to particle surface, spraying and drying properties, and flow rate)
- The fluid bed (geometry, spraying system, airflow rate and temperature, continuous/batch)

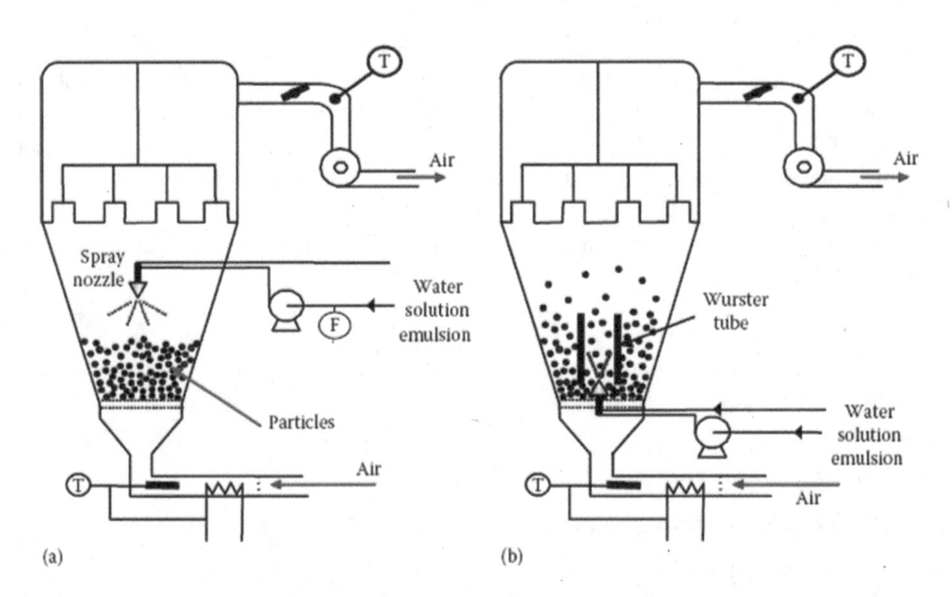

FIGURE 16.12 Coating of solid particles in fluid bed. (a) Top spray; (b) Wurster bottom mode.

TABLE 16.3

Coating Operation: Optimization Parameters for Solid Products

Product	Coating Agent	Chosen Process
Physical state	Solution or powder	Chosen equipment
Composition	Composition	Continuous, non-continuous
Water content, fat content	Fat content	
Shape	Water or solvent content (modified drying time)	Static, dynamic (agitation)
Dimensions		Volume
Density	Solubility in water of coating constituents	Temperature range
Texture (resistance, stickiness)		
Presence of fines	Melting point or gel temperature	Flow rate
Surface state	Density	Residence time
	Viscosity, flowability limit	
Hygroscopicity	Size (if powder particles)	Energy flow rate
Surface tension	Necessary pretreatment (mix, emulsion)	Air temperature and RH
Temperature (for easy dispersion, absorption)		Coating recycling
	Surface tension	
Production flow rate	Flow rate	

Another method to coat fine particles (~20 μm) with aroma may consist of forming a suspension of particles in a liquid containing the coating agent (lipid), saturated with CO_2 at a supercritical pressure. Then, after spraying the suspension, followed by decompression and expansion, microcapsules may be collected in the gas (Perrut, 2003; Gouin, 2004). The main advantages are the low temperature and the absence of water.

16.4 Examples of Studies Using Complementary Processes

The encapsulated aroma has been protected by one process, but the final product may need further additional treatment to improve or modify the final end use properties.

For example, in the case of powders, the controlled agglomeration of particles may contribute to improving the dissolution rate (porous agglomerate), to increasing the size of particles, to changing the density or improving the mixing with other particles, or to adding a protection (coating) to delay and control the release (Figure 16.13).

16.4.1 Vegetable Oil Encapsulation Using Spray-Drying and Fluid Bed Agglomeration

The objective was to encapsulate a vegetable oil (representing a model for an aroma compound) into a powder. The oil (ISIO4) (VO) was to be at a concentration of 5% TS in a matrix made of maltodextrin (MD) (DE 12) and acacia gum (AG) (3/2 MD/AG). The composition of the matrix was defined in previous research work on the encapsulation of aroma by spray-drying (Bhandari et al., 1992; Turchiuli et al., 2005; Fuchs et al., 2006).

The desired end use properties (brief) for the powder with 5% aroma were the following:

- A low water content to avoid sticking during processing and storage and to obtain stability
- High solubility in water for complete dissolution
- Controlled size, size distribution, and density facilitating good mixing with other powders without fine particles (a source of dust)
- Good flowability for easy handling, dosage, and transport
- Protection regarding oxidation during storage

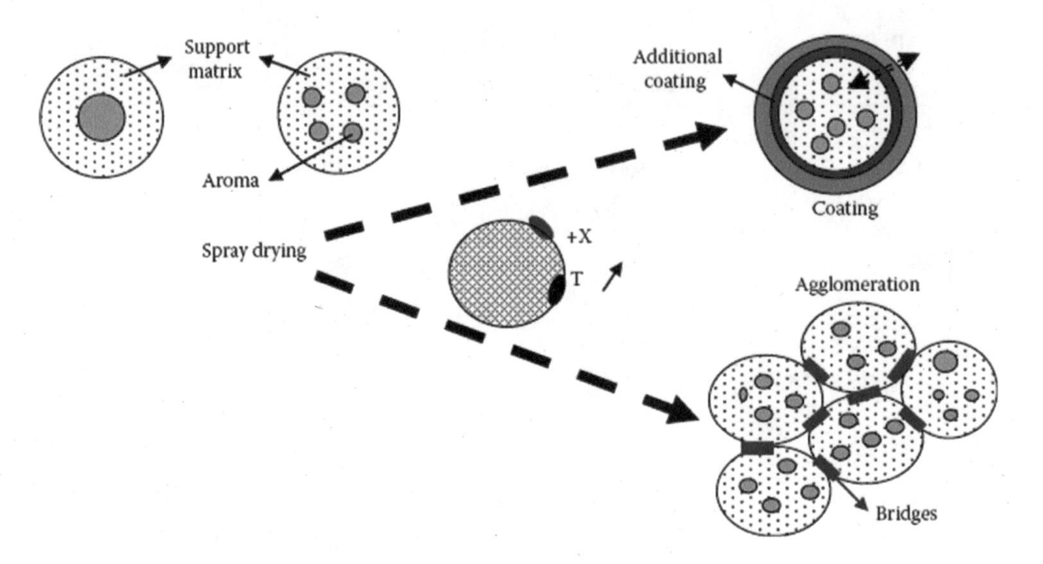

FIGURE 16.13 Complementary processes for aroma encapsulated powders: spray-drying, fluid bed coating/agglomeration (X, coating; T, temperature).

Three processes were tested at pilot scale (Figure 16.14):

- (A) Spray-drying of a formulated MD/AG/VO emulsion: a continuous process, with a powder yield of 65% and a final formulated powder with small mean size (30 μm)
- (B) Agglomeration of spray-dried fine powder in a fluid bed: a batch process, with a yield of 89%, giving porous agglomerates (200 μm) with low mechanical resistance but good wettability
- (C) Direct agglomeration in a fluid bed of MD powder sprayed with formulated AG/VO emulsion: a batch process, with a yield of 89%, giving porous agglomerates (240 μm) with both good mechanical resistance and good flowability

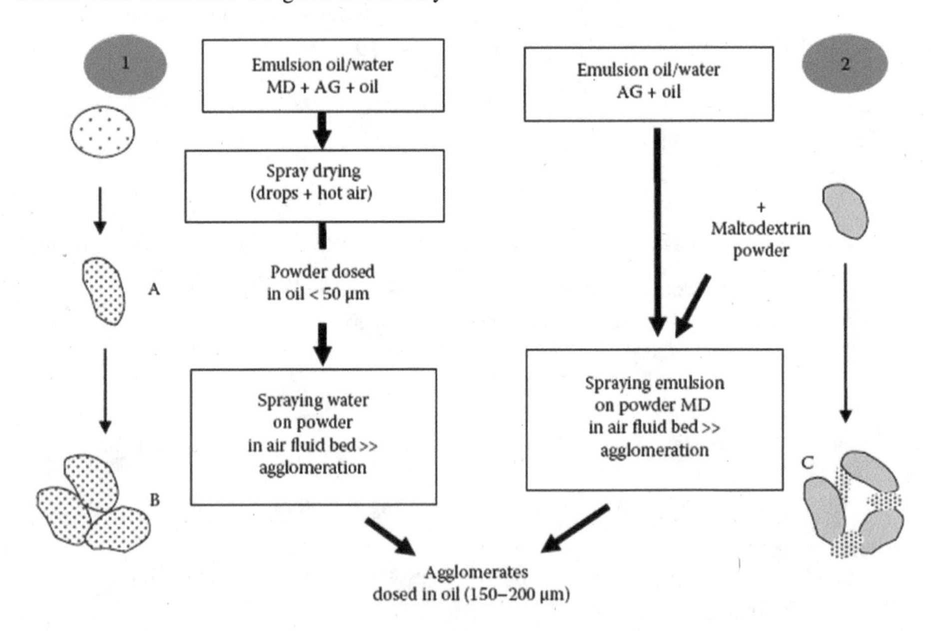

FIGURE 16.14 Two processing methods for dry emulsion preparation: three powders, A, B, and C.. AG, acacia gum; MD, maltodextrin.

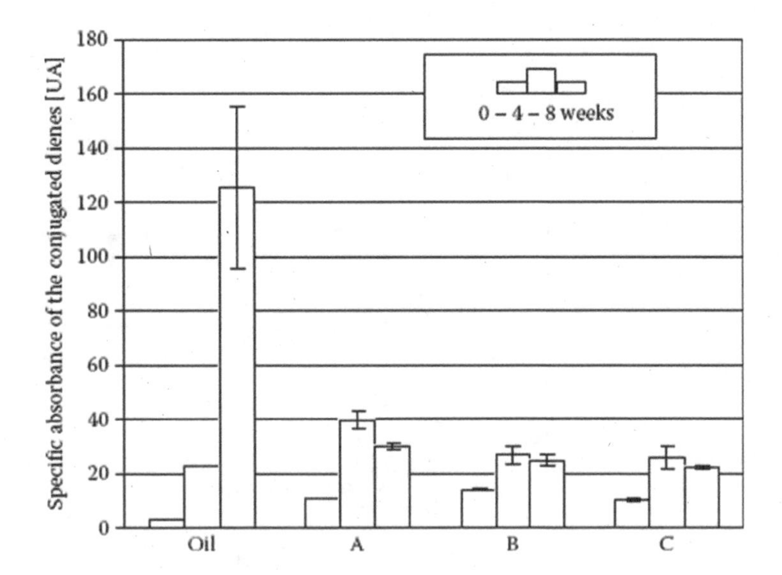

FIGURE 16.15 Oxidation tests (accelerated 60°C; conjugated dienes in extracted oil [234 nm]; 0, 4, 8 weeks). A, spray-dried powder; B, agglomerated A; C, MD agglomerated with emulsion.

The three powders were well dosed with oil (>4.4 g/100 g TS), with low losses of oil during the different processes (>88% of initial oil in the powder), and showing good protection of oil against oxidation (Figure 16.15). The oil on the particle surfaces was less than 2% of total oil for sprayed and agglomerated powders and 6% for the maltodextrin agglomerated with emulsion.

In summary, the tested encapsulation method is applicable for other components and different concentrations (aromas, antioxidants, or other oil substances). Other possible supports may be proposed to replace maltodextrin and AG, such as modified starch (Buffo et al., 2002), proteins, gums, or protective agents.

16.4.2 Aroma Encapsulation into Powder for Chewing Gum

The main industrial objective was to replace liquid concentrated aroma by a powder with the same sensorial characteristics as liquid (aroma profile, intensity, etc.) to be produced at industrial scale. This meant protection and stability of the aroma (during processing, storage, and use) with a final controlled release during chewing (i.e., by temperature, humidity, and mechanical stress) (Turchiuli and Dumoulin, 2013a) (Figure 16.16).

It was proposed to prepare aroma powders with ~20% aroma in TS with different particle size, structure, and distribution of aroma within the solid matrix.

The liquid aroma to consider was a complex mixture (M2). For simplification, a representative model mixture (M1) with three main molecules T, C, and F, with different volatilities T > C > F (M1 representing 30% M2), was first used:

$$T\ (\text{ester})\ 96.57\% w/w + C\ (\text{aldehyde})\ 0.93\% + F(\text{lactone})\ 2.5\%$$

The encapsulation of M1 was studied in a matrix made of maltodextrin DE12 and AG (ratio MD/AG = 3/2).

First, an aqueous emulsion (<5 μm, 40% TS) was spray-dried to obtain a dry emulsion (SD) (20% aroma, 80% MD/AG). Then, some trials were done to agglomerate the spray-dried particles in a fluid bed by spraying water on the surface of the fluidized particles (SDA). Finally, in the fluid bed, the agglomerates were coated by spraying the emulsion on them (SDAC). In the three cases, the total composition of the final powder was preserved.

At a pilot scale, the powder yield for the three studied processes was superior to 75%, with good physical properties for the powders.

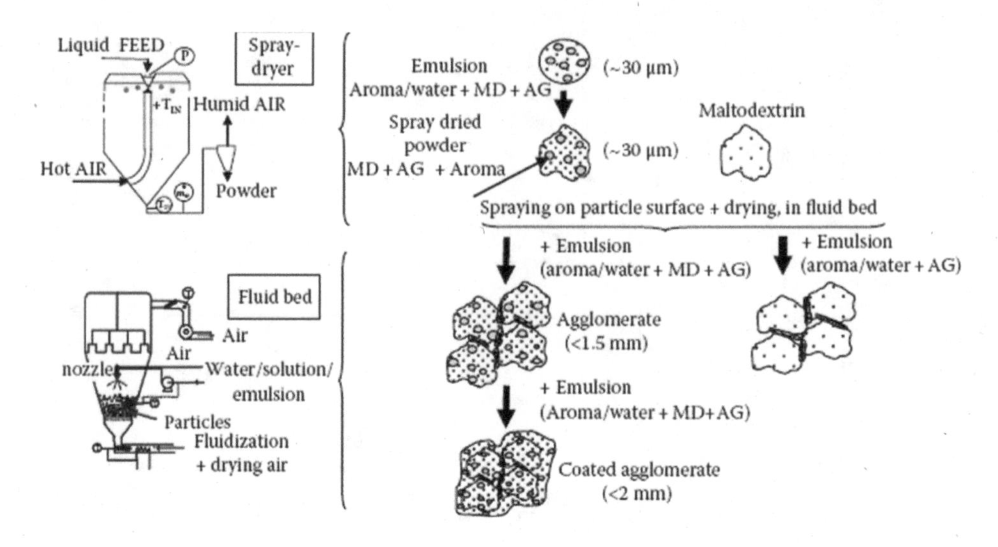

FIGURE 16.16 Aroma in chewing gum powder. AG, acacia gum; MD, maltodextrin.

The aroma concentration in dry matter was 15% in the spray-dried powder and agglomerates (compared with 20% in the initial emulsion). It was less than 9–15% for the coated agglomerates, showing some aroma loss.

Tests for aroma perception in a gum paste (0.6% w/w aroma) to compare with liquid aroma were conducted by the aroma producer. The perception was good for SD and SDA powders. For the coated agglomerates, there were changes in aroma perception. Probably due to the long coating process, it means that in this case, more aroma protection is needed in the coating, modifying the final global composition.

A spray-dried powder was prepared at a larger scale (×10) using the real aroma M2, and the results of sensorial tests were encouraging.

16.4.3 Delayed Release of Flavor Material

First, a base powder is prepared containing the flavor ingredient encapsulated in a matrix (water soluble or partially hydrophilic). The initial aqueous emulsion may be formed with gelatin, AG, a plasticizer (glycerol), and an emulsifier. Then, it is dried as a thin layer by drum drying. Second, the powder is ground to get a high-density powder (no holes), which will be further coated (in a fluid bed) with one or two layers of a water-insoluble material (i.e., polyvinylacetate, shellac, or zein). The main objective is to prevent the flavor ingredient from migrating/diffusing into the product (food or toothpaste). The release will occur at a given temperature by breaking, chewing, and brushing, delivering a high perception of flavors. Several different ways are described in the work by Merrit et al. (1985).

A double-encapsulated flavor powder (spray-dried) is prepared by a secondary fat coating process, with better resistance to moisture and oxygen than a single-encapsulated powder (Cho and Park, 2002).

16.4.4 Production of Soluble Beverage Powder

One example is given of encapsulating an aroma/flavor for a beverage (Liu and Rushmore, 1996).

An oil-in-water emulsion was prepared from coffee aroma incorporated into coffee/vegetable oil (5–20% by weight) and water-soluble coffee solids forming the aqueous continuous phase (50–75% TS). Individual drops of emulsion (nozzle/N_2, 0.4–1 mm; core of coffee oil) were sprayed onto soluble coffee powder (in a fluid bed and pan coater), with little modification of the water content of the powder (<4% by weight). The capsules were attached to the coffee powder surface (0.1–1% aroma by weight) and were able to release the aroma only by dissolution in the hot water cup. The film-forming agent may be another support such as maltodextrin, AG, carbohydrates, tea or cocoa solids, or vegetable derivatives. And other flavors may be used, for example, for instant soups.

16.5 Conclusion

For the aroma encapsulation, a wide variety of possible supports and techniques exists to prepare a stable product with controlled aroma composition, taking into account the different steps of production, storage, and final use.

The choice for an aroma encapsulation project will be decided according to the initial objectives, with a multidisciplinary cooperation between the aroma and support suppliers and the process and chemical engineers, according to the objectives of properties for the final product, the release conditions (Table 16.4). They have to exchange their complementary knowledge and competencies and find a compromise between the sciences, the choice, and the limits of both formulation and process to reach the objectives of both industry and consumers.

TABLE 16.4

Summary of Techniques for Aroma Encapsulation in Food Products with Some Examples of Use

Operation Initial → Final State	Characteristics (% = AC in Support)	Size (μm) and Use
Emulsion		
L	Oil/water	(0.1–1) Drinks, dressings before drying
Spray-drying		
L → S		(5–200)
Hot air	Carbohydrates, proteins, gums (10–50%)	Citrus fruits Cinnamon, butter aroma
Cold air	Fats with low melting point	Pastry
Coacervation		(5–5000)
Emulsion + polymerization (+ drying) L → L/S	(80–90%)	Citrus fruits, raspberry Vanilla, mint, coffee, onion
Inclusion		
L → L/S	In cyclodextrins (6–15%)	(10–100) Spices, onion, garlic Remove bitterness
Extrusion		
Dispersion in matrix + extrusion in medium (for surface washing) L/P → S	(8–20%)	(1–1000) Drinks Cakes, desserts
Coextrusion		
Two concentric streams (heart + wall) and hardening L → S	(90%)	(> 500)
Fluid bed coating		
Solid particles + coating S + L → S		(>100)
Dry coating (mix)		
Mixing sugar + AC + … S → S	Lipid, cellulose, proteins, starch	Sugar + vanilla
Cocrystallization		
Supersaturated sugar solution S + L → S	(10–25%)	(5–25) Drinks

AC, active component; L, liquid; P, paste; S, solid.

From the beginning of the project, the following general questions must be considered in the following order—with some possible comeback at any moment:

- What will be the functions that the encapsulated ingredients (aroma or oil) provide to the final product? What is the final physical state for use in liquid/solid/paste, and in what environment?
- What are the particle size, density, and stability requirements for the encapsulated ingredient (i.e., if mixed, mouthfeel factor)?
- What is the optimal concentration of the active material in the microcapsule/final product?
- By what mechanism will the aroma be released from the microcapsule?

From the answers to the previous questions, it is possible to define some possible materials that can be used as support related to the desired protection and final use and release, within the constraints of cost and legislation (security, etc.).

Considering all these answers, one or more processes will be proposed with the following questions for each:

- What processing conditions must the encapsulated ingredient resist before releasing its content? What is the best time to introduce the aroma into the process? What are the risks of losses of some aroma constituents, modifying the final aroma profile?
- What will be the resulting possible lifetime, with which packaging, for which storage conditions?

The final choice for the whole process will be made after research with pilot experiments, followed by possible scale-up to industrial trials (batch or continuous), with the necessary multiple analyses, according to the type of products to consider, for food, chemistry, and medicine. The questions of cost and security must be included throughout the chosen process. Then, a new tailored product will be born!

REFERENCES

Augustin M.A. and Hemar Y., 2009. Nano- and micro-structured assemblies for encapsulation of food ingredients. *Chemical Society Reviews*, 38, 902–912.

Baranauskiene R., Bylaite E., Zukauskaite J., and Venskutonis R.P., 2007. Flavor retention of peppermint essential oil spray-dried in modified starches during encapsulation and storage. *Journal of Agricultural and Food Chemistry*, 55, 3027–3036.

Baranauskiene R., Venkustonis P.R., Dewettinck K., and Verhé R., 2006. Properties of oregano, citronella and marjoram flavors encapsulated into milk protein-based matrices. *Food Research International*, 39, 413–425.

Beristain C.I., Vasquez A., Garcia H.S., and Vernon-Carter E.J., 1996. Encapsulation of orange peel oil by co-crystallization. *LWT*, 29, 645–647.

Bhandari B., D'Arcy B., and Young G., 2001. Flavour retention during high temperature short time extrusion cooking process: A review. *International Journal of Food Science and Technology*, 36, 453–461.

Bhandari B.R., D'Arcy B.R., and Le Thi Bich L., 1998. Lemon oil to β-cyclodextrin ratio effect on the inclusion efficiency of β-cyclodextrin and the retention of oil volatiles in the complex. *Journal of Agricultural and Food Chemistry*, 46, 1494–1499.

Bhandari B.R., D'Arcy B.R., and Padukka I., 1999. Encapsulation of lemon oil by paste method using beta-cyclodextrin: Encapsulation efficiency and profile of oil volatiles. *Journal of Agricultural and Food Chemistry*, 47, 5194–5197.

Bhandari B.R., Dumoulin E.D., Richard H.M.J., Noleau I., and Lebert A., 1992. Flavor encapsulation by spray drying: Application to citral and linatyl acetate. *Journal of Food Science*, 57(1), 217–221.

Boh B., Staresinic M., and Sumiga B., October 5–8, 2011. Synthesis and applications of scented microcapsules in textile products. In: *XIX International Conference on Bioencapsulation*, Amboise, France.

Bouquerand P.E., Dardelle G., and Erni P., 2012. Chapter 18: An industry perspective on the advantages and disadvantages of different flavour delivery systems. In: *Encapsulation Technologies and Delivery Systems for Food Ingredients and Nutraceuticals*, Eds. N. Garti and D.J. Clements. Woodhead Publishing, Oxford, pp. 453–487.

Buffo R., Probst K., Zehentbauer G., Luo Z., and Reineccius G.A., 2002. Effects of agglomeration on the properties of spray-dried encapsulated flavours. *Flavour and Fragrance Journal*, 17, 292–299.

Buffo R. and Reineccius G.A., 2001. Comparison among assorted drying processes for the encapsulation of flavors. *Perfumer and Flavorist*, 26, 58–67.

Chen A.C., Veiga M.F., and Rizzuto A.B., November 1988. Co-crystallization: An encapsulation process. *Food Technology*, 42(11), 87–90.

Cho Y.H. and Park J., 2002. Characteristics of double-encapsulated Fflavor powder prepared by secondary fat coating process. *Food Chemistry and Toxicology*, 67, 968–972.

Cho Y.H., Shim H.K., and Park J., 2003. Encapsulation of fish oil by an enzymatic gelation process using transglutaminase cross-linked proteins. *Journal of Food Science*, 68(9), 2717–2723.

Clark J. and Shen C., 2004. Fast flavor release coating for confectionery. Patent WO/2004/077956.

Coumans W.J., Kerkhof P.J., and Bruin S., 1994. Theoretical and practical aspects of aroma retention in spray drying and freeze drying. *Drying Technology*, 12(1&2), 99–149.

De Kruif C.G., Weinbreck F., and De Vries R., 2004. Complex coacervation of proteins and anionic polysaccharides. *Current Opinion in Colloid and Interface Science*, 9, 340–349.

De Roos K.B., 2003. Effect of texture and microstructure on flavour retention and release. *International Dairy Journal*, 13, 593–605.

De Vos P., Faas M.M., Spasojevic M., and Sikkema J., 2010. Encapsulation for preservation of functionality and targeted delivery of bioactive food components. *International Dairy Journal*, 20, 292–302.

Drusch S., Regier M., and Bruhn M., 2012. Chapter 7: Recent advances in the microencapsulation of oils high in polyunsaturated fatty acids. In: *Novel Technologies in Food Science*, Eds. A. McElhatton and PJ. do Amaral Sobral. Springer, New York, pp. 159–181.

Dumoulin E. and Bimbenet J.J., 1998. Chapter 10: Spray drying and quality changes. In: *The Properties of Water in Foods. ISOPOW 6*, Ed. D.S. Ried. Blackie Academic and Professional, London, pp. 209–232.

Dziezak J.D., April 1988. Microencapsulation and encapsulated ingredients. *Food Technology*, 42(4), 136–151.

Edris A. and Bergnstähl B., 2001. Encapsulation of orange oil in a spray dried double emulsion. *Nahrung/ Food*, 45(2), 133–137.

Finney J., Buffo R., and Reineccius G.A., 2002. Effects of type atomization and processing temperatures on the physical properties and stability of spray-dried flavors. *Journal of Food Science: Food Engineering and Physical Properties*, 67, 1108–1114.

Fuchs M., Turchiuli C., Bohin M., Cuvelier M.E., Ordonnaud C., Peyrat-Maillard M.N., and Dumoulin E., 2006. Encapsulation of oil in powder using spray drying and fluidised bed agglomeration. *Journal of Food Engineering*, 75, 27–35.

Gianfrancesco A., Turchiuli C., Flick D., and Dumoulin E., 2010. CFD modeling and simulation of maltodextrin solutions spray drying to control stickiness. *Food and Bioprocess Technology*, 3, 946–955.

Gibbs B.F, Kermasha S., Alli I., and Mulligan C.N., 1999. Encapsulation in the food industry: A review. *International Journal of Food Sciences and Nutrition*, 50, 213–224.

Gouin S., 2004. Microencapsulation: Industrial appraisal of existing technologies and trends. *Trends in Food Science and Technology*, 15, 330–347.

Guignon B., Duquenoy A., and Dumoulin E., 2002. Fluid bed encapsulation of particles and practice. *Drying Technology*, 20(2), 419–447.

Guignon B., Regalado E., Duquenoy A., and Dumoulin E. 2003. Helping to choose operating parameters for a coating fluid bed process. *Powder Technology*, 130, 193–198.

Jafari S.M., 2009. Encapsulation of nano-emulsions by spray drying. PhD, Pub LAP Lambert, Saarbrücken, Germany.

Jafari S.M., Assadpoor E., He Y., and Bhandari B., 2008. Encapsulation efficiency of food flavours and oils during spray drying. *Drying Technology*, 26(7), 816–835.

Jafari S.M., He Y., and Bhandari B., 2007a. Encapsulation of nanoparticles of d-limonene by spray drying: Role of emulsifiers and emulsifying techniques. *Drying Technology*, 25, 1069–1079.

Jafari S.M., He Y., and Bhandari B., 2007b. Role of particle size on the encapsulation efficiency of oils during spray drying. *Drying Technology*, 25, 1081–1089.

Jayasundera M., Adhikari B., Aldred P., and Ghandi A., 2009. Surface modification of spray dried food and emulsion powders with surface-active proteins: A review. *Journal of Food Engineering*, 93, 266–277.

Karbowiak T., Debeaufort F., and Voilley A., Avril–Mai 9–17, 2007. Les emballages comestibles: Nature, fonctionnalité et utilisations. *Industries Alimentaires Agricoles*, 124(4/5), 9–17.

Kerkhof PJ.A.M. and Thijssen H.A.C., 1974. Retention of aroma components in extractive drying of aqueous carbohydrate solutions. *Journal of Food Technology*, 9, 415–423.

Kim S.S., Han Y.J., Hwang T.J., Roh H.J., Hahm T.S., Chung M.S., and Shin S.G., 2001. Microencapsulation of ascorbic acid in sucrose and lactose by cocrystallization. *Food Science and Biotechnology*, 10(2), 101–107.

King C.J., 1995. Spray drying: Retention of volatile compounds revisited. *Drying Technology*, 13(5–7), 1221–1240.

King C.J., Kieckbusch T.G., and Greenwald C.G., 1984. Food-quality factors in spray drying. In: *Advances in Drying*, Ed. A.S Mujumdar. Hemisphere Publisher Corporation, New York, Vol. 3, pp. 70–120.

Krishnan S., Kshirsagar A.C., and Singhal R.S., 2005. The use of gum arabic and modified starch in the microencapsulation of a food flavoring agent. *Carbohydrate Polymers*, 62, 309–315.

Lakkis J.M., 2007. *Encapsulation and Controlled Release Technologies in Food Systems*, Ed. J.M. Lakkis. Blackwell Publisher, Ames, IA, 240p.

Leclercq S., Harlander K.R., and Reineccius G., 2009. Formation and characterization of microcapsules by complex coacervation with liquid or solid aroma cores. *Flavour Fragrance Journal*, 24, 17–24.

Liu R.T. and Rushmore D.F., 1996. Process for making encapsulated sensory agents. US Patent 5,580,593.

Liu X.D., Atarashi T., Furuta T., Yoshii H., Aishima S., Ohkawara M., and Linko P., 2001. Microencapsulation of emulsified hydrophobic flavors by spray drying. *Drying Technology*, 19(7), 1361–1374.

Madene A., Jacquot M., Scher J., and Desobry S., 2006. Flavour encapsulation and controlled release: A review. *International Journal of Food Science and Technology*, 41, 1–21.

Marion J.P. and Audrin A., October 10–24, 1988. L'encapsulation d'arômes en images. *RIA*, 411, 41–46.

Masters K., 1985. *Spray Drying*, 4th edn., Longman Scientific and Technical, John Wiley & Sons, New York, 696p.

McNamee B.F., O'Riordan E.D., and O'Sulivan M., 2001. Effect of partial replacement of gum arabic with carbohydrates on its microencapsulation properties. *Journal of Agriculture and Food Chemistry*, 49, 3385–3388.

Merrit C.G., Wingerd W.H., and Keller D.J., 1985. Encapsulated flavorant material, method for its preparation, and food and other composition incorporating same. US Patent 4,515,769.

Onwulata C. (Ed.), 2005. *Encapsulated and Powdered Foods*. CRC, Taylor & Francis, Boca Raton, FL, 514p.

Parris N., Cooke P.H., and Hicks K.B., 2005. Encapsulation of essential oils in zein nanospherical particles. *Journal of Agriculture and Food Chemistry*, 53, 4788–4792.

Penbunditkul P., Yoshii H., Ruktanonchai U., Charinpanitkul T., and Soottitantawat A., 2012. The loss of OSA- modified starch emulsifier property during the high-pressure homogeniser and encapsulation of multi-flavour bergamot oil by spray drying. *International Journal of Food Science and Technology*, 47, 2325–2333.

Perrut M., 2003. Method for encapsulating fine solid particles in the form of microcapsules. Brevet FR2811913, US2003157183.

Pisecky J., 1997. *Handbook of Milk Powder Manufacture*. Niro A/S/, Copenhagen, 261p.

Qi Z.H. and Xu A., 1999. Starch base ingredients for flavour encapsulation. *Cereal Foods World*, 44(7), 460–465.

Re M.I., 1998. Mcroencapsulation by spray drying. *Drying Technology*, 16(6), 1195–1236.

Reineccius G.A., 2004. The spray drying of food flavors. *Drying Technology*, 22(6), 1289–1324.

Richard H. and Multon J.J., coord. 1992. *Les arômes alimentaires*. Tech. & Doc. Lavoisier, Paris, 438p.

Risch S.J. and Reineccius G.A. (Eds.), 1988. *Flavor Encapsulation*. ACS Symposium Series 370, Washington, DC, 202p.

Risch S.J. and Reineccius G.A. (Eds.), 1995. *Encapsulation and Controlled Release of Food Ingredients*. ACS Symposium Series 590, American Chemical Society, Washington, DC, 214p.

Rosenberg M., Kopelman I.J., and Talmon Y., 1990. Factors affecting retention in spray drying microencapsulation of volatile materials. *Journal of Agricultural and Food Chemistry*, 38, 1288–1294.

Rulkens W.H. and Thijssen H.A.C., 1972. The retention of organic volatiles in spray drying aqueous carbohydrate solutions. *Journal of Food Technology*, 7, 95–105.

Schlameus W., 1995. Centrifugal extrusion encapsulation. In: *Encapsulation and Controlled Release of Food Ingredients*, Eds. S.J. Risch and G.A. Reineccius. ACS Symposium Series 590, American Chemical Soc., Washington DC, pp. 96–103.

Schubert H., Engel R., and Kempa L., 2009. Chapter 1: Principles of structured food emulsion: Novel formulations and trends. In: *Global Issues and Food Science and Technology*, Eds. G. Barbosa Canovas et al. Elsevier Inc., Amsterdam, pp. 3–20.

Senoussi A., Bhandari B., Dumoulin E., and Berk Z., 1994. Flavour retention in different methods of spray drying. In: *Developments in Food Engineering*, Eds. T. Yano, R. Matsuno, and K. Nakamura. Blackie Academic and Professional, London, pp. 433–435.

Senoussi A., Dumoulin E., and Berk Z. 1995. Retention of diacetyl in milk during spray-drying and storage. *Journal of Food Science*, 60(5), 894–897, 905.

Shahidi F. and Han X., 1993. Encapsulation of food ingredients. *Critical Reviews in Food Science and Nutrition*, 33(6), 501–547.

Sheu T.Y. and Rosenberg M., 1995. Microencapsulation by spray drying ethyl caprylate in whey protein and carbohydrate wall systems. *Journal of Food Science*, 60(1), 98–103.

Soottitantawat A., Bigeard H., Yoshii H., Furuta T., Ohgawara M., and Linko P., 2005a. Influence of emulsion and powder size on the stability of encapsulated D-limonene by spray drying. *Innovative Food Science and Emerging Technologies*, 6, 107–114.

Soottitantawat A., Takayama K., Okamura K., Muranaka D., Yoshii H., and Furuta T., 2005b. Microencapsulation of L-menthol by spray drying and its release characteristics. *Innovative Food Science and Emerging Technologies*, 6, 163–170.

Soottitantawat A., Yoshii H., Furuta T., Ohkawara M., and Linko P., 2003. Microencapsulation by spray drying: Influence of emulsion size on the retention of volatile compounds. *Journal of Food Science*, 68, 2256–2262.

Turchiuli C. and Dumoulin E., 2013a. Chapter 14: Aroma encapsulation in powder by spray drying, and fluid bed agglomeration and coating. In: *Advances in Food Process Engineering Research and Applications*, Eds. S. Yanniotis, P. Taoukis, N.G. Stoforos, and T. Karathanos. Springer, New York, pp. 255–265.

Turchiuli C., Fuchs M., Bohin M., Cuvelier M.E., Ordonnaud C., Peyrat-Maillard M.N., and Dumoulin E., 2005. Oil encapsulation by spray drying and fluidised bed agglomeration. *Innovative Food Science and Emerging Technologies*, 6, 29–35.

Turchiuli C., Jimenez Munguia M., Hernandez Sanchez M., Cortes Ferre H., and Dumoulin E. 2014. Use of different supports for oil encapsulation in powder by spray drying. *Powder Technology*, 255, 103–108.

Turchiuli C., Lemarie N., Cuvelier M.E. and Dumoulin E., 2013b. Production of fine emulsions at pilot scale for oil compounds encapsulation. *Journal of Food Engineering*, 115, 452–458.

Van Lengerich B.H., 2003. Embedding and encapsulation of controlled release particles. Patent EP 1342548.

Vandamme T., Poncelet D., and Subra-Paternault P., 2007. *Microencapsulation. Des sciences aux technologies.* Ed. Tec & Doc, Lavoisier, Paris, 355p.

Vega C. and Roos Y.H., 2006. Invited review: Spray-dried dairy and dairy-like emulsions. Compositional considerations. *Journal of Dairy Science*, 89, 383–401.

Whelehan M. and Marison I.W. October 5–8, 2011. Microencapsulation by dripping and jet break-up. In: *XIX International Conference on Bioencapsulation*, Amboise, France.

Yeo Y., Bellas E., Firestone W., Langer R., and Kohane D.S., 2005. Complex coacervates for thermally sensitive controlled release of flavor compounds. *Journal of Agriculture and Food Chemistry*, 53, 7518–7525.

Yuliani S., Torley P.J., D'Arcy B., Nicholson T., and Bhandari B., 2006. Extrusion of mixtures of starch and d-limonene encapsulated with β-cyclodextrine. *Food Research International*, 39, 318–331.

Zilberboim R., Kopelman I.J., and Talmon Y., 1986. Microencapsulation by a dehydrating liquid: Retention of paprika oleoresin and aromatic esters. *Journal of Food Science*, 51(5), 1301–1310.

17

Encapsulation of Flavors, Nutraceuticals, and Antibacterials

Stéphane Desobry and Frédéric Debeaufort

CONTENTS

17.1 Introduction

Edible coatings provide a physical barrier against mass transport from the environment to food and from food to the environment, as shown in the previous chapters, and these barrier properties are important for passive protection of food. Consumers are asking today for better food safety and for higher nutritional and flavor properties. In recent years, active packaging has been developed to extend food shelf life by increasing coatings' positive effect. For example, more activity can be provided to edible coatings by adding active compounds, such as flavors, antibacterials, or nutraceuticals. Active compounds can be incorporated directly into the edible polymer matrix or can be encapsulated to better protect their activity and properties. The Eastern countries are the most advanced in this particular domain, North America follows, and Europe is now developing more and more active material solutions.

Active compound stability in different foods has been of increasing interest due to its relationship with food quality and acceptability. Manufacturing and storage processes, packaging materials, and ingredients in foods often cause modifications in overall flavor and nutritional value by reducing an active compound's activity or intensity or by producing newly formed components. Many factors, such as physicochemical properties, concentration, and interaction of active molecules with food components, affect the resulting quality. As a result, it is beneficial to encapsulate active ingredients prior to use.

Encapsulation is the technique by which one material or mixture of materials is coated with or entrapped within another material or system. The coated material is called the *active* or *core material*, and the coating material is called the *shell*, *wall material*, *carrier*, *matrix*, or *encapsulant*. The development of microencapsulation products started in the 1950s with research into pressure-sensitive coatings for the manufacture of carbonless copying paper (Green and Scheicher, 1955). Encapsulation technology is now well developed and accepted within the pharmaceutical, chemical, cosmetic, food, and printing industries. In food products, fats and oils, aroma compounds and oleoresins, vitamins, minerals, colorants, and enzymes have been encapsulated, while in coating films, oils, aroma compounds, antimicrobials, and enzymes have been encapsulated.

Encapsulation processes are receiving more and more interest in food applications, and recently, an evolution from micro- to nanoencapsulation has been observed (Shimoni, 2009). The main interest in reducing the size of the capsules was to limit the impact of capsule addition on the physical properties of food. Laboratories and industrial development teams have shown the high efficiency of this size reduction (Imran et al., 2010a, b). This evolution is nevertheless limited by the suspicion of possible toxicity of some nanoparticles for the human body after inhalation, skin permeation, and ingestion (Bouwmeester et al., 2009). Nanoparticle absorption through cell membranes may oxidize cellular compounds or cause inflammation to develop. This risk exists in cases of both free nanocapsules and nanocapsules incorporated into edible films or coatings. A lot of research is required now to better understand the real effect of nanocapsules incorporated into foods. In particular, the effects of solid and "soft" nanoparticles should be compared more precisely, because some studies have reported a negligible effect of soft and organic compounds easily degradable by the cell compared with solid mineral particles.

These studies will make possible the development of clear legislation in the countries that use nanoparticles. While waiting for this, the industry is not very confident about this particular technology, and only a few applications of nanoparticle incorporation in an edible matrix exist.

In this chapter, after describing the encapsulation matrix and techniques, two parts will be presented to cover the main aspects of encapsulation in edible films and coating. The first concerns volatile compound encapsulation, and the second concerns non-volatile molecules such as nutraceuticals and antimicrobials.

17.2 Capsule Matrices

The encapsulation of biomolecules can be achieved using two main methods. The first consists of making capsules in which the compound is included as a core or entrapped in a polymeric matrix (Figure 17.1 a, b).

These encapsulation methods are used for a wide range of applications. The protective efficiency of the two systems is very different, with much more efficient barrier action in the core system compared with easier release and lower cost in the matrix system.

The second method consists of developing films or coatings in which the biomolecules are directly included and entrapped just as in a matrix but on a larger scale (Figure 17.1c). In this matrix film system, the film or coating contains aroma, nutraceuticals, and active compounds. The release is completely controlled by the molecular diffusivity.

To increase the potentiality of the films, more and more applications combine the two previous methods (Figure 17.1a through c) to obtain film/capsule systems as presented in Figure 17.1d. In this structured film or coating, the amount of volatile compound can be reduced thanks to the optimized preservation, low release from the capsules, and low transfer through the film. These systems allow negligible oxidation of the active compounds due to reduced oxygen diffusivity.

The matrix material has to be chosen to face the high number of performance requirements placed on microcapsules when a limited number of encapsulating materials and methods exist. Since the late 1980s, they have been prepared from a large range of materials including proteins, carbohydrates, lipids, and gums (Brazel, 1999). Each group of materials has certain advantages and disadvantages, and these materials are combined to produce high-quality systems. The functional characteristics of an effective wall material used to encapsulate active components must

- Be inert regarding the encapsulated molecules
- Allow complete elimination of solvent used for matrix or capsule formation
- Provide maximum protection of the active ingredient against external factors
- Produce a stable system before the solidification of capsules within a film
- Release the active compounds at the time and the place desired

The characteristics of major wall materials used for encapsulation are reported in the following sections.

17.2.1 Carbohydrates

Carbohydrates are used extensively in spray-dried encapsulations of food ingredients as the encapsulating support—that is, the wall material or carrier. The ability of carbohydrates to form a protective matrix,

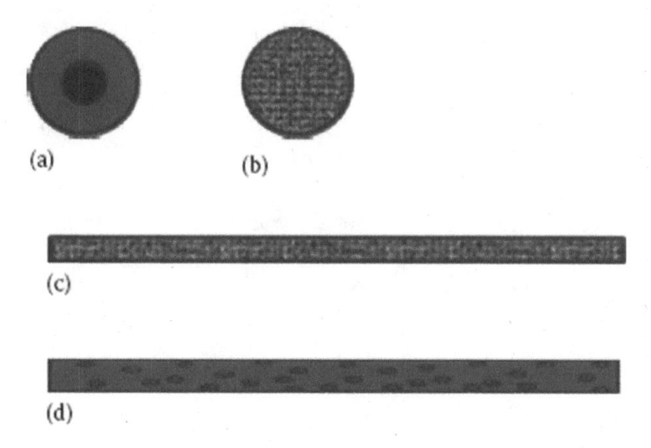

FIGURE 17.1 (a) Core, (b) matrix in nano-/microcapsule system, (c) matrix film, and (d) film or coating with nano-/microcapsules entrapped.

complemented by their diversity and low cost, makes them the first choice for encapsulation. The main limit for their application is their high water sensitivity and fast hydration when in contact with water.

The preferred carbohydrate matrices are starch and starch-based ingredients (modified starches, maltodextrins, and β-cyclodextrins). Maltodextrins possess matrix-forming properties important in a wall system. In selecting the wall materials for encapsulation, maltodextrin is a good compromise between cost and effectiveness, as it is bland in flavor, has low viscosity at high solid ratios, and is available in different average molecular weights. Gums and thickeners are generally bland or tasteless, but they can have a pronounced effect on the taste and flavor of foods. In general, hydrocolloids decrease sweetness, with much of the effect being attributed to viscosity and hindered diffusion. Gum arabic (GA) is also an excellent encapsulating material. Its solubility, low viscosity, emulsification characteristics, and good retention of active compounds make it very versatile for most encapsulation methods, such as spray-drying. Its application is nevertheless limited due to its high cost compared with other carbohydrates, and its availability and cost are subject to high variation.

17.2.2 Proteins

Although food hydrocolloids are widely used as microencapsulants, food proteins (i.e., sodium caseinate, whey protein isolates (WPIs), and soy protein isolates) have a good ability to produce efficient encapsulation matrices. WPIs provide a good barrier against oxidation and an effective basis for microencapsulation by spray-drying. In combination with maltodextrins and corn syrup solids, whey proteins have been reported to be one of the most effective encapsulation materials during spray-drying. In such a system, whey proteins served as emulsifying and film-forming agents, while the carbohydrates acted as a matrix-forming material.

Protein-based materials such as polypeptone, soy protein, milk proteins, and gelatin derivatives are able to form stable emulsions with hydrophobic compounds. However, their solubility in cold water, the potential to react with carbonyls, and their high cost limit potential applications.

17.3 Encapsulation Methods

Encapsulation is accomplished by a large variety of methods (Madene et al., 2006). The two major industrial processes are spray-drying and extrusion; however, freeze-drying, coacervation, and adsorption techniques are also widely used in the case of heat-sensitive compounds.

17.3.1 Spray-Drying

Spray-drying is the commercial process most widely used in large-scale production of encapsulated molecules. The merits of the process have ensured its dominance, including availability of equipment, low process cost, wide choice of carrier solids, good retention of volatiles, good stability of the finished product, and large-scale production in continuous mode. The production of encapsulated powders by spray-drying involves the formation of a stable emulsion in which the wall material acts as a stabilizer. When core materials of limited water solubility are encapsulated by spray-drying, the resulting capsules are of matrix-type structure. As such, the core has been shown to be organized in small wall material–coated droplets embedded in the wall matrix.

The main disadvantage of spray-drying is that some low–boiling point aromatics or heat-sensitive molecules can be lost, and some core material may also be on the surface of the capsule, where it is subject to oxidation. Another problem is that the product is a very fine powder, typically in the range of 10–100 μm in diameter, which needs further processing to make agglomerates that are more readily soluble for liquid consumption.

17.3.2 Freeze-Drying

The freeze-drying technique is one of the most useful processes for drying thermosensitive substances. This technique is based on low-temperature dehydration under vacuum, avoiding water phase transition and oxidation. The dried mixture has a porous structure and must be ground, resulting in heterogeneous particles. This drying technique is less attractive than others, because its cost is up to 50 times higher than spray-drying. Freeze-drying, nevertheless, gives excellent preservation results and is adapted to high-value encapsulated molecules.

17.3.3 Spray-Cooling and Spray-Chilling

The loss of heat-sensitive material during the spray-drying process has led to a number of alternative methods for dehydration of sprayed microcapsules. Spray-cooling and spray-chilling are similar to spray-drying; the core material is dispersed in a liquefied coating or wall material and atomized at low temperature.

17.3.4 Extrusion

Encapsulation via extrusion is slightly more expensive compared with spray-drying. The principal advantage of the extrusion method is the high density of the produced matrix, allowing great stability of encapsulated materials against oxidation. In extrusion, there are two steps. First, the feed is extruded without cooking. Second, the raw ingredients are cooked by the combined action of heat, mechanical shearing, and pressure.

17.3.5 Coacervation

Coacervation consists in separating from the solution the colloidal particles, which agglomerate into a separate liquid phase called a *coacervate*. Coacervation can be simple or complex. Simple coacervation involves only one type of polymer with the addition of strongly hydrophilic agents to the colloidal solution. For complex coacervation, two or more types of polymer are used. Active molecules are entrapped in the matrix during coacervate formation by adjusting precisely the ratio between the matrix polymer and the entrapped molecule (Figure 17.2).

17.4 Controlled Release

Controlled release is a method by which one or more active agents or ingredients are made available at a desired site and time and at a specific rate. For matrix systems encapsulating an active compound, release

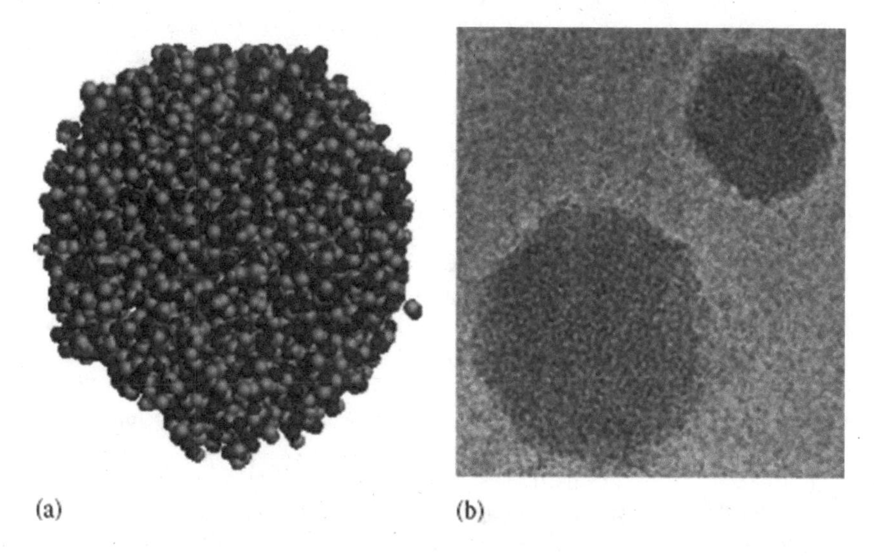

FIGURE 17.2 Gum arabic/β-lactoglobulin coacervates: (a) molecular model and (b) scanning electron micrograph.

depends on diffusion process, particle type and geometry, and controlled degradation or dissolution of matrix material.

Compared with molecules added to food, the advantages of controlled release are that the active ingredients are released at controlled rates over a prolonged period of time; loss of ingredients during processing, cooking, and even digestive molecular destruction can be strongly reduced; and molecule bioavailability can be increased for nanocapsules.

17.4.1 Release Controlled by Diffusion

Diffusion is the main regulation phenomenon for controlled release in a stable matrix. The diffusion process is well known, thanks to thousands of experimental studies related to mass transfer through films, coatings, and capsules. The macromolecular network density of the matrix and the molecular weight of the encapsulated compounds are the key factors for release controlled by diffusion. The chemical potential difference at each side of the matrix is the major driving force influencing diffusion. The principal steps in the release of a compound from a matrix system are diffusion of the active agent to matrix surface; component partition between matrix and food; and finally, transport away from the matrix surface.

17.4.2 Release Controlled by Matrix Degradation

The release of an active compound from a matrix-type delivery system may be controlled by a combination of diffusion and erosion. Heterogeneous erosion occurs when degradation is confined to a thin layer at the delivery system surface, whereas homogeneous erosion is a result of degradation occurring at a uniform rate throughout the polymer matrix.

17.4.3 Release Controlled by Swelling

In swelling-controlled systems, the molecule dispersed in a polymeric matrix is unable to diffuse to any significant extent within the matrix. When the matrix polymer is placed in a thermodynamically compatible medium, the polymer swells due to absorption of fluid from the medium. In the swollen matrix, the encapsulated molecule is able to diffuse due to larger intermolecular spaces.

17.4.4 Release Controlled by Melting

This mechanism of release involves the melting of the capsule wall to release the active material. It is readily accomplished in the food industry because there are numerous materials of low melting point that are approved for food use (lipids, modified lipids, or waxes). In such applications, coated particles are stored at temperatures well below the melting point of the coating and then heated above this temperature during preparation, cooking, and consumption.

17.5 Edible Films for Volatile Molecules (Flavors, Essential Oils)

Even if most research deals with interactions between packaging and volatile compounds in the case of beverages, similar problems occur for viscous liquids, gels, and solid foods. One possible way suggested to lower interactions between aroma compounds and plastics is to retain flavor molecules inside the food product by using an additional barrier. This could be an edible film or coating or a thin layer that has a high selectivity against aroma transfer and can be eaten along with the protected food (Miller and Krochta, 1997; Debeaufort et al., 2002). This entails some kind of macroencapsulation of the product. The main application is using only the barrier properties of the coating to retain aroma within the food; however, the film or coating could be used as a carrier or support for flavors at the surface of the product. So, edible films serve many purposes, including the production of a dry free-flowing flavor (most flavors are liquids), protection of the flavoring from interaction with the food or deleterious reactions such as oxidation, confinement during storage, and finally, controlled release (Reineccius, 2009). The degree to which the edible film meets these requirements depends on the process used to form the film around the flavoring and the film composition.

These volatile compounds will be released rapidly when the consumer tastes the product. Several products are already on the market using this technology for flavoring. For instance, there is a roasted peanut with a curry-flavored coating that is instantaneously dissolved in the mouth and gives immediately the perception of the Indian spice. Another example designed for children is a multi-sugar-coated sweet in which each layer of the coating contains different tastes and flavors separated by arabic gum or hydrocolloid layers to prevent the migration of aroma compounds from one layer to another. For this application, the diffusivity of volatile compounds should be very low and they should have a high affinity for the coating, which should be highly soluble in the mouth. As previously outlined, edible films and coatings can deliver and maintain desirable concentrations of color, flavor, spiciness, sweetness, saltiness, and so forth. Several commercial films, especially Japanese pullulan-based films, are available in a variety of colors, with spices and seasonings included (Guilbert and Gontard, 2005). For instance, (Laohakunjit and Kerdchoechuen 2007) coated milled rice with sorbitol–rice starch coatings containing 25% natural pandan leaf extract (*Pandanus amaryllifolius* Roxb.). The rice starch coating containing pandan extract allowed production of jasmine-flavored rice after cooking. Recently, Origami Foods (Pleasantown, California) commercialized vegetable and fruit edible films as alternatives to the seaweed sheets (nori) traditionally used for sushi and other Asian cuisine. They are made from broccoli, tomato, carrot, mango, apple, peach, pear, as well as a variety of other fruit and vegetable products, and they can contain spices, seasonings, colorants, flavors, vitamins, and other beneficial plant-derived compounds (Martin-Belloso et al., 2009).

17.5.1 Flavor Compounds and Matrices Involved in Edible Films and Coatings

Edible packaging materials are mainly composed of a film-forming substance that provides cohesiveness to the matrix (a continuous network) or a barrier substance that lowers impermeability. These are usually polysaccharides, proteins, or lipids used alone or as mixtures. Only a few substances have simultaneously good film-forming and barrier properties, such as, for example, wheat gluten–based films, which have satisfactory mechanical resistance and very low oxygen permeability. Moreover, the permeability of D-limonene (one of the main compounds of citrus flavor) in whey protein–based films is lower than in ethylene vinyl alcohol (EVOH) or polyvinylidene chloride (PVDC) films (Fayoux et al., 1997a,b; Miller et al., 1998). Because edible packaging containing volatile compounds is consumed along with the food, its edibility and safety are essential. For this reason, flavor compounds can be used to obtain active edible packaging, because they can act as antimicrobials, antioxidants, and flavoring agents. In particular, essential oils can be added to edible films and coatings to modify flavor, aroma, and odor as well as to introduce antimicrobial properties (Sánchez-González et al., 2010a).

Very few published data exist on the incorporation of plant essential oils into edible films and coatings. Essential oils are regarded as alternatives to chemical preservatives, and their use in foods meets the demands of consumers for minimally processed natural products, as reviewed by (Burt 2004). Vanillin has been used recently as a bacteriostatic rather than a bactericidal agent in fresh-cut apples (Rupasinghe et al., 2006). Essential oils have also been evaluated for their ability to protect food against pathogenic bacteria in contaminated apple juice (Friedman et al., 2004; Raybaudi-Massilia et al., 2006) and other foods, and they are used as flavoring agents in baked goods, sweets, ice cream, beverages, and chewing gum (Fenaroli, 1995; Burt, 2004). Sánchez-González et al. (2009, 2010a,b) introduced tea tree and bergamot essential oils in chitosan or hydroxypropylmethylcellulose (HPMC) edible films at a range of 0% to 3% (w/w) in the film-forming suspension for antimicrobial properties. They displayed antimicrobial efficiency at the higher concentration of bergamot on *Penicillium italicum* and at the lowest concentration of tea tree oil on *Listeria monocytogenes*. However, no information was given on the concentration in the film after drying or on the sensory impact.

(McHugh et al. 1996) developed the first edible films made from fruit purees, which were shown to be a promising tool for improving the quality and extending the shelf life of minimally processed fruit. (Rojas-Grau et al. 2006) recently investigated the effect of plant essential oils on the antimicrobial and physical properties of apple puree edible films. Alginate–apple puree films containing plant essential oils were further explored as edible coatings by (Rojas-Grau et al. 2007) with the aim of studying the effect of lemongrass, oregano oil, and vanillin on native psychrophilic aerobic bacteria, yeasts, molds, and inoculated *Listeria innocua* in fresh-cut "Fuji" apples. Coatings with essential oils seemed to effectively inhibit the growth of *L. innocua* inoculated on apple pieces as well as psychrophilic aerobic bacteria, yeasts, and molds. In some cases, essential oils such as thymol and carvacrol were added to a bio-based coating as antimicrobial agents and not as flavoring compounds (Ben Arfa et al., 2007; Del Nobile et al., 2008). (Ponce et al. 2008) used oleoresins containing both volatile and non-volatile compounds extracted from oregano, rosemary, olive, capsicum, garlic, onion, and cranberries in edible films based on sodium caseinate and carboxymethylcellulose and chitosan. These authors revealed that the use of chitosan enriched with rosemary and olive did not introduce deleterious effects on the sensorial acceptability of squash. Chitosan enriched with rosemary and olive improved the antioxidant protection of the minimally processed squash, offering a great advantage in the prevention of browning reactions that typically result in quality loss in fruits and vegetables. These coatings provide both antioxidant and antimicrobial properties of coatings at 1% oleoresin content without too much sensory disturbance.

Gelatin- and chitosan-based edible films with clove essential oil incorporated were tested for antimicrobial activity against six selected microorganisms: *Pseudomonas fluorescens, Shewanella putrefaciens, Photobacterium phosphoreum, L. innocua, Escherichia coli*, and *Lactobacillus acidophilus*. The clove-containing film inhibited all these microorganisms irrespective of the film matrix (Gómez-Estaca et al., 2010). The effect on the microorganisms during this period was in accordance with biochemical indexes of quality, indicating the viability of these films for fish preservation.

These authors also tested on 18 bacterial strains other essential oils: fennel (*Foeniculum vulgare* Miller), cypress (*Cupressus sempervirens* L.), lavender (*Lavandula angustifolia*), thyme (*Thymus vulgaris* L.), herb-of-the-cross (*Verbena officinalis* L.), pine (*Pinus sylvestris*), and rosemary (*Rosmarinus officinalis*). Antioxidant properties as well as light barrier properties of gelatin-based edible films containing oregano or rosemary aqueous extracts have been assessed by (Gómez-Estaca et al. 2009). The interaction between essential oil polyphenols and the protein was found to be more extensive when tuna-skin gelatin was employed. However, this did not clearly affect the antioxidant properties of the films, although it could affect diffusion of phenolic compounds in the essential oil from film to food. The light barrier properties were improved by the addition of oregano or rosemary extracts, irrespective of the type of gelatin employed. The shelf life of cold-smoked sardine (*Sardina pilchardus*) was improved also by gelatin-based films using, singly or in combination, high pressure (300 MPa/20 °C/15 min) and films enriched by adding an extract of oregano (*Origanum vulgare*) or rosemary (*Rosmarinus officinalis*) or by adding chitosan (Gomez-Estaca et al., 2007). Gelatin seems to be a good way of encapsulating both antimicrobial and antioxidant volatile essential oils. Seaweed extracts, such as alginates and carrageenans, have been extensively studied by Hambleton et al. (2008, 2009a,b, 2010, 2012) and (Fabra et al. 2008, 2009). These authors revealed that both carrageenans and alginates are able to be used as films or coatings with encapsulated volatile compounds such as aroma or essential oils. The addition of lipids such as acetylated monoglycerides, beeswax, and oleic acid, used alone or as mixture, allows moisture transfer to be reduced but affects the ability of the coatings to retain and release volatile compounds. Lipids could have positive or negative influences depending on the polarity of the volatile compounds. Zein-based monolayer and multilayer films were loaded with spelt bran and thymol (35% w/w) to obtain edible polymeric materials. Various composite systems were developed to control thymol release (Mastromatteo et al., 2009). (Madene et al. 2006) described the process for encapsulation of sensitive volatile compounds. Encapsulation can be employed to retain aroma in a food product during storage, protect the flavor from undesirable interactions with food, minimize flavor/flavor interactions, guard against light-induced reactions and/or oxidation, increase flavor shelf life, and allow controlled release. The incorporation of small amounts of flavors into foods can greatly influence the finished product quality, cost, and consumer satisfaction. The food industry is continuously developing ingredients, processing methods, and packaging materials to improve flavor preservation and delivery. The stability of the matrices is an important condition to preserve the properties of the flavor materials. Many factors, such as the kind of wall material, the ratio of the core material to the wall material, the encapsulation method, and storage conditions affect the antioxidative stability of the encapsulated flavor. According to the technological process used, the matrices of encapsulation will present various shapes (films, spheres, or irregular particles), various structures (porous or compact), and various physical structures (vitreous or crystalline dehydrated solid, or rubbery matrix) that will influence the diffusion of flavor properties or external substances (oxygen or solvent) and the stability of the product during its storage.

17.5.2 Retention and Release of Volatile Compounds in Edible Films and Coatings

Food matrix ingredients, among them food proteins, have little flavor of their own but are known to bind and trap aroma compounds. Depending on the nature and strength of the binding, aroma release in the gas phase will be more or less decreased. The mechanism of flavor binding is dependent on the role of the protein structure as well as the type of flavor compound (aldehyde, alcohol, ketone, and ester) involved in the binding process (Heng et al., 2004). The most extensively studied proteins are milk proteins, known for their emulsifying properties. For instance, affinity for β-lactoglobulin increases with hydrophobic chain length or overall hydrophobicity of flavor compounds, except for terpenes, the main constituents of essential oils. However, it was not possible to find a simple explanation for the binding strength of aroma compounds from different chemical classes. Among the other food proteins, β-lactalbumin, caseins, bovine serum albumin, and soy proteins have been studied to a lesser extent for their binding properties toward flavor compounds. β-Lactalbumin was found to bind ketones and aldehydes but with a poor flavor-binding capacity compared with other whey proteins. This explains why whey protein films are often desired for flavor or essential oil encapsulation. Moreover, whey proteins have both film-forming properties and emulsifying capacity.

Carbohydrates can have a measurable influence on the release and perception of flavors. Carbohydrates change the volatility of compounds relative to water, but the effect depends on the interaction between the particular volatile molecule and the particular carbohydrate. As a general rule, carbohydrates, especially polysaccharides, decrease the volatility of compounds relative to water by a small to moderate amount as a result of molecular interactions. However, some carbohydrates, especially the monosaccharides and disaccharides, exhibit a salting-out effect, causing an increase in volatility relative to water (Godshall, 1997).

Lipids are rather homogeneous, hydrophobic, and non-polar materials, existing in aqueous media in the form of distinct regions. In these systems, flavor compounds are distributed between the lipid and the aqueous phases following the physical laws of partition (Solms et al., 1973). Lipids are carriers and release modulators of aroma, but they are also flavor precursors. Lipid oxidation as well as lipolysis generates numerous short-chain compounds, to which we are extremely sensitive because of their low levels of olfactory detection. The global effect of lipids on aroma compound release is to decrease their volatility. Generally, aroma compounds are hydrophobic and show greater affinity for the lipidic phase than for the aqueous or vapor phase. The influence of the physicochemical properties of both the flavor compound and the fat content has been studied by the determination of air–oil partition coefficients. Solid fat content and crystallinity of fat improves the controlled release by the reduction of diffusion kinetic. However, increased temperature induces melting of the solid fat content, which is unfavorable to flavor retention.

A food product with active molecules on the surface and a food product coated with an edible film with active volatile molecules are compared in Figure 17.3. In food products, surface contamination and oxidation are most probable and need to be prevented (Han, 2002).

In both systems, with and without edible film, the food layers that do not contain active agents (flavoring, antimicrobial, and antioxidant) have initially a very large volume compared with the volume of thin films, coatings, or food surfaces where active volatile molecules were dispersed. Because of the almost infinite volume of food layer without active molecules, migration of active molecules from the surface into the food will be favored. If the agent is sprayed onto the surface of food, the initial surface concentration will be very high and will begin to decrease due to dissolution and diffusion of the agent toward the center of the food (Marcuzzo et al., 2011). Therefore, the solubility (or partition coefficient) and diffusion coefficient (diffusivity) of the agent in a food are very important characteristics to maintain the surface concentration needed to hinder microbial growth, oxidation, or flavor loss during the expected shelf life. Otherwise, the active molecule concentration in the surface layer will be reduced or depleted. The release rate must be controlled to prevent the early depletion of active molecules due to fast migration. The application of edible coatings or films with active molecules could allow a very low

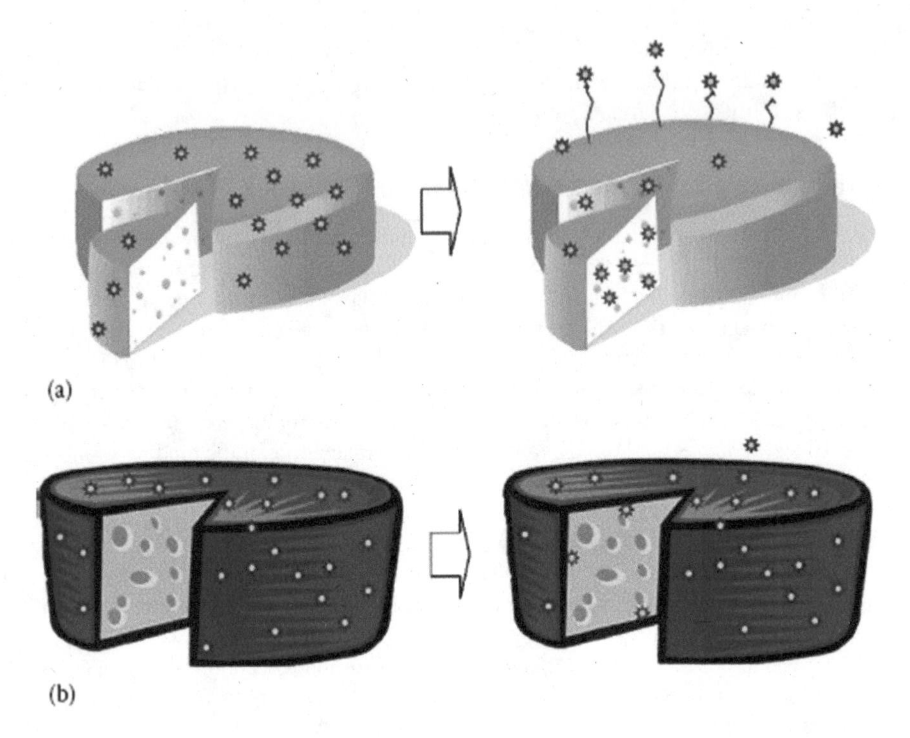

FIGURE 17.3 Retention and controlled release of aroma compounds added at the food surface or entrapped in edible coatings/films. (a) Deposition of volatile active compound on food surface and (b) food coated with an edible film encapsulating the volatile active compounds.

diffusion rate of active molecules into the food to maintain their efficiency on the surface. Moreover, the effectiveness of the surface function could allow a lower active molecule concentration in the food. If the active molecule dispersed in the edible film or coating is an aroma compound, the factors that influence flavor release have to be considered. The release rate of the volatile agent from the packaging system is highly dependent on the volatility, which relates to the chemical interaction between the volatile agent and packaging materials. The absorption rate of headspace volatiles into a food surface is related to the composition of the food, as the ingredients undergo chemical interactions with the gaseous agents. Because most volatile agents are generally lipophilic, the lipid content of the food is an important factor in headspace concentration. The use of volatile antimicrobial agents has many advantages. This system can be used effectively for highly porous, powdered, shredded, irregularly shaped, and particulate foods, such as ground beef, shredded cheese, small fruits, and mixed vegetables (Han, 2002).

The ability of edible films to entrap or to retain flavor compounds during film storage and above all, during film processing is of importance. If flavors are lost during the drying of a hydrocolloid-based film-forming liquid (solution, emulsion, or suspension), the resultant aroma will become unbalanced. The more volatile or the less interactive volatile compounds will be preferentially lost during the film/coating process, inducing a loss of the fresh aroma notes. Using emulsions or suspensions with hydrocolloids interacting strongly with the aroma compounds tends to reduce the aroma loss during drying. When edible films are enriched with essential oils, the drying temperatures usually employed to form the edible coating are high enough to volatilize a high percentage of the aromatic components. (Sánchez-González 2010a) showed that D-limonene (the main component of bergamot essential oil) loss during drying ranged from 39% to 99% when added from 0.5% to 3% (w/w) in chitosan films. Moreover, (Monedero et al. 2010) found that losses during drying are tremendously greater than during film storage. Soy protein–beeswax–oleic acid emulsified films, dried at moderate temperature (20 °C, 45% relative humidity (RH), lost up to 98% of encapsulated *n*-hexanal, but this could be reduced by using a greater amount of beeswax. Beeswax particles decreased by half the aroma diffusivity in the protein-based matrix. However, the greater the loss during drying, the faster is the release to the vapor phase. Most of the remaining *n*-hexanal

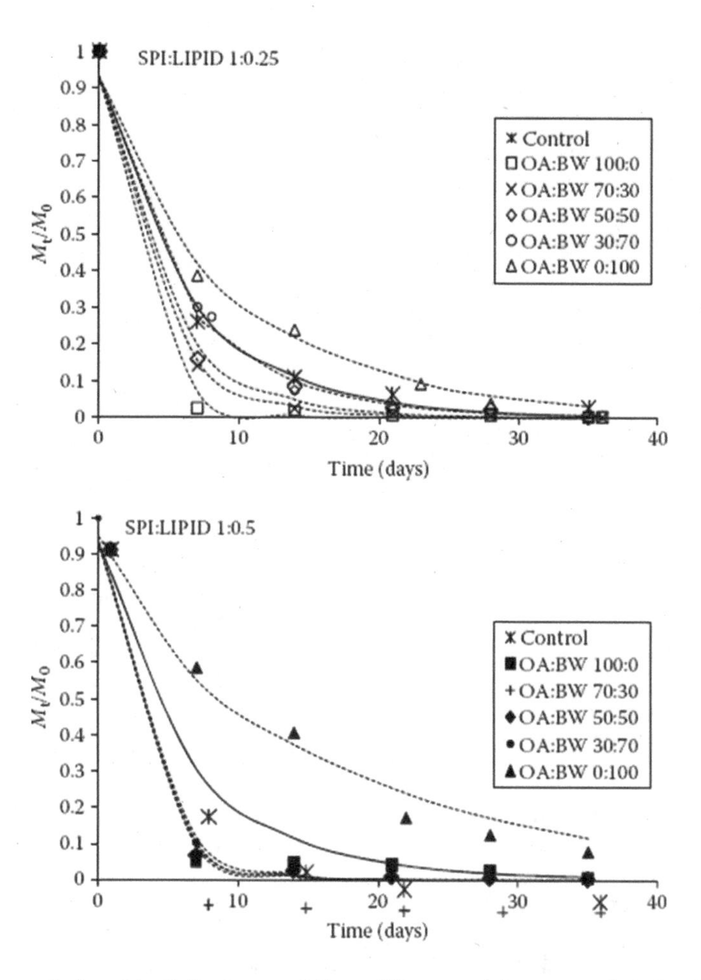

FIGURE 17.4 *n*-Hexanal release kinetic for the control film and films containing soy protein isolate (SPI) combined with oleic acid (OA) and beeswax (BW) in SPI:lipid ratios 1:0.25 and 1:0.5 (experimental data: symbols, and fitted model: lines). M_t/M_0 is the ratio of volatile compounds in the film (M_t, amount of Hexanal in the film at time t; M_0, initial amount of Hexanal in the film. (From Monedero, M. et al., *J. Food Eng.*, 100, 128, 2010. With permission.)

in the film after drying was lost after 35 days' storage in an open ventilated chamber (Figure 17.4). Similar trends were observed by (Tunc and Duman 2011) when the montmorillonite (nanoclay) content in chitosan films increased. Solid particles such as beeswax or montmorillonite probably reduce the diffusivity of the volatile compound because of increasing "tortuosity" and then delay of the aroma loss by the films. A significant increase of the thymol release rate with an increase of bran concentration was observed by (Mastromatteo et al. 2009). So, in this case, the incorporation of nanoparticles (spelt bran) had the opposite effect and was unfavorable for volatile compound retention.

The advantage of substituting essential oils or aroma compounds for the corresponding food-grade oleoresins could lie in the introduction of other, non-volatile components positively affecting food quality (Ponce et al., 2008).

Flavor retention by the film matrix during processing depends as much on the temperature as on the film or coating composition. On the one hand, temperature increases the aroma volatility, according to Henry's law (Buttery et al., 1971), and the diffusivity. On the other hand, many film constituents enable reduced volatility and diffusivity. Polyols used as plasticizers in film formulation are very good for aroma support and are able to significantly reduce flavor release. Acacia gum provides the same advantages for flavor retention and could also strengthen the film's mechanical properties. Recent studies showed that the addition of nanoparticles such as nanoclays is significant in improving the retention efficiency of chitosan-based edible films (Tunc and Duman, 2011). Usually, the mass partition coefficient K_{mass} (ratio

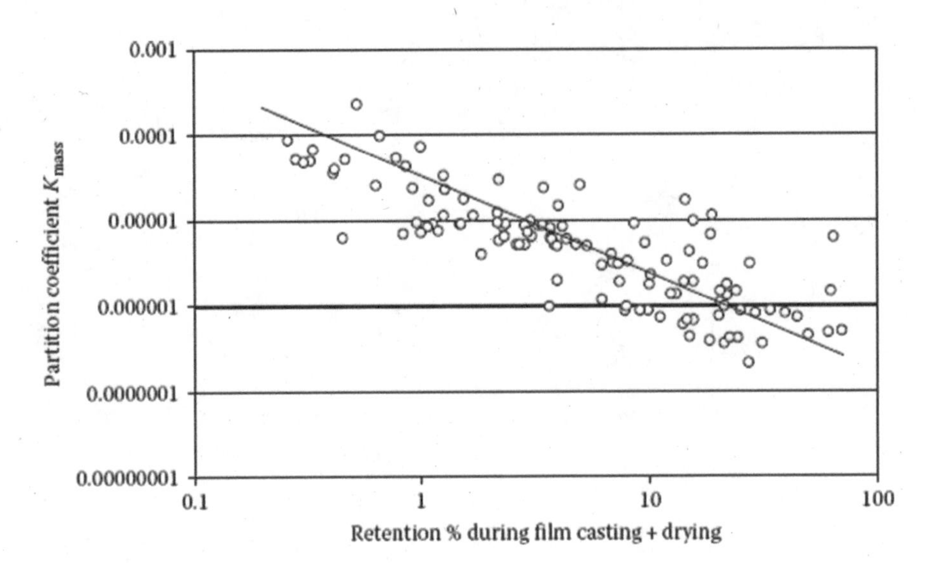

FIGURE 17.5 Relationship between carvacrol retention during film processing and air/film partition coefficient during film storage. (From Kurek, M. and Debeaufort, F., Development of an antimicrobial coating for packaging films: Physico-chemical and microbiological approaches, Doctoral school intermediate PhD report, University of Burgundy, Dijon, France, 2010.)

between aroma concentration in air and aroma concentration in film) is inversely proportional to the retention capacity during film processing.

As displayed in Figure 17.5, the partition coefficient (during film storage) can be related to carvacrol retention during film processing (casting and drying) by chitosan-based films (more than 200 recipes in a 25–80 °C drying temperature range) at various humidities (0–98% RH).

The release of various aroma compounds (ethyl esters, methylketones, and alcohols), from either carrageenan-based, carrageenan–acetylated monoglycerides emulsion–based, or acetylated monoglycerides–based films strongly differs (Marcuzzo et al., 2010). In lipid films, the aroma compound release is more affected by factors related to diffusivity, whereas in carrageenan emulsified films, the affinity between the volatile compounds and the polymer preponderantly influences sorption phenomena and thus, the release. Carrageenan films resulted in possible encapsulating matrices: they display better performance in the retention of more polar aroma compounds than pure lipid or emulsified films. Carrageenan films were able to retain volatile compounds during film processing and released them gradually with time.

The surrounding medium of the encapsulated aroma, such as polysaccharides, proteins, lipids, and salts, could play an important role in release in liquid media and then, aroma compound retention by the film matrix. Different behaviors of aroma compound volatility in food products have been observed in the presence of salt or sucrose molecules, observing that some of them presented a “salting out” effect (favoring release by volatilization), others an opposite “salting in” effect, and yet others no modification (Lubbers et al., 1998; Van Ruth et al., 2002). Similar effects could be observed in the release of the aroma compounds encapsulated in films. While the “salting out” effect should accelerate the release, the “salting in” could decrease the rate of the release.

The effects of aqueous media (containing 0.9% NaCl) and temperature (25 °C and 37 °C) on the release of encapsulated aroma compounds (*n*-hexanal and D-limonene) in ι-carrageenan-based films with and without lipid have been studied by (Fabra et al. 2011). D-Limonene was released quickly from water at higher temperatures. However, no effect of temperature was observed on *n*-hexanal release in water. Only between 40% and 56% of the hexanal was released, depending on film composition, while D-limonene was completely retained in the fat material (Table 17.1). The presence of salt in the liquid release medium was significantly and tremendously favorable to the aroma retention by carrageenan-based

TABLE 17.1

Percentage of *n*-Hexanal and D-Limonene Retained in the Film at the Equilibrium of Release Kinetics in Liquid Media

Type of Film	Aqueous Medium	Retention of n-Hexanal (%)[a]	Retention of D-Limonene (%)
Carrageenans without lipid	Water	9.3 (1.6)	6.2 (1.7)
	0.9% NaCl	57 (3)	79 (2)
Carrageenans with lipid	Water	11 (1)	7 (3)
	0.9% NaCl	44 (2)[e]	100

[a] Mean value (standard deviation).

films. Retention was 4 to 15 times higher according to aroma compound hydrophobicity and the presence of fat particles in the film matrix.

In a similar way, (Sánchez-González et al. 2011) studied chitosan films enriched with different concentrations of bergamot oil and the migration of D-limonene (the major oil component) into five liquid simulated foods (water, 10% ethanol, 50% ethanol, 95% ethanol, and isooctane). D-Limonene migration (release) was significant in 95% ethanol, whereas in the other food simulants, its release remained less impressive. Composite films remained intact with isooctane chitosan-bergamot oil (CH-BO), and no release of limonene was observed. The polarity of the simulant and the migrant seems to be a key factor to explain these results. However, in the reports of both Sánchez-González (2010) and (Fabra et al. 2011), chitosan- or carrageenan-based films immersed in liquid media probably swelled, and some of their soluble components (plasticizers) probably migrated into the contacting liquid. Because the film integrity was not maintained, the release phenomenon could not be attributed only to the partition (solubility) and diffusion coefficient of the aroma compounds. Other parameters must also be taken into account to better explain and describe the release phenomenon of aroma compounds in contact with liquid or gel foods, such as swelling, partial dissolution of film components, counterdiffusion of other solutes from food to film, and osmotic gradients.

Volatile compound retention by edible films is complex, and many factors could affect it and interact. Figure 17.6 is a simplified summary view of the main parameters influencing the flavor release from edible films.

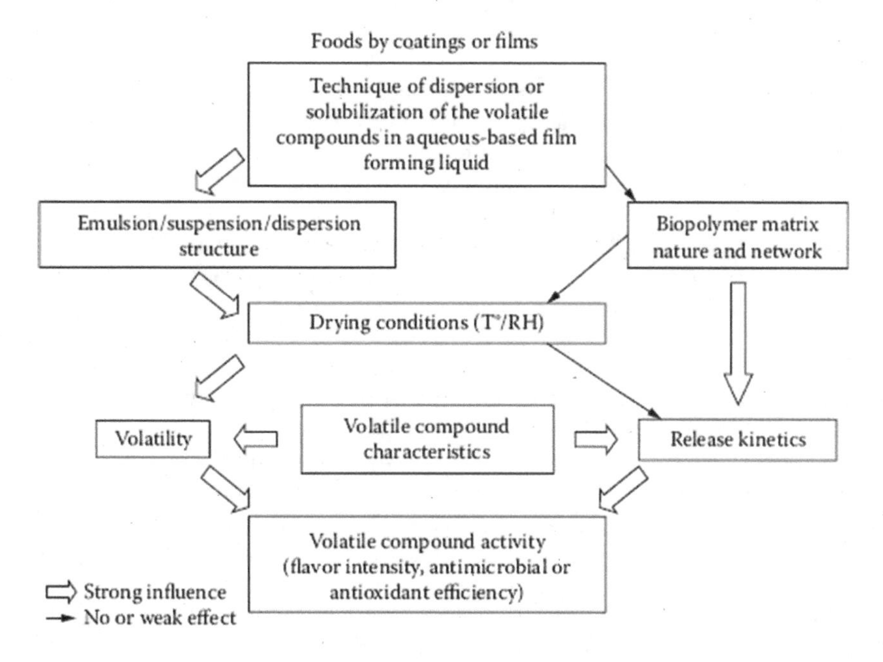

FIGURE 17.6 Overall mechanism of flavor release in air from edible films: main influencing factors.

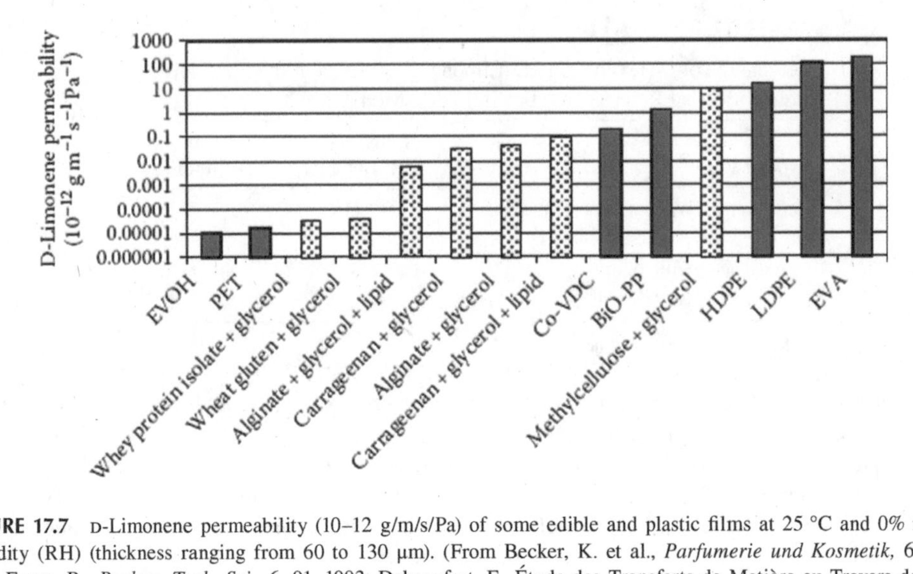

FIGURE 17.7 D-Limonene permeability (10–12 g/m/s/Pa) of some edible and plastic films at 25 °C and 0% relative humidity (RH) (thickness ranging from 60 to 130 µm). (From Becker, K. et al., *Parfumerie und Kosmetik,* 68, 268, 1987; Franz, R., *Packag. Tech. Sci.,* 6, 91, 1993; Debeaufort, F., Étude des Transferts de Matière au Travers de Films d'Emballages: Perméation de l'Eau et de Substances d'Arôme en Relation avec les Propriétés Physico-chimiques des Films Comestibles, PhD Dissertation, ENS.BANA, Université de Bourgogne, Dijon, France, 1994; Debeaufort, F. and Voilley, A., *J. Agr. Food Chem.,* 42, 2871, 1994; Debeaufort, F. and Voilley, A., *Cellulose,* 2, 1, 1995; Kobayashi, M. et al., *J. Food Sci.,* 60, 205, 1995; Paik, J.S. and Writer, M.S. *J. Agr. Food Chem.,* 43, 175, 1995; Miller, K.S. and Krochta, J.M., *Trends Food Sci. Technol.,* 8, 228, 1997; Miller, K.S. et al., *J. Food Sci.,* 63, 244, 1998; Quezada-Gallo, J.A. et al., *J. Agr. Food Chem.,* 47, 108, 1999; Quezada-Gallo, A. et al., *New Developments in the Chemistry of Packaging Materials,* ACS Books, Dallas, TX, 1999; Quezada-Gallo, J.A. et al., Influence de la Structure et de la Composition de Réseaux Macromoléculaires sur les Transferts de Molécules Volatiles (eau et Arômes). Application aux Emballages Comestibles et Plastiques, PhD Dissertation, Université de Dijon, Dijon, France, 1999; Hambleton, A. et al., *Biomacromolecules,* 9(3), 1058, 1999; Hambleton, A. et al., *Food Hydrocolloids,* 23(8), 2116, 2009; Hambleton, A. et al., *J. Food Eng.,* 93, 80, 2009. With permission.) Bio-PP, bioriented polypropylene; Co-VDC, polyvinylidene chloride copolymer; EVA, ethylene vinyl acetate; EVOH, ethylene vinyl alcohol; HDPE and LDPE, high- and low-density polyethylene; PET, polyethylene terephthalate.

17.5.3 Barrier Performance of Edible Films and Coatings against Aroma Compounds

D-Limonene is the most studied flavor compound. It is considered as the typical aroma probe of citrus juices and soda beverages, and many data are available in the literature. Figure 17.7 gives a comparison between the D-limonene permeabilities of edible and plastic films. Polysaccharide-based films are of the same order as heat-sealable polymers such as polyolefins, while protein-based films and coatings have barrier properties as good as the best plastics. The permeability of wheat gluten and glycerol or whey protein and glycerol films is about 10^{-17} g/m/s/Pa, and it is from 10^{-14} to 10^{-16} g/m/s/Pa in the case of polysaccharides and other protein-based films (Miller et al., 1998; Quezada-Gallo, 1999; Hambleton et al., 2010).

Although D-limonene is probably the most studied flavor compound in the field of polymer permeability, it is not often used in the flavoring of solid food products. Debeaufort and Voilley (1995) studied the mass transfer through edible films of several other molecules commonly found in cheese, fruits, and dairy products. Table 17.2 displays the permeability of some flavor compounds through edible films. It seems that the permeability is always much lower (from 100 to 100,000 times) through protein-based than through carbohydrate films. These results allow them to be considered for use as a protective layer against scalping and permeation through classically used plastic films. Edible barriers are able to retain food flavors within the food matrix due to very low permeability. However, because of a lack of aroma permeability data, no trend or general rules could be given about the barrier efficiency of edible barriers. Interactions between film constituents and aroma compounds have a great impact on the permeability mechanism. Very few papers have focused on the aroma transfers related to interactions, sorption, and diffusion phenomena. It seems that chemical affinities between flavor compounds and film components have more influence on the permeability value than the classical sorption–diffusion models. Aroma induces significant changes in the physical-chemical and structural properties of

TABLE 17.2

Permeability (p) to Aroma Compounds of Several Edible Films at 25 °C

Edible Film Composition	**Aroma Compounds**	**P (10^{-12} g m^{-1} s^{-1} Pa^{-1})**
Methylcellulose + glycerol	1-Octen-3-ol	122
	2-Pentanone	19
	2-Heptanone	39
	2-Octanone	338
	2-Nonanone	420
	Ethyl acetate	128
	Ethyl butyrate	119
	Ethyl isobutyrate	106
	Ethyl hexanoate	668
	D-Limonene	10
ι-Carrageenan + glycerol	Ethyl acetate	0.046
	Ethyl butyrate	1.25
	Ethyl hexanoate	<0.014
	2-Hexanone	0.196
	1-Hexanol	<0.0011
	Cis-3-hexenol	<0.012
	n-Hexanal	0.0189
	D-Limonene	0.0291
ι-Carrageenan + glycerol + acetylated monoglycerides + beeswax (emulsion)	Ethyl acetate	0.022
	Ethyl butyrate	0.33
	Ethyl hexanoate	<0.014
	2-Hexanone	0.111
	1-Hexanol	<0.0011
	Cis-3-hexenol	<0.017
	n-Hexanal	<0.00005
	D-Limonene	0.0803
ι-Carrageenan + glycerol + oleic acid + beeswax (emulsion)	Ethyl acetate	0.83
	Ethyl butyrate	0.75
	Ethyl hexanoate	194
	2-Hexanone	<0.0011
	1-Hexanol	1.12
	Cis-3-hexenol	<0.0012
Sodium alginate + glycerol	n-Hexanal	0.0134
	D-Limonene	0.00034
Sodium alginate + glycerol + acetylated monoglycerides + beeswax (emulsion)	Ethyl butyrate	0.033
	Ethyl hexanoate	134.2
	2-Hexanone	0.094
	n-Hexanal	<0.00005
	D-Limonene	0.000053
Wheat gluten + glycerol	1-Octen-3-ol	4.6
	2-Pentanone	0.12
	2-Heptanone	0.50
	2-Octanone	<0.005
	Ethyl acetate	0.059
	Ethyl butyrate	0.670
	Ethyl isobutyrate	0.04

(*Continued*)

TABLE 17.2 (CONTINUED)

Permeability (*p*) to Aroma Compounds of Several Edible Films at 25 °C

	Aroma Compounds	**P (10^{-12} g m^{-1} s^{-1} Pa^{-1})**
	Ethyl hexanoate	<0.005
	D-Limonene	0.00004
Sodium caseinate + glycerol	Ethyl acetate	0.006
	Ethyl butyrate	0.19
	Ethyl hexanoate	<0.0004
Sodium caseinate + glycerol + oleic acid (emulsion)	Ethyl acetate	36.1
	Ethyl butyrate	1732.7
	Ethyl hexanoate	1632.1
	2-Hexanone	1.85
	1-Hexanol	3116
	Cis-3-hexenol	404
Sodium caseinate + glycerol + oleic acid + beeswax (emulsion)	Ethyl acetate	0.73
	Ethyl butyrate	134.7
	Ethyl hexanoate	<0.0004
	2-Hexanone	18.7
	1-Hexanol	238.8
	Cis-3-hexenol	639.1
Sodium caseinate + glycerol + beeswax (emulsion)	Ethyl acetate	0.0061
	Ethyl butyrate	0.08
	Ethyl hexanoate	<0.00045
	2-Hexanone	<0.00003
	1-Hexanol	<0.005
	Cis-3-hexenol	<0.0004
Whey protein isolate + glycerol	D-Limonene	0.0003

Source: Debeaufort, F., Étude des Transferts de Matière au Travers de Films d'Emballages: Perméation de l'Eau et de Substances d'Arôme en Relation avec les Propriétés Physico-chimiques des Films Comestibles, PhD Dissertation, ENS.BANA, Université de Bourgogne, Dijon, France, 1994; Debeaufort, F. and Voilley, A., *J. Agr. Food Chem.*, 42, 2871, 1994; Debeaufort, F. and Voilley, A., *Cellulose*, 2, 1, 1995; Kobayashi, M. et al., *J. Food Sci.*, 60, 205, 1995; Miller, K.S. and Krochta, J.M., *Trends Food Sci. Technol.*, 8, 228, 1997; Miller, K.S. et al., *J. Food Sci.*, 63, 244, 1998; Quezada-Gallo, A. et al., *New Developments in the Chemistry of Packaging Materials*, ACS Books, Dallas, TX, 1999; Quezada-Gallo, J.A. et al., *J. Agr. Food Chem.*, 47, 108, 1999; Quezada-Gallo, J.A., Influence de la Structure et de la Composition de Réseaux Macromoléculaires sur les Transferts de Molécules Volatiles (eau et Arômes). Application aux Emballages Comestibles et Plastiques, PhD Dissertation, Université de Dijon, Dijon, France, 1999; Hambleton, A. et al., *Biomacromolecules*, 9(3), 1058, 2008; Hambleton, A. et al., *Food Hydrocolloids*, 23(8), 2116, 2009; Hambleton, A. et al., *J. Food Eng.*, 93, 80, 2009; Fabra, M.J. et al., *Carbohydr. Polym.*, 76, 325, 2009; Fabra, M.J. et al., *Biomacromolecules*, 9(5), 9, 1406, 2008.

polysaccharide- or protein-based edible films when entrapped or when transferred as described in the following.

17.5.4 Protection of Flavor Compounds against Chemical Degradation by Edible Films and Coatings

The most common problem occurring during flavor storage is deterioration due to oxidation. Any flavoring containing citrus oils or oils based on aldehydes is susceptible to oxidative reactions and the development of off-flavors. Thus, an important function of an edible film is protection of the flavoring from oxygen.

Choosing an encapsulating agent with adequate barrier properties (to oxygen) will lead to more stable dry flavors. Because D-limonene is sensitive to oxidative degradation, several studies were focused on the possibility of encapsulating this aroma compound. Different ingredients were chosen as wall materials, such as starch, maltodextrin, and GA (Wyler and Solms, 1982; Anandaraman and Reineccius, 1986; Bertolini et al., 2001). Generally, lemon oils are added to food in the form of water-in-oil emulsions. (Djordevic et al. 2008) studied the possibility of stabilizing oil-in-water emulsions with whey protein isolate (WPI) instead of GA to inhibit D-limonene degradation. The formation of the limonene oxidation products limonene oxide and carvone was lower in the WPI- than in the GA-stabilized emulsions. This was in agreement with (Kim and Morr 1996), who found that limonene oxide, carvone, and carveol formation in microencapsulated orange oil was lower in emulsions stabilized with WPI and soy protein isolate than in emulsions stabilized with GA. Among wall materials that can be applied to preserve aroma compounds, κ-carrageenans and wheat gluten were selected for their useful gas barrier properties. Hambleton et al. (2008) showed that carrageenan-based edible film prevents the oxidation of *n*-hexanal to hexanoic acid in oxidative conditions. (Marcuzzo et al. 2011) also showed that both carrageenan- and wheat gluten–based edible films seem to prevent D-limonene oxidation. Wheat gluten–based film protected D-limonene from degradative reactions, and the increase in carvone release was probably due to oxidation in the headspace once D-limonene was released and not within the matrix, as was confirmed by Fourier transform infrared (FTIR) analysis of edible film containing the aroma compounds.

Oxygen permeability is, then, a key factor for improved efficiency of film and coating matrices to protect aroma compounds encapsulated or initially present in coated food products. Table 17.3 gives the oxygen permeability of the more studied edible films and coatings. Protein-based edible films are more efficient barriers against oxygen transfer and are more suitable for flavor protection. Protein also interacts to a greater extent with flavor compounds. But, the oxygen protection strongly varies with the moisture level. Increasing water activity or relative humidity tends to increase the oxygen permeability from 10 to 10^5 times. For example, the oxygen permeability of collagen film rises from 6.6×10^{-19} to 14.68×10^{-15} g/m/s/Pa when water activity increases from 0 to 0.93, but this behavior is also observed for water-sensitive plastic films such as EVOH or nylon. The plasticization of food macromolecules by water increases molecular mobility, which favors both oxygen diffusivity and aroma release and oxidation.

17.5.5 Effect of Volatile Compounds on Structural and Physical-Chemical Properties of Edible Films and Coatings

Incorporating volatile active compounds may significantly affect the structural organization of the film-forming substance network and consequently, its physical-chemical properties.

17.5.5.1 Impact on Film Appearance and Transparency

The incorporation of aroma compounds or essential oils always decreases the transparency and gloss of edible films made of carbohydrate or protein because of emulsion structure formation during drying. The solubility of most aroma compounds and essential oils is much lower than the amount added to film-forming suspensions. All the authors who have measured the optical properties of films containing essential oils or flavor compounds observed this behavior, regardless of the nature of the matrix (chitosan, HPMC, methylcellulose, soy protein isolate, WPI, carrageenans, or sodium alginate) or the volatile compounds (tea tree oil, bergamot, cloves, ginger, onion, carvacrol, thymol, etc.) (Pranoto et al., 2005a; (Fabra et al., 2008, 2009, 2011; Hambleton et al., 2009a,b, 2010; Sanchez Gonzalez et al., 2009, 2010a,b; Atarés et al., 2010; Tunc and Duman, 2011). Changes in appearance could generally be related to the microstructure of the film containing the volatile compound.

TABLE 17.3

Oxygen Permeability of Edible Films and Coatings Compared with Common Plastic Films (10^{-15} g/m/s/Pa)

Film	Oxygen Permeability	T (°C)	a_w
Low-density polyethylene	16.050	23	0
Polyesters	0.192	23	0
Ethylene vinyl alcohol	0.003	23	0
Polyvinylidene chloride	0.0066	23	0
Methylcellulose + glycerol	8.352	30	0
Carrageenan + glycerol	0.72	25	0
Carrageenan + glycerol + GBS	0.86	25	0
Alginate + glycerol	9.4	25	0
Alginate + glycerol + GBS	2.7	25	0
Chitosan + glycerol	0.009	25	0
Starch alginate	0.087	23	0
Beeswax	7.68	25	0
Carnauba wax	1.296	25	0
Collagen	0.00066	25	0
Zein + glycerol	0.56	38	0
Wheat gluten + glycerol	0.016	25	0
Soy protein isolate + glycerol	0.1	25	0
Casein + glycerol	0.0115	25	0
Casein + sorbitol	0.0132	25	0
Peanut protein + glycerol	0.034	23	0
Fish myofibrillar protein + glycerol	0.031	25	0
Low-density polyethylene	30.85	23	0.5
Polyesters	0.192	23	1
Ethylene vinyl alcohol	0.096	23	0.95
Polyvinylidene chloride	0.084	23	0.95
Pectin + glycerol	21.44	25	0.96
Starch + glycerol	17.36	25	1
Hydroxypropylmethylcellulose	4.88	25	0.50
Chitosan + glycerol	7.552	25	0.93
Collagen	0.48	25	0.63
Collagen	14.68	25	0.93
Wheat gluten + glycerol	20.64	25	0.95
Whey protein + glycerol	33	23	0.34
Whey protein + glycerol	435	23	0.56
Fish myofibrillar protein + glycerol	27.93	25	0.93

Source: Gennadios, A., *Protein-Based Edible Films and Coatings*, CRC Press, Boca Raton, FL, 2002; With kind permission from Springer Science+Business Media: *Edible Films and Coatings for Food Applications*, 2009, Embuscado, M.E. and Huber, K.C.

a_w, water availability; GBS, Grindsted Barrier System.

17.5.5.2 Changes in Film Microstructure Induced by Volatile Compound Encapsulation

(Hambleton et al. 2009a) showed that the incorporation of 3% of *n*-hexanal in alginate matrices induced a more homogeneous structure, as observed by an environmental scanning electron microscope (ESEM). In contrast, *n*-hexanal provoked a more heterogeneous distribution of emulsified fat (beeswax + acetylated monoglycerides) in the film cross section. This was attributed to the competition between the emulsifier (glycerol monostearate) and *n*-hexanal, which is amphipolar. The second behavior was also observed for carrageenan-based films with encapsulated *n*-hexanal (Hambleton et al., 2009b). Sánchez-González et al. (2009) observed a more heterogeneous structure with increasing tea tree oil content in HPMC films. The essential oil was at a concentration higher than the solubility limit of the volatile in the film-forming suspension, which resulted in an emulsion of tea tree oil droplets in the HPMC network. While a continuous structure was observed for the HPMC film, the presence of tea tree oil caused discontinuities associated with the formation of two phases in the matrix: lipid droplets embedded in a continuous polymer network. Lipid droplets, whose number increased with the tea tree oil concentration, were homogeneously distributed across the film. This reveals that very little creaming occurred during the film drying, probably due to the highly viscous effect of HPMC.

17.5.5.3 Influence of Volatile Compound Encapsulation on Film Mechanical Properties

The oil droplets or discontinuities usually induce a loss of mechanical properties, such as a decrease of the tensile strength and Young (elastic) modulus, as given in Table 17.4.

The addition of tea tree oil in the 0.5%–3% concentration range caused a significant decrease in the elastic modulus and in tensile strength at break of HPMC films, although with no significant effect on deformation at break (Sánchez-González et al., 2009). As elongation did not increase, the loss of mechanical properties cannot be attributed to a plasticization of HPMC by tea tree oil. This coincides with the results reported by other authors when adding essential oil to a chitosan matrix (Pranoto et al., 2005a,b; Zivanovic et al., 2005) and is in agreement with the effect of the structural discontinuities provoked by the incorporation of the oil on the mechanical behavior. These discontinuities reduced the film's resistance to fracture. Bergamot oil (0.5%) added to chitosan films reduced the tensile strength and elongation by two and three times, respectively (Sánchez-González et al., 2010a). Cinnamon oil seemed to have some plasticizing effect on soy protein isolate–based films, making them more extensible as the oil content increased (Atarés et al., 2010). On the contrary, cinnamon oil may have caused some degree of rearrangement in the protein network, thus strengthening it and increasing the film's resistance to elongation. This effect was not observed when ginger oil was added. Films with ginger oil were less resistant and less elastic than those with cinnamon oil ($p < 0.01$). The discontinuities in the protein matrix may imply a decrease in the deformability of the films with ginger oil, because these reach the break point at lower deformation. This tendency could be explained by the fact that lipids are unable to form a cohesive and continuous matrix. Very different behavior was observed by Hambleton et al. (2012) for carrageenan films incorporating *n*-hexanal. In fact, the addition of incorporated *n*-hexanal tends to increase the elastic modulus and tensile strength. This is probably due to the stabilizing effect of the aroma compound on the film matrix. *n*-Hexanal interacts with κ-carrageenan's lateral chains and plays a stabilizing role in the interface due to its amphipolar character, leading to a much more homogeneous structure that increases the film's stiffness. Nevertheless, *n*-hexanal does not affect the film's capacity to stretch, as the elongation percentages are not significantly different from the same film without *n*-hexanal. The addition of incorporated *n*-hexanal has the same effect as the presence of fat material; it reduces the elastic modulus and tensile strength, contrary to ι-carrageenan-based film. Incorporated *n*-hexanal weakly interacts with the sodium alginate, because this type of film has a well-organized structure in an egg-box model, stabilized by divalent ions to form stronger gels and thus, stronger films. But, *n*-hexanal interacts with the other components in the film, such as glycerol, which being a polyol, has a great affinity for flavors of this type. These interactions lead to a reduction of stiffness and to the film's resistance to elongation. The presence of both *n*-hexanal and fat material has a significant effect, reducing elastic modulus and tensile strength more than in other types of films, probably because the aroma compound interacts primarily with the fat material when added to the film. Film plasticization can occur during aroma transfer, and this depends on the aroma concentration, as observed by (Quezada-Gallo et al. 1999a, b) for the permeation of 2-heptanone and 2-pentanone through methylcellulose-based films. The transfer rate of 2-heptanone strongly increased for an aroma gradient higher than 10 μg/ml. Mechanical film properties changed on exposure to flavor concentrations higher than 10 μg/ml. 2-Heptanone and 2-pentanone increased film elongation, suggesting that polymer plasticization occurred. The ketone group of the

TABLE 17.4

Influence of Volatile Compound Incorporation on Mechanical Properties of Edible Films

Film Matrix	Volatile Compounds	Elongation (%)	Tensile Strength (MPa)	Elastic Modulus (MPa)
Soy protein isolate	/	4	11	412
	Cinnamon oil 1%	7.5	14.1	354
	Ginger oil 1%	3	8	340
Chitosan	/	22	113	2182
	Bergamot oil 3%	1.7	22	682
	Tea tree oil 2%	8	54	653
Alginate	/	4.1	66.1	/
	Garlic oil 0.4%	2.7	38.7	/
Sodium alginate	/	4.7	55	3280
	n-Hexanal 1%	1.5	28	2247
Sodium alginate + GBS (emulsion)	/	2.1	31	2320
	n-hexanal 1%	1.9	20	1605
Carrageenan	/	1.2	8.8	95
	n-hexanal 1%	1.6	15	1259
Carrageenan + GBS (emulsion)	/	2.6	10	927
	n-hexanal 1%	2.4	10	751
HPMC	/	0.1	59	1697
	Tea tree oil 40%	0.11	42	956

Source: Atarés, L. et al., *J. Food Eng.*, 99, 384, 2010; Sánchez-González, L. et al., *J. Food Eng.*, 98, 443, 2010b; Sánchez-González, L. et al., *Food Hydrocolloids*, 23, 2102, 2009; Pranoto, Y. et al., *LWT—Food Sci. Technol.*, 38(8), 859, 2005; Hambleton, A. et al., *Food Chem.*, 2012.

flavor compound interacts with hydroxyl groups of methylcellulose. 2-Heptanone forms weak hydrogen bonds with methylcellulose. This likely widens the spaces among the polymer chains, resulting in swelling and plasticization of the film network, which decreases the mechanical properties and enhances the transfer of volatiles and water vapor.

17.5.5.4 Impact of Flavor Compound Encapsulation on Barrier Properties of Film

Both microstructure and mechanical properties are affected by flavor compound or essential oil encapsulation, including the permeability of films to other volatile compounds such as water vapor, oxygen, or other aroma compounds. This suggests that the choice of matrix for the encapsulation of a volatile compound on the basis of its oxygen barrier performance, for instance, could not be counted on because of permeability changes due to encapsulation. When 1% *n*-hexanal was encapsulated, the oxygen permeability of carrageenan–glycerol and carrageenan–glycerol–lipid edible films increased by 15% and 100%, respectively (Hambleton et al., 2008). The incorporation of 1% *n*-hexanal in a sodium alginate film induced a doubling in oxygen permeability and a 10-fold increase when the sodium alginate film contained lipid emulsions (Hambleton et al., 2009a). On the contrary, (Rojas-Grau et al. 2007) did not show any change in the oxygen permeability of alginate–apple puree film when oregano, carvacrol, lemongrass oil, citral, cinnamon oil, or cinnamaldehyde was added in a range of 0.1–0.5%. In these cases, the low content of encapsulated volatile compounds did not disturb the network structure.

The effect of RH on flavor transfer rates is probably as important as the effect of temperature. Moisture increases the mass transfer rates of gases and vapors through hydrophilic biopolymer films. (Miller et al. 1998) observed substantial increases of D-limonene permeability through whey protein films when RH increased, increasing 2–20-fold when RH varied from 40% to 80%. Quezada-Gallo (1999) observed similar behavior for the permeability of 2-pentanone, 2-heptanone, and ethyl esters through both methylcellulose and wheat gluten films. This behavior was attributed to plasticization of the biopolymer network by water. But if the effect of water on the permeability of volatile compounds is well known and seems obvious, the effect of aroma compounds or essential oils on the water vapor permeability (WVP) is quite a bit more complicated, as displayed in Table 17.5.

The WVP of soy protein isolate or alginate-based films without lipid emulsions tends to increase when essential oils or aroma compounds are encapsulated, whereas that of films based on sodium alginate plus lipid, carrageenan, chitosan, or HPMC decreases. (Atarés et al. 2010) showed that the addition of small proportions of ginger and cinnamon essential oils resulted in a reduction in the water vapor barrier properties of soy protein isolate films. This could be due to the interactions of oil components with some protein tails, which could promote a decrease in the hydrophobic character of the protein matrix. Because the amount of oil incorporated is very low, the lipid discontinuities seem not to be relevant to increasing the tortuosity factor for transfer of water molecules, responsible for the reduction of WVP, although a further increase in cinnamon oil content resulted in a WVP decrease. The effectiveness of cinnamon oil, as compared with ginger oil, in reducing WVP at a fixed protein-to-oil ratio suggests that the former remains partially integrated in the protein network of the dry films. The difficulties in integrating the essential oils or aromas in a hydrophilic network may be due to matrix disruptions and creation of void spaces at the protein–essential oil interface. Therefore, it cannot be assumed that the WVP of edible films is reduced simply by adding a hydrophobic component such as an aroma or essential oil to the formulation, but the impact of lipid addition on the microstructure of the emulsified film is a determining

TABLE 17.5

Influence of Volatile Compound Incorporation on Water Vapor Permeability of Edible Films

Film Matrix	Volatile Compounds	WVP (10^{-10} g/m/s/Pa)	T (°C)	ΔRH (%)
Soy protein isolate	/	1.38	25	33–53
	Cinnamon oil 1%	1.52	25	33–53
	Ginger oil 1%	1.89	25	33–53
Chitosan	/	12.4	25	100–54
	Bergamot oil 3%	6.5	25	100–54
	Tea tree oil 2%	7.4	25	100–54
Alginate	/	2.35	25	0–50
	Garlic oil 0.4%	72.64	25	0–50
Sodium alginate	/	2.13	25	30–84
	n-Hexanal 1%	2.96	25	30–84
Sodium alginate + GBS (emulsion)	/	1.68	25	30–84
	n-Hexanal 1%	1.21	25	30–84
Carrageenan	/	23.5	25	30–84
	n-Hexanal 1%	22.2	25	30–84
Carrageenan + GBS (emulsion)	/	29	25	30–84
	n-Hexanal 1%	25.3	25	30–84
HPMC	/	8.2	25	100–54
	Tea tree oil 40%	5.5	25	100–54

Source: Atarés, L. et al., *J. Food Eng.*, 99, 384, 2010; Sánchez-González, L. et al., *Food Hydrocolloids*, 23, 2102, 2009; Sánchez-González, L. et al., *J. Food Eng.*, 98, 443, 2010; Pranoto, Y. et al., *Food Res. Int.*, 38, 267, 2005; Hambleton, A. et al., *Food Chem.*, 2012. With permission.

factor in water barrier efficiency. In the case of HPMC, chitosan, or carrageenan, the previous explanation for increasing the hydrophobicity does not fit, because WVP increased with essential oil content. The WVP values showed a significant decrease in line with the increase in tea tree oil concentration, following a linear trend and reaching a maximum WVP reduction of about 40% with the incorporation of 2% of the essential oil (Sánchez-González et al., 2010a,b). This behavior is expected, as an increase in the hydrophobic compound fraction usually leads to an improvement in the water barrier properties of films, as was previously reported for essential oil addition in chitosan films (Zivanovic et al., 2005).

The incorporation of *n*-hexanal in an ι-carrageenan film induced a twofold decrease in the permeability of ethyl acetate, ethyl butyrate, and 2-hexanone and a sixfold decrease for D-limonene. The encapsulated *n*-hexanal interacted with CH_2OH or sulfated groups of ι-carrageenan lateral chains, inducing lower permeability, as shown by Hambleton et al. (2008, 2010). When the film contained a lipid dispersed in the carrageenan matrix, the behavior of the aroma compound permeability varied. A decrease was observed by 30% for 2-hexanone, twofold for ethyl acetate, and 100-fold for d-limonene. Conversely, a decrease occurred by 15-fold for ethyl butyrate due to reduced solubility of ethyl butyrate in the hydrophilic ι-carrageenan matrix. The higher the log K, the greater the hydrophobicity of the aroma compound. In contrast, it promotes interactions with the fat. Ethyl butyrate is, then, more retained in fat globules. Films with fat are also less uniform, the fat globule particle size increases, and there are fewer open spaces in the film structure to facilitate the aroma diffusivity. Therefore, the aroma compound transfer is limited. Alginate–lipid emulsion films exhibit a significant increase in ethyl butyrate (×2000), ethyl hexanoate (×10,000), and D-limonene permeability (×300), whereas permeability drops fourfold for 2-hexanone. The permeability of aroma compounds is much more complex than that of gases or water vapor. It depends on the solubility of the volatile compounds in the edible film; their vapor pressure, volatility, hydrophobicity (often expressed as the partition coefficient between water and octanol), solubility (governed by the chemical nature of both film and volatile compound), molecular mobility of the matrix, and volatile diffusion coefficient; and many external parameters such as temperature, pressure, moisture levels, and so forth. Therefore, previous observations of the addition of aroma or essential oils to edible films cannot easily predict what would happen in other situations or conditions, because the physical-chemical conditions have not always been reported.

17.6 Edible Films and Coatings for Non-Volatile Molecule Encapsulation (Peptides, Polyunsaturated Fatty Acids, Antioxidants, and Antibacterials)

Over the last few years, consumer demand for foods of natural origin, high quality, elevated safety, longer shelf life, and fresh taste and appearance has been strongly increasing. Minimally processed and easy-to-eat foods are also requested. Currently, there is an escalating tendency to employ environmentally friendly packaging materials with the intention of substituting for non-degradable materials, thus reducing the environmental impact resulting from waste accumulation. Addressing the environmental issues and concurrently, extending the shelf life and food quality while reducing packaging waste has catalyzed the exploration of new bio-based packaging materials such as edible and biodegradable films. One of the approaches is to use renewable biopolymers such as polysaccharides, proteins, gums, lipids, and their derivatives, from animal and plant origin as described previously, not only to form capsules but also to form films and coatings able to encapsulate particular biomolecules. Such biodegradable/edible packagings/coatings not only ensure food safety but at the same time, are a good source of nutrition.

17.6.1 Nutraceuticals

In parallel with natural foods and wrapping materials, consumer demand is more and more focused on healthy foods to provide more than the basic needs. The encapsulation of nutraceuticals is at present a very hot topic. A great number of research articles are reporting on matrix and encapsulation processes. Milk proteins are largely used for encapsulation and controlled delivery using micelle structures or any other nanostructure (Semo et al., 2007; Livney, 2010). These proteins are often conjugated with polysaccharides to improve the encapsulation and controlled release properties. Maltodextrins and other modified starches are also largely used as encapsulation matrices due to their low cost and ease of use. Cyclodextrins show high potentiality for encapsulation but are limited in use by their industrial cost.

Several nutraceutical molecules can be incorporated into edible coatings, such as vitamins, peptides, polyunsaturated fatty acids (PUFA), or antioxidants to increase the nutritional value of food. The main problem with incorporating nutraceuticals into food is related to stability during storage. These reactive molecules rapidly lose their activity due to oxidation or other chemical reactions. Edible films and coatings are increasingly used to protect these active biomolecules from contact with foods. When incorporated in coatings or encapsulated, their bioactive effect is preserved, and nanoencapsulation could even increase a molecule's bioavailability. All the techniques cited earlier are commonly used for nutraceutical encapsulation with good efficiency. The main research topics are now the controlled release of the nutrient and also, and this is the more complex problem, concerns about the bioavailability and the focused release of the active compound (Chen et al., 2006; Kosaraju et al., 2006; Gonnet et al., 2010; Nair et al., 2010). As an example, the release of PUFA in the brain to limit Alzheimer's disease is a very hot research topic. Polyphenols are used as active compounds to reduce oxidative stress. (Fang and Bandhari 2010) reviewed research on the application of polyphenols. The unpleasant taste of most phenolic compounds can be completely masked by encapsulation. The common technologies for the encapsulation of polyphenols are spray-drying, coacervation, liposome entrapment, inclusion complexion, cocrystallization, nanoencapsulation, freeze-drying, yeast encapsulation, and the use of emulsions. In parallel with the development of preservation techniques, advanced research is developing flavonoid functionalization to improve their nutraceutical use. Common research on simultaneous functionalization and encapsulation should be developed to accelerate commercialization.

17.6.2 Antibacterials

Considering antibacterial molecules, postprocess contamination caused by product mishandling and faulty packaging is responsible for about two-thirds of all microbiologically related recalls in the United States, with most of these recalls originating from contamination of ready-to-eat food products. Antimicrobial agents, sometimes called food preservatives, are components that hinder the growth of microorganisms. According to the definition used by the Commission of the European Communities, preservatives are substances that extend the shelf life of foodstuffs by protecting them against deterioration caused by microorganisms (EU Directive 95/2/EC). Similar rules are applied in the United States, where the U.S. Food and Drug Administration (FDA) defines a preservative as any chemical that when added to food, tends to prevent or retard deterioration. Antimicrobials are used in food to control natural spoilage and to prevent or control the growth of microorganisms, including pathogenic microorganisms (da Silva Malheiros et al., 2010a,b; Drulis-Kawa and Dorotkiewicz, 2010).

Natural antimicrobials can be defined as substances produced by living organisms in their fight with other organisms for space and their competition for nutrients. The main sources of these compounds are plants (secondary metabolites in essential oils and phytoalexins), microorganisms (bacteriocins and organic acids), and animals (lysozyme from eggs and lactoferrins from milk). Across the various sources, the same types of active compounds can be encountered (e.g., enzymes, peptides, and organic acids). Reducing the need for antibiotics, controlling microbial contamination in food, improving shelf life extension technologies to eliminate undesirable pathogens, delaying microbial spoilage, decreasing the development of antibiotic resistance by pathogenic microorganisms, and strengthening immune cells in humans are some of the benefits (Tajkarimi et al., 2010). Most approved food antimicrobials have limited application due to pH or food component interactions. They are amphiphilic and can solubilize or be bound by lipids or hydrophobic proteins in foods, making them less available to inhibit microorganisms in the food product.

The term *bacteriocin* is mostly used to describe the small, heat-stable cationic peptides synthesized by Gram-positive bacteria, namely, lactic acid bacteria (LAB), which display a wider spectrum of inhibition

(Cotter et al., 2005). The bacteriocins produced by LAB offer several desirable properties that make them suitable for food preservation. They are generally recognized as safe (GRAS) substances, are inactive and non-toxic to eukaryotic cells, become inactivated by digestive proteases, and have little influence on the gut microbiota. Bacteriocins are usually pH and heat tolerant, and they have a relatively broad antimicrobial spectrum against many food-borne pathogenic and spoilage bacteria.

Nisin has been increasingly used as a biopreservative for direct incorporation in food as well as in active/edible films. Nisin effectively inhibits Gram-positive bacteria and outgrowth spores of *Bacillus* and *Clostridium*. If nisin is efficient against particular microorganisms, its activity rapidly decreases as it is hydrolyzed in the food product, and bacterial inactivation stops. All studies showed a restarting of bacterial growth a few days after nisin incorporation. Encapsulation and controlled release is an efficient way to avoid this resumption of bacterial growth, because active nisin is slowly delivered into the food or onto the food surface. In a recent study on edible films (Sebti et al., 2007) using HPMC/chitosan and incorporating pure nisin, the author evaluated the effect of nisin on the physical characteristics of films.

17.6.3 Edible Materials for Nutraceuticals or Antibacterial Molecule Encapsulation

The matrix used for nutraceuticals or antimicrobial molecules is of prime importance to allow good preservation or controlled release of these active compounds.

17.6.3.1 HPMC

Cellulose-based materials are widely used, as they offer advantages such as edibility, biocompatibility, barrier properties, and aesthetic appearance as well as being non-toxic and non-polluting and having low cost (Vasconez et al., 2009). HPMC edible films are attractive for food applications, because HPMC is a readily available non-ionic edible plant derivative shown to form transparent, odorless, tasteless, oil-resistant, water-soluble films with very efficient oxygen, carbon dioxide, aroma, and lipid barriers but with moderate resistance to water vapor transport. HPMC is used in the food industry as an emulsifier, film former, protective colloid, stabilizer, suspending agent, and thickener. HPMC is approved for food use by the FDA (21 CFR 172.874) and the EU (EU, 1995); its safety in food use has been affirmed by the Joint Food and Agriculture Organization (FAO)/World Health Organization (WHO) Expert Committee on Food Additives (JECFA) (Burdock, 2007). The tensile strength of HPMC films is high, and their flexibility is neither too high nor too fragile, which makes them suitable for edible coating purposes (Imran et al., 2010b).

17.6.3.2 Polyacetic Acid (PLA)

As a GRAS and biodegradable material, and also because of its biosorbability and biocompatible properties in the human body, PLA and its copolymers (especially polyglycolic acid) attracted pharmaceutical and medical scientists and researchers as carriers for releasing various drugs and agents, such as bupivacaine and many others. In the food domain, little research has been done on the suitability of PLA as an active packaging polymer. PLA is a new corn-derived polymer and needs time to be an accepted and effective active packaging material in the market.

(Van Aardt et al. 2007) studied the release of antioxidants from loaded poly (lactide-co-glycolide) (PLGA) (50:50) films with 2% α-tocopherol, and a combination of 1% butylated hydroxytoluene (BHT) and 1% butylated hydroxyanisole (BHA), into water, oil (food stimulant: Miglyol 812), and milk products at 4 °C and 25 °C in the presence and absence of light. They concluded that in a water medium, PLGA (50:50) showed hydrolytic degradation of the polymer and release of BHT into the water. In Miglyol 812, no degradation or antioxidant release took place even after 8 weeks at 25 °C. Milkfat was stabilized to some extent when light-exposed dry whole milk and dry buttermilk were exposed to antioxidant-loaded PLGA (50:50). They also suggested the potential use of degradable polymers as a unique active packaging option for the sustained delivery of antioxidants, which could be of benefit to the dairy industry by limiting the oxidation of high-fat dairy products, such as ice cream mixes.

17.6.4 Plasticizers

Plasticizers impressively affect the physical properties of biopolymer films (Zhang and Han, 2008). The plasticizer helps to decrease the inherent brittleness of films by reducing intermolecular forces, increasing the mobility of polymer chains, decreasing the glass transition temperature of these materials, and improving their flexibility (Zhang and Han, 2008; Galdeano et al., 2009). Thus, it is important to study the effects of the commonly used polyol glycerol on the homogeneous dispersion of Nisaplin® (nisin, salt, and milk solid) for the formation of composite active films of improved quality. However, plasticizers generally cause increased water permeability, so they must be added at a certain level to obtain a film with the desired flexibility, thickness, and transparency without a significant decrease of mechanical strength and barrier properties to mass transfer (Möller et al., 2004; Jongjareonrak et al., 2006; Brindle and Krochta, 2008).

17.6.5 Molecular Diffusion from Film and Coating to Food Surface

Migration from capsules included in films or coatings can be represented as shown in Figure 17.8.

Diffusion and erosion lead to slow release of the antibacterials, allowing food stabilization for a longer period than by adding the antibacterials directly into the food. Quantitative measurement of the rate at which a diffusion process occurs is usually expressed in terms of diffusivity (also called the *diffusion coefficient*), expressed in square meters per second. The classical theory used to model the diffusion process is based on Fick's law (Crank, 1975; Stannet, 1978). Diffusion in a homogeneous medium is based on the assumption that the rate of transfer, R, of a migrant passing perpendicularly through the unit area of a section is proportional to the concentration gradient between the two sides of the packaging:

$$R = \frac{dM}{Adt} = -D\frac{dC}{dx}$$

where

- D is the diffusion coefficient (m^2/s)
- A is the film area (m^2)
- D is a function of the local diffusant concentration, C (g/m)
- t is time (s)
- x is the thickness of the film or coating (m)

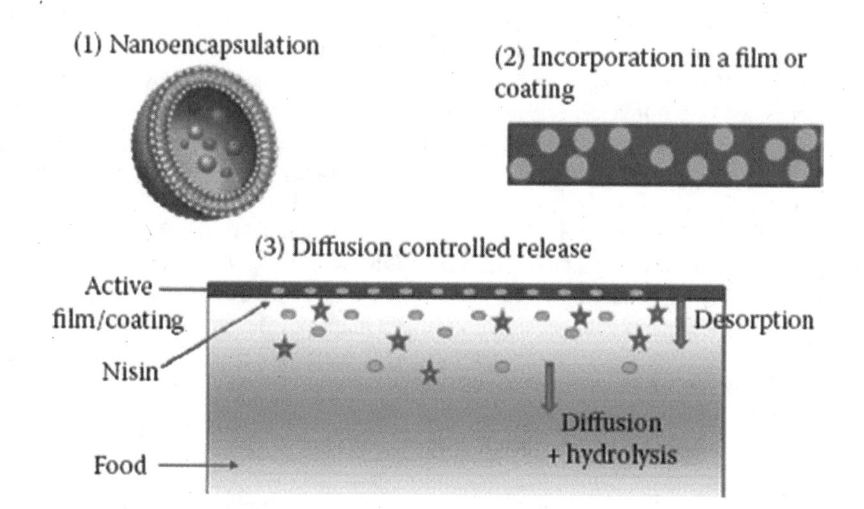

FIGURE 17.8 Nanoencapsulation, film/coating inclusion, and controlled release into food.

The quantity of package components that may migrate from a packaging material into liquid or solid food depends on the chemical and physical properties of the food and the polymer. Various factors, such as migrant concentration, molecular weight, solubility, diffusivity, partition coefficient between polymer and food, time, temperature, polymer and food composition, and structures (density, crystallinity, and chain branching), are the main controlling factors in migration.

Legally, polymers for packaging are regulated through global or specific migration levels. Global migration measures the total amount of all compounds migrating into food simulants independently of migrant composition. Specific migration concerns a given migrant. Several studies have measured global and specific migration from packaging materials to foods (Baner, 1991; Jamshidian et al., 2010). In the case of active films, the global migration limit does not apply, because active compound migration is required. Regulations are now more precise about the possible activity of the packaging and coating.

17.6.6 Chemical Structure of Migrant and Matrix State

The chemical structure of an encapsulated molecule is an important parameter that can influence the partition coefficient and then, the controlled release into a food. Alcohols and short-chained esters have higher partition coefficients in an oil/polymer system than in a water/polymer system. Several studies have attempted to model the relationship between the encapsulated molecule, the composition of the food, and the partition coefficient (Arab Tehrany and Desobry, 2004). It is also known that the matrix crystallinity and glass transition of the matrix are key factors for an efficient controlled release of an active compound. A controlled transition from the glassy to the rubbery state (temperature and water activity) leads to the best system for good food preservation. A lot of work still has to be done to achieve perfect control of the release of an active compound.

17.7 Conclusion

A hot topic in the functional food and pharmaceutical industry is the efficient encapsulation of high-value-added ingredients, such as PUFA, flavors, vitamins, and health-promoting ingredients, in relation to improved functionalities. In addition, because many of the most popular nutritional ingredients on the market today have unpleasant sensory characteristics, keeping the objectionable flavors out of products can sometimes be as important as keeping the enjoyable flavors in. Numerous developments have been made in the field of encapsulated food flavors. This is because of several favorable properties of the encapsulated form of flavors: ease in handling and mixing; stability against air, light, and evaporation; masking of undesirable tastes and aromas; and delivery of ingredients at the desired stage and at specifically targeted release sites. Advances in the development of new wall materials and microencapsulation methods have paved the way for value-added ingredients of higher quality and consistency, enhanced performance, and improved prices. Each encapsulation process, generally developed to solve a particular problem encountered by product development, presents advantages and disadvantages. The relationships among problems, capabilities, and encapsulation methods have been presented. Microencapsulation by spray-drying is the most economical and flexible way for the food industry to encapsulate ingredients. Thus, this technology is now becoming available to satisfy the increasingly specialized needs of the market. In addition, the fluid-bed process is a promising encapsulation technique for large-scale production of flavor powders to be applied in the food industry. The choice of an appropriate technique of encapsulation depends on the properties of the compounds, the degree of stability required during storage and processing, the properties of the food components, the specific release properties required, the maximum obtainable molecule load in the powder, and the production cost.

REFERENCES

Anandaraman, S., Reineccius, G.A. 1986. Stability of encapsulated orange peel oil. *Food Technology*, 11: 88–93.

Arab Tehrany, E., Desobry, S. 2004. Partition coefficient in food/packaging systems. *Food Additives and Contaminants*, 21(12): 1186–1202.

Atarés, L., De Jesús, C., Talens, P., Chiralt, A. 2010. Characterization of SPI-based edible films incorporated with cinnamon or ginger essential oils. *Journal of Food Engineering*, 99: 384–391.

Baner, A.L. 1991. Prediction of solute partition coefficients between polyolefins and alcohols using the regular solution theory and group contribution methods. *Industrial and Engineering Chemistry Research*, 30(7): 1506–1515.

Becker, K., Koszinowski, J., Piringer, O. 1987. Permeation von Riech- und Aromastoffen Durch Polyolefine. *Parfurmerie und Kosmetik*, 68: 268–278.

Ben Arfa, A., Chrakabandhu, Y., Preziosi-Belloy, L., Chalier, P., Gontard, N. 2007. Coating papers with soy protein isolates as inclusion matrix of carvacrol. *Food Research International*, 40: 22–32.

Bertolini, A.C., Siani, A.C., Grosso, C.R.F. 2001. Stability of monoterpenes encapsulated in gum arabic by spray-drying. *Journal of Agricultural and Food Chemistry*, 49: 780–785.

Bouwmeester, H., Dekkers, S., Noordam, M.Y., Hagens, W.I., Bulder, A.S., de Heer, C., ten Voorde, S., Wijnhoven, S., Marvin, H., Sips, A. 2009. Review of health safety aspects of nanotechnologies in food production. *Regulatory Toxicology and Pharmacology*, 53(1): 52–62.

Brazel, C.S. (1999). Microencapsulation: Offering solution for the food industry. *Cereal Foods World*, 44: 388–393.

Brindle, L.P., Krochta, J.M. 2008. Physical properties of whey protein hydroxypropyl methylcellulose blend edible films. *Journal of Food Science*, 73(9): 446–454.

Burdock, G.A. 2007. Safety assessment of hydroxypropyl methylcellulose as a food ingredient. *Food and Chemical Toxicology*, 45(12): 2341–2351.

Burt, S., 2004. Essential oils: Their antibacterial properties and potential applications in foods—A review. *International Journal of Food Microbiology*, 94: 223–253.

Buttery, R.G., Bomben, J.L., Guadagni, D.G., Ling, L.C. 1971. Volatilities of organic flavor compounds in foods. *Journal of Agricultural and Food Chemistry*, 19(6): 1045–1048.

Chen, L., Remondetto, G.E., Subirade, M. 2006. Food protein-based materials as nutraceutical delivery systems. *Trends in Food Science and Technology*, 17(5): 272–283.

Cotter, P.D., Hill, C., Ross, P.R. 2005. Bacteriocins: Developing innate immunity for food. *Nature Reviews Microbiology*, 3(10): 777–788.

Crank, J. 1975. *Mathematics of Diffusion*, 2nd edn. Clavedon Press, Oxford.

Da Silva Malheiros, P., Joner Daroit, D., Brandelli, A. 2010b. Food applications of liposome-encapsulated antimicrobial peptides. *Trends in Food Science and Technology*, 21(6): 284–292.

Da Silva Malheiros, P., Joner Daroit, D., Pesce da Silveira, N., Brandelli, A. 2010a. Effect of nanovesicle-encapsulated nisin on growth of *Listeria monocytogenes* in milk. *Food Microbiology*, 27(1): 175–178.

Debeaufort, F. 1994. Étude des Transferts de Matière au Travers de Films d'Emballages: Perméation de l'Eau et de Substances d'Arôme en Relation avec les Propriétés Physico-chimiques des Films Comestibles. PhD Dissertation, ENS.BANA, Université de Bourgogne, Dijon, France.

Debeaufort, F., Quezada-Gallo, J.A., Voilley, A. 2002. Chapter 24: Edible films and coatings as aroma barrier. In *Protein-Based Edible Films and Coatings*, Gennadios, A. (ed.). CRC Press, Boca Raton, FL, pp. 579–600.

Debeaufort, F., Tesson, N., Voilley, A. 1995. Aroma compounds and water vapour permeability of edible films and polymeric packagings. In *Food and Packaging Materials—Chemical Interactions*, Ackermann, P., Jägerstad, M. and Ohlsson, T. (eds.). The Royal Society of Chemistry, Cambridge, pp. 169–175.

Debeaufort, F., Voilley, A. 1994. Aroma compound and water vapor permeability of edible films and polymeric packagings, *Journal of Agricultural and Food Chemistry*, 42: 2871–2875.

Debeaufort, F., Voilley, A. 1995. Methylcellulose-based edible films and coatings: 1. Effect of plasticizer content on water and 1-octen-3-ol sorption and transport. *Cellulose*, 2: 1–10.

Del Nobile, M.A., Conte, A., Incoronato, A.L., Panza, O. 2008. Antimicrobial efficacy and release kinetics of thymol from zein films. *Journal of Food Engineering*, 89: 57–63.

Djordevic, D., Cercaci, L., Alamed, J., McClements, D.J., Decker, E.A. (2008). Chemical and physical stability of protein- and gum arabic- stabilized oil-in-water emulsions containing limonene. *Journal of Food Science*, 73: C167–C172.

Drulis-Kawa, Z., Dorotkiewicz-Jach, A. 2010. Liposomes as delivery systems for antibiotics. *International Journal of Pharmaceutics*, 387(1–2): 187–198.

Embuscado, M.E., Huber, K.C. 2009. *Edible Films and Coatings for Food Applications*. Springer Science, New York, p. 403.

Fabra, M.J., Chambin, O., Assifaoui, A., Debeaufort, F. 2011. Influence of temperature and salt concentration on the release in liquid media of aroma compounds encapsulated in edible films. *Journal of Food Engineering*, 108(1): 30–36.

Fabra, M.J., Hambleton, A., Talens, P., Debeaufort, F., Chiralt, A., Voilley, A. 2008. Aroma barrier properties of sodium caseinate-based edible films. *Biomacromolecules*, 9(5): 1406–1410.

Fabra, M.J., Hambleton, A., Talens, P., Debeaufort, F., Chiralt, A., Voilley, A. 2009. Influence of interactions on the water and aroma permeabilities of iota-carrageenan-oleic acid-beeswax edible films used for flavour encapsulation. *Carbohydrate Polymers*, 76: 325–332.

Fang, Z., Bhandari, B. 2010. Encapsulation of polyphenols—A review. *Trends in Food Science and Technology*, 21(10): 510–523.

Fayoux, S., Seuvre, A.M., Voilley, A. 1997a. Aroma transfers in and through plastic packagings: Orange juice and d-limonene. A review. Part 1: Orange juice aroma sorption. *Packaging Technology Science*, 10: 69–82.

Fayoux, S., Seuvre, A.M., Voilley, A. 1997b. Aroma transfers in and through plastic packagings: Orange juice and d-limonene. A review. Part 2: Overall sorption mechanism and parameter—A literature survey. *Packaging Technology Science*, 10: 145–160.

Fenaroli, G. (ed.). 1995. *Fenaroli's Handbook of Flavor Ingredients*. CRC Press, Boca Raton, FL.

Franz, R. 1993. Permeation of volatile organic compounds across polymer films. Part I: Development of a sensitive test method suitable for high barrier packaging films at very low permeant vapour pressures. *Packaging Technology Science*, 6: 91.

Friedman, M., Henika, P.R., Levin, C.E., Mandrell, R.E. 2004. Antibacterial activities of plant essential oils and their components against *Escherichia coli* O157:H7 and *Salmonella enterica* in apple juice. *Journal of Agricultural and Food Chemistry*, 52: 6042–6048.

Galdeano, M.C., Mali, S., Grossmann, M.V.E., Yamashita, F., Garcia, M.A. 2009. Effects of plasticizers on the properties of oat starch films. *Materials Science and Engineering C*, 29(2): 532–538.

Gennadios, A. 2002. *Protein-Based Edible Films and Coatings*. CRC Press, Boca Raton, FL, p. 639.

Godshall, M.A. 1997. How carbohydrate influence food flavor. *Food Technology*, 51(1): 63–67.

Gómez-Estaca, J., López de Lacey, A., López-Caballero, M.E., Gomez-Guillén, M.C., Montero, P. 2010. Biodegradable gelatin–chitosan films incorporated with essential oils as antimicrobial agents for fish preservation. *Food Microbiology*, 27(7): 889–896.

Gómez-Estaca, J., Montero, P., Fernández-Martí, F., Alemán, A., Gómez-Guillén, M.C. 2009. Physical and chemical properties of tuna-skin and bovine-hide gelatin films with added aqueous oregano and rosemary extracts. *Food Hydrocolloids*, 23(5): 1334–1341.

Gomez-Estaca, J., Montero, P., Giménez, B., Gómez-Guillén, M.C. 2007. Effect of functional edible films and high pressure processing on microbial and oxidative spoilage in cold-smoked sardine (*Sardina pilchardus*). *Food Chemistry*, 105(2): 511–520.

Gonnet, M., Lethuaut, L., Boury, F. 2010. New trends in encapsulation of liposoluble vitamins. *Journal of Controlled Release*, 146(3): 276–290.

Green, B.K., Scheicher, L. 1955. Pressure sensitive record materials. US Patent number 2, 217, 507, Ncr C.

Guilbert, S., Gontard, N. 2005. Agro-polymers for edible and biodegradable films: Review of agricultural polymeric materials, physical and mechanical characteristics. In *Innovations in Food Packaging*, Han, J.H. (ed.). Elsevier Academic Press, Oxford, pp. 263–276.

Hambleton, A., Debeaufort, F., Beney, L., Karbowiak, T., Voilley, A. 2008. Protection of active aroma compound against moisture and oxygen by encapsulation in biopolymeric emulsion-based edible films. *Biomacromolecules*, 9(3): 1058–1063.

Hambleton, A., Debeaufort, F., Bonnotte, A. Voilley, A. 2009b. Influence of alginate emulsion-based films structure on its barrier properties and on its protection of microencapsulated aroma compound. *Food Hydrocolloids*, 23(8): 2116–2124.

Hambleton, A., Fabra, M.J. Debeaufort, F. Brun-Dury, C. Voilley, A. 2009a. Interface and aroma barrier properties of iota-carrageenan emulsion-based films used for encapsulation of active food compounds. *Journal of Food Engineering*, 93: 80–88.

Hambleton, A., Perpiñan-Saiz, N., Fabra, M.J., Voilley, A., Debeaufort, F. 2012. The Schroeder paradox or how the state of water affects the moisture transfers through edible films. *Food Chemistry*, 132(4): 1629–2230.

Hambleton, A., Voilley, A., Debeaufort, F. 2010. Transport parameters for aroma compounds through ι-carrageenan and sodium alginate-based edible films. *Food Hydrocolloids*, doi:10.1016/j.foodhyd.2010.10.010

Han, J. 2002. Protein-based edible films and coatings carrying antimicrobial agents. In *Protein-Based Films and Coatings*, Gennadios, A. (ed.). CRC Press, Boca Raton, FL, pp. 485–500.

Heng, L., Van Koningsveld, G.A., Gruppen, H., Van Boekel, M., Vincken, J.P., Roozen, J.P., Voragen, A.G. 2004. Protein-flavour interactions in relation to development of novel protein foods. *Trends Food Science and Technology*, 15(3): 217–224.

Imran, M., El-Fahmy, S., Revol-Junelles, A.M., Desobry, S. 2010a. Cellulose derivative based active coatings: Effects of nisin and plasticizer on physico-chemical and antimicrobial properties of hydroxypropyl methylcellulose films. *Carbohydrate Polymers*, 81: 219–225.

Imran, M., Revol-Junelles, A.-M., Martyn, A., Tehrany, E.A., Jacquot, M., Linder, M., Desobry, S. 2010b. Active food packaging evolution: Transformation from micro- to nanotechnology. *Critical Reviews in Food Science and Nutrition*, 50(9): 799–821.

Jamshidian, M., Arab Tehrany, E., Imran, M., Jacquot, M., Desobry, S. 2010. PLA: Production, application and controlled release. *Comprehensive Reviews: Food Science and Food Safety*, 9(5): 552–571.

Jongjareonrak, A., Benjakul, S., Visessanguan, W., Tanaka, M. 2006. Effects of plasticizers on the properties of edible films from skin gelatin of bigeye snapper and brownstripe red snapper. *European Food Research and Technology*, 222(3–4): 229–235.

Kim, Y.D., Moor, C.V. 1996. Microencapsulation properties of gum arabic and several food proteins: Spray-dried orange oil emulsion particles. *Journal of Agricultural Food Chemistry*, 44(5): 1314–1320.

Kobayashi, M., Kanno, T., Hanada, K., Osanai, S.I. 1995. Permeability and diffusivity of d-limonene vapor in polymeric sealant films. *Journal of Food Science*, 60: 205–209.

Kosaraju, S.L., D'ath, L., Lawrence, A. 2006. Preparation and characterisation of chitosan microspheres for antioxidant delivery. *Carbohydrate Polymers*, 64(2): 163–167.

Kurek, M., Debeaufort, F. 2010, Development of an antimicrobial coating for packaging films: Physicochemical and microbiological approaches. Doctoral school intermediate PhD report, University of Burgundy, Dijon, France.

Laohakunjit, N., Kerdchoechuen, O. 2007. Aroma enrichment and the change during storage of non-aromatic milled rice coated with extracted natural flavour. *Food Chemistry*, 101: 339–344.

Livney, Y.D. 2010. Milk proteins as vehicles for bioactives. *Current Opinion in Colloid and Interface Science*, 15(1–2): 73–83.

Lubbers, S., Landy, P., Voilley, A. 1998. Retention and release of aroma compounds in foods containing proteins. *Food Technology*, 52(5): 68–74, 208–214.

Madene, A., Jacquot, M., Scher, J., Desobry, S. 2006. Flavour encapsulation and controlled release—A review. *International Journal of Food Science and Technology*, 41(1): 1–21.

Marcuzzo, E., Debeaufort, F., Hambleton, A., Sensidoni, A., Tat, L., Beney, L., Voilley, A. 2011. Encapsulation of aroma compounds in biopolymeric emulsion-based edible films to prevent oxidation. *Food Research International.*

Marcuzzo, E., Sensidoni, A., Debeaufort, F., Voilley, A. 2010. Encapsulation of aroma compounds in biopolymeric emulsion based edible films to control flavour release. *Carbohydrate Polymers*, 80(3): 984–988.

Martin-Belloso, O., Rojas-Grau, M.A., Soliva-Fortuny, R. 2009. Delivery of flavour and active ingredients using edible films and coatings. In *Edible Films and Coatings for Food Applications*, Embuscado, M.E. and Huber, K.C. (eds.). Springer Science, New York, pp. 295–313.

Mastromatteo, M., Barbuzzi, G., Conte, A., Del Nobile, M.A. 2009. Controlled release of thymol from zein based film. *Innovative Food Science and Emerging Technologies*, 10(2): 222–227.

McHugh, T.H., Huxsoll, C.C., Krochta, J.M. 1996. Permeability properties of fruit puree edible films. *Journal of Food Science*, 61: 88–91.

Miller, K.S., Krochta, J.M. 1997. Oxygen and aroma barrier properties of edible films: A review. *Trends in Food Science and Technology*, 8: 228–237.

Miller, K.S., Upadhyaya, S.K., Krochta, J.M. 1998. Permeability of d-limonene in whey protein films. *Journal of Food Science*, 63: 244–247.

Möller, H., Grelier, S., Pardon, P., Coma, V. 2004. Antimicrobial and physicochemical properties of chitosan-HPMC-based films. *Journal of Agricultural and Food Chemistry*, 52(21): 6585–6591.

Monedero, M., Hambleton, A., Talens, P., Debeaufort, F., Chiralt, A., Voilley, A. 2010. Study of the retention and release of n-hexanal from soy protein isolate-lipid composite films. *Journal of Food Engineering*, 100: 128–133.

Nair, H.B., Sung, B., Yadav, V.R., Kannappan, R., Chaturvedi, M.M., Aggarwal, B.B. 2010. Delivery of anti-inflammatory nutraceuticals by nanoparticles for the prevention and treatment of cancer. *Biochemical Pharmacology*, 80(12): 1833–1843.

Paik, J.S., Writer, M.S., 1995. Prediction of flavor sorption using the Flory-Huggins equation. *Journal of Agriculture and Food Chemistry*, 43: 175–178.

Ponce, A.G., Roura, S.I., del Valle, C.E., Moreira, M.R. 2008. Antimicrobial and antioxidant activities of edible coatings enriched with natural plant extracts: In vitro and in vivo studies. *Postharvest Biology and Technology*, 49: 294–300.

Pranoto, Y., Rakshit, S.K., Salokhe, V.M. 2005b. Enhancing antimicrobial activity of chitosan films by incorporating garlic oil, potassium sorbate and nisin. *LWT—Food Science and Technology*, 38(8): 859–865.

Pranoto, Y., Salokhe, V.M., Rakshit, S.K. 2005a. Physical and antibacterial properties of alginate-based edible film incorporated with garlic oil. *Food Research International*, 38: 267–272.

Quezada-Gallo, A., Debeaufort, F., Voilley, A. 1999b. Mechanism of aroma transport through edible and plastic packagings. In *New Developments in the Chemistry of Packaging Materials*, Rish, S. (ed.). ACS Books, Dallas, TX, pp. 125–140.

Quezada-Gallo, J.A. 1999. Influence de la Structure et de la Composition de Réseaux Macromoléculaires sur les Transferts de Molécules Volatiles (eau et Arômes). Application aux Emballages Comestibles et Plastiques. PhD Dissertation, Université de Dijon, Dijon, France.

Quezada-Gallo, J.A., Debeaufort, F., Voilley, A. 1999a. Interactions between aroma and edible films. 1. Permeability of methylcellulose and polyethylene films to methyl ketones. *Journal of Agriculture and Food Chemistry*, 47: 108–113.

Raybaudi-Massilia, R., Mosqueda-Melgar, J., Martin-Belloso, O. 2006. Antimicrobial activity of essential oils on *Salmonella Enteritidis*, *Escherichia coli*, and *Listeria innocua* in fruit juices. *Journal of Food Protection*, 69: 1579–1586.

Reineccius, G. 2009. Edible films and coatings for flavour encapsulation. In *Edible Films and Coatings for Food Applications*, Embuscado, M.E. and Huber, K.C. (eds.). Springer Science, New York, pp. 269–294.

Rojas-Grau, M., Raybaudi-Massilia, R.M., Soliva-Fortuny, R.S., Avena-Bustillos, R.J., McHugh, T.H., Martin-Belloso, O. 2007. Apple puree-alginate edible coating as carrier of antimicrobial agents to prolong shelf-life of fresh-cut apples. *Postharvest Biology and Technology*, 45: 254–264.

Rojas-Grau, M.A., Avena-Bustillos, R., Friedman, M., Henika, P., Martin-Belloso, O., McHugh, T., 2006. Mechanical, barrier and antimicrobial properties of apple puree edible films containing plant essential oils. *Journal of Agricultural and Food Chemistry*, 54: 9262–9267.

Rupasinghe, H.P., Boulter-Bitzer, J., Ahn, T., Odumeru, J. 2006. Vanillin inhibits pathogenic and spoilage microorganisms in vitro and aerobic microbial growth in fresh-cut apples. *Food Research International*, 39: 575–580.

Sánchez-González, L. 2010. Caracterizacion y aplicacion de recubrimientos antimicrobianos a base de polisacaridos y aceites esenciales. PhD Dissertation, Universidad Politecnica de Valencia, Valencia, Spain, p. 309.

Sánchez-González, L., Chafer, M., Chiralt, A., Gonzalez-Martinez, C. 2010a. Physical properties of edible chitosan films containing bergamot essential oil and their inhibitory action on *Penicillium italicum*. *Carbohydrate Polymers*, 82: 277–283.

Sánchez-González, L., Cháfer, M., González-Martínez, C., Chiralt, A., Desobry, S. 2011. Study of the release of limonene present in chitosan films enriched with bergamot oil in food simulants. *Journal of Food Engineering*, 105(1), 138–143.

Sánchez-González, L., Gonzalez-Martinez, C., Chiralt, A., Chafer, M. 2010b. Physical and antimicrobial properties of chitosan–tea tree essential oil composite films. *Journal of Food Engineering*, 98: 443–452.

Sánchez-González, L., Vargas, M., Gonzalez-Martinez, C., Chiralt, A., Chafer, M. 2009. Characterization of edible films based on hydroxypropylmethylcellulose and tea tree essential oil. *Food Hydrocolloids*, 23: 2102–2109.

Sebti, I., Chollet, E., Degraeve, P., Noel, C., Peyrol, E. 2007. Water sensitivity, antimicrobial, and physicochemical analyses of edible films based on HPMC and/or chitosan. *Journal of Agriculture and Food Chemistry*, 55(3): 693–699.

Semo, E., Kesselman, E., Danino, D., Livney, Y.D. 2007. Casein micelle as a natural nano-capsular vehicle for nutraceuticals. *Food Hydrocolloids*, 21(5–6): 936–942.

Shimoni, E. 2009. Nanotechnology for foods: Delivery systems. In *Global Issues in Food Science and Technology*, Gustavo, B.-C., Alan, M., David, L., Walter, S., Ken, B. and Paul, C. Eds., Academic Press, San Diego, CA, pp. 411–424.

Solms, J., Osman-Ismail, F., Beyler, M. 1973. The interaction of volatiles with food components. *Canadian Institute of Food Science and Technology Journal*, 6: A10–A16.

Stannet. 1978. The transport of gases in synthetic polymeric membranes—An historic perspective. *Journal of Membrane Science*, 3: 97–115.

Tajkarimi, M.M., Ibrahim, S.A., Cliver, D.O. 2010. Antimicrobial herb and spice compounds in food. *Food Control*, 21(9): 1199–1218.

Tunc, S., Duman, O. 2011. Preparation of active antimicrobial methyl cellulose/carvacrol/montmorillonite nanocomposite film and investigation of carvacrol release. *LWT—Food Science and Technology*, 44: 465–472.

Van Aardt, M., Duncan, S.E., Marcy, J.E., Long, T.E., O'Keefe, S.F., Sims, S.R. 2007. Release of antioxidants from poly(lactide-co-glycolide) films into dry milk products and food simulating liquids. *International Journal of Food Science and Technology*, 42(11): 1327–1337.

van Ruth, S.M., King, C., Giannouli, P. 2002. Influence of lipid fraction, emulsifier fraction, and mean particle diameter of oil-in-water emulsions on the release of 20 aroma compounds. *Journal of Agricultural and Food Chemistry*, 50(8): 2365–2371.

Vasconez, M.B., Flores, S.K., Campos, C.A., Alvarado, J., Gerschenson, L.N. 2009. Antimicrobial activity and physical properties of chitosan-tapioca starch based edible films and coatings. *Food Research International*, 42(7): 762–769.

Wyler, L., Solms, J. 1982. Starch flavour complexes III. Stability of dried starch-flavor complexes and other dried flavour preparations. *Lebensmittel-Wissenschaft und Technologie*, 15(2): 93–97.

Zhang, Y., Han, J.H. 2008. Sorption isotherm and plasticization effect of moisture and plasticizers in pea starch film. *Journal of Food Science*, 73(7): 313–324.

Zivanovic, S., Chi, S., Draughon, A.F. 2005. Antimicrobial activity of chitosan films enriched with essential oils. *Journal of Food Science*, 70(1): 1145–1151.

18

Microencapsulation of Phase Change Materials

Jessica Giro-Paloma, Mònica Martínez, A. Inés Fernández, and Luisa F. Cabeza

CONTENTS

18.1 Introduction

Energy storage is an important area of research, as it is one of the key issues to improve the efficiency and the economic feasibility of energy systems. There are different types of energy storage[1,2]:

1. Mechanical energy storage, which includes gravitational energy storage, compressed air energy storage, and flywheels. The first two can be used for large- utility-scale energy storage, and flywheels are appropriate for intermediate storage.
2. Electrical storage, which can be used through a battery when it is charged by connecting it to a source of direct electric current. When it is discharged, the stored chemical energy is converted into electrical energy.
3. Thermal energy storage (TES),[3–5] which is stored as a change in the internal energy of a material as sensible or latent heat.
   a. Sensible heat storage (SHS): see Equation 18.1. The thermal energy (Q) is stored by raising the temperature of a solid or liquid, using the heat capacity (C_p at constant pressure) and the change in temperature of the material during the charging/discharging process. T_f is the final temperature after heating, T_i is the initial temperature (before heating), and m is the mass.

$$Q = \int_{T_i}^{T_f} mC_p \mathrm{d}T = mC_p\left(T_f - T_i\right) \tag{18.1}$$

   b. Latent heat storage (LHS)[6,7]: see Equation 18.2. Latent heat storage is based on the heat absorption or release when a storage material undergoes a phase change from solid to liquid, or liquid to gas (or vice versa); Δh_m is the enthalpy of fusion, dT is the temperature difference, a_m is the fraction melted, m is the mass of material, C_{sp} is the average specific heat between T_i and T_m (J/kg/K), and C_{lp} is the average specific heat between T_m and T_f (J/kg/K).

$$\begin{aligned} Q &= \int_{T_i}^{T_m} mC_p \mathrm{d}T + ma_m\Delta h_m + \int_{T_m}^{T_f} mC_p \mathrm{d}T + mC_p\left(T_f - T_i\right) \\ &= m\left[C_{sp}\left(T_m - T_i\right) + a_m\Delta h_m + C_{lp}\left(T_f - T_m\right)\right] \end{aligned} \tag{18.2}$$

4. Thermochemical energy storage relies on the energy absorbed and released in breaking and reforming molecular bonds in a completely reversible chemical reaction. See Equation 18.3, where a_r is the fractioned reacted, m is the mass, and Δh_r is the endothermic heat of reaction.

$$Q = a_r m \Delta h_r \tag{18.3}$$

Materials used in latent heat storage are known as phase change materials (PCMs). The requirements for a PCM to be used as latent heat storage material are

- *Thermophysical properties*: Suitable phase-transition temperature; completely reversible freeze/melt cycle; low vapor pressure (P_v); high density (ρ); large change in enthalpy (ΔH), that is, high latent heat of transition; large specific heat capacity (C_p); good heat transfer, which means large thermal conductivity (k); no subcooling; small volume change (ΔV)
- *Kinetic properties*: High crystallization rate
- *Chemical properties*: Small volume pressure, low vapor pressure, compatibility with other materials, long-term chemical stability, non-toxic, no fire hazard
- *Economical*: Low price, recyclable, abundant

These materials have been studied for the last 40 years, and the most employed are hydrated salts, paraffin waxes, fatty acids, and eutectic mixtures. There are several applications for each PCM described in the literature, such as in textiles, [8] packaging, food transport, medical therapies, buildings,[9,10] industry, etc. The possible incorporation of PCMs in building materials has attracted the attention of researchers to investigate their ability to reduce energy consumption,[11] because improving the energy efficiency in buildings improves the total energy efficiency of a society, and it has important benefits for the economy[12] and the environment. Nevertheless, it is crucial to select the proper PCM depending on the final application, although it has been shown that in the catering sector, biomedical products, and the construction area, it is possible to modify the phase change temperature of the PCM by modifying its weight ratio, creating a binary core material, as Ma et al.[13,14] propose in their studies.

One of the applications for PCMs is in the field of *food conservation*, as in the studies by Lu and Tassou[15] and Oró et al.[16] Following in this area, Oró et al.[17] studied the effect of placing PCM outside a freezer, and the results showed a good ice cream quality. Moreover, drinks containers also can be mixed with these types of materials to store energy and release it when needed. As an example of this, Oró et al.[18] studied chilly bins to enhance thermal performance. Another application is *refrigeration*,[19] in which PCMs can be applied to refrigeration systems by solar cooling, as in the study of Gil et al.[20] It is possible to use PCMs in cold storage applications, as Oró et al.[21] evaluated the compatibility of PCMs with some metallic materials. Besides the studies on metal/PCM compatibility, Castellón et al.[22] evaluated plastic/PCM compatibility, concluding that low-density polypropylene (LDPE) and polypropylene (PP) show the worst encapsulation behavior compared with high-density polypropylene (HDPE). For this reason, the hardness and Young's modulus evaluation with two different experimental procedures for different type of polymers by applying nanoindentation was considered interesting, as Giro-Paloma et al.[23] studied. Subcooling is an ordinary problem in PCM cooling applications. It happens when a material is cooled and the crystallization does not start at the freezing temperature.[24] Huang et al.[25] and Günther et al.[26] evaluated the subcooling in PCM emulsions, remarking that in phase change slurry (PCS) applications, the changed nucleation and solidification behavior is critical. Hence, PCS are another important application widely employed in active pumping systems and can be used for refrigeration,[27] air conditioning,[28–30] and cold storage[31] applications. Taking into account *domestic heating applications* and *domestic hot water*,[32–35] there are some investigations studying the way to evaluate the domestic demand and improve the possible losses by using a tank. Another important application in the PCM area is *solar thermal energy storage*, because it is used for space heating, power generation, and other applications.[36] The attention received by this application is due to the large storage capacity and the isothermal nature of the storage process.[37] The main disadvantage is that it is accessible only for about 2000 h/year in most places. Hence, it is essential to study alternative methodologies to store solar thermal energy for the off hours. For this reason, PCM use is a key point in storing thermal energy with the associated reversible heat transfer, as Farid et al.[38] showed in their paper, where PCMs have to be encapsulated in these systems to prevent the large drop in heat transfer rates during the melting and solidification process.

An important drawback that researchers have found when PCMs are mixed with or incorporated into other materials is leakage. Because of this, most times, it is necessary to use them in a microencapsulated form to avoid leakage. Microencapsulation consists of enclosing a PCM in a microscopic polymer container. There are many advantages to microencapsulating PCMs, such as increasing the heat transfer area, reducing the PCM's reactivity toward the outside environment, and controlling the changes in the storage material volume as phase change occurs. This methodology is solving several problems, but at the same time, it has to be improved with more investigation in this field. Some of the aspects to be improved are related to the mechanical properties of the microcapsules, because when they are used in active systems, they break when the slurries are pumped and the microcapsules collide among them. In the case of passive systems,[39] the principal problem is the compression force against the microcapsules when they are mixed with building materials such as mortar or concrete and integrated into a building wall.[40] To avoid the fracture of the microcapsules, it is important to evaluate the whole shell/core system.

Besides, due to leakage in the liquid state when mixing the PCM with building construction materials,[7] and because of its thermal reliability, researchers have investigated the possibility of making a polymeric container for the paraffin waxes, using encapsulation technology and producing capsules named *encapsulated phase change material*.[41] When the sizes of these capsules are in the micrometer range, they are called *microencapsulated phase change material* (MPCM), which consists of a shell and a core made of the PCM. At this point, Kuznik et al.[42] revealed three relevant parameters for evaluating the MPCM quality, which are the mean diameter, the thickness of the shell, and the PCM mass percentage compared with the total mass of the capsule. The microcapsule skin has to be strong enough to hold the force generated in the system due to the volumetric changes during the phase change process of the PCM and due to the forces in the whole system. Moreover, the thickness of the coating material has to ensure the efficiency of the encapsulated PCM. Therefore, after checking that the PCM chosen is a good material to store energy, it is essential to model the system.[43]

18.2 Types of Phase Change Materials

The three types of PCM are organic, inorganic, and eutectic PCMs, as described in Table 18.1.

Cabeza et al.[44] listed numerous PCMs in their review, specifying their type, melting temperature, heat of fusion, thermal conductivity, density, and source. Figure 18.1 represents some PCMs that are in use, plotting the latent heat of fusion (kJ/kg) against the melting point (°C). They are classified by their application in buildings, including commercial buildings (black color).[45]

18.2.1 Organic PCMs

This type of PCM is divided into paraffin compounds and non-paraffin compounds (fatty acids mostly). The principal advantages of organic PCM are their chemical and thermal stability; they are non-corrosive, they are recyclable, and they have no subcooling. On the other hand, the disadvantages are their flammability, low thermal conductivity (k), and low phase change enthalpy.

18.2.1.1 Paraffin Compounds

These are tasteless, odorless, white, translucent, solid hydrocarbons. The source of paraffin wax is a by-product of petroleum refinery. Typically, waxes are produced as extracted residues during the dewaxing of lubricant oil. They consist of carbon (C) and hydrogen (H) atoms joined by single bonds with the general formula: C_nH_{2n+2}, where n is the number of carbons (C). Hydrocarbons with more than 17 C atoms

TABLE 18.1

Classification of Phase Change Materials (PCMs)

Organic	Inorganic	Eutectic
Paraffin compounds	Salt hydrates	Organic–organic
Fatty acids	Metals	Inorganic–inorganic
	Salts	Organic–inorganic

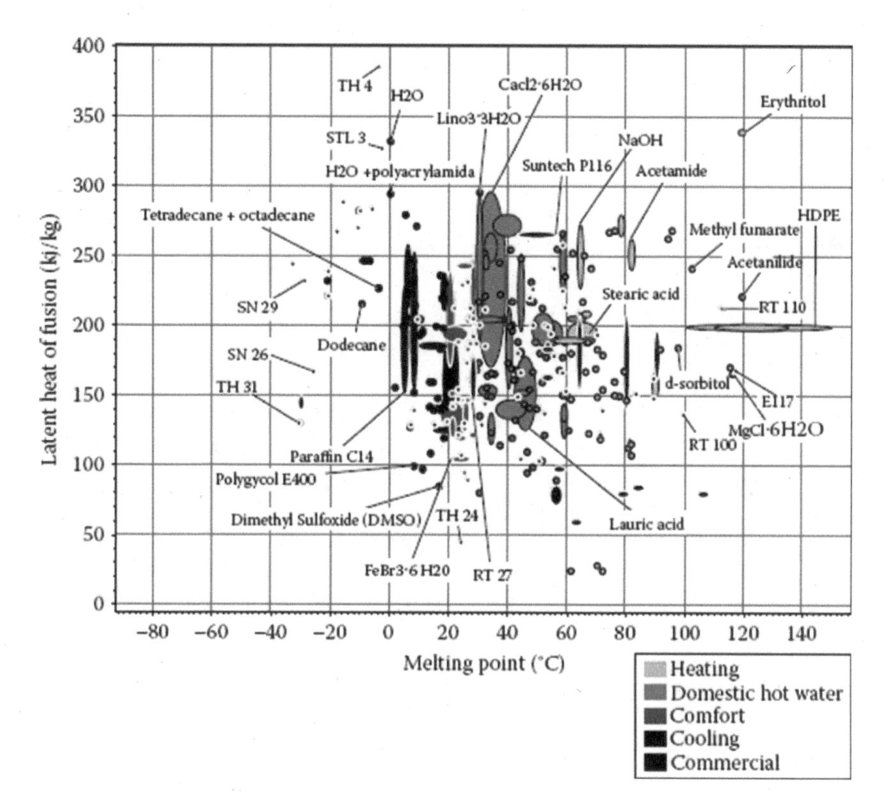

FIGURE 18.1 Plot of PCM latent heat of fusion (kJ/kg) versus melting point (°C). PCMs are classified by their application in buildings.

per molecule are waxy solids at room temperature. They are considered as paraffin wax when the number of carbon atoms is in the range of 20–40. Paraffin wax is used in paper coating, candle manufacture, protective sealants for food products and beverages, glass-cleaning preparations, floor polishing, and stoppers for acid bottles. In Table 18.2, some paraffin waxes used in TES systems are listed, detailing the number of C, the molecular weight, the melting point, and the latent heat of each one.[46] Moreover, in Figure 18.2, the melting point (°C) and the latent heat of fusion (kJ/kg) of some paraffin waxes used as PCMs in building applications are represented.

18.2.1.2 Fatty Acids

These are carboxylic acids with a long hydrocarbon chain, with the general formula $CH_3(CH_2)_{2n}COOH$, where n is between 4 and 14 C atoms. Fatty acids are non-toxic, with low corrosion activity and color, and chemically and thermally stable. Moreover, they can be obtained from natural products, not derived from fossil fuels. Fatty acids can be saturated (single bonds) or unsaturated (with one or more double bonds). Unsaturated fatty acids have lower melting points due to their molecular structure, which allows closer molecular interactions. Moreover, unsaturated fatty acids have *cis* and *trans* chain configurations, in which the intermolecular interactions are weaker than in saturated ones. Besides, different fatty acids can be mixed to design PCMs with different melting temperatures.

18.2.2 Inorganic PCMs

The two main subgroups for inorganic PCMs are the salt hydrates, salts, and the metals. Inorganic PCMs have fewer advantages than organic ones. The main advantage is that this type of PCM has higher phase change enthalpy, but the corrosion, subcooling, phase segregation, lack of thermal stability, and phase separation are the most important disadvantages.

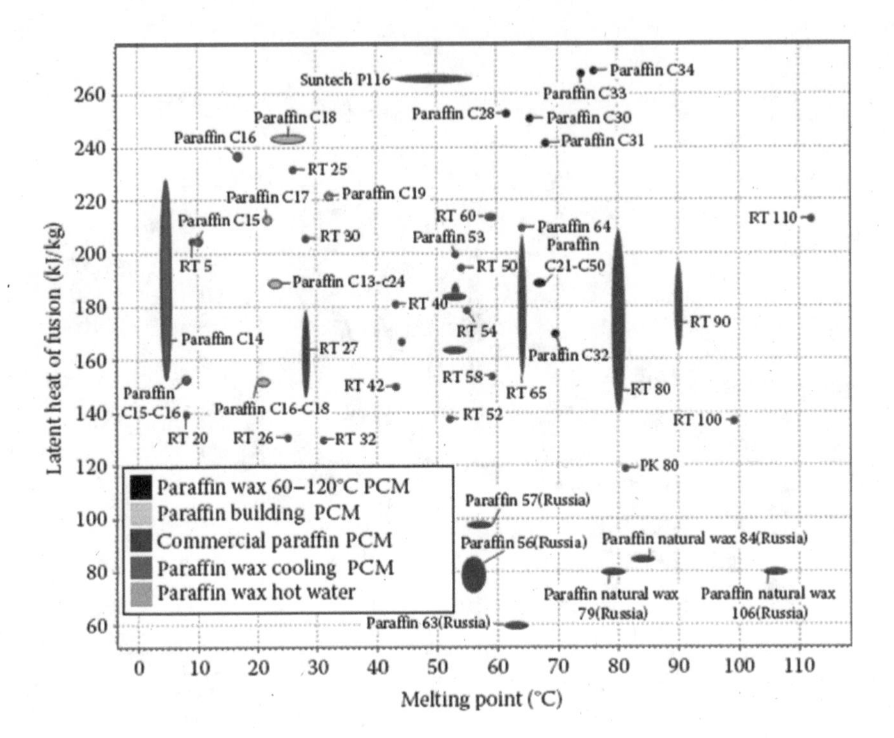

FIGURE 18.2 Paraffinic PCMs classified according to application in buildings by their melting point (°C) and their latent heat of fusion (kJ/kg).

18.2.2.1 Salt Hydrates

These have water molecules combined in a definite ratio as part of the crystal, with the general formula $M{\bullet}nH_2O$, where M is the salt. They are solid at room temperature, and when the melting point is reached, the salt starts to dissolve in the crystal water. There is a situation that will decrease the storage capacity of the salt, known as *incongruent melting*, where sometimes its own water is insufficient to dissolve all the salt. As a consequence, the salt can precipitate at the bottom, and only the salt dissolved in the solution will be involved in the phase change. A good solution to this problem is to build a container to prevent loss of water.

18.2.2.2 Salts

These types of inorganic PCM are those for which the working temperature ranges between 120 and 600 °C, such as nitrates ($NaNO_3$, KNO_3), hydroxides (KOH), chlorides (NaCl, $MgCl_2$), carbonates (Na_2CO_3, K_2CO_3), and fluorides (KF).[9,44]

18.2.2.3 Metals

This subcategory of inorganic PCMs includes the low-melting metals and metal eutectics. They are good candidates to consider when the volume is a key point in a system (because of their high heat of fusion per unit volume) and when the required temperature of the phase change has to be quite high. This type of PCM has high thermal conductivity, low specific heat, and low vapor pressure.

18.2.3 Eutectics

These are a combination of chemical compounds or elements that have a single chemical composition and solidify at a lower temperature than any other composition fabricated of the same ingredients. The

TABLE 18.2
Paraffin Waxes

	Number of Carbons	Molecular Weight (g/mol)	Melting Point (°C)	Latent Heat (J/g)
Heptane	7	100	−90.55	141
Octane	8	114	−56.75	181
Nonane	9	128	−53.45	170
Decane	10	142	−29.65	202
Undecane	11	156	−25.55	177
Dodecane	12	170	−9.55	216
Tridecane	13	184	−5.35	196
Tetradecane	14	198	5.75	227
Pentadecane	15	212	9.95	207
Hexadecane	16	226	18.15	236
Heptadecane	17	240	21.95	214
Octadecane	18	254	28.15	244
Nonadecane	19	268	32.05	222
Eicosane	20	282	36.65	248
Heneicosane	21	296	40.25	213
Docosane	22	310	44.05	252
Tricosane	23	324	47.55	234
Tetracosane	24	338	64.85	255
Pentacosane	25	352	50.65	238
Hexacosane	26	366	53.55	250
Heptacosane	27	380	56.35	235
Octacosane	28	394	58.75	254
Nonacosane	29	408	63.25	239
Triacontane	30	422	65.45	252

combinations can be organic–organic, inorganic–inorganic, or organic–inorganic.[1] There are numerous eutectic mixtures suitable to be used as PCMs, and they are usually preferred in cooling applications.

Of the different PCMs described, paraffin waxes are the most used PCMs in MPCM systems, because of their latent heat and TES capacity, their abundance, their low cost, and the large number of applications. Furthermore, these materials are very stable after several charging/discharging cycles,[47] and for this reason, the thermal reliability using paraffin waxes as PCMs is satisfactory for MPCM[48] systems. Moreover, it is well known that they exhibit a slow thermal response due to their relatively low thermal conductivity, and this low thermal conductivity can be increased by introducing some additives, such as graphite powder and metal particles.[49] The most commonly employed paraffin wax PCM is *n*-octadecane (often used in building applications[50–52]). Moreover, *n*-hexadecane, *n*-docosane, n-eicosane, *n*-nonadecane, and fatty acids are widely employed as PCMs in MPCM systems.

18.3 Materials Used to Microencapsulate PCMs

There are a lot of studies about the shells of microcapsules. The shell of an MPCM has to be strong enough to stand all the forces they have to support in an active or passive system. There are many possibilities for a shell, such as polymeric, glass,[50] or metallic,[53] but we will focus on the polymeric ones. The most frequently employed material is the thermoplastic acrylic polymer[54–56] polymethylmethacrylate (PMMA),[57–61,70] because it has good mechanical properties and good protection against the environment, and it is cheap and inert. Moreover, BASF® has a variety of products with this polymeric shell for MCPM. In fact, there are some studies taking into account Micronal® sample, as Castellón et al.[62] and Tzvetkov et al.[63] ones. PMMA has good compatibility with a wide variety of PCMs; for example, with fatty acids[71,72] such as stearic acid, palmitic acid, myristic acid, and lauric acid (see Table 18.3). Alkan and Sarı[64] studied their use for latent heat thermal energy storage (LHTES).

Another material often used as a shell in an MPCM system is melamine–formaldehyde resin (MF). Su et al.[65] studied the influence of temperature on deformation for HDPE, polystyrene (PS), and polyurethane (PUR), concluding that the yield point of MPCM decreases with increasing temperature. Moreover, SiO_2 is also a very common non-polymeric shell used in MPCM systems.[66–68]

18.4 Types of Microencapsulated Phase Change Materials

Microcapsules with good mechanical resistance are crucial to allow reversible liquid–solid–liquid phase transitions and to protect the PCM during the whole product life. Microencapsulation is a process of

TABLE 18.3

Most Employed PCM/Shell Combinations

	PCM						
			Fatty Acids				Mixture Fatty Acids
Shell	**Paraffin**	***n*-Octadecane**	**Palmitic Acid**	**Stearic Acid**	**Lauric Acid**	**Myristic Acid**	
Polymethyl methacrylate (PMMA)	[60, 62, 69]	[51, 118, 135]	[64]	[64, 91]	[64]	[64]	
Melamine–formaldehyde (MF) resin	[65, 80]	[106]					[96]
SiO_2	[66–68, 93]	[114]			[92]		
PMMA+SiO_2	[70]						
Methyl methacrylate (MMA)		[55]					
Polyethyl methacrylate (PEMA)		[51]					
Styrene–methyl methacrylate	[86]						
Urea–formaldehyde (UF) resin		[134]					[96]
β-naphthol–formaldehyde							[96]
High-density polyethylene (HDPE)	[68, 71]				[71]		
HDPE/wood flour (HDPE/WF)	[49]						
Polycarbonate (PC)				[72]			
Polyvinyl chloride (PVC)	[71]				[71]		
Polyethylene terephthalate (PET)	[134]	[134]					
Polystyrene (PS)	[8]	[105]					
Polyurethane (PUR)	[71]	[134]			[71]		
Polyvinyl acetate (PVA)	[71]				[71]		
Styrene–butadiene–styrene (SBS)	[71]				[71]		[96]
Titania		[53]					
AlOOH (Boehmite)			[104]				

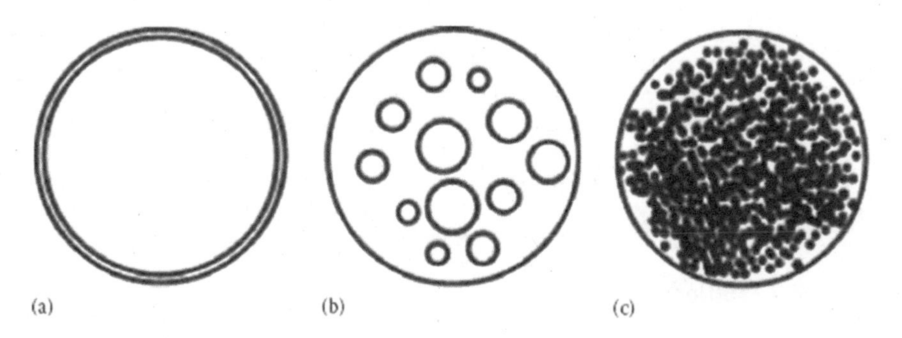

FIGURE 18.3 Types of microcapsule: (a) mononuclear, (b) polynuclear, and (c) matrix encapsulation.

enclosing particles of micrometer size contained in an inert shell to isolate and protect them from the external environment. The shell/core combination is the main point in the fabrication of these microcapsules; the shell's purpose is to protect the core, and the core's role is to contain the PCM. The shell can be permeable, semi-permeable, or impermeable, and the core can be in the gas, liquid, or solid state. The shape can be spherical or irregular. A suitable shell material compatible with the core is required. Furthermore, the MPCM description depends on the core material and also on the formation of the shell. Figure 18.3 shows the different types of MPCM, which are as follows:

- Mononuclear microcapsules contain the shell around the core.
- Polynuclear capsules have many cores enclosed within the shell.
- In matrix encapsulation, the core material is distributed homogeneously into the shell material.

Three main steps are required for microencapsulation preparation: to form the shell around the PCM, ensure that there is no leakage, and guarantee that no desired materials are included in the system core/shell MPCM.

The inclusion of waxes into the microcapsules makes it possible to increase the heat transfer and to control the volume changes when the phase change is happening.

There are numerous studies on the development of MPCM with different shell/core combinations. Table 18.3 shows the most frequently employed combinations of shell and PCM. Besides, other shell/PCM combinations can be possible, as Table 18.4 describes.

Moreover, there are additional possibilities, with PMMA as a capsule with hexadecane,[58] heptadecane,[73] or docosane[54]; methylmethacrylic acid (MMA) with *n*-pentadecane as PCM[61]; MF resin with *n*-dodecanol[74]; polyurea as shell and butyl stearate as PCM[75]; and SiO_2, and nonadecane.[76] More possibilities are PS with RT-31[77]; aminoplastics involving $C_{16}H_{33}Br$,[78] nickel with zinc[36]; stainless steel as shell and a eutectic as PCM[36]; and the combination of cryosol with RT-6, RT-10, and RT-20[30] as PCMs.

18.5 Technologies Used to Encapsulate Organic PCMs

The microencapsulation process includes two main steps: the emulsification step and the formation of the capsules. The emulsification step determines the size, and the size distribution of the microcapsules may be influenced both by physical parameters, such as the apparatus configuration, the stirring rate, the temperature, and the volume ratio of the two phases, and also by the physico-chemical properties, such as the interfacial tension, the viscosities, the densities, and the chemical compositions of the two phases.

The formation of microcapsules is greatly affected by the surfactant, which influences not only the mean diameter but also the stability of the dispersion. The surfactants used in the system have two roles: the first to reduce the interfacial tension between the oil and aqueous phases, allowing the formation of smaller microcapsules, and the other to prevent coalescence by adsorption on the oil–water interface, thereby forming a layer around the oil droplets. The synthesis of a core/shell particle or other possible morphologies is mainly governed by kinetic and thermodynamic factors.

The most frequently employed methods to encapsulate the PCM are chemical, physico-chemical, and physico-mechanical methodologies. Emulsion and *in situ* polymerization are the two most used as chemical methodologies. Among physico-chemical methodologies, sol-gel and coacervation are the most employed methods, and finally, for the physico-mechanical methodologies, the one-step method is the most used. Some examples of these methodologies are shown in Table 18.5.

18.5.1 Chemical Methodology

There are different techniques, such as *in situ*, emulsion, suspension, interfacial, and dispersion polymerization. The monomers polymerize around droplets of an emulsion and form a solid polymeric wall. In the case of MPCM, this technique is extensively used, as reported in the literature.

18.5.1.1 In Situ Polymerization

In this methodology, the PCM emulsion has to be prepared, and then the synthesis of the prepolymer solution has to be carried out by mixing two monomers and water.[79] This prepolymer has to be added to the emulsion in droplet form while the emulsion is agitated for a fixed time. The microcapsules are obtained after cooling and filtering the emulsion. Finally, the MPCM has to be dried. This microencapsulation methodology can start either with monomers or with commercial prepolymers. It is suggested to add modifying agents[80] to improve the mechanical properties. For instance, Boh et al.[81] used a modifier, with MF prepolymers as wall materials and styrene–maleic acid anhydride copolymers as modifying/emulsifying agents. Moreover, Boh and Šumiga[82] defined *in situ* polymerization as the procedure whereby monomers or precondensates are added only to the aqueous phase of the emulsion. Another study using this technique is that of Yang et al.,[83] in which they evaluated the best shell to encapsulate tetradecane as PCM with PS, PMMA, polyethylmethacrylate (PEMA), and polyvinyl acetate (PVAc) as PCS. After this study, Yang et al.[84] used the same PCM contained in different shell materials, acrylonitrile–styrene copolymer (AS), acrylonitrile–styrene–butadiene copolymer (ABS), and polycarbonate (PC), concluding that all three shell materials could be used to microencapsulate *n*-tetradecane. Using this same PCM, but in this case with urea and formaldehyde, Fang et al.[85] concluded that the *n*-tetradecane encapsulation is efficient enough with good thermal stability and attractive for thermal energy storage and heat transfer applications. Besides, Chen et al.[66] synthesized paraffin/SiO_2 MPCMs by *in situ* polymerization.

18.5.1.2 Emulsion Polymerization

This method consists of adding the polymer into an oiled system with an emulsifier. This can be performed in a chemical, thermal, or enzymatic way. The last step is washing the emulsion, eliminating the oil to isolate the microcapsules. Sarı et al.[73] employed this methodology to prepare MPCM with PMMA as shell and *n*-heptadecane as PCM for TES.

18.5.1.3 Suspension Polymerization

This polymerization is governed by multiple simultaneous mechanisms, such as particle coalescence and breakup, secondary nucleation, and the diffusion of monomer to the interface, as Sánchez-Silva et al.[86] revealed in their study. The collective effect of these mechanisms confers the size, the structure, and the surface properties on the microcapsules. A method based on a free radical polymerization suspension process to fabricate non-polar MPCM was developed by Sánchez et al.[87] Also, these authors[88] studied the influence of the temperature in a reaction, the stirring rate, and the mass ratio of paraffin to styrene on the thermal properties of MPCM. More studies using this process were done by Borreguero et al.,[89] who

TABLE 18.4

Other PCM/Shell Combinations

	PCM						
Shell	***006E*-Tetradecane**	***n*-Octacosane**	**Tetradecane**	**Eicosane**	**Polyethylene Glycol**	**Butyl Stearate + Paraffin**	**RT 27**
Polymethyl methacrylate (PMMA)		[57, 134]	[83]	[58]			
Polymethyl methacrylate-co- divinylbenzene (P(MMA-co-DVB))						[14]	
Polyethyl methacrylate (PEMA)			[83]				
Acrylate						[56]	
Urea–formaldehyde (UF) resin	[85]						
High-density polyethylene (HDPE)					[71]		
Polycarbonate (PC)	[84]						
Polyvinyl chloride (PVC)					[71]		
Polystyrene (PS)			[83]				[100]
Polyurethane (PUR)					[71]		
Polyurea + PUR						[13]	
Low-density polyethylene ethyl + vinyl acetate (LDPE + EVA)							[100]
Polyvinyl acetate (PVA)			[83]		[71]		
Acrylonitrile–styrene–butadiene (ABS)	[84]						
Acrylonitrile–styrene copolymer (AS)	[84]						
Styrene–butadiene–styrene (SBS)					[71]		
Polysiloxane				[121]			
Gelatin/gum arabic (G/AG)							[103]
Sterilized gelatin/gum arabic (SG/AG)							[103]
Agar-agar/gum arabic (AA/AG)							[103]

TABLE 18.5
List of Types of Polymerization with References

Type of Microencapsulation Technique	References
Coacervation technique	[65, 103]
Emulsion polymerization	[51, 54, 57–59, 73, 83]
In situ cross-linking by coemulsification	[121]
In situ polymerization method	[13, 74, 80, 85, 104]
Interfacial polycondensation	[61, 114]
One-step method	[99]
Sol-gel method	[66, 68, 92, 93]
Spay-drying	[100]
Suspension-like polymerization	[55, 56, 77, 86]

considered two main steps: a continuous one with deionized water and the stabilizer (polyvinyl-pyrrolidone [PVP]) and a discontinuous one containing the styrene monomer, paraffin wax, and benzoyl peroxide.

18.5.1.4 Interfacial Polycondensation

This involves the addition of an organic phase (containing poly-functional monomers and/or oligomers) into an aqueous phase (containing a mixture of emulsifiers and protective colloid stabilizers) along with the encapsulating material. The presence of cross-linked materials influences the morphology of the external microcapsule surface.[90] The major advantages of this technique are the high reaction speed and the low penetrability of the products. As an example of this procedure, Liang et al.[75] used interfacial polymerization to prepare MPCM of butyl stearate as a PCM in a polyurea system. The SiO_2/paraffin nanoparticle study by Li et al.[67] using a polycondensation technique led to the conclusion that this methodology could also be used to fabricate other organic and inorganic PCMs with different core/shell compositions.

18.5.1.5 Dispersion Polymerization

In this type of polymerization, it is important to study the effects of parameters such as initiator, monomer, and stabilizer concentration and the reaction time on the characteristics of the final microcapsules. In this method, the inherent simplicity of the single-step process makes the procedure useful. Typical examples are alcohol, alcohol–ether, and alcohol–water mixtures. By ultraviolet (UV) photoinitiated dispersion polymerization, Wang et al.[91] prepared a composite composed of a stearic acid PCM in a PMMA shell.

18.5.2 Physico-Chemical Methodology

These techniques include sol-gel encapsulation, the coacervation, and the supercritical CO_2–assisted methods.

18.5.2.1 Sol-Gel Encapsulation or Core Templating

Fang et al.[92] described the preparation of a form-stable lauric acid/silicon dioxide composite PCM by sol-gel encapsulation. Furthermore, Li et al.[93] prepared a shape-stabilized paraffin/silicon dioxide composite PCM with the same procedure. Also, Tang et al.[94] enhanced the thermal conductivity of polyethylene glycol (PEG)/SiO_2 via *in situ* chemical reduction of $CuSO_4$ through an ultrasound-assisted sol-gel process, concluding that the Cu/PEG/SiO_2 hybrid material had excellent thermal stability and a good, stable performance. However, this process technology does not allow obtaining a polymeric shell sufficiently tight to prevent the diffusion of small water molecules during the phase changes of a salt hydrate used as a PCM.[95]

18.5.2.2 Coacervation

This is a phenomenon that is exploited in colloid systems, whereby macromolecular colloid-rich coacervate droplets surround dispersed microcapsule cores and generate a viscous microcapsule wall, solidifying with cross-linking agents. Hence, the polymeric solute is separated in the form of small liquid droplets, forming the coacervate. Then, it surrounds the insoluble particles dispersed into a liquid. These droplets slowly unite and form a continuous cover around the core. For obtaining longer-lifetime microcapsules, this procedure has to be finished by adding the polymer twice. In the course of this mechanism, a thin microcapsule shell is obtained. Besides, the compatibility and impermeability are enhanced (a lower speed of polymer deposition increases impermeability), providing more stability to the microcapsules and preserving the size and the spherical shape. The texture is smoother, and the form is more regular, compared with one-step coacervation, whereby rougher, coarser, and more porous microcapsules with many protrusions are obtained. This methodology is useful when hydrosoluble polymers are needed to create the MPCM. Some researchers use this technique, such as Özonur et al.,[96] concluding that a gelatin + gum arabic mixture was the best wall material for microencapsulating coco fatty acid mixtures compared with urea–formaldehyde resin, MF resin, and β-naphthol–formaldehyde. Figure 18.4 shows a scheme of the coacervation technique.

18.5.2.3 Supercritical CO_2–Assisted

This method is a good alternative to conventional processes, because it is an effective synthetic method, which has gained interest for the synthesis of polymeric composites. As an example, Haldorai et al.[97] presented an overview of the synthesis of polymer–inorganic filler nanocomposites in supercritical CO_2.

18.5.3 Physico-Mechanical Methodology

This methodology includes the one-step method and spray-drying encapsulation. Neither of these techniques is capable of producing microcapsules smaller than 100 μm.[98]

18.5.3.1 One-Step Method

The advantages of using this method are the easy scale-up and that there is no need for a stabilizing agent due to self-stabilization. As an example, Jin et al.[99] used it without surfactants or dispersants or acids/bases for stabilizing capsules via an oil/water emulsion. It allows tuning of the size and polydispersity of the capsules and the use of nonadecane as core material.

18.5.3.2 Spray-Drying

This consists of the preparation of an emulsion, dispersing the PCM material in a concentrated solution to form the capsule, until the preferred size of the MPCM is obtained. This emulsion is pulverized into

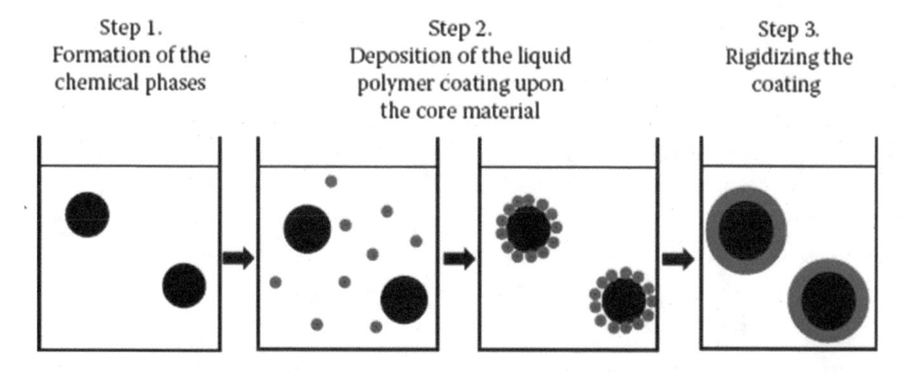

FIGURE 18.4 Steps for the coacervation technique.

droplets. Then, it has to be dried to obtain the MPCM. Borreguero et al.[100] use this methodology, mixing the two polymers LDPE and ethylene–vinyl acetate (EVA) with the purpose of creating the polymeric shell. These two polymers have a chemically similar structure as well as low density, versatility, and low cost. The technique involves the atomization of a homogeneous liquid stream in a drying chamber, where the solvent is evaporated and solid particles are obtained, and is suitable for heat-sensitive materials. Besides, Hawlader et al.[101] and Fei et al.[53] applied this technique to produce another type of microcapsules, using different polymeric shells and PCM cores. This methodology is also useful when hydrosoluble polymers are needed to create the MPCM.[95]

In Table 18.5, several references with the type of polymerization used to perform their experiments are summarized.

18.6 MPCM Characterization Techniques

Several techniques to characterize the microparticles are described. To analyze the shape and size of samples, particle size distribution (PSD), optical microscopy (OM), scanning electron microscopy (SEM), and transmission electron microscopy (TEM) are the most often employed. When the crystallographic phases of MPCM are studied, X-ray diffraction (XRD), wide angle X-ray scattering (WAXS), and low angle laser light scattering (LLALS) are techniques used to characterize them. The way to describe the physical properties of an MPCM in a fluid (PCS) is by using viscometry (η) and density (ρ) techniques. Also, conductivity (k) and flammability are parameters to take into account. To study the thermophysical properties, the methods most often used are differential scanning calorimetry (DSC) and thermogravimetric analysis (TGA). Fourier transform infrared spectroscopy (FTIR) is used to analyze the chemical properties of the MPCM. Moreover, to extract the mechanical parameters of MPCM, atomic force microscopy (AFM) is the technique to be used. Finally, several MPCM studies describe the cycling tests and further characterization of the cycled microcapsules with any of the abovementioned techniques.

18.6.1 Particle Size Distribution (PSD)

This technique defines the relative amount of particles according to size. An exhaustive preparation of the sample is not necessary. A curve plotting the percentage in number or the percentage in volume against the size is obtained. The results obtained must be estimated considering the size as well as the transparency of the MPCM. To calculate the approximation, the Mie and Fraunhofer models are available.[102] For this reason, to decide which of them fits better with the sample, it is necessary to evaluate the MPCM by SEM, verifying the results from the laser diffraction. The Mie model fits better for homogeneous and spherical particles, opaque or transparent, and with diameters below 30 μm, and the refractive index of the substance and the absorption are needed to calculate the size of the MPCM. To use the Fraunhofer model, opaque particles bigger than 30 μm are needed. A lot of MPCM studies include the use of this technique in their results. For instance, Yu et al.[74] used this technique to study the diameter distribution of MPCM prepared with different mass ratios of emulsifier to PCM.

18.6.2 Optical Microscopy (OM)

This technique supplies images of the microspheres with high-quality resolution. Sánchez et al.[8] developed thermoregulating textiles using MPCM with paraffin in the core and evaluated their morphology, fixation, and durability using a transmitted light and reflection mode. Moreover, Bayés-García et al.[103] characterized gelatin/gum arabic (G/AG) MPCM by using a thermo-optical microscope at different temperatures.

18.6.3 Scanning Electron Microscopy (SEM)

This technique is very useful to study the MPCM size and shape. Moreover, the preparation of the sample is very simple and fast. The SEM uses a focused beam of high-energy electrons to generate a variety of signals at the surface of solid specimens. The signals, which derive from electron–sample interactions, reveal information about the sample, including the texture, chemical composition, crystalline structures, and orientation of materials. Data are collected over a selected area of the surface of the sample, and a two-dimensional image is generated that displays spatial variations in these properties.

18.6.4 Transmission Electron Microscopy (TEM)

This technique is used when the MPCM is in the nanometer size range. The specimen must have a low density, allowing the electrons to travel through the sample. There are different ways to prepare the material: it can be cut into very thin slices either by fixing it in plastic or by working with it as frozen material. Pan et al.[104] studied nanostructures that were prepared through the *in situ* interfacial polycondensation method.

18.6.5 X-ray Diffraction (XRD)

This technique is used to analyze the microencapsulated paraffin crystalloid phase, and it is used to guarantee the PCM encapsulation. PS shell nanocapsules were analyzed by Fang et al.[105], and Fang et al.[68] synthesized and characterized MPCM paraffin composites with SiO_2 shells.

18.6.6 Wide Angle X-ray Scattering (WAXS) and Low Angle Laser Light Scattering (LALLS)

These techniques are used to determine the degree of crystallinity of a PCM and the crystallographic forms of the MPCM. Zhang and Wang[106] analyzed the WAXS patterns of *n*-octadecane/resorcinol-modified MF MPCM. Besides, Sánchez-Silva et al.[86] have used the LALLS technique to characterize styrene–methyl methacrylate copolymer shell MPCM.

18.6.7 Viscosity Measurement (η), Density (ρ), and Conductivity (k)

The measurement of PCM viscosity is a key point in flowage, taking into account that the higher the temperature, the lower the viscosity. Also, it is important to measure it at different temperatures with a rotation viscometer. An increase in the viscosity will give rise to an increase in pump energy consumption, which will counteract the positive effect to some extent.

Another important property to consider in a PCS system is the density (ρ_{PCS}). The definition of ρ_{PCS} is expressed by the sum of the products of the weight fraction x_i and the density ρ_i of each component, as Equation 18.4 defines, and it can be obtained by the use of a specific gravimeter or a pycnometer by determining the volume. Toppi and Mazzarella[107] proposed to calculate the properties of a composite material as a function of its composition instead of needing to determine the thermal properties on every occasion the composition changes.

$$\rho_{PCS} = \frac{1}{\sum_i \frac{x_i}{\rho_i}} \tag{18.4}$$

An extra property very often used to characterize MPCM and PCS is the thermal conductivity. It is calculated in different ways, as Youssef et al.[108] reported in a review. One mode to calculate it is by Maxwell's[109] relation, expressed in Equation 18.5, where k_{MPCM} is represented as the thermal conductivity of the microcapsule, k_{cont} is the thermal conductivity of the content, and c_{MPCM} is the volume fraction of the MPCM:

$$k_{MPCM} = k_{cont} \frac{2 + {k_{MPCM}}/{k_{cont}} + 2c_{MPCM}\left({k_{MPCM}}/{k_{cont}} - 1\right)}{2 + {k_{MPCM}}/{k_{cont}} + c_{MPCM}\left({k_{MPCM}}/{k_{cont}} - 1\right)} \tag{18.5}$$

TABLE 18.6
Techniques Used

References	FTIR	TGA	SEM	TEM	PSD	DSC	Encapsulation Efficiency	OM	η	ρ	*k*	Cycling	WAXS	XRD	LALLS	Elasticity	AFM	UV	Microindentation
[8]	×	–	×	–	–	×	–	×	–	–	–	–	–	–	–	–	–	–	–
[13]	×	×	×	–	×	–	–	–	–	–	–	×	–	–					
[14]	×	×	×	–	×	×	–	–	–	–	–	×	–	–	–	–	–	–	–
[27]	–	–	–	–	–	×	–	×	–	–	–	–	–	–	–	–	–	–	–
[30]	–	–	–	–	×	×	–	–	×	×	–	–	–	–	–	–	–	–	–
[36]	–	–	–	–	–	×	–	–	–	–	–	×	–	–	–	–	–	–	–
[40]	×	×	×		×	×													
[49]	–	–	×	–	–	×	–	–	–	–	–	×	–	–	–	–	–	–	–
[51]	×	×	×	×	×	×	–	–	–	–	–	–	–	×	–	–	–	–	–
[53]	×		×			×								×				×	
[54]	×	×	×	–	×	×	–	–	–	–	–	×	–	–	–	–	–	–	–
[55]	×	×	×	–	–	×	–	–		–	–	–	–	–	–	–	–	–	–
[56]	×	×	×	–	–	×	–	–	–	–	–	×	–	–	–	–	–	–	–
[57]	×	×	×		×	×						×							
[58]	×	×	×	–	×	×	–	–	–	–	–	×	–	–	–	–	–	–	–
[59]	×	–	×	–	×	×	–	–	–	–	–	–	–	–	–	–	–	–	–
[60]	×	–	×	–	×	×	–	×	–	–	–	–	–	–	–	–	–	–	–
[61]	–	–	–	–	×	×	–	–	–	–	–	×	–	–	–	×	–	–	–
[62]	–	–	–	–	–	–	–	–	–	–	×	–	–	–	–	–	–	–	–
[63]	–	–	×	–	–	×	–	×	×	–	–	–	–	–	–	–	–	–	–
[65]	–	–	×	–	–	–	–	–	–	–	–	–	–	–	–	–	–	–	–
[66]	×	×	×	–	–	×	–	–	–	–	–	–	–	×	–	–	–	–	–
[67]	×	×	×	×	–	–	–	–	–	–	–	×	–	×	–	–	–	–	–
[68]	×	×	×	–	–	×	–	–	–	–	–	–	–	×	–	–	–	–	–
[69]	×	×	×	–	–	×	–	–	–	–	–	–	–	–	–	–	–	–	–
[70]	–	–	×	×	–	×	–	–	–	–	–	–	–	–	–	–	–	×	–
[71]	×	–	–	–	–	×	–	–	–	–	–	–	–	–	–	–	–	–	–
[72]	×	–	×	–	–	×	–	–	–	–	–	×	–	–	–	–	–	–	–
[73]	×	×	×	–	×	×	–	–	–	–	–	×	–	–	–	–	–	–	–
[74]	×	×	×	–	×	×	×	–	–	–	–	–	–	–	–	–	–	–	–
[75]	×	–	–	–	×	×	–	×	–	–	–	–	–	–	–	–	–	–	–

(Continued)

TABLE 18.6 (CONTINUED)

Techniques Used

References	FTIR	TGA	SEM	TEM	PSD	DSC	Encapsulation Efficiency	OM	η	ρ	*k*	Cycling	WAXS	XRD	LALLS	Elasticity	AFM	UV	Microindentation
[79]	–	×	×	–	–	–	–	–	–	–	–	–	–	×	–	–	–	–	–
[80]		–	×	–	–	–	–	–	–	–	–	–	–	–	–	–	–	–	–
[83]	×	–	–	×	×	×	–	–	×	–	–	–	–	–	–	–	–	–	–
[84]	×	–	×	–	×	×	×	–	–	–	–	–	–	–	–	–	–	–	–
[85]	×	×	×	–	–	×	–	–	–	–	–	–	–	–	–	–	–	–	–
[86]	×	×	×	–	×	×	–	×	–	–	–	–	–	–	×	–	–	–	–
[91]	×	–	×	–	–	×	–	–	–	–	–	×	–	–	–	–	–	–	–
[92]	×	×	×	–	–	×	–	–	–	–	–	–	–	–	–	–	–	–	–
[93]	×	×	×	–	–	×	–	–	–	–	–	–	–	–	–	–	–	–	–
[94]	×	×	×	–	–	×	–	–	–	–	×	×	–	×		–	–	–	–
[96]	×	–	–	–	–	–	–	×	–	–	–	×	–	–	–	–	–	–	–
[98]	–	×	×	–	×	×	×	–	×	×	×	×	–	–	–	–	–	–	–
[99]	–	×	×	–	–	×	–	–	–	–	–	–	–	–	–	–	–	–	–
[100]	–	–	×	–	–	×	–	–	–	–	–	×	–	–	–	–	×	–	–
[103]	–	×	×	–	×	×	–	×	–	–	–		–	–	–	–	–	–	–
[104]	×	×	–	×	–	×	–	–	–	–	–	–	–	–	–	–	–	–	–
[105]	×	×	–	×	×	×	–	–	–	–	–	–	–	×	–	–	–	–	–
[106]	×	–	×	–	×	×	–	×	–	–	–	–	×	–	–	–	–	–	–
[108]	–			–	×	×	–	–	–	×	×		–	–	–	–	–	–	–
[111]	–	×	×	–	–	×	–	–	–	–	–	–	–	–	–	–	–	–	–
[113]	×	×	×	–	×	×	–	–	–	–	–	–	×	–	–	–	–	–	–
[121]	–	–	×	–	–	×	–	×	–	–	–	×	–	–	–	–	–	–	–
[122]	×	×	×	–	–	×	–	–	×	–	×	×	–	–	–	–	–	–	–
[123]	–	–	–	–	×	×	–	–	×		–	–	–	–	–	–	–	–	–
[124]	–	–	×	–	–	×	–	–	×	×	–	–	–	–	–	–	–	–	–
[125]	–	–	×	–	×	×	–	–	–	–	–	–	–	–	–	–	–	–	×
[132]	–	–	×	–	–	×	–	–	–	–	–	–	–	–	–	–	–	–	–
[139]	–	–	–	–	–	–	–	–	–	×	×	–	–	–	–	–	–	–	–

The study of these three explained properties (η, ρ, k) in a PCS was evaluated by Wang and Niu[110]; the density was calculated by weighted fraction of the densities of PCM, the coating material, and the water, based on the mass and energy balance, and the thermal conductivity of the PCS was calculated based on a composite sphere approach.

18.6.8 Flammability

As the lower thermal stability and inflammability properties have been severely restricted, especially in building applications,[6] studies of the flammability properties of PCMs have high importance in numerous applications. To evaluate the combustion properties of polymer materials, the cone calorimeter is one of the most effective bench-scale methods. One flammability composite studied by Cai et al.[111] evaluated PCM based on paraffin/HDPE. A proposed solution to reduce the flammability is to introduce flame retardants, as studied by Sittisart and Farid[112]. As microencapsulating PCMs avoids leakage, it will be expected that materials containing MPCM will have better fire-resistant behavior than those containing PCM without microencapsulation.

18.6.9 Differential Scanning Calorimetry (DSC)

This is the most widely employed thermal analysis technique. It measures endothermic and exothermic transitions as a function of the temperature, measures heat capacities, and can detect glass, fusion, crystallization, and oxidation transitions. It provides the melting and solidifying enthalpy as well as the melting and solidifying temperature of a sample, giving an idea of whether the material has a good capacity to store energy. Zhang et al.[113] used this calorimetric technique to characterize two PCMs suitable for applications in concentrating solar power systems.

18.6.10 Thermogravimetric Analysis (TGA)

This technique is widely employed, and it measures the amount and rate of change in the weight of a material as a function of temperature under a controlled atmosphere. The measurements are used primarily to evaluate the thermal stability. The technique can characterize materials that exhibit weight loss or gain due to decomposition, oxidation, or dehydration. As an example, Zhang et al.[114] evaluated the step thermal degradation and the thermal reliability of a silica/*n*-octadecane MPCM.

18.6.11 Fourier Transform Infrared Spectroscopy (FTIR)

The infrared segment of the electromagnetic spectrum is separated into three regions: the near (14000–4000 cm^{-1}), the mid (4000–400 cm^{-1}), and the far infrared (400–10 cm^{-1}). This FTIR mid region is used in MPCM studies to evaluate the shell material of the microcapsule and also to study its feasible degradation. It can be performed using attenuated total reflectance (ATR) or with KBr pellets,[115] without and with further preparation, respectively. Infrared spectroscopy uses the fact that the functional groups present in the molecules absorb some specific resonant frequencies depending on their structure. An example of a study using this technique is that by Zou et al.[116] to characterize the shell of the MPCM *n*-hexadecane in polyurea.

18.6.12 Atomic Force Microscopy (AFM)

This is a surface technique that has the ability to obtain topographic images of surfaces with nanometric resolution and can also be used as a nanomanipulator to move and test the surface of the samples in a variety of ways, such as electrically, magnetically, and mechanically. AFM has been proposed to assess the mechanical properties of microcapsules. It is based on the measurement of deformation under well-defined stress. The microcapsule maximum force and the total deformation of an individual microcapsule[100] have to be measured by increasing the applied force until the typical force–displacement curves are obtained. It is necessary to repeat this procedure several times (a minimum of three times), because polymeric shells usually do not provide repeatable results. A single microcapsule might not reflect the actual strength when microcapsules are piled together.[117] Moreover, AFM can extract the deformation and the Young's modulus histogram, taking into account a selected small area on the top of the microcapsule. Giro-Paloma et al.[118] used it to evaluate the maximum force that Micronal® DS 5001 MPCM can stand before their breakage at three different temperatures: 25, 45, and 80 °C. The main conclusion was that the applied load required to break the sample was not constant and depended on the working temperature. In this line of research, using the AFM technique, Giro-Paloma et al.[119] estimated the highest force that the Micronal® DS 5007 PCS can hold on the top of the microsphere.

18.6.13 Cycling

The thermal cycling test exposes the thermal reliability and chemical stability of a microcapsule. This experiment is an important analysis to estimate the quality of the microcapsules so as to guarantee no alteration of their geometrical shape after numerous cycles.[120] Cycling can be performed on organic and inorganic PCMs, as studied by Shukla et al.[121], who concluded that although the studied inorganic PCMs were not found to be appropriate materials after some cycles, for the studied organic PCMs, after 1000 thermal cycles, gradual changes in the melting temperature as well as the latent heat of fusion were observed. Fortuniak et al.[122] evaluated 50 fusion–crystallization cycle tests in a DSC device. So, this experiment can be performed in a DSC or using a thermocycler. If the number of cycles increases to 5000, the study of Sari et al.[57], who conducted the thermal cycling test of a PMMA shell and *n*-octacosane as PCM, is relevant. Besides, when the cycles are finished, the samples have to be evaluated by other techniques to estimate the possible PCM degradation, for instance by FTIR, as Sari and Biçer[123] evaluated. In most studies, the thermal reliability was studied following the thermal properties, and an example is the study by Sari and Biçer[123] of some fatty acid composites for building materials.

In Table 18.6, some studies using the different techniques discussed here are listed.

18.7 MPCM Applications

There are several applications for MPCMs,[3] reflected in many studies of these materials for storing energy. The first industrial application of microcapsules was introduced at the end of the 1950s in the production of pressure-sensitive copying papers. Since then, this technology has been enhanced, modified, and adapted for different reasons and uses. Consequently, there has been a rapid growth in patent applications, reflecting industrial research and development, as well as an increase in the number of articles and papers. Also, microcapsules have been used in the graphic and printing industries, in food and cosmetic products, for pharmaceutical and medical purposes, and in agricultural formulations. Furthermore, they are used in the chemical, textile, and construction materials industries, biotechnology, photography, electronics, and waste treatment.

The applications of MPCMs can be divided into two major groups: thermal protection or inertia and storage. The main difference between these relates to the thermal conductivity of the substance. Zalba et al.[3] explain that storage systems with low thermal conductivity can generate a true problem, as there can be enough energy stored but insufficient capacity to dispose of this energy quickly enough.

One of the applications for MPCMs is in *refrigeration systems*.[128] PCSs are another important application to take into account.[124–126] When an MPCM is suspended in a carrier fluid, mainly water[129] or with other substances such as glycerol,[130] a PCS is created, enhancing the heat transfer to the MPCM. These substances are extensively employed in active pumping systems,[131] and they can be used for refrigeration,[110] heat exchangers,[132] heating,[133] ventilation, air conditioning (HVAC),[134] and solar energy[135] applications. PCSs not only act as an energy storage device but also as a heat transport system, as Salunkhe and Shembekar[136] explained in their review. The way of pumping the slurries in an active system is very important,[131] because it is known that the microcapsules can be broken when they collide with each other. Moreover, Zhang and Niu[137] studied numerically the influences of microparticles and phase change in fluid (pure water), PCS, and MPCM.

TABLE 18.7
Results of Some Studies on MPCM Properties

References	MPCM	Latent Heat (J/g)		Phase Change (°C) Peak Temperature		Diameter (μm)	Shape MPCM	Emulsifier	Cycles	PCM Content (%)
[8]	PS + paraffin wax	104.7		40	45	5.5	Tubular	–	–	–
[13]	Polyurea/polyurethane/ butylstearate + paraffin	87.6		27.68		5–15	Spherical	–	500	63.7
[14]	PMMA-co-DVB	135	28	35	5–10	spherical	–	50	50–85	
[25]	RT-20 emulsion	21.2	–	0.2–12.5		–	–	–		
[49]	HDPE Wood + paraffin	45.9	44.3	16.7	7.9	–	Spherical	–	100	–
[51]	PMMA + n-octadecane	198.5	197.1	29.2	33.6	0.05–0.3	Spherical	89.5	–	–
	PEMA + n-octadecane	208.7	205.9	31.1	32.3	0.06–0.36	Spherical	–	–	–
[53]	Titania + n-octadecane	92–97	–			–	–	–		
[54]	PMMA + docosane	54.6	–48.7	41	40.6	0.16	Spherical	–	–	–
[55]	BMA–MMA	173.7	174.4	–		–	Spherical	–	–	77.3
[56]	Acrylate/butyl stearate + paraffin	85		–		10–30	Spherical		500	48–68
[57]	PMMA + n-octacosane	86.4	88.5	50.6	53.2	0.25	Spherical	–	5000	
[58]	PMMA + eicosane	35.2	34.9	54.2	87.5	0.7	Spherical	–	5000	
[59]	PMMA + n-hexadecane	68.89 and 145.61		–		0.22–1.050	Spherical	–	–	29.04–61.4
[60]	PMMA + paraffin	106.9	112.3	55.8	50.1	0.21	Spherical	66	3000	
[61]	MMA + n-pentadecane	107	97	10	9.5	650–760	Spherical	–	20	–
[62]	Micronal (BASF)	100		26		5	Spherical	–	–	
[64]	PMMA + SA	187	190	67	66	–	–	–	–	80
[64]	PMMA + PA	173	175	60	59	–	–	–	–	80
[64]	PMMA + MA	166	168	51	50.7	–	–	–	–	80
[64]	PMMA + LA	149	151	41	41.5	–	–	–	–	80
[66]	SiO_2 + paraffin	156.86	144.09	57.96	55.78	40–60	Spherical	82.2	–	–
[67]	Paraffin + SiO_2	45.5	43.8	56.5	45.5	0.2–0.5	Quasi-spherical	31.7	30	–

(Continued)

TABLE 18.7 (CONTINUED)

Results of Some Studies on MPCM Properties

References	MPCM	Latent Heat (J/g)			Phase Change (°C) Peak Temperature			Diameter (μm)	Shape MPCM	Emulsifier	Cycles	PCM Content (%)
[68]	SiO_2 + paraffin	130.82		93.04	57.84		57.01	8–15	Spherical	–	–	–
[69]	PMMA+ paraffin		101			24–33		0.5–2	Spherical	–	–	61.2
[72]	Polycarbonate + stearic acid	91.4		96.8	60.2		51.2	0.5	Spherical	–	1000	–
[73]	SiO_2 + heptadecane	81.5		84.7	18.2		18.4	0.26	Spherical	–	5000	–
[74]	MF + n-dodecanol		187.5			21.5		30.6	Irregular spherical	93.1		–
[75]	Polyurea + butyl stearate		80			29		20-35	Spherical	–	–	–
[77]	PS + RT-31		75.7			31.56		4	Spherical	–	–	–
[78]	$C_{16}H_{33}Br$ + amino plastics		137			14.3		4.3	8.2	–		–
[83]	PMMA + PCM	66.26		60.62	2.95		5.97	5 to 30	–	–	–	–
	PEMA + PCM	80.62		65.35	3.19		5.68	6 to 30	Spherical			
[84]	Acrylonitrile–styrene copolymer + n-tetradecane		142.3			10		<1	Spherical	84.6	–	–
	ABS + n-tetradecane		107.1			10		<1	Spherical	66.3	–	–
	PC + n-tetradecane		49.5			10		<1	Spherical	30.6	–	–
[85]	Urea/formaldehyde + n-tetradecane		134.16		–			0.1			–	60
[86]	Melamine–formaldehyde + paraffin		87.5			45		380	Spherical	–	–	–
[91]	Acrylonitrile–styrene + n-tetradecane	60.4		50.6	92.1		95.9	2–3	Spherical	–	500	–
[92]	SiO_2 + polyethylene glycol	117.21		90	44.78		40.33	–	–	–	–	–
[93]	SiO_2 + paraffin	35.8		32.51	57.8		56.85	–	Spherical	–	–	–
[94]	Cu/PEG/SiO_2		110.2			–		–	–	–	–	–
[99]	Silica + nonadecane-		124.7			–		27	Spherical		–	–

(*Continued*)

TABLE 18.7 (CONTINUED)
Results of Some Studies on MPCM Properties

References	MPCM	Latent Heat (J/g)		Phase Change (°C) Peak Temperature		Diameter (μm)	Shape MPCM	Emulsifier	Cycles	PCM Content (%)
[100]	LDPE/EVA + RT27	98.14		25.42	31.15	3.5	Spherical	63	3000	49.32
	Polystyrene + RT27	96.14		26.12	30.28	360	Spherical			48.61
[103]	Gelatin/gum arabic + RT-27	–		–		9	Spherical	–	–	–
	Sterilized gelatin/gum arabic + RT-27	79		25.15	28.15	12	Spherical	49	–	–
	Agar-agar/gum arabic + RT-27	78		26.35	29.35	4.3	Spherical	48	–	–
[104]	AlOOH + palmitic acid	19		12.7		0.2	Spherical	–	–	–
[105]	Polystyrene + n-octadecane	124.4		–		0.100–0.123	Spherical	–	–	–
[106]	MF + n-octadecane	122.5	132.2	26.75	20.33	12.45	Spherical	88	–	–
[111]	HDPE + paraffin	96.79		37	55	300	Lamellar			
[113]	Silica + n-octadecane					18.72	Spherical			
[121]	Polysiloxane + n-eicosane	139		37.4	30	22.9	Spherical	–	80	–
[122]	Galactitol hexa myristate (GHM)	61	121	39	46	–	–	–	1000	–
	Galactitol hexa laurate (GHL) esters									
[124]	PCS	90–100		28			Spherical	–	–	–
[125]	Micronal® DS 5008	135		23		11.2	Spherical	–	–	–
[133]	BASF slurries	–		Up to 70		2–8	–	–	–	–

DVB: Divinylbenzene.
BMA: butyl-methacrylate.
SA: stearic acid.
PA: palmitic acid.
MA: myristic acid.
LA: lauric acid.

Besides the *air conditioning*[128] applications, *heat exchangers* are also a possible application. Furthermore, as was mentioned before, the introduction of PCMs in *building constructive solutions*, such as in floors, [138,139] walls, and ceilings, is very common because of the interest in evaluating the consequences when PCMs are incorporated in passive systems.

18.8 Data

In Table 18.7, a summary of some data reported in various studies is found. As a main conclusion, it is important to remark that it is very interesting to measure the same sample several times to estimate the repeatability. As can be seen in Table 18.7, there is a wide dispersion of values. For this reason, to ensure the reproducibility of results, it is essential to specify the experimental procedure, the instruments used, and the conditions applied in the measurement.

18.9 Conclusions

Thermal energy storage is studied today as a very good technology for energy efficiency in many applications. Latent heat storage is one of the technologies used, in which PCMs are employed. Since PCMs can leak when included in a system, they are always encapsulated, and microencapsulation is sometimes considered, producing the so-called MPCMs.

The most frequently used MPCMs have paraffin wax as core material and PMMA as a shell. This type is the one commercially available MPCM. Nowadays, there are several studies of organic, eutectic, and inorganic salts as PCM in MPCM. Emulsion polymerization and *in situ* polymerization are the two methods most often employed to create MPCMs. Moreover, DSC is the technique most often employed to measure the main thermophysical properties of MPCMs.

Acknowledgments

The work is partially funded by the Spanish government (ENE2011-28269-C03-02 and ENE2011-22722). The authors would like to thank the Catalan Government for the quality accreditation given to their research group GREA (2009 SGR 534) and research group DIOPMA (2009 SGR 645). The research leading to these results has received funding from the European Union's Seventh Framework Programme (FP7/2007-2013) under grant agreement n° PIRSES-GA-2013-610692 (INNOSTORAGE).

REFERENCES

1. Sharma, A.; Tyagi, V.V.; Chen, C.R.; Buddhi, D. Review on thermal energy storage with phase change materials and applications. *Renew Sust Energ Rev* 13 (2009): 318–345.
2. McCormack, S.; Griffiths, P. Phase change materials. A primer for architects and engineers. TU0802 next generation cost effective phase change materials for increased energy efficiency in renewable energy systems in buildings (NeCoEPCM) (2013): Available at: https://www.researchgate.net/profile/Philip_Griffiths2/publication/261284791_Phase_Change_Materials_A_primer_for_Architects_and_Engineers/links/0c960533bfcef5d050000000/Phase-Change-Materials-A-primer-for-Architects-and-Engineers.pdf
3. Zalba, B.; Marín, J.M.; Cabeza, L.F.; Mehling, H. Review on thermal energy storage with phase change: Materials, heat transfer analysis and applications. *Appl Therm Eng* 23 (2003): 251–283.
4. Hamdan, M.A.; Elwerr, F.A. Thermal energy storage using phase change material. *Sol Energ* 56(2) (1996): 183–189.
5. Kabbaraa, M.J.; Abdallaha, N.B. Experimental investigation on phase change material based thermal energy storage unit. *Procedia Comput Sci* 19 (2013): 694–701.
6. Khudhair, A.M.; Farid, M.M. A review on energy conservation in building applications with thermal storage by latent heat using phase change materials. *Energ Convers Manage* 45 (2004): 263–275.

7. Pasupathy, A.; Velraj, R.; Seeniraj, R.V. Phase change material-based building architecture for thermal management in residential and commercial establishments. *Renew Sust Energ Rev* 12 (2008): 39–64.
8. Sánchez, P.; Sánchez-Fernández, M.V.; Romero, A.; Rodríguez, J.F.; Sánchez-Silva, L. Development of thermo-regulating textiles using paraffin wax microcapsules. *Thermochim Acta* 498 (2010): 16–21.
9. Baetens, R.; Petter Jelle, B.; Gustavsen, A. Phase change materials for building applications: A state-of-the-art review. *Energ Buildings* 42 (2010): 1361–1368.
10. Tyagi, V.V.; Buddhi, D. PCM thermal storage in buildings: A state of art. *Renew Sust Energ Rev* 11 (2007): 1146–1166.
11. Ling, T.-C.; Poon, C.-S. Use of phase change materials for thermal energy storage in concrete: An overview. *Constr Build Mater* 46 (2013): 55–62.
12. Zhang, D.; Li, Z.; Zhou, J.; Wu, K. Development of thermal energy storage concrete. *Cement Concrete Res* 34 (2004): 927–934.
13. Ma, Y.; Chu, X.; Tang, G.; Yao, Y. Adjusting phase change temperature of microcapsules by regulating their core compositions. *Mater Lett* 82 (2012): 39–41.
14. Ma, Y.; Chu, X.; Li, W.; Tang, G. Preparation and characterization of poly(methyl methacrylate-co-divinylbenzene) microcapsules containing phase change temperature adjustable binary core materials. *Sol Energ* 86 (2012): 2056–2066.
15. Lu, W.; Tassou, S.A. Characterization and experimental investigation of phase change materials for chilled food refrigerated cabinet applications. *Appl Energ* 112 (2013): 1376–1382.
16. Oró, E.; Miró, L.; Farid, M.M.; Cabeza, L.F. Thermal analysis of a low temperature storage unit using phase change materials without refrigeration system. *Int J Refrig* 35 (2012): 1709–1714.
17. Oró, E.; de Gracia, A.; Cabeza, L.F. Active phase change material package for thermal protection of ice cream containers. *Int J Refrig* 36 (2013): 102–109.
18. Oró, E.; Cabeza, L.F.; Farid, M.M. Experimental and numerical analysis of a chilly bin incorporating phase change material. *Appl Therm Eng* 58 (2013): 61–67.
19. Oró, E.; Gil, A.; Miró, L.; Peiró, G.; Álvarez, S.; Cabeza, L.F. Thermal energy storage implementation using phase change materials for solar cooling and refrigeration applications. *Energ Procedia* 30 (2012): 947–956.
20. Gil, A.; Oró, E.; Miró, L.; Peiró, G.; Ruiz, A.; Salmerón, J.M.; Cabeza, L.F. Experimental analysis of hydroquinone used as phase change material (PCM) to be applied in solar cooling refrigeration. *Int J Refrig* 39 (2014): 95–103.
21. Oró, E.; Miró, L.; Barreneche, C.; Martorell, I.; Farid, M.M.; Cabeza, L.F. Corrosion of metal and polymer containers for use in PCM cold storage. *Appl Energ* 109 (2013): 449–453.
22. Castellón, C.; Martorell, I.; Cabeza, L.F.; Fernández, A.I.; Manich, A.M. Compatibility of plastic with phase change materials (PCM). *Int J Energy Res* 35 (2011): 765–771.
23. Giro-Paloma, J.; Roa, J.J.; Diez-Pascual, A.M.; Rayón, E.; Flores, A.; Martínez, M.; Chimenos, J.M.; Fernández, A.I. Depth-sensing indentation applied to polymers: A comparison between standard methods of analysis in relation to the nature of the materials. *Eur Polym J* 49 (2013): 4047–4053.
24. Günther, E.; Schmid, T.; Mehling, H.; Hiebler, S.; Huang, L. Subcooling in hexadecane emulsions. *Int J Refrig* 33 (2010): 1605–1611.
25. Huang, L.; Günther, E.; Doetsch, C.; Mehling, H. Subcooling in PCM emulsions—Part 1: Experimental. *Thermochim Acta* 509 (2010): 93–99.
26. Günther, E.; Huang, L.; Mehling, H.; Dötsch, C. Subcooling in PCM emulsions—Part 2: Interpretation in terms of nucleation theory. *Thermochim Acta* 522 (2011): 199–204.
27. Huang, L.; Petermann, M.; Doetsch, C. Evaluation of paraffin/water emulsion as a phase change slurry for cooling applications. *Energy* 34 (2009): 1145–1155.
28. Abduljalil, A.A.; Mat, S.B.; Sopian, K.; Sulaiman, M.Y.; Lim, C.H.; Abdulrahman, T. Review of thermal energy storage for air conditioning systems. *Renew Sust Energ Rev* 16 (2012): 5802–5819.
29. Domínguez, M.; García, C. Aprovechamiento de los Materiales de Cambio de Fase (PCM) en la Climatizacion. *Informatión Tecnológica* 20(4) (2009): 107–115.
30. Huang, L.; Doetsch, C.; Pollerberg, C. Low temperature paraffin phase change emulsions. *Int J Refrig* 33 (2010): 1583–1589.
31. Augood, P.C.; Newborough, M.; Highgate, D.J. Thermal behaviour of phase-change slurries incorporating hydrated hydrophilic polymeric particles. *Exp Therm Fluid Sci* 25 (2001): 457–468.

32. Cabeza, L.F.; Ibáñez, M.; Solé, C.; Roca, J.; Nogués, M. Experimentation with a water tank including a PCM module. *Sol Energ Mat Sol C* 90 (2006): 1273–1282.
33. Jordan, U.; Vajen, K. Realistic domestic hot-water profiles in different time scales. IEA SHC. Task 26: Solar combisystems (2001).
34. Mehling, H.; Cabeza, L.F.; Hippeli, S.; Hiebler, S. PCM-module to improve hot water heat stores with stratification. *Renew Energ* 28 (2003): 699–6711.
35. Haillot, D.; Nepveu, F.; Goetz, V.; Py, X.; Benabdelkarim, M. High performance storage composite for the enhancement of solar domestic hot water systems Part 2: Numerical system analysis. *Sol Energ* 86 (2012): 64–77.
36. Zhao, W.; Neti, S.; Oztekin, A. Heat transfer analysis of encapsulated phase change materials. *Appl Therm Eng* 50 (2013): 143–151.
37. Joulin, A.; Younsic, Z.; Zalewski, L.; Lassue, S.; Rousse, D.R.; Cavrot, J.-P. Experimental and numerical investigation of a phase change material: Thermal-energy storage and release. *Appl Energ* 88 (2011): 2454–2462.
38. Farid, M.M.; Khudhair, A.M.; Razack, S.A.K.; Al-Hallaj, S. A review on phase change energy storage: Materials and applications. *Energ Convers Manage* 45 (2004): 1597–1615.
39. Soares, N.; Costa, J.J.; Gaspar, A.R.; Santos, P. Review of passive PCM latent heat thermal energy storage systems towards buildings' energy efficiency. *Energ Buildings* 59 (2013): 82–103.
40. Tyagi, V.V.; Kaushik, S.C.; Tyagi, S.K.; Akiyama, T. Development of phase change materials based microencapsulated technology for buildings: A review. *Renew Sust Energ Rev* 15 (2011): 1373–1391.
41. Zheng, Y.; Zhao, W.; Sabol, J.C.; Tuzla, K.; Neti, S.; Oztekin, A.; Chen, J.C. Encapsulated phase change materials for energy storage—Characterization by calorimetry. *Sol Energ* 87 (2013): 117–126.
42. Kuznik, F.; David, D.; Johannes, K.; Roux, J.-J. A review on phase change materials integrated in building walls. *Renew Sust Energ Rev* 15 (2011): 379–391.
43. Dutil, Y.; Rousse, D.; Lassue, S.; Zalewski, L.; Joulin, J.; Virgone, A.; Kuznik, F. et al. Modeling phase change materials behavior in building applications: Comments on material characterization and model validation. *Renew Energ* 61 (2014): 132–135.
44. Cabeza, L.F.; Castell, A.; Barreneche, C.; de Gracia, A.; Fernández, A.I. Materials used as PCM in thermal energy storage in buildings: A review. *Renew Sust Energ Rev* 15 (2011): 1675–1695.
45. Barreneche, C.; Navarro, H.; Serrano, S.; Cabeza, L.F.; Fernández, A.I. New database on phase change materials for thermal energy storage in buildings to help PCM selection. ISES Solar World Congress, Cancun, Mexico, 2013.
46. Mochane, M.J. Polymer encapsulated paraffin wax to be used as phase change material for energy storage. Thesis, Master of Science, 2011.
47. Sharma, S.D.; Buddhi, D.; Sawhney, R.L. Accelerated thermal cycle test of latent heat-storage materials. *Sol Energ* 66 (1999): 483–490.
48. Silakhori, M.; Naghavi, M.S.; Metselaar, H.S.C.; Mahliạ, T.M.I.; Mehrali, H.; Fauzi, M. Accelerated thermal cycling test of microencapsulated paraffin wax/polyaniline made by simple preparation method for solar thermal energy storage. *Materials* 6 (2013): 1608–1620.
49. Li, J.; Xue, P.; Ding, W.; Han, J.; Sun, G. Micro-encapsulated paraffin/high-density polyethylene/wood flour composite as form-stable phase change material for thermal energy storage. *Sol Energ Mat Sol C* 93 (2009): 1761–1767.
50. Tan, F.L.; Hosseinizadeh, S.F.; Khodadadi, J.M.; Fan, L. Experimental and computational study of constrained melting of phase change materials (PCM) inside a spherical capsule. *Int J Heat Mass Tran* 52 (2009): 3464–3472.
51. Zhang, G.H.; Bon, S.A.F; Zhao, C.Y. Synthesis, characterization and thermal properties of novel nanoencapsulated phase change materials for thermal energy storage. *Sol Energ* 86 (2012): 1149–1154.
52. Zhang, X.X.; Tao, X.M.; Yick, K.L.; Wang, X.C. Structure and thermal stability of microencapsulated phase-change materials. *Colloid Poly Sci* 282(4) (2004): 330–336.
53. Fei, B.; Lu, H.; Qi, K.; Shi, H.; Liu, T.; Li, X.; Xin, J.H. Multi-functional microcapsules produced by aerosol reaction. *J Aerosol Sci* 39 (2008): 1089–1098.
54. Alkan, C.; Sari, A.; Karaipekli, A.; Uzun, O. Preparation, characterization, and thermal properties of microencapsulated phase change material for thermal energy storage. *Sol Energ Mat Sol C* 93 (2009): 143–147.

55. Qiu, X.; Li, W.; Song, G.; Chu, X.; Tang, G. Microencapsulated *n*-octadecane with different methylmethacrylate-based copolymer shells as phase change materials for thermal energy storage. *Energy* 46 (2012): 188–199.
56. Ma, Y.; Chu, X.; Tang, G.; Yao, Y. Synthesis and thermal properties of acrylate-based polymer shell microcapsules with binary core as phase change materials. *Mater Lett* 91 (2013): 133–135.
57. Sari, A.; Alkan, C.; Karaipekli, A.; Uzun, O. Microencapsulated *n*-octacosane as phase change material for thermal energy storage. *Sol Energ* 83 (2009): 1757–1763.
58. Alkan, C.; Sari, A.; Karaipekli, A. Preparation, thermal properties and thermal reliability of microencapsulated n-eicosane as novel phase change material for thermal energy storage. *Energ Convers Manage* 52 (2011): 687–692.
59. Alay, S.; Alkan, C.; Göde, F. Synthesis and characterization of poly(methyl methacrylate)/*n*-hexadecane microcapsules using different cross-linkers and their application to some fabrics. *Thermochim Acta* 518 (2011): 1–8.
60. Wang, Y.; Shi, H.; Xia, T.D.; Zhang, T.; Feng, H.X. Fabrication and performances of microencapsulated paraffin composites with polymethylmethacrylate shell based on ultraviolet irradiation-initiated. *Mater Chem Phys* 135 (2012): 181–187.
61. Taguchi, Y.; Yokoyama, H.; Kado, H.; Tanaka, M. Preparation of PCM microcapsules by using oil absorbable polymer particles. *Colloid Surf A* 301 (2007): 41–47.
62. Castellón, C.; Medrano, M.; Roca, J.; Cabeza, L.F.; Navarro, M.E.; Fernández, A.I.; Lazaro, A.; Zalba, B. Effect of microencapsulated phase change material in sandwich panels. *Renew Energ* 35 (2010): 2370–2374.
63. Tzvetkov, G.; Graf, B.; Wiegner, R.; Raabe, J.; Quitmann, C.; Fink, R. Soft X-ray spectromicroscopy of phase-change microcapsules. *Micron* 39 (2008): 275–279.
64. Alkan, C.; Sari, A. Fatty acid/poly(methyl methacrylate) (PMMA) blends as form-stable phase change materials for latent heat thermal energy storage. *Sol Energ* 82 (2008): 118–124.
65. Su, J.-F.; Wang, X.-Y.; Dong, H. Influence of temperature on the deformation behaviours of melamine-formaldehyde microcapsules containing phase change material. *Mater Lett* 84 (2012): 158–161.
66. Chen, Z.; Cao, L.; Fang, G.; Shan, F. Synthesis and characterization of microencapsulated paraffin microcapsules as shape-stabilized thermal energy storage materials. *Nanosc Microsce Therm* 17(2) (2013): 112–123.
67. Li, B.; Liu, T.; Hu, L.; Wang, Y.; Gao, L. Fabrication and properties of microencapsulated paraf-fin@SiO2 phase change composite for thermal energy storage. *ACS Sustainable Chem Eng J* 1 (2013): 374–380.
68. Fang, G.; Chen, Z.; Li, H. Synthesis and properties of microencapsulated paraffin composites with SiO2 shell as thermal energy storage materials. *Chem Eng J* 163 (2010): 154–159.
69. Ma, S.; Song, G.; Li, W.; Fan, P.; Tang, G. UV irradiation-initiated MMA polymerization to prepare microcapsules containing phase change paraffin. *Sol Energ Mater Sol C* 94 (2010): 1643–1647.
70. Wu, X.; Wang, Y.; Zhu, P.; Sun, R.; Yu, S.; Du, R. Using UV-vis spectrum to investigate the phase transition process of PMMA–SiO2@paraffin microcapsules with copper-chelating as the ion probe. *Mater Lett* 65 (2011): 705–707.
71. Kenisarin, M.M.; Kenisarina, K.M. Form-stable phase change materials for thermal energy storage. *Renew Sust Energ Rev* 16 (2012): 1999–2040.
72. Zhang, T.; Wang, Y.; Shi, H.; Yang, W. Fabrication and performances of new kind microencapsulated phase change material based on stearic acid core and polycarbonate shell. *Energ Convers Manage* 64 (2012): 1–7.
73. Sari, A.; Alkan, C.; Karaipekli, A. Preparation, characterization and thermal properties of PMMA/n-heptadecane microcapsules as novel solid-liquid microPCM for thermal energy storage. *Appl Energ* 87 (2010): 1529–1534.
74. Yu, F.; Chen, Z.-H.; Zeng, X.-R. Preparation, characterization and thermal properties of microPCMs containing ra-dodecanol by using different types of styrene-maleic anhydride as emulsifier. *Colloid Polym Sci* 287 (2009): 549–560.
75. Liang, C.; Lingling, X.; Honbo, S.; Zhibin, Z. Microencapsulation of butyl stearate as a phase change material by interfacial polycondensation in a polyurea system. *Energ Corner Manag* 50 (2009): 723–729.
76. Zuidam, N.J.; Shimoni, E. Overview of microencapsulates for use in food products or processes and methods to make them. In: Zuidam, N.J.; Nedovic, V. A., eds. *Encapsulation Technologies for Food Active Ingredients and Food Processing*. Springer, Dordrecht; 2009, pp. 3–31.

77. Sánchez-Silva, L.; Rodríguez, J.F.; Sanchez, P. Influence of different suspension stabilizers on the preparation of Rubitherm RT31 microcapsules. *Colloid Surf A* 390 (2011): 62–66.
78. Zeng, R.; Wang, X.; Chen, B.; Zhang, Y.; Niu, J.; Wang, X.; Hongfa D. Heat transfer characteristics of microencapsulated phase change material slurry in laminar flow under constant heat flux. *Appl Energ* 86 (2009): 2661–2670.
79. Choi, J.K.; Lee, J.G.; Kim, J.H.; Yang, H.S. Preparation of microcapsules containing phase change materials as heat transfer media by in-situ polymerization. *J Irad Eng Chem* 7 (2001): 358–362.
80. Šumiga, B.; Knez, E.; Vrtacnik, M.; Ferk-Savec, V.; Staresinic, M.; Boh, B. Production of melamine-formaldehyde PCM microcapsules with ammonia scavenger used for residual formaldehyde reduction. *Acta Chim Slov* 58 (2011): 14–25.
81. Boh, B.; Knez, E.; Staresinic, M. Microencapsulation of higher hydrocarbon phase change materials by *in situ* polymerization. *J Microeracapsul* 22(7) (2005): 715–735.
82. Boh, B.; Šumiga, B. Microencapsulation technology and its applications in building construction materials. *RMZ—Mater Georavirora* 55(3) (2008): 329–344.
83. Yang, R.; Xu, H.; Zhang, Y. Preparation, physical property and thermal physical property of phase change microcapsule slurry and phase change emulsion. *Sol Eraerg Mat Sol* C 80 (2003): 405–416.
84. Yang, R.; Zhang, Y.; Wang, X.; Zhang, Y.; Zhang, Q. Preparation of *n*-tetradecane-containing microcapsules with different shell materials by phase separation method. *Sol Eraerg Mat Sol* C 93 (2009): 1817–1822.
85. Fang, G.; Li, H.; Yang, F.; Liua, X.; Wu, S. Preparation and characterization of nano-encapsulated ra-tetradecane as phase change material for thermal energy storage. *Chem Erag* J 153 (2009): 217–221.
86. Sánchez-Silva, L.; Rodríguez, J.F.; Romero, A.; Borreguero, A.M.; Carmona, M.; Sánchez, P. Microencapsulation of PCMs with a styrene-methyl methacrylate copolymer shell by suspension-like polymerization. *Chem Erag* J 157 (2010): 216–222.
87. Sánchez, L.; Sánchez, P.; de Lucas, A.; Carmona, M.; Rodríguez, J. Microencapsulation of PCMs with a polystyrene shell. *Colloid Polym Sci* 285 (2007): 1377–1385.
88. Sánchez, L.; Sánchez, P.; Carmona, M.; de Lucas, A.; Rodríguez, J.F. Influence of operation conditions on the microencapsulation of PCMs by means of suspension-like polymerization. *Colloid Polym Sci* 286 (2008): 1019–1027.
89. Borreguero, A.M.; Carmona, M.; Sanchez, M.L.; Valverde, J.L.; Rodriguez, J.F. Improvement of the thermal behavior of gypsum blocks by the incorporation of microcapsules containing PCMs obtained by suspension polymerization with an optimal core/coating mass ratio. *Appl Therm Erag* 30 (2010): 1164–1169.
90. Pascu, O.; Garcia-Valls, R.; Giamberini, M. Interfacial polymerization of an epoxy resin and carboxylic acids for the synthesis of microcapsules. *Poly Irat* 57 (2008): 995–1006.
91. Wang, Y.; Dong X.T.; Xia Feng, H.; Zhang, H. Stearic acid/polymethylmethacrylate composite as form-stable phase change materials for latent heat thermal energy storage. *Reraew Eraerg* 36 (2011): 1814–1820.
92. Fang, G.; Li, H.; Liu, X. Preparation and properties of lauric acid/silicon dioxide composites as form-stable phase change materials for thermal energy storage. *Mater Chem Phys* 122 (2010): 533–536.
93. Li, H.; Fang, G.; Liu, X. Synthesis of shape-stabilized paraffin/silicon dioxide composites as phase change material for thermal energy storage. *J Mater Sci* 45 (2010): 1672–1676.
94. Tang, B.; Qiu, M.; Zhang, S. Thermal conductivity enhancement of PEG/SiO2 composite PCM by *ira situ* Cu doping. *Sol Eraerg Mat Sol C* 105 (2012): 242–248.
95. Fabien, S. The manufacture of microencapsulated thermal energy storage compounds suitable for smart textile. Univ Lille Nord de France, Ensait, France, September 15, 2011.
96. Özonur, Y.; Mazman, M.; Paksoy, H.Ö.; Evliya, H. Microencapsulation of coco fatty acid mixture for thermal energy storage with phase change material. *Iratl J Eraerg Res* 30(10) (2006): 741–749.
97. Haldorai, Y.; Shim, J.-J.; Lim, K.T. Synthesis of polymer-inorganic filler nanocomposites in supercritical CO2. *J Supercrit Fluid* 71 (2012): 45–63.
98. Zhao, C.Y.; Zhang, G.H. Review on microencapsulated phase change materials (MEPCMs): Fabrication, characterization and applications. *Renew Sust Energ Rev* 15 (2011): 3813–3832.
99. Jin, Y.; Lee, W.; Musina, Z.; Ding, Y. A one-step method for producing microencapsulated phase change materials. *Particuology* 8(6) (2010): 588–590.

100. Borreguero, A.M.; Valverde, J.L.; Rodriguez, J.F.; Barber, A.H.; Cubillo, J.J.; Carmona, M. Synthesis and characterization of microcapsules containing Rubitherm® RT27 obtained by spray drying. *Chem Eng J* 166 (2011): 384–390.
101. Hawlader, M.N.A.; Uddin, M.S.; Kihn, M.M. Microencapsulated PCM thermal energy storage system. *Appl Energ* 74 (2003): 195–202.
102. de Boer, G.B.J.; de Weerd, C.; Thoenes, D.; Goossens, H.W.J. Laser diffraction spectrometry: Fraunhofer diffraction versus Mie scattering. *Part Syst Charact* 4 (1987): 9–14.
103. Bayés-García, L.; Ventolà, L.; Cordobilla, R.; Benages, R.; Calvet, T.; Cuevas-Diarte, M.A. Phase change materials (PCM) microcapsules with different shell compositions: Preparation, characterization and thermal stability. *Sol Energ Mat Sol C* 94 (2010): 1235–1240.
104. Pan, L.; Tao, Q.; Zhang, S.; Wang, S.; Zhang, J.; Wang, S.; Wang, Z.; Zhang, Z. Preparation, characterization and thermal properties of micro-encapsulated phase change materials. *Sol Energ Mat Sol C* 98 (2012) : 66–70.
105. Fang, Y.; Kuang, S.; Gao, X.; Zhang, Z. Preparation and characterization of novel nanoencapsulated phase change materials. *Energ Convers Manage* 49 (2008): 3704–3707.
106. Zhang, H.; Wang, X. Fabrication and performances of microencapsulated phase change materials based on *n*-octadecane core and resorcinol-modified melamine-formaldehyde shell. *Colloid Surf A* 332 (2009): 129–138.
107. Toppi, T.; Mazzarella, L. Gypsum based composite materials with micro-encapsulated PCM: Experimental correlations for thermal properties estimation on the basis of the composition. *Energ Buildings* 57 (2013): 227–236.
108. Youssef, Z.; Delahaye, A.; Huang, L.; Trinquet, F.; Fournaison, L.; Pollerberg, C.; Doetsch, C. State of the art on phase change material slurries. *Energ Convers Manage* 65 (2013): 120–132.
109. Maxwell, J.C. *A Treatise on Electricity and Magnetism*. Dover, New York; 1954, pp. 1–440.
110. Wang, X.; Niu, J. Heat transfer of microencapsulated PCM slurry flow in a circular tube. *AIChE J* 54(4) (2008): 1110–1120.
111. Cai, Y.; Wei, Q.; Huang, F.; Lin, S.; Chen, F.; Gao, W. Thermal stability, latent heat and flame retardant properties of the thermal energy storage phase change materials based on paraffin/high density polyethylene composites. *Renew Energ* 34 (2009): 2117–2123.
112. Sittisart, P.; Farid, M.M. Fire retardants for phase change materials. *Appl Energ* 88 (2011): 3140–3145.
113. Zhang, H.; Sun, S.; Wang, X.; Wu, D. Fabrication of microencapsulated phase change materials based on *n*-octadecane core and silica shell through interfacial polycondensation. *Colloid Surf A* 389 (2011): 104–117.
114. Alkan, C. Enthalpy of melting and solidification of sulfonated paraffins as phase change materials for termal energy storage. *Thermochim Acta* 451 (2006): 126–130.
115. Zou, G.-L.; Lan, X.-Z.; Tan, Z.-C.; Sun, L.-Z.; Zhang, T. Microencapsulation of n-hexadecane as a phase change material in polyurea. *Acta Phys Chim Sin* 20(1) (2004): 90–93.
116. Su, J.F.;Wang, S.B.; Zhang, Y.Y.; Huang, Z. Physicochemical properties and mechanical characters of methanol modified melamine–formaldehyde (MMF) shell microPCMs containing paraffin. *Colloid Polym Sci* 289 (2011): 111–119.
117. Giro-Paloma, J.; Oncins, G.; Barreneche, C.; Martínez, M.; Fernandez, A.I.; Cabeza, L.F. Physicochemical and mechanical properties of microencapsulated phase change material. *Appl Energ* 109 (2013): 441–448.
118. Giro-Paloma, J.; Barreneche, C.; Delgado, M.; Martínez, M.; Fernández, A.I.; Cabeza, L.F. Physicochemical and thermal study of a MPCM of PMMA shell and paraffin wax as a core. *Energy Procedia* 48 (2014): 347–354 (2014 International Conference on Solar Heating and Cooling for Buildings and Industry).
119. Hawlader, M.N.A.; Uddin, M.S.; Zhu, H.J. Encapsulated phase change materials for thermal energy storage: Experiments and simulation. *Int J Energ Res* 26 (2002): 159–171.
120. Shukla, A.; Buddhi, D.; Sawhney, R.L. Thermal cycling test of new selected inorganic and organic phase change materials. *Renew Energ* 33 (2008): 2606–2614.
121. Fortuniak, W.; Slomkowski, S.; Chojnowski, J.; Kurjata, J.; Tracz, A.; Mizerska, U. Synthesis of a paraffin phase change material microencapsulated in a siloxane polymer. *Colloid Polym Sci* 291 (2013): 725–733.

122. Sari, A.; Biçer, A. Thermal energy storage properties and thermal reliability of some fatty acid esters/building material composites as novel form-stable PCMs. *Sol Energ Mat Sol C* 101 (2012): 114–122.
123. Vorbeck, L.; Gschwander, S.; Thiel, P.; Lüdemann, B.; Schossig, P. Pilot application of phase change slurry in a 5 m3 storage. *Appl Energ* 109 (2013): 538–543.
124. Gschwander, S.; Schossig, P.; Henning, H.M. Micro-encapsulated paraffin in phase-change slurries. *Sol Energ Mat Sol C* 89 (2005): 307–315.
125. Rahman, A.; Dickinson, M.; Farid, M.M. Microindentation of microencapsulated phase change materials. *Adv Mat Res* 275 (2011): 85–88.
126. Zhang, P.; Ma, Z.W. An overview of fundamental studies and applications of phase change material slurries to secondary loop refrigeration and air conditioning systems. *Renew Sust Energ Rev* 16 (2012): 5021–5058.
127. Baronetto, S.; Serale, G.; Goia, F.; Perino, M. Numerical model of a slurry PCM-based solar thermal collector. *Proceedings of the 8th International Symposium on Heating, Ventilation and Air Conditioning.* Lecture Notes in Electrical Engineering 263. Springer, Heidelberg; 2014, pp. 13–20.
128. Hideo, I.; Yanlai, Z.; Akihiko, H.; Naoto, H. Numerical simulation of natural convection of latent heat phase-change-material microcapsulate slurry packed in a horizontal rectangular enclosure heated from below and cooled from above. *Heat Mass Transfer* 43 (2007): 459–470.
129. Lu, W.; Tassou, S.A. Experimental study of the thermal characteristics of phase change slurries for active cooling. *Appl Energ* 91(1) (2012): 366–374.
130. Delgado, M.; Lazaro, A.; Mazo, J.; Zalba, B. Review on phase change material emulsions and microencapsulated phase change material slurries: Materials, heat transfer studies and applications. *Renew Sust Energ Rev* 16 (2012): 253–273.
131. Alvarado, J.L.; Marsh, C.; Sohn, C.; Phetteplace, G.; Newell, T. Thermal performance of microencapsulated phase change material slurry in turbulent flow under constant heat flux. *Int J Heat Mass Tran* 50 (2007): 1938–1952.
132. Zhang, P.; Ma, Z.W.; Wang, R.Z. An overview of phase change material slurries: MPCS and CHS. *Renew Sust Energ Rev* 14 (2010): 598–614.
133. Huang, M.J.; Eames, P.C.; McCormack, S.; Griffiths, P.; Hewitt, N.J. Microencapsulated phase change slurries for thermal energy storage in a residential solar energy system. *Renew Energ* 36 (2011): 2932–2939.
134. Salunkhe, P.B.; Shembekar, P.S. A review on effect of phase change material encapsulation on the thermal performance of a system. *Renew Sust Energ Rev* 16 (2012): 5603–5616.
135. Zhang, S; Niu, J. Experimental investigation of effects of super-cooling on microencapsulated phasechange material (MPCM) slurry thermal storage capacities. *Sol Energ Mat Sol C* 94 (6) (2010): 1038–1048.
136. Lin, C.C.; Yu, K.P.; Zhao, P.; Lee, G.W.M. Evaluation of impact factors on VOC emissions and concentrations from wooden flooring based on chamber tests. *Build Environ* 44 (2009): 525–533.
137. Schossig, P.; Henning, H.M.; Gschwander, S.; Haussmann, T. Micro-encapsulated phase change materials integrated into construction materials. *Sol Energ Mat Sol C* 89 (2005): 297–306.
138. Kuznik, F.; Virgone, J.; Roux, J.J. Energetic efficiency of room wall containing PCM wallboard: A full-scale experimental investigation. *Energ Buildings* 40 (2008): 148–156.
139. Griffiths, P.W.; Eames, P.C. Performance of chilled ceiling panels using phase change materials slurries as the heat transport medium. *Appl Therm Eng* 27 (2007): 1756–1760.

19

Encapsulation of Bioactive Compounds

Francesco Donsì, Mariarenata Sessa, and Giovanna Ferrari

CONTENTS

19.1 Introduction

The scientific and industrial interest in functional foods, in which bioactive compounds are incorporated, has been significantly reinforced in recent years by the increasing demand from consumers for health promotion and disease prevention through diet and nutrition.

The main bioactive compounds for food functionalization include several classes of compounds that have been discovered to positively affect health. They can be classified as phytochemicals, micronutrients, dietary fibers, prebiotics, and probiotics. In particular, prebiotic compounds and probiotic microorganisms, due to their contribution to regulating intestinal flora, have recently gained significant attention, because the human biota is considered to affect several body functions directly and indirectly. Despite being not single chemical molecules but living organisms, probiotics are generally classified among the bioactive compounds, and therefore, their encapsulation is also treated in this chapter.

The incorporation of bioactive compounds in foods is challenged by significant technological hurdles, which are related to the desired *in-product* as well as *in-body* behavior.[1,2]

The *in-product* behavior is affected by the following:

- The efficient dispersion in the food matrix and compatibility with it
- The reactivity of bioactive compounds, which in turn, is responsible for the high degradation rate and interaction with other food components
- The complex environment of food matrices, with different interfaces between aqueous and lipid phases, where bioactive compounds may adsorb

- The intense treatment conditions to which food is subjected because of processing, preservation, or preparation (high or low temperatures, pH extremes, intense shearing, etc.)

The *in-body* behavior is affected by the following:

- The release from the food matrix, preferably triggered by environmental changes, such as pH (chewing, gastrointestinal tract), temperature (body temperature, cooking), mechanical shear (chewing, mastication), enzymes (gastrointestinal tract), and the addition of moisture (dissolution, chewing)
- The fate of the bioactive compounds during gastric and intestinal digestion
- The bioavailability of the active compounds, taking into account uptake by epithelial cells, absorption in the bloodstream, and reaching the target sites

To control and regulate in-product as well as in-body behavior, the bioactive compounds need a suitable encapsulation system, which is able not only to promote their efficient dispersion and protection during most intense phases of processing, preservation, or preparation but also to control their release during mastication (taste masking or enhancing) and gastrointestinal digestion, enhancing their bioaccessibility and bioavailability.[1]

In particular, controlled release of the bioactive compounds plays a double role: preserving the bioactive compounds in product and maximizing their release in the intestinal tract to promote their uptake.

Encapsulation involves the immobilization of the bioactive compounds within a capsule that completely embeds them or in which they are dispersed. The particles obtained on encapsulation may have a size ranging from a few nanometers to a few millimeters, with smaller sizes preferred to increase the specific surface for the release of the encapsulated compounds (also known as *payload*) as well as the interaction with biological tissues.[2]

Taking into account the requirements in terms of physicochemical stability, mean droplet size, controlled or triggered release, and food compatibility and ease of production at industrial scale, the main technological and scientific efforts toward the development of efficient encapsulation systems for the food industry in recent years aimed to

- Develop novel capsule architectures with a compartmentalized structure able to host bioactive compounds with different interfacial properties[3]
- Exploit the functionality of existing food-grade ingredients to replace artificial molecules with limited food acceptability[4]
- Design novel production processes, with low cost and high productivity, to produce novel delivery systems compatible with the food industry at an industrial scale[1]

It must be remarked that because of the health-beneficial properties of bioactive compounds, the interest in their encapsulation extends not only to the food and beverage industry but also to the pharmaceutical, nutraceutical, and agricultural industries.

19.2 Bioactive Compounds in Foods

In recent years, a strong correlation between nutrition and chronic diseases, such as diabetes, obesity, cardiovascular diseases, hypertension, some types of cancer, osteoporosis, and dental diseases, has become progressively more evident. The increased incidence of these chronic illnesses, as well as their appearance in life earlier than before, which are considered to be the main consequence of increasing consumption of high-fat and energy-dense foods, regular intake of so-called junk foods and low fruit and vegetable consumption, sedentary lifestyles, and insufficient physical activity, impose a growing burden on health care systems, as recorded by the World Health Organization.[5]

Good nutrition is important for maintaining good health and promoting socio-economic development. Epidemiological studies have shown a positive relationship between dietary intake of whole grains, fruits, vegetables, fish, and fermented milk products and health status[6–9] because of their content of bioactive compounds such as dietary fibers, phytosterols, carotenoids, peptides, bioactive lipids, and probiotics.

Bioactive compounds, as defined by Kitts,[10] are "extranutritional" constituents that are naturally present in small quantities in foodstuffs of both plant and animal origin. In fact, these compounds cannot be considered nutrients in the classical sense of the term, which are capable of developing, growing, and keeping an organism alive, but can be defined as substances that can modulate several important biological activities and functions of the human body. The biological activities modulated by the intake of bioactive compounds are different, including antioxidant and anti-inflammatory activity, modulation of detoxification enzymes, stimulation of the immune system, antibacterial and antiviral activity, antiproliferative and proapoptotic activity, etc.[11] However, studies that highlight their beneficial effects on human health are not yet conclusive and are generally focused on specific compounds. In fact, many aspects related to their bioavailability, metabolism, and interaction with the food matrix are still unclear.[12] In addition, some of these substances, which are reported to exhibit either a positive or a negative effect on health depending on the amount taken, are contained in foods whose consumption should not be promoted, such as alcoholic beverages.

Nevertheless, consumers' interest in the relationship between diet and health has increased the demand for functional foods, fostering a continuous advancement in the related science and technology. Many bioactive compounds are highly lipophilic, resulting in poor absorption and limited bioavailability; others are chemically unstable once extracted from the animal or plant source tissue. It has become apparent that current difficulties associated with the inclusion of bioactives in food matrices are one of the major problems that manufacturers struggle with when developing functional foods. There is a pressing need within the food industry for edible delivery systems to encapsulate, protect, and release bioactive compounds.[13]

The most important types of bioactive compounds can be classified as phytochemicals, micronutrients, dietary fibers, prebiotics, and probiotics, which are briefly discussed in the following sections.

19.2.1 Phytochemicals

Phytochemicals are chemical compounds that naturally occur in plants. They are responsible for some organoleptic properties, such as the deep purple of blueberries and the smell of garlic. They are nonessential nutrients, meaning that they are not required by the human body for sustaining life but have protective or disease-preventive properties.[14] It is well known that the production of these chemicals by plants has a protective purpose; recent research has demonstrated that phytochemicals can also protect humans against several diseases.[15–17]

Without specific knowledge of their cellular actions or mechanisms, phytochemicals have been considered as drugs for millennia. For example, Hippocrates considered that willow tree leaves could relieve fever. During the nineteenth and twentieth centuries, the main strategy of scientists for the cure of illnesses was to discover the active ingredients, which had medicinal or pesticidal properties. Examples of these discoveries include salicylic acid, morphine, and pyrethroids (pesticides). During the 1980s, many laboratories started to identify phytochemicals in plants that might be used as medicines. At the same time, other scientists conducted epidemiological studies to determine the relationship between the consumption of certain phytochemicals and human health.[18] Although scientific evidence supports the health-promoting functions of phytochemicals, their beneficial effects are often lost due to their poor solubility in aqueous and lipid phases, their physicochemical instability under food processing conditions (temperature, light, oxygen, and interaction with food matrix ingredients) and in the gastrointestinal tract (pH, enzymes, and presence of other nutrients), as well as insufficient gastric residence time and low permeability within the gut, limiting their activity and potential benefits.[19] Today, scientific research is focused on the study of protective mechanisms able to maintain the active molecular form until the time of consumption and to deliver it to the target sites within the organism[20] through suitable delivery systems.

TABLE 19.1
Food Sources, Beneficial Effects, and Technological Hurdles Related to Food Incorporation of Phytochemicals

Class	Examples	Food Sources	Beneficial Effects	Technological Hurdles for Food Incorporation	References
Phenolic compounds					
Polyphenols					
Flavonoids	Quercetin	Capers, red onion, lovage, green tea, apple, broccoli	Antioxidant activity, anti-inflammatory activity, bronchodilator, reduces the release of histamine	Poor solubility in aqueous phase and low chemical stability	[21,22]
	Catechins	Cocoa/chocolate, tea, red wine, berries, peach, apple, pear, apricot, vinegar	Anti-atherosclerotic effect, inhibit the oxidation of low-density lipoproteins, cancer chemopreventive activity	Sensitive to oxidation, light, and pH, astringent and bitter taste, slightly soluble in water	[23,24]
Stilbenoids	Resveratrol	Red wine, red grapes, peanuts, cocoa/ chocolate	Anti-inflammatory activity, beneficial cardiovascular effects, anti-atherosclerotic effects, antioxidant activity, chemoprotective advantages	Low solubility in aqueous and lipid phases, easily degraded by sunlight, and susceptible to react with dissolved O_2, bitter taste	[25–27]
Curcuminoids	Curcumin	Indian spice turmeric	Antioxidant activity, chemopreventive and anticarcinogenic effects, anti-inflammatory activity	Scarce solubility in water and oil phases, unstable at neutral–basic pH values	[28–30]
Aromatic acids					
Phenolic acids	Gallic acid	Blackberry, mango, chocolate, raspberry, vinegar, wine	Antifungal and antiviral properties, antioxidant activity, used as remote astringent in cases of internal hemorrhage	Sensitive to temperature, oxidation, light, and pH	[31, 32]
Hydroxycinnamic acids	Caffeic acid	Coffee, barley grain, argan oil	Antioxidant activity, immunomodulatory and anti-inflammatory activity	Low stability in UV irradiation and O_2 presence, low aqueous solubility, and bitter taste	[33, 34]
Terpenes					
Carotenoids	β-carotene	Carrots, pumpkins, sweet potatoes, cantaloupe, mango, papayas	Protection against photooxidative stress and prevention against skin damage	Susceptible to light, oxygen, and autooxidation	[35, 36]

(*Continued*)

TABLE 19.1 (CONTINUED)

Food Sources, Beneficial Effects, and Technological Hurdles Related to Food Incorporation of Phytochemicals

Class	Examples	Food Sources	Beneficial Effects	Technological Hurdles for Food Incorporation	References
	Lycopene	Tomatoes, watermelon, papaya	Potential preventive and/or therapeutic effect in prostate cancer		[37–40]
	Lutein	Collard greens, kale, spinach, eggs, asparagus, broccoli, carrots	Supports maintenance of eye health		[41]
Monoterpenes	Limonene	Essential oils of citrus fruit	Antimicrobial activity, potential chemopreventive agent	Chemically reactive species	[42,43]
Lipids	Omega-3, -6, and -9 fatty acids	Vegetable oils, nuts, fish, eggs	Reduce blood triglyceride levels, decrease risks of cardiovascular diseases	Vulnerable to oxidation and rancidity	[44–46]
Organosulfides	Allicin, diallylsulfide, allylmethyl-trisulfide	Garlic, leek, onion	Enhance detoxification of undesirable compounds; support maintenance of heart, immune, and digestive health	Chemically unstable and offensive odor	[47,48]

Table 19.1 presents a list of phytochemicals, divided into different classes, with their beneficial effects, food sources, and the technological hurdles related to their incorporation in food matrices.

Polyphenols are a heterogeneous group of natural compounds, particularly known for their beneficial effects on human health. In nature, polyphenols are produced by the secondary metabolism of plants, where they have different roles in relation to their various chemical characteristics:

- Defense against herbivores, imparting an unpleasant taste, and pathogens (phytoalexins)
- Mechanical support (lignins)
- Barrier against microbial invasion
- Attraction of pollinators by improving pigmentation (anthocyanins)

The polyphenols are characterized by possessing at least one aromatic ring, to which one or more hydroxyl groups are bound. More than 8000 different structures of polyphenols have been identified, ranging from simple, low–molecular weight compounds with a single aromatic ring up to highly polymerized compounds with very high molecular weight. Polyphenols can be classified into two groups, flavonoids and non-flavonoids (phenolic acids, stilbenes, and lignans), as a function of the number of phenolic rings and of the structural elements bound to these rings.[49–51]

The intake of polyphenols in the human diet varies in relation to the type, quantity, and quality of the plant foods consumed. Not only fruits and vegetables, but also tea, red wine, cocoa, and derivatives, are particularly rich in polyphenols. However, the cooking process considerably reduces the polyphenol content of food.[52,53]

Polyphenols such as curcumin, quercetin, and resveratrol are examples of naturally occurring phytochemicals with proved antioxidant and anti-inflammatory activity,[54,55] beneficial cardiovascular effects,[49,51,56] anti-atherosclerotic effects,[57,58] and chemopreventive and anticarcinogenic effects.[59–61] An abundance of mechanistic information has become available on how polyphenols derived from dietary sources, which have putative chemopreventive properties, interfere with tumor promotion and progression.[62] The effect of these bioactive compounds on the organism is influenced by their bioavailability, namely, their ability to be effectively absorbed by the human body. However, polyphenols have poor bioavailability and are rapidly metabolized by the human body, losing their potential beneficial effects.[50,63]

To solve these problems and ensure that the polyphenols retain their beneficial properties even after ingestion, it is necessary to develop encapsulation systems that protect these compounds once extracted from plants from interaction with atmospheric agents (light, oxygen, and temperature), during the production, transformation, storage, and cooking of foods in which they are incorporated, and finally, during the digestion process, to reach the target sites in a chemically stable form that is able to exert beneficial effects.[26,27]

Carotenoids contribute to the yellow and red colors of many foods. Carotenoids containing oxygen are known as xanthophylls (e.g., lutein and zeaxanthin), while those without oxygen are known as carotenes (e.g., lycopene and β-carotene). The carotenoids have been reported to exhibit several potential health benefits: for example, lutein and zeaxanthin contribute to decreasing age-related macular degeneration and cataracts,[41] and lycopene contributes to decreasing the risk of prostate cancer.[64] In their endogenous form in foods, carotenoids are generally stable. However, when incorporated as food additives, carotenoids are relatively unstable, because they are susceptible to light, oxygen, and autooxidation.[65] Consequently, the dispersion of carotenoids into a food matrix can result in their rapid degradation.[38] An additional challenge to using carotenoids as ingredients in functional foods is their high melting point, making them crystalline on food storage.

Bioactive lipids, such as omega-3, -6, and -9 fatty acids, are unsaturated fatty acids. Their important physiological role has been attributed to their ability to decrease the risks of cardiovascular disease,[44] diseases induced by immune response disorders (e.g., type 2 diabetes, inflammatory bowel diseases, and rheumatoid arthritis),[45,66] and mental disorders,[67,68] as well as to benefit infant development.[69–71] The growing list of disorders positively affected by bioactive lipids strongly suggests that large portions of the population would benefit from increased consumption of omega-3, -6 and -9 fatty acids, making them an excellent candidate for incorporation into functional foods.

However, numerous challenges exist in the production, transportation, and storage of fatty acid fortified functional foods, since these lipids are extremely susceptible to oxidative deterioration. The encapsulation of bioactive lipids has been found to be an excellent method for their stabilization.[46]

19.2.2 Micronutrients

Micronutrients are different from macronutrients (such as carbohydrates, protein, and fat) because they are necessary only in very small quantities. Nevertheless, micronutrients are essential for good health, and a deficiency can cause serious health problems. Micronutrients include dietary minerals, which are important for the smooth operation of all biological functions and cellular activity. Therefore, their consumption is associated with the reduction of the risk of certain types of diseases. The most important micronutrients are the following:

- *Calcium,* which reduces the risk of osteoporosis and of other disorders related to its deficiency in the diet, has a protective role against hypertension and certain types of cancer.[72,73]
- *Potassium,* which reduces the risk of high blood pressure and stroke,[74] in combination with a low sodium diet.
- *Magnesium,* which supports the maintenance of normal muscle and nerve functions and immune and bone health.[75]
- *Iron,* which helps the human body to produce red blood cells and lymphocytes.[76]
- *Chromium,* which reduces hyperglycemia and maintains the correct level of cholesterol and glucose in the blood.[77,78]

The minerals are used to fortify various types of products, especially bread and baked goods, breakfast cereals, dairy products, and vegetables. Moreover, they are often mixed with some vitamins and homogeneously distributed in the products. However, in some formulations, it is necessary to adopt measures to protect the micronutrients by some factors that may cause them to be lost or decrease their bioavailability, such as the use of microencapsulation techniques.[79–81]

19.2.3 Dietary Fibers

Dietary fibers represent a group of carbohydrates, which can be isolated from different plant sources, that are resistant to hydrolysis by enzymes of the gastrointestinal tract. They are divided into two groups: water-soluble fibers, among which the main examples are the β-glucans and arabinoxylans, and water-insoluble fibers, which include lignins, celluloses, and hemicelluloses.

Dietary fibers exert both functional and metabolic effects, which make them important components of the diet. In addition to the increase of satiety and improvement of bowel function and disorders associated with it (constipation and diverticulosis), the intake of fiber in food has been related to a reduction in the risk of major chronic diseases, such as cardiovascular diseases,[7] thanks to the reduction of cholesterol,[82] of glucose in the blood, and of insulin.[83]

Among the water-soluble fibers, the most important group is represented by β-glucans, which have important positive effects on heart diseases, reducing the levels of cholesterol and glucose in the blood, and which are able to promote the growth of lactobacilli and bifidobacteria present in the gastrointestinal tract. The β-glucans are present in all cereals, but barley and oats contain the highest amounts.

Fibers are used as ingredients in many processed foods, such as bread, pasta, breakfast cereals, yogurt, and meat products. They may be either mixed in the form of concentrated isolated products, such as concentrates of oat fiber or introduced into the formulations through the addition of flour or other ingredients that contain them.

The main difficulty related to their incorporation in food matrices derives from the consideration that dietary fibers have a high capacity to bind water, thus determining a reduction of the final volume of the product (in the case of bakery products) and in general, an alteration of the organoleptic properties of the food. For this reason, encapsulation seems a promising technique to improve the inclusion of dietary fibers in functional foods without modifying the sensory characteristics, such as color, appearance, flavor, and taste, of the food itself.

19.2.4 Prebiotics and Probiotics

Prebiotics and probiotics are an area of growing scientific interest for the development and production of functional foods.[84] The terms *prebiotics* and *probiotics* themselves reveal important implications for human health that may result from their use and that are related to the intestinal microbiota. In fact, the target organ of the action of these "ingredients" is the intestines, but indirectly, the whole body is the real beneficiary of their effects. The function of prebiotics and probiotics is to promote the proliferation and balance of the bacterial composition that constitutes the intestinal ecosystem. The intestinal microbiota is made of hundreds of different bacterial species, whose multiple metabolic activities affect the state of health of the host.

Under psycho-physical, dietary, and environmental stress conditions, or after medication intake, there is an imbalance of microflora that makes the body susceptible to attack by pathogens. A proper diet is one of the main factors that influence the qualitative and quantitative composition of intestinal microflora. The most common approach involves the consumption of traditional foods, such as yogurt and fermented milk, which essentially contain probiotics, defined as live microorganisms, which can positively affect the host by improving its intestinal microbial balance.

The intestinal microflora, consisting of populations of lactobacilli and bifidobacteria, beneficial microorganisms, and a high number of bacteria, such as clostridia and enteropathogens, capable of causing health risks, has the primary role of recovering energy through the fermentation processes that metabolize specific substrates, especially carbohydrates, peptides, proteins, and some lipids, that are not digested in the upper part of the intestine.

Therefore, health benefits from the consumption of prebiotics and probiotics are observed when

- The composition of the intestinal flora is influenced, with an increase in the amount of probiotic bacteria in the gut.
- The development of a limited number of probiotic bacteria is selectively stimulated in the colon through the fermentation of non-digestible substrates due to the intake of prebiotic ingredients

19.2.4.1 Prebiotics

Prebiotics are not live microorganisms but non-digestible food ingredients, which when administered in adequate amounts, bring benefit to the consumer due to their ability to selectively promote the growth and/or the activity of one or more bacteria already present in the gastrointestinal tract or ingested together with the prebiotics.[85,86] To be defined as prebiotic, a substance must be resistant to attack by hydrochloric acid in the stomach and to the hydrolytic and enzymatic processes that occur in the duodenum, acting as a substrate for fermentation by intestinal bacteria, selectively stimulating their growth and/or activity.

Therefore, the effect of a prebiotic is essentially indirect, because it acts selectively as a substrate for one or for a limited number of microorganisms, causing a modification of the intestinal microflora; it is not the prebiotic itself but rather, the changes that it promotes in the composition of the intestinal microflora that will bring the expected benefits.[87]

As defined by Wang,[88] an ingredient can be classified as prebiotic if (1) it is resistant to digestion in the upper gut tract, (2) it can be fermented by intestinal microbiota, (3) it brings beneficial effects to the host health, (4) it selectively stimulates the activity of probiotics, and (5) it is stable to food processing treatments.

Moreover, prebiotics can be incorporated into food matrices if they are chemically stable during food processing, under conditions of low pH and high temperatures, and to Maillard reactions.

Most prebiotics are non-digestible oligosaccharides; they are obtained by extraction from plants (e.g., inulin from chicory), possibly followed by enzymatic hydrolysis (e.g., oligofructose from inulin), or by synthesis from mono- or disaccharides.[89] Among all the prebiotics, inulin and oligosaccharides are certainly the most studied and have been recognized throughout the world as dietary fibers . The most important prebiotics are presented in Table 19.2.

Prebiotics are increasingly used to achieve a double action: an improvement of the organoleptic properties of the food and a better balance of the nutritional benefits.[100]

The use of inulin and non-digestible oligosaccharides improves the taste and texture of food to which they are added. In addition, these fibers are readily fermentable by specific bacteria such as lactobacilli and bifidus bacteria, resulting in an increase in their population with the simultaneous production of short-chain fatty acids.[101] These fatty acids, especially butyrate, acetate, and propionate, provide metabolic energy for the host.

19.2.4.2 Probiotics

Probiotics are defined by the World Health Organization as "live microorganisms that, when administered in adequate amounts, confer a health benefit on the host."[102] This definition refers to non-pathogenic microorganisms present in foods or added to them and "excludes references to biotherapeutic agents and beneficial microorganisms not used in the food industry."

The main probiotic preparations available on the market are known as lactic acid bacteria (LAB), for the most part represented by lactobacilli and bifidobacteria, which are important constituents, normally present, of the gastrointestinal microflora, and which produce lactic acid as a major metabolite. Moreover, certain yeasts and bacilli are also counted as probiotics. Table 19.3 reports the microorganisms recognized as probiotics and used in the production of functional foods.

The bacteria normally used in the production of yogurt, in particular *Streptococcus thermophilus* and *Lactobacillus bulgaricus*, are not expected to survive and overcome the intestinal tract and therefore, are not considered probiotics.[103]

TABLE 19.2

Food Sources and Beneficial Effects of the Most Important Prebiotics

Prebiotics	Food Sources	Beneficial Effects	References
Inulin	Chicory, banana, onion, garlic	Promotes colonic health, decreases amount of cholesterol and triglycerides	[90,91]
Lactulose	Milk	Beneficial effects on digestive health	[92]
Isomaltooligosaccharides (IMO)	Wheat, barley, corn, pulses, oats, tapioca, rice, potato, and other starch sources	Reduce flatulence (i.e., generating less gas), low glycemic index, and anticaries activities	[93–95]
Xylooligosaccharides (XOS)	Bamboo shoots, milk, and honey	Anti-allergy, anti-infection and anti-inflammatory properties, immunomodulatory, and antimicrobial activity	
Soybean oligosaccharides (SOS)	Soybeans	Prevent constipation due to the production of short-chain fatty acids, improve absorption of calcium and other minerals, reduce risk of colon cancer	[96]
Fructooligosaccharides (FOS)	Banana, onion, chicory root, garlic, asparagus, wheat, and leeks	Promote calcium absorption, provide some energy to the body	[97]
Galactooligosaccharides (GOS)	Bovine milk	Excellent source for health-promoting bacteria, support the intestinal immune system, improve mineral absorption	[98,99]

There is much scientific evidence, supported by clinical studies, on the efficacy of probiotics in the prevention and treatment of gastrointestinal disorders and respiratory and urogenital diseases.[104] Many microbial strains with probiotic properties are able not only to restore the intestinal microbial balance but also to impart other beneficial effects on health, associated with the production of acids and bacteriocins and with competition with pathogenic microorganisms. Among these, the main effects are the reduction of the level of cholesterol in the blood, the reduction of fecal enzymes with potentially mutagenic activity that can induce the onset of tumors, the reduction of lactose intolerance, the increase of the response of the immune system, the increase of calcium absorption, and the synthesis of vitamins.[105]

As evidenced in Figure 19.1, probiotics have been shown to work by the following mechanisms[106]:

- *Competition for nutrients:* Within the gut, beneficial and pathogenic microorganisms use the same types of nutrients, creating a general competition between bacteria for these nutrients. When a probiotic is administered, there is an overall reduction in the nutrients available to pathogenic bacteria, and consequently, this minimizes the numbers of pathogenic microorganisms.

TABLE 19.3

Microorganisms Used as Probiotics for the Production of Functional Foods

Lactobacilli	Bifidobacteria	Other Microorganisms
L. acidophilus	*B. animalis*	*Enterococcus faecium*
L. casei	*B. breve*	*Bacillus subtilis*
L. johnsonii	*B. infantis*	*Escherichia coli*
L. reuteri	*B. longum*	*Saccharomyces boulardii*
L. rhamnosus	*B. adolescentis*	*Clostridium butyricum*
L. salivarius	*B. lactis*	
L. plantarum, L. crispatus	*B. bifidum*	

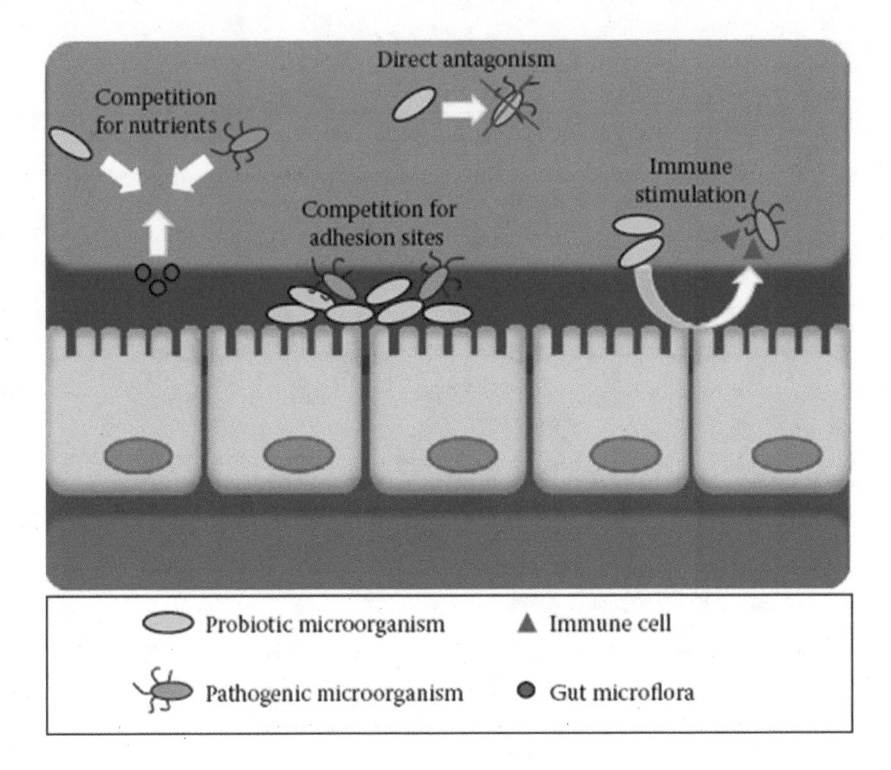

FIGURE 19.1 Mechanisms of action of probiotic microorganisms in the gut.

- *Competition for adhesion sites:* Beneficial bacteria can attach to the gut wall and form colonies at various sites throughout the gut. This prevents pathogenic bacteria from gaining a foothold, resulting in their expulsion from the body.
- *Lactic acid production:* Probiotics produce lactic acid, which acts to reduce the gut pH, inhibiting the growth of pathogenic bacteria, which prefer a more alkaline environment.
- *Effect on immunity:* Probiotics have been shown to increase the levels of cell-signaling chemicals and the effectiveness of infection-fighting cells (white blood cells).

The probiotics used for the production of functional foods are incorporated as a supplement or added in advance to facilitate fermentation processes during the preparation of the food itself. It appears evident that the physicochemical properties of a food used as a carrier to deliver the probiotics within the gastrointestinal tract are an important factor that determines the survival and thus, the potential beneficial effect of the probiotic.[107] Fat content, concentration and type of proteins and sugars, oxygen level, pH, and storage temperature of the product are some factors that affect the growth and survival of the probiotic. By formulating the food to achieve an appropriate pH value and a high buffer capacity, it is possible to increase the pH of the gastric tract and in this way, improve the stability of the probiotic.[108]

Therefore, to achieve beneficial effects, it is necessary not only to ensure a high level of viable microorganisms in the food (at least 10^7 colony forming units [CFU]/g) but also to ensure the protection of the probiotics during the production, storage, and consumption of the food product. Moreover, it is necessary to protect them from the action of gastrointestinal acids and enzymes, from adhesion to the intestinal epithelium, and from attack by antibiotics.[109]

Although probiotic cultures do not tend to markedly change the sensory properties of the products to which they are added, in many cases, consumers have found products fermented with *L. delbrueckii* subsp. *bulgaricus* too acidic and with a strong flavor of acetaldehyde (the typical flavor of yogurt). For the production of probiotic products, it is necessary to develop probiotic cultures that do not alter the organoleptic properties but are able to enhance the flavors of the foods to which they are added.[110]

Probiotics are mainly added to dairy products, but in recent years, thanks to the use of different technologies, the food industry has begun trying to add probiotics to different types of foods, such as beverages,[111] baking products,[112,113] ice cream,[114–116] and chocolate.[117]

To ensure the viability and stability of probiotics without altering the organoleptic properties of food after their inclusion in food matrices, different strategies have been developed, ranging from the use of particular technologies, such as microencapsulation, that protect bacteria from unfavorable conditions that may occur during the preparation, production, and storage of the product by enclosing them in special coatings, to the use of special substrates that can be used selectively by probiotic microorganisms for their growth. Such substrates are represented by prebiotic ingredients, and growing interest is developing in the synbiotic use of probiotics and prebiotics for the functionalization of foods.[118]

19.3 Encapsulation Systems

Encapsulation systems for bioactive compounds may take different configurations and architectures depending on the materials, fabrication process, and dispersion medium as well as interaction with the payload. However, despite an ultimate classification of encapsulation systems being difficult, Figure 19.2 tries to provide approximate categories,[2,119] identifying two broad classes of systems, one pertaining to the matrix type and the other to the core and shell type.

In the *matrix type*, the payload is well dispersed within the encapsulation material, more or less homogeneously distributed, and in general, it is present also on the surface of the matrix. In contrast, *core-shell*-type systems are characterized by an external shell completely covering a core containing the bioactive compounds, which therefore, are not in direct contact with the external environment.

As shown in Figure 19.2, core-shell systems encompass different architectures, the simplest of which is single core/single shell (also known as *reservoir type*), and the most complex comprise several cores

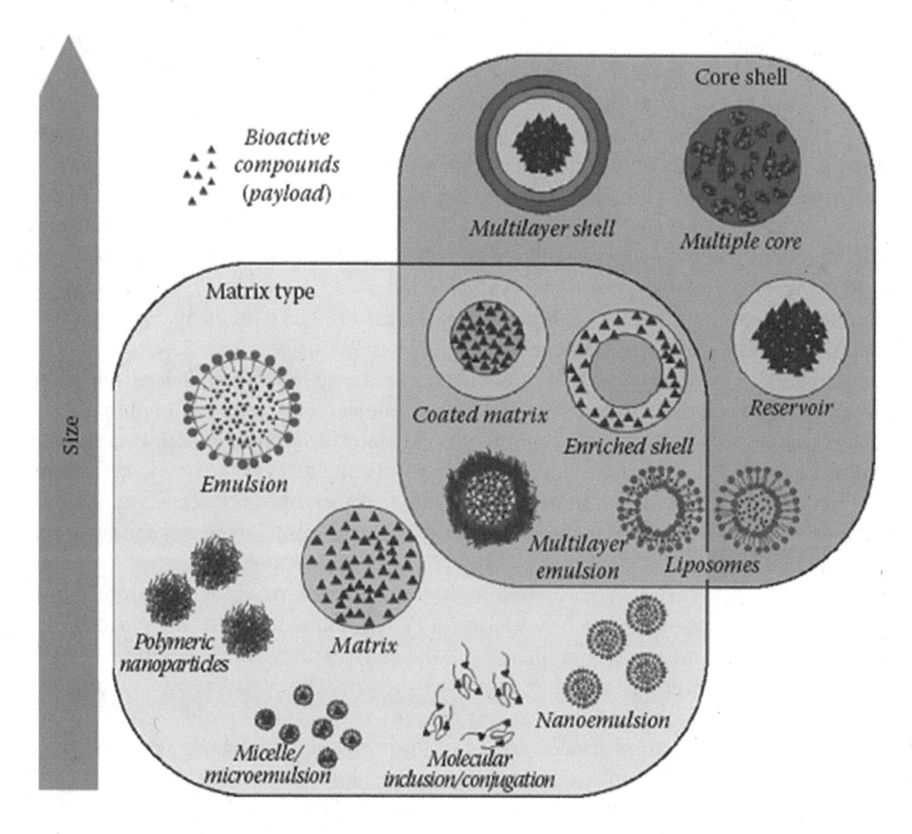

FIGURE 19.2 Classification of the main encapsulation systems for bioactive compounds.

and/or a multiple-layer shell. Ideally, the shell material should be able to limit the diffusion of the payload *in product*, while once *in body*, it should facilitate its release when needed. Therefore, waxes, fats, and proteins are primarily used as shell materials.[120,121] Because of their structural organization, with a higher degree of complexity than the matrix type, core-shell systems are generally characterized by larger mean particle sizes and a more complex fabrication process, whose key step is the formation of the shell layer, which implies higher processing costs.[122,123]

In contrast, matrix-type systems are less expensive to fabricate but are not suitable for taste masking and are characterized by a lower loading capability.

Molecular inclusion or conjugation complexes, micelles, microemulsions, polymeric particles, and emulsions and nanoemulsions can all be classified as matrix-type encapsulation systems.

For example, at the molecular level, micellization and molecular inclusion complexation represent some of the simplest approaches to fabricate matrix-type systems.

Micelles result from the spontaneous aggregation of amphiphilic molecules (emulsifier or surfactant) in a solvent at a concentration level above the critical micelle concentration (CMC). If the surfactant concentration remains above the CMC, micelles are thermodynamically stabilized against disassembly.[124]

The use of micelles in foods is limited by the low loading capacity of bioactive compounds and the amount of surfactants used, which may significantly impact on the organoleptic properties of the end product, as well as by their potential metastability with respect to the solubility of bioactive compounds (dynamic formation and rupture of micelle structures), which may cause precipitation on dilution or crystallization over long storage times.[125]

Inclusion complexes are instead based on the steric and hydrophobic interaction of the hydrophobic polysaccharide chain with hydrophobic molecules. For example, amylase was reported to be able to form a complex with conjugated linoleic acid due to its helical molecular structure, whose moieties can form steric and hydrophobic interactions with compounds such as free fatty acids.[126] However, inclusion complexation is more efficient if the polysaccharide structure is a ring of well-defined size, such as for cyclodextrins, which are naturally occurring cyclic oligosaccharides derived from starch, with six, seven, or eight glucose residues assembled in a ring, characterized by external hydrophilic moieties and internal hydrophobic moieties.[127] Thanks to their architecture, cyclodextrins can accommodate in the internal cavity non-polar molecules that can fit their specific cavity. The inclusion of bioactive molecules in cyclodextrins may significantly increase their water solubility,[127] as shown for flavonols, such as quercetin and myricetin, that were inclusion complexed with β-cyclodextrins[128] as well as for lycopene in α-, β-, and γ-cyclodexrins.[129] Inclusion complexation in cyclodextrins finds application in deodorizing processes or in the taste masking of bitter molecules. The encapsulated compounds, which are efficiently protected in the presence of oxygen and radiation, are released under high-moisture and low-temperature conditions.[130] β-cyclodextrins have been approved for food use.[131] However, it must be highlighted that the use of cyclodextrins is limited by the high cost of the material as well as by the high ratio between the encapsulant and the encapsulated material.

At a scale larger than molecular but still in the nanoscale range (<100 nm), *microemulsions* are self-assembling encapsulation systems, consisting of inner droplets stabilized by amphiphilic molecules, generating hydrophilic and hydrophobic regions of large interfacial area, which enable them to host different guest molecules such as food additives, nutraceuticals, aromas, cosmetic compounds, active ingredients, and drugs.[132] They differ from emulsions and nanoemulsions because they are thermodynamically stable systems, with characteristic dimensions ranging from 5 to 50 nm and typical emulsifier to oil ratio larger than 1.[133]

Depending on the properties of the amphiphilic molecules used, of the temperature, and of the oil and water fractions, oil-in-water (O/W) or water-in-oil (W/O) microemulsions can be formed, but O/W emulsions have the highest potential for application to the delivery of bioactive compounds in food products.[3,132] In comparison with emulsions, microemulsions require the use of relatively large amounts of surfactant,[134] with the consequence of their loading capacity being significantly lower.[125]

With a typical size ranging from nanometric (<100 nm) to submicrometric (<1 μm), *biopolymeric particles and nanoparticles*, made of proteins or polysaccharides, thanks to their excellent compatibility with foods, are able to efficiently encapsulate, protect, and deliver bioactive compounds, forming different structures, such as random coils, sheets, or rods, around the bioactive molecules.[135,136] The most suitable biopolymers for incorporation into foods include (1) proteins, such as whey proteins, casein, gelatin, soy protein, and zein, and (2) polysaccharides, such as starch, cellulose, and other hydrocolloids,[137] with the particle formulation depending on the desired particle functionality (size, morphology, charge, permeability, and environmental stability), end product compatibility, and general in-product behavior as well as on release properties and in-body behavior.

Remarkably, the steric and electrical characteristics of the biopolymeric particles, which are controlled by biopolymer properties and assembly conditions, enable not only the control of physicochemical stability but also the interaction with other species or food ingredients with opposite charge, as well as the interaction with biological surfaces (i.e., mucoadhesiveness).[137]

In general, the fabrication of biopolymer particles is based on the induced aggregation of homogeneous or heterogeneous molecules, controlling their intrinsic properties as well as the environmental conditions (i.e., temperature, pH, and ionic strength). In particular, the self-assembly of proteins is induced by physical interactions due to van der Waals, electrostatic, and hydrophobic forces and hydrogen bonding,[138] while the self-assembly of polysaccharides results from van der Waals and electrostatic forces as well as hydrogen bonding.[139] Following spontaneous association, a consolidation step of the particle structures, such as covalent bonding, is desired to prevent the formation of larger structures or of a gelling network.

Emulsions and nanoemulsions are heterogeneous systems consisting of two immiscible liquids, with one liquid phase being dispersed as droplets into another continuous liquid phase and stabilized by an appropriate emulsifier. In particular, nanoemulsions are characterized by a nanometric size (<100 nm), while emulsions are in the submicrometric and micrometric range.

O/W emulsions, which are of prevalent interest for encapsulating bioactive compounds and delivering them into food systems, are composed of oil droplets dispersed in an aqueous medium and are stabilized by a food-grade surfactant or biopolymeric layer, whose properties control the interfacial behavior (charge, thickness and droplet size, and rheology) as well as the response to environmental stresses (pH, ionic strength, temperature, and enzyme activity) of the encapsulation system.[136]

Unlike microemulsions, emulsions are kinetically stable, requiring energy to be formed, and are subject to several instability phenomena, including coalescence and gravitational separation. The most common approach to reduce their instability is to reduce their mean droplet size to such values that Brownian motion effects dominate over gravitational forces.[133] Therefore, nanoemulsions are highly stable to gravitational separation and show a lower tendency to droplet aggregation than conventional emulsions, because the strength of the net attractive forces acting between droplets usually decreases with decreasing droplet diameters.[133]

The encapsulation of bioactive compounds in O/W emulsions enables their efficient incorporation in foods. Depending on the properties of the bioactive compounds and in particular, on their lipo- or hydrophilicity, the payload localization within an O/W emulsion may change from a prevalent entrapment within the inner oil phase (bioactive-enriched core) to a prevalent concentration in the outer stabilizer film (bioactive-enriched shell). Remarkably, the localization of the bioactive compounds influences their stability, release, and bioavailability.

Similarly to emulsions, *solid lipid particles* consist of emulsifier-coated lipid droplets dispersed within an aqueous phase, with the main difference that the lipid phase is in the solid or semi-solid state.

In comparison with emulsions, solid lipid particles exhibit increased chemical protection against degradation, higher encapsulation efficiency (>90%), and better controlled release due to the immobilization of the encapsulated bioactive compound in the solid lipid matrix.[140]

In addition, when the lipid phase consists of fats of different types and with different properties as well as crystallization kinetics, the encapsulation system is characterized by voids and defects of the solid fat crystalline structure with high loading capability. In contrast, when the lipid phase consists of a more ordered crystalline structure, made of a single type of fats, less space will be available for the bioactive molecules, reducing the loading capability.[141] Therefore, typically, two or more lipids with different melting points are used, such as mixtures of purified triglycerides, waxes, or fatty acids.[142] The main limitation to the use of solid lipid particles is in their fabrication, which is based on high-energy emulsification processes at temperatures above the melting point of the lipids and requires fine control of the lipid crystallization, which is significantly dependent on the temperature history of the system, on the presence of impurities in the lipid phase, and on droplet size.[143]

Nanostructure lipid carriers indicate lipid particles with a disperse phase made of a mixture of solid and liquid lipids. Due to the decreased melting point of the lipid phase, such systems can be produced at lower temperatures, reducing the extent of degradation of thermolabile compounds.[141]

Liposomes are spherical bilayers made of amphiphilic molecules, such as phospholipids, which are characterized by an inner hydrophilic water domain physically separated from the bulk of water. Due to their structure, liposomes may enclose bioactive molecules of hydrophilic nature in the inner hydrophilic domain as well as molecules of lipophilic nature in the bilayer. Liposome fabrication is based on the spontaneous association of amphiphilic molecules in a lamellar phase, induced by the proper selection of solvent, concentration, and temperature conditions, followed by its dispersion by intense mechanical disruption or the use of solvent evaporation techniques. Liposomes can be structured in (1) single bilayers, forming unilamellar vesicles, (2) several concentric bilayers, forming multilamellar vesicles, or (3) non-concentric bilayers, forming multivesicular vesicles.[143] The use of liposomes in the food industry is limited by the high cost of pure lecithins, which are the most suitable food-grade amphiphilic molecules, the low encapsulation efficiency, and the complicated and costly fabrication process.

Matrix-type systems in which an additional coating layer is applied, are defined as *coated matrix type*.[2] They are usually fabricated in two-step processes, consisting of the formation of the matrix-type system (i.e., spray-drying or emulsification) followed by the formation of a coating layer (i.e., fluidized bed coating or layer-by-layer deposition).

In the case of emulsions, the deposition of a *multiple layer* of emulsifiers and/or polyelectrolytes may positively affect the absorption properties in the gastrointestinal tract[144] and significantly improve emulsion stability by increasing the packing density at the interface.[145] For example, the adsorption of layers of biopolymers, such as chitosan, on lipid-stabilized emulsion droplets was reported not only to improve the physical stability against thermal processing, freeze–thaw cycling, and drying, as well as the chemical stability against lipid oxidation,[146] but also to enable controlled or triggered release.[147] Moreover, alternating layers of proteins and hydrocolloids are used, based on electrostatic interactions, to realize multiple-layer systems, with improved stability to extremes of temperature, pH, and ionic strength.[148]

It must be remarked that in practice, many of these encapsulation systems are not spherical in shape, and therefore, the terms *microcapsules* and *nanocapsules* are used in a general way to denote either irregular or spherical shapes.[149]

The type of encapsulation system, together with the properties of the encapsulating material, significantly affects the release properties of the payload. In Figure 19.3, a schematic representation of the release profiles of the main classes of encapsulation types is shown.

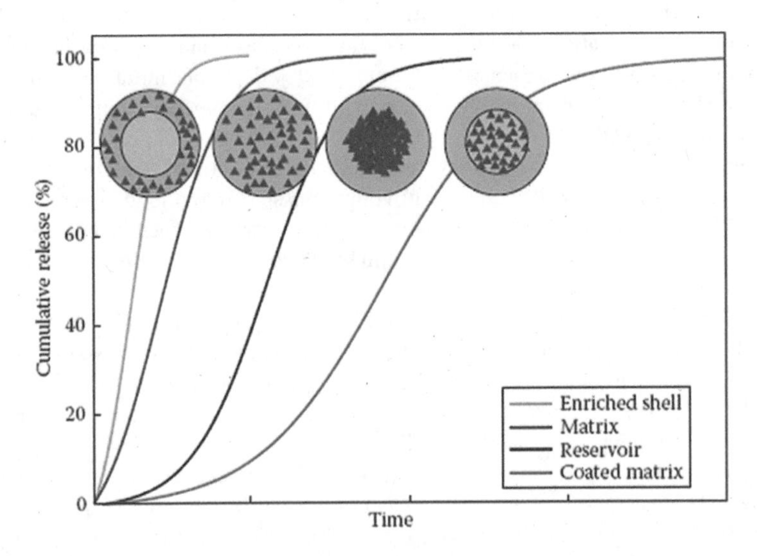

FIGURE 19.3 Schematic representation of cumulative release profiles of the main encapsulation architectures (reservoir type, matrix type, coated matrix type).

In general, faster release kinetics are expected from those systems where the bioactive compounds are in close contact with the release medium, which occurs in particular for enriched shell and matrix-type systems. In contrast, when a shell layer is present, the release may be significantly slowed down, depending on the mechanism of permeabilization of the external shell. Therefore, it is possible to trigger the payload release by opportunely triggering the degradation of the external shell under specific environmental conditions.

In addition to the architecture of the encapsulation system, the other factors affecting the release kinetics of the encapsulated bioactive compounds can be classified into chemical and morphological factors. The chemical factors comprise the size of the bioactive molecules and the properties of the encapsulant material, such as its thermal stability, amorphous or crystalline structure, glass transition properties, melting point, phase transition, and ionic charge. The morphological factors include size distribution, shape, coating uniformity, moisture content, physico-chemical stability, and hygroscopicity.[2]

Tables 19.4 and 19.5 provide a survey of recent applications of micrometric and submicrometric encapsulation systems for bioactive compounds.

In particular, Table 19.4 provides examples of core-shell and matrix-type systems, with details on the type of bioactive compounds, the formulation of the shell or the matrix, the fabrication technique, and the desired application. In Table 19.4, liposomes are also included, whose size, especially for food applications, is in the micrometric range, but they can be fabricated also at smaller sizes.

Table 19.5 gives examples of encapsulation systems of submicrometric scale, providing the same details as Table 19.4.

19.4 Encapsulation Techniques

The fabrication of encapsulation systems may be carried out through mechanical processes, in general based on a top-down approach to the disruption of larger systems into homogeneously sized particles or droplets with desired properties, or by physicochemical processes, based on a bottom-up approach to the assembling of molecular building blocks into structured systems (i.e., micelles, microemulsions, and some biopolymeric nanoparticles), as well as by a mixed approach, whereby molecular assembly and comminution processes are combined together (i.e., liposomes, multilayer emulsions, and some biopolymeric nanoparticles).

Different techniques are used at industrial level, including spray-drying and spray-chilling, fluid bed coating, melt injection and melt infusion, extrusion, coacervation, and crystallization, as well as the production of emulsions, particles, and liposomes.[189,190]

Figure 19.4 shows a classification of the main encapsulation techniques as a function of the type of capsules that can be produced and of their typical size. Interestingly, smaller-size systems are usually matrix type, such as nanoemulsions, micelles, microemulsions, and molecular complexes. Only in the case of biopolymeric particles is it possible to attain a certain degree of structural complexity at submicrometric size. In contrast, for larger-size systems, different architectures are possible through mechanical technologies, such as atomization and extrusion, together with mechanical or physico-chemical coating approaches.

19.4.1 Mechanical Processes

Mechanical processes for the production of encapsulation systems are in general based on the disruption of a dispersion of the bioactive compounds in the encapsulant material into small particles or droplets and their stabilization by drying, phase transition, or deposition of molecular layers on their surface.

Spray-drying is a relatively simple and inexpensive drying process that can be advantageously used to encapsulate bioactive and aroma compounds in foods.[191,192] In the spray-drying process, the payload is preliminarily finely dispersed or homogenized in a highly concentrated (up to 30% by weight) biopolymeric aqueous solution, containing, for example, starches, succinylated starches, cellulosics, gelatin, gums, and proteins, or in organic solvent solutions, containing, for example, poly(lactic-co-glycolic acid) (PLGA), ethyl cellulose, or acrylates, with the main requirement being the ability to form a glassy

TABLE 19.4

Encapsulation of Bioactive Compounds at the Microscale

Bioactives	Formulation	Fabrication	Application	References
Core-shell systems				
Jasmine essential oil	Gelatin and gum arabic	Complex coacervation	Preservation at high temperature	[150]
Vanilla oil	Chitosan	Complex coacervation	Controlled release and thermostability in spice industry	[151]
Sucralose	Gelatin and gum arabic	Double emulsion and coacervation	Preservation	[152]
Ascorbic acid	Gelatin and gum arabic	Double emulsion and coacervation	Preservation	[153]
Riboflavin	Whey protein/alginate gel	Cold gelation	Controlled release in beverages	[154]
Matrix systems				
Lycopene	Gelatin and gum arabic	Complex coacervation	Preservation in cake making	[155]
Natural vitamins	Gum arabic	Spray-drying	Preservation	[156]
Blackcurrant polyphenols	Maltodextrins and inulin	Spray-drying	Protection of antioxidant activity	[157]
Bilberry extract	Whey proteins	Emulsion templating and hot gelation	Preservation	[158]
Bayberry extracts	Ethyl cellulose	Phase separation	Protection of antioxidant activity	[159]
Pomegranate polyphenols	Maltodextrin or soybean protein isolates	Spray-drying	Preservation at high temperature	[160]
Cactus pear extracts	Maltodextrins and inulin	Spray-drying	Preservation at high temperature	[161]
Liposomes				
Peptides, bacteriocins	Different phospholipids	Encapsulation in the inner aqueous core	Biopreservatives	[162]
Curcumin	Soy lecithin	Encapsulation in the phospholipid bilayer	Oral carrier (in rats)	[163]
Enzymes	Unsaturated soybean phospholipid	Encapsulation in the inner aqueous core	Alleviate bitterness in protein hydrolysates	[164]
Vitamin C	Soy lecithin	Encapsulation in the inner aqueous core	Increased vitamin protection and controlled release	[165]
Gallic acid	Nopal mucilage	Spray-drying	Preservation	[166]

material on drying. Eventually, another emulsifier might be added to the solution to improve payload dispersion. The dispersion is then sprayed in a drying chamber, forming fine droplets, which are rapidly dried on contact with a cocurrent or countercurrent flow of hot gas, ultimately forming small micrometric droplets, which are collected in a cyclone or in a filter cloth. The payload of the produced microcapsules is dispersed in the matrix of the encapsulant material. The encapsulation systems obtained by spray-drying are in general capsules of matrix type, whose properties are mainly affected by the properties of the emulsion and by the process conditions. However, especially for vitamins, aroma compounds, and probiotics, the high temperatures required for complete solvent evaporation may induce significant thermal damage or volatilization.[193,194]

Spray-chilling represents an alternative to spray-drying, used to prevent the volatilization or degradation of thermolabile food additives. Spray-chilling is also based on the preliminary dispersion of the payload in a solution, which can be made of low–molecular weight polymers, resins, hydrogenated vegetable

TABLE 19.5

Encapsulation of Bioactive Compounds at the Submicrometric Scale

Bioactives	Formulation	Fabrication	Application	References
Nanoemulsions				
β-carotene	Medium-chain triglyceride oil and Tween 20–80	Emulsification via high-pressure homogenization	Food functionalization	[167]
Curcumin	Medium-chain triglycerols and Tween 20	Emulsification via high-pressure homogenization	Food functionalization	[168]
D-Limonene	Palm oil and soy lecithin	Emulsification via hot high-pressure homogenization	Antimicrobial activity in foods	[42]
Molecular inclusion/conjugation				
Iron	Casein and whey protein isolates	Electrostatic bonds	Food functionalization	[169]
Resveratrol	β-lactoglobulin	Complexation	Increase of dispersibility in water and photostability	[170]
Conjugated linoleic acid	Amylase	Inclusion complex via hydrophobic interactions	Protection against oxidation and dispersion in aqueous phase	[126]
Quercetin	Chitosan	Complexation followed by ionic gelation	Increased dispersibility in water and bioavailability	[171]
Quercetin and myricetin	HydrP-β-cyclodextrins	Inclusion complexation	Increase of solubility and inhibition of oxidation	[128]
Kaempferol, quercetin, and myricetin	HP-β-cyclodextrins	Inclusion complexation	Increase of solubility and inhibition of oxidation	[172]
Curcumin	HP-β-, M-β-, and HP-γ-cyclodextrins	Inclusion complexation	Preservation of antioxidant activity	[173]
Curcumin	Hydrophobically modified starch	Hydrophobic interaction	Increased dispersibility in water and anticarcinogenic activity	[174]
Micelles				
Curcumin	Casein micelles	Hydrophobic interactions	Food functionalization	[175]
Vitamins D2, D3	Casein micelles	Hydrophobic interactions	Food functionalization	[176, 177]
Naringenin	Polyvinylpyrrolidone	Solvent evaporation method	Improved dissolution rate, high physicochemical stability	[178]

(*Continued*)

TABLE 19.5 (CONTINUED)

Encapsulation of Bioactive Compounds at the Submicrometric Scale

Bioactives	Formulation	Fabrication	Application	References
Quercetin	Pluronic F68 + lecithin	Solvent evaporation method, high-pressure homogenization	Improved dissolution rate	[179]
Citral	Tween 80	Micellization	Preservation of flavor in carbonated beverages	[180]
Phytosterols	Water/propylene glycol/ R(+)-limonene/ethanol/ Tween 60	Microemulsion	Solubilization and dispersion in food systems	[181]
Lycopene	Water/propylene glycol/ R(+)-limonene/ethanol/ Tween 60	Microemulsion	Solubilization and dispersion in food systems	[182]
Curcumin	So Tween 20/Glycerol monooleate/medium-chain triglycerides/water	Microemulsion	Solubilization and dispersion in food systems	[183]
Multiple-layer nanoemulsions				
β-Carotene	Lactoferrin and beta-lactoglobulin	Multilayer emulsions	Preservation	[184]
Strawberry flavors	Pea protein isolate and pectin	Spray-dried multilayer emulsions	Flavor preservation	[185]
Citral	Chitosan and ε-polylysine	Layer-by-layer emulsions	Flavor preservation	[186]
β-Carotene	Chitosan and soybean soluble polysaccharides	Two-stage deposition on emulsion by homogenization	Preservation	[187]
ω-3 Fatty acids	Lecithin and chitosan	Spray-dried multilayer emulsions	Increasing oxidative stability	[188]

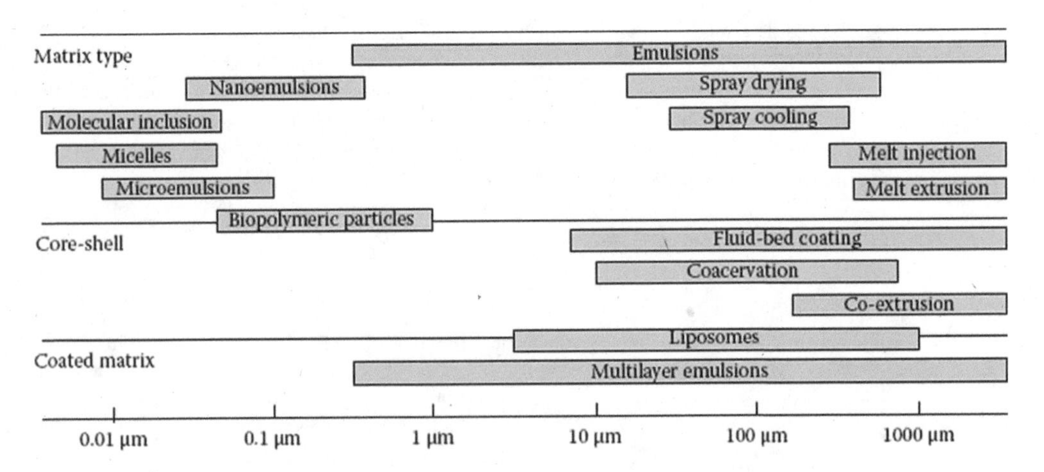

FIGURE 19.4 Main encapsulation techniques classified as a function of the prevalent architecture and mean particle size of the encapsulation system.

oils, or waxes, and its atomization, with the capsule consolidation being based not on dehydration but on glass transition or crystallization of the encapsulant material on rapid chilling in a cooled gas flow.[195,196]

Fluid bed coating is based on the deposition of a shell layer on preformed particles (e.g., by spray-drying), therefore constituting the final step of production of a core-shell or a coated-matrix-type architecture. The particles containing the payload are fluidized, and the coating material is sprayed over them at high pressure,[197] forming a shell layer that is then dried by solvent evaporation or crystallization. The coating material usually consists of starches, dextrins, protein derivatives, molasses, lipids, and waxes. Prior to spraying, the coating material is either melted or dispersed in a suitable solvent that can be easily evaporated, forming a viscous system with an enhanced tendency to deposition and adhesion on the payload particles. The air or gas flow through the fluidized bed serves to chill the molten material or to evaporate the solvent, causing its consolidation in a shell layer.[198] Fluid bed coating is an inexpensive process, which is often associated with spray-drying.[199]

Melt extrusion is a process used to encapsulate the bioactive compounds in a glassy, impermeable, dense matrix through the extrusion of a suitable dispersion of the payload in a molten material, which is immediately chilled to induce glassy transition. Typical materials used include fats, fatty acids, mono-, di- , and triglycerides, waxes, polyethylene glycols, and other commercial products, such as Shellac. The extrusion conditions determine the mean particle size and morphology of the encapsulation system.[196,200] *Melt injection* is based on the same concept as melt extrusion, with the main difference that extrusion is carried out through a filter within a cooling or dehydrating medium.[196,200]

Coextrusion technology is instead used to produce core-shell particles and is based on extrusion through a concentric nozzle, with the payload dispersion being extruded through the inner nozzle and the wall materials being extruded from the outer nozzle. Due to its ability to fabricate in a simple and robust process encapsulation systems with multiple coating layers (using extrusion nozzles with multiple concentricity), coextrusion is used when a slow and controlled release of the payload, as well as taste masking, is desired.[201] Particle consolidation occurs either via chilling and glass transition of the wall material, via gelation, or via evaporation of the solvent. In the first case, the same coating materials as for melt extrusion are used, while in the other cases, viscous polymer solutions are used, based on proteins and polysaccharides, gums, and other commercial polymers. The main disadvantages of melt extrusion and coextrusion processes are the high temperatures and the high shear rates attained in the extruder.[202]

The top-down approaches to the production of submicrometric particulate systems are mainly based on comminution by mechanical size reduction techniques, which require minimal use of chemical additives.[4]

Media milling is a process based on the action of grinding media, which are usually balls or beads of hard materials, which are rotated at a very high speed to generate strong shear forces to disintegrate larger particles into nanoparticles.[203] The milling chamber is loaded not only with the powder material to be comminuted and with the grinding media, but also with a dispersion medium (e.g., water) and a stabilizer, to prevent aggregation or coalescence phenomena. The efficiency of the milling process mainly depends on number and type (i.e., hardness) of the grinding media, the ratio of bioactive compound and stabilizer, the milling time, the milling speed, and the process temperature.[204]

Colloid milling is instead based on a high-speed rotor/stator system, in which the processed material is ground, dispersed, or emulsified as a consequence of its exposure to intense shear stresses, friction, and high-frequency vibrations. Colloid mills are constituted by smooth or toothed rotors and stators, gear-rim dispersion machines, and intensive mixers.[205]

The main disadvantages of media and colloid milling are the difficulty of removal of residual grinding media from the final product and the loss of bioactive compounds due to adhesion to the inner surface of the milling chamber. However, they represent low-cost and easily scalable methods of comminution of solid particles, with wide industrial application.[204]

High-pressure homogenization is another mechanical disruption process for the production of submicrometric emulsions, suspensions, and dispersions, with significant advantages in terms of ease of operation, industrial scalability, reproducibility, and high throughput.[205,206] The process consists in the mechanical disruption of the disperse phase by application of high-intensity fluid-mechanical stresses as a consequence of the flow of the continuous phase under high pressure (50–400 MPa) through a specifically designed homogenization chamber. The homogenization chamber can be realized in different geometries, ranging from a simple orifice plate to colliding jets and radial diffuser assemblies.[1,207–211]

19.4.2 Physico-Chemical Processes

Submicrometric systems can also be produced through physico-chemical processes, in general based on bottom-up approaches to the spontaneous association of molecules or larger building blocks around the bioactive compounds, driven by the balance of attractive and repulsive forces tending to thermodynamic equilibrium. The intensity of the involved forces can be controlled by environmental factors, such as temperature, concentration, pH, and ionic strength of the system,[212] with the required mechanical energy being limited to system agitation.[4] Since the encapsulation systems formed through bottom-up approaches exist at the thermodynamic equilibrium, any environmental modification would contribute to their disassembly, therefore limiting their use and incorporation in real food systems, unless a stabilization step is added through physical or chemical changes, such as rapid quench cooling, sudden solvent dilution or evaporation, or chemical reactions.

In general, the fabrication of encapsulation systems via bottom-up approaches is based on the dissolution of the bioactive compound in a suitable organic solvent followed by its precipitation through a non-solvent addition in the presence of stabilizers, such as surfactants or hydrocolloids.[4,213] Eventually, a polymer soluble in the internal solvent but insoluble in the external one may be added to the system.[4]

Phase inversion methods are based on spontaneous O/W nanoemulsion formation induced by controlling the interfacial behavior, from predominantly lipophilic to predominantly hydrophilic, of the surfactants at the O/W interface in response to changes in system composition or environmental conditions.[214,215]

For non-ionic surfactants, the change in surfactant behavior can be achieved by changing the temperature of the system[215]: at low temperatures, the volume of the hydrated surfactant head group is larger than that of the hydrophobic tail group, thus favoring the formation of O/W emulsions, while at high temperatures, the volume of the dehydrated surfactant head group is smaller than that of the hydrophobic tail group, thus favoring the formation of W/O emulsions.[214] During cooling, the system crosses a point of minimal surface tension, which is referred to as the *phase inversion temperature*,[215] around which the formation of nanometric droplets is promoted.[216] The change of surfactant behavior can also be achieved through environmental changes, such as salt concentration or pH value,[215] as well as surfactant concentration.[214]

Due to the formulation requirements of phase inversion methods, the nanoemulsions are highly unstable to coalescence when conditions closer to the phase inversion are attained, such as on incorporation in real foods or during high-temperature treatments.[214]

The *phase separation method* is based on the spontaneous solvent-in-water emulsification induced by the addition of an aqueous solution containing a surfactant to a polar organic solvent, such as acetone or methylene chloride, containing the bioactive compound and a lipophilic polymer, due to the affinity between the water and the polar solvent. Further water addition is subsequently used to cause the diffusion of the solvent out of the emulsion droplets, inducing the precipitation of the lipophilic polymer and the encapsulation of the bioactive compound.[4,217]

Coacervation is a relatively simple approach to the fabrication of core-shell materials, exploiting the phase separation method. It is based on the electrostatic interaction of the components of an emulsion to create microcapsules that are resistant to moisture and heating.[121] Basically, coacervation is a phase separation process in aqueous phase, which begins with the dispersion or emulsification of the payload in an aqueous solution of a primary biopolymer, such as gum arabic or gelatin.[119] Subsequently, the system is mixed with another aqueous solution of a secondary encapsulant biopolymer with opposite charge to the primary one to promote the formation of complexes that are not soluble in the original aqueous phase.[218] The mean droplet size of the capsules can be controlled through the pH and temperature of the original solution but also depends on the properties of the encapsulated compounds as well as of the encapsulant agent.[121] The capsules formed by the coacervation method can be further consolidated by inducing polymer gelation through the addition of minerals and changing the temperature of the system[219] or inducing polymer precipitation by changes of pH, temperature, or electrolyte composition.[4,136] Because of the structural complexity of coacervate systems, their typical sizes are in the micrometric range.

The *layer-by-layer method* enables the fabrication of multiple-layer polyelectrolyte capsules and emulsions.[220] It consists of the spontaneous deposition of biopolymer stabilizing layers at the interface of particles or emulsion droplets, driven by the electrostatic attraction between the templating interface and the forming monolayers. In particular, the spontaneous deposition of a first layer of charged polyelectrolyte occurs around primary systems with opposite charge, producing a secondary particulate system coated with a two-layer interface. In addition, each deposited layer not only fully compensates the charge of the previous templating layer but also imparts an uncompensated counter-charge, generating a charge reversal. Once a monolayer is formed, the surface is saturated with polyelectrolytes, thus preventing further adsorption. As a consequence, the deposition of multiple biopolymer layers is possible,[146] improving the stability to environmental stresses in comparison with conventional single-layer systems.[221]

19.5 Encapsulation of Probiotics

The main techniques of encapsulation of probiotic cells are based on spray-drying, emulsification, and extrusion, which lead to matrix-type systems, as well as coextrusion and fluid bed coating, which instead lead to core-shell systems. In both cases, the typical size of the encapsulation system is between 1 and 5 μm in diameter, being dictated by the size of the microbial cells to be encapsulated.[222]

The design of encapsulation systems for probiotics is primarily aimed at preserving their viability as far as possible, protecting them from adverse environments and ensuring their release in the specific part of the intestine to which the probiotics are targeted.[223] In particular, the physico-chemical properties of the capsules are the controlling factors for cell viability and targeted release. For example, capsules with shells that are insoluble in water are required to ensure physical stability in most of the food matrices and in the initial regions of the gastrointestinal tract.

Different materials are used to encapsulate probiotic cells, such as vegetable and animal proteins, starch and its derivatives, and maltodextrins. However, other biopolymers are also used.[222]

For example, *calcium and sodium alginates* are polysaccharides derived from algae, which are frequently used to encapsulate probiotics because of their lack of toxicity, biocompatibility, low cost, and ease of use.[224] On drying, they form a porous structure, which is not resistant to acidic environments and therefore, is not suitable to ensure adequate stability in the gastric tract. However, enhanced stability can be achieved by blending calcium alginate with other biopolymers or by further coating alginate capsules with a layer of insoluble polymers.

Gellan or xanthan gums are polysaccharides of microbial origin, obtained from *Pseudomonas elodea* and *Xanthomonas campestris*, respectively. A mixture of these was reported to be extremely suitable to encapsulate probiotics in systems with high resistance in acidic environments.[222]

Kappa-carrageenan is a natural polymer, widely used in the food industry, which is extremely compatible with microbial cells, ensuring high viability after the encapsulation process. However, the resulting gel structures have limited physical stability in the stress conditions experimented in food transformation,[225] requiring it to be blended with other polymers.

Chitosan is a linear polysaccharide, which is typically used to form coating layers on preformed systems. For example, probiotics encapsulated in an alginate matrix and further coated with a layer of chitosan are characterized by improved protection in the gastrointestinal tract, which enables them to reach the colon with high viability.[226] The main disadvantage of chitosan is an observed inhibitory effect on LAB.

Cellulose acetate phthalate is a polymer typically used as an enteric coating[227] because it is not soluble in water at acidic pH but only at pH > 6.

Matrix-type systems to encapsulate probiotics use spray-drying or spray-freeze-drying. Spray-drying ensures a rapid and low-cost process, which suits different industrial applications. However, its wide use is limited by the high temperatures required in the drying chamber, which significantly affect cell viability. The use of protective materials, such as granular starches, soluble fibers, or trehalose, which are thermal protector agents, is recommended to limit thermal damage to probiotic cells.

Moreover, an additional coating layer deposited in a fluid bed coating system integrated with the spray-dryer can improve the capsules' resistance in the gastrointestinal tract.[228]

Spray-freeze-drying instead combines the characteristics of the freeze-drying process with those of spray-drying. The probiotic dispersion is sprayed in a chilled vapor of a cryogenic liquid, causing the formation of a dispersion of frozen droplets, which are subsequently spray-dried,[228,229] with the advantage of a narrow particle size distribution, especially if compared with a freeze-dried microbial dispersion, followed by mechanical comminution, short freeze-drying times, and high cell viability. However, the main drawback of this technique is represented by the high process costs, which are between 30 and 50 times higher than spray-drying.[230]

The emulsification of probiotics enables microbial cells to be encapsulated in small-diameter capsules, maintaining high viability. However, the encapsulated cells remain in the aqueous phase, with consequent limitation to their use as additives to liquid products. Typical coating materials are alginates, carrageenan, gellan, and xanthan gums, but also whey proteins have been used to improve compatibility with milk products as well as to overcome the regulations that restrict the use of excipients of microbial origin.[231]

Extrusion is a simple and low-cost process of encapsulation with core-shell architecture, which is able to preserve probiotic cell viability due to the limited use of harmful solvents and the small stresses exerted on microbial cells. However, its use on a large scale is limited by the slow process of capsule fabrication. In contrast, due to its easier scalability, fluid bed coating is more widely used in the encapsulation of probiotic cells.

Table 19.6 reports some details, such as type of microbial cells, encapsulant materials, and encapsulation process, about the encapsulation systems for probiotics described by recent literature to provide some indication of the main research trends.

From the analysis of the data reported in Table 19.6, it is evident that the consolidation of the outer layer of the encapsulation system, carried out through gelation or cross-linking processes, is a fundamental step not only toward the improvement of viability preservation but also toward the controlled release of the probiotics in the gastrointestinal tract.

Therefore, the adequate selection of the encapsulant materials, of the process of encapsulation, and of the consolidation method enables enhanced protection of the payload even during intense transformation processes, including extremes of temperature and pH, as well as high shear stresses.

19.6 Conclusions and Perspectives

The encapsulation of bioactive compounds represents an efficient approach, widely used in the food and nutraceutical industry, to promote the homogeneous dispersion of bioactive compounds in the end product, to protect them from interaction with other ingredients and from degradation during end product transformation, storage, and preparation, and to control their release where needed.

Currently, different architectures of encapsulation systems are used to tailor their properties to the requirements of the end product. For example, matrix-type systems, in general fabricated through simple and inexpensive processes, are indicated for sustained release applications. In addition, due to their simplicity, matrix systems can also be fabricated with smaller particle sizes, in the submicrometric range. In contrast, core-shell systems, with a higher structural complexity that is reflected in multiple-stage fabrication processes, are especially suitable for the enhanced protection of the bioactive compounds as well as for a triggered release in the intestinal tract. However, their complexity also limits their typical size, which usually is in the micrometric range.

Future trends are toward using the different encapsulation systems to prepare food additives and comparing their stability and performance when introduced into real foods.

This could answer two different needs, one related to the extension of the use of nutraceutical compounds in foodstuffs to increase the content of nutrients with beneficial effects on human health, and the other related to the design of novel foods tailored to fulfill peculiar dietetic requirements; that is, to overcome specific problems related to the lack of essential nutrients in the diet of the population of certain countries and/or to answer the demand for personalized nutrition of groups of the population, such as young or elderly people.

TABLE 19.6

Review of Recently Investigated Encapsulation Systems for Probiotics with Details on Encapsulant Materials, Type of Probiotic Cells, Encapsulation Process, and Viability

Encapsulant Material	Microorganism	Technique	Viability	References
Microcapsules of calcium alginate with an external coating of whey proteins	*Lactobacillus plantarum* 299v (1) *Lactobacillus plantarum* 800 (2) *Lactobacillus plantarum* CIP A159 (3)	Freeze-drying	Reduction after 60 min at 37°C (simulated gastric fluid) for coated capsules: 2.3 log CFU/g for (1), 4.2 log CFU/g for (2), and 3.3 log CFU/g for (3). Reduction after 60 min at 37°C (simulated gastric fluid) for uncoated capsules: 7.8 log CFU/g for (1), 8.4 log CFU/g for (2), and 8.2 log CFU/g for (3). After 180 min at 37°C (simulated intestinal fluid): only bacteria in coated capsules survive.	[232]
Casein-based microcapsules produced by enzymatic gelation with transglutaminase	*Lactobacillus* FI9 *Bifidobacterium* BB12	Emulsification + freeze-drying	*Lactobacillus* F19 survives in significantly higher numbers in the encapsulated form compared with unencapsulated cells; encapsulation improves survival of *Bifidobacterium* BB12 during storage to a maximum of 90 days in all conditions tested; for *Lactobacillus* F19, no positive effect of the encapsulation during storage was found.	[235]
Microcapsules of calcium alginate with a coating of sodium alginate	Lactobacillus acidophilus (1) Lactobacillus rhamnosus (2)	Emulsification + gelation	After incubation in simulated gastric fluid (60 min) and simulated intestinal fluid (pH 7.25, 2 h), the number of surviving cells encapsulated into double layer–coated alginate microspheres corresponds to 6.5 log CFU/ml for (1) and 7.6 log CFU/ml for (2), while the corresponding results are 2.3 (1) and 2.0 (2) log CFU/ml for unencapsulated cells.	[235]
Capsules of sodium alginate or amidated pectin and their combination	*Lactobacillus casei*	Extrusion	Pectin and alginate induce a sort of synergic effect in the consolidation of the trapping matrix. Capsules produced with 2–3% pectin combined with 0.5% alginate tend to display optimal counts of *L. casei*, which were above the therapeutic requirement of 10^7 CFU/g in yogurt after 20 days of storage at 4°C.	[235]
Microcapsules of sodium alginate and corn starch	*Lactobacillus acidophilus* LAI	Emulsification	The microorganisms survived better in the encapsulated form at high temperatures and at high salt concentrations. The unencapsulated cells were completely destroyed at 90°C, whereas the microencapsulated cells were reduced by 4.14 log cycles. After 3 h incubation in simulated intestinal fluid, the unencapsulated and encapsulated cells registered 5.47 and 2.16 log cycle reduction, respectively.	[235]

(*Continued*)

TABLE 19.6 (CONTINUED)

Review of Recently Investigated Encapsulation Systems for Probiotics with Details on Encapsulant Materials, Type of Probiotic Cells, Encapsulation Process, and Viability

Encapsulant Material	Microorganism	Technique	Viability	References
Capsules of maltodextrin	Lactobacillus acidophilus (1) Lactobacillus rhamnosus (2)	Spray-drying	Maximum survival at 100°C: 81.17%; maximum survival at 130°C: 55%.	[236]
Microcapsules of whey proteins gelled with transglutaminase	*Bifidobacterium bifidum* F-35	Gelation vs. spray-drying	The gelation technique enables the production of bigger and denser capsules, which degrade more slowly in simulated gastric fluids and provide better protection for the cells compared with the capsules produced by spray-drying.	[237]
Capsules of trehalose consolidated with monosodium glutamate	*Lactobacillus rhamnosus* GG	Spray-drying	Spray-drying of *L. rhamnosus* in trehalose resulted in a survival rate of 69%, but the addition of glutamate has significantly increased the survival rate to 80.8%.	[238]

The approach described is particularly valuable considering that the strong interaction between nutrition and health is today receiving renewed attention due to the awareness of researchers, medical doctors, and public administrators responsible for health care that there is a pressing need to prevent illness, reduce the number of hospitalizations, and deal with the problem of the aging of the population of developed countries to reduce the costs and at the same time, ensure the survival of the public health care management system.

Moreover, the introduction to the market of novel food products whose consumption is beneficial to prevent health risks that are available to a wide range of consumers, not only limited to niche markets due to the elevated costs of industrial transformation, could also sustain the agro-food industry of developed countries in the very aggressive globalized scenario.

REFERENCES

1. Donsì, F.; Sessa, M.; Ferrari, G., Nanometric-size delivery systems for bioactive compounds for the nutraceutical and food industries. In *Bio-Nanotechnology: A Revolution in Food, Biomedical and Health Sciences*, Bagchi, D., Bagchi, M, Moriyama, H., Shahidi, F., (Eds.) John Wiley & Sons, Oxford, 2013, p. 619.
2. Lakkis, J. M., Introduction. In *Encapsulation and Controlled Release Technologies in Food Systems*, Lakkis, J. M., (Ed.) Blackwell Publishing Ltd, Oxford, 2007, pp. 1–12.
3. Garti, N.; Yuli-Amar, I., Micro- and nano-emulsions for delivery of functional food ingredients. In *Delivery and Controlled Release of Bioactives in Foods and Nutraceuticals*, Garti, N., (Ed.) Woodhead Publishing Limited, Cambridge, 2008, pp. 149–183.
4. Acosta, E., Bioavailability of nanoparticles in nutrient and nutraceutical delivery. *Current Opinion in Colloid and Interface Science* (2009): *14*, 3–15.
5. WHO; FAO Diet, *Nutrition and the Prevention of Chronic Diseases: Report of a Joint WHO/FAO Expert Consultation.* World Health Organization, Geneva, Switzerland, 2003.
6. Kris-Etherton, P. M.; Hecker, K. D.; Bonanome, A.; Coval, S. M.; Binkoski, A. E.; Hilpert, K. F.; Griel, A. E.; Etherton, T. D., Bioactive compounds in foods: Their role in the prevention of cardiovascular disease and cancer. *American Journal of Medicine* (2002): *113*, 71–88.

7. Satija, A.; Hu, F. B., Cardiovascular benefits of dietary fiber. *Current Atherosclerosis Reports* (2012): *14*, 505–514.
8. Zamora-Ros, R.; Fedirko, V.; Trichopoulou, A.; Gonzalez, C. A.; Bamia, C.; Trepo, E.; Noethlings, U. et al., Dietary flavonoid, lignan and antioxidant capacity and risk of hepatocellular carcinoma in the European prospective investigation into cancer and nutrition study. *International Journal of Cancer* (2013): *133*, 2429–2443.
9. Ford, D. W.; Jensen, G. L.; Hartman, T. J.; Wray, L.; Smiciklas-Wright, H., Association between dietary quality and mortality in older adults: A review of the epidemiological evidence. *Journal of Nutrition in Gerontology and Geriatrics* (2013): *32*, 85–105.
10. Kitts, D. D., Bioactive substances in food—Identification and potential uses. *Canadian Journal of Physiology and Pharmacology* (1994): *72*, 423–434.
11. Kris-Etherton, P. M.; Lefevre, M.; Beecher, G. R.; Gross, M. D.; Keen, C. L.; Etherton, T. D., Bioactive compounds in nutrition and health-research methodologies for establishing biological function: The antioxidant and anti-inflammatory effects of flavonoids on atherosclerosis. *Annual Review of Nutrition* (2004): *24*, 511–538.
12. Carrato, B.; Sanzini, E., Biologically-active phytochemicals in vegetable food. *Annali dell'Istituto superiore di sanita* (2005): *41*, 7–16.
13. McClements, D. J.; Decker, E. A.; Weiss, J., Emulsion-based delivery systems for lipophilioc bioactive components. *Journal of Food Science* (2007): *72*, R109–R124.
14. Kanazawa, K., Bioavailability of non-nutrients for preventing lifestyle-related diseases. *Trends in Food Science and Technology* (2011): *22*, 655–659.
15. Gillespie, S.; Gavins, F. N. E., Phytochemicals: Countering risk factors and pathological responses associated with ischaemia reperfusion injury. *Pharmacology and Therapeutics* (2013): *138*, 38–45.
16. Su, Z.-Y.; Shu, L.; Khor, T. O.; Lee, J. H.; Fuentes, F.; Kong, A.-N. T., A Perspective on dietary phytochemicals and cancer chemoprevention: Oxidative stress, Nrf2, and epigenomics. *Natural Products in Cancer Prevention and Therapy* (2013): *329*, 133–162.
17. Dao, C. A.; Patel, K. D.; Neto, C. C., Phytochemicals from the fruit and foliage of Cranberry (Vaccinium macrocarpon)—Potential benefits for human health. *Emerging Trends in Dietary Components for Preventing and Combating Disease* (2012): *1093*, 79–94.
18. Wildman, R. E. C., *Handbook of Nutraceuticals and Functional Foods*. CRC Press, New York, 2001.
19. Bell, L. N., Stability testing of nutraceuticals and functional foods. In *Handbook of Nutraceuticals and Functional Foods*, Wildman, R. E. C., (Ed.) CRC Press, New York, 2001, pp. 501–516.
20. Wang, X.; Jiang, Y.; Huang, Q., Encapsulation technologies for preserving and controlling the release of enzymes and phytochemicals. In *Encapsulation and Controlled Release Technologies in Food Systems*, Lakkis, J. M., (Ed.) Wiley-Blackwell, Hoboken, NJ, 2007.
21. Wang, H. K., The therapeutic potential of flavonoids. *Expert Opinion on Investigational Drugs* (2000): *9*, 2103–2119.
22. Lamson, D. W.; Brignall, M. S., Antioxidants and cancer, Part 3: Quercetin. *Alternative Medicine Review: A Journal of Clinical Therapeutic* (2000): *5*, 196–208.
23. Kielhorn, S.; Thorngate, J. H., Oral sensations associated with the flavan-3-ols (+)-catechin and (–)-epicatechin. *Food Quality and Preference* (1999): *10*, 109–116.
24. Siess, M. H.; Le Bon, A. M.; Canivenc-Lavier, M. C.; Suschetet, M., Mechanisms involved in the chemoprevention of flavonoids. *Biofactors* (2000): *12*, 193–199.
25. Jang, M. S.; Cai, E. N.; Udeani, G. O.; Slowing, K. V.; Thomas, C. F.; Beecher, C. W. W.; Fong, H. H. S. et al., Cancer chemopreventive activity of resveratrol, a natural product derived from grapes. *Science* (1997): *275*, 218–220.
26. Sessa, M.; Tsao, R.; Liu, R.; Ferrari, G.; Donsi, F., Evaluation of the stability and antioxidant activity of nanoencapsulated resveratrol during in vitro digestion. *Journal of Agricultural and Food Chemistry* (2011): *59*, 12352–12360.
27. Sessa, M.; Balestrieri, M. L.; Ferrari, G.; Servillo, L.; Castaldo, D.; D'Onofrio, N.; Donsi, F.; Tsao, R., Bioavailability of encapsulated resveratrol into nanoemulsion-based delivery systems. *Food Chemistry* (2014): *147*, 42–50.
28. Chauhan, D. P., Chemotherapeutic potential of curcumin for colorectal cancer. *Current Pharmaceutical Design* (2002): *8*, 1695–1706.

29. Donsi, F.; Wang, Y.; Li, J.; Huang, Q., Preparation of curcumin sub-micrometer dispersions by high-pressure homogenization. *Journal of Agricultural and Food Chemistry* (2010): *58*, 2848–2853.
30. Sharma, D.; Sukumar, S., Big punches come in nanosizes for chemoprevention. *Cancer Prevention Research* (2013): *6*, 1007–1010.
31. Umadevi, S.; Gopi, V.; Vellaichamy, E., Inhibitory effect of gallic acid on advanced glycation end products induced up-regulation of inflammatory cytokines and matrix proteins in H9C2 (2–1) cells. *Cardiovascular Toxicology* (2013): *13*, 396–405.
32. Kratz, J. M.; Andrighetti-Frohner, C. R.; Leal, P. C.; Nunes, R. J.; Yunes, R. A.; Trybala, E.; Bergstrom, T.; Monte Barardi, C. R.; Oliveira Simoes, C. M., Evaluation of anti-HSV-2 activity of gallic acid and pentyl gallate. *Biological and Pharmaceutical Bulletin* (2008): *31*, 903–907.
33. Fathi, M.; Mirlohi, M.; Varshosaz, J.; Madani, G., Novel caffeic acid nanocarrier: Production, characterization, and release modeling. *Journal of Nanomaterials* (2013): Article ID 434632.
34. Variyar, P. S.; Ahmad, R.; Bhat, R.; Niyas, Z.; Sharma, A., Flavoring components of raw monsooned arabica coffee and their changes during radiation processing. *Journal of Agricultural and Food Chemistry* (2003): *51*, 7945–7950.
35. Cao-Hoang, L.; Fougere, R.; Wache, Y., Increase in stability and change in supramolecular structure of beta-carotene through encapsulation into polylactic acid nanoparticles. *Food Chemistry* (2011): *124*, 42–49.
36. Biesalski, H. K.; Obermueller-Jevic, U. C., UV light, beta-carotene and human skin—Beneficial and potentially harmful effects. *Archives of Biochemistry and Biophysics* (2001): *389*, 1–6.
37. Arab, L.; Steck, S., Lycopene and cardiovascular disease. *American Journal of Clinical Nutrition* (2000): *71*, 1691S–1695S.
38. Ribeiro, H. S.; Ax, K.; Schubert, H., Stability of lycopene emulsions in food systems. *Journal of Food Science* (2003): *68*, 2730–2734.
39. Kucuk, O.; Sarkar, F. H.; Djuric, Z.; Sakr, W.; Pollak, M. N.; Khachik, F.; Banerjee, M.; Bertram, J. S.; Wood, D. P., Effects of lycopene supplementation in patients with localized prostate cancer. *Experimental Biology and Medicine* (2002): *227*, 881–885.
40. Krinsky, N. I., Overview of lycopene, carotenoids, and disease prevention. *Proceedings of the Society for Experimental Biology and Medicine* (1998): *218*, 95–97.
41. Stringham, J. M.; Bovier, E. R.; Wong, J. C.; Hammond, B. R., Jr., The Influence of dietary lutein and zeaxanthin on visual performance. *Journal of Food Science* (2010): *75*, R24–R29.
42. Donsi, F.; Sessa, M.; Ferrari, G., Nanoencapsulation of essential oils to enhance their antimicrobial activity in foods. *Journal of Biotechnology* (2010): *150*, S67–S67.
43. Crowell, P. L., Prevention and therapy of cancer by dietary monoterpenes. *Journal of Nutrition* (1999): *129*, 775S–778S.
44. Lluis, L.; Taltavull, N.; Munoz-Cortes, M.; Sanchez-Martos, V.; Romeu, M.; Giralt, M.; Molinar-Toribio, E. et al., Protective effect of the omega-3 polyunsaturated fatty acids: Eicosapentaenoic acid/docosahexaenoic acid 1:1 ratio on cardiovascular disease risk markers in rats. *Lipids in Health and Disease* (2013): *12*, 140.
45. Galgani, J. E.; Uauy, R. D.; Aguirre, C. A.; Diaz, E. O., Effect of the dietary fat quality on insulin sensitivity. *British Journal of Nutrition* (2008): *100*, 471–479.
46. Garg, M.; Mishra, D.; Agashe, H.; Jain, N. K., Ethinylestradiol-loaded ultraflexible liposomes: Pharmacokinetics and pharmacodynamics. *Journal of Pharmacy and Pharmacology* (2006): *58*, 459–468.
47. Fukushima, S.; Takada, N.; Hori, T.; Wanibuchi, H., Cancer prevention by organosulfur compounds from garlic and onion. *Journal of Cellular Biochemistry* (1997): *67*, 100–105.
48. Hirsch, K.; Danilenko, M.; Giat, J.; Miron, T.; Rabinkov, A.; Wilchek, M.; Mirelman, D.; Levy, J.; Sharoni, Y., Effect of purified allicin, the major ingredient of freshly crushed garlic, on cancer cell proliferation. *Nutrition and Cancer-an International Journal* (2000): *38*, 245–254.
49. Curin, Y.; Andriantsitohaina, R., Polyphenols as potential therapeutical agents against cardiovascular diseases. *Pharmacological Reports* (2005): *57*, 97–107.
50. D'Archivio, M.; Filesi, C.; Vari, R.; Scazzocchio, B.; Masella, R., Bioavailability of the polyphenols: Status and controversies. *International Journal of Molecular Sciences* (2010): *11*, 1321–1342.
51. Habauzit, V.; Morand, C., Evidence for a protective effect of polyphenols-containing foods on cardiovascular health: An update for clinicians. *Therapeutic Advances in Chronic Disease* (2012): *3*, 87–106.

52. Kita, A.; Bakowska-Barczak, A.; Hamouz, K.; Kulakowska, K.; Lisinska, G., The effect of frying on anthocyanin stability and antioxidant activity of crisps from red- and purple-fleshed potatoes (Solanum tuberosum L.). *Journal of Food Composition and Analysis* (2013): *32*, 169–175.
53. Perla, V.; Holm, D. G.; Jayanty, S. S., Effects of cooking methods on polyphenols, pigments and antioxidant activity in potato tubers. *Lwt-Food Science and Technology* (2012): *45*, 161–171.
54. Tipoe, G. L.; Leung, T.-M.; Hung, M.-W.; Fung, M.-L., Green tea polyphenols as an anti-oxidant and anti-inflammatory agent for cardiovascular protection. *Cardiovascular and Hematological Disorders Drug Targets* (2007): *7*, 135–144.
55. Bognar, E.; Sarszegi, Z.; Szabo, A.; Debreceni, B.; Kalman, N.; Tucsek, Z.; Sumegi, B.; Gallyas, F., Jr., Antioxidant and anti-inflammatory effects in RAW264.7 macrophages of malvidin, a major red wine polyphenol. *Plos One* (2013): *8*, e65355.
56. Quinones, M.; Miguel, M.; Aleixandre, A., Beneficial effects of polyphenols on cardiovascular disease. *Pharmacological Research* (2013): *68*, 125–131.
57. Widmer, R. J.; Freund, M. A.; Flammer, A. J.; Sexton, J.; Lennon, R.; Romani, A.; Mulinacci, N.; Vinceri, F. F.; Lerman, L. O.; Lerman, A., Beneficial effects of polyphenol-rich olive oil in patients with early atherosclerosis. *European Journal of Nutrition* (2013): *52*, 1223–1231.
58. Kurosawa, T.; Itoh, F.; Nozaki, A.; Nakano, Y.; Katsuda, S. I.; Osakabe, N.; Tsubone, H.; Kondo, K.; Itakura, H., Suppressive effect of cocoa powder on atherosclerosis in Kurosawa and Kusanagi-hypercholesterolemic rabbits. *Journal of Atherosclerosis and Thrombosis* (2005): *12*, 20–28.
59. Stoner, G. D.; Mukhtar, H., Polyphenols as cancer chemopreventive agents. *Journal of Cellular Biochemistry. Supplement* (1995): *22*, 169–180.
60. Araujo, J. R.; Goncalves, P.; Martel, F., Chemopreventive effect of dietary polyphenols in colorectal cancer cell lines. *Nutrition Research* (2011): *31*, 77–87.
61. Giftson, J. S.; Jayanthi, S.; Nalini, N., Chemopreventive efficacy of gallic acid, an antioxidant and anticarcinogenic polyphenol, against 1,2-dimethyl hydrazine induced rat colon carcinogenesis. *Investigational New Drugs* (2010): *28*, 251–259.
62. D'Incalci, M.; Steward, W. P.; Gescher, A. J., Use of cancer chemopreventive phytochemicals as antineoplastic agents. *Lancet Oncology* (2005): *6*, 899–904.
63. Hu, M., Commentary: Bioavailability of flavonoids and polyphenols: Call to arms. *Molecular Pharmaceutics* (2007): *4*, 803–806.
64. Basu, A.; Imrhan, V., Tomatoes versus lycopene in oxidative stress and carcinogenesis: Conclusions from clinical trials. *European Journal of Clinical Nutrition* (2007): *61*, 295–303.
65. Xianquan, S.; Shi, J.; Kakuda, Y.; Yueming, J., Stability of lycopene during food processing and storage. *Journal of Medicinal Food* (2005): *8*, 413–422.
66. Calder, P. C., Omega-3 fatty acids and inflammatory processes. *Nutrients* (2010): *2*, 355–374.
67. Mozurkewich, E.; Chilimigras, J.; Klemens, C.; Keeton, K.; Allbaugh, L.; Hamilton, S.; Berman, D.; Vazquez, D.; Marcus, S.; Djuric, Z.; Vahratian, A., The mothers, Omega-3 and mental health study. *BMC Pregnancy and Childbirth* (2011): *11*, 46.
68. Goren, J. L.; Tewksbury, A. T., The use of omega-3 fatty acids in mental illness. *Journal of Pharmacy Practice* (2011): *24*, 452–471.
69. Seida, J. C.; Mager, D. R.; Hartling, L.; Vandermeer, B.; Turner, J. M., Parenteral omega-3 fatty acid lipid emulsions for children with intestinal failure and other conditions: A systematic review. *Journal of Parenteral and Enteral Nutrition* (2013): *37*, 44–55.
70. Shek, L. P.; Chong, M. F.-F.; Lim, J. Y.; Soh, S.-E.; Chong, Y.-S., Role of dietary long-chain polyunsaturated fatty acids in infant allergies and respiratory diseases. *Clinical and Developmental Immunology* (2012): *2012*, 730568–730568.
71. Huffman, S. L.; Harika, R. K.; Eilander, A.; Osendarp, S. J. M., Essential fats: How do they affect growth and development of infants and young children in developing countries? A literature review. *Maternal and Child Nutrition* (2011): *7*, 44–65.
72. Major, G. C.; Alarie, F.; Dore, J.; Phouttama, S.; Tremblay, A., Supplementation with calcium plus vitamin D enhances the beneficial effect of weight loss on plasma lipid and lipoprotein concentrations. *American Journal of Clinical Nutrition* (2007): *85*, 54–59.
73. Abbott, R. D.; Curb, J. D.; Rodriguez, B. L.; Sharp, D. S.; Burchfiel, C. M.; Yano, K., Effect of dietary calcium and milk consumption on risk of thromboembolic stroke in older middle-aged men—The Honolulu Heart Program. *Stroke* (1996): *27*, 813–818.

74. Aburto, N. J.; Hanson, S.; Gutierrez, H.; Hooper, L.; Elliott, P.; Cappuccio, F. P., Effect of increased potassium intake on cardiovascular risk factors and disease: Systematic review and meta-analyses. *British Medical Journal* (2013): *346*, f1378.
75. Cherbuin, N.; Kumar, R.; Sachdev, P. S.; Anstey, K. J., Dietary mineral intake and risk of mild cognitive impairment: The PATH through life project. *Frontiers in Aging Neuroscience* (2014): *6*, 4.
76. Domellof, M.; Thorsdottir, I.; Thorstensen, K., Health effects of different dietary iron intakes: A systematic literature review for the 5th Nordic Nutrition Recommendations. *Food and Nutrition Research* (2013): *57*, 21667.
77. Cekic, V.; Vasovic, V.; Jakovljevic, V.; Lalosevic, D.; Cabo, I.; Mikov, M.; Sabo, A., Effect of chromium enriched fermentation product of barley and brewer's yeast and its combination with rosiglitazone on experimentally induced hyperglycaemia in mice. *Srpski Arhiv Za Celokupno Lekarstvo* (2011): *139*, 610–618.
78. Liu, J.; Bao, W.; Jiang, M.; Zhang, Y.; Zhang, X.; Liu, L., Chromium, selenium, and zinc multimineral enriched yeast supplementation ameliorates diabetes symptom in streptozocin-induced mice. *Biological Trace Element Research* (2012): *146*, 236–245.
79. Nicolae, A.; Gabaldon Hernandez, J. A.; Martinez San Martin, A., Evaluation of the calcium yield in the microencapsulation process. *Romanian Biotechnological Letters* (2013): *18*, 8685–8688.
80. Li, Y. O.; Diosady, L. L., Microencapsulation and its application in micronutrient fortification through "engineered" staple foods. *Agro Food Industry Hi-Tech* (2012): *23*, 18–21.
81. Wegmuller, R.; Zimmermann, M. B.; Buhr, V. G.; Windhab, E. J.; Hurrell, R. E., Development, stability, and sensory testing of microcapsules containing iron, iodine, and vitamin a for use in food fortification. *Journal of Food Science* (2006): *71*, S181–S187.
82. Brown, L.; Rosner, B.; Willett, W. W.; Sacks, F. M., Cholesterol-lowering effects of dietary fiber: A meta-analysis. *American Journal of Clinical Nutrition* (1999): *69*, 30–42.
83. Salmeron, J.; Ascherio, A.; Rimm, E. B.; Colditz, G. A.; Spiegelman, D.; Jenkins, D. J.; Stampfer, M. J.; Wing, A. L.; Willett, W. C., Dietary fiber, glycemic load, and risk of NIDDM in men. *Diabetes Care* (1997): *20*, 545–550.
84. Figueroa-Gonzalez, I.; Quijano, G.; Ramirez, G.; Cruz-Guerrero, A., Probiotics and prebiotics—Perspectives and challenges. *Journal of the Science of Food and Agriculture* (2011): *91*, 1341–1348.
85. Van Loo, J. A. E., Prebiotics promote good health: The basis, the potential, and the emerging evidence. *Journal of Clinical Gastroenterology* (2004): *38*, S70–S75.
86. Roberfroid, M., Prebiotics: The concept revisited. *Journal of Nutrition* (2007): *137*, 830S–837S.
87. Teitelbaum, J. E.; Walker, W. A., Nutritional impact of pre- and probiotics as protective gastrointestinal organisms. *Annual Review of Nutrition* (2002): *22*, 107–138.
88. Wang, Y., Prebiotics: Present and future in food science and technology. *Food Research International* (2009): *42*, 8–12.
89. Crittenden, R. G.; Playne, M. J., Production, properties and applications of food-grade oligosaccharides. *Trends in Food Science and Technology* (1996): *7*, 353–361.
90. Komninou, D.; Ayonote, A.; Richie, J. P.; Rigas, B., Insulin resistance and its contribution to colon carcinogenesis. *Experimental Biology and Medicine* (2003): *228*, 396–405.
91. Kaur, N.; Gupta, A. K., Applications of inulin and oligofructose in health and nutrition. *Journal of Biosciences* (2002): *27*, 703–714.
92. Ballongue, J.; Schumann, C.; Quignon, P., Effects of lactulose and lactitol on colonic microflora and enzymatic activity. *Scandinavian Journal of Gastroenterology. Supplement* (1997): *222*, 41–44.
93. Minami, T.; Miki, T.; Fujiwara, T.; Kawabata, S.; Izumitani, A.; Ooshima, T.; Sobue, S.; Hamada, S., Caries-inducing activity of isomaltooligosugar (IMOS) in in vitro and rat experiments. *Shoni shikagaku zasshi. The Japanese Journal of Pedodontics* (1989): *27*, 1010–1017.
94. Mussatto, S. I.; Mancilha, I. M., Non-digestible oligosaccharides: A review. *Carbohydrate Polymers* (2007): *68*, 587–597.
95. Tateyama, I.; Hashii, K.; Johno, I.; Iino, T.; Hirai, K.; Suwa, Y.; Kiso, Y., Effect of xylooligosaccharide intake on severe constipation in pregnant women. *Journal of Nutritional Science and Vitaminology* (2005): *51*, 445–448.
96. Espinosa-Martos, I.; Ruperez, P., Soybean oligosaccharides. Potential as new ingredients in functional food. *Nutricion Hospitalaria* (2006): *21*, 92–96.

97. Dominguez, A. L.; Rodrigues, L. R.; Lima, N. M.; Teixeira, J. A., An overview of the recent developments on fructooligosaccharide production and applications. *Food and Bioprocess Technology* (2014): *7*, 324–337.
98. Tomomatsu, H., Health-effects of oligosaccharides. *Food Technology* (1994): *48*, 53–53.
99. Bodera, P., Influence of prebiotics on the human immune system (GALT). *Recent Patents on Inflammation and Allergy Drug Discovery* (2008): *2*, 149–153.
100. Franck, A., Technological functionality of inulin and oligofructose. *British Journal of Nutrition* (2002): *87*, S287–S291.
101. Nelson, A. L., Properties of high-fiber ingredients. *Cereal Foods World* (2001): *46*, 93–97.
102. Sanders, M. E., Probiotics: Definition, sources, selection, and uses. *Clinical Infectious Diseases* (2008): *46*, S58–S61.
103. Senok, A. C.; Ismaeel, A. Y.; Botta, G. A., Probiotics: Facts and myths. *Clinical Microbiology and Infection* (2005): *11*, 958–966.
104. Gardiner, G. E.; Bouchier, P.; O'Sullivan, E.; Kelly, J.; Collins, J. K.; Fitzgerald, G.; Ross, R. P.; Stanton, C., A spray-dried culture for probiotic Cheddar cheese manufacture. *International Dairy Journal* (2002): *12*, 749–756.
105. Scholz-Ahrens, K. E.; Ade, P.; Marten, B.; Weber, P.; Timm, W.; Asil, Y.; Glueer, C.-C.; Schrezenmeir, J., Prebiotics, probiotics, and synbiotics affect mineral absorption, bone mineral content, and bone structure. *Journal of Nutrition* (2007): *137*, 838S–846S.
106. Hemarajata, P.; Versalovic, J., Effects of probiotics on gut microbiota: Mechanisms of intestinal immunomodulation and neuromodulation. *Therapeutic Advances in Gastroenterology* (2013): *6*, 39–51.
107. Shah, N. P., Functional cultures and health benefits. *International Dairy Journal* (2007): *17*, 1262–1277.
108. Kailasapathy, K.; Harmstorf, I.; Phillips, M., Survival of Lactobacillus acidophilus and Bifidobacterium animalis ssp lactis in stirred fruit yogurts. *LWT-Food Science and Technology* (2008): *41*, 1317–1322.
109. Ranadheera, R. D. C. S.; Baines, S. K.; Adams, M. C., Importance of food in probiotic efficacy. *Food Research International* (2010): *43*, 1–7.
110. Granato, D.; Branco, G. F.; Cruz, A. G.; Fonseca Faria, J. d. A.; Shah, N. P., Probiotic dairy products as functional foods. *Comprehensive Reviews in Food Science and Food Safety* (2010): *9*, 455–470.
111. Shah, N. P.; Ding, W. K.; Fallourd, M. J.; Leyer, G., Improving the stability of probiotic bacteria in model fruit juices using vitamins and antioxidants. *Journal of Food Science* (2010): *75*, M278–M282.
112. Zhang, L.; Huang, S.; Ananingsih, V. K.; Zhou, W.; Chen, X. D., A study on Bifidobacterium lactis Bb12 viability in bread during baking. *Journal of Food Engineering* (2014): *122*, 33–37.
113. Cote, J.; Dion, J.; Burguiere, P.; Casavant, L.; Van Eijk, J., Probiotics in bread and baked products: A new product category. *Cereal Foods World* (2013): *58*, 293–296.
114. Mohammadi, R.; Mortazavian, A. M.; Khosrokhavar, R.; da Cruz, A. G., Probiotic ice cream: Viability of probiotic bacteria and sensory properties. *Annals of Microbiology* (2011): *61*, 411–424.
115. Cruz, A. G.; Antunes, A. E. C.; Sousa, A. L. O. P.; Faria, J. A. F.; Saad, S. M. I., Ice-cream as a probiotic food carrier. *Food Research International* (2009): *42*, 1233–1239.
116. Kailasapathy, K.; Sultana, K., Survival and beta-D-galactosidase activity of encapsulated and free Lactobacillus acidophilus and Bifidobacterium lactis in ice-cream. *Australian Journal of Dairy Technology* (2003): *58*, 223–227.
117. Possemiers, S.; Marzorati, M.; Verstraete, W.; Van de Wiele, T., Bacteria and chocolate: A successful combination for probiotic delivery. *International Journal of Food Microbiology* (2010): *141*, 97–103.
118. Ziemer, C. J.; Gibson, G. R., An overview of probiotics, prebiotics and synbiotics in the functional food concept: Perspectives and future strategies. *International Dairy Journal* (1998): *8*, 473–479.
119. Jyothi, N. V. N.; Prasanna, P. M.; Sakarkar, S. N.; Prabha, K. S.; Ramaiah, P. S.; Srawan, G. Y., Microencapsulation techniques, factors influencing encapsulation efficiency. *Journal of Microencapsulation* (2010): *27*, 187–197.
120. Ezhilarasi, P. N.; Karthik, P.; Chhanwal, N.; Anandharamakrishnan, C., Nanoencapsulation techniques for food bioactive components: A review. *Food and Bioprocess Technology* (2013): *6*, 628–647.
121. Dong, Q.-Y.; Chen, M.-Y.; Xin, Y.; Qin, X.-Y.; Cheng, Z.; Shi, L.-E.; Tang, Z.-X., Alginate-based and protein-based materials for probiotics encapsulation: A review. *International Journal of Food Science and Technology* (2013): *48*, 1339–1351.
122. Shewan, H. M.; Stokes, J. R., Review of techniques to manufacture micro-hydrogel particles for the food industry and their applications. *Journal of Food Engineering* (2013): *119*, 781–792.

123. Oxley, J. D., Coextrusion for food ingredients and nutraceutical encapsulation: Principles and technology. In: *Woodhead Publishing Series in Food Science, Technology and Nutrition, Encapsulation Technologies and Delivery Systems for Food Ingredients and Nutraceuticals*, Garti. N and Julian McClements, D., (Eds.) Woodhead Publishing (2012): 131–150.
124. Letchford, K.; Burt, H., A review of the formation and classification of amphiphilic block copolymer nanoparticulate structures: Micelles, nanospheres, nanocapsules and polymersomes. *European Journal of Pharmaceutics and Biopharmaceutics* (2007): *65*, 259–269.
125. Narang, A. S.; Delmarre, D.; Gao, D., Stable drug encapsulation in micelles and microemulsions. *International Journal of Pharmaceutics* (2007): *345*, 9–25.
126. Yang, Y.; Gu, Z. B.; Zhang, G. Y., Delivery of bioactive conjugated linoleic acid with self-assembled amylose-CLA complex. *Journal of Agricultural and Food Chemistry* (2009): *57*, 7125–7130.
127. Fang, Z. X.; Bhandari, B., Encapsulation of polyphenols—A review. *Trends in Food Science and Technology* (2010): *21*, 510–523.
128. Lucas-Abellan, C.; Fortea, I.; Gabaldon, J. A.; Nunez-Delicado, E., Encapsulation of quercetin and myricetin in cyclodextrins at acidic pH. *Journal of Agricultural and Food Chemistry* (2008): *56*, 255–259.
129. Patricia Blanch, G.; Luisa Ruiz Del Castillo, M.; Del Mar Caja, M.; Perez-Mendez, M.; Sanchez-Cortes, S., Stabilization of all-trans-lycopene from tomato by encapsulation using cyclodextrins. *Food Chemistry* (2007): *105*, 1335–1341.
130. Zeller, B. L.; Saleeb, F. Z.; Ludescher, R. D., Trends in development of porous carbohydrate food ingredients for use in flavor encapsulation. *Trends in Food Science and Technology* (1998): *9*, 389–394.
131. Astray, G.; Gonzalez-Barreiro, C.; Mejuto, J. C.; Rial-Otero, R.; Simal-Gandara, J., A review on the use of cyclodextrins in foods. *Food Hydrocolloids* (2009): *23*, 1631–1640.
132. Spernath, A.; Aserin, A., Microemulsions as carriers for drugs and nutraceuticals. *Advances in Colloid and Interface Science* (2006): *128*, 47–64.
133. McClements, J. D., *Food Emulsions: Principles, Practices and Techniques*, 2nd edn. CRC Press, Boca Raton, FL, 1999.
134. de Campo, L.; Yaghmur, A.; Garti, N.; Leser, M. E.; Folmer, B.; Glatter, O., Five-component food-grade microemulsions: Structural characterization by SANS. *Journal of Colloid and Interface Science* (2004): *274*, 251–267.
135. Mezzenga, R.; Schurtenberger, P.; Burbidge, A.; Michel, M., Understanding foods as soft materials. *Nature Materials* (2005): *4*, 729–740.
136. Chen, L. Y.; Remondetto, G. E.; Subirade, M., Food protein-based materials as nutraceutical delivery systems. *Trends in Food Science and Technology* (2006): *17*, 272–283.
137. Jones, O. G.; McClements, D. J., Functional biopolymer particles: Design, fabrication, and applications. *Comprehensive Reviews in Food Science and Food Safety* (2010): *9*, 374–397.
138. Damodaran, S., Protein stabilization of emulsions and foams. *Journal of Food Science* (2005): *70*, R54–R66.
139. Rinaudo, M., Main properties and current applications of some polysaccharides as biomaterials. *Polymer International* (2008): *57*, 397–430.
140. Fathi, M.; Mozafari, M. R.; Mohebbi, M., Nanoencapsulation of food ingredients using lipid based delivery systems. *Trends in Food Science and Technology* (2012): *23*, 13–27.
141. Iqbal, M. A.; Md, S.; Sahni, J. K.; Baboota, S.; Dang, S.; Ali, J., Nanostructured lipid carriers system: Recent advances in drug delivery. *Journal of Drug Targeting* (2012): *20*, 813–830.
142. Schubert, M. A.; Muller-Goymann, C. C., Characterisation of surface-modified solid lipid nanoparticles (SLN): Influence of lecithin and nonionic emulsifier. *European Journal of Pharmaceutics and Biopharmaceutics* (2005): *61*, 77–86.
143. Sagalowicz, L.; Leser, M. E., Delivery systems for liquid food products. *Current Opinion in Colloid and Interface Science* (2010): *15*, 61–72.
144. McClements, D. J.; Li, Y., Structured emulsion-based delivery systems: Controlling the digestion and release of lipophilic food components. *Advances in Colloid and Interface Science* (2010): *159*, 213–228.

145. Wackerbarth, H.; Schon, P.; Bindrich, U., Preparation and characterization of multilayer coated microdroplets: Droplet deformation simultaneously probed by atomic force spectroscopy and optical detection. *Langmuir* (2009): *25*, 2636–2640.
146. Grigoriev, D. O.; Miller, R., Mono- and multilayer covered drops as carriers. *Current Opinion in Colloid and Interface Science* (2009): *14*, 48–59.
147. Ogawa, S.; Decker, E. A.; McClements, D. J., Production and characterization of O/W emulsions containing droplets stabilized by lecithin-chitosan-pectin mutilayered membranes. *Journal of Agricultural and Food Chemistry* (2004): *52*, 3595–3600.
148. Gu, Y. S.; Decker, E. A.; McClements, D. J., Application of multi-component biopolymer layers to improve the freeze-thaw stability of oil-in-water emulsions: beta-Lactoglobulin-iota-carrageenan-gela tin. *Journal of Food Engineering* (2007): *80*, 1246–1254.
149. Arshady, R., In the name of particle formation. *Colloids and Surfaces A-Physicochemical and Engineering Aspects* (1999): *153*, 325–333.
150. Lv, Y.; Yang, F.; Li, X. Y.; Zhang, X. M.; Abbas, S., Formation of heat-resistant nanocapsules of jasmine essential oil via gelatin/gum arabic based complex coacervation. *Food Hydrocolloids* (2014): *35*, 305–314.
151. Yang, Z. M.; Peng, Z.; Li, J. H.; Li, S. D.; Kong, L. X.; Li, P. W.; Wang, Q. H., Development and evaluation of novel flavour microcapsules containing vanilla oil using complex coacervation approach. *Food Chemistry* (2014): *145*, 272–277.
152. Rocha-Selmi, G. A.; Theodoro, A. C.; Thomazini, M.; Bolini, H. M. A.; Favaro-Trindade, C. S., Double emulsion stage prior to complex coacervation process for microencapsulation of sweetener sucralose. *Journal of Food Engineering* (2013): *119*, 28–32.
153. Comunian, T. A.; Thomazini, M.; Alves, A. J. G.; de Matos, F. E.; Balieiro, J. C. D.; Favaro-Trindade, C. S., Microencapsulation of ascorbic acid by complex coacervation: Protection and controlled release. *Food Research International* (2013): *52*, 373–379.
154. Wichchukit, S.; Oztop, M. H.; McCarthy, M. J.; McCarthy, K. L., Whey protein/alginate beads as carriers of a bioactive component. *Food Hydrocolloids* (2013): *33*, 66–73.
155. Rocha-Selmi, G. A.; Favaro-Trindade, C. S.; Grosso, C. R. F., Morphology, stability, and application of lycopene microcapsules produced by complex coacervation. *Journal of Chemistry* (2013): Article ID 982603.
156. Romo-Hualde, A.; Yetano-Cunchillos, A. I.; Gonzalez-Ferrero, C.; Saiz-Abajo, M. J.; Gonzalez-Navarro, C. J., Supercritical fluid extraction and microencapsulation of bioactive compounds from red pepper (*Capsicum annum* L.) by-products. *Food Chemistry* (2012): *133*, 1045–1049.
157. Bakowska-Barczak, A. M.; Kolodziejczyk, P. P., Black currant polyphenols: Their storage stability and microencapsulation. *Industrial Crops and Products* (2011): *34*, 1301–1309.
158. Betz, M.; Kulozik, U., Microencapsulation of bioactive bilberry anthocyanins by means of whey protein gels. *11th International Congress on Engineering and Food (Icef11)* (2011): *1*, 2047–2056.
159. Zheng, L. Q.; Ding, Z. S.; Zhang, M.; Sun, J. C., Microencapsulation of bayberry polyphenols by ethyl cellulose: Preparation and characterization. *Journal of Food Engineering* (2011): *104*, 89–95.
160. Robert, P.; Gorena, T.; Romero, N.; Sepulveda, E.; Chavez, J.; Saenz, C., Encapsulation of polyphenols and anthocyanins from pomegranate (Punica granatum) by spray drying. *International Journal of Food Science and Technology* (2010): *45*, 1386–1394.
161. Saenz, C.; Tapia, S.; Chavez, J.; Robert, P., Microencapsulation by spray drying of bioactive compounds from cactus pear (Opuntia ficus-indica). *Food Chemistry* (2009): *114*, 616–622.
162. Malheiros, P. D.; Daroit, D. J.; Brandelli, A., Food applications of liposome-encapsulated antimicrobial peptides. *Trends in Food Science and Technology* (2010): *21*, 284–292.
163. Takahashi, M.; Uechi, S.; Takara, K.; Asikin, Y.; Wada, K., Evaluation of an oral carrier system in rats: Bioavailability and antioxidant properties of liposome-encapsulated curcumin. *Journal of Agricultural and Food Chemistry* (2009): *57*, 9141–9146.
164. Nongonierma, A. B.; Abrlova, M.; Fenelon, M. A.; Kilcawley, K. N., Evaluation of two food grade proliposomes to encapsulate an extract of a commercial enzyme preparation by microfluidization. *Journal of Agricultural and Food Chemistry* (2009): *57*, 3291–3297.
165. Kirby, C. J.; Whittle, C. J.; Rigby, N.; Coxon, D. T.; Law, B. A., Stabilization of ascorbic-acid by microencapsulation in liposomes. *International Journal of Food Science and Technology* (1991): *26*, 437–449.

166. Medina-Torres, L.; Garcia-Cruz, E. E.; Calderas, F.; Laredo, R. F. G.; Sanchez-Olivares, G.; Gallegos-Infante, J. A.; Rocha-Guzman, N. E.; Rodriguez-Ramirez, J., Microencapsulation by spray drying of gallic acid with nopal mucilage (Opuntia ficus indica). *LWT-Food Science and Technology* (2013): *50*, 642–650.
167. Yuan, Y.; Gao, Y.; Zhao, J.; Mao, L., Characterization and stability evaluation of b-carotene nanoemulsions prepared by high pressure homogenization under various emulsifying conditions. *Food Research International* (2008): *41*, 61–68.
168. Wang, X.; Jiang, Y.; Wang, Y. W.; Huang, M. T.; Ho, C. T.; Huang, Q., Enhancing anti-inflammation activity of curcumin through o/w nanoemulsions. *Food Chemistry* (2008): *108*, 419–424.
169. Sugiarto, M.; Ye, A.; Singh, H., Characterisation of binding of iron to sodium caseinate and whey protein isolate. *Food Chemistry* (2009): *114*, 1007–1013.
170. Liang, L.; Tajmir-Riahi, H. A.; Subirade, M., Interaction of beta-Lactoglobulin with resveratrol and its biological implications. *Biomacromolecules* (2008): *9*, 50–56.
171. Zhang, Y. Y.; Yang, Y.; Tang, K.; Hu, X.; Zou, G. L., Physicochemical characterization and antioxidant activity of quercetin-loaded chitosan nanoparticles. *Journal of Applied Polymer Science* (2008): *107*, 891–897.
172. Mercader-Ros, M. T.; Lucas-Abellan, C.; Fortea, M. I.; Gabaldon, J. A.; Nunez-Delicado, E., Effect of HP-beta-cyclodextrins complexation on the antioxidant activity of flavonols. *Food Chemistry* (2010): *118*, 769–773.
173. Tomren, M. A.; Masson, M.; Loftsson, T.; Tonnesen, H. H., Studies on curcumin and curcuminoids XXXI. Symmetric and asymmetric curcuminoids: Stability, activity and complexation with cyclodextrin. *International Journal of Pharmaceutics* (2007): *338*, 27–34.
174. Yu, H. L.; Huang, Q. R., Enhanced in vitro anti-cancer activity of curcumin encapsulated in hydrophobically modified starch. *Food Chemistry* (2010): *119*, 669–674.
175. Sahu, A.; Kasoju, N.; Bora, U., Fluorescence study of the curcumin-casein micelle complexation and its application as a drug nanocarrier to cancer cells. *Biomacromolecules* (2008): *9*, 2905–2912.
176. Forrest, S. A.; Yada, R. Y.; Rousseau, D., Interactions of vitamin D-3 with bovine beta-lactoglobulin A and beta-casein. *Journal of Agricultural and Food Chemistry* (2005): *53*, 8003–8009.
177. Semo, E.; Kesselman, E.; Danino, D.; Livney, Y. D., Casein micelle as a natural nano-capsular vehicle for nutraceuticals. *Food Hydrocolloids* (2007): *21*, 936–942.
178. Kanaze, F. I.; Kokkalou, E.; Niopas, I.; Barmpalexis, P.; Georgarakis, E.; Bikiaris, D., Dissolution rate and stability study of flavanone aglycones, naringenin and hesperetin, by drug delivery systems based on polyvinylpyrrolidone (PVP) nanodispersions. *Drug Development and Industrial Pharmacy* (2010): *36*, 292–301.
179. Gao, L.; Liu, G. Y.; Wang, X. Q.; Liu, F.; Xu, Y. F.; Ma, J., Preparation of a chemically stable quercetin formulation using nanosuspension technology. *International Journal of Pharmaceutics* (2011): *404*, 231–237.
180. Choi, S. J.; Decker, E. A.; Henson, L.; Popplewell, L. M.; McClements, D. J., Inhibition of citral degradation in model beverage emulsions using micelles and reverse micelles. *Food Chemistry* (2010): *122*, 111–116.
181. Garti, N.; Yaghmur, A.; Leser, M. E.; Clement, V.; Watzke, H. J., Improved oil solubilization in oil/water food grade microemulsions in the presence of polyols and ethanol. *Journal of Agricultural and Food Chemistry* (2001): *49*, 2552–2562.
182. Spernath, A.; Yaghmur, A.; Aserin, A.; Hoffman, R. E.; Garti, N., Food-grade microemulsions based on nonionic emulsifiers: Media to enhance lycopene solubilization. *Journal of Agricultural and Food Chemistry* (2002): *50*, 6917–6922.
183. Lin, C.C.; Lin, H.Y.; Chi, M.H.; Shen, C.M.; Chen, H.W.; Yang, W.J.; Lee, M.H., Preparation of curcumin microemulsions with food-grade soybean oil/lecithin and their cytotoxicity on the HepG2 cell line. *Food Chemistry* (2014): 154, 282–290.
184. Mao, Y. Y.; Dubot, M.; Xiao, H.; McClements, D. J., Interfacial engineering using mixed protein systems: Emulsion-based delivery systems for encapsulation and stabilization of beta-carotene. *Journal of Agricultural and Food Chemistry* (2013): *61*, 5163–5169.
185. Gharsallaoui, A.; Roudaut, G.; Beney, L.; Chambin, O.; Voilley, A.; Saurel, R., Properties of spray-dried food flavours microencapsulated with two-layered membranes: Roles of interfacial interactions and water. *Food Chemistry* (2012): *132*, 1713–1720.

186. Yang, X. Q.; Tian, H. X.; Ho, C. T.; Huang, Q. R., Stability of citral in emulsions coated with cationic biopolymer layers. *Journal of Agricultural and Food Chemistry* (2012): *60*, 402–409.
187. Hou, Z. Q.; Gao, Y. X.; Yuan, F.; Liu, Y. W.; Li, C. L.; Xu, D. X., Investigation into the physicochemical stability and rheological properties of beta-carotene emulsion stabilized by soybean soluble polysaccharides and chitosan. *Journal of Agricultural and Food Chemistry* (2010): *58*, 8604–8611.
188. Shaw, L. A.; McClements, D. J.; Decker, E. A., Spray-dried multilayered emulsions as a delivery method for omega-3 fatty acids into food systems. *Journal of Agricultural and Food Chemistry* (2007): *55*, 3112–3119.
189. Parada, J.; Aguilera, J. M., Food microstructure affects the bioavailability of several nutrients. *Journal of Food Science* (2007): *72*, R21–R32.
190. de Vos, P.; Faas, M. M.; Spasojevic, M.; Sikkema, J., Encapsulation for preservation of functionality and targeted delivery of bioactive food components. *International Dairy Journal* (2010): *20*, 292–302.
191. Ubbink, J.; Kruger, J., Physical approaches for the delivery of active ingredients in foods. *Trends in Food Science and Technology* (2006): *17*, 244–254.
192. Champagne, C. P.; Fustier, P., Microencapsulation for the improved delivery of bioactive compounds into foods. *Current Opinion in Biotechnology* (2007): *18*, 184–190.
193. Fuchs, M.; Turchiuli, C.; Bohin, M.; Cuvelier, M. E.; Ordonnaud, C.; Peyrat-Maillard, M. N.; Dumoulin, E., Encapsulation of oil in powder using spray drying and fluidised bed agglomeration. *Journal of Food Engineering* (2006): *75*, 27–35.
194. Gharsallaoui, A.; Roudaut, G.; Chambin, O.; Voilley, A.; Saurel, R., Applications of spray-drying in microencapsulation of food ingredients: An overview. *Food Research International* (2007): *40*, 1107–1121.
195. Okuro, P. K.; de Matos Junior, F. E.; Favaro-Trindade, C. S., Technological challenges for spray chilling encapsulation of functional food ingredients. *Food Technology and Biotechnology* (2013): *51*, 171–182.
196. Nedovic, V.; Kalusevic, A.; Manojlovic, V.; Levic, S.; Bugarski, B., An overview of encapsulation technologies for food applications. *11th International Congress on Engineering and Food (ICEF11)* (2011): *1*, 1806–1815.
197. Takei, N.; Unosawa, K.; Matsumoto, S., Effect of the spray-drying process on the properties of coated films in fluidized bed granular coaters. *Advanced Powder Technology* (2002): *13*, 333–342.
198. Lopez-Rubio, A.; Gavara, R.; Lagaron, J. A., Bioactive packaging: Turning foods into healthier foods through biomaterials. *Trends in Food Science and Technology* (2006): *17*, 567–575.
199. Barbosa-Canovas, G. V.; Juliano, P., Physical and chemical properties of food powders. *Encapsulated and Powdered Foods* (2005): *146*, 39–71.
200. Vaz, C. M.; van Doeveren, P.; Reis, R. L.; Cunha, A. M., Soy matrix drug delivery systems obtained by melt-processing techniques. *Biomacromolecules* (2003): *4*, 1520–1529.
201. Henrist, D.; Van Bortel, L.; Lefebvre, R. A.; Remon, J. P., In vitro and in vivo evaluation of starch-based hot stage extruded double matrix systems. *Journal of Controlled Release* (2001): *75*, 391–400.
202. Breitenbach, J., Melt extrusion: From process to drug delivery technology. *European Journal of Pharmaceutics and Biopharmaceutics* (2002): *54*, 107–117.
203. Merisko-Liversidge, E.; Liversidge, G. G.; Cooper, E. R., Nanosizing: A formulation approach for poorly-water-soluble compounds. *European Journal of Pharmaceutical Sciences* (2003): *18*, 113–120.
204. Singh, S. K.; Srinivasan, K. K.; Gowthamarajan, K.; Singare, D. S.; Prakash, D.; Gaikwad, N. B., Investigation of preparation parameters of nanosuspension by top-down media milling to improve the dissolution of poorly water-soluble glyburide. *European Journal of Pharmaceutics and Biopharmaceutics* (2011): *78*, 441–446.
205. Schultz, S.; Wagner, G.; Urban, K.; Ulrich, J., High-pressure homogenization as a process for emulsion formation. *Chemical Engineering and Technology* (2004): *27*, 361–368.
206. Liedtke, S.; Wissing, S.; Muller, R. H.; Mader, K., Influence of high pressure homogenisation equipment on nanodispersions characteristics. *International Journal of Pharmaceutics* (2000): *196*, 183–185.
207. Stang, M.; Schuchmann, H.; Schubert, H., Emulsification in high-pressure homogenizers. *Engineering in Life Sciences* (2001): *1*, 151–157.
208. Donsi, F.; Annunziata, M.; Vincensi, M.; Ferrari, G., Design of nanoemulsion-based delivery systems of natural antimicrobials: Effect of the emulsifier. *Journal of Biotechnology* (2012): *159*, 342–350.

209. Donsi, F.; Sessa, M.; Ferrari, G., Effect of emulsifier type and disruption chamber geometry on the fabrication of food nanoemulsions by high pressure homogenization. *Industrial and Engineering Chemistry Research* (2012): *51*, 7606–7618.
210. Donsi, F.; Sessa, M.; Mediouni, H.; Mgaidi, A.; Ferrari, G., Encapsulation of bioactive compounds in nanoemulsion-based delivery systems. *Procedia Food Science* (2011): *1*, 1666–1671.
211. Donsì, F.; Ferrari, G.; Maresca, P., High-pressure homogenization for food sanitization. In *Global Issues in Food Science and Technology*, Barbosa-Cánovas, G. V.; Mortimer, A.; Lineback, D.; Spiess, W.; Buckle, K., (Eds.) Academic Press, Burlington, MA, 2009.
212. Sanguansri, P.; Augustin, M. A., Nanoscale materials development—A food industry perspective. *Trends in Food Science and Technology* (2006): *17*, 547–556.
213. Gao, Y.; Li, Z. G.; Sun, M.; Guo, C. Y.; Yu, A. H.; Xi, Y. W.; Cui, J.; Lou, H. X.; Zhai, G. X., Preparation and characterization of intravenously injectable curcumin nanosuspension. *Drug Delivery* (2011): *18*, 131–142.
214. Rao, J. J.; McClements, D. J., Stabilization of phase inversion temperature nanoemulsions by surfactant displacement. *Journal of Agricultural and Food Chemistry* (2010): *58*, 7059–7066.
215. Fernandez, P.; Andre, V.; Rieger, J.; Kuhnle, A., Nano-emulsion formation by emulsion phase inversion. *Colloids and Surfaces A-Physicochemical and Engineering Aspects* (2004): *251*, 53–58.
216. Roger, K.; Cabane, B.; Olsson, U., Formation of 10–100 nm size-controlled emulsions through a subPIT cycle. *Langmuir* (2010): *26*, 3860–3867.
217. Horn, D.; Rieger, J., Organic nanoparticles in the aqueous phase—Theory, experiment, and use. *Angewandte Chemie-International Edition* (2001): *40*, 4331–4361.
218. Singh, S. S.; Siddhanta, A. K.; Meena, R.; Prasad, K.; Bandyopadhyay, S.; Bohidar, H. B., Intermolecular complexation and phase separation in aqueous solutions of oppositely charged biopolymers. *International Journal of Biological Macromolecules* (2007): *41*, 185–192.
219. Li, B. Z.; Wang, L. J.; Li, D.; Bhandari, B.; Li, S. J.; Lan, Y. B.; Chen, X. D.; Mao, Z. H., Fabrication of starch-based microparticles by an emulsification-crosslinking method. *Journal of Food Engineering* : *92*, 250–254.
220. Yang, W. J.; Trau, D.; Renneberg, R.; Yu, N. T.; Caruso, F., Layer-by-layer construction of novel bio-functional fluorescent microparticles for immunoassay applications. *Journal of Colloid and Interface Science* (2001): *234*, 356–362.
221. Guzey, D.; McClements, D. J., Formation, stability and properties of multilayer emulsions for application in the food industry. *Advances in Colloid and Interface Science* (2006): *128*, 227–248.
222. Burgain, J.; Gaiani, C.; Linder, M.; Scher, J., Encapsulation of probiotic living cells: From laboratory scale to industrial applications. *Journal of Food Engineering* (2011) *104*, 467–483.
223. Picot, A.; Lacroix, C., Encapsulation of bifidobacteria in whey protein-based microcapsules and survival in simulated gastrointestinal conditions and in yoghurt. *International Dairy Journal* (2004): *14*, 505–515.
224. Krasaekoopt, W.; Bhandari, B.; Deeth, H., Evaluation of encapsulation techniques of probiotics for yoghurt. *International Dairy Journal* (2003): *13*, 3–13.
225. Chen, M. J.; Chen, K. N., Application of probiotic encapsulation in dairy products. In *Encapsulation and Controlled Release Technologies in Food System*, Lakkis, J. M., (Ed.) Blackwell Publishing Ltd, Oxford, 2007, pp. 83–107.
226. Chavarri, M.; Maranon, I.; Ares, R.; Ibanez, F. C.; Marzo, F.; del Carmen Villaran, M., Microencapsulation of a probiotic and prebiotic in alginate-chitosan capsules improves survival in simulated gastro-intestinal conditions. *International Journal of Food Microbiology* (2010): *142*, 185–189.
227. Mortazavian, A. M.; Aziz, A.; Ehsani, M. R.; Razavi, S. H.; Mousavi, S. M.; Sohrabvandi, S.; Reinheimer, J. A., Survival of encapsulated probiotic bacteria in Iranian yogurt drink (Doogh) after the product exposure to simulated gastrointestinal conditions. *Milchwissenschaft-Milk Science International* (2008): *63*, 427–429.
228. Semyonov, D.; Ramon, O.; Kaplun, Z.; Levin-Brener, L.; Gurevich, N.; Shimoni, E., Microencapsulation of Lactobacillus paracasei by spray freeze-drying. *Food Research International* (2010): *43*, 193–202.
229. Wang, Z. L.; Finlay, W. H.; Peppler, M. S.; Sweeney, L. G., Powder formation by atmospheric spray-freeze-drying. *Powder Technology* (2006): *170*, 45–52.

230. Zuidam, N. J.; Shimoni, E., Overview of Microencapsulates for use in food products or processes and methods to make them. In: *Encapsulation Technologies for Active Food Ingredients and Food Processing*, Zuidam, N., Nedovic, V., (Eds). Springer, New York (2010): 3–29.
231. Heidebach, T.; Foerst, P.; Kulozik, U., Influence of casein-based microencapsulation on freeze-drying and storage of probiotic cells. *Journal of Food Engineering* (2010): *98*, 309–316.
232. Gbassi, G. K.; Vandamme, T.; Ennahar, S.; Marchioni, E., Microencapsulation of Lactobacillus plantarum spp in an alginate matrix coated with whey proteins. *International Journal of Food Microbiology* (2009): *129*, 103–105.
233. Mokarram, R. R.; Mortazavi, S. A.; Najafi, M. B. H.; Shahidi, F., The influence of multi stage alginate coating on survivability of potential probiotic bacteria in simulated gastric and intestinal juice. *Food Research International* (2009): *42*, 1040–1045.
234. Sandoval-Castilla, O.; Lobato-Calleros, C.; Garcia-Galindo, H. S.; Alvarez-Ramirez, J.; Vernon-Carter, E. J., Textural properties of alginate-pectin beads and survivability of entrapped Lb. casei in simulated gastrointestinal conditions and in yoghurt. *Food Research International* (2010): *43*, 111–117.
235. Sabikhi, L.; Babu, R.; Thompkinson, D. K.; Kapila, S., Resistance of microencapsulated Lactobacillus acidophilus LA1 to processing treatments and simulated gut conditions. *Food and Bioprocess Technology* (2010): *3*, 586–593.
236. Anekella, K.; Orsat, V., Optimization of microencapsulation of probiotics in raspberry juice by spray drying. *LWT-Food Science and Technology* (2013): *50*, 17–24.
237. Zou, Q.; Liu, X.; Zhao, J.; Tian, F.; Zhang, H.-p.; Zhang, H.; Chen, W., Microencapsulation of Bifidobacterium bifidum F-35 in whey protein-based microcapsules by transglutaminase-induced gelation. *Journal of Food Science* (2012): *77*, M270–M277.
238. Sunny-Roberts, E. O.; Knorr, D., Cellular injuries on spray-dried Lactobacillus rhamnosus GG and its stability during food storage. *Nutrition and Food Science* (2011): *41*, 191–200.

20

Cell Immobilization Technologies for Applications in Alcoholic Beverages

Argyro Bekatorou, Stavros Plessas, and Athanasios Mallouchos

CONTENTS

20.1 Introduction

Whole cell immobilization is defined as "the physical confinement or localization of intact cells to a certain region of space with preservation of some desired catalytic activity."[1,2] Immobilization resembles the conditions in which microbial cells are found in nature, taking into account that they are usually found adhering to and growing on different kinds of surfaces. The considerable research and industrial interest in the use of immobilized cells for alcoholic and other food- and fuel-related fermentation applications is due to the numerous advantages that such technologies offer compared with conventional free cell systems. Specifically, the advantages of immobilized cells for alcoholic beverages production include (1) achievement of higher cell densities in the bioreactors, and therefore increased substrate uptake, higher productivities, and shorter process times, (2) protection against shear forces and stress (pH, temperature, substrate concentration and end-product inhibition, presence of heavy metals, etc.), leading to extended operational stability of the biocatalyst, (3) feasibility of continuous processing (Figure 20.1), easy product recovery, and reusability of the biocatalyst, (4) feasibility of low-temperature fermentation, which can lead to improved product quality, (5) reduction of secondary fermentation (maturation) times, (6) reduced contamination risk due to the higher cell densities and increased fermentation activity, and (7) reduction of investment and energy costs due to the construction of smaller bioreactors, fewer separation and filtration requirements, and higher productivity.[2–6]

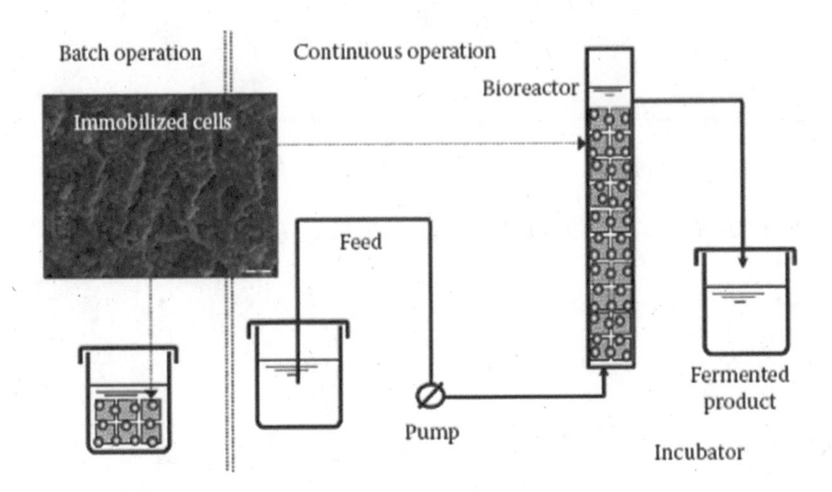

FIGURE 20.1 Batch and continuous fermentation by immobilized cells.

Various materials have been proposed as carriers for cell immobilization in wine, beer, cider, distillates, and ethanol production in various process designs and bioreactor configurations. Carriers include organic and inorganic natural or synthesized products as well as waste materials. Organic materials can be synthetically made (e.g., synthetic polymers) or extracted from natural sources (e.g., cellulose, polymeric hydrogels, etc.). Natural carriers can be used with minimal or no pretreatment and include products such as wood pieces, parts of fruit, etc., or wastes and by-products of food-grade purity, such as sawdust, spent grains, crop residues, etc. A carrier is suitable for cell immobilization for use in the production of alcoholic beverages when various prerequisites are satisfied, such as (1) good stability against enzymes, temperature, solvents, pressure, and shearing forces, (2) large surface area, (3) porous structure or functional groups for cells to adhere to, (4) ease of handling and regeneration, (5) ability to protect cells and extend their viability and activity, (6) easy substrate and product transfer, and (7) an immobilization technique that is easy, cost-effective, and suitable for scale-up. Last but not least, a suitable cell immobilization carrier should not leave residues that affect product quality and should be readily accepted by consumers.[2,4]

Nevertheless, industrial use of immobilized cells is still limited and it will depend on the development of processes that can be readily scaled up.[2] Data available in the literature on materials and techniques used for viable cell immobilization for application in alcoholic beverages and food-grade ethanol production are highlighted and discussed.

20.2 Cell Immobilization Techniques

Cell immobilization has been used in the production of all types of alcoholic beverages and various other biotechnological processes, and therefore, many such techniques have been developed, which can be grouped into the following four major categories:[1,2,7–9] (1) immobilization on a solid carrier surface, (2) immobilization by entrapment in a porous matrix, (3) carrier-free immobilization, and (4) containment behind barriers.

20.2.1 Immobilization on a Solid Carrier Surface

Cell immobilization on a solid carrier can be done by physical adsorption due to electrostatic forces or by covalent binding between the cell membrane and the carrier. Immobilization can also be done by growth of the cells into natural cavities on a surface and therefore, containment of the cells due to entrapment and/or a combination with electrostatic and other weak forces (Figure 20.2). The strength with which the cells are bonded to the carrier as well as the depth of the biofilm varies depending on the cell strain

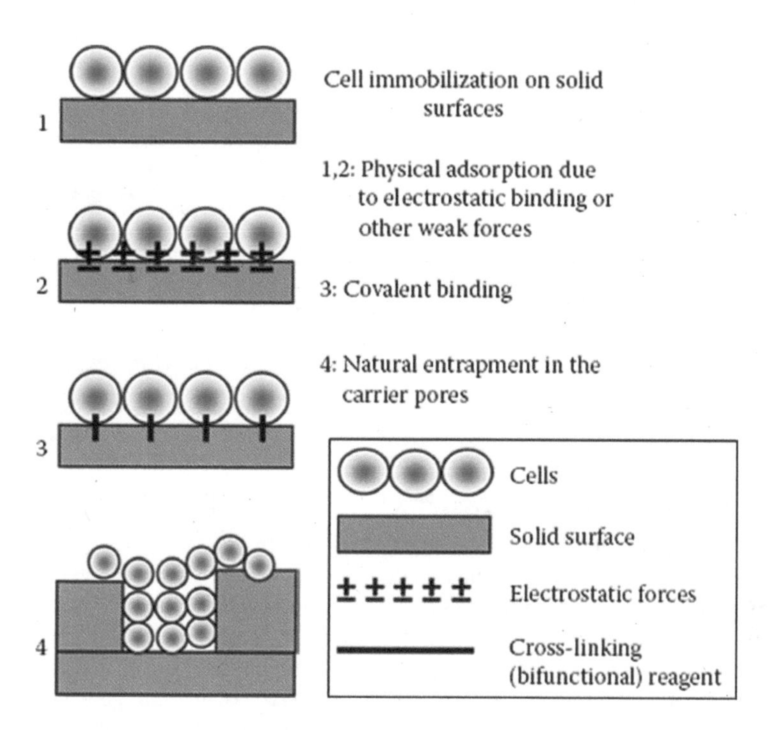

FIGURE 20.2 Basic types of cell immobilization on solid carrier surfaces. (Adapted from Kourkoutas, Y. et al., *Food Microbiol.*, 21(4), 377, 2004.)

and the nature of the carrier surface.[2] In this type of immobilization, the adhered viable cells may grow and escape, depending on the conditions applied, and therefore, may be in equilibrium with free cells in suspension. Such immobilized cell systems have been extensively used mainly due to the ease of the immobilization technique, which in some cases can be achieved by simple contact of a cell suspension with the carrier for a small time period (Figure 20.3). Examples of solid carriers used in this type of immobilization are cellulosic materials such as diethylaminoethyl (DEAE)-cellulose, wood, sawdust, delignified sawdust, cereal bran, etc., and inorganic materials such as polygorskite, montmorillonite, hydromica, porous porcelain, porous glass, pumice stone, etc. Solid materials such as glass or cellulose can also be treated with various chemicals (polycations, chitosan, etc.) to improve their characteristics as cell binding carriers.[2]

Cellulosic materials are very popular as immobilization carriers for alcoholic beverage production due to their food-grade purity, low cost, and availability all year round. Delignified cellulosic materials (DCMs) have been successfully used as carriers for the development of immobilized cell biocatalysts for use in various bioprocesses related to the food and fuel industries, such as alcoholic and lactic acid fermentations for the production of alcoholic beverages and dairy products.[2,10,11] Nano/microporous cellulose (NMC) prepared after the removal of lignin from wood cellulose was found suitable for the development of "cold pasteurization" processes, acting as a biofilter for cell removal. It was also used successfully as a biocatalyst in food fermentations, acting both as a cell immobilization carrier and as a promoter of biochemical reactions, even at extremely low temperatures.[10] The cumulative surface area of the NMC pores was found to be 0.8 to 0.89 m^2/g as indicated by porosimetry analysis. This surface is relatively small compared with other porous materials such as γ-alumina; however, using a natural organic material is attractive from the point of view that it is safer for bioprocess applications and is better accepted by consumers. The NMC/immobilized yeast biocatalyst increased the fermentation rate and was more effective at lower temperatures compared with free cells. Furthermore, the activation energy (E_a) of fermentation was found to be 28% lower than that of free cells, indicating that it is an excellent material to promote the catalytic action of cells for alcoholic fermentation.

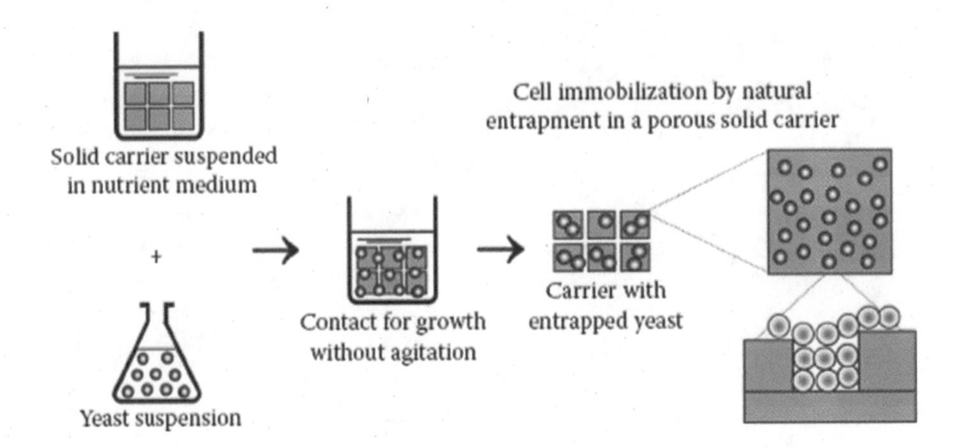

FIGURE 20.3 A simple technique for cell immobilization by natural entrapment on a porous solid carrier.

Other natural materials, including agroindustrial wastes and by-products, used as carriers for this type of viable cell immobilization are dried gluten pellets, brewer's spent grains, whole cereal grains, composite biocatalysts of cellulosic materials and gels, pieces and parts of dried or fresh fruit, grape stems and grape skins, cork pieces, olive pits, natural sponges, etc. All these materials were used successfully in the production of alcoholic beverages at research level (wine making and malolactic fermentation [MLF], brewing, distillates, and whey-based alcoholic drinks), showing good potential for larger-scale application due to advantages such as their low cost, easy industrial preparation, good operational stability, improved quality of the beverages produced compared with free cells, and easier acceptance by consumers. To facilitate commercialization for industrial or home-scale fermentations, research on the development of active dried ready-to-use immobilized biocatalysts was carried out, mainly aiming to optimize freeze-drying or simpler, mild, and cost-effective thermal drying.[2] For example, no protecting medium was needed for the freezing and freeze-drying of *Saccharomyces cerevisiae* cells immobilized on DCM and gluten pellets. The immobilized biocatalysts retained their viability during storage and showed high productivity and stability for glucose, wine, and beer fermentation,[12–15] leading to products of similar quality to those produced by fresh immobilized cells and of improved quality compared with free cells.

Inorganic materials are to some extent advantageous for use as yeast immobilization carriers for alcoholic fermentation, because they are usually abundant and cheap materials that can improve productivity and in most cases, product flavor. They can withstand shear forces; they can be easily recovered and reused; and therefore, they can facilitate the development of large-scale continuous operations. Porous inorganic materials, such as γ-alumina pellets and the mineral kissiris (a cheap, porous volcanic rock found in Greece, which contains mainly 70% SiO_2), ceramic chamotte (clay), hydroxylapatite, etc., have been evaluated as carriers for yeast immobilization.[2,16] They were found suitable for ambient and low-temperature fermentation, increasing both ethanol productivity and the biocatalytic stability of the yeast and leading to products of improved aroma. However, their composition (which liberates mineral residues in the process medium) limits their applications related to food. On the other hand, they are suitable for distillates, potable and fuel-grade alcohol production, biodiesel, or other non-food purposes.[2]

20.2.2 Immobilization by Entrapment in a Porous Matrix

The entrapment of cells in a porous matrix is a more definite type of immobilization that does not depend on the cell properties.[9] In this type of immobilization, either the cells are allowed to penetrate into a porous matrix until their mobility is obstructed by the presence of other cells or the porous material is formed *in situ* in a culture of cells (Figure 20.4). Both entrapment methods are based on the inclusion of cells in a rigid network, which allows mass transfer of nutrients and metabolites.

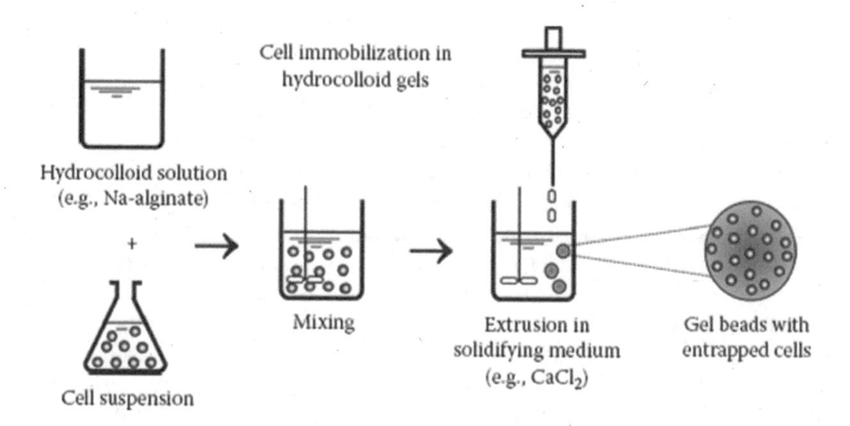

FIGURE 20.4 Common extrusion technique for generation of immobilized cells/polysaccharide gel beads.

At research level, this approach for cell immobilization has been by far the most popular for various applications, including the production of alcoholic beverages, with the main advantage being that it is a simple technique that proceeds under very mild conditions and is therefore compatible with most living cells. More frequently used than any other materials for this application are hydrogels of natural polysaccharides found widely in nature as constituents of cell walls of plants, crustaceans, or insects, such as alginate salts, cellulose, κ-carrageenan, agar, pectic acid, and chitosan, or synthetic polymeric matrixes such as polyacrylamide. Other examples include gelatin, collagen, and polyvinyl alcohol (PVA).[2,9,17–19]

The entrapment of cells in Ca alginate beads is the most widely used and studied technique for immobilizing living cells, although it has some limitations, such as (1) low stability (e.g., disruption by substances that have a high affinity for Ca^{2+}, such as phosphate or citrate), (2) high porosity, which although it allows easy mass transfer, may lead to leakage of nutrients and limit their usefulness to cells, and (3) wide pore size distribution, which makes controlled release difficult.[19] Cell growth in the gel depends initially on diffusion limitations (substrates, metabolites, and oxygen), affected by the porosity of the material, and later, on the impact of the accumulated cell mass. Due to these limitations, cells near the outer surface may behave differently compared with cells inside the beads; they may multiply and be released from the beads.[4] In that case, the fermentation system would comprise immobilized cells and free cells in suspension. To avoid this problem, double-layer beads have been developed, in which the external layer prevents the cells from escaping.[2,20] Composites of alginate gels with other organic or inorganic materials have also been developed for beverage fermentations with more than one entrapped microbial species. For example, alginate beads were used to entrap *Saccharomyces cerevisiae* or *Oenococcus oeni* cells, functioning as a protective environment for the deposition of silica gel membranes.[21] The biocatalyst was evaluated for both alcoholic fermentation of glucose and MLF. Bacterial cellulose–alginate (BCA) composite sponge was used as a yeast carrier for ethanol production.[22] The biocatalyst had an asymmetric structure, with a thin and dense outer layer covering a macroporous interior, which was effective for yeast immobilization. Kregiel et al.[16] immobilized yeasts in foamed alginate gels to study the induced alterations in growth, vitality, and metabolic activity as well as the potential for fermentation applications. The scale-up of ethanol production from molasses using yeast immobilized in alginate-based microporous and mesoporous zeolite composites was also attempted.[23]

The enumeration of biomass entrapped in a gel matrix is critical for the application of biotechnological processes using viable immobilized cells. Such methods usually are gravimetric or include the determination of proteins, DNA, NADH, and ATP, which are expressed as biomass concentration. For example, a reliable, accurate, and rapid luminometric method (ATP determination) was developed for estimating the active biomass of brewing, wine-making, and ethanol-producing yeast strains immobilized in alginate, pectate, and κ-carrageenan hydrogels.[24]

20.2.3 Carrier-Free Immobilization

Carrier-free immobilization can be done naturally using self-aggregation by flocculation or artificially induced using cross-linking agents. Cell flocculation has been defined by many authors as an aggregation of cells to form a larger unit or the property of cells in suspensions to adhere in clumps and sediment rapidly.[2,25] Flocculation can be considered as an immobilization technique, since the large size of the aggregates facilitates their potential use in bioreactors, such as packed bed, fluidized bed, and continuous stirred tank reactors. It is the most simple and inexpensive immobilization method, although it is very difficult to predict and control, which is essential for maximizing bioreactor efficiency. The natural flocculation characteristics of yeasts are affected by many factors, such the genetic characteristics of the strain, the structure and surface charges of the cell wall, the growth phase, the process temperature, the medium pH and composition, etc.[8] Weak flocculation ability may result in cell washout of the bioreactor, resulting in low cell concentration and therefore, insufficient fermentation rates. Artificial flocculating agents or cross-linkers can be used to enhance aggregation in cell cultures that do not naturally flocculate. On the other hand, high flocculation activity may result in low concentrations of active cells due to nutrient diffusion limitations to the cells in the core of large aggregates.[8] Yeast flocculation is a property of major importance for the brewing industry, as it affects fermentation productivity and beer quality in addition to yeast removal and recovery. The importance of the flocculation properties of *S. cerevisiae* for alcoholic beverage production and the mechanisms and factors affecting them has been widely reviewed.[2,8,25,26]

A new trend in carrier-free immobilization is the coimmobilization of cells in the form of biocapsules. In this immobilization technique, the immobilization matrix is provided by one of the microorganisms to be immobilized without the need for an external support and the associated costs. For example, a special procedure to immobilize *S. cerevisiae* on the filamentous fungus *Penicillium chrysogenum* (generally regarded as safe [GRAS]) was developed via a simple, inexpensive, natural coimmobilization process for use in ethanol production and wine making.[27–29] The resulting biocapsules were spherical, hollow, smooth, elastic, and strong and were used with no loss of integrity during the fermentation processes. The fungus died during the fermentation process, remaining as an inert yeast carrier, facilitating reuse of the biocapsules.

20.2.4 Mechanical Containment behind a Barrier

Containment of cells behind a barrier can be achieved either by the use of semipermeable membranes to isolate the cells from the bulk liquid, or by entrapment of cells in a microcapsule, or by cell immobilization on the interaction surface of two immiscible liquids. The cells can be immobilized on the membrane (as in the case of biosensors), or they can be allowed to grow into a void enclosed by the membrane (as in the case of membrane reactor systems). However, growth must be controlled to prevent excessive buildup of biomass, which could cause membrane rupture.[2,8,9] This type of immobilization is ideal when cell-free product and minimum transfer of compounds are required,[17] and it is widely used in cell recycling and continuous processes.[30,31] In wine making, constructions using yeasts confined by microfiltration membranes have been developed and are available in the market, such as the "Millispark" cartridge, which was developed for the secondary fermentation of sparkling wine inside the bottle.[2] The major disadvantages of membrane immobilization techniques are mass transfer limitations (supply of oxygen and nutrients to the cells and the removal of carbon dioxide) and possible membrane plugging caused by cell growth.[9,32,33]

An important technique for cell entrapment behind a barrier is microencapsulation, which involves the inclusion of cells in polymeric microspheres in the 1 to 1000 μm size range. This technique has numerous advantages, such as higher cell densities and increased cell survival and productivity.[34] Microencapsulation has found application in various biotechnological processes, such as encapsulation of probiotics for food production, development of encapsulated biocatalysts for fermentation processes, and environmental bioremediation. The microspheres have a larger specific surface area to allow good diffusion of nutrients and metabolites. Additionally, microencapsulation allows easy separation of cells during fermentation processes and minimizes cell washout, but the microspheres must be mechanically strong to withstand shear forces and other destructive conditions such as exposure to acids, gases, and solvents.[34]

20.3 Effect of Immobilization on Microbial Cells

Alterations in cell growth, physiology, and metabolic activity may be induced by cell immobilization, such as effects on the activation of energetic metabolism; targeted protein expression to support the altered metabolic behavior of immobilized cells; altered growth rates; increased substrate uptake and product yield; increase in storage polysaccharides; altered yield of fermentation by-products, including flavor compounds; higher intracellular pH values; increased tolerance against toxic and inhibitory compounds; increased hydrolytic enzyme activities; modifications in the nucleic acid contents, etc. Parameters that have been considered responsible for these alterations include mass transfer limitations, disturbances in the growth pattern, surface tension and osmotic pressure effects, reduced water activity, cell-to-cell communication, changes in the cell morphology, altered membrane permeability, etc.[2,5,35–37] Generally, it is considered that it is the microenvironment inside the immobilization matrix that affects the physiology and metabolic behavior of yeasts and not the nature of the matrix.[38] For example, the insufficient space for growth in the immobilization matrix was considered to be responsible for differences in the morphology between free and immobilized yeast cells, while the increased viability and activity of the immobilized cells during storage have been attributed to the protective effect of the matrix.[36,37,39] Differences in intracellular pH values observed between free and immobilized yeast were attributed to the increased permeability of the cell membranes to protons, higher ATP consumption, increased glucose uptake, and increased glycolytic activity.[2] The increased ethanol tolerance and lower substrate inhibition of immobilized yeasts were attributed to the protective effect of the matrix or to modified structural features that affect permeability, such as the fatty acid composition of cell membranes. Additionally, osmotic stress caused by the immobilization techniques was found to lead to intracellular production of pressure-regulating compounds such as polyols, which lead to decreased water activity and consequently, higher tolerance to toxic compounds.[2,36,37] Finally, due to these reasons, cells immobilized in various types of carriers showed enhanced viability and stability during freezing, freeze-drying, and thermal drying, which was exploited for the development of ready-to-use dried biocatalysts for commercial distribution regarding alcoholic fermentation applications.[12–15,40–42]

20.3.1 Effect of Immobilization on Product Flavor

The flavor of fermented foods depends highly on the metabolic activity of the culture used. Especially, amino acid and lipid metabolism in yeasts makes a crucial contribution to flavor, because it is linked to the production of flavor-active compounds such as esters, alcohols, carbonyl compounds, fatty acids, and non-volatile components.[2,43] The increased ester and decreased fusel alcohol (mainly amyl alcohols) formation that has been observed during fermentations using immobilized cells, or the improved ratios of esters to higher alcohols, especially at low temperatures, is considered to have a great impact on beverage quality and technology.[2,44,45]

The aroma of wine is the result of a complex combination of varietal and fermentation- and maturation-derived compounds that give each wine its distinctive character. A considerable fraction of these aroma-related compounds is produced during the primary fermentation of grape must. These compounds are mainly acetate esters of higher alcohols, ethyl esters of fatty acids, higher alcohols, fatty acids, ketones and aldehydes, and sulfur compounds.[45,46] Among these, esters are considered particularly important for wine aroma due to their characteristic fruity and floral odors and because their concentrations usually exceed their odor threshold values in wines. Higher alcohols have higher threshold values and contribute mainly to the complexity of wine aroma at concentrations lower than 400 mg/l, above which they are considered off-flavors.[47] Among carbonyl compounds, acetaldehyde, smelling like green apple, freshly cut grass, or green leaves, has a great impact on wine quality. In most wines, above threshold values, it is considered an off-odor. Finally, among fatty acids, the ones that mainly affect wine aroma are acetic, propanoic, butanoic, 3-methylbutanoic, hexanoic, octanoic, and decanoic acids.[48] The concentrations of these compounds are affected by any factor that affects the progress of fermentation, such as the yeast strain used, the immobilization technique, fermentation temperature, composition of grape must, dissolved oxygen levels, etc. Most of the research works investigating the effect of fermentation by immobilized cells on wine aroma focus on the analysis of the major volatile compounds in wines (acetaldehyde, ethyl acetate, propanol, isobutanol, and amyl alcohols) as well as minor volatile compounds with low threshold values, using various gas chromatography (GC) techniques. To evaluate the effect of cell immobilization and other process parameters on wine flavor, the analytical results are compared with sensory evaluations by expert panelists.[45]

Beer is produced by a number of complex biochemical (germination, enzyme hydrolysis of carbohydrates and proteins, etc.) and technological processes (malting, kilning, mashing, fermentation, and maturation), all of which affect the flavor of the final product. The fermentation stage alone contributes more than 600 flavor-active compounds to beer.[49] The flavor of beer is affected by the same groups of compounds as in wine and other fermented products, along with carbon dioxide, ethanol, and glycerol, which control the overall effect of the minor constituents. Amino acid metabolism is also a key to the formation of the mentioned compounds, and since it is affected by immobilization technology, as mentioned earlier, this technology has become interesting for controlling or altering flavor, leading to the production of beers with characteristic flavor profiles.[2,50,51] A few very important changes also occur during the maturation (lagering) stage of the brewing process. One of the key compounds in beer maturation is diacetyl, a compound with an undesired butter flavor above threshold values. Diacetyl is produced from α-acetolactate by an oxidative non-enzymatic reaction and is slowly converted to the flavorless derivatives acetoin and 2,3-butanediol by yeast metabolism during the maturation stage. This process, known as *diacetyl rest*, is time consuming and energy demanding, since it must take place at very low temperatures (~0 °C) to avoid degradation of product quality.[52] The combination of yeast immobilization and low-temperature fermentation can reduce primary fermentation time, accelerate the removal of diacetyl (maturation), and improve overall product quality. Therefore, it is has been reported as a promising technological strategy for significant reduction of production costs.[2,18,53]

20.3.2 Low-Temperature Fermentation with Immobilized Cells

A large number of studies report the selection or improvement of psychrophilic, psychrotolerant, or cold-adapted yeasts for low-temperature alcoholic beverages or ethanol production, and many more describe the development of low-temperature fermentation processes employing immobilized cell systems. Metabolic and physiological changes in yeast are not only induced in psychrophilic species evolved in cold environments but are also common during growth or fermentation processes at low temperatures. *S. cerevisiae* is naturally found in environments, such as the surface of fruit, that can be subjected to low temperatures. In alcoholic fermentation processes, these yeasts can be exposed to temperatures around 10 to 12 °C, while industrial strains may be stored at very low temperatures (4 °C), at which viability is maintained but growth is restricted.[54] In *S. cerevisiae*, low temperatures induce the expression of genes that display a cold-sensitivity phenotype, including the induction of fatty acid desaturases, proteins involved in pre-rRNA processing and ribosome biogenesis, specific amino acid–rich cell wall proteins, altered nitrogen metabolism, changes in the membrane fatty acids, alterations in aroma-related biochemical reactions, etc.[54]

Regarding alcoholic beverage production, low-temperature fermentation possesses a number of advantages, such as the production of beers of superior quality and the ability to ferment low-acidity musts to produce more malic and succinic acid, glycerol, and β-phenylethanol and less acetic acid.[54,55] However, these processes are not commonly used due to the increased risk of stuck and sluggish fermentations. The use of immobilized yeasts has facilitated the development of productive low-temperature fermentation processes, as demonstrated by numerous published works.[2,18,54] The greatest impact of such technologies is considered to be the improved flavor of the products, which was mainly attributed to the better ratios of esters to higher alcohols on total volatiles. Specifically, wines produced at low temperatures have aromas with more fruity notes due to increased synthesis or reduced conversion of esters. The use of immobilized cells for very low-temperature fermentation (below 10 °C) led to wines of improved aroma due to this effect.[2,43,45,56–60] The use of natural food-grade supports for cell immobilization, such as DCM and gluten pellets, proved to be effective for low-temperature wine making, with significantly improved fermentation productivity and product quality compared with free cells. To facilitate commercialization of such biocatalysts, freeze-drying techniques were also evaluated for the production of ready-to-use active dry formulations.[15]

Beer, on the other hand, is produced by more complex biochemical and technological processes, all of which affect its flavor. Yeast amino acid metabolism, a key to the development of beer flavor as described earlier, is affected by process temperature and use of cell immobilization techniques. Therefore, technologies based on these features as well as other process conditions and strain selection have been developed to control beer flavor.[2,51] The combination of immobilized yeast and low-temperature primary fermentation was found to produce beers with low levels of diacetyl, therefore indicating the potential for low-cost industrial application, since maturation is a highly energy-consuming process.[2] Finally, Perpete and Collin[61] showed that during alcohol-free beer production, the enzymatic reduction of *worty* flavor (caused by Strecker aldehydes) by brewer's yeast was improved by cold contact fermentation.

20.4 Wine Making by Immobilized Cells

Although research on wine making by immobilized cells is extensive, industrial applications are still limited, mainly due to the strong traditional character of this product and consumer susceptibility. The objectives of using immobilized cells in wine making are to improve fermentation productivity, reduce maturation time, reduce production cost and installation size, improve flavor, and produce novel products of distinct characters.[2] A variety of materials and techniques have been proposed for wine yeast immobilization, involving organic, inorganic, natural, or synthesized immobilization carriers that may be used with or without modifications to optimize their characteristics.

Inorganic materials that have been proposed for yeast immobilization for batch and continuous primary fermentation in wine making, leading to significant improvement of process kinetics even at extremely low temperatures (0–10 °C), include kissiris,[56,62,63] γ-alumina,[63,64] polygorskite, montmorillonite, hydromica, porous porcelain, porous glass, glass pellets covered with a layer of alginates, etc.[2,65] Despite their good attributes, the use of such biocatalysts is limited by the possibility to transfer undesirable residues in the final product (e.g., Al in the case of γ-alumina or kissiris). Efforts to remove such residues have also been reported, while wines produced by these techniques can be used as raw materials for distillates production, since mineral residues do not distil.[64,66] Apart from safety and quality, consumer acceptance is another main drawback when using inorganic materials in wine making.

To avoid the presence of undesirable residues in wine released by inorganic materials, food-grade organic carriers were also evaluated for wine making, mainly polysaccharide hydrogels. Ca alginate gels were considered suitable for laboratory- and pilot-scale fermentation and under real vinification conditions, with interesting results regarding the formation of glycerol and secondary fermentation by-products.[2] Nevertheless, their use does not offer a good industrial choice because of their high cost and low stability, which leads to cell washout and release of residues in wine. Practical applications of alginates concern mainly secondary fermentation inside the bottle in sparkling wine production for easy clarification and removal of cells.[65,67,68] Another application is the treatment of sluggish and stuck fermentations.[69]

Natural supports for yeast immobilization, such as DCM,[44] gluten pellets,[57] and cork pieces,[70] were found to be effective for both ambient and low-temperature wine making with significantly increased fermentation rates compared with free cells. The produced wines were of improved quality, which was demonstrated by both sensorial tests and chemical analysis of aroma volatiles.[45] These materials are of food-grade purity, cheap, abundant, and easy to prepare industrially. They can be easily accepted by consumers, and compared with other natural supports such as fruit pieces, they have higher operational and mechanical stability. Freeze-dried biocatalysts consisting of yeast cells immobilized on gluten pellets or DCM were also evaluated for producing wines of similar quality to those made by fresh immobilized cells.[15]

The idea of using fruit parts and pieces as yeast immobilization supports for wine making was also attractive; because fruits are the natural habitats of yeasts, they can be easily accepted by consumers, and can affect flavor, leading to new types of wines. Cells immobilized on apple, pear, and quince pieces,[2] fresh sugarcane pieces,[71] dried raisin berries,[72] and grape skins[73] were successfully used for batch and continuous wine making at ambient and extremely low temperatures. Raisin berries and grape skins are fully compatible with wine, and consumer acceptance is not an issue. Although the use of fruit in high-volume bioreactors is problematic, such materials could still be employed in low-capacity processes due to the fine quality of the produced wines, which can add value to the product.

In search of natural supports for yeast immobilization suitable for wine making, starch-containing materials such as potatoes,[74] whole wheat,[42] corn,[75] barley grains,[76] and starch gels[77] were evaluated, leading to increased productivity even at extremely low temperatures compared with free cells. Finally, for economical and sustainable food-grade wine biocatalysts, various agroindustrial by-products and residues have been proposed as wine yeast carriers, including brewer's spent grains,[58] watermelon rind pieces,[78] grape pomace,[79] grape seeds, skins, stems, and corncobs,[80] providing all the discussed advantages in wine making, with the main drawback usually a negative effect on product color.

Other applications of immobilization techniques in wine making include the use of immobilized yeasts for color or acidity correction and immobilized enzymes for aroma release from bonded precursors. For example, chitin, chitosan, diethyl-amino-ethyl chitosan, and acrylic beads have been used for enzyme immobilization (e.g., β-glucosidase, α-arabinosidase, and α-rhamnosidase) to enhance the aroma of wines, musts, fruit juices, and beverages through the breakage of glycosidic linkages of rhamnose with aromatic compounds (e.g., monoterpenes and norisoprenoids) present in glycosidic form.[81–83] Yeasts immobilized in κ-carrageenate and alginate gels were used effectively for color correction treatments as well as to delay accelerated browning during storage of sherry pale white wines.[84,85] An *S. cerevisiae* stain immobilized in double-layer alginate–chitosan beads was used for bioreduction of the volatile acidity of acidic wines with a volatile acidity higher than 1.44 g/l acetic acid with no detrimental impact on wine aroma.[86] Finally, a few applications of membrane technology in wine making have been reported. Takaya et al.[87] studied the efficiency of two membrane bioreactor systems for continuous dry wine making: a single-vessel bioreactor in which cells were entrapped by a cross-flow-type microfilter and a two-vessel system consisting of a continuous stirred tank reactor and a membrane bioreactor. The double vessel was found more suitable for dry wine making.

20.4.1 Malolactic Fermentation of Wine by Immobilized Cells

MLF is a secondary process that usually occurs in red wines, or wines with high acidity, during the maturation period. During MLF, L-malic acid is converted to L-lactic acid and carbon dioxide by bacteria such as *Leuconostoc*, *Lactobacillus*, and *Pediococcus* spp. Most bacteria convert malic acid to lactic acid with the intermediate formation of pyruvic acid, while *O. oeni* (previously classified as *Leuconostoc oenos*) expresses the malolactic enzyme to directly convert malic acid in a one-step reaction. *O. oeni* has been extensively studied for the controlled MLF of wine due to its higher tolerance to ethanol, low pH, and SO_2.[2,88,89] Yeasts such as *Schizosaccharomyces pombe* and *Saccharomyces* strains can also convert malic acid through a maloethanolic-type fermentation.[90] Lactic acid is less acidic than malic acid, and as a consequence, MLF leads to improvement of the sensory properties and biological stability of the wines. Additionally, the production of various other by-products of the MLF reaction may affect wine flavor positively.

MLF may occur in bottled wines that have not been adequately preserved with SO_2 by transformation of residual sugars by wild microflora, causing undesirable turbidity and development of off-flavors. The use of selected immobilized lactic acid bacteria can accelerate controlled MLF by higher cell densities and increased tolerance to inhibitory wine constituents, leading to flavor improvements. Immobilization can also facilitate the reuse of the MLF biocatalyst and the application of continuous processing.[2] Most of the initial efforts to conduct controlled MLF in wine involved the use of immobilized *O. oeni* in alginate gels.[65] Later works investigated the use of a variety of other carriers and bioreactor designs to optimize immobilized cell systems for practical MLF applications. Carriers such as polyacrylamide, κ-carrageenan, silica gel, pactate gels, chitosan, PVA (LentiKats®), positively charged cellulose sponge, composites of alginates with organo-silica or charcoal, polyurethane foams, gel-like membranes, and even oak chips were used as immobilization supports for various species with satisfactory results.[91–96]

The use of immobilized bacteria in must inoculated with free yeast or a coimmobilized bacteria and yeast biocatalyst can allow simultaneous alcoholic fermentation and MLF. An integrated wine-making process, including sequential alcoholic fermentations and MLF operated continuously, was developed by Genisheva et al.[97] *S. cerevisiae* cells immobilized either on grape stems or on grape skins, and *O. oeni* cells immobilized on grape skins only, were employed for a high-yield production of dry white wine having a good physico-chemical quality and 67% reduced malic acid concentration. Similar results were obtained by the use of three cultures (two *S. cerevisiae* spp. and *L. delbrueckii*) immobilized in alginate gel beads packed in near-horizontal acrylic columns.[98] The use of immobilized *O. oeni* on DCM also led to improvements of MLF in wine making,[99] and this effort was further enhanced by the development of a two-layer composite biocatalyst for simultaneous alcoholic fermentation and MLF.[100] The biocatalyst consisted of DCM with entrapped *O. oeni* cells covered with starch gel containing an alcohol-resistant and cryotolerant *S. cerevisiae* strain. The significance of such composite biocatalysts is the feasibility of two or three bioprocesses in the same bioreactor, thus reducing production cost in the food industry.[100]

Most of these efforts claimed possible industrial application of the immobilized biocatalysts due to their high operational stability, faster conversion of malic acid compared with free cells, and increased tolerance to SO_2 and ethanol inhibition. Understanding the nature and factors that affect MLF, which is a very complex process, as well as its effect on flavor formation is crucial for optimizing wine technology and applying immobilization techniques at full scale.

20.5 Brewing by Immobilized Cells

In brewing, research has also focused on cell immobilization techniques to facilitate continuous processing, reduce maturation time, and produce alcohol-free beer. Brewing requires long fermentation times, large-scale fermentation, maturation at very low temperature, and large storage capacity, and it is therefore a highly energy-consuming process.[2,101,102] The rapid maturation of beer has been attempted by employing immobilization techniques in batch and continuous processes with claimed potential industrial application. Another advantage of immobilization is the reduced need for filtration, since the concentration of free cells in the product is usually very low.[2] Nevertheless, continuous beer fermentation has not seen significant industrial application, because process characteristics such as simplicity of design, low investment costs, flexible operation, effective process control, and good product quality have not yet been achieved. The application of effective, cheap, and sustainable carrier materials for yeast immobilization could significantly lower the investment costs of continuous fermentation systems, given that the correct sensory characteristics are achieved in the short time typical of such systems.[102,103] The research efforts to optimize brewing with immobilized cells are extensive, and several materials have been proposed as yeast immobilization carriers, such as polysaccharide gels (mainly alginates), polyethylene film, polyethylene rings, PVA, polyvinyl chloride (PVC), DCM, gluten pellets, DEAE-cellulose, waste materials such as wood sawdust, spent grains, and corncobs, and a few inorganic materials. In the following, the most recent of these efforts are highlighted.

A complete continuous beer fermentation system consisting of a main fermentation reactor (gas-lift) and a maturation reactor (packed bed) containing yeast immobilized on spent grains and corncobs, respectively, was evaluated.[103] It was found that by fine tuning of process parameters (residence time and aeration), it was possible to adjust the flavor profile of the final product, which was of a regular quality according to consumer evaluation and comparisons of analytical and sensorial profiles. Continuous, primary bottom beer fermentation maintaining stable and high-level yeast activity during long-term operation was also evaluated in a bioreactor filled with liquid non-absorbent carrier particles, resulting in improved fermentation performance.[104] Two innovative brewing processes, high-gravity batch and complete continuous beer fermentation systems, were studied, showing a significant influence of variables such as concentration and temperature on the ethanol yield and consequently, on the productivity of the high-gravity batch process. The technological feasibility of continuous production of beer based on yeast immobilization on cheap alternative carriers (delignified spent grains and corncob cylinders) was also demonstrated. The influence of process parameters on the fermentation performance and quality of the obtained beers was studied by sensorial analysis. No significant difference in the degree of acceptance between the obtained products and some traditional market brands was observed.[105] Optimization of process parameters and monitoring of flavor formation were studied for continuous brewing with yeasts immobilized on spent grains at 7 to 15 °C in a bubble column reactor with high-gravity all-malt wort (15°Plato). As the fermentation temperature was increased, the degree of fermentation, the rate of sugar consumption, the ethanol volumetric productivity, the consumption of free amino nitrogen, and the ratio of higher alcohols to esters increased.[106] The continuous fermentation of more concentrated worts (16.6 and 18.5°Plato) resulted in beers with unbalanced flavor profiles due to excessive ethyl acetate formation.[107] The potential of application of the non-aggressive LentiKat® (R) technique for brewer's yeast immobilization on PVA was assessed.[108] High cell loads achieved by this procedure and the immobilization procedure had no adverse effect on cell viability. The immobilized cells exhibited high fermentation activity in both laboratory- and pilot-scale fermentations in three successive gas-lift bioreactors, indicating good potential of immobilized cells for the development of continuous primary beer fermentation.

Several methods have also been developed to meet the increasing demand for alcohol-free beer over the last decade, due to health issues, safety in the workplace or during driving, and strict social regulations. These methods include alcohol removal from the product or limited fermentation of wort. In the case of limited fermentation, production is most efficient when immobilized cells are employed.[2,109] However, non-alcoholic beer suffers from flavor defects as well as improper body and foaming properties. Therefore, the production of alcohol-free beer with satisfactory sensory characteristics has recently given rise to increased technological and economic interest.[109] Such systems have already been successfully applied. Van Iersel et al.[110] used a system for the production of non-alcoholic beer by limited fermentation, achieved by low temperatures and anaerobic conditions, using immobilized *S. cerevisiae* in a packed bed reactor. Ethanol contents lower than 0.08% were obtained, while the production of esters and alcohols was stimulated. Flavor formation and cell physiology during alcohol-free beer production by limited fermentation in a controlled down-flow packed bed using immobilized *S. cerevisiae* on polystyrene coated with DEAE-cellulose were also investigated.[110] The system was characterized as highly controllable, and optimal flavor was achieved by the introduction of regular aerobic periods to stimulate yeast growth and temperature variations to control the growth rate and flavor formation.[111] To optimize laboratory-scale continuous alcohol-free beer production, experiments were carried out using real wort and a mimicking model medium in a continuously operating gas-lift reactor with brewing yeast immobilized on spent grains. The results also suggested that the process parameters represent a powerful tool in controlling the degree of fermentation and flavor formation by the immobilized biocatalyst.[112] Finally, the influence of production strains (bottom-fermenting *S. pastorianus* and *S. cerevisiae* with disruption in the KGD2 gene), carrier materials (spent grains and corncobs), bioreactor designs (packed bed and gas-lift), and mixing regimes (ideally mixed and plug flow) on the formation of flavor-active compounds was demonstrated during alcohol-free beer production.[113]

Among polysaccharide hydrogels, alginate gels are the most extensively studied supports for brewer's yeast immobilization.[2] A two-stage reactor system was proposed for continuous secondary fermentation of wort at laboratory scale using immobilized yeast:[114] an up-flow gas-lift bioreactor for main fermentation, and column packed bed reactors with yeast entrapped in three different polysaccharide hydrogels (Ca alginate, Ca pectate, and κ-carrageenan). All three carriers were found suitable for continuous secondary fermentation of green beers produced by continuous main fermentation. Patkova et al.[115] used Ca alginate–entrapped yeast to ferment high-gravity wort in half the time needed for fermentation by free yeast. They also observed that when the original wort gravity was increased, the specific rate of ethanol production remained constant, and the viability did not fall below 95% of living cells, confirming protection of the immobilized cells against osmotic stress. The influence of immobilized yeasts on fermentation parameters and beer quality using a continuous gas-lift bioreactor system with brewer's yeast entrapped in Ca pectate or κ-carrageenan was evaluated. The produced beers had suitable flavor, with low levels of diacetyl, an optimal ratio of higher alcohols to esters, and a maximal specific rate of sugar use.[50] The feasibility to explore the potential uses of immobilization on sensorial characteristics of stout beer (color, flavor, the headspace compounds) was evaluated in batch fermentation using yeast microencapsulated in alginate. Fermentation by free and immobilized yeasts showed no significant difference for all process variables of interest. The profile of headspace compounds was different, perhaps because of changes in yeast behavior and the presence of secondary metabolites.[116]

As in the case of wine making, biocatalysts prepared by immobilization of the alcohol-resistant and cryotolerant strain *S. cerevisiae* AXAZ-1 on DCM and gluten pellets were found to be suitable for batch and continuous fermentation of wort at ambient and low temperatures.[117] The immobilized yeast showed important operational stability with no decrease of activity even at very low temperatures (below 5 °C). Batch fermentations at various temperatures were faster than those of free cells and those usual in commercial brewing, while beer produced by the immobilized yeast contained lower amounts of diacetyl and polyphenols as well as lower bitterness and pH compared with beer produced by free cells. The fruity aroma of beers obtained at low temperatures was attributed to improved ratios of higher alcohols to esters. The use of these immobilized biocatalysts was considered advantageous for brewing, and their possible commercialization in an active freeze-dried form was evaluated.[13,40] The freeze-dried biocatalysts retained their viability during long fermentation periods (13–14 months), and the produced beers were clear with low concentrations of suspended cells and lower diacetyl contents compared with beers produced by free cells. Bekatorou et al.[118] used the same *S. cerevisiae* strain immobilized on dried figs for batch beer fermentations at low and room temperatures (3–20 °C) with similar results. The produced green beers had a fine clarity and were sweet and smooth with a special fruity fig-like aroma and taste, clearly distinct from other commercial products. The safety, low cost, and consumer acceptance of such natural carriers are unquestionable.[2] Finally, a biocatalyst prepared by immobilization of the strain on delignified brewer's spent grains was used for brewing at very low temperatures, also resulting in beers with fine clarity, excellent quality, and mature character after the end of primary fermentation.[119] Fermentation times were low (only 20 days at 0 °C) with high ethanol and beer productivity and low vicinal diketone, DMS, and amyl alcohol concentrations. GC and gas chromatography-mass spectrometry (GC-MS) analysis showed significant quantitative differences in the composition of aroma volatiles, revealing an impact of the fermentation temperature on sensory properties.

A few inorganic materials have been used as immobilization supports for brewing, including porous ceramics, diatomaceous silica (*kieselguhr*), porous brick pieces, and porous glass. Porous glass beads were used for the development of continuous brewing processes for rapid maturation of beer.[52,120] The fermentation times were reduced by half compared with the conventional batch processes, and the proposed technology was demonstrated to be feasible for application in brewing if combined with a heat treatment system to reduce the relatively high diketone content. Finally, a novel material, foam ceramic, with an enormous surface area and good mechanical properties, was applied as a yeast immobilization carrier for continuous secondary fermentation of beer.[121] The results indicated that dilution rate was the primary factor for reducing diacetyl, which could be promptly decreased to a permissible level after about 6 h. Therefore, the secondary fermentation cycle for beer maturation, and the associated costs, could be cut significantly.

Scale-up processes and industrial applications of immobilized cell systems for brewing have been reported and reviewed by various authors.[2,52,101,22]

20.6 Production of Ethanol and Distillates by Immobilized Cells

The requirement for ethanol as an additive in the beverage industries is high, and so is the pursuit of development of efficient bioethanol production systems, including the use of immobilized cells. Research on food-grade alcohol production usually focuses on controlling the formation of volatile by-products, since their concentration in the fermented broths is critical for the production of good-quality distillates and alcoholic beverages. The nature of the immobilization carrier does not usually affect the distillate composition, since non-volatile constituents do not distil. Therefore, the requirement for food-grade purity carriers is not essential due to the employment of the distillation step. Immobilized microbes such as conventional, psychrotolerant, cold-adapted, thermotolerant, alcohol-resistant, and genetically modified yeasts (e.g., *S. cerevisiae, S. diastaticus, Kluyveromyces marxianus, Candida* spp., etc.), and bacteria (e.g., *Zymomonas mobilis*) have been evaluated for food- and fuel-grade ethanol production. These works involved various types of immobilization carriers, as described earlier, synthetic substrates (e.g., glucose media), or waste effluents (e.g., molasses and cheese whey), in various process designs, to evaluate the use of these biocatalysts for bioethanol production.[2,54]

Alcohol can be produced by single, double, or fractional distillation of the fermented liquids. The quality of the distillates depends on the quality of the raw materials, which is generally considered to be good in the case of low-temperature fermentations by immobilized cells due to the improved profiles of volatile compounds, as discussed earlier. In the production of distillates, inorganic materials such as γ-alumina pellets and kissiris can be very useful as immobilization carriers, since they are cheap, abundant, and stable and can be easily reused. Moreover, mineral residues in the fermented liquid are not an issue, since they do not distil. For that reason, batch and continuous low-temperature fermentation of grape must using the cryotolerant and alcohol-resistant strain *S. cerevisiae* AXAZ-1 immobilized on γ-alumina and kissiris was successfully performed for improved production of wine-based distillates.[64,66] Other natural materials proposed as immobilization carriers for the strain were DCM,[10] gluten pellets,[2] orange peel,[123] and brewer's spent grains,[124] used for the alcoholic fermentation of glucose media or molasses at ambient, low, and high temperatures (40 °C). Kopsahelis et al.[125] further proposed an integrated cost-effective system for the continuous alcoholic fermentation of sterilized and non-sterilized molasses at 30 to 40 °C with the same yeast strain immobilized on brewer's spent grains in two types of bioreactor, a multistage fixed-bed tower (MFBT) and a packed bed reactor (PB). The MFBT bioreactor gave better results regarding ethanol concentration, productivity, and conversion. Higher ethanol productivity was obtained in the case of non-sterilized molasses, with no contamination observed during 32 days of continuous operation, which was considered particularly interesting for industrial application.

In the same manner, ethanol production from various single and mixed sugar substrates and industrial effluents was reported, using yeasts immobilized on natural materials such as corncobs,[126] maize stem ground tissue,[127] carob pod,[128] pine wood chips,[129] and woven cotton fiber,[130] as well as in the form of yeast biocapsules.[27] Other research groups studied the effect of composite materials or chemically modified immobilization carriers for efficient bioethanol production. For example, the effect of derivatization was evaluated for the development of delignified corncob grits derivatized with 2-(diethylamino)ethyl chloride hydrochloride as a carrier to immobilize *S. cerevisiae*. The biocatalyst produced ethanol under optimized derivatization and adsorption conditions between yeast cells and the DEAE-corncobs.[131] Similarly, high bioethanol productivity was reported by *S. bayanus* immobilized onto cross-linked graft carboxymethylcellulose-g-poly(N-vinyl-2-pyrrolidone) copolymer beads,[132] and was increased when the percentage of N-vinyl-2-pyrrolidone in copolymer was increased. Efficient bioethanol production from glucose–xylose mixtures was also obtained using *Scheffersomyces stipitis*, free or immobilized in silica–hydrogel films and in cocultures with *S. cerevisiae*.[133] A BCA sponge fabricated by a freeze-drying process was successfully used as a yeast cell carrier for ethanol fermentation,[22] exhibiting several advantageous properties, such as high porosity, appropriate pore size, strong hydrophilicity, and high mechanical, chemical, and thermal stability. The scale-up of ethanol production from molasses using yeast immobilized in alginate–mesoporous zeolite composite carriers exhibited much shorter fermentation times, higher ethanol productivity, and better operational durability, indicating potential for commercial applications.[23]

The impact of high temperature (30 −45°C) on ethanol fermentation by immobilized *K. marxianus* was evaluated by Du Le et al.[134] and Eiadpum et al.[135] The immobilized yeast demonstrated faster sugar assimilation and a higher ethanol level in the fermentation broth in comparison with the free yeast. The different response of the free and immobilized yeast to thermal stress was attributed to the observed changes in the cell membrane fatty acid composition.

Finally, various attempts have been made to optimize alcoholic fermentation processes using immobilized *Z. mobilis* cells. Altuntas and Ozcelik[136] performed continuous ethanol production in a stirred bioreactor using coimmobilized amyloglucosidase and *Z. mobilis* cells for simultaneous saccharification of starch and fermentation to ethanol. Bioethanol production from cane molasses was also studied using a *Z. mobilis* strain entrapped in *Luffa cylindrica* L. sponge discs and Ca alginate gel beads.[137] The immobilization carriers were found to be equally good for ethanol production, but *Luffa* was considered advantageous due to its lower cost, non-destructive nature, and lack of environmental hazard.

20.7 Cider Making by Immobilized Cells

Cider production is a complex process consisting of the alcoholic fermentation step by yeasts followed by MLF by lactic acid bacteria. Traditional cider is produced by natural fermentation of apple juice by the apple's wild microflora, leading to an unstable product of variable quality. Recent studies have been focusing on the use of selected starter cultures and novel technologies, including the use of immobilization techniques, to increase productivity, accelerate maturation, and improve cider quality and stability. The immobilized cell systems used for cider production usually involve coimmobilized species in the same bioreactor for a one-step process or a series of bioreactors containing different species to conduct sequential alcoholic fermentation and MLF.[2] The proposed carrier materials, microbial species, and process designs are generally similar to those applied in wine making. Although most of the research efforts proved to be efficient for both main and secondary fermentation, further research is needed for application at industrial scale to produce ciders with improved and controlled flavor profiles (Figure 20.5).

20.8 Production of Cheese Whey–Based Beverages by Immobilized Cells

Whey is the main liquid waste of the dairy industries and is produced in large quantities worldwide. Its high organic load makes disposal in biological treatment plants impossible, while discarding it imposes a serious environmental threat. Partial exploitation of whey includes its use as animal feed, as a raw material for protein supplement formulations, and as an industrial food additive. The cost-effective microbial conversion of whey into products of added value is very important from both economic and environmental points of view.[138,139] Recent advances include the use of selected cultures, single or mixed, including thermotolerant and genetically modified strains (e.g., to express β-galactosidase activity), in appropriate process designs to produce or recover products of added value from whey, such as single-cell protein (SCP), organic acids, enzymes, lactose, and ethanol.

The bioconversion of whey into ethanol has been mainly attempted using immobilized *K. marxianus* strains in various types of bioreactor system.[140–143] The proposed processes enabled high ethanol yields and productivities from whey. The natural mixed culture kefir, consisting of lactose-fermenting yeasts

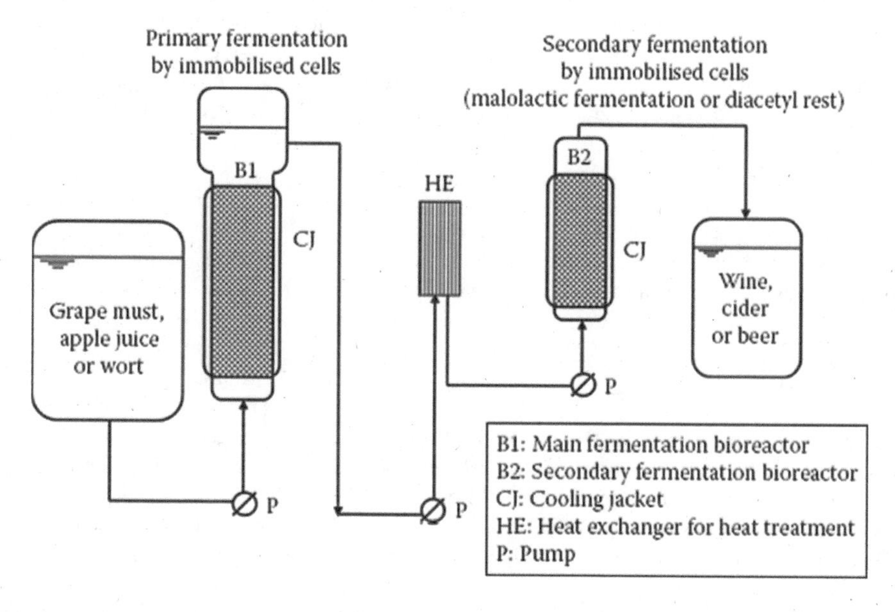

FIGURE 20.5 A schematic process design involving two immobilized cell bioreactors for sequential primary and secondary fermentation for wine, cider, or beer production. (Adapted from Kourkoutas, Y. et al., *Food Microbiol.*, 21(4), 377, 2004.)

and bacteria, has also been used to conduct alcoholic and lactic acid fermentation of whey, depending on the process conditions applied. Kefir immobilized on DCM was found to be suitable for continuous whey fermentation supplemented with 1% raisin extract or molasses.[144] An industrial scale-up process of whey alcoholic fermentation, promoted by raisin extracts, was successfully developed using suspended kefir cells.[138] The fermented whey could be exploited as a raw material to produce kefir-like whey-based drinks and food-grade and fuel alcohol. The development of this technology was supported by the easily precipitated biomass in the form of about 1 mm aggregates, leading to the avoidance of centrifugal separators, which are high-cost equipment.[138] Recent advances in lactic acid production by microbial fermentation processes, including cell immobilization and cell recycling techniques, have been reviewed.[6,145]

20.9 Conclusions

A huge number of immobilization carriers and techniques have been proposed by various research groups for the production of alcoholic beverages due to the numerous advantages associated with such techniques (e.g., increased fermentation rates and yields, high cell densities in the bioreactors, reduced risk of contamination, feasibility of continuous processing, biocatalyst recycling, easier product separation, etc.). However, for full-scale application, features such as the product quality, mechanical strength, and operational stability of the biocatalyst, installation and production costs, flavor control, and consumer acceptance still need to be optimized.

REFERENCES

1. Karel S.F., Libicki S.B. and Robertson C.R. The immobilization of whole cells-engineering principles. *Chemical Engineering Science* 40 (8) (1985): 1321–1354.
2. Kourkoutas Y., Bekatorou A., Banat I.M., Marchant R. and Koutinas A.A. Immobilization technologies and support materials suitable in alcohol beverages production: A review. *Food Microbiology* 21 (4) (2004): 377–397.
3. Mussatto S.I., Dragone G., Guimaraes P.M.R., Silva J.P.A., Carneiro L.M., Roberto I.C., Vicente A., Domingues L. and Teixeira J.A. Technological trends, global market, and challenges of bio-ethanol production. *Biotechnology Advances* 28 (6) (2010): 817–830.
4. Freeman A. and Lilly M.D. Effect of processing parameters on the feasibility and operational stability of immobilized viable microbial cells. *Enzyme and Microbial Technology* 23 (5) (1998): 335–345.
5. Junter G.-A. and Jouenne T. 2.36—Immobilized viable cell biocatalysts: A paradoxical development. In *Comprehensive Biotechnology*, 2nd edn; M. Moo-Young (Ed.); Oxford: Pergamon, 2 (2011): 491–505.
6. Kosseva M.R. Immobilization of microbial cells in food fermentation processes. *Food and Bioprocess Technology* 4 (6) (2011): 1089–1118.
7. Verbelen P.J., De S.D.P., Delvaux F., Verstrepen K.J. and Delvaux F.R. Immobilized yeast cell systems for continuous fermentation applications. *Biotechnology Letters* 28 (19) (2006): 1515–1525.
8. Pilkington P.H., Margaritis A., Mensour N.A. and Russell I. Fundamentals of immobilized yeast cells for continuous beer fermentation: A review. *Journal of the Institute of Brewing* 104 (1) (1998): 19–31.
9. Scott C.D. Immobilized cells: A review of recent literature. *Enzyme and Microbial Technology* 9 (2) (1987): 66–72.
10. Koutinas A.A., Sypsas V., Kandylis P., Michelis A., Bekatorou A., Kourkoutas Y., Kordulis C. et al. Nano-tubular cellulose for bioprocess technology development. *Plos One* 7 (4) (2012): e34350.
11. Koutinas A.A., Papapostolou H., Dimitrellou D., Kopsahelis N., Katechaki E., Bekatorou A. and Bosnea L.A. Whey valorisation: A complete and novel technology development for dairy industry starter culture production. *Bioresource Technology* 100 (15) (2009): 3734–3739.
12. Bekatorou A., Koutinas A.A., Kaliafas A. and Kanellaki M. Freeze-dried *Saccharomyces cerevisiae* cells immobilized on gluten pellets for glucose fermentation. *Process Biochemistry* 36 (6) (2001): 549–557.
13. Bekatorou A., Koutinas A. A., Psarianos K. and Kanellaki M. Low-temperature brewing by freeze-dried immobilized cells on gluten pellets. *Journal of Agricultural and Food Chemistry* 49 (1) (2001): 373–377.

14. Iconomopoulou M., Kanellaki M., Psarianos K. and Koutinas A.A. Delignified cellulosic material supported biocatalyst as freeze-dried product in alcoholic fermentation. *Journal of Agricultural and Food Chemistry* 68 (3) (2000): 958–961.
15. Iconomopoulou M., Psarianos K., Kanellaki M. and Koutinas A.A. Low temperature and ambient temperature wine making using freeze-dried immobilized cells on gluten pellets. *Process Biochemistry* 37 (7) (2002): 707–717.
16. Kregiel D., Berlowska J. and Ambroziak W. Adhesion of yeast cells to different porous supports, stability of cell-carrier systems and formation of volatile by-products. *World Journal of Microbiology and Biotechnology* 28 (12) (2012): 3399–3408.
17. Park J.K. and Chang H.N. Microencapsulation of microbial cells. *Biotechnology Advances* 18 (4) (2000): 303–319.
18. Willaert R. and Nedovic V. Primary beer fermentation by immobilised yeast–A review on flavour formation and control strategies. *Journal of Chemical Technology and Biotechnology* 81 (8) (2006): 1353–1367.
19. Smidsrod O. and Skjak-Braek G. Alginate as immobilization matrix for cells. *Trends in Biotechnology* 8 (3) (1990): 71–78.
20. Tanaka H., Irie S. and Ochi H. A novel immobilization method for prevention of cell leakage from the gel matrix. *Journal of Fermentation and Bioengineering* 68 (3) (1989): 216–219.
21. Callone E., Campostrini R., Carturan G., Cavazza A. and Guzzon R. Immobilization of yeast and bacteria cells in alginate microbeads coated with silica membranes: Procedures, physico-chemical features and bioactivity. *Journal of Materials Chemistry* 18 (40) (2008): 4839–4848.
22. Kirdponpattara S. and Phisalaphong M. Bacterial cellulose-alginate composite sponge as a yeast cell carrier for ethanol production. *Biochemical Engineering Journal* 77 (2013): 103–109.
23. Zheng C., Sun X., Li L. and Guan N. Scaling up of ethanol production from sugar molasses using yeast immobilized with alginate-based MCM-41 mesoporous zeolite composite carrier. *Bioresource Technology* 115 (2012): 208–214.
24. Navratil M., Domeny Z., Hronsky V., Sturdic E., Smogrovicova D. and Gemeiner P. Use of bioluminometry for determination of active yeast biomass immobilized in ionotropic hydrogels. *Analytical Biochemistry* 284 (2) (2000): 394–400.
25. Zhao X.-Q. and Bai F.-W. Yeast flocculation: New story in fuel ethanol production. *Biotechnology Advances* 27 (6) (2009): 849–856.
26. Jin Y.-L. and Speers A.R. Flocculation of Saccharomyces cerevisiae. *Food Research International* 31 (6–7) (1998): 421–440.
27. Garcia-Martinez T., Puig-Pujol A., Peinado R.A., Moreno J. and Mauricio J.C. Potential use of wine yeasts immobilized on *Penicillium chrysogenum* for ethanol production. *Journal of Chemical Technology and Biotechnology* 87 (3) (2012): 351–359.
28. Garcia-Martinez T., Lopez de Lerma N., Moreno J., Peinado R.A., Carmen M.M. and Mauricio J.C. Sweet wine production by two osmotolerant Saccharomyces cerevisiae strains. *Journal of Food Science* 78 (6) (2013): M874–M879.
29. Lopez de Lerma N., Garcia-Martinez T., Moreno J., Mauricio J.C. and Peinado R.A. Volatile composition of partially fermented wines elaborated from sun dried Pedro Ximenez grapes. *Food Chemistry* 135 (4) (2012): 2445–2452.
30. Lebeau T., Jouenne T. and Junter G.-A. Simultaneous fermentation of glucose and xylose by pure and mixed cultures of *Saccharomyces cerevisiae* and Candida shehatae immobilized in a two-chambered bioreactor. *Enzyme and Microbial Technology* 21 (1997): 265–272.
31. Kargupta K., Siddhartha D. and Sanyal S.K. Analysis of the performance of a continuous membrane bioreactor with cell recycling during ethanol fermentation. *Biochemical Engineering Journal* 1 (1) (1998): 31–37.
32. Lebeau T., Jouenne T. and Junter G.-A. Diffusion of sugars and alcohols through composite membrane structures immobilising viable yeast cells. *Enzyme and Microbial Technology* 22 (1998): 434–438.
33. Gryta M. The assessment of microorganism growth in the membrane distillation system. *Desalination* 142 (1) (2002): 79–88.
34. Rathore S., Desai M.P., Liew C.V., Chan L.W. and Heng P.W.S. Microencapsulation of microbial cells. *Journal of Food Engineering* 116 (2) (2013): 369–381.
35. Westrin B.A. and Axelsson A. Diffusion in gels containing immobilized cells: A critical review. *Biotechnology and Bioengineering* 38 (5) (1991): 439–446.

36. Junter G.-A., Coquet L., Vilain S. and Jouenne T. Immobilized-cell physiology: Current data and the potentialities of proteomics. *Enzyme and Microbial Technology* 31 (2002): 201–212.
37. Junter G.-A. and Jouenne T. Immobilized viable microbial cells: From the process to the proteome... or the cart before the horse. *Biotechnology Advances* 22 (8) (2004): 633–658.
38. Jamai L., Sendide , K., Ettayebi , K. Errachidi F., Hamdouni-Alami O., Tahri-Jouti M.A., McDermott T. and Ettayebi M. Physiological difference during ethanol fermentation between calcium alginate-immobilized Candida tropicalis and Saccharomyces cerevisiae. *FEMS Microbiology Letters* 204 (2) (2001): 375–379.
39. Melzoch K., Rychtera M. and Habova V. Effect of immobilization upon the properties and behavior of Saccharomyces cerevisiae cells. *Journal of Biotechnology* 32 (1) (1994): 59–65.
40. Bekatorou A., Soupioni M.J., Koutinas A.A. and Kanellaki M. Low-temperature brewing by freez-edried immobilized cells. *Applied Biochemistry and Biotechnology* 97 (2) (2002): 105–121.
41. Tsaousi K., Velli A., Akarepis F., Bosnea L, Drouza C., Koutinas A.A. and Bekatorou A. Low-temperature winemaking by thermally dried immobilized yeast on delignified brewer's spent grains. *Food Technology and Biotechnology* 49 (3) (2011): 379–384.
42. Kandylis P., Manousi M.E., Bekatorou A. and Koutinas A.A. Freeze-dried wheat supported biocatalyst for low temperature wine making. *LWT—Food Science and Technology* 43 (10) (2010): 1485–1493.
43. Mallouchos A., Komaitis M., Koutinas A.A. and Kanellaki M. Wine fermentations by immobilized and free cells at different temperatures. Effect of immobilization and temperature on volatile by-products. *Food Chemistry* 80 (1) (2003): 109–113.
44. Bardi E. and Koutinas A.A. Immobilization of yeast on delignified cellulosic material for room temperature and low-temperature wine making. *Journal of Agricultural and Food Chemistry* 42 (1) (1994): 221–226.
45. Mallouchos A. and Argyro B. Wine fermentations by immobilized cells. Effect on wine aroma. In *Microbial Implication for Safe and Qualitative Food Products*. Psarianos, C. and Kourkoutas, C. Eds. Kerala: Research Signpost, 2008.
46. Schreier P. and Jennings W.G. Flavor composition of wines: A review. *Critical Reviews in Food Science and Nutrition* 12 (1) (1979): 59–111.
47. Vidrih R. and Hribar J. Synthesis of higher alcohols during cider processing. *Food Chemistry* 67 (3) (1999): 287–294.
48. Etievant X.P. *Volatile Compounds in Foods and Beverages*. New York: Henk Maarse, 1991.
49. Russell I. and Stewart G.G. Contribution of yeast and immobilization technology to flavor development in fermenting beverages. *Food Technology* 46 (11) (1992): 146.
50. Smogrovicova D. Chapter 23—Formation of beer volatile compounds at different fermentation temperatures using immobilized yeasts. *Flavour Science* (2014): 129–131.
51. Branyik T., Vicente A. A., Dostalek P. and Teixeira J.A. A review of flavour formation in continuous beer fermentations. *Journal of the Institute of Brewing* 114 (1) (2008): 3–13.
52. Yamauchi Y., Okamoto T., Murayama H., Kajino K., Amikura T., Hiratsu H., Nagara A., Kamiya T. and Inoue T. Rapid maturation of beer using an immobilized yeast bioreactor. 1. Heat conversion of alpha-acetolactate. *Journal of Biotechnology* 38 (2) (1995): 101–108.
53. Moll M. Fermentation and maturation of beer with immobilised yeasts. *Journal of the Institute of Brewing* 112 (4) (2006): 346–346.
54. Kanellaki M., Bekatorou A. and Koutinas A.A. Low-temperature production of wine, beer and distillates using cold-adapted yeasts. In *Cold-Adapted Yeasts: Biodiversity, Adaptation Strategies and Biotechnological Significance*, Buzzini, P. and Margesin, R. Eds. pp. 417–439. Heidelberg: Springer, 2014.
55. Kanellaki M. and Koutinas A.A. Low temperature fermentation by cold-adapted and immobilized yeast cells. In *Biotechnological Applications of Cold-Adapted Organisms*, Margesin, R. and Schinner, F. Eds. pp. 117–145. Berlin, Heidelberg: Springer-Verlag, 1999.
56. Bakoyianis V., Kana K., Kalliafas A. and Koutinas A.A. Low-temperature continuous wine-making by kissiris-supported biocatalyst: Volatile by-products. *Journal of Agricultural and Food Chemistry* 41 (3) (1993): 465–468.
57. Bardi E.P., Bakoyianis V., Koutinas A.A. and Kanellaki M. Room temperature and low temperature wine making using yeast immobilized on gluten pellets. *Process Biochemistry* 31 (5) (1996): 425–430.

58. Mallouchos A., Loukatos P., Bekatorou A., Koutinas A.A. and Komaitis M. Ambient and low temperature winemaking by immobilized cells on brewer's spent grains: Effect on volatile composition. *Food Chemistry* 104 (3) (2007): 918–927.
59. Kourkoutas Y., Kanellaki M., Koutinas A.A. and Tzia C. Effect of fermentation conditions and immobilization supports on the wine-making. *Journal of Food Engineering* 69 (2005): 115–123.
60. Yajima M. and Yokotsuka K. Volatile compound formation in white wines fermented using immobilized and free yeast. *American Journal of Enology and Viticulture* 52 (3) (2001): 210–218.
61. Perpete P. and Collin S. How to improve the enzymatic worty flavour reduction in a cold contact fermentation. *Food Chemistry* 70 (4) (2000): 457–462.
62. Bakoyianis V., Kanellaki M., Kalliafas A. and Koutinas A.A. Low temperature wine-making by immobilized cells on mineral kissiris. *Journal of Agricultural and Food Chemistry* 40 (7) (1992): 1293–1296.
63. Bakoyianis V., Koutinas A. A., Aggelopoulos K. and Kanellaki M. Comparative study of kissiris, γ-alumina and Ca-alginates as supports of cells for batch and continuous wine making at low temperatures. *Journal of Agricultural and Food Chemistry* 45 (12) (1997): 4884–4888.
64. Loukatos P., Kiaris M., Ligas I., Bourgos G., Kanellaki M., Komaitis M. and Koutinas A.A. Continuous wine making by γ-alumina-supported biocatalyst. *Applied Biochemistry and Biotechnology* 89 (1) (1999): 1–13.
65. Colagrande O., Silva A. and Fumi M.D. Recent applications of biotechnology in wine production. *Biotechnology Progress* 10 (1) (1994): 2–18.
66. Loukatos P., Kanellaki M., Komaitis M., Athanasiadis I. and Koutinas A.A. A new technological approach proposed for distillate production using immobilized cells. *Journal of Bioscience and Bioengineering* 95 (1) (2003): 35–39.
67. Busova K., Magyar I. and Janky F. Effect of immobilized yeasts on the quality of bottle-fermented sparkling wine. *Acta Alimentaria* 23 (1) (1994): 9–23.
68. Torresi S., Frangipane M.T. and Anelli G. Biotechnologies in sparkling wine production. Interesting approaches for quality improvement: A review. *Food Chemistry* 129 (3) (2011): 1232–1241.
69. Silva S., Ramon P.F., Silva P., Maria de Fatima T. and Strehaiano P. Use of encapsulated yeast for the treatment of stuck and sluggish fermentations. *Journal International Des Sciences De La Vigne Et Du Vin* 36 (3) (2002): 161–168.
70. Tsakiris A., Kandylis P., Bekatorou A., Kourkoutas Y. and Koutinas A.A. Dry red wine-making using yeast immobilized on cork pieces. *Applied Biochemistry and Biotechnology* 162 (5) (2010): 1316–1326.
71. Reddy V.L., Reddy P.L., Wee Y.-J. and Reddy O.V.S. Production and characterization of wine with sugarcane piece immobilized yeast biocatalyst. *Food and Bioprocess Technology* 4 (1) (2011): 142–148.
72. Tsakiris A., Bekatorou A., Koutinas A.A., Marchant R. and B.I.M.Immobilization of yeast on dried raisin berries for use in dry white wine making. *Food Chemistry* 87 (2004): 11–15.
73. Mallouchos A., Reppa P., Aggelis G., Kanellaki M., Koutinas A.A. and Komaitis M. Grape skins as a natural support for yeast immobilization. *Biotechnology Letters* 24 (16) (2002): 1331–1335.
74. Kandylis P. and Koutinas A.A. Extremely low temperature fermentations of grape must by potatoes supported yeast-strain AXAZ-1. A contribution is performed to catalysis of alcoholic fermentation. *Journal of Agricultural and Food Chemistry* 56 (9) (2008): 3317–3327.
75. Kandylis P., Mantzari A., Koutinas A.A. and Kookos I.K. Modelling of low temperature wine-making, using immobilized cells. *Food Chemistry* 133 (4) (2012): 1341–1348.
76. Kandylis P., Dimitrellou D. and Koutinas A.A. Winemaking by barley supported yeast cells. *Food Chemistry* 130 (2) (2012): 425–431.
77. Kandylis P., Goula A. and Koutinas A.A. Corn starch gel for yeast cell entrapment a view for catalysis of wine fermentation. *Journal of Agricultural and Food Chemistry* 56 (24) (2008): 12037–12045.
78. Reddy V.L., Reddy H.K.Y., Reddy P.A.L. and Reddy V.S.O. Wine production by novel yeast biocatalyst prepared by immobilization on watermelon (Citrullus vulgaris) rind pieces and characterization of volatile compounds. *Process Biochemistry* 43 (7) (2008): 748–752.
79. Genisheva Z., Macedo S., Mussatto S.I., Teixeira J.A. and Oliveira J.M. Production of white wine by Saccharomyces cerevisiae immobilized on grape pomace. *Journal of the Institute of Brewing* 118 (2) (2012): 163–173.
80. Genisheva Z., Mussatto S.I., Oliveira J.M. and Teixeira J.A. Evaluating the potential of wine-making residues and corn cobs as support materials for cell immobilization for ethanol production. *Industrial Crops and Products* 34 (1) (2011): 979–985.

81. Spagna G., Barbagallo R.N., Casarini D. and Pifferi P.G. A novel chitosan derivative to immobilize a-L-rhamnopyranosidase from *Aspergillus niger* for application in beverage technologies. *Enzyme and Microbial Technology* 28 (2001): 427–438.
82. Gonzalez-Pombo P., Farina L., Carrau F., Batista-Viera F. and Brena B.M. Aroma enhancement in wines using co-immobilized. *Aspergillus niger* glycosidases. *Food Chemistry* 143 (2014): 185–191.
83. Gallifuoco A., D'Ercole L., Alfani F., Cantarella M., Spagna G. and Pifferi P.G. On the use of chitosan-immobilized p-glucosidase in wine-making: Kinetics and enzyme inhibition. *Process Biochemistry* 33 (2) (1998): 163–168.
84. Merida J., Lopez-Toledano A. and Medina M. Immobilized yeasts in kappa-carragenate to prevent browning in white wines. *European Food Research and Technology* 225 (2) (2007): 279–286.
85. Lopez-Toledano A., Merida J. and Medina M. Colour correction in white wines by use of immobilized yeasts on kappa-carragenate and alginate gels. *European Food Research and Technology* 225 (5–6) (2007): 879–885.
86. Vilela A., Schuller D., Mendes-Faia A. and Corte-Real M. Reduction of volatile acidity of acidic wines by immobilized Saccharomyces cerevisiae cells. *Applied Microbiology and Biotechnology* 97 (11) (2012): 4991–5000.
87. Takaya M., Matsumoto N. and Yanase H. Characterization of membrane bioreactor for dry wine production. *Journal of Bioscience and Bioengineering* 93 (2) (2002): 240–244.
88. Versari A., Parpinello G.P. and Cattaneo M. Leuconostoc oenos and malolactic fermentation in wine: A review. *Journal of Industrial Microbiology and Biotechnology* 23 (1999): 447–455.
89. Zhang D. and Lovitt R.W. Strategies for enhanced malolactic fermentation in wine and cider maturation. *Journal of Chemical Technology and Biotechnology* 81 (7) (2006): 1130–1140.
90. Redzepovic S., Orlic S., Madjak A., Kozina B., Volschenk H. and Viljoen-Bloom M. Differential malic acid degradation by selected strains of Saccharomyces during alcoholic fermentation. *International Journal of Food Microbiology* 83 (1) (2003): 49–61.
91. Kosseva M., Beschkov V., Kennedy J.F. and Lloyd L.L. Malolactic fermentation in Chardonnay wine by immobilized *Lactobacillus casei* cells. *Process Biochemistry* 33 (8) (1998): 793–797.
92. Rodriguez-Nogales M.J., Vila-Crespo J. and Fernandez-Fernandez E. Immobilization of *Oenococcus oeni* in lentikats (R) to develop malolactic fermentation in wines. *Biotechnology Progress* 29 (1) (2013): 60–65.
93. Maicas S. The use of alternative technologies to develop malolactic fermentation in wine. *Applied Microbiology and Biotechnology* 56 (2001): 35–39.
94. Iorio G., Catapano G., Drioli E., Rossi M. and Rella R. Malic enzyme immobilization in continuous capillary membrane reactors. *Journal of Membrane Science* 22 (1985): 317–324.
95. Hong S.-K., Lee H.-J., Park H.-J., Hong Y.-A., Rhee I.-K., Lee W.-H., Choi S.-H., Lee O.-S. and Park H.-D. Degradation of malic acid in wine by immobilized Issatchenkia orientalis cells with oriental oak charcoal and alginate. *Letters in Applied Microbiology* 50 (5) (2010): 522–529.
96. Guzzon R., Carturan G., Krieger-Weber S. and Cavazza A. Use of organo-silica immobilized bacteria produced in a pilot scale plant to induce malolactic fermentation in wines that contain lysozyme. *Annals of Microbiology* 62 (1) (2012): 381–390.
97. Genisheva Z., Mota A., Mussatto S.I., Oliveira J.M. and Teixeira J.A. Integrated continuous winemaking process involving sequential alcoholic and malolactic fermentations with immobilized cells. *Process Biochemistry* 49 (1) (2014): 1–9.
98. Aaron R.T., Davis R.C., Hamdy M.K. and Toledo R.T. Continuous alcohol/malolactic fermentation of grape must in a bioreactor system using immobilized cells. *Journal of Rapid Methods and Automation in Microbiology* 12 (2) (2004): 127–148.
99. Agouridis N., Kopsahelis N., Plessas S., Koutinas A.A. and Kanellaki M. *Oenococcus oeni* cells immobilized on delignified cellulosic material for malolactic fermentation of wine. *Bioresource Technology* 99 (18) (2008): 9017–9020.
100. Servetas I., Berbegal C., Camacho N., Bekatorou A., Ferrer S., Nigam P., Drouza C. and Koutinas A.A. Saccharomyces cerevisiae and *Oenococcus oeni* immobilized in different layers of a cellulose/starch gel composite for simultaneous alcoholic and malolactic wine fermentations. *Process Biochemistry* 48 (9) (2013): 1279–1284.
101. Masschelein C.A., Ryder D.S. and Simon J.-P. Immobilized cell technology in beer production. *Critical Reviews in Biotechnology* 14 (1994): 155–177.

102. Branyik T., Silva D.P., Baszczynski M., Lehnert R. and Almeida e Silva J.B. A review of methods of low alcohol and alcohol-free beer production. *Journal of Food Engineering* 108 (4) (2012): 493–506.
103. Branyik T., Silva D.P., Vicente A.A., Lehnert R., Silva J.B., Almeida E., Dostalek P. and Teixeira J.A. Continuous immobilized yeast reactor system for complete beer fermentation using spent grains and corncobs as carrier materials. *Journal of Industrial Microbiology and Biotechnology* 33 (12) (2006): 1010–1018.
104. Inoue T. and Mizuno A. Preliminary study for the development of a long-life, continuous, primary fermentation system for beer brewing. *Journal of the American Society of Brewing Chemists* 66 (2) (2008): 80–87.
105. Silva D.P., Branyik T., Dragone G., Vicente A.A., Teixeira J.A. and Almeida e Silva J.B. High gravity batch and continuous processes for beer production: Evaluation of fermentation performance and beer quality. *Chemical Papers* 62 (1) (2008): 34–41.
106. Dragone G., Mussatto S.I. and Almeida e Silva J.B. Influence of temperature on continuous high gravity brewing with yeasts immobilized on spent grains. *European Food Research and Technology* 228 (2) (2007): 257–264.
107. Dragone G., Mussatto S.I. and Almeida e Silva J.B. High gravity brewing by continuous process using immobilised yeast: Effect of wort original gravity on fermentation performance. *Journal of the Institute of Brewing* 113 (4) (2007): 391–398.
108. Bezbradica D., Obradovic B., Leskosek-Cukalovic I., Bugarski B. and Nedovic V. Immobilization of yeast cells in PVA particles for beer fermentation. *Process Biochemistry* 42 (9) (2007): 1348–1351.
109. Sohrabvandi S., Mousavi S.M., Razavi S.H., Mortazavian A.M. and Rezaei K. Alcohol-free beer: Methods of production, sensorial defects, and healthful effects. *Food Reviews International* 26 (4) (2010): 335–352.
110. Van Iersel M.F.M., Van Dieren B., Rombouts F.M. and Abee T. Flavor formation and cell physiology during the production of alcohol-free beer with immobilized Saccharomyces cerevisiae. *Enzyme and Microbial Technology* 24 (7) (1999): 407–411.
111. Van Iersel M.F.M., Brouwer P.E., Rombouts F.M. and Abee T. Influence of yeast immobilization on fermentation and aldehyde reduction during the production of alcohol free beer. *Enzyme and Microbial Technology* 26 (8) (2000): 602–607.
112. Lehnert R., Novak P., Macieira F., Kurec M., Teixeira J.A. and Tomas B. Optimisation of lab-scale continuous alcohol-free beer production. *Czech Journal of Food Sciences* 27 (4) (2009): 267–275.
113. Mota A., Novak P., Macieira F., Vicente A. A., Teixeira J.A., Smogrovicova D. and Branyik T. Formation of flavor-active compounds during continuous alcohol-free beer production: The influence of yeast strain, reactor configuration, and carrier type. *Journal of the American Society of Brewing Chemists* 69 (1) (2011): 1–7.
114. Domeny Z., Smogrovicova D., Gemeiner P., Sturdik E., Patkova J. and Malovikova A. Continuous secondary fermentation using immobilized yeast. *Biotechnology Letters* 20 (11) (1998): 1041–1045.
115. Patkova J., Smogrovicova D., Domeny Z. and Bafrncova P. Very high gravity wort fermentation by immobilized yeast. *Biotechnology Letters* 22 (14) (2000): 1173–1177.
116. Almonacid S.F., Najera A.L., Young M.E., Simpson R.J. and Acevedo C.A. A comparative study of stout beer batch fermentation using free and microencapsulated yeasts. *Food and Bioprocess Technology* 5 (1) (2012): 750–758.
117. Bardi E., Koutinas A. A., Soupioni M. and Kanellaki M. Immobilization of yeast on delignified cellulosic material for low temperature brewing. *Journal of Agricultural and Food Chemistry* 44 (2) (1996): 463–467.
118. Bekatorou A., Sarellas A., Ternan N.G., Mallouchos A., Komaitis M., Koutinas A.A. and Kanellaki M. Low-temperature brewing using yeast immobilized on dried figs. *Journal of Agricultural and Food Chemistry* 50 (25) (2002): 7249–7257.
119. Kopsahelis N., Kanellaki M. and Bekatorou A. Low temperature brewing using cells immobilized on brewer's spent grains. *Food Chemistry* 104 (2) (2007): 480–488.
120. Tata M., Bower P., Bromberg S., Duncombe D., Fehring J., Lau V., Ryder D. and Stassi P. Immobilized yeast bioreactor systems for continuous beer fermentation. *Biotechnology Progress* 15 (1) (1999): 105–113.
121. Cheng J., Liu J., Shao H. and Qiu Y. Continuous secondary fermentation of beer by yeast immobilized on the foam ceramic. *Research Journal of Biotechnology* 2 (3) (2007): 40–42.

122. Mensour N.A., Margaritis A., Briens C.L., Pilkington H. and Russell I. Developments in the brewing industry using immobilized yeast cell bioreactor systems. *Journal of the Institute of Brewing* 103 (6) (1997): 363–370.
123. Plessas S., Bekatorou A., Koutinas A.A., Soupioni M., Banat I.M. and Marchant R. Use of Saccharomyces cerevisiae cells immobilized on orange peel as biocatalyst for alcoholic fermentation. *Bioresource Technology* 98 (4) (2007): 860–865.
124. Kopsahelis N., Agouridis N., Bekatorou A. and Kanellaki M. Comparative study of spent grains and delignified spent grains as yeast supports for alcohol production from molasses. *Bioresource Technology* 98 (7) (2007): 1440–1447.
125. Kopsahelis N., Papachronopoulos A., Bosnea L., Bekatorou A., Tzia C. and Kanellaki M. Alcohol production from sterilized and non-sterilized molasses by Saccharomyces cerevisiae immobilized on brewer's spent grains in two types of continuous bioreactor systems. *Biomass and Bioenergy* 46 (2012): 809–809.
126. Laopaiboon L. and Laopaiboon P. Ethanol production from sweet sorghum juice in repeated-batch fermentation by Saccharomyces cerevisiae immobilized on corncob. *World Journal of Microbiology and Biotechnology* 28 (2) (2012): 559–566.
127. Razmovski R. and Vucurovic V. Bioethanol production from sugar beet molasses and thick juice using Saccharomyces cerevisiae immobilized on maize stem ground tissue. *Fuel* 92 (1) (2012): 1–8.
128. Yatmaz E., Turhan I. and Karhan M. Optimization of ethanol production from carob pod extract using immobilized *Saccharomyces cerevisiae* cells in a stirred tank bioreactor. *Bioresource Technology* 135 (2013): 365–371.
129. Dhabhai R., Chaurasia S.P. and Dalai A.K. Efficient bioethanol production from glucose-xylose mixtures using co-culture of Saccharomyces cerevisiae immobilized on canadian pine wood chips and free Pichia stipitis. *Journal of Biobased Materials and Bioenergy* 6 (5) (2012): 594–600.
130. Chen Y., Liu Q., Zhou T., Li B., Yao S., Wu J. and Ying H. Ethanol production by repeated batch and continuous fermentations by Saccharomyces cerevisiae immobilized in a fibrous bed bioreactor. *Journal of Microbiology and Biotechnology* 23 (4) (2013): 511–517.
131. Lee S.-E., Lee C.G., Kang D.H., Lee H.-Y. and Jung K.-H. Preparation of corncob grits as a carrier for immobilizing yeast cells for ethanol production *Journal of Microbiology and Biotechnology* 22 (12) (2012): 1673–1680.
132. Gokgoz M. and Yigitoglu M. High productivity bioethanol fermentation by immobilized Saccharomyces bayanus onto carboxymethylcellulose-g-poly(N-vinyl-2-pyrrolidone) beads. *Artificial Cells Nanomedicine and Biotechnology* 41 (2) (2013): 137–143.
133. De Bari I., De Canio P., Cuna D., Liuzzi F., Capece A. and Romano P. Bioethanol production from mixed sugars by Scheffersomyces stipitis free and immobilized cells, and co-cultures with *Saccharomyces cerevisiae. New Biotechnology* 30 (6) (2013): 591–597.
134. Du Le H., Pornthap T. and Van Viet M.L. Impact of high temperature on ethanol fermentation by *Kluyveromyces marxianus* immobilized on banana leaf sheath pieces. *Applied Biochemistry and Biotechnology* 171 (3) (2013): 806–816.
135. Eiadpum A., Limtong S. and Phisalaphong M. High-temperature ethanol fermentation by immobilized coculture of *Kluyveromyces marxianus* and *Saccharomyces cerevisiae. Journal of Bioscience and Bioengineering* 114 (3) (2012): 325–329.
136. Altuntas E.G. and Ozcelik F. Ethanol production from starch by co-immobilized amyloglucosidase—Zymomonas mobilis cells in a continuously-stirred bioreactor. *Biotechnology and Biotechnological Equipment* 27 (1) (2013): 3506–3512.
137. Behera S., Mohanty R.C. and Ray R.C. Ethanol fermentation of sugarcane molasses by *Zymomonas mobilis* MTCC 92 immobilized in *Luffa cylindrica* L. sponge discs and Ca-alginate matrices. *Brazilian Journal of Microbiology* 43 (4) (2012): 1499–1507.
138. Koutinas A.A., Athanasiadis I., Bekatorou A., Psarianos C., Kanellaki M., Agouridis N. and Blekas G. Kefir-yeast technology: Industrial scale-up of alcoholic fermentation of whey, promoted by raisin extracts, using kefir-yeast granular biomass. *Enzyme and Microbial Technology* 41 (5) (2007): 576–582.
139. Koutinas A.A., Bekatorou A., Nigam P., Banat I.M. and Marchant R. Whey utilization and SCP production. In *Advances in Cheese Whey Utilization*, Ma Esperanza Cerdan, Ma Isabel Gonzalez-Siso and Manuel Bacerra, Eds. pp. 147–161. Kerala: Research Signpost, 2008.

140. Dale C.M., Eagger A. and Okos M.R. Osmotic inhibition of free and immobilized *K. marxianus* anaerobic growth and ethanol productivity in whey permeate concentrate. *Process Biochemistry* 29 (7) (1994): 535–544.
141. Gabardo S., Rech R., Zachia A. and Marco A. Performance of different immobilized-cell systems to efficiently produce ethanol from whey: Fluidized batch, packed-bed and fluidized continuous bioreactors. *Journal of Chemical Technology and Biotechnology* 87 (8) (2012): 1194–1201.
142. Kourkoutas Y., Dimitropoulou S., Kanellaki M., Marchant R., Nigam P., Banat I.M. and Koutinas A.A. High-temperature alcoholic fermentation of whey using *Kluyveromyces marxianus* IMB3 yeast immobilized on delignified cellulosic material. *Bioresource Technology* 82 (2002): 177–181.
143. Ozmihci S. and Kargi F. Fermentation of cheese whey powder solution to ethanol in a packed-column bioreactor: Effects of feed sugar concentration. *Journal of Chemical Technology and Biotechnology* 84 (1) (2009): 106–111.
144. Kourkoutas Y., Psarianos C., Koutinas A.A., Kanellaki M., Banat I.M. and Marchant R. Continuous whey fermentation using kefir yeast immobilized on delignified cellulosic material. *Journal of Agricultural and Food Chemistry* 50 (9) (2002): 2543–2547.
145. Abdel-Rahman M.A., Tashiro Y. and Sonomoto K. Recent advances in lactic acid production by microbial fermentation processes. *Biotechnology Advances* 31 (6) (2013): 877–902.

Index